Compendium of Trace Metals and Marine Biota

Volume 1: Plants and Invertebrates

Compendium of Trace Metals and Marine Biota

Volume 1: Plants and Invertebrates

by
Ronald Eisler

ELSEVIER

Amsterdam • Boston • Heidelberg • London • New York • Oxford
Paris • San Diego • San Francisco • Singapore • Sydney • Tokyo

Elsevier
The Boulevard, Langford Lane, Kidlington, Oxford OX5 1GB, UK
Radarweg 29, PO Box 211, 1000 AE Amsterdam, The Netherlands

First edition 2010

Copyright © 2010 Elsevier BV. All rights reserved

No part of this publication may be reproduced, stored in a retrieval system or transmitted in any form or by any means electronic, mechanical, photocopying, recording or otherwise without the prior written permission of the publisher

Permissions may be sought directly from Elsevier's Science & Technology Rights Department in Oxford, UK: phone (+44) (0) 1865 843830; fax (+44) (0) 1865 853333; email: permissions@elsevier.com. Alternatively you can submit your request online by visiting the Elsevier web site at http://elsevier.com/locate/permissions, and selecting *Obtaining permission to use Elsevier material*

Notice
No responsibility is assumed by the publisher for any injury and/or damage to persons or property as a matter of products liability, negligence or otherwise, or from any use or operation of any methods, products, instructions or ideas contained in the material herein.

British Library Cataloguing in Publication Data
A catalogue record for this book is available from the British Library

Library of Congress Cataloging-in-Publication Data
A catalog record for this book is available from the Library of Congress

ISBN (Set): 978-0-444-53439-2
ISBN (Vol 1): 978-0-444-53436-1

For information on all Elsevier publications
visit our website at books.elsevier.com

Printed and bound in Great Britain
10 11 12 10 9 8 7 6 5 4 3 2 1

Working together to grow
libraries in developing countries

www.elsevier.com | www.bookaid.org | www.sabre.org

ELSEVIER   BOOK AID International   Sabre Foundation

This volume is dedicated to the memory of my mentors:

Yohay Bin-Nun
Lauren R. Donaldson
Clarence P. Idyll
Morton I. Grossman
Frank G. Lowman
James E. Lynch
Lionel A. Walford

Contents

Acknowledgments .... *xvii*
Books by Ronald Eisler .... *xix*
About the Author .... *xxi*
List of Tables – Vol. 1 .... *xxiii*

Chapter 1 Introduction 1
    1.1   Literature Cited 6

Chapter 2 Algae and Macrophytes 7
    2.1   Aluminum 7
    2.2   Americium 8
    2.3   Antimony 8
    2.4   Arsenic 9
    2.5   Barium 14
    2.6   Beryllium 14
    2.7   Bismuth 15
    2.8   Boron 15
    2.9   Cadmium 16
    2.10 Cerium 22
    2.11 Cesium 22
    2.12 Chromium 23
    2.13 Cobalt 26
    2.14 Copper 28
    2.15 Gallium 36
    2.16 Germanium 37
    2.17 Gold 37
    2.18 Iron 38
    2.19 Lead 42
    2.20 Lithium 48
    2.21 Manganese 48
    2.22 Mercury 51
    2.23 Molybdenum 55

2.24 Nickel 56
2.25 Plutonium 60
2.26 Radium 61
2.27 Rhenium 61
2.28 Rubidium 61
2.29 Ruthenium 61
2.30 Selenium 62
2.31 Silicon 62
2.32 Silver 64
2.33 Strontium 66
2.34 Technetium 67
2.35 Tin 67
2.36 Titanium 69
2.37 Tungsten 69
2.38 Uranium 69
2.39 Vanadium 70
2.40 Yttrium 71
2.41 Zinc 72
2.42 Zirconium 79
2.43 Literature Cited 80

Chapter 3 Protists 99
3.1 Aluminum 99
3.2 Arsenic 99
3.3 Barium 101
3.4 Cadmium 101
3.5 Cesium 102
3.6 Chromium 102
3.7 Cobalt 102
3.8 Copper 102
3.9 Gallium 103
3.10 Iron 103
3.11 Lead 103
3.12 Manganese 104
3.13 Mercury 104
3.14 Nickel 107
3.15 Silver 107
3.16 Strontium 108
3.17 Tin 108
3.18 Titanium 109
3.19 Zinc 109
3.20 Literature Cited 109

Chapter 4 Sponges 115
    4.1   Aluminum 115
    4.2   Antimony 115
    4.3   Arsenic 115
    4.4   Barium 118
    4.5   Cadmium 118
    4.6   Cobalt 119
    4.7   Copper 119
    4.8   Gallium 119
    4.9   Iron 119
    4.10 Lead 120
    4.11 Manganese 120
    4.12 Mercury 120
    4.13 Nickel 120
    4.14 Plutonium 120
    4.15 Ruthenium 120
    4.16 Selenium 120
    4.17 Silver 121
    4.18 Strontium 121
    4.19 Vanadium 121
    4.20 Zinc 121
    4.21 Literature Cited 121

Chapter 5 Coelenterates 123
    5.1   Aluminum 123
    5.2   Antimony 123
    5.3   Arsenic 123
    5.4   Barium 124
    5.5   Bismuth 125
    5.6   Boron 125
    5.7   Cadmium 126
    5.8   Cesium 127
    5.9   Chromium 127
    5.10 Cobalt 127
    5.11 Copper 127
    5.12 Gallium 129
    5.13 Iron 131
    5.14 Lead 131
    5.15 Lithium 133
    5.16 Manganese 133
    5.17 Mercury 134
    5.18 Molybdenum 134

5.19 Nickel..134
5.20 Radium..134
5.21 Rubidium..134
5.22 Ruthenium..134
5.23 Scandium..135
5.24 Silicon...135
5.25 Silver...135
5.26 Strontium..137
5.27 Thorium..137
5.28 Tin...137
5.29 Titanium...137
5.30 Uranium...139
5.31 Vanadium...140
5.32 Zinc...140
5.33 Literature Cited..141

Chapter 6 Molluscs 143

6.1 Aluminum...144
6.2 Americium..147
6.3 Antimony..148
6.4 Arsenic..150
6.5 Barium..155
6.6 Bismuth..157
6.7 Boron..157
6.8 Cadmium..157
6.9 Cerium..175
6.10 Cesium..175
6.11 Chromium..177
6.12 Cobalt...186
6.13 Copper..192
6.14 Europium..224
6.15 Gallium...224
6.16 Gold..224
6.17 Hafnium...226
6.18 Iron...226
6.19 Lanthanum...237
6.20 Lead..237
6.21 Lithium...255
6.22 Manganese...255
6.23 Mercury..266
6.24 Molybdenum..280
6.25 Neptunium...282
6.26 Nickel...282
6.27 Niobium...292

6.28	Plutonium	292
6.29	Polonium	293
6.30	Protactinium	293
6.31	Radium	293
6.32	Rhenium	294
6.33	Rubidium	294
6.34	Ruthenium	295
6.35	Samarium	295
6.36	Scandium	295
6.37	Selenium	295
6.38	Silicon	299
6.39	Silver	299
6.40	Strontium	312
6.41	Tantalum	312
6.42	Technetium	312
6.43	Tellurium	315
6.44	Terbium	315
6.45	Thallium	315
6.46	Thorium	316
6.47	Tin	316
6.48	Titanium	329
6.49	Tungsten	329
6.50	Uranium	329
6.51	Vanadium	329
6.52	Ytterbium	329
6.53	Zinc	329
6.54	Zirconium	363
6.55	Literature Cited	363

Chapter 7 Crustaceans 399

7.1	Aluminum	399
7.2	Antimony	399
7.3	Arsenic	399
7.4	Barium	406
7.5	Beryllium	407
7.6	Bismuth	407
7.7	Boron	407
7.8	Cadmium	407
7.9	Cerium	417
7.10	Cesium	418
7.11	Chromium	418
7.12	Cobalt	423
7.13	Copper	424
7.14	Dysprosium	438
7.15	Erbium	438

7.16	Europium	439
7.17	Gadolinium	439
7.18	Gallium	439
7.19	Germanium	440
7.20	Gold	440
7.21	Holmium	440
7.22	Iron	440
7.23	Lanthanum	443
7.24	Lead	443
7.25	Lithium	448
7.26	Lutetium	448
7.27	Manganese	449
7.28	Mercury	452
7.29	Molybdenum	460
7.30	Neodymium	461
7.31	Neptunium	461
7.32	Nickel	461
7.33	Niobium	463
7.34	Plutonium	463
7.35	Polonium	464
7.36	Praseodymium	464
7.37	Protactinium	464
7.38	Radium	464
7.39	Rhenium	464
7.40	Rubidium	465
7.41	Ruthenium	465
7.42	Samarium	465
7.43	Scandium	465
7.44	Selenium	468
7.45	Silicon	469
7.46	Silver	469
7.47	Strontium	470
7.48	Terbium	472
7.49	Thallium	472
7.50	Thorium	472
7.51	Thulium	472
7.52	Tin	472
7.53	Titanium	476
7.54	Tungsten	476
7.55	Uranium	476
7.56	Vanadium	476
7.57	Ytterbium	476
7.58	Yttrium	477

7.59	Zinc	477
7.60	Zirconium	495
7.61	Literature Cited	495

Chapter 8 Insects 517

8.1	Cadmium	517
8.2	Chromium	517
8.3	Copper	519
8.4	Iron	520
8.5	Lead	520
8.6	Manganese	520
8.7	Nickel	520
8.8	Zinc	520
8.9	Literature Cited	520

Chapter 9 Chaetognaths 521

9.1	Arsenic	521
9.2	Barium	521
9.3	Boron	524
9.4	Cadmium	524
9.5	Chromium	524
9.6	Cobalt	524
9.7	Copper	524
9.8	Gallium	524
9.9	Iron	525
9.10	Lead	525
9.11	Lithium	525
9.12	Molybdenum	525
9.13	Nickel	525
9.14	Silver	525
9.15	Strontium	525
9.16	Tin	525
9.17	Titanium	526
9.18	Vanadium	526
9.19	Zinc	526
9.20	Zirconium	526
9.21	Literature Cited	526

Chapter 10 Annelids 527

10.1	Aluminum	527
10.2	Americium	527
10.3	Antimony	528
10.4	Arsenic	528
10.5	Barium	529
10.6	Cadmium	529

10.7	Cesium	532
10.8	Chromium	532
10.9	Cobalt	534
10.10	Copper	535
10.11	Iron	537
10.12	Lead	537
10.13	Manganese	537
10.14	Mercury	537
10.15	Nickel	540
10.16	Plutonium	540
10.17	Ruthenium	541
10.18	Samarium	541
10.19	Scandium	542
10.20	Selenium	542
10.21	Silver	542
10.22	Strontium	542
10.23	Tin	543
10.24	Zinc	543
10.25	Zirconium	547
10.26	Literature Cited	547

Chapter 11 Echinoderms *553*

11.1	Aluminum	553
11.2	Antimony	553
11.3	Arsenic	553
11.4	Barium	556
11.5	Boron	556
11.6	Cadmium	556
11.7	Cerium	558
11.8	Cesium	558
11.9	Chromium	559
11.10	Cobalt	559
11.11	Copper	561
11.12	Europium	563
11.13	Gallium	564
11.14	Iron	564
11.15	Lead	565
11.16	Manganese	566
11.17	Mercury	568
11.18	Molybdenum	569
11.19	Nickel	569
11.20	Plutonium	569
11.21	Rubidium	571
11.22	Scandium	571

11.23	Selenium	571
11.24	Silver	571
11.25	Strontium	573
11.26	Technetium	575
11.27	Tin	575
11.28	Zinc	575
11.29	Literature Cited	578

Chapter 12 Tunicates 583

12.1	Antimony	583
12.2	Arsenic	583
12.3	Barium	585
12.4	Boron	585
12.5	Cadmium	585
12.6	Cesium	585
12.7	Chromium	586
12.8	Cobalt	587
12.9	Copper	587
12.10	Iron	589
12.11	Lead	589
12.12	Manganese	589
12.13	Mercury	589
12.14	Nickel	589
12.15	Niobium	589
12.16	Rubidium	591
12.17	Ruthenium	592
12.18	Scandium	592
12.19	Selenium	592
12.20	Silver	592
12.21	Strontium	592
12.22	Tin	593
12.23	Titanium	593
12.24	Vanadium	593
12.25	Zinc	595
12.26	Literature Cited	596

Chapter 13 Concluding Remarks 599

13.1	General	599
13.2	Breadth of Coverage	600
13.3	Depth of Coverage	600
13.4	Literature Cited	602

Acknowledgments

Early work on this project was conducted at research libraries of the United States Environmental Protection Agency, the United States Department of the Interior, and the National Library of Medicine. During the past several years, all work was conducted at the National Agricultural Library (NAL) of the United States Department of Agriculture, located in Beltsville, Maryland. I thank the librarians and staff of the NAL for their assistance in procuring needed research materials.

Books by Ronald Eisler

2010. *Compendium of Trace Metals and Marine Biota. Volume 2: Vertebrates*, Elsevier, Amsterdam, 522 pp.

2007. *Eisler's Encyclopedia of Environmentally Hazardous Priority Chemicals*, Elsevier, Amsterdam, 950 pp.

2006. *Mercury Hazards to Living Organisms*, CRC Press, Boca Raton, Florida, 312 pp.

2004. *Biogeochemical, Health, and Ecotoxicological Perspectives on Gold and Gold Mining*, CRC Press, Boca Raton, Florida, 356 pp.

2000. *Handbook of Chemical Risk Assessment: Health Hazards to Humans, Plants, and Animals. Vol. 1, Metals; Vol. 2, Organics; Vol.3, Metalloids, Radiation, Cumulative Index to Chemicals and Species*, Lewis Publishers, Boca Raton, Florida, 1903 pp.

1981. *Trace Metal Concentrations in Marine Organisms*, Pergamon Press, Elmsford, New York, 687 pp.

About the Author

Ronald Eisler received the B.A. degree from New York University, and the M.S. and Ph.D. degrees from the University of Washington. As a federal research scientist, he served with the United States Department of the Interior in the Territory of Alaska (Egegik), New Jersey (Highlands), Maryland (Laurel), and Washington, DC; the United States Environmental Protection Agency in Rhode Island (Narragansett); and the United States Army Medical Service Corps in Colorado (Denver). In addition to federal service, he was a research assistant at the University of Miami Marine Laboratory (Coral Gables, Florida), a radiochemist at the Laboratory of Radiation Ecology at the University of Washington (Seattle), an aquatic biologist at the New York State Department of Environmental Conservation (Raybrook), and the senior science advisor to the American Fisheries Society (Bethesda, Maryland). Dr. Eisler participated in research and monitoring studies in the Pacific Northwest, the Territory of Alaska, Colorado, the Marshall and Marianas Islands, all along the eastern seaboard of the United States Atlantic coast, the Adirondacks region of New York, the Gulf of Aqaba in the Red Sea, and the Gulf of Mexico. Since 1955 he has authored more than 150 technical articles—including several books and 16 book chapters—mainly on contaminant hazards to plants, animals, and human health, with emphasis on trace metals. He has held a number of adjunct professor appointments and taught for extended periods at the Graduate School of Oceanography of the University of Rhode Island, and the Department of Biology of American University in Washington, DC. He also served as Visiting Professor and Resident Director of Hebrew University's Marine Biology Laboratory in Eilat, Israel. In retirement, he actively consults and writes on chemical risk assessment.

Eisler resides in Potomac, Maryland, with his wife, Jeannette, a teacher of French and Spanish.

List of Tables – Vol. 1

Table 1.1: Symbol, Atomic Number, and Atomic Weight of the Known Elements
Table 2.1: Aluminum Concentrations in Field Collections of Algae and Macrophytes
Table 2.2: Antimony Concentrations in Field Collections of Algae and Macrophytes
Table 2.3: Arsenic Concentrations in Field Collections of Algae and Macrophytes
Table 2.4: Barium Concentrations in Field Collections of Algae and Macrophytes
Table 2.5: Beryllium, Bismuth, and Boron Concentrations in Field Collections of Algae and Macrophytes
Table 2.6: Cadmium Concentrations in Field Collections of Algae and Macrophytes
Table 2.7: Cesium Concentrations in Field Collections of Algae and Macrophytes
Table 2.8: Chromium Concentrations in Field Collections of Algae and Macrophytes
Table 2.9: Cobalt Concentrations in Field Collections of Algae and Macrophytes
Table 2.10: Copper Concentrations in Field Collections of Algae and Macrophytes
Table 2.11: Gallium, Germanium, and Gold Concentrations in Field Collections of Algae and Macrophytes
Table 2.12: Iron Concentrations in Field Collections of Algae and Macrophytes
Table 2.13: Lead Concentrations in Field Collections of Algae and Macrophytes
Table 2.14: Lithium Concentrations in Field Collections of Algae and Macrophytes
Table 2.15: Manganese Concentrations in Field Collections of Algae and Macrophytes
Table 2.16: Mercury Concentrations in Field Collections of Algae and Macrophytes
Table 2.17: Molybdenum Concentrations in Field Collections of Algae and Macrophytes
Table 2.18: Nickel Concentrations in Field Collections of Algae and Macrophytes
Table 2.19: Rubidium Concentrations in Field Collections of Algae and Macrophytes
Table 2.20: Selenium and Silicon Concentrations in Field Collections of Algae and Macrophytes
Table 2.21: Silver Concentrations in Field Collections of Algae and Macrophytes

Table 2.22: Strontium Concentrations in Field Collections of Algae and Macrophytes
Table 2.23: Tin Concentrations in Field Collections of Algae and Macrophytes
Table 2.24: Titanium Concentrations in Field Collections of Algae and Macrophytes
Table 2.25: Vanadium Concentrations in Field Collections of Algae and Macrophytes
Table 2.26: Zinc Concentrations in Field Collections of Algae and Macrophytes
Table 3.1: Trace Element Concentrations in Field Collections of Protists
Table 4.1: Trace Element Concentrations in Field Collections of Sponges
Table 5.1: Aluminum, Antimony, and Arsenic Concentrations in Field Collections of Coelenterates
Table 5.2: Barium, Bismuth, and Boron Concentrations in Field Collections of Coelenterates
Table 5.3: Cadmium Concentrations in Field Collections of Coelenterates
Table 5.4: Chromium and Cobalt Concentrations in Field Collections of Coelenterates
Table 5.5: Copper Concentrations in Field Collections of Coelenterates
Table 5.6: Iron Concentrations in Field Collections of Coelenterates
Table 5.7: Lead and Lithium Concentrations in Field Collections of Coelenterates
Table 5.8: Manganese and Mercury Concentrations in Field Collections of Coelenterates
Table 5.9: Molybdenum and Nickel Concentrations in Field Collections of Coelenterates
Table 5.10: Scandium, Silicon, and Silver Concentrations in Field Collections of Coelenterates
Table 5.11: Strontium, Thorium, Tin, and Titanium Concentrations in Field Collections of Coelenterates
Table 5.12: Uranium and Vanadium Concentrations in Field Collections of Coelenterates
Table 5.13: Zinc Concentrations in Field Collections of Coelenterates
Table 6.1: Aluminum Concentrations in Field Collections of Molluscs
Table 6.2: Antimony Concentrations in Field Collections of Molluscs
Table 6.3: Arsenic Concentrations in Field Collections of Molluscs
Table 6.4: Barium, Bismuth, and Boron Concentrations in Field Collections of Molluscs
Table 6.5: Cadmium Concentrations in Field Collections of Molluscs
Table 6.6: Cesium Concentrations in Field Collections of Molluscs
Table 6.7: Chromium Concentrations in Field Collections of Molluscs
Table 6.8: Cobalt Concentrations in Field Collections of Molluscs
Table 6.9: Copper Concentrations in Field Collections of Molluscs

Table 6.10: Europium, Gallium, Gold, and Hafnium Concentrations in Field Collections of Molluscs
Table 6.11: Iron Concentrations in Field Collections of Molluscs
Table 6.12: Lead Concentrations in Field Collections of Molluscs
Table 6.13: Manganese Concentrations in Field Collections of Molluscs
Table 6.14: Mercury Concentrations in Field Collections of Molluscs
Table 6.15: Molybdenum Concentrations in Field Collections of Molluscs
Table 6.16: Nickel Concentrations in Field Collections of Molluscs
Table 6.17: Rubidium Concentrations in Field Collections of Molluscs
Table 6.18: Scandium Concentrations in Field Collections of Molluscs
Table 6.19: Selenium Concentrations in Field Collections of Molluscs
Table 6.20: Silver Concentrations in Field Collections of Molluscs
Table 6.21: Strontium Concentrations in Field Collections of Molluscs
Table 6.22: Thorium, Tin, Titanium, Tungsten, and Uranium Concentrations in Field Collections of Molluscs
Table 6.23: Vanadium Concentrations in Field Collections of Molluscs
Table 6.24: Zinc Concentrations in Field Collections of Molluscs
Table 7.1: Aluminum Concentrations in Field Collections of Crustaceans
Table 7.2: Antimony Concentrations in Field Collections of Crustaceans
Table 7.3: Arsenic Concentrations in Field Collections of Crustaceans
Table 7.4: Barium, Beryllium, and Boron Concentrations in Field Collections of Crustaceans
Table 7.5: Cadmium Concentrations in Field Collections of Crustaceans
Table 7.6: Cerium and Cesium Concentrations in Field Collections of Crustaceans
Table 7.7: Chromium Concentrations in Field Collections of Crustaceans
Table 7.8: Cobalt Concentrations in Field Collections of Crustaceans
Table 7.9: Copper Concentrations in Field Collections of Crustaceans
Table 7.10: Europium, Gallium, Germanium, and Gold Concentrations in Field Collections of Crustaceans
Table 7.11: Iron Concentrations in Field Collections of Crustaceans
Table 7.12: Lead Concentrations in Field Collections of Crustaceans
Table 7.13: Manganese Concentrations in Field Collections of Crustaceans
Table 7.14: Mercury Concentrations in Field Collections of Crustaceans

Table 7.15: Molybdenum Concentrations in Field Collections of Crustaceans
Table 7.16: Nickel Concentrations in Field Collections of Crustaceans
Table 7.17: Scandium, Selenium, and Silver Concentrations in Field Collections of Crustaceans
Table 7.18: Strontium Concentrations in Field Collections of Crustaceans
Table 7.19: Tin, Titanium, and Tungsten Concentrations in Field Collections of Crustaceans
Table 7.20: Vanadium Concentrations in Field Collections of Crustaceans
Table 7.21: Zinc Concentrations in Field Collections of Crustaceans
Table 8.1: Trace Element Concentrations in Field Collections of Marine Insects
Table 9.1: Trace Element Concentrations in Field Collections of Chaetognaths
Table 10.1: Antimony, Arsenic, Barium, and Cadmium Concentrations in Field Collections of Annelids
Table 10.2: Chromium, Cobalt, and Copper Concentrations in Field Collections of Annelids
Table 10.3: Iron, Lead, and Manganese in Field Collections of Annelids
Table 10.4: Mercury and Nickel Concentrations in Field Collections of Annelids
Table 10.5: Samarium, Scandium, Silver, and Strontium in Field Collections of Annelids
Table 10.6: Zinc Concentrations in Field Collections of Annelids
Table 11.1: Aluminum Concentrations in Field Collections of Echinoderms
Table 11.2: Antimony Concentrations in Field Collections of Echinoderms
Table 11.3: Arsenic Concentrations in Field Collections of Echinoderms
Table 11.4: Cadmium Concentrations in Field Collections of Echinoderms
Table 11.5: Chromium Concentrations in Field Collections of Echinoderms
Table 11.6: Cobalt Concentrations in Field Collections of Echinoderms
Table 11.7: Copper Concentrations in Field Collections of Echinoderms
Table 11.8: Iron Concentrations in Field Collections of Echinoderms
Table 11.9: Lead Concentrations in Field Collections of Echinoderms
Table 11.10: Manganese Concentrations in Field Collections of Echinoderms
Table 11.11: Mercury Concentrations in Field Collections of Echinoderms
Table 11.12: Nickel Concentrations in Field Collections of Echinoderms
Table 11.13: Rubidium Concentrations in Field Collections of Echinoderms
Table 11.14: Selenium Concentrations in Field Collections of Echinoderms
Table 11.15: Silver Concentrations in Field Collections of Echinoderms
Table 11.16: Strontium and Tin Concentrations in Field Collections of Echinoderms

Table 11.17: Zinc Concentrations in Field Collections of Echinoderms
Table 12.1: Antimony, Arsenic, Barium, Boron, and Cadmium Concentrations in Field Collections of Tunicates
Table 12.2: Cesium, Chromium, and Cobalt Concentrations in Field Collections of Tunicates
Table 12.3: Copper, Iron, and Lead Concentrations in Field Collections of Tunicates
Table 12.4: Manganese, Mercury, Nickel, and Niobium Concentrations in Field Collections of Tunicates
Table 12.5: Rubidium, Scandium, Selenium, and Silver Concentrations in Field Collections of Tunicates
Table 12.6: Strontium, Tin, and Titanium Concentrations in Field Collections of Tunicates
Table 12.7: Vanadium Concentrations in Field Collections of Tunicates
Table 12.8: Zinc Concentrations in Field Collections of Tunicates
Table 13.1: Trace Metals and Marine Biota: Breadth of Coverage
Table 13.2: Trace Metals and Marine Biota: Depth of Coverage

CHAPTER 1

Introduction

The first major attempt to systematically summarize all that was known about trace metal and metalloid content of marine biota was Vinogradov's classic *The Elementary Chemical Composition of Marine Organisms* (Vinogradov, 1953) and the last was *Trace Metal Concentrations in Marine Organisms* (Eisler, 1981). At that time, I recommended major revision in about 25 years owing to a growing technical literature and to a greater availability of atomic absorption spectroscopy and newer analytical techniques that accurately measure trace metal concentrations in tissues of marine plants and animals at biologically significant levels.

This volume on plants and invertebrates—and the companion volume on marine vertebrates (Eisler, 2009)—has two main objectives. The first is to summarize the available world literature on trace metal and metalloid concentrations in tissues of representative field populations of marine, estuarine, and oceanic algae and macrophytes, bacteria and protozoans, sponges, coelenterates, molluscs, arthropods, chaetognaths, annelids, echinoderms, and tunicates, and their significance to organism health and to their consumers. The database in this subject area alone has more than doubled in the past several decades. Information on the following elements are presented: aluminum, americium, antimony, arsenic, barium, beryllium, bismuth, boron, cadmium, cerium, cesium, chromium, cobalt, copper, dysprosium, erbium, gadolinium, gallium, germanium, gold, hafnium, holmium, iron, lanthanum, lead, lithium, lutetium, manganese, mercury, molybdenum, nickel, niobium, plutonium, polonium, praseodymium, protactinium, radium, rhenium, rubidium, ruthenium, samarium, scandium, selenium, silicon, silver, strontium, tantalum, technetium, tellurium, terbium, thallium, thorium, thulium, tin, titanium, tungsten, uranium, vanadium, ytterbium, yttrium, zinc, and zirconium (Table 1.1). Information on sodium, potassium, calcium, and magnesium were especially abundant, but these were excluded as their concentrations in marine organisms were almost always in excess of the 100.0 mg/kg dry weight limit that I set arbitrarily as a trace concentration.

The second objective is to synthesize selected information on biological, chemical, and physical factors known to modify uptake, retention, and translocation of each element by representative groups of marine plants and invertebrates under field and laboratory

Table 1.1: Symbol, Atomic Number, and Atomic Weight of the Known Elements

Element	Symbol	Atomic Number	Atomic Weight
Actinium	Ac	89	227[a]
Aluminum	Al	13	26.98
Americium	Am	95	243[a]
Antimony	Sb	51	121.76
Argon	Ar	18	39.95
Arsenic	As	33	74.92
Astatine	At	85	210[a]
Barium	Ba	56	137.33
Berkelium	Bk	97	247[a]
Beryllium	Be	4	9.01
Bismuth	Bi	83	208.98
Bohrium	Bh	197	264[a]
Boron	B	5	10.81
Bromine	Br	35	79.90
Cadmium	Cd	48	112.41
Calcium	Ca	20	40.01
Californium	Cf	98	251[a]
Carbon	C	6	12.01
Cerium	Ce	58	140.12
Cesium	Cs	55	132.90
Chlorine	Cl	17	35.45
Chromium	Cr	24	52.00
Cobalt	Co	27	58.93
Copper	Cu	29	63.54
Curium	Cm	96	247[a]
Dubnium	Db	105	262[a]
Dysprosium	Dy	66	162.50
Einsteinium	Es	99	252[a]

(Continues)

Table 1.1: Cont'd

Element	Symbol	Atomic Number	Atomic Weight
Erbium	Er	68	167.26
Europium	Eu	63	151.96
Fermium	Fm	100	257[a]
Fluorine	F	9	19.00
Francium	Fr	87	223[a]
Gadolinium	Gd	64	157.25
Gallium	Ga	31	69.72
Germanium	Ge	32	72.61
Gold	Au	79	196.97
Hafnium	Hf	72	178.49
Hassium	Hs	108	269[a]
Helium	He	2	4.00
Holmium	Ho	67	164.93
Hydrogen	H	1	1.01
Indium	In	49	114.82
Iodine	I	53	126.90
Iridium	Ir	77	192.22
Iron	Fe	26	55.85
Krypton	Kr	36	83.80
Lanthanum	La	57	138.91
Lawrencium	Lr	103	262[a]
Lead	Pb	82	207.20
Lithium	Li	3	6.94
Lutetium	Lu	71	174.97
Magnesium	Mg	12	24.31
Manganese	Mn	25	54.94
Meitnerium	Mt	109	268[a]
Mendelevium	Md	101	258[a]

(Continues)

Table 1.1: Cont'd

Element	Symbol	Atomic Number	Atomic Weight
Mercury	Hg	80	200.59
Molybdenum	Mo	42	95.94
Neodymium	Nd	60	144.24
Neon	Ne	10	20.18
Neptunium	Np	93	237[a]
Nickel	Ni	28	58.69
Niobium	Nb	41	92.91
Nitrogen	N	7	14.01
Nobelium	No	102	259[a]
Osmium	Os	76	190.23
Oxygen	O	8	16.00
Palladium	Pd	46	106.42
Phosphorus	P	15	30.97
Platinum	Pt	78	195.08
Plutonium	Pu	94	244[a]
Polonium	Po	84	210[a]
Potassium	K	19	39.10
Praseodymium	Pr	59	140.91
Promethium	Pm	61	145[a]
Protactinium	Pa	91	231.04
Radium	Ra	88	226[a]
Radon	Rn	86	222[a]
Rhenium	Re	75	186.21
Rhodium	Rh	45	102.91
Rubidium	Rb	37	85.47
Ruthenium	Ru	44	101.07
Rutherfordium	Rf	104	261[a]
Samarium	Sm	62	150.36

(*Continues*)

Table 1.1: Cont'd

Element	Symbol	Atomic Number	Atomic Weight
Scandium	Sc	21	44.96
Seaborgium	Sg	106	266[a]
Selenium	Se	34	78.96
Silicon	Si	14	28.09
Silver	Ag	47	107.87
Sodium	Na	11	23.00
Strontium	Sr	38	87.62
Sulfur	S	16	32.07
Tantalum	Ta	73	180.95
Technetium	Tc	43	98[a]
Tellurium	Te	52	127.6
Terbium	Tb	65	158.92
Thallium	Tl	81	204.38
Thorium	Th	90	232.04
Thulium	Tm	69	168.93
Tin	Sn	50	118.71
Titanium	Ti	22	47.88
Tungsten	W	74	183.84
Ununbium	Unb	112	277[a]
Ununennium	Uue	119	–
Ununhexium	Uuh	116	–
Ununnilium	Uun	110	269[a]
Ununoctiium	Uuo	118	–[b]
Ununpentium	Uup	115	–[b]
Ununquadium	Uuq	114	–[b]
Ununseptium	Uus	117	–[b]
Ununtriium	Uut	113	–[b]
Unununium	Unn	111	272[a]

(Continues)

Table 1.1: Cont'd

Element	Symbol	Atomic Number	Atomic Weight
Uranium	U	92	238.03
Vanadium	V	23	50.94
Xenon	Xe	54	131.29
Ytterbium	Yb	70	173.04
Yttrium	Y	39	88.91
Zinc	Zn	30	65.39
Zirconium	Zr	40	91.22

[a]Most stable or best known isotope.
[b]Yet to be reported.

conditions. Recognition of the importance of these modifiers and their accompanying interactions is essential to the understanding of metals kinetics in marine systems and to the interpretation of baseline residue data in marine biota. In many cases, the relation between concentrations of these elements in tissues had little relation to concentrations of the same element in the organisms's immediate biogeophysical environment including sediments, sediment interstitial waters, diet, and water column. The reasons for this are explored, and their role examined in formulation of proposed criteria by regulatory agencies to protect sensitive natural resources and their consumers. It is emphasized that major changes are now being recorded in global climate extremes and in the amounts of metals, metalloids, and other contaminants discharged into the biosphere as a result of human activities; these changes, and others, may ultimately render obsolete certain terms—now used liberally in this volume—such as "controls," "reference site," and "environmentally pristine area."

The organization of this book is similar to that of my earlier work on trace metal concentrations in marine biota (Eisler, 1981): chapters are arranged in evolutionary order starting with most primitive to most advanced; within each chapter, metals and metalloids are arranged in alphabetical order; and, finally, all concentrations are listed in milligrams per kilogram (mg/kg = parts per million) on a fresh weight (FW), dry weight (DW), or ash weight (AW) basis. In all tables, concentrations shown in parentheses represent the range of values documented; others, the means.

1.1 Literature Cited

Eisler, R., 1981. Trace Metal Concentrations in Marine Organisms. Pergamon, Elmsford, NY, 687 pp.
Eisler, R., 2009. Compendium of Trace Metals and Marine Biota. Volume 2: Vertebrates. Elsevier, Amsterdam.
Vinogradov, A.P., 1953. The elementary chemical composition of marine organisms. Sears Foundation for Marine Research, Memoir 2. Yale University, New Haven, CT, 647 pp.

CHAPTER 2
Algae and Macrophytes

This group is ubiquitous in aquatic environments with distribution usually limited only by available light and nutrients. It is generally acknowledged that photosynthetic plants form the base of marine food pyramids and that highest biomass is linked to our most productive fisheries, especially biomass of unicellular algae. Continuing interest in the role of algae and macrophytes as indicators of trace metal contaminants, their role in transferring these substances to other environmental compartments, and their potential in removing toxic metals from impacted ecosystems, is reflected by a growing technical literature on this subject.

2.1 Aluminum

Aluminum is common in the aquatic environment, but is comparatively harmless to biota under circumneutral or alkaline conditions typical of marine ecosystems (Sparling and Lowe, 1996). Relatively high concentrations of aluminum were evident in marine plants, with many species containing more than 200.0 mg Al/kg dry weight (DW) whole plant and some more than 3000.0 mg Al/kg DW (Table 2.1). More than 90% of the total aluminum was associated with water insoluble and water soluble organic fractions from the cell wall (Bohm, 1972).

Laboratory studies with the euryhaline alga *Suaeda maritima* showed that aluminum accumulations were lower and growth better among saline-reared algae (Yeo and Flowers, 1977). In saline waters, growth was stimulated at 1.0 mg Al/L; after 14 days of exposure, roots contained 0.75 mg Al/kg DW and shoots 0.24 mg Al/kg DW. In nonsaline waters, growth was inhibited under the same dose-time regimen; roots contained 1.59 mg Al/kg DW after 14 days and shoots 0.24 mg Al/kg DW. Higher concentrations of 4.0-16.0 mg Al/L medium inhibited growth, regardless of ambient salinity. After 14 days, *Suaeda* exposed in saline waters to 16.0 mg Al/L had 4.1 and 0.4 mg Al/kg DW in roots and shoots, respectively. *Suaeda* immersed for 14 days in fresh water containing 16.0 mg Al/L had elevated levels of 36.7 mg Al/kg DW in roots, and 3.7 mg Al/kg DW in shoots (Yeo and Flowers, 1977).

Table 2.1: Aluminum Concentrations in Field Collections of Algae and Macrophytes

Organism	Concentration	Reference[a]
Algae, whole	34.0-4200.0 DW	1, 2
Algae, whole, 15 species		
12 species	118.0-490.0 DW	3
3 species	330.0-1750.0 DW	3
Alga, *Enteromorpha* sp.		
Chesapeake Bay, Maryland	22,830.0 DW	6
San Francisco Bay, California	16,020.0 DW	6
Phytoplankton; organic fractions vs. siliceous frustules		
No titanium group	110.0 DW vs. 620.0 DW	4
Titanium group	444.0 DW vs. 2550.0 DW	4
Strontium-concentrated group	38.0 DW vs. 80.0-110.0 DW	4
Sargassum spp., whole	33.0-13,450.0 DW	7
Seaweeds, whole, 30 species	533.0 (57.0-3290.0) DW	5

Values are in mg Al/kg dry weight (DW).
[a] 1, Culkin and Riley, 1958; 2, Tijoe et al., 1977; 3, Riley and Roth, 1971; 4, Martin and Knauer, 1973; 5, Yamamoto et al., 1976; 6, as quoted in Sparling and Lowe, 1996; 7, Trefry and Presley, 1976.

2.2 Americium

Accumulation of americium in marine vegetation is similar to that of plutonium; however, some marine plants, such as *Sargassum* and *Fucus*, selectively accumulate plutonium over americium, while others, such as *Desmaidetta*, *Daisia*, and *Chondrus*, accumulate americium over plutonium (Livingston and Bowen, 1976). The reasons for this are not known.

2.3 Antimony

Algae reportedly concentrate the antimony content of the surrounding seawater by factors that range from 110 to 340 (Van As et al., 1973). However, the highest antimony levels recorded in whole marine plants were 0.47 mg Sb/kg on a fresh weight (FW) basis and 2.5 mg/kg on a DW basis (Table 2.2).

Antimony trioxide is at least six times more toxic to freshwater algae than marine algae. For example, the marine alga *Skeletonema costatum* had normal chlorophyll *a* production and normal growth during 96-h exposure to antimony trioxide concentrations less than

Table 2.2: Antimony Concentrations in Field Collections of Algae and Macrophytes

Organism	Concentration	Reference[a]
Algae, whole; 6 species	0.02-0.08 DW	1
Macrophytes; whole; 5 species	0.08-0.47 FW	2
Macrophytes; whole; 3 species	0.07-0.22 FW	5
Macrophytes; whole; 4 spp.; Strait of Magellan, Chile; 2000-2001	<0.2-1.98 DW	6

Seaweeds
  Whole, 5 species
  Whole
	
	0.08-0.24 FW
	0.05-2.5 DW
		
		3
		4

Values are in mg Sb/kg fresh weight (FW) or dry weight (DW).
[a]1, Leatherland and Burton, 1974; 2, Van As et al., 1973; 3, Van As et al., 1975; 4, Lunde, 1970; 5, Strohal et al., 1975; 6, Astorga-Espana et al., 2008.

4.2 mg Sb/L, whereas adverse effects in these variables were observed in freshwater algae during 96-h exposure to 0.01-0.63 mg Sb/L (USEPA, 1989).

2.4 Arsenic

Field collections of algae and macrophytes show considerable variation in arsenic content, with known modifiers that include proximity to arsenic mining activities, arsenic valence state, fractions analyzed, and geographic location (Table 2.3). Arsenosugars were the most abundant arsenic species in 8 of 10 species of Adriatic Sea littoral zone algae sampled in 2004; however, *Cystoseira* and *Ceramium* (the two species that contained the highest concentrations of total arsenic at 17.6 and 28.1 mg/kg FW, respectively) contained high amounts of mainly inorganic arsenic (Slejkovec et al., 2006). Total arsenic concentrations in the Antarctic region macroalgae ranged from 5.8 mg/kg DW in *Myriogramme* sp. to 152.0 mg/kg DW in *Himantothallus grandfolius* (Farias et al., 2007); in general, total arsenic concentrations in brown algae were higher than those in red algae, and this confirms observations in non-Antarctic species of macroalgae (Almela et al., 2006; Cullen and Reimer, 1989).

Arsenate reduction to arsenite in seawater is dependent on pentavalent phosphorus in solution and algal biomass (Johnson and Burke, 1978). During algal growth, as phosphate is depleted and the P^{5+}/As^{5+} ratio decreases, arsenate reduction occurs at an increasing rate. The resultant arsenite (As^{3+}), after an initial peak, is rapidly oxidized to arsenate, suggesting the biological catalysis of oxidation as well as mediation of arsenate reduction. Although most investigators agree that arsenite is more toxic than arsenate to higher organisms, it seems

Table 2.3: Arsenic Concentrations in Field Collections of Algae and Macrophytes

Organism	Concentration	Reference[a]
Adriatic Sea; 2004; littoral zone; total arsenic		
Ulva sp.	1.4 FW	14
Enteromorpha sp.	1.4 FW	14
Padina sp.	1.9 FW	14
Gelidium spp.	1.7-2.4 FW	14
Fucus sp.	8.4-12.1 FW	14
Cystoseira sp.	17.6 FW	14
Ceramium sp.	28.1 FW	14
Algae, whole		
Green	0.5-5.0 DW	1
Brown	Max. 30.0 DW	1
11 species	2.0-58.0 DW	2
Various species	10.0-100.0 DW	3
Algae; whole; 2 species; near arsenic waste outfall vs. sites 5-300 km distant	17.2 DW vs. 9.8-12.1 DW	7
Antarctic region macroalgae; whole; summer 2002; total arsenic vs. inorganic arsenic		
Brown algae; 6 species	33.0-152.0 DW vs. 0.25-0.84 DW	15
Red algae; 3 species	5.8-28.0 DW vs. 0.12-0.55 DW	15
Argentina; 3 sites; summer 2002; *Porphyra columbina* vs. *Ulva* sp.; whole	22.9-33.8 DW vs. 3.0-5.6 DW	12
Seaweed, *Chondrus crispus*	5.2 DW	1
Seaweed, *Fucus* sp.; whole; arsenic-contaminated site at Seal Harbour, Nova Scotia vs. reference site		
Total arsenic	23.0-43.0 FW vs. 6.0-10.0 FW	11
Inorganic arsenic		
Trivalent arsenic	3.1 FW vs. 0.31 FW	11
Pentavalent arsenic	10.1 FW vs. 0.62 FW	11
Bioaccessible fraction	18.0-29.0 FW vs. 0.7-1.0 FW	11
Alga, *Fucus* sp.; oil vs. fatty acid	6.0-27.0 FW vs. 5.0-6.0 FW	1
Brown alga, *Fucus vesiculosus*		
Whole	94.0 DW	1
Whole	35.2-80.0 DW	4
Oil	155.0-221.0 DW	1

(Continues)

Table 2.3: Cont'd

Organism	Concentration	Reference[a]
Fatty acid	8.0-36.0 DW	1
Shoalgrass, *Halodule wrightii*; rhizomes; south Texas, Lower Laguna Madre; 1986-1987	12.2 (2.0-25.0) DW	8
Brown alga, *Laminaria digitata*		
Whole	94.0 DW	1
Whole	42.0-50.0 DW	4
Oil	155.0-221.0 DW	1
Fatty acid	8.0-36.0 DW	1
Alga, *Laminaria hyperborea*; total arsenic vs. organic arsenic	142.0 DW vs. 139.0 DW	2
Seaweed, *Padina* sp.; Guam; June 1998-1999; fronds	5.3-32.2 DW; max. 38.1 DW	9
Seagrass, *Posidonia oceanica*; Mediterranean Sea, Spain; 2001-2003; leaves	0.69 (0.2-1.6) DW	13
Sargassum weed, *Sargassum fluvitans*		
Total arsenic	19.5 FW	5
As^{3+}	1.8 FW	5
As^{5+}	17.7 FW	5
Organoarsenicals	0.2 FW	5
Seaweed, *Sargassum* sp.		
Total arsenic	4.1-8.7 FW	6
As^{3+}	0.14-0.35 FW	6
As^{5+}	1.9-7.3 FW	6
Organoarsenicals	Max. 0.1 FW	6
Seaweeds		
Whole	3.8-93.8 DW	1
Whole	10.0-109.0 DW	2
Oil fraction	5.7-221.0 FW	2
Eelgrass, *Zostera marina*; Oregon		
Leaf	4.6 DW	10
Rhizome	4.8 DW	10
Root	12.0 DW	10

Values are in mg As/kg fresh weight (FW) or dry weight (DW).
[a] 1, NAS, 1977; 2, Eisler, 1981; 3, Pershagen and Vahter, 1979; 4, Jenkins, 1980a; 5, Woolson, 1975; 6, NRCC, 1978; 7, Penrose et al., 1975; 8, Custer and Mitchell, 1993; 9, Denton et al., 2006; 10, Kaldy, 2006; 11, Koch et al., 2007; 12, Perez et al., 2007; 13, Fourqurean et al., 2007; 14, Slejkovec et al., 2006; 15, Farias et al., 2007.

that arsenate has a more profound effect than arsenite on growth and morphology of marine algae. It is suggested that marine algae erect a barrier against the absorption of arsenite, but not arsenate; within the cell, arsenate could be reduced to the comparatively toxic arsenite (Johnson and Burke, 1978). Thus, *Tetraselmis chui* and *Hymenomonas carterae* cultured in media containing various concentrations of arsenate or arsenite took up arsenate that was partially released by both species; differences between rates of uptake and release of arsenicals suggest that arsenate undergoes chemical changes after incorporation into algal cells (Bottino et al., 1978).

Methylated arsenicals such as monomethyl arsonate (MMA) and dimethylarsinate (DMA) are about 1000 times less toxic than inorganic arsenites and arsenates to *Chlorella salina*; uptake of organoarsenicals is similarly reduced (Karadjova et al., 2008). Seawater enrichment with phosphate—up to 1.3 mg P/L—reduced both intracellular content and toxicity due to inorganic trivalent and pentavalent species, although MMA and DMA were unaffected. The release of inorganic arsenate and arsenite, together with bioreduction of arsenate and subsequent methylation of inorganic arsenite, may be a detoxification mechanism for this alga (Karadjova et al., 2008).

In Charlotte Harbor, Florida, a region that has become phosphate enriched due to agricultural activity, virtually all of the arsenic taken up by phytoplankton was biomethylated and returned to the estuary, usually as methylarsonic acid and dimethylarsenic acid (Froelich et al., 1985). The ability of marine phytoplankton to methylate arsenic and release the products to a surrounding environment varies between species and even within a particular species in relation to their possession of necessary methylating enzymes (Sanders, 1985). The process involved in detoxifying arsenate after its absorption by phytoplankton are not firmly established, but seem to be nearly identical in all plants, suggesting a similar evolutionary development (Benson, 1984). Like phosphates and sulfates, arsenate may be fixed with ADP, reduced to the arsonous level, and successfully methylated and adenosylated, ultimately producing the 5-dimethylarsenosoribosyl derivatives accumulating in algae (Benson, 1984). The former kelp industry is considered a probable cause of present day arsenic contamination in remote Scottish islands (Riekie et al., 2006). Burning seaweed to produce kelp was in years past a significant industry in remote coastal areas of Scotland and elsewhere, and may have caused arsenic contamination of soils to 10.7 mg As/kg soil versus 1.7 mg As/kg in reference areas. From 1764 to 1772, an estimated 4900 metric tons of kelp was produced commercially in Scotland, and in 1770-1820 another 2500 tons of kelp was produced on Orkney. Seaweeds containing up to 100.0 mg As/kg DW were burned, transforming low-toxicity arsenosugars to highly toxic inorganic forms, mainly arsenate (Riekie et al., 2006).

The ability of marine phytoplankton to accumulate high concentrations of inorganic arsenicals and transform them to methylated arsenicals that are later efficiently transferred in the food chain is well documented (Benson, 1984; Freeman, 1985; Froelich et al., 1985;

Irgolic et al., 1977; Maeda et al., 1985; Matsuto et al., 1984; Norin et al., 1985; Sanders, 1985; Yamaoka and Takimura, 1986). In a three-step food chain consisting of *Dunaliella marina* (alga), brine shrimp *Artemia salina* (grazer), and a carnivorous shrimp *Lysmata seticaudata*, organic forms of arsenic are derived from an *in vivo* synthesis by *Dunaliella* and efficiently transferred along the food chain (Wrench et al., 1979). Arsenic is taken up from seawater by a variety of organisms, but there is no evidence of magnification along the aquatic food chain (Hallacher et al., 1985; Lindsay and Sanders, 1990; NAS, 1977; NRCC, 1978; Woolson, 1975). In a marine ecosystem based on the alga *Fucus vesiculosus*, arsenate at 7.5 mg As^{5+}/L was taken up by all biota. After 3 months, arsenic was accumulated most efficiently by *Fucus* (120.0 mg/kg DW in apical fronds) and filamentous algal species; little or no accumulation occurred in invertebrates, although arsenic seemed to be retained by gastropods and mussels (Rosemarin et al., 1985). Microcosms of a Delaware cordgrass (*Spartina alterniflora*), salt marsh exposed to elevated levels of As^{5+}, showed that virtually all arsenic was incorporated into plant tissue or strongly sorbed to cell surfaces (Sanders and Osman, 1985). In general, arsenic concentrations are significantly higher in marine plants and animals than in freshwater counterparts because of their ability to accumulate arsenic from seawater or food sources, and not because of localized arsenic pollution (Eisler, 2000; Maher, 1985). The great majority of arsenic in marine biota exists as water-soluble and lipid-soluble organoarsenicals that include arsenolipids, arsenosugars, arsenocholine, arsenobetaine [$(CH_3)_3AsCH_2COOH$], MMA [$CH_3AsO(OH)_2$], DMA [$(CH_3)_2AsO(OH)$], and other forms (Edmonds et al., 1993). Arsenosugars (arsenobetaine precursors) are the dominant arsenic species in brown kelp, *Ecklonia radiata*, and certain molluscs and crustaceans (Eisler, 2000).

Adverse effects of arsenites are reported in marine plants at 0.019-0.58 mg As^{3+}/L and of arsenates at 0.013-100.0 mg As^{5+}/L. Among the most sensitive of all marine species of plants and animals tested were three species of algae, which had reduced growth during exposure to 0.019-0.022 mg As^{3+}/L (USEPA, 1985a).

Arrested spore development was documented in the red alga *Plumaria elegans* 7 days after exposure to 0.58 mg As^{3+}/L for 18 h (Sanders, 1986). Normal sexual reproduction was documented in the red alga *Champia parvula* at concentrations <0.065 mg As^{3+}/L; however, sexual reproduction ceased at 0.095 mg As^{3+}/L and death occurred at 0.3 mg As^{3+}/L (Thursby and Steele, 1984). For *Champia* and pentavalent arsenic, normal growth—but no sexual reproduction—was observed at 10.0 mg As^{5+}/L (Thursby and Steele, 1984). Chronic studies with mass cultures of natural phytoplankton communities exposed to low levels of arsenate (0.001-0.015 mg As^{5+}/L) showed that As^{5+} differentially inhibits certain plants, causing a marked change in species composition, succession, and predator-prey relations; the significance of these changes on carbon transfer between trophic levels is unknown (Sanders, 1986; Sanders and Cibik, 1985). Growth inhibition occurred in *S. costatum* at

0.013 mg As^{5+}/L (USEPA, 1985a) and reduced chlorophyll *a* production was observed in *Thalassiosira aestivales* at 0.075 mg As^{5+}/L (USEPA, 1985a). *Sargassum* communities seemed normal during exposure to 0.028 mg As^{5+}/L (Blake and Johnson, 1976). Results from laboratory studies with five species of euryhaline algae grown in freshwater or in seawater show that all five species synthesize fat-soluble and water-soluble organoarsenicals from both arsenate and arsenite. Arsenic concentration factors over seawater among the five species ranged from 200 to 3000, with accumulations highest in lipid phases (Lunde, 1973).

2.5 Barium

The maximum barium concentrations in whole marine plants were 12.0 mg Ba/kg FW and 287.0 mg/kg DW (Table 2.4). The highest concentration of 287.0 mg Ba/kg DW was measured in a phytoplankton organic fraction containing high strontium content (Martin and Knauer, 1973), emphasizing the importance of interactions.

2.6 Beryllium

Beryllium concentrations in marine algae ranged from a low of 1.1 mg Be/kg DW in *Dunaliella tertiolecta* to 8.4 mg Be/kg DW in *Heteromastix longifillis* (Riley and Roth, 1971; Table 2.5).

Table 2.4: Barium Concentrations in Field Collections of Algae and Macrophytes

Organism	Concentration	Reference[a]
Algae, whole, 15 species		
9 species	34.5-76.0 DW	1
5 species	80.0-145.0 DW	1
1 species	262.0 DW	1
Algae, whole		
Chlorophycae	0.09-0.44 FW	2
Phaeophyceae	0.05-0.38 FW	2
Rhodophyceae	0.05-0.36 FW	2
Phytoplankton, organic reactions		
No titanium group	33.0 DW	3
With titanium group	19.0 DW	3
Strontium-concentrated group	287.0 DW	3
Seaweeds, whole, 5 species		
May	13.0-44.0 DW	4
June	16.0-64.0 DW; 3.1-12.0 FW	4

Values are mg Ba/kg fresh weight (FW) or dry weight (DW).
[a] 1, Riley and Roth, 1971; 2, Mauchline and Templeton, 1966; 3, Martin and Knauer, 1973; 4, Black and Mitchell, 1952.

Table 2.5: Beryllium, Bismuth, and Boron Concentrations in Field Collections of Algae and Macrophytes

Element and Organism	Concentration	Reference[a]
Beryllium		
Algae, whole, 14 species		
6 species	1.1-2.0 DW	1
4 species	2.1-3.0 DW	1
2 species	3.1-4.0 DW	1
2 species	5.7-8.4 DW	1
Bismuth		
Algae, whole, 6 species	2.0-7.7 DW	1
Boron		
Seaweeds, whole, 41 species	106.0 (16.0-319.0) DW	2
Algae	4.0-120.0 DW	3
Argentina; summer 2002; 3 sites; *Porphyra columbina* vs. *Ulva* sp.; whole	12.6-16.5 DW vs. 37.9-58.3 DW	4

Values are in mg element/kg dry weight (DW).
[a] 1, Riley and Roth, 1971; 2, Yamamoto et al., 1971; 3, Jenkins, 1980b; 4, Perez et al., 2007.

2.7 Bismuth

Bismuth concentrations in field collections of marine algae ranged from 2.0 mg Bi/kg DW in *Monochrysis lutheri* to 7.7 mg Bi/kg DW in *Chlamydomonas* sp. (Riley and Roth, 1971; Table 2.5).

2.8 Boron

Boron content among 41 species of seaweeds collected from Wakayama Prefecture, Japan, ranged from 16.0 to 319.0 mg B/kg DW (Table 2.5).

Boron is an essential trace element for the diatom *Cylindrotheca fusiformis* (Lewin and Chen, 1976) and probably other marine plants. Boron interactions with water temperature, nitrates, phosphates, and silicates are important governors of phytoplankton growth (Rao, 1981). At 30.0 mg B/L, at low temperatures and high concentrations of nitrates, phosphates, and silicates, marine phytoplankton show increased primary production and carbon assimilation. At 30.0 mg B/L, under conditions of high temperatures and low nutrients, photosynthesis is inhibited up to 62% (Rao, 1981). *Cylindrotheca* grown in boron-deficient medium of 0.02 mg B/L had a generation time of 90 h compared to only 11 h when cultured in complete medium containing 0.5 mg B/L (Lewin and Chen, 1976). Boron-deficient *Cylindrotheca* accumulate rubidium, phenolic compounds, nitrates, and phosphates, and show increased activity of various enzymes, especially glucose-6-phosphate dehydrogenase; however, respiratory adjustment is negligible until nutrient stress becomes irreversible in about 48 h (Smyth and Dugger, 1980, 1981).

Phytoplankton are comparatively resistant to high boron concentrations, and no measurable effects were observed during exposure to media containing 5.0-10.0 mg B/L; however, growth inhibition of 25% was documented at 50.0 mg B/L, and about 40% were killed at 100.0 mg B/L (Antia and Cheng, 1975).

2.9 Cadmium

The grossly elevated levels of 22.0-220.0 mg Cd/kg DW in algae (Table 2.6) were almost always directly associated with cadmium pollution from human activities.

The ability of algae and macrophytes to accumulate cadmium from ambient seawater is well documented. Concentration factors ranged from 4200 to 11,000 for field collections (Melhuus et al., 1978) and from 11 to 36 in laboratory studies with radiocadmium (Hiyama and Shimizu, 1964) and stable cadmium (Sivalingam, 1978b). Marine algae accumulated more cadmium at lower salinities than at higher salinities; algae bound only free cadmium ions, and these were inversely related to the concentration of suspended particulate matter (Favero et al., 1996). *Ulva reticulata* showed a 36-fold increase in 48 h over seawater containing 50.0 mg Cd/L (Sivalingam, 1978b). Others have shown that diatoms held in seawater containing 1.0 mg Cd/L for 12 days had 8300.0 mg Cd/kg FW (Cossa, 1976), *S. alterniflora* seedlings reared in media containing 100.0 mg Cd/L for 8 weeks contained 94.0 mg Cd/kg FW (Dunstan et al., 1975), and 85-95% uptake was measured among two algal species held in 1.8 mg Cd/L for 24 h (Maclean et al., 1972). Up to 80% of accumulated cadmium was retained for at least 13 days in *Ulva* sp. (Ueda et al., 1978). Sporophytes of the kelp, *Laminaria saccharina*, when placed in cadmium-seawater solutions begin to accumulate cadmium almost immediately without apparent regulation (Markham et al., 1980). Saturation was not reached within 6 days suggesting the potential for high accumulations. Slower growing plants and slower growing regions of the thallus, such as the stipe, holdfast, and distal blade regions, accumulated more cadmium than did other portions of the plant (Markham et al., 1980). The sea rush, *Juncus maritimus*, a saltmarsh plant, accumulated cadmium in roots continuously throughout the year in the Douro River estuary in Portugal, and is considered a major photostabilizer in cadmium regulation in that area (Almeida et al., 2006).

Cadmium-copper antagonism in seaweeds inhabiting coastal Chile affected by copper mine wastes is reported (Andrade et al., 2006). High copper accumulations in various species of

Table 2.6: Cadmium Concentrations in Field Collections of Algae and Macrophytes

Organism	Concentration	Reference[a]
Algae, whole		
Chlorophycae; 5 spp.	8.3-16.3 DW	1
Phaeophyceae; 4 spp.	5.6-13.8 DW	1
Rhodophyceae, 10 spp.	3.7-13.2 DW	1
Rio Tinto estuary, Spain	Max. 7.4 DW	2
West Norway coast	Max. 20.0 DW	20
Israel; Mediterranean Sea coast; 2 spp.	0.9-2.1 DW	3
United Kingdom, Solant region; 2 spp.	0.3-0.4 DW	4
Argentina; 3 sites; summer 2002; *Porphyra columbina* vs. *Ulva* sp.; whole	2.8-3.5 DW vs. 0.2-1.0 DW	27
Brown alga, *Ascophyllum nodosum*; whole		
Sorfjorden, Norway	6.0-15.0 DW	9
Eikhamrane, Norway	3.5-5.7 DW	10
Flak, Norway	<1.0 DW	10
Transferred from Eikhamrane to Flak for 120 days	<1.0-4.0 DW	10
Menai Straits, United Kingdom	1.8 DW	11
Dulas Bay, United Kingdom	1.5 DW	11
Lofoten, Norway	<0.7 DW	12
Trondheimsfjord, Norway	<0.7-1.0 DW	12
Hardangerfjord, Norway	0.7-16.0 DW	12
Coast of Ireland; 1999-2000	0.1-0.6 DW	33
Green alga, *Cladophora* sp.; 2000-2003; Baltic Sea		
Gulf of Gdansk	0.29 (0.08-0.62) DW	28
Vistula Lagoon	<0.1 DW	28
Reference site	0.19 DW; max. 0.39 DW	28
Green alga, *Enteromorpha* sp.; whole; 2000-2003; Gulf of Gdansk, Baltic Sea	0.44 (0.03-1.08) DW	24
Alga, *Skeletonema costatum*, whole; water content 0.01 mg Cd/L	4.0 DW	17
Fucus distichus, whole	0.7-1.8 DW	13

(Continues)

Table 2.6: Cont'd

Organism	Concentration	Reference[a]
Fucus serratus, whole		
Sorfjorden, Norway	22.0-29.0 DW	9
Portland, Dorset, United Kingdom	0.3 DW	4
Severn estuary, United Kingdom	53.0 DW	9
Fucus vesiculosus, whole		
Sorfjorden, Norway	8.6-10.6 DW	9
Tamar estuary, United Kingdom	4.8 (1.8-9.0) DW	14
Menai Straits, United Kingdom	2.1 DW	11
Dulas Bay, United Kingdom	1.8 DW	11
Severn estuary, United Kingdom	220.0 DW	15
Irish Sea	1.4 DW	16
Halophytes; 2 spp.; leaves, Tamil Nadu, India; anuary 2007	<0.31 DW	38
Macroalgae (6 spp). vs. phytoplankton; Gulf of California, Mexico; 1998-1999; whole	0.13-0.87 DW vs. 0.27 DW	25
Macroalgae; 3 spp.; epiphytic on estuarine mangrove aerial roots; Sydney region, Australia; sites	Not detectable	29
Mangroves; Gulf of California, Mexico; 1998-1999; twigs vs. leaves		
Rhizophora mangle	0.59 vs. 0.17 DW	25
Avicennia germinans	0.12 DW vs. 0.1 DW	25
Laguncularia racemosa	0.29 DW vs. 0.25 DW	25
Mexico; Gulf of California; 1999-2000		
Microalgae, *Coscinodiscus centralis*; whole	0.27 DW	35
Macroalgae; *Gracilaria* sp. vs. *Polysiphonia* sp.; whole	0.23 DW vs. 0.87 DW	35
Mangroves; leaves		35
Rhizophora mangle	0.17 DW	35
Avicennia germinans	0.1 DW	35
Laguncularia racemosa	0.25 DW	35
Seaweed, *Padina* sp.; Guam; 1998-1999; fronds	0.07-0.26 DW	21

(Continues)

Table 2.6: Cont'd

Organism	Concentration	Reference[a]
Macrophyte, *Posidonia oceanica*		
Italy		
Sheath	0.6-2.0 DW	23
Rhizome	0.8-2.4 DW	23
Mediterranean Sea; blade; summers 2004-2005	2.1-5.4 DW	31
Corsica, France; summer 2003; adult leaves; blades vs. sheaths	2.8 (1.5-4.0) DW vs. 1.0 (0.7-1.7) DW	34, 36
Mangrove, *Rhizophora mangle*; Natal, Brazil; summer 2001; leaves	1.1-1.4 DW	3
Saltmarsh grass, *Spartina alterniflora*		
Whole	0.6 DW	18
Leaves and stalks	0.5 DW	17, 19
Seaweed, *Ulva lactuca*		
Gulf of California, Mexico; 2002; whole		
Summer	1.3-1.5 DW	26
Winter	2.6 DW	26
India; 2004-2005; industrial area	38.1 (21.7-59.6) DW	32
Seaweeds		
Australia, 3 spp.	0.03-85.0 DW	6
Japan, 5 spp.	0.1-0.3 DW	5
Korea		
17 spp.	0.02-1.5 DW	7
20 spp.	0.2-3.4 DW	8
Eelgrass, *Zostera marina*		
Baja California, Mexico; March 2000; reference site vs. salt mining site; whole	1.6 (0.03-3.4) DW vs. 6.2 (0.008-66.0) DW	37
Oregon		
Leaf	1.7 DW	22
Rhizome	0.8 DW	22
Root	6.4 DW	22

Values are in mg Cd/kg dry weight (DW) or fresh weight (FW).

[a] 1, Sivalingam, 1978b; 2, Stenner and Nickless, 1975; 3, Roth and Hornung, 1977; 4, Leatherland and Burton, 1974; 5, Ishibashi et al., 1964a; 6, Harris et al., 1979; 7, Kim, 1972; 8, Pak et al., 1977; 9, Melhuus et al., 1978; 10, Myklestad et al., 1978; 11, Foster, 1976; 12, Haug et al., 1974; 13, Bohn, 1979; 14, Bryan and Uysal, 1978; 15, Butterworth et al., 1972; 16, Preston et al., 1972; 17, Windom et al., 1976; 18, Dunstan et al., 1975; 19, Windom, 1975; 20, Stenner and Nickless, 1974; 21, Denton et al., 2006; 22, Kaldy, 2006; 23, Baroli et al., 2001; 24, Zbikowski et al., 2006; 25, Ruelas-Inzunza and Paez-Osuna, 2006; 26, Orduna-Rojas and Longoria-Espinoza, 2006; 27, Perez et al., 2007; 28, Zbikowski et al., 2007; 29, Melville and Pulkownik, 2007; 30, Silva et al., 2006; 31, Lafabrie et al., 2007b; 32, Kamala-Kannan et al., 2008; 33, Morrison et al., 2008; 34, Lafabrie et al., 2008a; 35, Ruelas-Inzunza and Paez-Osuna, 2008; 36, Lafabrie et al., 2008b; 37, Marcias-Zamora et al., 2008; 38, Agoramoorthy et al., 2008.

seaweeds collected in 2003-2004 appear to suppress cadmium burdens, and seaweeds from copper-contaminated sites had lower burdens (0.1-1.9 mg Cd/kg DW) than did reference sites (2.5-17.8 mg Cd/kg DW). Transplants of *Lessonia nigrescens* from a reference site to a copper-contaminated site resulted in cadmium depuration; however, cadmium loss occurs more slowly than copper uptake (Andrade et al., 2006).

Studies on accumulation of cadmium by eelgrass, *Zostera marina*, under laboratory conditions, show active transport across cell membranes, as opposed to passive diffusion (Faraday and Churchill, 1979). Most, if not all, of the cadmium uptake occurred through the roots, and this was directly related to cadmium concentration and exposure time. Cadmium uptake by the root rhizomes was similar at dissolved oxygen concentrations of 6% or 90% saturation. In separate studies, cadmium could also accumulate in *Zostera* leaves, with increasing translocation to the root rhizomes over time, where it remained. It was concluded that *Z. marina* is not important in cycling of cadmium from sediments into the water column, but may be very important in removing cadmium from the water column and returning it to the sediments via translocation to the root rhizomes. Maximum cadmium concentrations recorded in roots rhizomes and leaves after immersion for 72 h in 1.0 mg Cd/L were 48.0 mg Cd/kg DW in roots and 94.0 mg Cd/kg DW in leaves (Faraday and Churchill, 1979).

C. salina, a euryhaline alga, can accumulate up to 142.0 mg Cd/kg FW after exposure for 14 days to 0.1 mg Cd/L in fresh water; however, cadmium uptake is drastically reduced at lower ambient cadmium concentrations and at salinities of 5 ppt, which was the lowest test salinity examined (Wong et al., 1979). Salt marsh plants accumulated higher concentrations of cadmium from contaminated soils under oxidizing conditions: however, well-oxidized soils amended with up to 50.0 mg Cd/kg were associated with reduced yields (Gambrell et al., 1977). Algae can also accumulate cadmium from sewage sludge as shown in 120-day exposure studies with *Thalassiosira testudinum* (Montgomery et al., 1978). Seagrass, *T. testudinum*, exposed to graded concentrations of cadmium for 96 h took up cadmium in proportion to the dose, with concentrations highest in green blades (max. 150.0 mg Cd/kg DW), roots and rhizomes (17.8 mg/Cd kg DW), and live sheaths (8.0 mg/kg DW); all tissues had increases in thiol-containing compounds as a response to cadmium exposure (Alvarez-Legorreta et al., 2008).

Below ground portions of two species of saltmarsh plants collected from the Tagus estuary, Portugal, in March 2004, had elevated concentrations of cadmium when compared to aerial portions (Reboreda and Cacador, 2007). Cadmium values in *Spartina maritima*, in mg Cd/m^2 DW, were 19.3 for roots, 0.8 for stems, and 0.2 for leaves. For *Halimione portulacoides*, these values were 22.4 for roots, 2.3 in stems, and 0.7 in leaves. Authors conclude that areas colonized by *Spartina* are effective sinks for cadmium in the estuary and those colonized mainly by *Halimione* are more effective sources (Reboreda and Cacador, 2007).

Sensitivity of algae to cadmium varied considerably. Growth inhibition in the diatom *Phaeodactylum tricornutum* occurred at 5.4 mg Cd/L in 72 h; however, growth stimulation was recorded at 0.16-1.0 mg Cd/L (Horvatic and Persic, 2007), although this requires verification. Growth inhibition was evident at cadmium concentrations in excess of 0.45 mg Cd/L over a 10-day period for *Pelvetia* sp., *Fucus* spp., and *Ascophyllum nodosum* (Stromgren, 1980). At lower concentrations, growth was significantly enhanced in *Pelvetia* and *Ascophyllum* (Stromgren, 1980). Berland et al. (1976) with reports that concentrations of 0.05-1.0 mg Cd/L killed 18 species of algae in 17 days, with growth inhibition reported in the range of 0.005-0.25 mg Cd/L. Concentrations of 0.01-0.025 mg Cd/L depressed the growth of *P. tricornutum* (Cossa, 1976) and *S. costatum* (Berland et al., 1977). However, no deleterious effects were observed at 11.2 mg Cd/L and lower among four species grown in batch culture (Bentley-Mowat and Reid, 1977), and among *Spartina* seedlings held in 100.0 mg Cd/L for 8 weeks (Dunstan et al., 1975); concentrations up to 100.0 mg Cd/L had no measurable effect on oxygen uptake of most species of algae tested (Mills and Colwell, 1977). But concentrations of 112.0 mg Cd/L inhibited photosynthesis within 15 min in the four species of alga tested (Overnell, 1976). It is probable that some of this variability in cadmium sensitivity was attributable to biological availability. For example, chelated cadmium had no demonstrable effect on *P. tricornutum* (Cossa, 1976). Macroalgae, *Enteromorpha prolifera* and *Enteromorpha linza* stressed by 0.1 or 0.2 mg Cd/L react differently (Malea et al., 2006). In *E. linza*, glutathione pools increased, but were susceptible to Cd-induced oxidative stress. In *E. prolifera*, cadmium was actively detoxified by phytochelatins (Malea et al., 2006).

Red alga, *Audouinella saviana*, was exposed to lethal concentrations of cadmium for up to 15 days (Talarico, 2002). Among survivors, thalli of treated algae—when compared to controls—had abnormal plasmalemna, enhanced protein and carbohydrate biosynthesis, and smooth and thick cell walls. Chloroplasts did not accumulate cadmium and were the least affected organelles. Cadmium toxicity is reduced in this cadmium-tolerant species by formation of polysaccharides and structural proteins as a cell wall barrier blocking Cd^{2+} ions in the cell wall by negatively charged groups, cytoplasmic ion detoxification by chelating proteins and cytoplasmic sequestering, and by accumulation of metal complexes into large vacuoles (Talarico, 2002).

Mixtures of cadmium and zinc ions were more effective in the inhibition of growth of *Thalassiosira pseudonana*, and of clone Skel-5 of *S. costatum* than either metal alone (Braek et al., 1980). However, cadmium-zinc mixtures were less than additive in producing growth inhibition of *P. tricornutum* and *S. costatum*, clone Skel-0. The relative resistance of the latter two species is attributed to competition for uptake sites by different chemical forms of cadmium and zinc. In the most resistant alga, *P. tricornutum*, sorption of cadmium as well as zinc was dependant on magnesium concentration of the medium. It was concluded that high levels of nontoxic divalent cations, such as calcium, magnesium, and strontium in seawater, compete with the same uptake sites as toxic divalent cations such as cadmium, copper, and zinc, and may, therefore, protect against metal poisoning in marine organisms (Braek et al., 1980).

2.10 Cerium

Radiocerium-144 was concentrated from the medium by six species of algae by factors of 300-3000 within 30 min and by factors of 2000-4500 after 24 h (Rice and Willis, 1959); similar results are reported by Ancellin and Vilquin (1968). Dead algae continue to adsorb ^{144}Ce, as was the case for *Cystoseira barbata* (Polikarpov, 1961), suggesting that residue levels in algae dead on collection should be viewed with reservation (Eisler, 1981).

Uptake and retention of cerium by algae is variable. Nondividing cells of *Nitzchia closterium* kept in the dark continue to accumulate ^{144}Ce for as long as 3 days, and particulate cerium became associated with the cells more rapidly than did ionic cerium (Rice and Willis, 1959); further, chelating compounds added to the medium reduced uptake of radiocerium by *Nitzchia* (Rice and Willis, 1959), and other species (Hiyama and Shimizu, 1964). Loss rate of cerium from algae is variable; about 25% in 5 h on transfer to cerium-free media; this effect was accelerated when the transfer medium contained chelating agents (Rice and Willis, 1959). The greatest loss of ^{144}Ce was from biological dilution as a result of cell division; this loss amounted to 50% of the contained activity of the cells at each division (Rice and Willis, 1959).

2.11 Cesium

The highest concentration of stable cesium recorded in a marine plant is 0.071 mg/kg DW in whole *Sargassum thunbergii* collected off the coast of Japan (Ishii et al., 1978). On a DW basis, cesium values in other species of marine plants ranged from 0.031 to 0.061 mg/kg (Table 2.7).

Table 2.7: Cesium Concentrations in Field Collections of Algae and Macrophytes

Organism	Concentration	Reference[a]
Algae, whole, 6 species	0.037-0.061 DW	1
Alga, *Hizakia fusiforme*, whole; varying lengths		
up to 10 cm	0.031 DW	1
10.1-20 cm	0.036 DW	1
20.1-30 cm	0.051 DW	1
Sargassum spp., whole, 5 species	0.037-0.071 DW	1
Seaweeds, whole, 5 species	0.003-0.023 FW	2, 3

Values are in mg Cs/kg fresh weight (FW) or dry weight (DW).
[a] 1, Ishii et al., 1978; 2, Van As et al., 1973; 3, Van As et al., 1975.

Radiocesium accumulation rates by marine plants vary significantly among species. Concentration factors up to 2 are listed with equilibrium reached in 24 h (Hiyama and Shimizu, 1964), up to 3 (Ueda et al., 1978), up to 6 (Ichikawa, 1961), 4 to 30 (Gutknecht, 1965), up to 30 (Ishii et al., 1978), up to 40 (Ancellin and Vilquin, 1968), 8 to 70 (Van As et al., 1973), and 2740 to 3360 in algulose and alginic acid fractions (Ryndina, 1976). Accumulations were higher in fresh than in decaying samples of *Cystoseira* and other species; however, the opposite was observed for *C. barbata* (Polikarpov, 1961).

Light, nitrogen, oxygen, dissolved cesium concentrations, and chemical form affect cesium dynamics in seaweeds (Gutknecht, 1965). Uptake and loss of ^{137}Cs by seaweeds were stimulated by light, although some activity occurred in the dark. Previous exposure to light stimulated radiocesium-137 uptake in the dark; exposure to nitrogen in the dark inhibited both influx and efflux of ^{137}Cs. Neither darkness nor anoxia caused appreciable loss of ^{137}Cs. Radiocesium was not extensively adsorbed by cell walls or killed tissues. Influx and efflux of cesium were proportional, respectively, to the external and internal radiocesium concentrations. Finally, intracellular cesium appears to be largely ionic, with cesium movements closely related to metabolism (Gutknecht, 1965).

2.12 Chromium

High concentrations of chromium were recorded in marine algae and other biota near electroplating plants, tanneries, oil drilling operations, sewage outfalls, drift cooling towers, dump sites, and other sources of chromium-containing wastes that were being discharged into the environment (Eisler, 2000c). Marine plants can accumulate chromium from seawater by factors ranging from 5890 to 10,930 (El Ati-Hellal et al., 2007; Van As et al., 1973), from solutions containing 50.0 mg Cr/L by a factor of 18 in 49 h (Sivalingam, 1978a), and from sewage sludge containing 25.0 mg Cr/kg DW by a factor of 2.2 (Montgomery et al., 1978). The unusually high chromium concentrations observed among some species of algae and macrophytes from Puerto Rico (Bernhard and Zattera, 1975) and from Narragansett Bay, Rhode Island, United States (Phelps et al., 1975), were almost certainly due to chromium wastes discharged into the Bay from electroplaters in the case of Rhode Island and from other anthropogenic sources in Puerto Rico. A similar situation probably exists wherever grossly elevated chromium levels are documented in marine algae and macrophytes (Table 2.8). Accidental contamination of field-collected samples is another source of elevated chromium residues that has been investigated (Martin and Knauer, 1973). They found that chromium contamination of phytoplankton samples collected under field conditions comes from metal particles in samples (up to 150.0 mg Cr/kg), rust from stainless hydrowire (up to 460.0 mg Cr/kg), and from hull paint (up to 1.0 mg/kg). Dead biomass of the brown seaweed *Ecklonia* sp. can reduce the toxic hexavalent chromium ion to the less-toxic trivalent chromium ion; the rate of reduction was most rapid at pH < 3.0 (Park et al., 2007).

Table 2.8: Chromium Concentrations in Field Collections of Algae and Macrophytes

Organism	Concentration	Reference[a]
Algae, whole		
Sea of Japan, 12 spp.	1.0-140.0 DW	1
United Kingdom, 11 spp.	2.8-30.0 DW	2
Monaco, 4 spp.	0.4-4.2 DW	3, 4
Soviet Union, 4 spp.	6.7-13.4 DW	5
Egypt; Mex Bay, Alexandria; 2 pp.	1.9-3.5 DW	17
Puerto Rico, 18 spp.	0.4-110.0 DW	6
Argentina; summer 2002; 3 sites; *Porphyra columbina* vs. *Ulva* sp.; whole,	0.65-0.73 DW vs. 0.84-1.14 DW	21
Knotted wrack, *Ascophyllum nodosum*, whole		
Norway	4.0 DW	7
England	1.1-10.0 DW	8, 9
Coast of Ireland; 1999-2000	0.05-1.1 DW	26
Mangrove, *Avicennia marina*; pneumatophores; Queensland, Australia	<1.0-4.0 DW	29
Shoalgrass, *Halodule wrighti*; rhizomes; Texas; Laguna Madre; 1986-1987	13.0 (5.0-149.0) DW	15
Macroalgae; epiphytic on estuarine mangrove aerial roots; Sydney, Australia		
Caloglossa leprieurii	39.0 DW	22
Caloglossa nipae	9.0 DW	22
Bostrychia sp.	61.0 DW	22
Seaweed, *Patina* sp.; Guam; 1998-1999; ronds	0.6-14.0 DW	18
Macrophyte, *Phragmites* sp,; New Jersey; Hackensack River wetlands vs. reference site; August-November 1991		
Roots	240.0 DW vs. 65.0 DW	16
Shoots	0.3 DW vs. 0.2 DW	16
Phytoplankton, whole; Narragansett Bay, Rhode Island	4.3-73.3 DW	10

(*Continues*)

Table 2.8: Cont'd

Organism	Concentration	Reference[a]
Seagrass, *Posidonia oceanica*		
Blades; Mediterranean Sea; summers 2004-2005; 3 sites	0.2-1.3 DW	24
Corsica, France; summer 2003; adult leaves; blades vs. sheaths	0.32 (0.15-1.1) DW vs. 0.84 (0.2-3.8) DW	27, 28
Seaweeds, whole		
Japan, 44 spp.	0.1-2.5 DW	11
Korea, 20 spp.	0.7-7.4 DW	12
United Kingdom, 5 spp.		
May	0.7-1.8 DW	13
June	0.7-3.7 DW; 0.14-0.68 FW	13
Cordgrass, *Spartina* spp., whole		
Reference areas	2.3-3.1 DW	14
Areas treated with sewage-amended sludge at 10,300 mg Cr/m^2 over a 7-year period	31.0-44.0 DW	14
Macroalgae; Tunisia; 2 spp.; Mediterranean Sea coast; whole	2.2-17.0 DW	20
Salt marsh plants; Venice lagoon; June-December 2001; aerial portions; maximum concentration; *Phragmites australis* vs. *Bolboschoenus maritimus*	120.0 DW vs. 28.0 DW	26
Seaweed, *Ulva lactuca*; whole; India; 2004-2005	22.7 (10.5-45.7) DW	25
Eelgrass, *Zostera marina*; Oregon		
Leaf	7.2 DW	19
Rhizome	7.4 DW	19
Root	38.0 DW	19

Values are in mg Cr/kg fresh weight (FW), dry weight (DW), or ash weight (AW).
[a]1, Gryzhankova et al., 1973; 2, Riley and Roth, 1971; 3, Fukai, 1965; 4, Fukai and Broquet, 1965; 5, Khristoforova et al., 1976; 6, Bernhard and Zattera, 1975; 7, Lande, 1977; 8, Foster, 1976; 9, Bryan and Uysal, 1978; 10, Phelps et al., 1975; 11, Ishibashi and Yamamoto, 1960; 12, Pak et al., 1977; 13, Black and Mitchell, 1952; 14, Giblin et al., 1980; 15, Custer and Mitchell, 1993; 16, Hall and Pulliam, 1995; 17, Dahab et al., 1990; 18, Denton et al., 2006; 19, Kaldy, 2006; 20, El Ati-Hellal et al., 2007; 21, Perez et al., 2007; 22, Melville and Pulkownik, 2007; 23, Bragato et al., 2006; 24, Lafabrie et al., 2007b; 25, Kamala-Kannan et al., 2008; 26, Morrison et al., 2008; 27, Lafabrie et al., 2008a; 28, Lafabrie et al., 2008b; 29, Preda and Cox, 2002.

Chromium concentrations in blades of the seagrass, *Posidonia oceanica*, were positively correlated with chromium concentrations in the surrounding sediments (Lafabrie et al., 2007b). Although chromium is abundant in primary producers, there is little evidence of biomagnification through marine food chains consisting of herbivores and carnivores (Osterberg et al., 1964). In an experimental food chain consisting of phytoplankton, brine shrimp, postlarval fish, and adult fish, chromium was successfully transferred through each trophic level, with concentrations declining up the food chain (Baptist and Lewis, 1969). Based on laboratory studies, it was concluded that chromium uptake from diet was usually more efficient than uptake from seawater (Baptist and Lewis, 1969).

Among sensitive species of marine algae, concentrations of 0.01 mg/L of hexavalent chromium (Cr^{6+}) partially inhibited growth of *Olisthodiscus leucas*. All cultures, including those in which growth was inhibited, contained viable, active (>75%) cells at the end of 10 days. Inhibitory effects were reversed by chelators such as EDTA (Mahoney, 1982), suggesting that naturally occurring ligands and sequestering agents in seawater may alleviate the toxicity of Cr^{6+}, and perhaps other metals. In the giant kelp, *Macrocystis pyrifera*, photosynthesis was inhibited 20% in 5 days at 1.0 mg/L of Cr^{6+}, and 50% in 4 days at 5.0 mg/L (USEPA, 1980d); this kelp seems to be one of the more chromium-resistant plants. Euryhaline species of algae tested show increasing resistance to hexavalent chromium at increasing salinities (Wong and Trevors, 1988). Less than 0.05 mg Cr^{6+}/L is proposed as safe for marine biota (USEPA, 1999).

2.13 Cobalt

Cobalt concentrations in algal field collections seldom exceed 15.0 mg/kg DW, with most values notably lower (Table 2.9). The average cobalt content in 38 species of seaweeds from Japan was 0.8 mg/kg DW, with values ranging from 0.1 for *Eisenia bicyclis* to 4.5 mg/kg DW for *S. thunbergii* (Ishibashi et al., 1964c). Differences in cobalt content of Japanese seaweeds were not attributable to season of collection or habitat. Green seaweeds were comparatively high in cobalt when compared to red seaweeds. Most brown seaweeds were low in cobalt, but some species such as *S. thunbergii*, *S. hemiphyllum*, and *Padina arborescens* contained high cobalt residues. Seaweeds with high cobalt concentrations also contain high iron concentrations and vice versa (Ishibashi et al., 1964c), and this could account for some of the observed variation.

Six species of macrophytes collected from the southwest coast of India between 1999 and 2001 had elevated concentrations in summer monsoon and hot seasons (up to 10.1 mg Co/kg DW), with little difference between species in ability to accumulate cobalt from seawater (Kalesh and Nair, 2006). Cobalt concentrations in the seagrass, *P. oceanica*, from the Mediterranean Sea in summers of 2004-2005 were positively correlated with cobalt concentrations in the surrounding sediments, suggesting that this species may be useful as a biomonitor for cobalt and other trace metals (Lafabrie et al., 2007b).

Table 2.9: Cobalt Concentrations in Field Collections of Algae and Macrophytes

Organism	Concentration	Reference[a]
Algae, whole		
Sea of Japan		
Reds and browns, 9 spp.	2.0-147.0 DW	1
Greens, 3 spp.	2.0-31.0 DW	1
Iceland, 10 spp.	0.3-11.9 DW	2
Goa, India, 33 spp.	1.3-15.2 DW	3
Canada, Atlantic Provinces, 11 spp.	0.1-6.3 DW	4
Algae and macrophytes; whole, 10 spp.	0.1-0.4 DW	5
Argentina; summer 2004; 3 sites; *Porphyra columbina* vs. *Ulva* sp.; whole	0.38-0.55 DW vs. 0.27-0.73 DW	15
Seaweed, *Ascophyllum nodosum*; whole; Ireland; 1999-2000	0.11-0.92 DW	17
Fucus vesiculosus; whole; United Kingdom	4.8 (1.8-9.0) DW	6
Macrophytes; 6 spp.; whole; India; 1999-2001	0.8-10.1 DW	14
Marsh plants; whole; Texas, Harbour Island, 14 spp.	0.06-0.065 DW	7
Porphyra spp., whole	0.5-2.6 DW	2
Seagrass, *Posidonia oceanica*, blades		
Mediterranean Sea; summers 2004-2005; 3 sites	1.7-12.1 DW	16
Corsica, France; summer 2003; adult leaves; blades vs. sheaths	3.8 (1.8-7.7) DW vs. 0.2 (0.1-0.3) DW	18
Seaweeds. Whole		
Korea; 20 spp.	0.9-2.4 DW	8
Norway	0.1-5.2 DW	9
South Africa; 7 spp.	0.006-0.075 FW	10, 11
Sea of Japan; 38 spp.	0.8 (0.1-4.5) DW	12
Irish Sea; United Kingdom; 5 spp.		
May	0.7-1.4 DW	13
June	0.6-2.0 DW; 0.1-0.4 FW	13

Values are in mg Co/kg fresh weight (FW) or dry weight (DW).
[a] 1, Gryzhankova et al., 1973; 2, Munda, 1978; 3, Agadi et al., 1978; 4, Young and Langille, 1958; 5, Ishii et al., 1978; 6, Bryan and Uysal, 1978; 7, Lytle et al., 1973; 8, Pak et al., 1977; 9, Lunde, 1970; 10, Van As et al., 1973; 11, Van As et al., 1975; 12, Ishibashi et al., 1964c; 13, Black and Mitchell, 1952; 14, Kalesh and Nair, 2006; 15, Perez et al., 2007; 16, Lafabrie et al., 2007b; 17, Morrison et al., 2008; 18, Lafabrie et al., 2008a.

Bioconcentration of cobalt from ambient seawater under laboratory conditions ranged from 15 (Ichikawa, 1961) to 136 (Sivalingam, 1978a); the latter value was derived from immersion for 48 h of *U. reticulata* in solutions containing 50.0 mg Co/L. Under field conditions, cobalt bioconcentration was variously estimated for different species at 215-1600 (Van As et al., 1973), 2000 (Ishii et al., 1978), and 990 and 9710 in algulose and barium-complexed fucoidan fractions, respectively, for *Cystoseira* spp. (Ryndina, 1976).

Cobalt is considered an essential trace element for most organisms, including algae. Normal growth was evident in *Cricosphaera carterae* at concentrations less than 0.6 mg Co/L; however, higher concentrations up to 6.0 mg Co/L inhibited cell division, and still higher concentrations up to 12.0 mg Co/L drastically reduced protein synthesis (Blankenship and Wilbur, 1975). In the diatom, *P. tricornutum*, growth stimulation occurred at 0.08-12.5 mg Co/L, and growth inhibition at 19.6 mg Co/L (Horvatic and Persic, 2007).

2.14 Copper

Large variability in copper concentrations of marine plants is due, in part, to proximity to heavily populated or industrialized areas, fraction or tissue analyzed, copper concentration of the surrounding seawater, season of collection, geographic area, interspecies variations, and age of the plant (Almeida et al., 2006; Eisler, 1981, 2000g; Kalesh and Nair, 2006; Table 2.10). Algae reared or found in high-copper environments can tolerate additional levels of copper more readily than algae from low-copper environments. It is not clear if these differences are adaptive or genetic. In one case, *Ectocarpus siliculosus* collected from the hulls of ocean-going freighters—that had been treated with copper-based antifouling paints—were an order of magnitude more resistant to dissolved $CuCl_2$ over a 5-week period than were conspecifics collected from an uncontaminated rocky shore (Russell and Morris, 1970). Copper tolerance in various strains of *E. siliculosus* is attributed to a copper-exclusion mechanism (Hall et al., 1979). Changes in the cell membrane, possibly in combination with intracellular changes, are the most probable causes of the exclusion mechanism rather than extracellular releases of organic matter (Hall et al., 1979).

Copper is readily accumulated from seawater by algae and macrophytes by factors estimated at 2060 for *Enteromorpha* spp. (El Ati-Hellal et al., 2007), 6000-20,000 for *A. nodosum* (Foster, 1976; Melhuus et al., 1978), 2000 for *Sargassum* (Ishii et al., 1978), 1070-27,000 for *F. vesiculosus* (Bryan and Hummerstone, 1973; El Ati-Hellal et al., 2007; Foster, 1976; Seeliger and Edwards, 1977), and 4700-200,000 for other species (Kim and Won, 1974; Seeliger and Edwards, 1977; Yoshimura et al., 1976). Of all marine organisms examined, algae—together with certain bivalve molluscs—have the greatest ability to accumulate copper from the surrounding environment (Eisler, 1979). Copper in sewage sludge is also readily mobilized by marine plants (Montgomery et al., 1978). Field collections of marine algae may be subjected to contamination. For example, copper contamination of

Table 2.10: Copper Concentrations in Field Collections of Algae and Macrophytes

Organism	Concentration	Reference[a]
Alga, *Ascophyllum nodosum*, whole		
England: polluted bay vs. reference site	68.0 (46.0-96.0) DW vs. 12.0 (6.0-18.0) DW	1
Norway; polluted fjord vs. reference ite	45.0-240.0 DW vs. 5.5 (4.0-8.0) DW	1
Algae and macrophytes, whole		
Canada; Atlantic Provinces; 11 spp.	6.0-62.0 DW	2
Iceland, 10 spp.	1.8-8.8 DW	3
Goa, India; 39 spp.	3.2-80.4 DW	4, 5
Irish Sea	5.6-86.0 DW	6
Sea of Japan; 12 spp.	2.0-8.0 DW	7
Norway; west coast	Max. 170.0 DW	8
Spain; Rio Tinto estuary	Max. 26.0 DW	8
Israel; Mediterranean Sea coast; 2 spp.	2.9-7.6 DW	9
United Kingdom; 13 spp.	25.0-210.0 DW	10
Argentina; summer 2002; 3 sites; *Porphyra columbina* vs. *Ulva* sp.; whole	1.9-6.9 DW vs. 1.7-3.8 DW	32
Chile; 2003-2004; seaweeds		
Copper-contaminated site from copper mine wastes; 4 spp.	93.5-1609.2 DW	37
Reference site; 8 spp.	1.7-27.1 DW	37
Ulva compressa; reference site vs. copper-contaminated site	4.2-6.1 DW vs. 93.5-790.6 DW	37
Mangrove, *Avicennia marina*		
Hong Kong		
Sediment	13.0 DW	38
Root	13.0 DW	38
Leaf	16.0 DW	38
Western Australia		
Sediment	16.0 DW	39
Root	18.0 DW	39
Leaf	7.3 DW	39
Shenzhen, China		
Sediment	36.0 DW	40, 41
Root	13.0-14.0 DW	40, 41
Leaf	5.0-5.2 DW	40, 41
Southeast Australia		
Sediment	61.0 DW	42
Root	101.0 DW	42
Leaf	9.0 DW	42
Pneumatophore	6.0-12.0 DW	45

(*Continues*)

Table 2.10: Cont'd

Organism	Concentration	Reference[a]
Green alga, *Cladophora* sp.; 2000-2003; Baltic Sea		
Gulf of Gdansk	5.3 (1.1-11.2) DW	33
Vistula Lagoon	8.4 DW; max. 11.4 DW	33
Reference site	5.1 DW; max. 7.2 DW	33
Chlorophyceae, whole, 14 spp.	9.2-27.7 DW	11, 12
Green alga, *Enteromorpha* sp.; whole; 2000-2003; Gulf of Gdansk, Baltic Sea	4.9 (1.8-11.6) DW	26
Fucus serratus		
Near desalinization plant outfall vs. reference site		
Tips		
May	5.9 DW vs. 6.7 DW	13
August	204.0 DW vs. 4.3 DW	13
Thallus		
May	17.0 DW vs. 6.7 DW	13
August	104.0 DW vs. 4.7 DW	13
Fucus spiralis; near desalination plant outfall vs. reference site; August		
Tips	231.0 DW vs. 2.8 DW	13
Thallus	231.0 DW vs. 3.8 DW	13
Microalgae, *Haslea ostrearia*		
Adsorbed onto cell surface	0.06 DW	47
Incorporated into cell; soluble vs. insoluble	0.017 DW vs. 0.023 DW	47
India; Tamil Nadu; leaves; January 2007		
Mangroves; 8 spp.	24.1 (8.1-95.1) DW	46
Halophytes; 5 spp.	14.4 (7.8-32.2) DW	46
Macroalgae; Tunisia; Mediterranean Sea coast; 2 spp.; whole	1.3-8.2 DW	29
Macroalgae (6 spp.) vs. phytoplankton; Gulf of California, Mexico; 1998-1999; whole	7.5-24.6 DW vs. 30.0 DW	27
Macroalgae; epiphytic on estuarine mangrove aerial roots; Sydney region, Australia		
Caloglossa lepriaurii	200.0 DW	31
Caloglossa nipae	34.0 DW	31
Bostrychia sp.	86.0 DW	31

(Continues)

Table 2.10: Cont'd

Organism	Concentration	Reference[a]
Mangroves; Gulf of California, Mexico; 1998-1999; twigs vs. leaves		
Rhizophora mangle	3.0 DW vs. 7.4 DW	27
Avicennia germinans	18.1 DW vs. 4.6 DW	27
Laguncularia racemosa	5.0 DW vs. 2.3 DW	27
New Jersey, United States; Raritan Bay; water content 0.002 mg Cu/L vs. 0.022 mg Cu/L		
Blidingia minima	2.0 DW vs. 160.0 DW	14
Enteromorpha linza	17.0 DW vs. 77.0 DW	14
Fucus vesiculosus	8.0 DW vs. 48.0 DW	14
Ulva spp.	14.0 DW vs. 38.0 DW	14
Macrophytes; 6 spp.; whole; India; 1999-2001	0.6-14.6 DW	30
Phaeophyceae, whole, 23 spp.	4.5-49.7 DW	11, 12
Brown alga, *Pelvetia canaliculata*; whole		
Norway	35.0 DW	1
Scotland	5.0-16.0 DW	1
Macrophyte, *Posidonia oceanica*; Italy		
Sheath	6.0-17.0 DW	25
Rhizome	5.4-15.3 DW	25
Mangrove, *Rhizophora mangle*; Natal, Brazil; summer 2001; leaves	1.1-2.7 DW	35
Mangrove, *Rhizophora stylosa*		
Western Australia		
Sediment	14.0 DW	39
Root	6.5 DW	39
Leaf	3.7 DW	39
Yingluo Bay, China		
Sediment	19.0 DW	43
Root	1.1 DW	43
Leaf	0.6 DW	43
Rhodophyceae, whole, 29 spp.	2.0-13.4 DW	7,11,12
Saltmarsh macrophytes; 3 spp.; Yangtze River estuary, China; July 2005		
Above ground portions	30.0 DW	34
Below ground portions	100.0 DW	34

(*Continues*)

Table 2.10: Cont'd

Organism	Concentration	Reference[a]
Salt marsh plants; Venice lagoon; 2001; aerial portions		
Phragmites australis	Max. 14.0 DW	36
Bolboschoenus maritimus	Max. 5.0 DW	36
Sargassum spp., whole		
Japan	Max. 10.0 DW	15
Texas	4.1 (2.8-7.0) DW	16
Korea	5.7 (3.6-13.1) DW	17
Seaweed, *Ulva lactuca*; Gulf of California, Mexico; 2002; whole		
Summer	23.0 DW	28
Winter	44.0 DW	28
Seaweeds, whole		
Korea, 17 spp.	0.9-17.0 DW; 0.2-4.0 FW	18
Korea, 20 spp.	5.0-54.0 DW	19
Norway	6.0-63.0 DW	20
United Kingdom, 5 spp.; May vs. June	4.0-7.0 DW vs. 6.0-31.0 DW	21
Eelgrass, *Zostera marina*		
Baja California, Mexico; March 2000; whole; reference site vs. salt mining site	2.9 (1.3-14.0) DW vs. 4.0 (3.8-15.0) DW	44
Oregon		
Leaf	10.0 DW	24
Rhizome	8.3 DW	24
Root	47.0 DW	24
Eelgrass, *Zostera* spp.		
Denmark; 1979-80; metals-contaminated site vs. reference site		
Leaves	9.0-13.0 DW vs. 5.0-6.0 DW	22
Roots	27.4 DW vs. 6.0-7.0 DW	1
Portugal and Spain; metals-contaminated site vs. reference site	Max. 1350.0 DW vs. 9.0-36.0 DW	23

Values are in mg Cu/kg fresh weight (FW) or dry weight (DW).

[a] 1, Jenkins, 1980b; 2, Young and Langille, 1958; 3, Munda, 1978; 4, Agadi et al., 1978; 5, Zingde et al., 1976; 6, Culkin and Riley, 1958; 7, Gryzhankova et al., 1973; 8, Stenner and Nickless, 1974; 9, Roth and Hornung, 1977; 10, Riley and Roth, 1971; 11, Sivalingam, 1978b; 12, Ishibashi et al., 1962; 13, Romeril, 1977; 14, Seeliger and Edwards, 1977; 15, Ishii et al., 1978; 16, Horowitz and Presley, 1977; 17, Kim and Won, 1974; 18, Kim, 1972; 19, Pak et al., 1977; 20, Lunde, 1970; 21, Black and Mitchell, 1952; 22, Brix and Lyngby, 1982; 23, Stenner and Nickless, 1975; 24, Kaldy, 2006; 25, Baroli et al., 2001; 26, Zbikowski et al., 2006; 27, Ruelas-Inzunza and Paez-Osuna, 2006; 28, Orduna-Rojas and Longoria-Espinoza, 2006; 29, El Ati-Hellal et al., 2007; 30, Kalesh and Nair, 2006; 31,Melville and Pulkownik, 2007; 32, Perez et al., 2007; 33, Zbikowski et al., 2007; 34, Quan et al., 2007; 35, Silva et al., 2006; 36, Bragato et al., 2006; 37, Andrade et al., 2006; 38, Chen et al., 2003; 39, Alongi et al., 2003; 40, Peng et al., 1997; 41, Zheng and Lin, 1996; 42, MacFarlane et al., 2003; 43, Zheng et al., 1997; 44, Marcias-Zamora et al., 2008; 45, Preda and Cox, 2002; 46, Agoramoorthy et al., 2008; 47, Amiard-Triquet et al., 2006.

phytoplankton samples by metal particles in samples, rust from stainless and nonstainless hydrowire, hull paint, and open ocean tar balls are all documented (Martin and Knauer, 1973).

Bioconcentration and biomagnification of copper occurs in the food chain of diatom (*S. costatum*) to clams (*Donax cuneatus*) to prawns (*Penaeus indicus*). All species accumulate copper from the medium, and clams and shrimp from the diet. Maximum concentrations after exposure to 0.2 mg Cu/L and diets for 10 days, in mg Cu/kg FW, are 2.8 in whole diatoms, 13.6 in clam soft parts, and 33.9 in whole shrimps (Rao and Govindarajan, 1992). Transfer of copper from wood treated with chromated copper arsenate (CCA) occurs in estuarine algae (*Ulva* sp., *Enteromorpha* sp.), American oysters, mud snails, and fiddler crabs (Weis and Weis, 1992). Algae from CCA-treated lumber show elevated concentrations of copper when compared to reference sites (Weis and Weis, 1992). Bioavailability of copper determined from metal speciation in diet permits an accurate prediction of trophic transfer in some cases (Amiard-Triquet et al., 2006). In the case of the microalga *Haslea ostrearia*, which was exposed to 0.03 mg Cu/L until exponential growth was achieved and then fed to Pacific oysters, *Crassostrea gigas*, for 3 weeks, only the exchangeable copper was available, that is, copper adsorbed onto cell surface and soluble copper incorporated into the cell. Copper-loaded *Haslea* had 0.23 mg Cu/kg DW adsorbed onto the cell surface, 0.085 mg Cu/kg DW in incorporated soluble fractions, and 0.035 mg Cu/kg DW as insoluble copper within the cell. Thus, 90% of the copper in *Haslea* was exchangeable and similar to the 93% stored by oysters (Amiard-Triquet et al., 2006). In the case of copper-loaded diatoms, *Tetraselmis suecica*, only half (21% of the potentially available copper of 0.067 mg Cu/kg DW) was incorporated by oysters (Amiard-Triquet et al., 2006).

Below ground portions of two species of saltmarsh plants collected from the Tagus estuary, Portugal, in March 2004 had elevated concentrations of copper when compared to aerial portions (Reboreda and Cacador, 2007). For *S. maritima*, copper concentrations, in mg Cu/m^2 DW, were 230.1 for roots, 1.8 for stems, and 0.8 for leaves. For *H. portulacoides*, these values were 660.9 for roots, 3.9 for stems, and 1.8 for leaves, suggesting that *Spartina* is a more effective sink for copper and *Halimione* a more effective source (Reboreda and Cacador, 2007). Saltmarsh plants, including *H. portulacoides*, can increase the potential mobility of metals, especially copper, from sediments to pore water and sometimes to the water column when rhizosediments are resuspended (Almeida et al., 2008). High transfer of copper from *Halimione*-colonized sediments to pore waters occurs with attendant risks to biota; however, strong copper-complexing organic ligands in water and elutriates reduce metal bioavailability. In areas not colonized by *Halimione*, sediments remove copper and other metals to pore waters (Almeida et al., 2008). Although mangroves are a taxonomically diverse group, copper accumulation and partitioning is similar across genera and families (MacFarlane et al., 2007). Patterns of copper accumulation are also similar for salt-secreting and nonsecreting species. Copper is accumulated in roots to concentrations similar to adjacent sediments, and copper concentrations in leaves are usually less than half that of roots. Leaf copper concentrations decreased as environmental copper concentrations increased, suggesting regulation. A similar case is made for zinc, another essential metal (MacFarlane et al., 2007).

Copper residue data in photosynthetic organisms should be interpreted with the knowledge that numerous intrinsic and extrinsic variables significantly affect copper uptake, as well as growth and survival. Thus, increasing temperatures were positively associated with copper uptake and biomass in *Coccochloris elegans*, *D. tertiolecta*, and *S. costatum* (Mandelli, 1969). Growth rate and copper content of *T. pseudonana* and *Nannochloris atomus* were altered independently of medium copper concentrations by varying pH (Sunda and Guillard, 1976). Chelators reduced the availability of copper to algae (Davey et al., 1973; Erickson, 1972; Erickson et al., 1970; Morris and Russell, 1974). In one instance, the presence of decomposed natural plankton and detritus decreased toxicity of copper to *T. pseudonana*, probably by complexation or chelation of the copper to a biologically unavailable form (Erickson, 1972). In another instance, effects of potentially inhibitory levels of cupric ion in seawater on marine algae can be reduced or eliminated depending on the degree of copper complexation by natural organic ligands (Sunda and Guillard, 1976). The anionic constituent must also be considered when evaluating algal copper kinetics, with differences observable in chloride and sulfate salts (Nielsen and Wium-Anderson, 1970). High concentrations of iron or citrate in seawater both reportedly reduced the biocidal properties of copper to algae (Nielsen and Kamp-Nielsen, 1970), presumably by complexation.

In laboratory studies, thalli of two species of benthic red algae, *Ceramium pedicellatum* and *Neogardhiella baileyi* accumulated copper from media containing 0.0014, 0.0064, or 0.011 mg Cu/L by factors ranging from 4540 and 6864 (Seeliger and Edwards, 1979). Up to 22% of the copper was excreted in organic matter, but less than 10% was released in the inorganic form. Dead algae released 80-90% of the copper in organic fractions, both dissolved and particulate. Copper was thus present to a large extent in both dissolved and detrital fractions, components which are readily consumed by marine heterotrophs, and only to a small extent in dissolved inorganic form (Seeliger and Edwards, 1979). Copper tolerance of the diatom *N. closterium* was measured after copper acclimatization to 0.005 or 0.025 mg Cu/L for 100 days; there were no significant changes over a 72-h period in growth rate or tolerance to zinc or copper when compared to nonacclimatized algae (Johnson et al., 2007). Seagrass (*Heterozostera tasmanica*) held in seawater containing 0.042 mg Cu/L for several weeks had 2700.0 mg Cu/kg DW; sea grasses in media containing 0.0003 mg Cu/L had 2.5 mg Cu/kg DW, and intermediate values are reported for 0.01 mg Cu/L (306.0-564.0 mg/kg DW) and 0.02 mg Cu/L (1280.0 mg Cu/kg DW; Ahsanullah and Williams, 1991). Brown alga, *Sargassum linearifolium*, experimentally contaminated in the laboratory with 0.3 mg Cu/L for 24 h and transferred to the field, had greatly reduced colonial epifauna taxa (Roberts et al., 2006).

Ionic copper (Cu^{2+}) is phytotoxic. Copper was the most toxic metal tested to five species of fucoid algae, followed by ionic mercury, zinc, lead, and cadmium, in that order (Stromgren, 1980). The biocidal properties of copper were evident among complex metal mixtures. In one study using the diatom *Thalassiosira aestivalis*, effects of a mixture of 10 metals (copper, zinc, nickel, chromium, lead, cadmium, mercury, arsenic, selenium, and antimony) were investigated at metals concentrations expected to occur in a moderately polluted estuary. Diatom growth was not inhibited under laboratory conditions. But growth was inhibited when all metals concentrations were increased by a factor of 5 or 10. Growth inhibition was due solely to copper with normal growth evident when copper was deleted from the mixture (Thomas et al., 1980). In another study, three species of algae were grown *in situ* in Norwegian fjords contaminated to various degrees with copper and other metals in solution including cadmium, zinc, lead, and mercury. In every case, growth was significantly reduced and metals uptake increased in the most heavily contaminated fjord when compared to a less-polluted fjord, and especially to reference sites (Eide and Jensen, 1979). It is probable that copper was the primary agent responsible for the observed deleterious effects. Copper toxicity in a diatom, *Thalassiosira weissflogii*, is predictable, in part, by the intracellular copper and its subcellular distribution as compared with the currently used free Cu^{2+} ion activity model (Miao and Wang, 2007). Antifouling paints containing mixtures of zinc pyrithione and copper (as a replacement for tributyltin, TBT) were more than additive in toxicity during 96-h survival tests than either alone, and this should be considered in developing copper and zinc water quality criteria (Bao et al., 2008). Increased production of phytochelatins in the diatom *P. tricornutum* is a useful indicator of free copper in the medium, suggesting that phytochelatins can serve as biomarkers of exposure to the bioavailable copper fraction (Morelli and Fantozzi, 2008). A linear dose-response relation between metal exposure—including cadmium, lead, and especially copper—and phytochelatin synthesis was found in seawater enriched with known amounts of metal (Morelli and Fantozzi, 2008).

Copper sensitivity of 13 species of marine microalgae—as judged by 50% inhibition of growth rate in 72 h—ranged from 0.0006 mg Cu/L for *Minutocellus polymorphus*, a centric diatom, to 0.53 mg/L for *D. tertiolecta*, a green algae; values for most other species tested were less than 0.015 mg Cu/L (Levy et al., 2007). Interspecies sensitivity to copper was attributed to differences in uptake rates across the plasma membrane, differences in internal binding mechanisms, and variability in detoxification mechanisms between species (Levy et al., 2007). However, use of internal copper concentrations and net uptake rates alone cannot satisfactorily account for differences in species sensitivity to copper, suggesting that cellular detoxification mechanisms merit additional research (Levy et al., 2008). Considerable variability in sensitivity to copper was documented among algal species. For example, exposure of algae to very low sublethal concentrations of copper (0.00003 mg/L) increases their sensitivity toward additional copper challenge and to cadmium salts (Visviki and Rachlin, 1994a). Adverse effects on growth, survival, and development were documented for sensitive species of algae at nominal copper concentrations of 0.005-0.01 mg Cu/L (Davey, 1976; Jensen et al., 1976; Saifullah, 1978; Saward et al., 1975; USEPA, 1980a), and at 0.011-0.05 mg/L (Braek et al., 1976; Gnassi-Barelli et al., 1978; Havens, 1994; Hopkins and Kain, 1971; Pace et al., 1977; Sunda and Guillard, 1976; Visviki and Rachlin, 1994a). Growth inhibition (50% in 96 h) is reported for *Chlorella pyrenoidosa* at 0.51 (0.39-0.63) mg Cu/L (Wang et al., 2007). Among resistant species, adverse effects were evident at comparatively high nominal copper concentrations of greater than 0.1-100.0 mg Cu/L (as quoted in Abalde et al., 1995; Coppellotti, 1989; Eisler, 1981; Visviki and Rachlin, 1994b). Minor effects on lipid metabolism of *Dunaliella salina* was reported at 0.000031 mg Cu/L after exposure for 8 months (Visviki and Rachlin, 1994a), but this requires verification. A proposed criterion to protect algal life is less than 0.0031 mg Cu/L (El Ati-Hellal et al., 2007).

2.15 Gallium

Gallium concentrations in algae and seaweeds ranged from 0.01 to 0.64 mg/kg DW, and up to 20.0 on an ash weight (AW) basis (Table 2.11). The role of gallium in plant physiology is imperfectly understood.

Table 2.11: Gallium, Germanium, and Gold Concentrations in Field Collections of Algae and Macrophytes

Element and Organism	Concentration	Reference[a]
Gallium		
Algae, whole, Irish Sea	0.01 DW	1
Diatoms, whole, Black Sea	20.0 AW	2
Seaweeds, whole, Japan, 30 spp.	0.14 (0.02-0.64) DW	3
Germanium		
Sargassum sp., whole, Gulf of Mexico	0.003-0.032 FW	4
Gold		
Seaweeds, whole, Japan		
Ulva sp.	0.015-0.093 AW	5
Ulva sp. and *Porphyra* sp.	0.021-0.035 DW	6

Values are in mg element/kg fresh weight (FW), dry weight (DW), or ash weight (AW).
[a]1, Culkin and Riley, 1958; 2, Vinogradova and Koual'skiy, 1962; 3, Yamamoto et al., 1976; 4, Johnson and Braman, 1975; 5, Fukai and Meinke, 1959; 6, Fukai and Meinke, 1962.

2.16 Germanium

Germanium concentrations in *Sargassum* collected from the Gulf of Mexico ranged from 0.003 to 0.032 mg/kg FW (Table 2.11).

Marine diatoms under laboratory conditions were more sensitive to germanium salts than were other groups of phytoplanktons (Azam et al., 1973; Lewin, 1966; Thomas and Dodson, 1974). In one study, four species of diatoms accumulated radiogermanium from their growth media and incorporated up to 80% into the silica of cell walls (Azam et al., 1973). Uptake and incorporation of germanic acid was dependent upon the Ge/Si ratio; at a ratio of 0.01, no inhibition of growth or silicic acid uptake was observed (Lewin, 1966). At a Ge/Si ratio of 0.1, growth and silicic acid uptake in diatoms was reduced by 95%. Lewin (1966) concluded that high levels of germanium interfered with silicate utilization by diatoms. Thomas and Dodson (1974) demonstrated that 0.5 and 1.0 mg Ge/L as $Ge(OH)_4$ interfered with diatom photosynthesis, with degree of inhibition proportional to $Si(OH)_4$ concentrations in the medium; however, at 5.0 and 10.0 mg Ge/L, inhibition was independent of $Si(OH)_4$ concentration.

2.17 Gold

Unlike certain species of terrestrial plants that routinely accumulate gold to concentrations greater than 0.1 mg Au/kg DW (Eisler, 2004), maximum values recorded in seaweeds from Japan were 0.035 mg/kg on a DW basis and 0.093 mg/kg on an AW basis (Table 2.11). Gold is not an essential element for living organisms (Brown and Smith, 1980).

Trivalent gold is considered the most toxic gold species tested. Trivalent gold as tetrachloroaurate ($AuCl_4^-$) depressed chlorophyll concentrations, photosynthetic rates, and thiol levels at concentrations greater than 0.0985 mg Au^{3+}/L over a 21-day period in *Amphora coffeaeformis*, a marine diatom (Robinson et al., 1997). Cells were able to recover at concentrations <0.985 mg Au^{3+}/L due to cellular and photoreduction of $AuCl_4^-$. Adverse effects were exacerbated by Cu^{2+}. Uptake of Au^{3+} by *Amphora* is not an energy-dependent process. At 0.394-0.985 mg Au^{3+}/L, only 30% of the total gold uptake after 24 h was internal. Robinson et al. (1997) concluded that algal cells, alive or dead, rapidly accumulate Au^{3+} and start to reduce it within 2 days to elemental gold (Au^o) and monovalent gold (Au^+).

Gold can be sequestered from acid solutions by dead biomass of the brown alga *Sargassum natans* and deposited in its elemental form (Kuyucak and Volesky, 1989). The cell wall of *Sargassum* was the major site for gold deposition, with carbonyl groups (C=O) playing a major role in binding, and *N*-containing groups a lesser role. Like activated carbon, the biomass of *S. natans* is extremely porous and this accounts, in part, for its ability to accumulate gold (Kuyucak and Volesky, 1989). Ground-dried seaweeds representing species of *Sargassum*, *Gracilaria*, *Eisenia*, and *Ulva*, accumulated gold from solution (Zhao et al., 1994). Seaweeds can remove 75-90% of gold within 60 min at pH 2 from solutions containing 5.0 mg Au^{3+}/L.

2.18 Iron

Open ocean marine diatoms are bioindicators of available iron originating from atmospheric dust. Depending on the physicochemical characteristics of the dust, phytoplankton growth is stimulated (Visser et al., 2003). In the laboratory, Arctic diatoms (*Actinocyclus* sp., *Thalassiosira* sp.) were exposed to road dust containing $FeCl_3$. Diatom growth was positively correlated with dissolution rate of iron in seawater and concentration of amorphous iron in dust (Visser et al., 2003).

Iron concentrations in marine algae and macrophytes are elevated (Table 2.12) when compared to other trace elements; substantial variability in iron content was evident, even among closely related species. The high iron content is probably due to several factors. First, is the established need of iron by marine plants for normal growth (Davies, 1970; Goldberg, 1952; Lewin and Chen, 1971; Smith, 1970). Davies (1970) studied the growth of *D. tertiolecta* under iron-limiting conditions and concluded that cells maintained in culture under iron-deficient conditions gradually adapt to that situation. Moreover, growth rate of

Table 2.12: Iron Concentrations in Field Collections of Algae and Macrophytes

Organism	Concentration	Reference[a]
Algae and macrophytes; whole		
Admiralty Bay, Antarctica; January 2004; 3 spp.	91.0-4450.0 DW	30
Irish Sea	34.0-4680.0 DW	1
Japan; 11 spp.	25.0-590.0 DW	2
Japan; 44 spp.	74.0-3410.0 DW	4
Sea of Japan; 12 spp.	82.0-886.0 DW	5
Goa, India; 33 spp.	130.0-1796.0 DW	6
Argentina; summer 2002; 3 sites; *Porphyra columbina* vs. *Ulva* sp.; whole	213.0-440.0 DW vs. 201.0-532.0 DW	34
Brown alga, *Ascophyllum nodosum*; whole		
Trondheimsfjord, Norway	52.0-291.0 DW	7
United Kingdom		
Menai Straits	86.0 DW	8
Dulas Bay	30.0 DW	8
Mangrove, *Avicennia marina*; Queensland, Australia; pneumatophores	184.0-4687.0 DW	37
Alga, *Chaetomorpha brychagona*; whole		
Near iron-ore tailings	7624.0-10,346.0 DW	9
Reference site	114.0 DW	9
Chlorophyceae; whole; 5 spp.	4735.0-13,816.0 DW	10

(*Continues*)

Table 2.12: Cont'd

Organism	Concentration	Reference[a]
Alga, *Enteromorpha cunita*; whole		
Near iron-ore tailings	2491.0-6772.0 DW	9
Reference site	143.0 DW	9
Fucus vesiculosus		
Whole; Norway	140.0 DW	7
Whole; United Kingdom		
Coastal waters	249.0 DW	11
Menai Straits	218.0 DW	8
Dulas Bay	75.0 DW	8
Tamar estuary	808.0 (401.0-1300.0) DW	12
Thallus	506.0-1920.0 DW	13
Halimeda spp., whole	24.0-27.6 DW	14, 15
Hizakia fusiforme; whole; varying lengths		
up to 10 cm	91.0 DW	2
10-20 cm	82.0 DW	2
20-30 cm	72.0 DW	2
India; Tamil Nadu; leaves; January 2007		
Mangroves; 8 spp.	193.6 (103.8-293.0) DW	38
Halophytes; 5 spp.	418.6 (311.2-977.3) DW	38
Mangroves; Gulf of California, Mexico; 998-1999; twigs vs. leaves		
Rhizophora mangle	41.0 DW vs. 115.0 DW	31
Avicennia germinans	77.0 DW vs. 171.0 DW	31
Laguncularia racemosa	28.8 DW vs. 97.0 DW	31
Marsh plants; whole; 14 spp.	47.0-200.0 DW	16
Macroalgae (6 spp.) vs. phytoplankton; Gulf of California, Mexico; 1998-1999; whole	318.0-2131.0 DW vs. 889.0 DW	31
Macroalgae; epiphytic on estuarine mangrove aerial roots; Sydney region, Australia		
Caloglossa leprieurii	23,000.0 DW	33
Caloglossa nipae	3100.0 DW	33
Bostrychia sp.	18,000.0 DW	33
Padina australis, whole	Max. 504.0 DW	17
Padina gymnospora, whole	Max. 4100.0 DW	18
Phaeophyceae, whole, 4 spp.	348.0-15,743.0 DW	10

(Continues)

Table 2.12: Cont'd

Organism	Concentration	Reference[a]
Phytoplankton		
Whole	624.0 (280.0-1290.0) DW	19
Organic fractions		
No titanium group	224.0 DW	20
Titanium group	1510.0 DW	20
Strontium-concentrated group	231.0 DW	20
Siliceous frustules		
No titanium group	220.0 DW	20
Titanium group	560.0 DW	20
Strontium-concentrated group	180.0-500.0 DW	20
Macrophyte, *Posidonia oceanica*		
Italy		
Sheath	0.5-4.8 DW	29
Rhizome	0.09-0.1 DW	29
Spain; Mediterranean Sea; leaves; 2001-2003	72.7 (31.1-167.7) DW	36
Mangrove, *Rhizophora mangle*; summer 2001; Natal, Brazil; leaves	76.7-120.0 DW	35
Rhodophyceae, whole, 9 spp.	486.0-2301.0 DW	10
Sargassum fulvellum, whole	130.0-265.0 DW; 1137.0-1950.0 AW	21
Sargassum horneri		
Whole	98.0 DW	2
Whole; varying lengths		
up to 50 cm	137.0 DW	2
50-100 cm	80.0 DW	2
100-150 cm	58.8 DW	2
Lamina	335.0 DW	2
Air bladder	61.0 DW	2
Stipe	41.0 DW	2
Receptacle	18.0 DW	2
Seaweeds, whole		
South Africa; 11 spp.	3.3-230.0 FW	22, 23
Coast of Japan; 38 spp.	52.0-3296.0 DW	21
Norway	33.0-931.0 DW	24
Japan; 30 spp.	501.0 DW (47.0-3310.0) DW	25
Korea; 20 spp.	148.0-2343.0 DW	26
May vs. June; 5 spp.	Max. 638.0 DW vs. Max. 3380.0 DW	27

(*Continues*)

Table 2.12: Cont'd

Organism	Concentration	Reference[a]
Spartina alterniflora		
Sprout	700.0-4800.0 DW	3
Mature plant	320.0-2600.0 DW	3
Dead plant	2250.0-10,500.0 DW	3
Leaves and stalks	750.0 DW	3
Seaweed, *Ulva lactuca*; Gulf of California, Mexico; 2002; whole		
August	3652.0 DW	32
Rest of year	500.0-1500.0 DW	32
Eelgrass, *Zostera marina*; Oregon		
Leaf	2422.0 DW	28
Rhizome	1584.0 DW	28
Root	11,933.0 DW	28

Values are in mg Fe/kg fresh weight (FW), dry weight (DW), or ash weight (AW).
[a]1, Culkin and Riley, 1958; 2, Ishii et al., 1978; 3, Williams and Murdock, 1969; 4, Ishibashi and Yamamoto, 1960; 5, Gryzhankova et al., 1973; 6, Agadi et al., 1978; 7, Lande, 1977; 8, Foster, 1976; 9, Wong et al., 1979; 10, Sivalingam, 1978b; 11, Preston et al., 1972; 12, Bryan and Uysal, 1978; 13, Bryan and Hummerstone, 1973; 14, Bohn, 1979; 15, Khristoforova and Bogdanova, 1980; 16, Lytle et al., 1973; 17, Pillai, 1956; 18, Stevenson and Ufret, 1966; 19, Horowitz and Presley, 1977; 20, Martin and Knauer, 1973; 21, Ishibashi et al., 1964a; 22, Van As et al., 1973; 23, Van As et al., 1975; 24, Lunde, 1970; 25, Yamamoto et al., 1976; 26, Pak et al., 1977; 27, Black and Mitchell, 1952; 28, Kaldy, 2006; 29, Baroli et al., 2001; 30, Santos et al., 2006; 31, Ruelas-Inzunza and Paez-Osuna, 2006; 32, Orduna-Rojas and Longoria-Espinoza, 2006; 33, Melville and Pulkownik, 2007; 34, Perez et al., 2007; 35, Silva et al., 2006; 36, Fourqurean et al., 2007; 37, Preda and Cox, 2002; 38, Agoramoorthy et al., 2008.

cells may be expressed mathematically in terms of the metabolic iron content of the cells, and that chlorophyll production by phytoplankton may be governed by the supply of organoiron complexes. Second, is the ability of most algal species to bioconcentrate iron from the surrounding environment. Typical concentration factors ranged from 810,000 for thallus of *F. vesiculosus* (Bryan and Hummerstone, 1973) to a low of 100-300 for *Ceramium* and *Enteromorpha* (Taylor and Odum, 1960); intermediate values for various species were also documented (Ichikawa, 1961; Ishii et al., 1978; Ryndina, 1976; Van As et al., 1973; Young, 1975; Yoshimura et al., 1976). Davies (1967) reports that iron is present in a particulate form, probably as hydrous ferric oxide, which can scavenge isotopes of other elements. The ability of a diatom *P. tricornutum* to adsorb iron from the medium at high ambient iron levels was documented (Davies, 1967). Third, is the possibility of accidental contamination of samples during collection. Martin and Knauer (1973) observed that iron contamination associated with field samples of phytoplankton included metal particles in samples, rust form stainless hydrowire, hull paint, and open ocean tar balls. Fourth, and most likely, iron contamination as a result of industrial and other anthropogenic activities may account for a significant amount of the variability observed in Table 2.12, especially among estuarine species. For example, the highest iron (and manganese) concentrations in salt marsh plants grown on contaminated soils occurred under acid-reducing conditions, and this is in accord with effects of reducing conditions on chemical speciation of iron and manganese and their bioavailability (Gambrell et al., 1977). In another study, thalli of marine algae from intertidal environments, especially *P. elegans* and *Cladophora rupestris*, retained about 29% of their FW as iron after shaking for 6 h in solutions containing 5000.0 mg iron ore dust/L. On this basis, Boney (1978) concluded that marine algae have potential as bioindicators of ore dust loading of seawater due to accidental spillage at ore unloading terminals. Bryan and Hummerstone (1973) state that other factors should be considered when interpreting the significance of iron loadings in marine plants, such as seasonal changes, location of the plant in the intertidal zone, age of the plant, and portion analyzed. Ishibashi et al. (1964b) showed that iron content in 38 species of dried seaweeds from Japan ranged from a low of 52.0 mg/kg in *Sargassum gigantifoleum* to 3296.0 mg/kg *Caulerpa okamurai*; however, these differences were not attributable to season or habitat. Green seaweeds usually contained the highest iron concentrations and red seaweeds the lowest; however in almost every case, high iron content was positively associated with high cobalt content, and low iron content with low cobalt loadings (Ishibashi et al., 1964b), suggesting a possible complexation mechanism.

2.19 Lead

Lead is not essential for photosynthetic organisms, and excessive amounts cause growth inhibition as well as a reduction in mitosis, photosynthesis, and water absorption (Demayo et al., 1982). Mangroves as a group tend to operate as excluder species for nonessential metals and as regulators of essential metals. Unlike essential metals, such as copper and zinc, the nonessential lead is excluded from leaf tissues regardless of environmental concentrations (MacFarlane et al., 2007; Table 2.13).

Many factors can modify lead concentrations in field-collected algae, including geographic location, proximity to anthropogenic lead sources, water lead concentrations, association with strontium and titanium, and season of collection (Eisler, 1981; Table 2.13). In addition, low pH and high phosphate levels of the surrounding medium may inhibit lead uptake in *Ulva lactuca* (Shiber and Washburn, 1978). Differences in concentrations of lead—as well as copper, iron, manganese, and zinc—among Caulerpaceae seaweeds from shallows of coral islands in the southwestern Pacific Ocean were associated with differences in naturally occurring geochemical properties of island waters (Khristoforova and Bogdanova, 1980). Lead concentrations in field-collected phytoplankton samples may be contaminated by sampling devices. Contamination from sampling is estimated to approach 1500.0 mg Pb/kg from metal particles, 660.0 mg/kg stainless hydrowire, and 485.0 mg/kg from hull paint (Martin and Knauer, 1973). In addition, tetramethyllead (TML) was 4-10 times more effective than tetraethyllead (TEL) salts in producing deleterious effects in *Dunaliella* spp. (Marchetti, 1978), and this is presumably reflected in lead-uptake kinetics.

Table 2.13: Lead Concentrations in Field Collections of Algae and Macrophytes

Organism	Concentration	Reference[a]
Algae, whole		
India, Goa; 33 spp.	usually 6.1-28.3 DW; max. 197.5 DW	1
Norway; west coast	Max. 1200.0 DW	2
Spain; Rio Tinto estuary	Max. 22.0 DW	2
United Kingdom; 11 spp.	8.1-56.6 DW	3
Argentina; summer 2002; 3 sites; *Porphyra columbina* vs. *Ulva* sp.; whole	<0.5 DW vs. 0.8-1.7 DW	23
Brown alga, *Ascophyllum nodosum*; whole		
Norway		
Eikhamrane	11.0-54.0 DW	4
Flak	<3.0 DW	4
Transferred from Eikhamrane to Flak for 120 days	<3.0-5.0 DW	4
Sorfjorden	10.4-85.0 DW	5
Hardangerfjord	<3.0-95.0 DW	6
United Kingdom		
Menai Straits	2.3 DW	7
Dulas Bay	2.2 DW	7
Irish coast; 1999-2000	0.12-2.1 DW	34
Mangrove, *Avicennia marina*		
Shenzhen, China		
Sediment	31.0-34.0 DW	28, 29
Root	3.4-3.5 DW	28, 29
Leaf	1.8-1.9 DW	28, 29
Hong Kong		
Sediment	33.0 DW	30
Root	15.0 DW	30
Leaf	8.0 DW	30
Southeast Australia		
Sediment	100.0 DW	31
Root	164.0 DW	31
Leaf	5.0 DW	31
Alga, *Blidingia minima*; whole; Raritan Bay, New Jersey, United States; water lead content of 0.002 mg Pb/L or 0.01 mg Pb/L		
0.002 mg/L	12.0 DW	8
0.01 mg/L	172.0 DW	8

(*Continues*)

Table 2.13: Cont'd

Organism	Concentration	Reference[a]
Chesapeake, Bay; submerged aquatic vegetation; 5 spp.; 1989-1981; whole	7.4 (0.5-30.0) DW	18
Green alga, *Cladophora* sp.; Baltic Sea; 2000-2003		
Gulf of Gdansk	5.1 (1.8-13.2) DW	24
Vistula Lagoon	8.4 DW; max. 11.4 DW	24
Reference site	4.3 DW; max. 6.2 DW	24
Alga, *Enteromorpha linza*; whole; Raritan Bay; water lead content of 0.002 or 0.01 mg Pb/L		
0.002 mg/L	18.0 DW	8
0.01 mg/L	68.0 DW	8
Chlorophyceae, whole, 5 spp.	13.0-58.0 DW	9
Green alga, *Enteromorpha* sp.; whole; 2000-2003; Gulf of Gdansk, Baltic Sea	3.8 (0.9-10.4) DW	20
Alga, *Fucus vesiculosus*; whole		
United Kingdom		
Severn estuary	8.5 DW	10
Dulas Bay	2.3 DW	7
Menai Straits	3.2 DW	7
Irish Sea	4.0 DW	11
Tamar estuary	26.5 (5.9-109.0) DW	12
Norway, Sorfjorden	27.5-163.3 DW	5
United States, Raritan Bay; water content of 0.002 or 0.01 mg Pb/L		
0.002 mg/L	8.0 DW	8
0.01 mg/L	38.0 DW	8
India; Tamil Nadu; leaves; January 2007		
Mangroves; 8 spp.	16.7 (2.2-27.3) DW	39
Halophytes; 5 spp.	12.6 (10.6-17.0) DW	39
Macroalgae; whole; 4 spp.; Strait of Magellan, Chile; 2000-2001	<0.5-11.2 DW	36
Macroalgae (6 spp.) vs. phytoplankton; Gulf of California, Mexico; 1998-1999; whole	0.29-3.1 DW vs. 23.0 DW	21

(*Continues*)

Table 2.13: Cont'd

Organism	Concentration	Reference[a]
Macroalgae; epiphytic on estuarine mangrove aerial roots; Sydney region, Australia		
Caloglossa leprieurii	257.0 DW	25
Caloglossa nipae	61.0 DW	25
Bostrychia sp.	264.0 DW	25
Mangroves; Gulf of California, Mexico; 1998-1999; twigs vs. leaves		
Rhizophora mangle	0.9 DW vs. 2.1 DW	21
Avicennia germinans	0.4 DW vs. 2.2 DW	21
Laguncularia racemosa	0.4 DW vs. 0.94 DW	21
Marsh plants, whole, 4 spp.; Texas, United States	0.83-4.2 DW	13
Mexico; Gulf of California; 1999-2000		
Microalgae, *Coscinodiscus centralis*; whole	2.3 DW	37
Macroalgae; *Gracilaria* sp. vs. *Polysiphonia* sp.; whole	4.9 DW vs. 3.1 DW	37
Mangroves; 3 spp.; leaves	0.9-2.2 DW	37
Phaeophyceae, whole, 4 spp.	5.2-49.7 DW	9
Macrophyte, *Posidonia oceanica*		
Blade; Mediterranean Sea; summers 2004-2005; 3 sites Italy	1.4-1.8 DW	27
Sheath	5.2-11.2 DW	19
Rhizome	0.8-2.4 DW	19
Corsica, France; summer 2003; adult leaves; blades vs. sheaths	2.2 (1.3-3.4) DW vs. 0.3 (0.2-0.5) DW	35
Mangrove, *Rhizophora stylosa*; Yingluo Bay, China		
Sediment	10.0 DW	32
Root	0.9 DW	32
Leaf	0.8 DW	32
Rhodophyceae, whole, 8 spp.	1.7-44.8 DW	9
Saltmarsh macrophytes; 3 spp.; July 2005; Yangtze River estuary, China		
Above ground portions	14.0 DW	26
Below ground portions	28.0 DW	26

(Continues)

Table 2.13: Cont'd

Organism	Concentration	Reference[a]
Seaweeds, whole		
Korea, 17 spp.	0.2-1.3 DW; 0.05-0.3 FW	14
Korea, 20 spp.	1.7-82.0 DW	15
May vs. June, 5 spp.	2.0-10.0 DW vs. 4.0-13.0 DW	16
Seaweed, *Ulva lactuca*		
Gulf of California, Mexico; 2002; whole		
August (low)	94.0 DW	22
February (high)	259.0 DW	22
India; whole; 2004-2005	11.6 (5.9-20.3) DW	33
Ulva spp., whole		
Puget Sound, Washington, United States	1.0-2.0 DW	17
Raritan Bay, New Jersey		
Water content 0.002 mg Pb/L	20.0 DW	8
Water content 0.01 mg Pb/L	76.0 DW	8
Eelgrass, *Zostera marina*; whole; Baja California, Mexico; March 2000; reference site vs. salt mining site	14.9 (2.6-46.0) DW vs. 8.8 (4.5-18.8) DW	38

Values are in mg Pb/kg dry weight (DW) or fresh weight (FW).
[a] 1, Agadi et al., 1978; 2, Stenner and Nickless, 1974; 3, Riley and Roth, 1971; 4, Myklestad et al., 1978; 5, Melhuus et al., 1978; 6, Haug et al., 1974; 7, Foster, 1976; 8, Seeliger and Edwards, 1977; 9, Sivalingam, 1978b; 10, Butterworth et al., 1972; 11, Preston et al., 1972; 12, Bryan and Uysal, 1978; 13, Lytle et al., 1973; 14, Kim, 1972; 15, Pak et al., 1977; 16, Black and Mitchell, 1952; 17, Schell and Nevissi, 1977; 18, Di Giulio and Scanlon, 1985; 19, Baroli et al., 2001; 20, Zbikowski et al., 2006; 21, Ruelas-Inzunza and Paez-Osuna, 2006; 22, Orduna-Rojas and Longoria-Espinoza, 2006; 23, Perez et al., 2007; 24, Zbikowski et al., 2007; 25, Melville and Pulkownik, 2007; 26, Quan et al., 2007; 27, Lafabrie et al., 2007b; 28, Tam et al., 1995; 29, Zheng and Lin, 1996; 30, Chen et al., 2003; 31, MacFarlane et al., 2003; 32, Zheng et al., 1997; 33, Kamala-Kannan et al., 2008; 34, Morrison et al., 2008; 35, Lafabrie et al., 2008a; 36, Astorga-Espana et al., 2008; 37, Ruelas-Inzunza and Paez-Osuna, 2008; 38, Marcias-Zamora et al., 2008; 39, Agoramoorthy et al., 2008.

High bioconcentration factors (BCFs) for lead from ambient seawater by marine plants is documented by several investigators. BCFs ranged from 13,000 to 82,000 for algae from Raritan Bay, New Jersey (Seeliger and Edwards, 1977), and from 1200 to 26,000 for algae from Sorfjorden, Norway (Melhuus et al., 1978). Under laboratory conditions, *U. reticulata* held in seawater for 48 h containing 50.0 mg Pb/L had a BCF of 124 (Sivalingam, 1978b). The ability of mangrove *Rhizophora mangle* and the macrophyte *Thallassium testinudum* to accumulate lead from sewage sludge is reported by Montgomery et al. (1978). After exposure for 120 days, mangrove roots contained up to 5.0 mg/kg more lead than controls; for *Thallassium* this value was 80.0 mg/kg.

Saltmarsh plants contribute significantly to lead budgets in estuaries (Reboreda and Cacador, 2007). Below ground portions of two species of saltmarsh plants collected from the Tagus estuary, Portugal, in March 2004 had elevated concentrations of lead when compared to aerial portions. Values for *S. maritima*, in mg Pb/m^2 DW, were 670.0 in roots, 7.9 in stems, and 3.0 in leaves. For *H. portulacoides*, these values were 1147.1 in roots, 22.5 in stems, and 7.7 in leaves (Reboreda and Cacador, 2007).

Differential sensitivity of algal species to lead is well documented. Growth rate and biomass of the comparatively sensitive *S. costatum* were inhibited by dissolved lead concentrations between 0.00005 and 0.01 mg/L (Rivkin, 1979; USEPA, 1985b). Growth enhancement was observed among five species of intertidal algae at concentrations between 0.045 and 0.81 mg Pb/L (Stromgren, 1980); higher lead concentrations were associated with growth inhibition over a 10-day period in *Pelvetia canaliculatum*, three species of *Fucus*, and *A. nodosum*. At 0.1-1.0 mg Pb/L, photosynthesis and respiration were significantly reduced in several species (Woolery and Lewin, 1976; Zavodnik, 1977). In another study, *D. salina* was relatively unaffected at 0.3 mg Pb/L, but showed reduced growth and other adverse effects at 0.9 mg Pb/L and higher (Pace et al., 1977). In the range 0.25-2.0 mg Pb/L, growth was inhibited in 18 species of unicellular algae (Berland et al., 1976) and in *D. tertiolecta* (Stewart, 1977a). At 0.5 mg Pb/L, *Tiffaniella snyderae* showed reduced growth and reduced cell division rate over 28 days, and at 10.0 mg Pb/L slower growth occurred for *Platythamnion* and *Plenosporium* (Stewart, 1977b). At more than 10.0 mg Pb/L, kelp *M. pyrifera* showed 50% inhibition of photosynthesis in 4 days (Clendenning and North, 1959). However, algal cultures of *Phaeodactylum*, *Tetraselmis*, *Cricosphaera*, and *Dunaliella* all grew well at <20.0 mg Pb/L with adverse effects at higher concentrations (Bentley-Mowat and Reid, 1977). At 25.0-100.0 mg Pb/L, reduced oxygen uptake was measured in three species of unicellular algae (Mills and Colwell, 1977), and at 100.0 mg Pb/L *S. alterniflora* seedlings failed to grow with most dying within 8 weeks (Dunstan et al., 1975). Organolead compounds were more toxic to *P. tricornutum* than were inorganic lead compounds, and organoethylleads were more toxic than organomethylleads (Maddock and Taylor, 1980). Concentrations of various lead compounds causing 50% mortality in 96 h were greater than 5.0 mg/L for inorganic lead, 1.3 mg/L for TML, 0.8 mg/L for trimethyllead, and 0.1 mg/L for both triethyllead and TEL (Maddock and Taylor, 1980). A similar pattern held for growth inhibition of *D. tertiolecta*: 50% inhibition in 96 h at 0.15 mg TEL/L versus 1.65 mg TML/L (USEPA, 1985b).

Studies on lead uptake and retention in *P. tricornutum* and *Platymonas subcordiformes* showed that both species accumulated lead from the medium at ambient concentrations of 0.02 mg Pb/L and higher (Schulz-Baldes and Lewin, 1976). In the first phase, usually completed within minutes after lead addition, cells of *Phaeodactylum* became saturated when the lead burden reached a remarkable 11,640.0 mg/kg DW. In the second phase, the lead content of *Platymonas* continued to rise slowly, whereas that of *Phaeodactylum* declined after 2-3 days. The addition of chelating agents, such as EDTA, inhibited lead uptake by

Phaeodactylum. In both species, the content of bound lead increased with increasing exposure time, suggesting that during prolonged exposure to lead solutions the metal ions are first adsorbed to the cell surface and then translocated to within the cell wall, to the plasma membrane, and eventually to the cytoplasm.

2.20 Lithium

Available data indicate that lithium concentrations in plant tissues ranged from 4.0 to 8.0 mg Li/kg DW (Table 2.14). Diminished rhizoid development was observed in eggs of *Fucus evanescens* subsequent to rearing in 500.0 mg Li/L as LiI or LiCl (Nakazawa, 1977).

2.21 Manganese

Early interest in manganese was stimulated by the observation that ^{54}Mn, a common fallout product from early nuclear tests, was preferentially accumulated by marine algae over other radioisotopes examined (Slowey et al., 1965). Stable manganese is also readily accumulated by algae from seawater. Manganese BCFs of 3900 are reported for *A. nodosum* (Foster, 1976), 4600-26,000 for *F. vesiculosus* (Foster, 1976), 10,000-200,000 for *Laminaria japonica* (Yoshimura et al., 1976), and 9000 for *Sargassum* spp. (Ishii et al., 1978). Similar concentration factors are for other algal species (Ichikawa, 1961; Ryndina, 1976; Van As et al., 1973). Manganese is an essential element for normal growth of *D. tertiolecta*, and probably other species, and adverse effects occur in *Dunaliella* when manganese concentrations in the growth medium were less than 0.1 mg Mn/L (Noro, 1978).

Manganese concentrations in field collections of marine algae collected worldwide ranged from 3.8 to 3421.0 Mn/kg DW (Table 2.15). The high variability between and among species are attributed to a variety of biotic and extrinsic modifiers, including accidental contamination during the sampling process (Martin and Knauer, 1973), season of collection (Bryan and Hummerstone, 1973; Pillai, 1956), plant part analyzed (Bryan and Hummerstone, 1973; Khristoforova et al., 1976), tidal fluctuations (Sanders, 1978),

Table 2.14: Lithium Concentrations in Field Collections of Algae and Macrophytes

Organism	Concentration	Reference[a]
Algae, whole, 3 spp.	4.0-6.0 DW	1
Diatoms, whole	150.0 AW	2
Laminaria cloustini; frond; sterile vs. sporing	6.0 DW vs. 4.0 DW	1
Laminaria digitata; frond vs. stipe	8.0 DW vs. 4.0 DW	1

Values are in mg Li/kg dry weight (DW) or ash weight (AW).
[a]1, Black and Mitchell, 1952; 2, Vinogradova and Koual'skiy, 1962.

Table 2.15: Manganese Concentrations in Field Collections of Algae and Macrophytes

Organism	Concentration	Reference[a]
Algae and macrophytes; whole		
Sea of Japan; 12 spp.	5.0-805.0 DW	1
Canada, Atlantic Provinces; 11 spp.	20.0-50.0 DW	2
United Kingdom; coastal waters; 15 spp.	3.8-73.0 DW	3
Goa, India; 33 spp.	25.0-3421.0 DW	4
Japan; 12 spp.	4.4-41.0 DW	5
East Indies; 19 spp.	23.7-316.0 DW	6
Iceland; 12 spp.	13.0-680.0 DW	7
Argentina; summer 2002; 3 sites; *Porphyra columbina* vs. *Ulva* sp.; whole	31.7-41.8 DW vs. 8.1–51.4 DW	25
Brown alga, *Ascophyllum nodosum*, whole; United Kingdom	16.0-21.0 DW	8
Mangrove, *Avicennia marina*; SE Australia; pneumatophores	4.0-163.0 DW	30
Green alga, *Enteromorpha* sp.; whole; 2000-2003; Gulf of Gdansk, Baltic Sea	172.7 (30.7-384.1) DW	23
Near iron-ore tailings site vs. reference site		
Chaetomorpha brychagona	392.0-592.0 DW vs. 57.0 DW	9
Enteromorpha crinita	96.0-226.0 DW vs. 4.0 DW	9
Green alga, *Cladophora* sp.; Baltic Sea; 2002-2003		
Gulf of Gdansk	298.0 (48.7-740.0) DW	26
Vistula Lagoon	1185.0 DW; max. 1930.0 DW	26
Reference site	253.0 DW; max. 428.0 DW	26
Fucus vesiculosus, whole; United Kingdom		
Menai straits	103.0 DW	
Dulas Bay	71.0 DW	8
Irish Sea	99.0 DW	10
Tamar estuary	204.0 (85.0-393.0) DW	11
India; Tamil Nadu; leaves; January 2007		
Mangroves; 8 spp.	168.7 (51.1-391.0) DW	31
Halophytes; 5 spp.	38.3 (11.7-88.6) DW	31

(*Continues*)

Table 2.15: Cont'd

Organism	Concentration	Reference[a]
Macroalgae (6 spp.) vs. Phytoplankton; Gulf of California, Mexico; 1998-1999; whole	29.0-120.0 DW vs. 289.0 DW	24
Macroalgae; epiphytic on estuarine mangrove aerial roots; Sydney region, Australia		
Caloglossa leprieurii	1400.0 DW	27
Caloglossa nipae	1600.0 DW	27
Bostrychia sp.	1800.0 DW	27
Mangroves; Gulf of California, Mexico; 1998-1999; twigs vs. leaves		
Rhizophora mangle	47.0 DW vs. 30.0 DW	24
Avicennia germinans	11.1 DW vs. 64.0 DW	24
Laguncularia racemosa	106.0 DW vs. 28.0 DW	24
Marsh plants; whole; Harbour Island, Texas; 14 spp.	9.2-170.0 DW	12
Puerto Rico; whole		
Padina gymnospora	99.0 DW	13
Thalassia testudinium	49.0 DW	13
Phytoplankton		
Whole		
Texas Continental Shelf	21.7 (4.4-41.5) DW	14
Monterey Bay, California	35.0 DW	15
Organic fractions vs. siliceous frustules		
No titanium group	6.1 DW vs. not detected	16
Titanium group	13.3 DW vs. 4.3 DW	16
Strontium- concentrated group	7.7 DW vs. 1.2-22.0 DW	16
Mangrove, *Rhizophora mangle*; summer 2001; Natal, Brazil; leaves	190.0-550.0 DW	28
Sargassum horneri		
Lamina	12.7 DW	5
Air bladder	7.0 DW	5
Stipe	4.3 DW	5
Receptacle	3.4 DW	5
Whole, varying lengths		
Up to 50 cm	9.9 DW	5
50-100 cm	8.4 DW	5
100-150 cm	8.7 DW	5

(Continues)

Table 2.15: Cont'd

Organism	Concentration	Reference[a]
Sargassum spp.; whole; south Texas Continental Shelf	31.4 (12.0-89.5) DW	14
Seaweeds; whole		
Korea; 20 spp.	15.0-191.0 DW	17
South Africa; 5 spp.	0.3-6.1 FW	18, 19
Norway	4.0-164.0 DW	20
United Kingdom; 5 spp.; May vs. June	22.0-155.0 DW vs. 36.0-121.0 DW, 8.9-24.0 FW	21
Spartina alterniflora		
Sprout	50.0-130.0 DW	22
Mature plant	30.0-95.0 DW	22
Dead plant	90.0-330.0 DW	22
Leaves and stalks	50.0 DW	22
Eelgrass, *Zostera marina*; whole; Baja California, Mexico; reference site vs. salt mining site	61.0 (7.2-476.0) DW vs. 40.0 (18.0-73.0) DW	29

Values are in mg Mn/kg fresh weight (FW) or dry weight (DW).
[a]1, Gryzhankova et al., 1973; 2, Young and Langille, 1958; 3, Riley and Roth, 1971; 4, Agadi et al., 1978; 5, Ishii et al., 1978; 6, Sivalingam, 1978b; 7, Munda, 1978; 8, Foster, 1976; 9, Wong et al., 1979; 10, Preston et al., 1972; 11, Bryan and Uysal, 1978; 12, Lytle et al., 1973; 13, Stevenson and Ufret, 1966; 14, Horowitz and Presley, 1977; 15, Knauer and Martin, 1973; 16, Martin and Knauer, 1973; 17, Pak et al., 1977; 18, Van As et al., 1973; 19, Van As et al., 1975; 20, Lunde, 1970; 21, Black and Mitchell, 1952; 22, Williams and Murdock, 1969; 23, Zbikowski et al., 2006; 24, Ruelas-Inzunza and Paez-Osuna, 2006; 25, Perez et al., 2007; 26, Zbikowski et al., 2007; 27, Melville and Pulkownik, 2007; 28, Silva et al., 2006; 29, Marcias-Zamora et al., 2008; 30, Preda and Cox, 2002; 31, Agoramoorthy et al., 2008.

abundance of lithogenic materials in the immediate environment (Lowman et al., 1967), and proximity to anthropogenic sources (Table 2.15). For example, tidal fluctuations of 0.6-0.8 m were stimulatory to manganese-loaded phytoplankton, but the reverse was observed at higher tidal fluctuations of 1.0-1.2 m (Sanders, 1978). Lowered manganese burdens and lowered ash content occur in various species of *Posidonia, Sargassum, Dictyota*, and *Dictyopterus* collected long distances from the mouths of rivers in Puerto Rico when compared to conspecifics collected nearer the river mouth; these differences were attributed to the relatively high concentrations of lithogenic materials in riverine discharges to the sea (Lowman et al., 1967).

2.22 Mercury

The highest mercury concentration recorded in marine plants is 51.5 mg/kg DW in leaves of the marine flowering plant, *P. oceanica*, collected from a sewer outfall in Marseille, France (Table 2.16; Augier et al., 1978b). Concentrations of total mercury in marine flora

Table 2.16: Mercury Concentrations in Field Collections of Algae and Macrophytes

Organism	Concentration	Reference[a]
Algae and macrophytes, whole		
Admiralty Bay, Antarctica; 2004; 3 spp.	0.024-0.047 DW	21
Malaysia, 26 spp.	Max. 0.35 DW	1
Korea, 17 spp.	0.02-0.52 DW	2
Aleutian Islands; 2003-2004; used as food; 4 spp.	0.001-0.005 FW; max. 0.039 FW	23
Brown alga, *Ascophyllum nodosum*, whole		
60 cm length vs. 100-140 cm length	0.07 FW vs. 0.11 FW	3
Eikhamrane, Norway vs. Flak, Norway	0.012-1.09 DW vs. 0.02-0.03 DW	4
Transplanted from Eikhamrane to Flak for 4 months	0.04-0.95 DW	4
Lofoten, Norway	0.05-0.08 DW	5
Trondheimsfjord, Norway	0.05-0.18 DW	5
Hardangerfjord, Norway	0.05-20.0 DW	5
Seaweeds, *Fucus* spp.; NW Portugal; September 2002		
Receptacles	0.012-0.061 DW	22
Stipes	0.028-0.221 DW	22
Holdfasts	0.029-0.287 DW	22
India; Tamil Nadu; January 2007; leaves		
Mangroves; 8 spp.	0.06 (0.03-0.07) DW	28
Halophytes; 5 spp.	0.43 (0.09-0.94) DW	28
Kelp, *Laminaria digitata*, whole		
Firth of Tay, Scotland	0.13 FW; 0.79 DW	6
Solent region, England	0.17 DW	7
Macroalgae; whole; 4 spp.; Strait of Magellan, Chile; 2000-2001	<0.001-0.02 DW	27
Phytoplankton, whole		
Antarctica, Terra Nova Bay; 1989-91	0.04 DW	8
Chesapeake Bay, Maryland	0.11-0.13 DW	9
North Atlantic Ocean; offshore	0.05 FW	10
Japan, Minamata Bay; 1974	Max. 0.32 DW	20
Spain and Portugal	Max. 1.2 DW	11
Norway, west coast	0.6-25.2 DW	12
Laver, *Porphyra umbilicalus*; whole; United Kingdom	0.5 FW; 2.4 DW	13

(*Continues*)

Table 2.16: Cont'd

Organism	Concentration	Reference[a]
Marine flowering plant, *Posidonia oceanica*		
Near sewer outfall; Marseille, France		
Rhizomes	2.5 DW	19
Leaves	51.5 DW	19
Roots	0.6 DW	19
Dead sheaths of shoots; mercury-contaminated site (Tuscany, Italy) vs. reference site (Corsica, France)		
1995	0.21 DW vs. 0.05 DW	24
1996	0.23 DW vs. 0.05 DW	24
1997	0.3 DW vs. 0.06 DW	24
2003	0.32 DW vs. 0.05 DW	24
Mediterranean Sea; 3 sites; summers 2004-2005; blades	0.05-0.13 DW	25
Corsica, France; summer 2003; adult leaves; blades vs. sheaths	0.05 (0.03-0.07) DW vs. 0.04 (0.01-0.05) DW	26
Saltmarsh grass, *Spartina alterniflora*; Brunswick, Georgia, United States		
Whole	Max. 1.4 DW	14
Leaves and stalks	0.2 DW	14, 15
Rhizomes	0.5-0.7 DW	16
Roots	0.7-8.7 DW	16
Base of stalk	0.6-1.2 DW	16
Stalk	0.4-1.1 DW	16
Leaves	0.4-1.1 DW	16
Alga, *Ulva* spp., whole		
Firth of Tay, Scotland	6.3 FW; 25.5 DW	6
Minamata Bay, Japan	Max. 14.0 DW	17
Puget Sound, Washington State	0.005-0.011 DW	18

Values are in mg Hg/kg fresh weight (FW) or dry weight (DW).
[a]1, Sivalingam, 1980; 2, Kim, 1972; 3, Augier et al., 1978a; 4, Myklestad et al., 1978; 5, Haug et al., 1974; 6, Jones et al., 1972; 7, Leatherland and Burton, 1974; 8, Bargagli et al., 1998; 9, Cocoros et al., 1973; 10, Greig et al., 1975; 11, Stenner and Nickless, 1975; 12, Skei et al., 1976; 13, Preston et al., 1972; 14, Windom et al., 1976; 15, Windom, 1975; 16, Windom, 1973; 17, Matida and Kumada, 1969; 18, Schell and Nevissi, 1977; 19, Augier et al., 1978b; 20, Nishimura and Kumagi, 1983; 21, Santos et al., 2006; 22, Cairrao et al., 2007; 23, Burger et al., 2007; 24, Lafabrie et al., 2007a; 25, Lafabrie et al., 2007b; 26, Lafabrie et al., 2008a; 27, Astorga-Espana et al., 2008; 28, Agoramoorthy et al., 2008.

were usually less than 1.0 mg/kg DW; higher concentrations were almost always associated with contamination from environmental releases of mercury due to human activities (Table 2.16; Eisler, 2006). In general, mercury concentrations were highest in marine plants collected from sewage lagoons, near chloralkali plants, and proximity to industrialized areas (Augier et al., 1978a; Chigbo et al., 1982; Eisler, 2006). The macrophyte *P. oceanica* is proposed for use in reconstructing past mercury contamination of coastal sites by measuring mercury concentrations in the dead sheaths of shoots and taking into account degradation of the sheaths (Lafabrie et al., 2007a,b).

A mercury budget for estuaries along the Georgia, United States, coast indicates that the dominant salt marsh plant *S. alterniflora* exerts a strong control on mercury migration (Gardner et al., 1975; Windom, 1973; Windom et al., 1976). Mercury entered the estuary mainly in solution, delivering about 1.5 mg of mercury annually to each square meter of salt marsh surface. Annual uptake of mercury by *Spartina* alone was about 0.7 mg/m^2 salt marsh. Mangrove vegetation plays a similar role in mercury cycling in the Florida Everglades (Lindberg and Harriss, 1974; Tripp and Harriss, 1976). Another salt marsh macrophyte, *H. portulacoides*, accumulates mercury in the below ground part of the plant (Valega et al., 2008). However, this fraction is mobile and able to return mercury to the sediment pool throughout the mineralization process, thus contributing to reactive mercury concentrations in pore waters (Valega et al., 2008).

Rapid accumulation of mercury, especially organomercury compounds, by various species of algae under controlled conditions is documented (Davies, 1974; Fang, 1973; Laumond et al., 1973). However, mercury uptake rates and accumulations by marine algae are modified by mercury concentration (Sick and Windom, 1975; Windom et al., 1976), duration of exposure (Sick and Windom, 1975), salinity of the medium (Gambrell et al., 1977), biological surface area (Sick and Windom, 1975), variability in mercury detoxification mechanisms (Davies, 1976), cell density (Delcourt and Mestre, 1978), and accumulation after death (Glooschenko, 1969). For example, *Croomonas salina* took up 1400.0 mg Hg/kg DW after 48 h in a solution containing 0.164 mg Hg/L (Parrish and Carr, 1976); *Chaetoceros galvestonensis* and *P. tricornutum* contained 7400.0 and 2400.0 mg Hg/kg DW, respectively, when cultured in media containing 0.1 mg Hg/L (*Chaetoceros*) or 0.05 mg Hg/L (*Phaeodactylum*) (Hannan et al., 1973); and *Isochrysis* sp. contained up to 1000.0 mg Hg/kg DW after exposure to 0.015 mg Hg/L (Davies, 1974). Salt marsh plants show enhanced mercury accumulation in roots at lower salinities and higher pH (Gambrell et al., 1977).

Phytotoxic effects of mercurials include reduced growth, developmental abnormalities, photosynthesis inhibition, and death. Effects in sensitive species, such as *Scripsiella faroense* (Kayser, 1976) and *Nitzchia* spp. (Saboski, 1977), were measured at 0.001 mg Hg^{2+}/L. In one case, reduced growth was documented at 0.001-0.005 mg Hg/L in natural phytoplankton from Saanach Inlet, British Columbia, using artificial ecosystems; however, recovery to above-control levels occurred in 21 days (Thomas et al., 1980). Sensitivity to mercurials was documented for many species of algae and macrophytes in the range 0.005-0.03 mg/L (Berland et al., 1976, 1977; Davies, 1976; Delcourt and Mestre, 1978; Harrison et al., 1978; Hopkins and Kain, 1971; Horvatic and Persic, 2007; Stromgren, 1980). It is emphasized that phytotoxic effects of mercury are always more pronounced when mercury is present as an organomercurial than in the inorganic form (Boney, 1971; Boney and Corner, 1959; Boney et al., 1959; Delcourt and Mestre, 1978; Eisler, 1987, 2006; Ukeles, 1962).

2.23 Molybdenum

All plants contain molybdenum and it is probably essential for growth (Schroeder et al., 1970), although the concentrations required for marine plants are not known with certainty. The highest molybdenum value recorded for any marine plant is 5.8 mg/kg DW (Table 2.17). Among 11 species of seaweeds collected from Palk Bay, Norway, molybdenum values were highest in August and highest among the Rhodophyceae (Lunde, 1970).

Molybdenum occurs naturally in seawater as molybdate ion, MoO_4^{2-}, at about 0.01 mg/L (Abbott, 1977). Despite the high concentrations of dissolved molybdenum in offshore waters, phytoplanktons from these locales contain extremely low molybdenum concentrations, almost typical of molybdenum-deficient terrestrial plants (Howarth and Cole, 1985).

Table 2.17: Molybdenum Concentrations in Field Collections of Algae and Macrophytes

Organism	Concentration	Reference[a]
Algae, whole; Canada; 11 spp.	0.23-1.36 DW	1
Argentina; summer 2002; 3 sites; Porphyra columbina vs. Ulva sp.; whole	1.1-2.0 DW vs. 0.54-0.67 DW	8
Marsh plants, whole; Texas, United States; 14 spp.	0.36-2.5 DW	2
Porphyra spp., whole; Japan	Max. 1.0 DW; Max. 17.0 AW	3, 4
Seaweeds, whole		
United Kingdom, 5 spp.	0.2-1.3 DW; 0.04-0.2 FW	5
Norway, 11 spp.	0.3-5.8 DW	6, 7

Values are in mg Mo/kg fresh weight (FW), dry weight (DW) or ash weight (AW).
[a]1, Young and Langille, 1958; 2, Lytle et al., 1973; 3, Fukai and Meinke, 1959; 4, Fukai and Meinke, 1962; 5, Black and Mitchell, 1952; 6, Lunde, 1970; 7, Friberg et al., 1975; 8, Perez et al., 2007.

The low molybdenum levels are attributed to the high concentrations of sulfate in seawater; sulfate inhibits molybdate assimilation by phytoplankton, making it less available in seawater. As one result, nitrogen fixation and nitrate assimilation—processes that require molybdenum—may require greater energy expenditure in marine waters than in freshwaters and may explain, in part, why marine ecosystems are usually nitrogen limited and lakes are not (Howarth and Cole, 1985).

Laboratory studies on uptake of ^{99}Mo by *Fucus serratus* and *Laminaria digitata* show that *Fucus* accumulates twice as much molybdenum at 15°C than at 0°C, with light intensity having little apparent effect (Penot and Videau, 1975). *Laminaria*, under similar conditions, took up more than 5 times as much molybdenum as did *Fucus*; however, molybdenum uptake was reduced dramatically on addition of about 0.6 mg KCN/L (Penot and Videau, 1975). Experimentally increasing the ratio of sulfate to molybdate inhibits molybdate uptake by marine algae, slows nitrogen fixation rates, and slows the growth of organisms that use nitrate as a nitrogen source (Howarth and Cole, 1985).

Proposed molybdenum criteria to protect marine plants indicate that 0.0177 mg/L seawater and lower are associated with deficiency (Steeg et al., 1986; Vaishampayan, 1983), 0.000014 mg/L and higher are associated with high uptake (Short et al., 1971), and concentrations greater than 50.0 mg Mo/L are associated with growth reduction and other adverse effects (Eisler, 2000b; Sakaguchi et al., 1981).

2.24 Nickel

In general, nickel concentrations in plants, other living organisms, and abiotic materials are elevated in the vicinity of nickel smelters and refineries, nickel-cadmium battery plants, sewage outfalls, electroplating plants, coal ash disposal basins, and heavily populated areas (Eisler, 2000d). Nickel concentrations in marine algae and macrophytes from pollution-free locales, on a DW basis, ranged from 0.2 to 39.1 mg/kg, with most values less than 10.0 mg/kg (Table 2.18). High concentrations up to 60.3 mg Ni/kg DW blade were also reported for the seagrass, *P. oceanica*, and were positively correlated with sediment nickel concentrations (Lafabrie et al., 2007b). Algae reportedly bioconcentrate nickel from seawater by factors that range from 146 to over 1000 (Black and Mitchell, 1952; NAS, 1975; Sivalingam, 1978b). Accordingly, elevated levels of nickel were expected in algae in samples collected near heavily populated areas (Bernhard and Zattera, 1975), especially in the vicinity of sewage outfalls (Montgomery et al., 1978). For example, turtle grass, *Thalassia testudinium*, contained up to 45.0 mg Ni/kg DW in leaves over controls during exposure for 120 days in sewage sludge; for mangrove, *R. mangle*, this value was 10.0 mg Ni/kg DW higher in roots over conspecifics from a reference site (Montgomery et al., 1978). Concentrations of nickel in *Spartina* sp. roots from the vicinity of a discharge from a nickel-cadmium battery plant ranged from 30.0 to 500.0 mg/kg DW, and reflected sediment

Table 2.18: Nickel Concentrations in Field Collections of Algae and Macrophytes

Organism	Concentration	Reference[a]
Algae and macrophytes, whole		
Japan; coastal waters; 60 spp.	2.2-6.9 (0.2-31.0) DW	1, 21
India; Goa; 27 spp.	3.5-39.1 DW	2
Canada, Atlantic Ocean coast; 11 spp.	0.3 >2.0 DW	3
Sea of Japan; 12 spp.	5.0-31.0 DW	4
England; coastal waters; 14 spp.	2.7-10.3 DW	5
Israel; Mediterranean Sea coast; 2 spp.	5.2-5.8 DW	6
Nickel-contaminated areas vs. reference site	150.0 DW vs. <15.0 DW	20
Puerto Rico; 2 spp.	20.0-27.0 DW	7
Texas; Harbour Island; 14 spp.	0.2-2.6 DW	21
Argentina; summer 2002; 3 sites; *Porphyra columbina* vs. *Ulva* sp.; whole	0.7-1.5 DW vs. 1.0-4.1 DW	26
Brown alga, *Ascophyllum nodosum*, whole		
Trondheimsfjord, Norway	1.0-22.0 DW	8
Menai Straits; United Kingdom	22.0 DW	9
Dulas Bay; United Kingdom	4.6 DW	9
Nova Scotia	0.6 DW	19
Former Soviet Union	0.4 DW	19
Scotland	0.9 FW; 1.5-6.3 DW	19
Green alga, *Cladophora* sp.; Baltic Sea; 2000-2003		
Gulf of Gdansk	3.5 (0.8-9.2) DW	27
Vistula Lagoon	11.4 DW; max. 17.1 DW	27
Reference site	6.4 DW; max. 10.6 DW	27
Diatoms, whole, Black Sea	150.0 AW	10
Green alga, *Enteromorpha* sap.; whole; 2000-2003; Gulf of Gdansk, Baltic Sea	3.6 (0.1-7.1) DW	25
Florida; near sewage outfall; exposure for 120 days		
Turtle grass, *Thalassia testudinium*; leaves	45.0 DW	23
Mangrove, *Rhizophora mangle*; roots	10.0 DW	23

(Continues)

Table 2.18: Cont'd

Organism	Concentration	Reference[a]
Bladder wrack, *Fucus vesiculosus*, whole		
Norway	2.0-7.0 DW	8, 19
United Kingdom		
Tamar estuary	3.5 (0.7-4.8) DW	11
Dulas Bay	6.0 DW	9
Menai Straits	8.1 DW	9
Irish Sea	6.7 DW	12
England	1.2-29.6 DW	19
Greenland	0.6-2.3 DW	19
Nova Scotia	2.0 DW	19
Scotland	1.4 FW; 4.9 DW	19
Macroalgae; epiphytic on estuarine mangrove aerial roots; Sydney region, Australia		
Caloglossa leprieurii	10.0 DW	28
Caloglossa nipae	4.0 DW	28
Bostrychia sp.	5.0 DW	28
Marsh plants; whole		
Harbour Island, Texas; 14 spp.	0.21-2.6 DW	13
Macroalgae; whole; 4 spp.; Strait of Magellan; Chile; 2000-2001	<1.0-12.6 DW	33
Phytoplankton		
Organic fractions	1.9-7.8 DW	14
Whole, Puget Sound, Washington state	5.5 DW	15
South Texas coast	4.4 (2.2-7.5) DW	16
Seagrass, *Posidonia oceanica*		
Mediterranean Sea; summers 2004-2005; 3 sites; blades	27.5-60.3 DW	32
Corsica, France; summer 2003; adult leaves; blades vs. sheaths	24.4 (13.0-50.0) DW vs. 4.3 1.8-10.0 DW	31, 34
Mangrove, *Rhizophora mangle*; summer 2001; Natal, Brazil; leaves	0.24-0.58 DW	29
Salt marsh plants; Venice lagoon; 2001; aerial parts		
Phragmites australis	60.0 DW	30
Bolboschoenus maritimus	10.0 DW	30

(*Continues*)

Table 2.18: Cont'd

Organism	Concentration	Reference[a]
Sargassum, *Sargassum horneri*		
Whole, varying lengths		
Up to 50 cm	2.1 DW	1
50-100 cm	1.9 DW	1
100-150 cm	2.3 DW	1
Receptacle	2.0 DW	1
Lamina	2.9 DW	1
Air bladder	1.7 DW	1
Stipe	2.0 DW	1
Sargassum, *Sargassum* spp.; Gulf of Mexico; whole	0.9-15.6 DW	21
Sargassum weed, *Sargassum* sp.; Texas; outer continental shelf; whole	5.2 DW	16
Seaweeds, whole		
United Kingdom; 5 spp.	3.2-9.3 DW; 0.9-1.4 FW	17
Japan; Tomogashima; 38 spp.	2.8 (0.2-10.9) DW	18
Eelgrass, *Zostera marina*; Oregon		
Leaf	20.0 DW	24
Rhizome	25.0 DW	24
Root	40.0 DW	24
Algae, *Zygnema* sp.; Arctic Ocean; Spitsbergen, Svalbard; July-August 1988; whole	3.2 DW	22

Values are in mg Ni/kg fresh weight (FW), dry weight (DW), or ash weight.

[a]1, Ishii et al., 1978; 2, Agadi et al., 1978; 3, Young and Langille, 1958; 4, Gryzhankova et al., 1973; 5, Riley and Roth, 1971; 6, Roth and Hornung, 1977; 7, Stevenson and Ufret, 1966; 8, Lande, 1977; 9, Foster, 1976; 10, Vinogradova and Koual'skiy, 1962; 11, Bryan and Uysal, 1978; 12, Preston et al., 1972; 13, Lytle et al., 1973; 14, Martin and Knauer, 1973; 15, Laevastu and Thompson, 1956; 16, Horowitz and Presley, 1977; 17, Black and Mitchell, 1952; 18, Ishibashi et al., 1964a; 19, Jenkins, 1980b; 20, WHO, 1991; 21, Eisler, 1981; 22, Drbal et al., 1992; 23, Montgomery et al., 1978; 24, Kaldy, 2006; 25, Zbikowski et al., 2006; 26, Perez et al., 2007; 27, Zbikowski et al., 2007; 28, Melville and Pulkownik, 2007; 29, Silva et al., 2006; 30, Bragato et al., 2006; 31, Lafabrie et al., 2007b; 32, Lafabrie et al., 2008a; 33, Astorga-Espana et al., 2008; 34, Lafabrie et al., 2008b.

nickel concentrations that ranged up to 7000.0 mg/kg DW (Kniep et al., 1974). Dumping of sewage sludge in coastal tropical waters may result in elevated concentrations of nickel and other trace metals by turtle grass and other members of that ecosystem. The influence of airborne nickel species from automobile exhausts on coastal communities has not been addressed adequately.

Nickel is ubiquitous in the biosphere and is considered essential for the normal growth of many species of microorganisms and photosynthetic plants (Eisler, 2000d; USPHS, 1993, 1995). In higher plants, magnesium is the usual competitor for nickel in biological ion-exchange reactions; active binding sites are the carboxylic and hydroxycarboxylic groups fixed on the cell wall (Kasprzak, 1987).

There is little convincing evidence for the biomagnification of nickel in the food chain. Most authorities agree that nickel concentrations did not increase with ascending trophic levels of food chains and that predatory animals do not have higher concentrations (Chau and Kulikovsky-Cordeiro, 1995; Eisler, 2000d; Jenkins, 1980a; WHO, 1991). The potential for biomagnification exists because algae and macrophytes have comparatively elevated concentrations of nickel; however, animals seem to be able to regulate the nickel content of their tissues by controlled intake and increased excretion (Eisler, 2000d; Outridge and Scheuhammer, 1993).

The alga *Thalassiosira rotula* can accumulate as much as 90.0 mg Ni/kg DW from the ambient medium; however, growth was inhibited at 0.03 mg Ni/L and was toxic at 0.3 mg Ni/L (Dongmann and Nurnberg, 1982). For the diatom *Navicula pelliculosa*, 50% growth inhibition occurred in 14 days at 0.1 mg Ni/L (WHO, 1991), and reduced growth was reported in *P. tricornutum* at 1.0 mg Ni/L (USEPA, 1980b). Photosynthesis inhibition of 50% was reported in the comparatively resistant giant kelp, *M. pyrifera*, at 2.0 mg Ni/L (USEPA, 1980e). To protect marine plants and other biota, it is proposed that the 24-h average for total recoverable nickel should not exceed 0.0071 mg Ni/L and the maximum concentration should not exceed 0.14 mg/L at any time (USEPA, 1980e); however, this needs to be re-examined because 0.03 mg Ni/L adversely affects growth of marine diatoms (Dongmann and Nurnberg, 1982).

2.25 Plutonium

Marine algae and macrophytes concentrate radioplutonium isotopes from ambient seawater by factors up to 660 for dinoflagellates, 770 for the kelp *M. pyrifera*, 1080 for *Eisenia pyrifera*, and 1570 for *Enteromorpha* sp. (Pillai et al., 1964). Among Atlantic Ocean organisms examined for ^{239}Pu, Sargasso weed had the highest concentrations of radioplutonium on a FW basis, higher than mixed zooplankton, starfish *Asterias forbesi*, shell and meats of mussels and clams, shark liver and bone, and fish tissues, in that order (Wong, 1971). Marine algae may have potential in scavenging radioisotopes of plutonium and other fallout products, and heavy metals in general, from contaminated environments (Wong, 1971). It is alleged that there is negligible hazard to public health from the consumption of plutonium-contaminated algae, fish, or shellfish from the vicinity of the Windscale (United Kingdom) nuclear fuel reprocessing plant (Hetherington et al., 1976; Mitchell, 1977). *Porphyra* seaweed, as one instance, is processed into flour and subsequently sold as a baked product called laverbread. The radioactive dose to bone of people eating laverbread made from

Porphyra harvested near Windscale at a rate of 130 g daily was estimated at less than 7.5 millirems annually, which is considered a negligible risk (Hetherington et al., 1976; Mitchell, 1977).

2.26 Radium

Phytoplankton from various locations, including coastal Sweden, the Mediterranean Sea, and the Pacific Ocean contained about 6.0×10^{-7} mg Ra/kg calcium (Koczy and Titze, 1958). Radium, and possibly thorium, is concentrated by tropical marine algae through two mechanisms: (1) ion exchange or coprecipitation of the ion with the calcium carbonate matrix and (2) complex formation with either the protein nitrogen or some other component of he organic fraction (Edgington et al., 1970). For the Rhodophta and the highly calcified Chlorophyta, radium as well as uranium is concentrated by the first mechanism. For the Phaeophyta, radium is taken up by the second mechanism (Edgington et al., 1970).

2.27 Rhenium

Rhenium concentrations in marine plants ranged from 0.046 to 0.073 mg Re/kg AW in *Ulva* spp. collected from the coast of Japan, and from 0.011 to 0.016 mg Re/kg DW in *Ulva* spp. and *Porphyra* spp. from the same location (Fukai and Meinke, 1959, 1962).

2.28 Rubidium

Rubidium concentrations in marine photosynthetic organisms ranged from 11.0 to 250.0 mg Rb/kg DW (Table 2.19). Accumulation over ambient seawater by 40-fold is reported for field collections of *Sargassum* spp. from Japan (Ishii et al., 1978). Under laboratory conditions, *F. serratus* accumulated 1.7 mg Rb/kg DW; however, accumulation was reduced 50% in 5 h by 10^{-3} M dinitrophenol (Penot and Videau, 1975). *L. digitata* took up 1.0 mg Rb/kg DW in 2 h at 15°C, with uptake halved at 0°C (Penot and Videau, 1975). Up to 26% retention of radiorubidium is reported after 14 days for *Ulva pertussa* (Nakamura et al., 1977).

2.29 Ruthenium

Algae accumulated radioruthenium-106 from seawater solutions by factors of 100-200 times those of the medium (Ancellin et al., 1967; Ancellin and Vilquin, 1968; Ueda et al., 1978), and retained about 25% of the ^{106}Ru after 14 days (Nakamura et al., 1977). The chemical form in which ^{106}Ru is administered, especially its solubility, will significantly alter bioavailability to algae and other biota (Ancellin and Bovard, 1971). Uptake of ^{106}Ru by marine algae is a surface phenomenon associated mainly with extracellular polysaccharides (Jones, 1960). The presence of ferric hydroxide on sand and silt particles in the medium enhances algal accumulation of ruthenium (Jones, 1960).

Table 2.19: Rubidium Concentrations in Field Collections of Algae and Macrophytes

Organism	Concentration	Reference[a]
Algae and macrophytes, whole		
Coast of Japan; 12 spp.	11.0-52.0 DW	1
Coastal United Kingdom; 3 spp.	80.0-170.0 DW	2
Laminaria cloustini; frond		
Sterile	250.0 DW	2
Sporing	130.0 DW	2
Laminaria digitata		
Frond	240.0 DW	2
Stipe	240.0 DW	2

Values are in mg Rb/kg dry weight (DW).
[a]1, Ishii et al., 1978; 2, Black and Mitchell, 1952.

2.30 Selenium

Limited data show that selenium concentrations in field collections of algae and macrophytes range from 0.04 to 0.84 mg Se/kg DW (Table 2.20).

Studies on the metabolic transformation of inorganic selenium by *Tetraselmis tetrathele* and *Dunaliella minuta* demonstrate that the majority of accumulated selenium becomes associated with cellular protein, with the main selenium fraction integrated into the primary protein structure (Wrench, 1978). In marine algae, this may represent a form of storage prior to detoxification (Boisson et al., 1995).

Accumulation, transfer, and release of selenium by aquatic biota may affect the speciation and toxicity of dissolved selenium (Besser et al., 1993, 1994). Depletion of dissolved selenite and increased concentration of organoselenium compounds occur during seasonal peaks in phytoplankton abundance. For example, green algae previously exposed to inorganic radioselenium-75 produced increasing concentrations of organoselenium species during population blooms and crashes.

2.31 Silicon

Silicon is essential for the growth of many species of algae, especially diatoms (Borowitzka and Volcani, 1977; Morel et al., 1978). Silicon is readily accumulated from seawater by marine algae and macrophytes (Goering et al., 1977; Nelson and Goering, 1978; Yoshimura et al., 1976), and this may account, in part, for the elevated

Table 2.20: Selenium and Silicon Concentrations in Field Collections of Algae and Macrophytes

Element and Organism	Concentration	Reference[a]
Selenium		
Algae, whole	0.19-0.84 DW	1, 2
Argentina; summer 2002; 3 sites; *Porphyra columbina* vs. *Ulva* sp.; whole	1.1-1.7 DW vs. 0.5-0.7 DW	8

Seaweeds, whole
  Norway
  Seaweeds eaten by humans
	
	0.04-0.24 DW
	0.16-0.39 DW; 0.047 FW
		
		3
		6

Shoalgrass, *Halodule wrightii*; rhizomes; Texas, Lower Laguna Madre; 1986-1987	<0.2 DW	7
Silicon		
Halimeda opuntia, whole, Jamaica	2000.0 DW	4
Phytoplankton, whole, mostly diatoms, Japan	195,000.0 DW	5

Values are in mg element/kg fresh weight (FW) or dry weight (DW).
[a]1, Tijoe et al., 1977; 2, Chau and Riley, 1965; 3, Lunde, 1973; 4, Bohm, 1972; 5, Fujita, 1971; 6, Noda et al., 1979; 7, Custer and Mitchell, 1993; 8, Perez et al., 2007.

silicon levels in marine plants of 2000.0-195,000.0 mg Si/kg DW (Table 2.20). About 90% of the silicon in the cell wall of *Halimeda opuntia*, and presumably other algal species, was associated the water-soluble and water-insoluble organic fractions (Bohm, 1972).

Silicon is an important structural component of diatoms, silicoflagellates, some higher plants, sponges, and other organisms. It is necessary for DNA polymerase in diatoms. Silicified algal species are assumed to have an obligate concentration-dependent silicon requirement for growth, and transport systems for the accumulation of silicic acid in these species are well documented (Fuhrman et al., 1978). Phytoplankton growth in coastal waters is likely to become limited only when most of the silicate originally present has been removed in the course of diatom blooms (Paasche, 1973a,b). However, some nonsilicified algae, including *Platymonas* spp., can accumulate significant amounts of silicon without apparent requirement. Since silicon is a potentially limiting nutrient for diatoms, the removal of available silicon from the environment by organisms that do not require it could affect phytoplankton competition, succession, and eventually the cycling of silicon in the ocean (Fuhrman et al., 1978).

2.32 Silver

The highest silver concentration reported in field collections of marine photosynthetic organisms is 14.1 mg Ag/kg DW (Table 2.21). Silver accumulations in marine algae are due mainly to adsorption rather than active uptake (Ratte, 1999; USPHS, 1990). Dissolved silver speciation and bioavailability were important in determining silver uptake and retention by aquatic plants (Connell et al., 1991). Silver availability was controlled by the concentration of free silver ion (Ag^+) and the concentrations of other silver complexes, such as AgCl (Sanders and Abbe, 1989). Silver uptake by phytoplankton was rapid in proportion to silver concentration and inversely proportional to water salinity. The availability of free silver ion in marine environments is strongly controlled by salinity of the medium owing to the affinity of silver for the chloride ion (Sanders et al., 1991). Silver sorbs readily to phytoplankton and to suspended sediments. As salinity increases, the degree of sorption decreases. Nearly 80% of silver sorbed to suspended sediments at low salinities desorb at higher salinities; however, desorption

Table 2.21: Silver Concentrations in Field Collections of Algae and Macrophytes

Organism	Concentration	Reference[a]
Algae and macrophytes; whole; United Kingdom; 15 spp.	3.7-14.1 DW	1
Diatoms, whole, Black Sea	30.0 AW	2

Fucus vesiculosus, whole
  Irish Sea
  Tamar estuary
	
	0.35 DW
	0.23 (<0.1-0.36) DW
		
		3
		4

Macroalgae; whole; 4 spp.; Strait of Magellan, Chile; 2000-2001	<0.02-0.3 DW	9
Seaweed, *Padina* sp.; Guam; 1998-1999; fronds	Not detectable to 0.89 DW	8
Phytoplankton, organic fractions	0.2-0.6 DW	5

Seaweeds whole
  United Kingdom; 5 spp., May vs. June
  Arabian Sea; 4 spp.; whole; near Pakistan; 1987-88
	
	0.2-0.3 DW vs. 0.1-0.4 DW, 0.03-0.07 FW
	0.4-0.76 FW
		
		6
		
		7

Values are in mg Ag/kg fresh weight (FW), dry weight (DW), or ash weight (AW).
[a] 1, Riley and Roth, 1971; 2, Vinogradova and Koual'skiy, 1962; 3, Preston et al., 1972; 4, Bryan and Uysal, 1978; 5, Martin and Knauer, 1973; 6, Black and Mitchell, 1952; 7, Tariq et al., 1993; 8, Denton et al., 2006; 9, Astorga-Espana et al., 2008.

does not occur when silver is associated with phytoplankton. Thus, silver incorporation in or on cellular materials increases the retention of silver in the estuary, reducing the rate of transport (Sanders and Abbe, 1987, 1989). Diatoms, *Thalassiosira* spp., for example, readily accumulated silver from the medium. Once incorporated, silver was tightly bound to the cell membrane, even after the cells were mechanically disrupted (Connell et al., 1991).

High bioconcentration of silver from the ambient medium is documented for various species of marine plants. BCFs (mg Ag per kg FW organism/mg Ag per L medium) for silver and diatoms is 210 and for brown algae 240 (USEPA, 1980f). Recovery of silver from industrial and other waste outfalls using marine plants has not yet been evaluated. Marine algae held in solutions containing various concentrations of silver can accumulate the metal from the medium by factors of 13,000-66,000 at equilibrium (Fisher and Reinfelder, 1995; Fisher et al., 1984). Under laboratory conditions, algae held for 38 days in seawater containing radiosilver-110 concentrated ^{110}Ag by a factor of 2800 (Pouvreau and Amiard, 1974). Three species of marine algae subjected to 0.002 mg Ag/L for 24 h contained 27.8-58.6 mg Ag/kg DW at 10 ppt salinity, 16.4-33.4 mg/kg at 15 ppt salinity, and 9.8-25.2 mg Ag/kg DW at 20 ppt salinity (Sanders and Abbe, 1987).

Silver is comparatively phytotoxic to marine plants (Boney et al., 1959). Free silver ion is lethal to representative species of aquatic plants at concentrations of 0.0012-0.0049 mg/L; sublethal effects were documented at 0.0004-0.0006 mg/L (Eisler, 2000e). However, toxicity is reduced at increased salinities. For example, ionic silver produced 50% growth reduction in 5 days in the marine alga *Prorocentrum mariae-lebouriae* at 7.5 ppt and 0.0033 mg Ag/L; at 15.0-22.5 ppt and 30.0 ppt, these values were 0.0067 and 0082 mg Ag/L, respectively (Sanders and Abbe, 1989). Similar results are shown for *S. costatum* at 0.006-0.02 mg Ag/L (Sanders and Abbe, 1989). At 0.1-0.2 mg Ag/L, *Scenedesmus* sp. failed to grow (USEPA, 1980f).

Species composition and species succession in Chesapeake Bay phytoplankton communities were significantly altered in experimental ecosystems continuously stressed by low concentrations (0.0003-0.0006 mg/L) of silver (Sanders and Cibik, 1985; Sanders et al., 1990). At higher concentrations of 0.002-0.007 mg/L for 3-4 weeks, silver inputs caused disappearance of *Anacystis marina*, a mat-forming blue-green alga; increased dominance by *S. costatum*, a chain-forming centric diatom; and caused cell burdens of 8.6-43.7 mg Ag/kg DW (Sanders and Cibak, 1988).

The proposed silver criterion of 0.0023 mg total recoverable silver/L (maximum) at any time to protect marine life (USEPA, 1980f) needs to be reconsidered because phytoplankton species competition and succession are significantly altered at 0.0003-0.0006 mg total Ag/L, and some species of marine algae show extensive accumulations at 0.001-0.002 mg total Ag/L (Eisler, 2000e).

2.33 Strontium

The highest concentration of strontium recorded in field collections of algae and macrophytes was 7770.0 mg Sr/kg DW in the tropical calcareous alga *H. opuntia* (Table 2.22). Marked discrimination for strontium over calcium is recorded in several species of marine flora (Gogate et al., 1975; Weiss et al., 1976). This, and the observations that marine plants readily accumulate strontium from seawater (Hiyama and Shimizu, 1964; Ichikawa, 1961; Rice, 1956; Ryndina, 1976; Skipnes et al., 1975; Zlobin, 1966) and that there is an established need for strontium in normal algal metabolism (Weiss et al., 1976), may account, in part, for the elevated strontium content in *Halimeda*.

Marine plants were unable to discriminate between radiostrontium-90 and stable strontium (Parchevskii et al., 1965), but this pattern was not evident for strontium-89. For example, of 12 species of planktonic algae grown in media containing both strontium-89 and strontium-90, only *Carteria* and *Thoracomonas* took up ^{90}Sr, the others took up ^{89}Sr (Rice, 1956). In another case, an increase in stable strontium in seawater up to 160.0 mg Sr/L caused a threefold reduction in the uptake of ^{89}Sr by *F. serratus* (Zlobin, 1966). Polikarpov (1961) cautioned against use of dead algae in assessing strontium risk owing to loss of all accumulated strontium after death, as was documented for *C. barbata* and other species.

Table 2.22: Strontium Concentrations in Field Collections of Algae and Macrophytes

Organism	Concentration	Reference[a]
Algae and macrophytes, whole		
India; west coast; 9 spp.	19.0-646.0 DW; Max, 1458.0 DW	1
England; coastal waters; 10 spp.	5.8-39.0 DW	2
Irish Sea	0.02-6.2 FW	3
Diatoms; whole		
Mediterranean Sea	Max. 1500.0 AW	4
Black Sea	Max. 15,000.0 AW	5
Halimeda opuntia; whole; Jamaica	7700.0 DW	6
Phytoplankton; organic fractions		
No titanium group	147.0 DW	7
With titanium group	119.0 DW	7
Strontium-concentrated group	697.0 DW	7
Sargassum enerve, whole	2200.0 AW	4

Values are in mg Sr/kg fresh weight (FW), dry weight (DW), or ash weight (AW).
[a] 1, Gogate et al., 1975; 2, Riley and Roth, 1971; 3, Mauchline and Templeton, 1966; 4, Fukai et al., 1962; 5, Vinogradova and Koual'skiy, 1962; 6, Bohm, 1973; 7, Martin and Knauer, 1973.

Strontium availability probably influences the distribution of many macrophytes. Thus, mangrove vegetation comprising species of *Laguncularia*, *Conocarpus*, and *Rhizophora*—normally associated with brackish saline waters—were found in inland stations on Barbuda, West Indies, having no connection with the sea (Stoddart et al., 1973). Analysis of the inland waters showed higher concentrations of strontium and calcium, and lowered concentrations of sodium, potassium, and magnesium when compared with seawater having the same chloride content, with the higher strontium and calcium levels associated with changes in limestone composition.

2.34 Technetium

The brown seaweed *A. nodosum* was used to biomonitor the spatial distribution of technetium-99 in Wales (Oliver et al., 2006). During the period August 1999 to June 2000, eight sites containing *Ascophyllum* were selected 45-650 km from the Sellafield nuclear reprocessing facility, the putative source of the ^{99}Tc. *Ascophyllum* ^{99}Tc content decreased with increasing distance from the facility, being 202 times higher in June 2000 at the closest site than at the most distant site. Discharges of ^{99}Tc from the facility continue to decrease (Oliver et al., 2006).

2.35 Tin

Tin concentrations in field collections of marine algae and macrophytes ranged from 0.5 to 101.0 mg Sn/kg DW whole plant (Table 2.23), suggesting high bioconcentration potential over ambient seawater. BCFs were highest when ambient tin levels were <0.001 mg/L, when exposure times were comparatively lengthy, and when organism lipid content was elevated (Champ, 1986; Laughlin et al., 1986; Thain and Waldock, 1986; Thompson et al., 1985). BCFs for inorganic tin and marine algae were about 1900 (Eisler, 2000f). A BCF of 20,000 was reported after 7 days at 0.002 mg total Sn/L (Maguire et al., 1984). Partitioning or binding may control uptake of TBT by marine phytoplankton. A linear relation is documented for external concentrations of TBT compounds and cell burdens in *Nannochloris* sp. and *Chaetoceros gracilis*, but was not linear for *Isochrysis galbana* (Chiles et al., 1989).

Leaves of mangroves and halophytes collected on January 8, 2007 from a mangrove forest preserve in Tamil Nadu, India, had unusually high concentrations of total tin: up to 2516.7 mg/Sn kg DW (Table 2.23; Agoramoorthy et al., 2008). The reasons for this are unknown, but tin concentrations were positively correlated with iron and mercury (Agoramoorthy et al., 2008), and could have been a contaminant associated with these metals.

Table 2.23: Tin Concentrations in Field Collections of Algae and Macrophytes

Organism	Concentration	Reference[a]
Algae; whole; 12 spp.; United Kingdom	11.0-101.0 DW	1
Seaweeds, whole, 5 spp.	0.5-1.8 DW; 0.1-0.5 FW	2
Green alga, *Enteromorpha* spp.		
Inorganic tin	0.4 FW; 4.4 DW	3
Monomethyltins	0.5 FW	3
Dimethyltins	0.5 FW	3
Trimethyltins	<0.001 FW	3
Tetramethyltins	<0.001 FW	3
Monobutyltins	0.006 FW; 0.4 DW	3
Tributyltins	0.05 FW; 0.6 DW	3
India; Tamil Nadu; January 2007; leaves		
Mangroves; 8 spp.	1187.6 (354.2-1333.5) DW	5
Halophytes; 5 spp.	1871.2 (1655.8-2516.7) DW	5
Phytoplankton; Bohai Bay, North China; May-September, 2002		
Monobutyltins	0.007 FW	4
Dibutyltins	0.019 FW	4
Tributyltins	0.013 FW	4
Triphenyltins	0.001 FW	4

Values are in mg Sn/kg fresh weight (FW), dry weight (DW), and ash weight (AW).
[a] 1, Riley and Roth, 1971; 2, Black and Mitchell, 1952; 3, Donard et al., 1987; 4, Hu et al., 2006; 5, Agoramoorthy et al., 2008.

TBTs are highly toxic to aquatic plants and are present in some harbors where their release from antifouling paints is the putative source (Laughlin et al., 1986; USEPA, 1986; Walsh et al., 1985). The ability of algae to reduce various organotins to less-toxic metabolites that can be excreted rapidly seems to preclude biomagnification (Cardwell and Sheldon, 1986). Marine plants are important in the cycling of tin. Living algae are effective in immobilizing tin from seawater and regulating the formation and degradation of toxic methyltin compounds (Donard et al., 1987). Some species of algae, including two species of *Chlorella*, can degrade TBT when present in a high-growth medium (Tam et al., 2002; Tsang et al., 1999). Dead and decaying algae, however, accumulate inorganic and organic tin compounds, release them, and ultimately remove tin from the estuary to the atmosphere by formation of tetramethyltins (Donard et al., 1987; Tam et al., 2002).

Comparatively elevated concentrations of inorganic tin (>1.5 mg Sn/L) were associated with frustule abnormalities in *Nitzschia lierethrutti* after 14-day exposure (Saboski, 1977). In general, inorganic tins were always less toxic to marine flora than all organotins tested, and TBTs were the most toxic triorganotin compounds tested (Argese et al., 1998). In order of toxicity of triorganotins to marine plants, TBTs were the most toxic followed by tripropyltins, triphenyltins, trimethyltins, and tripentyltins (Eisler, 2000f). For algae, 50% growth inhibition in 72 h was reported for diatoms at 0.31 mg Sn (as inorganic tin)/L; for TBT compounds, growth inhibition occurred at 0.0001-0.0003 mg/L (Beaumont and Newman, 1986; Walsh et al., 1985), they inhibited photosynthesis at 0.003 mg/L (Hall and Pinkney, 1985) and death at 0.015 mg/L (Walsh et al., 1985). Resistance to tin is variable. A marine diatom, *T. pseudonana*, showed no adaptation or resistance to triphenyltins or TBTs (Walsh et al., 1985), but another diatom, *A. coffeaeformis*, was extremely resistant (Thomas and Robinson, 1986, 1987).

2.36 Titanium

Titanium concentrations in marine plants are variable and may reach as high as 10,000.0 mg Ti/kg DW in dinoflagellates under bloom conditions (Table 2.24; Collier, 1953).

Phytoplanktons in Monterey Bay, California, were analyzed for various metals based on three arbitrary groupings in organic fractions and frustules: titanium not detected and strontium not concentrated; titanium detected and strontium not concentrated; and strontium concentrated (Martin and Knauer, 1973). Concentrations of most metals were higher in the latter two groups, especially iron, manganese, aluminum, and titanium. It was speculated that these samples represented phytoplankton populations with reduced growth rate, and because of slow turnover rates sufficient time had elapsed for greater uptake of several metals. However, titanium is relatively abundant in many species of marine plants, especially in early summer (Table 2.24), and this does not support the scenario proposed by Martin and Knauer (1973).

2.37 Tungsten

On an AW basis, tungsten concentrations in *Ulva* spp. from Japanese waters ranged from 0.13 to 0.18 mg/kg whole plant (Fukai and Meinke, 1959). On a DW basis, *Porphyra* spp. and *Ulva* spp. from Japan contained between 0.029 and 0.042 mg W/kg whole plant (Fukai and Meinke, 1962).

2.38 Uranium

Accumulation of uranium salts in *Scenedesmus quadricauda* under laboratory conditions was highest at pH 6.0, 24°C, and algal biomass of 30 mg/L (Pribil and Marvan, 1976). Lowest uptake was observed at pH 4.0, 3°C, and biomass of 2290 mg/L; intermediate values were recorded for pH 8.0, 50°C, and biomass of 840 mg/L. At optimal pH and temperature, mean

Table 2.24: Titanium Concentrations in Field Collections of Algae and Macrophytes

Organism	Concentration	Reference[a]
Algae and macrophytes, whole		
Sea of Japan; 12 spp.	5.0-147.0 DW	1
United Kingdom; 15 spp.	7.8-99.5 DW	2
Diatoms, Black Sea	20,000.0 AW	3
Dinoflagellate, *Gymnodinium breve*	100.0-1000.0 DW	4
Phytoplankton		
Organic fraction		
No titanium group	not detected	5
With titanium group	27.0 DW	5
With strontium- concentrated group	not detected	5
Siliceous frustules		
No titanium group	115.0 DW	5
With titanium group	400.0 DW	5
With strontium- concentrated group	Max. 221.0 DW	5
Seaweeds		
Whole; 5 spp.		
May	20.0-38.0 DW	6
June	7.0-308.0 DW; 1.4-57.0 FW	6
Whole		
Brown	33.9 DW	7
Green	40.0 DW	7
Red	21.8 DW	7

Values are in mg Ti/kg fresh weight (FW), dry weight (DW), or ash weight (AW).
[a]1, Gryzhankova et al., 1973; 2, Riley and Roth, 1971; 3, Vinogradova and Koual'skiy, 1962; 4, Collier, 1953; 5, Martin and Knauer, 1973; 6, Black and Mitchell, 1952; 7, Yamamoto et al., 1970.

accumulations in *Scenedesmus* held in solutions containing less than 1.0 mg U/L with biomass greater than 500.0 mg/L for 6 h was 3300.0 mg U/kg DW; at higher nominal concentrations of 3.5-10.0 mg U/L, and lower initial biomass of less than 100.0 mg/L and uranium content was 100,500.0 mg/kg DW (Pribil and Marvan, 1976), suggesting that decontamination of uranium-impacted areas by selected species of algae should be given serious consideration.

2.39 Vanadium

Vanadium content of saline photosynthetic organisms ranged between 0.9 and 24.0 mg/kg DW (Table 2.25). Acute toxicity bioassays with vanadium as sodium metavanadate and

Table 2.25: Vanadium Concentrations in Field Collections of Algae and Macrophytes

Organism	Concentration	Reference[a]
Algae, whole		
Sea of Japan; 12 spp.	4.0-24.0 DW	1
Sea of Japan	2.8-3.4	2
Coastal waters; United Kingdom; 10 spp.	Max. 5.7 DW	3, 4
Argentina; summer 2002; 3 sites; *Porphyra columbina* vs. *Ulva* sp; whole	3.5-4.5 DW vs. 2.2-5.6 DW	9
Mangrove, *Avicennia marina*; Queensland, Australia; pneumatophores	<0.5-8.0 DW	11
Diatoms, whole, Black Sea	60.0 AW	5
Macroalgae; whole; 4 spp.; Strait of Magellan, Chile; 2000-2001	<1.2-11.3 DW	10
Seaweeds, whole		
3 spp.	Max. 3.1 DW; Max. 262.0 AW	6, 7
5 spp.		
May	1.5-2.6 DW	8
June	0.6-11.9 DW; 0.12-2.2 FW	8

Values are in mg V/kg fresh weight (FW), dry weight (DW), or ash weight (AW).
[a]1, Gryzhankova et al., 1973; 2, Yamamoto et al., 1970; 3, Chau and Riley, 1965; 4, Riley and Roth, 1971; 5, Vinogradova and Koual'skiy, 1962; 6, Fukai and Meinke, 1959; 7, Fukai and Meinke, 1962; 8, Black and Mitchell, 1952; 9, Perez et al., 2007; 10, Astorga-Espana et al., 2008; 11, Preda and Cox, 2002.

D. marina, *Prorocentrum micans*, and *Asterionella japonica* showed that algal populations were reduced by 50% in 9 days at nominal concentrations of 0.5-3.0 mg V/L (Miramand and Unsal, 1978). Algae were more sensitive to vanadium than were marine annelids (10.0 mg V/L), crabs (35.0 mg V/L), and mussels (65.0 mg V/L) during a similar period (Miramand and Unsal, 1978).

2.40 Yttrium

Yttrium accumulations in algae were apparently independent of yttrium concentrations in the medium (Parchevskii et al., 1965). Highest accumulations were consistently found in algulose fractions, followed by the fucoidan barium complex fraction, and the alginic acid fraction in the general ratio of 32:8:1 (Ryndina, 1976).

Selective discrimination of ^{90}Sr over ^{90}Y by *U. lactuca* occurred during exposure for 3 h (Hampson, 1967). Since ^{90}Sr has a physical half life of about 30 years and ^{90}Y only about 65 h, and this is unfortunate from the viewpoint of radiological damage control. The addition of 0.15 mg Y/L led to equilibrium within 3 h, with *Ulva* accumulating yttrium from the medium by a factor of about 500. Without added carrier, ^{90}Y remained on the outer surface of *Ulva*; however, with the addition of stable yttrium, ^{90}Y entered the cells in the region of protoplasmic inclusions (Hampson, 1967). Uptake of yttrium as well as strontium was enhanced as the pH of the medium decreased from 8.0 to 6.5 (Hampson, 1967).

The influence of chelating agents on ^{91}Y uptake by *Porphyra umbilicalis* was demonstrated by Vosjan (1969). *Porphyra* accumulated ^{91}Y from seawater to a concentration factor of 288 in 6 h. The accumulated yttrium was removed with seawater-EDTA washing, suggesting that yttrium is adsorbed onto outer layers of the thallus. When stable yttrium is added, the concentration factor decreases in concentrations greater than 0.008 mg Y/L owing to precipitation. From seawater containing low concentrations of EDTA, uptake of yttrium by *Porphyra* is negligible because the yttrium is chelated and biologically unavailable.

2.41 Zinc

Many species of marine algae and macrophytes contain more than 1000.0 mg Zn/kg DW (Table 2.26). These grossly elevated levels are usually associated with nearby industrial or domestic outfalls containing substantial amounts of zinc (Eisler, 1981, 2000a). Zinc residues from algae already dead on collection are of limited worth. For example, dead *Ulva lactosa* accumulate zinc more rapidly than live *Ulva* during a 6-h period (Gutknecht, 1965); a similar pattern was observed in *A. nodosum* (Skipnes et al., 1975). It was

Table 2.26: Zinc Concentrations in Field Collections of Algae and Macrophytes

Organism	Concentration	Reference[a]
Algae and macrophytes, whole		
Admiralty Bay, Antarctica; 2004; 3 spp.	27.7-39.6 DW	41
Canada, Atlantic Ocean coast; 11 spp.	35.0-97.0 DW	1
Iceland; 10 spp.	2.5-73.0 DW	2
Panama and Columbia	Max. 2150.0 AW	3
India, Goa; 33 spp.	Usually <25.0 DW; max. 203.9 DW	4
India; 11 spp.	10.8-80.0 DW	5
East Indies; 18 spp.	14.2-210.0 DW	6
Japan; 50 spp.	110.0-3950.0 FW	7
Norway, west coast	Max. 2370.0 DW	8
Spain, Rio Tinto estuary	Max. 160.0 DW	9

(*Continues*)

Table 2.26: Cont'd

Organism	Concentration	Reference[a]
United Kingdom; coastal waters; 14 spp.	75.0-410.0 DW	10
United States; Newport River estuary, North Carolina	33.0-120.0 DW	11
Argentina; summer 2002; 3 sites; *Porphyra columbina* vs. *Ulva* sp.; whole	21.1 DW vs.17.4-31.3 DW	47
Ascophyllum nodosum, whole		
Norway		
Flak	90.0-265.0 DW	12
Eikhamrane	825.0-2640.0 DW	12
Transferred from Eikhamrane to Flak for 4 months	70.0-1430.0 DW	12
Sorfjorden	2028.0-3700.0 DW	13
Lofoton	60.0-110.0 DW	14
Trondheimsfjord	66.0-640.0 DW	14
Hardangerfjord	240.0-3700.0 DW	14
United Kingdom		
Menai Straits	149.0 DW	15
Dulas Bay	199.0 DW	15
Coastal waters	60.0-103.0 DW	16
Mangrove, *Avicennia marina*		
Western Australia		
Sediment	34.0 DW	52
Root	16.0 DW	52
Leaf	14.0 DW	52
Hong Kong		
Sediment	55.0 DW	54
Root	16.0 DW	54
Leaf	15.0 DW	54
Shenzhen, China		
Sediment	99.0-106.0 DW	55,56
Root	70.0-79.0 DW	55,56
Leaf	23.0 DW	55,56
Southeast Australia		
Sediment	243.0 DW	57
Root	295.0 DW	57
Leaf	25.0 DW	57
Pneumatophore	29.0-60.0 DW	60

(Continues)

Table 2.26: Cont'd

Organism	Concentration	Reference[a]
Green alga, *Cladophora* sp.; Baltic Sea; 2000-2003		
Gulf of Gdansk	63.0 (21.7-146.4) DW	46
Vistula Lagoon	73.1 DW; max. 105.0 DW	46
Reference site	67.5 DW; max. 94.1 DW	46
Eisenia bicyclis, whole, Japan	39.0-127.0 DW	17, 18
Green alga, *Enteromorpha* sp.; whole; 2000-2003; Gulf of Gdansk, Baltic Sea	64.1 (13.6-175.9) DW	40
Fucus distichus; whole		
Distance from zinc ore deposit in km		
0.1	138.0 DW	19
1	88.0 DW	19
2	43.0 DW	19
7	33.0 DW	19
18	27.0 DW	19
Greenland		
Near Pb/Zn flotation mill outfall	300.0 FW	37
Reference site	8.0 FW	37
Fucus serratus, whole		
United Kingdom	70.0-600.0 DW	16, 20
Norway	3500.0-4400.0 DW	13
Fucus vesiculosus; whole		
Sorfjorden, Norway	2207.0-3090.0 DW	13
Tamar estuary, United Kingdom	379.0 (138.0-1330.0) DW	21
Irish Sea	171.0 DW	22
Wales, United Kingdom	216.0-507.0 DW	23
Severn estuary, United Kingdom	Max. 800.0 DW	24
Menai Straits, United Kingdom	116.0 DW	15
Dulas Bay, United Kingdom	306.0 DW	15
India; Tamil Nadu; January 2007; leaves		
Mangroves; 8 spp.	56.0 (9.3-116.9) DW	61
Halophytes; 5 spp.	52.9 (19.0-71.1) DW	61
Laminaria cloustini; frond		
Sterile, winter	117.0 DW	16
Sporing, winter	136.0 DW	16
Spring	76.0 DW	16

(Continues)

Table 2.26: Cont'd

Organism	Concentration	Reference[a]
Laminaria digitata		
Frond	64.0-99.0 DW	16
Stipe; winter	Not detectable	16
Stipe; spring	62.0 DW	16
Whole	4.0-10.0 FW	16
Macroalgae; whole; 4 spp.; Strait of Magellan, Chile; 2000-2001	14.1-79.0 DW	57
Macroalgae (6 spp.) vs. phytoplankton; Gulf of California, Mexico; 1998-1999; whole	26.2-103.0 DW vs. 117.0 DW	42
Macroalgae; epiphytic on estuarine mangrove aerial roots; Sydney, Australia		
Caloglossa leprieurii	407.0 DW	48
Caloglossa nipae	250.0 DW	48
Bostrychia sp.	394.0 DW	48
Macrocystis integrifolia		
Frond	97.0 (14.0-335.0) DW	25
Stipe	21.0 (10.0-34.0) DW	25
Macrophytes; 6 spp.; whole; India; 1999-2001	5.5-83.9 DW	45
Mangroves; Gulf of California, Mexico; 1998-1999; twigs vs. leaves		
Rhizophora mangle	10.1 DW vs. 8.7 DW	42
Avicennia germinans	11.0 DW vs. 64.0 DW	42
Laguncularia racemosa	11.0 DW vs. 15.0 DW	42
Mexico; Gulf of California; 1999-2000		
Microalgae, *Coscinodiscus centralis*; whole	117.0 DW	58
Macroalgae; *Gracilaria* sp. vs. *Polysiphonia* sp.; whole	36.0 DW vs. 34.0 DW	58
Mangroves; 3 spp.; leaves	8.7-21.0 DW	58
Mangrove, *Rhizophora mangle*; summer 2001; Natal, Brazil; leaves	4.2-6.9 DW	50
Marsh plants; whole; 14 spp.; Harbour Island, Texas	3.4-50.0 DW	26
Phytoplankton; whole		
Texas Continental Shelf	90.8 (42.0-158.0) DW	28
Japan	590.0 DW	29
Monterey Bay, California	750.0 DW	30

(*Continues*)

Table 2.26: Cont'd

Organism	Concentration	Reference[a]
Puerto Rico		
Cymodocea sp.	3.6-8.7 FW, 45.0-92.0 DW, 120.0-250.0 AW	27
Thalassia testudinum	5.5-12.0 FW, 35.0-82.0 DW, 75.0-200.0 AW	27
Udotea flabellum	4.9-800.0 FW, 16.0-1900.0 AW	27
Mangrove, *Rhizophora stylosa*		
Western Australia		
Sediment	29.0 DW	52
Root	15.0 DW	52
Leaf	6.6 DW	52
Yingluo Bay, China		
Sediment	47.0 DW	58
Root	6.2 DW	58
Leaf	5.9 DW	58
Saltmarsh macrophytes		
3 spp.; Yangtze River estuary; July 2005		
Above ground portions	110.0 DW	49
Below ground portions	160.0 DW	49
2 spp.; Venice lagoon; 2001; aerial portions		
Phragmites australis	Max. 14.0 DW	51
Bolboschoenus maritimus	Max. 7.5 DW	51
Seaweeds; whole		
Australia; 5 spp.; various distances from lead smelter outfall		
2.5-5.2 km	823.0 DW; max. 3540.0 DW	36
18.0-18.8 km	Max. 72.0 DW	36
South Africa; 7 spp.	1.2-12.0 FW	31
Norway	53.0-520.0 DW	32
Korea; 10 spp.	7.0-82.0 DW	33
Tunisia; 2 spp.	27.0-327.0 DW	44
Sorfjorden, Norway; whole		
Chorda sp.	1580.0-7350.0 DW	13
Enteromorpha sp.	95.0-6233.0 DW	13
Spartina alterniflora		
Sprout	11.0-20.0 DW	34
Mature plant	7.0-12.0 DW	34
Dead plant	14.0-37.0 DW	34

(*Continues*)

Table 2.26: Cont'd

Organism	Concentration	Reference[a]
Seaweed, *Ulva lactuca*; Gulf of California, Mexico; 2002; whole		
August (low)	98.0 DW	43
April (high)	547.0 DW	43
Eelgrass, *Zostera* spp., whole		
Australia	21.0-670.0 DW	35
Spain	Max. 1480.0 DW	9
United States, North Carolina	49.0 DW	11
Eelgrass, *Zostera marina*		
Leaf	Max. 195.0 DW	38
Rhizome	Max. 70.0 DW	38
Root	Max. 155.0 DW	38
Stem	Max. 85.0 DW	38
Baja California, Mexico; whole; March 2000; reference site vs. salt mining site	27.0 (1.7-221.0) DW vs. 38.0 (0.8-157.0) DW	59
Oregon		
Leaf	29.0 DW	39
Rhizome	18.0 DW	39
Root	52.0 DW	39

Values are in mg Zn/kg fresh weight (FW), dry weight (DW), or ash weight (AW).

[a]1, Young and Langille, 1958; 2, Munda, 1978; 3, Lowman et al., 1970; 4, Agadi et al., 1978; 5, Pillai, 1956; 6, Sivalingam, 1978b; 7, Ishibashi et al., 1965; 8, Stenner and Nickless, 1974; 9, Stenner and Nickless, 1975; 10, Riley and Roth, 1971; 11, Wolfe, 1974; 12, Myklestad et al., 1978; 13, Melhuus et al., 1978; 14, Haug et al., 1974; 15, Foster, 1976; 16, Black and Mitchell, 1952; 17, Ishibashi et al., 1965; 18, Ishii et al., 1978; 19, Bohn, 1979; 20, Leatherland and Burton, 1974; 21, Bryan and Uysal, 1978; 22, Preston et al., 1972; 23, Ireland, 1973; 24, Butterworth et al., 1972; 25, Wort, 1955; 26, Lytle et al., 1973; 27, Lowman et al., 1966; 28, Horowitz and Presley, 1977; 29, Fujita, 1972; 30, Knauer and Martin, 1973; 31, Van As et al., 1975; 32, Lunde, 1970; 33, Pak et al., 1977; 34, Williams and Murdock, 1969; 35, Harris et al., 1979; 36, Ward et al., 1986; 37, Loring and Asmund, 1989; 38, Brix and Lyngby, 1982; 39, Kaldy, 2006; 40, Zbikowski et al., 2006; 41, Santos et al., 2006; 42, Ruelas-Inzunza and Paez-Osuna, 2006; 43, Orduna-Rojas and Longoria-Espinoza, 2006; 44, El Ati-Hellal et al., 2007; 45, Kalesh and Nair, 2006; 46, Zbikowski et al., 2007; 47, Perez et al., 2007; 48, Melville and Pulkownik, 2007; 49, Quan et al., 2007; 50, Silva et al., 2006; 51, Bragato et al., 2006; 52, Alongi et al., 2003; 54, Chen et al., 2003; 55, Zheng and Lin, 1996; 56, Peng et al., 1997; 57, MacFarlane et al., 2003, 58, Zheng et al., 1997; 57, Astorga-Espana et al., 2008; 58, Ruelas-Inzunza and Paez-Osuna, 2008; 59, Marcias-Zamora et al., 2008; 60, Preda and Cox, 2002; 61, Agoramoorthy et al., 2008.

suggested that algal zinc-binding substances, presumably proteinaceous compounds (Parry and Hayward, 1973), are directly accessible to ambient zinc ions in seawater prior to death (Skipnes et al., 1975).

Zinc concentrations in field collections of marine algae and macrophytes are usually—but not always—at least several orders of magnitude higher than zinc concentrations of the surrounding seawater (Eisler, 1981). BCFs for *Enteromorpha* spp. and *F. vesiculosus* from the Mediterranean Sea near Tunisia were 9930 and 4980, respectively, and were attributed to harbor activities and urban discharges (El Ati-Hellal et al., 2007). Of 16 species of seaweeds analyzed, at least 3 did not markedly concentrate zinc over ambient seawater levels and only 7 species showed marked concentration (Saenko et al., 1976). In general, zinc concentrations in marine flora were elevated when seawater zinc concentrations were elevated; however, the relation was not linear. Marine flora, especially red and brown algae, are among the most effective marine zinc accumulators. Increasing accumulations of zinc in marine algae were associated with decreasing light intensity, decreasing pH, increasing water temperature, decreasing levels of DDT, and increasing oxygen; ionic zinc was accumulated more readily than other forms (Eisler, 1981). Mean BCFs for three species of marine algae were 1530 in 12 days, 4680 in 34 days, and 16,600 in 140 days (USEPA, 1980c). The maximum net daily accumulation rate for the alga *A. nodosum* was 1.3 mg Zn/kg FW for whole alga (Eisler, 1980). Season of collection is another important variable governing uptake: zinc concentrations in six species of macrophytes collected from the southwest coast of India between 1999 and 2001 were consistently highest during winter and summer monsoon seasons (Kalesh and Nair, 2006).

Zinc interacts with numerous chemicals, often producing greatly different patterns of accumulation, toxicity, and metabolism when compared to zinc alone. For example, cadmium-zinc interactions sometimes act to the organisms's advantage, and sometimes not, depending on nutritional status and other variables. In the case of marine algae, cadmium promotes the growth of zinc-limited phytoplankton (Price and Morel, 1990). Substitution of trace metals or metalloenzymes could be a common strategy for phytoplankton in trace-metal impoverished environments, such as the ocean, and could result in an effective limitation of phytoplankton growth by several bioactive elements. Thus, zinc-deficient diatoms *T. weissflogii* can grow at 90% of the maximum rate when supplied with cadmium, which substitutes for zinc in certain macromolecules; a similar case is made for cobalt, although less efficiently when compared to cadmium (Price and Morel, 1990).

Zinc is almost certainly required for normal growth and metabolism of all marine organisms, including the flora. Zinc deficiency in algae is associated with growth inhibition. For example, *T. weissflogii* grows best in the range 0.0007-0.065 mg Zn/L, but not at all at less than 0.0007 mg Zn/L (Anderson et al., 1978). In the diatom *P. tricornutum*, growth stimulation was observed at 0.08-0.31 mg Zn/L and growth inhibition at 5.4 mg Zn/L (Horvatic and Persic, 2007).

Adverse effects of zinc—including growth reduction, altered lipid metabolism, and photosynthesis inhibition—were reported in sensitive species of algae at 0.005-0.01 mg Zn/L for *Schroederella* and *Thalassiosira* and 0.05-0.1 mg Zn/L for *Amphidinium, Skeletonema,* and others (Braek et al., 1976; Eisler, 2000a; Malea et al., 2006; Spear, 1981; USEPA, 1980c, 1987). However, many species of algae are comparatively resistant to elevated zinc levels of 1.0-2.0 mg Zn/L, and higher (Eisler, 2000a; Stromgren, 1979; USEPA, 1980c, 1987). Specific effects of zinc on sensitive species of marine algae are documented at 0.0088-0.0095 mg/L (altered lipid metabolism and BCF of 10,770 in 140 days of brown macroalgae *F. serratus*; USEPA, 1987); at 0.015-0.025 mg/L (photosynthesis reduction in *Rhizoselenia* sp.; Spear, 1981); 0.019 mg/L (50% growth reduction in 48-96 h of the diatom *Schroederella schroederi*; USEPA, 1987); 0.02 mg/L (chlorophyll reduction of 65% in 2 days in the dinoflagellate *Glenodinium halli*; USEPA, 1987); 0.05-0.1 mg/L (growth reduced 23% in 15 days of the diatom *S. costatum*; USEPA, 1987); 0.065 mg/L (BCF of 255 in 6 days of green macroalgae *U. lactuca*; USEPA, 1987); 0.1 mg/L (growth inhibition of *Glenodinium splendens* in 38 days and 5 other species in 48 h; Spear, 1981; USEPA, 1987); 0.1 mg/L (growth stimulation of the diatom *Nitzschia longissima* during exposure for 1-5 days; USEPA, 1987); and 0.25 mg/L (decreased growth rate of brown alga *A. nodosum* in 10 days; USEPA, 1987).

Unlike algae, submerged macrophytes play a minor role in cycling zinc (Lyngby et al., 1982). Rooted aquatic macrophytes may participate in heavy metal cycling in the aquatic environment either as a source or as a sink; however, studies with eelgrass, *Z. marina*, show that zinc exchange between the water and sediment is negligible (Lyngby et al., 1982).

Although most species of marine algae and macrophytes are not adversely affected at less than 1.4 mg Zn/L (Eisler, 2000a), others show varying degrees of resistance. To protect sensitive species of algae and other marine life, the United States Environmental Protection Agency recommends that total recoverable zinc in seawater should average less than 0.056 mg/L and never exceed 0.17 mg/L (USEPA, 1980a); for acid-soluble zinc, these values are 0.086 and 0.095 mg/L, respectively (USEPA, 1987). More recently, saltwater zinc criteria were set at less than 0.081 mg Zn/L (USEPA, 1999). However, as shown earlier, several sensitive species of marine plants are adversely affected at zinc concentrations between 0.009 and 0.05 mg Zn/L, or significantly below the proposed criteria for marine life protection (Eisler, 2000a; Spear, 1981; USEPA, 1987).

2.42 Zirconium

Zirconium concentrations in *Gymnodinium brevis*, the dinoflagellate associated with "red tide" blooms in the Gulf of Mexico and elsewhere, contained 10.0-100.0 mg Zr/kg DW (Collier, 1953). Diatoms from the Black Sea contained up to 50.0 mg Zr/kg AW (Vinogradova and Koual'skiy, 1962). *U. pertussa* accumulated up to 41% of all ^{95}Zr after exposure for 14 days in solutions containing radiozirconium-95 (Nakamura et al., 1977).

2.43 Literature Cited

Abalde, J., Aid, A., Reisiz, S., Torres, E., Herrero, C., 1995. Response of the marine macroalga *Dunaliella tertiolecta* (Chlorophycea) to copper toxicity in short time experiments. Bull. Environ. Contam. Toxicol. 54, 317–324.

Abbott, O.J., 1977. The toxicity of ammonium molybdate to marine invertebrates. Mar. Pollut. Bull. 8, 204–205.

Agadi, V.V., Bhosle, N.B., Untawale, A.G., 1978. Metal concentration in some seaweeds of Goa (India). Botanica Marina 21, 247–250.

Agoramoorthy, G., Chen, F.A., Hsu, M.J., 2008. Threat of heavy metal pollution in halophytic and mangrove plants of Tamil Nadu, India. Environ. Pollut. 155, 320–326.

Ahsanullah, M., Williams, A.R., 1991. Sublethal effects and bioaccumulation of cadmium, chromium, copper and zinc in the marine amphipod *Allorchestes compressa*. Mar. Biol. 108, 59–65.

Almeida, C.M.R., Mucha, A.P., Vasconcelos, M.T.S.D., 2006. Variability of metal contents in the sea rush *Juncus maritimus*-estuarine sediment system through one year of plant's life. Mar. Environ. Res. 61, 424–438.

Almeida, C.M.R., Mucha, A.P., Bordalo, A.A., Vasconcelos, M.T.S.D., 2008. Influence of a salt marsh plant (*Halimione portulacoides*) on the concentrations and potential mobility of metals in sediments. Sci. Total Environ. 403, 188–195.

Almela, C., Clemente, M.J., Velez, D., Montoro, R., 2006. Total arsenic, lead and cadmium contents in edible seaweeds sold in Spain. Food Chem. Toxicol. 44, 1901–1908.

Alongi, D.M., Clough, B.F., Dixon, P., Tirendi, F., 2003. Nutrient partitioning and storage in arid-zone forests of the mangroves *Rhizophora stylosa* and *Avicennia marina*. Trees 17, 51–60.

Alvarez-Legorreta, T., Mendoza-Cozatl, D., Moreno-Sanchez, R., Gold-Bouchot, G., 2008. Thiol peptides induction in the seagrass *Thalassia testudinum* (Banks ex Konig) in response to cadmium exposure. Aquat. Toxicol. 86, 12–19.

Amiard-Triquet, C., Berthet, B., Joux, L., Perrein-Ettajani, H., 2006. Significance of physicochemical forms of storage in microalgae in predicting copper transfer to filter-feeding oysters (*Crassostrea gigas*). Environ. Toxicol. 21, 1–7.

Ancellin, J., Bovard, P., 1971. Observations concernant les contaminations experimentales et les contaminations "in situ" d'especes marine par le ruthenium 106. Rev. Int. Oceanogr. Med. 21, 85–92.

Ancellin, J., Vilquin, A., 1968. Nouvelles etudes de contaminations experimentales d'especes marines par le cesium 137, le ruthenium 106 et le cerium 144. Radioprotection 3, 185–213.

Ancellin, J., Bovard, P., Vilquin, A., 1967. In: New studies on experimental contamination of marine species by ruthenium-106. Actes du Congres Int. sur la Radioprotection du Milieu. Soc. Franc. de Radioprot, Toulouse, France, 14-16 January, 1967, pp. 213–214.

Anderson, M.A., Morel, F.M.M., Guillard, R.R.L., 1978. Growth limitation of a coastal diatom by low zinc ion activity. Nature 276, 70–71.

Andrade, S., Medina, M.H., Moffett, J.W., Correa, J.A., 2006. Cadmium-copper antagonism in seaweeds inhabiting coastal areas affected by copper mine waste disposals. Environ. Sci. Technol. 40, 4382–4387.

Antia, N.J., Cheng, J.Y., 1975. Culture studies on the effects from borate pollution on the growth of marine phytoplankters. J. Fish. Res. Board Can. 32, 2487–2494.

Argese, E., Bettiol, C., Ghirardini, A.V., Fasolo, M., Giurin, G., Ghetti, P.F., 1998. Comparison of in vitro submitochondrial particle and Microtox assays for determining the toxicity of organotin compounds. Environ. Toxicol. Chem. 17, 1005–1012.

Astorga-Espana, M.S., Calisto-Ulloa, N.C., Guerrero, S., 2008. Baseline concentrations of trace metals in macroalgae from the Strait of Magellan, Chile. Bull. Environ. Contam. Toxicol. 80, 97–101.

Augier, H., Gilles, G., Ramonda, G., 1978a. Studies on the mercury content in the thallus and in a commercial product of agricultural use of the brown algae *Ascophyllum nodosum* (Linne) Le Jois manufactured in Brittany. Botanica Marina 21, 41–416.

Augier, H., Gilles, G., Ramonda, G., 1978b. Recherche sur la pollution mercurielle du milieu maritime dans le region de Marseille (Mediterranee, France): Partie 1. Degre de contamination par le mercure de la phanerograme marine *Posidonia oceanica* delile a proximite du complexe portuaire et dans le zone de reject due grand collecteur d'egouts de la ville de Marseille. Environ. Pollut. 17, 269–285.

Azam, R., Hemmingsen, B.B., Volcani, B.E., 1973. Germanium incorporation into the silica of diatom cell walls. Arch. Microbiol. 92, 11–20.

Bao, V.W.W., Leung, K.M.Y., Kwok, K.W.H., Zhang, A.Q., Lui, G.C.S., 2008. Synergistic toxic effects of zinc pyrithione and copper to three marine species: implications on setting appropriate water quality criteria. Mar. Pollut. Bull. 57, 616–623.

Baptist, J.P., Lewis, C.W., 1969. Transfer of ^{65}Zn and ^{51}Cr through an estuarine food chain. In: Proceedings 2nd National Symposium on Radioecology. US Atomic Enerergy Community, Conference 670503, pp. 420–430.

Bargagli, R., Monaci, F., Sanchez-Hernandez, J.C., Cateni, D., 1998. Biomagnification of mercury in an Antarctic marine coastal food web. Mar. Ecol. Prog. Ser. 169, 65–76.

Baroli, M., Cristini, A., Cossu, A., De Falco, G., Gzale, V., Pergent-Martini, C., et al., 2001. Concentrations of trace metals (Cd, Cu, Fe, Pb) in *Posidonia oceanica* seagrass of Liscia Bay, Sardinia (Italy). In: Faranda, G.L., Guglielmo, G.L., Spezie, G. (Eds.), Mediterranean Ecosystems: Structures and Processes. Springer-Verlag, Milano, Italy, pp. 95–99.

Beaumont, A.R., Newman, P.B., 1986. Low levels of tributyltin reduce growth of marine micro-algae. Mar. Pollut. Bull. 17, 457–461.

Benson, A.A., 1984. Phytoplankton solved the arsenate-phosphate problem. In: Holm-Hansen, O., Bolis, R., Giles, R. (Eds.), Lecture Notes on Coastal and Estuarine Ecology. 8. Marine Phytoplankton and Productivity. Springer-Verlag, Berlin, pp. 55–59.

Bentley-Mowat, J.A., Reid, S.M., 1977. Survival of marine phytoplankton in high concentrations of heavy metals, and uptake of copper. J. Exp. Mar. Biol. Ecol. 16, 249–264.

Berland, D.R., Bonin, D.J., Kapkov, V.I., Maestrini, S.Y., Arlhac, D.P., 1976. Action toxique de quatre metaux lourds sur la croissance d'algues unicellulaires marines. CR Acad. Sci. Paris 282D, 633–636.

Berland, B.R., Bonin, D.J., Guerin-Ancey, O.J., Kapkov, V.I., Arlhac, D.P., 1977. Action de metaux lourds a des doses subletales sur les caracteristiques de la croisssance chez la diatomee. *Skeletonema costatum*. Mar. Biol. 42, 17–30.

Bernhard, M., Zattera, A., 1975. Major pollutants in the marine environment. In: Pearson, E.A., Frangipane, E.D. (Eds.), Marine Pollution and Marine Waste Disposal. Pergamon Press, Elmsford, NY, pp. 195–300.

Besser, J.M., Canfield, T.J., La Point, T.W., 1993. Bioaccumulation of organic and inorganic selenium in a laboratory food chain. Environ. Toxicol. Chem. 12, 57–72.

Besser, J.M., Huckins, J.N., Clark, R.C., 1994. Separation of selenium species released from Se-exposed algae. Chemosphere 29, 771–780.

Black, W.A.P., Mitchell, R.L., 1952. Trace elements in the common brown algae and in seawater. J. Mar. Biol. Assoc. UK 30, 575–584.

Blake, N.J., Johnson, D.L., 1976. Oxygen production-consumption of the pelagic *Sargassum* community in a flow-through system with arsenic additions. Deep Sea Res. 23, 773–778.

Blankenship, M.L., Wilbur, E.M., 1975. Cobalt effects on cell division and calcium uptake in the coccolithophorid *Cricosphaera carterae* (Haptophyceae). J. Phycol. 11, 211–219.

Bohm, E.L., 1972. Concentration and distribution of Al, Fe, and Si in the calcareous alga, *Halimeda opuntia* (L) (Chlorophyta, Udoteaceae). Int. Rev. Ges. Hydrobiol. 57, 631–636.

Bohm, E.L., 1973. Studies on the mineral content of calcareous algae. Bull. Mar. Sci. 23, 177–190.

Bohn, A., 1979. Trace metals in fucoid algae and purple sea urchins near a high Arctic lead/zinc ore deposit. Mar. Pollut. Bull. 10, 325–327.

Boisson, F., Gnassia-Barelli, M., Romeo, M., 1995. Toxicity and accumulation of selenite and selenate in the unicellular marine alga *Cricosphaera elongata*. Arch. Environ. Contam. Toxicol. 28, 487–493.

Boney, A.D., 1971. Sub-lethal effects of mercury on marine algae. Mar. Pollut. Bull. 2, 69–71.

Boney, A.D., 1978. Marine algae as collectors of iron ore dust. Mar. Pollut. Bull. 9, 175–180.

Boney, A.D., Corner, E.D.S., 1959. Application of toxic agents in the study of the ecological resistance of intertidal red algae. J. Mar. Biol. Assoc. UK 38, 267–275.

Boney, A.D., Corner, E.D.S., Sparrow, B.W.P., 1959. The effects of various poisons on the growth and viability of sporelings of the red alga *Plumaria elegans* (Bonnem). Biochem. Pharmacol. 2, 37–49.

Borowitzka, L.J., Volcani, B.E., 1977. Role of silicon in diatom metabolism. Arch. Microbiol. 112, 147–152.

Bottino, N.R., Newman, R.D., Cox, E.R., Stockton, R., Hoban, M., Zingaro, R.A., et al., 1978. The effects of arsenate and arsenite on the growth and morphology of the marine unicellular algae *Tetraselmis chui* (Chlorophyta) and *Hymenomonas carterae* (Chrysophyta). J. Exp. Mar. Biol. Ecol. 33, 153–168.

Braek, G.S., Jensen, A., Mohus, A., 1976. Heavy metal tolerance of marine phytoplankton. III. Combined effects of copper and zinc ions on cultures of four common species. J. Exp. Mar. Biol. Ecol. 25, 37–50.

Braek, G.S., Malnes, D., Jensen, A., 1980. Heavy metal tolerance of marine phytoplankton. IV. Combined effect of zinc and cadmium on growth and uptake in some marine diatoms. J. Exp. Mar. Biol. Ecol. 42, 39–54.

Bragato, C., Brix, H., Malagoli, M., 2006. Accumulation of nutrients and heavy metals in *Phragmites australis* (Cav.) Tri. ex Steudel and *Bolboschoenus maritimus* (L.) Palla in a constructed wetland of the Venice lagoon watershed. Environ. Pollut. 144, 967–975.

Brix, H., Lyngby, J.E., 1982. The distribution of cadmium, copper, lead, and zinc in eelgrass (*Zostera marina* L.). Sci. Total Environ. 24, 51–63.

Brown, D.H., Smith, W.E., 1980. The chemistry of the gold drugs used in the treatment of rheumatoid arthritis. Chem. Soc. Rev. 9, 217–240.

Bryan, G.W., Hummerstone, L.G., 1973. Brown seaweed as an indicator of heavy metals in estuaries in southwest England. J. Mar. Biol. Assoc. UK 53, 705–720.

Bryan, G.W., Uysal, H., 1978. Heavy metals in the burrowing bivalve *Scrobicularia plana* from the Tamar estuary in relation to environmental levels. J. Mar. Biol. Assoc. UK 58, 89–108.

Burger, J., Gochfeld, M., Jeitner, C., Burke, S., Stamm, T., Snigaroff, R., et al., 2007. Mercury levels and potential risk from subsistence foods from the Aleutians. Sci. Total Environ. 384, 93–105.

Butterworth, J., Lester, P., Nickless, G., 1972. Distribution of heavy metals in the Severn estuary. Mar. Pollut. Bull. 3, 72–74.

Cairrao, E., Pereira, M.J., Pastorinho, M.R., Morgado, F., Soares, A.M.V.M., Guilhermano, L., 2007. *Fucus* spp. as a mercury contamination bioindicator in coastal areas (northwestern Portugal). Bull. Environ. Contam. Toxicol. 79, 388–395.

Cardwell, R.D., Sheldon, A.W., 1986. A risk assessment concerning the fate and effects of tributyltins in the aquatic environment. In: Maton, G.L. (Ed.), Proceedings Oceans 86 Conference, Washington, DC, September 23-25, 1986, vol. 4, Organotin Symposium, pp. 1117–1129. Available from Marine Technology Society, Florida Avenue, Washington, DC.

Champ, M.A., 1986. Organotin symposium: introduction and review. In: Maton, G.L. (Ed.), Proceedings Oceans 86 Conference, Washington, DC, September 23-25, 1986, vol. 4, Organotin Symposium, 1093–1100. Available from Marine Technology Society, Florida Avenue, Washington, DC.

Chau, Y.K., Kulikovsky-Cordeiro, O.T.R., 1995. Occurrence of nickel in the Canadian environment. Environ. Res. 3, 95–120.

Chau, Y.K., Riley, J.P., 1965. The determination of selenium in sea water, silicates and marine organisms. Anal. Chim. Acta 33, 36–49.

Chen, X.Y., Tsang, E.P.K., Chan, A.L.W., 2003. Heavy metal contents in sediments, mangroves and bivalves from Ting Kok, Hong Kong. China Environ. Sci. 23, 480–484.

Chigbo, F.E., Smith, R.W., Shore, F.L., 1982. Uptake of arsenic, cadmium, lead and mercury from polluted waters by the water hyacinth *Eichornia crassipes*. Environ. Pollut. 27A, 31–36.

Chiles, T.C., Pendoley, P.D., Laughlin Jr., R.B., 1989. Mechanisms of tri-*n*-butyltin accumulation by marine phytoplankton. Can. J. Fish. Aquat. Sci. 46, 859–862.

Clendenning, K.A., North, W.J., 1959. Effects of wastes on the giant kelp, *Macrocystis pyrifera*. In: Pearson, E.A. (Ed.), Proceedings of the 1st Conference on Waste Disposal in the Marine Environment, Berkeley, CA, pp. 82–91.

Cocoros, G.P., Cahn, H., Siler, W., 1973. Mercury concentrations in fish, plankton and water from three western Atlantic estuaries. J. Fish Biol. 5, 641–647.

Collier, A., 1953. Titanium and zirconium in bloom of *Gymnodinium breve* Davis. Science 118, 328.

Connell, D.B., Sanders, J.G., Reidel, G.F., Abbe, G.R., 1991. Pathways of silver uptake and trophic transfer in estuarine organisms. Environ. Sci. Technol. 25, 921–924.

Coppellotti, O., 1989. Glutathione, cysteine, and acid-soluble thiol levels in *Euglena gracilis* cells exposed to copper and cadmium. Comp. Biochem. Physiol. 94C, 35–40.

Cossa, D., 1976. Sorption du cadmium par une population de la diatomee *Phaeodactylum tricornutum* en culture. Mar. Biol. 34, 163–167.

Culkin, F., Riley, J.P., 1958. The occurrence of gallium in marine organisms. J. Mar. Biol. Assoc. UK 37, 607–615.

Cullen, W., Reimer, K.J., 1989. Arsenic speciation in the environment. Chem. Rev. 89, 713–764.

Custer, T.W., Mitchell, C.A., 1993. Trace elements and organochlorines in the shoalgrass community of the lower Laguna Madre, Texas. Environ. Monit. Assess. 25, 235–246.

Dahab, O.A., Khalil, A.N., Halim, Y., 1990. Chromium fluxes through Mex Bay inshore waters. Mar. Pollut. Bull. 21, 68–73.

Davey, E.W., 1976. Potential roles of metal-ligands in the marine environment. In: Andrew, R.W., Hodson, P.V., Konesewigh, E.E. (Eds.), Toxicity to Biota of Metal Forms in Natural Water, Great Lakes Res. Adv. Bd., Stand. Comm. Sci. Basis Water Qual. Criteria, Inter. Jt. Comm. Res. Adv. Bd, pp. 197–209.

Davey, E.W., Morgan, M.J., Erickson, S.J., 1973. A biological measurement of the copper complexation capacity of seawater. Limnol. Oceanogr. 18, 993–997.

Davies, A.G., 1967. In: Studies on the accumulation of radio-iron by a marine diatom. Proceedings of an International Symposium on Radioecological Concentration Processes, Stockholm, 1966, pp. 983–991.

Davies, A.G., 1970. Iron, chelation and the growth of marine phytoplankton. I. Growth kinetics and chlorophyll production in cultures of the euryhaline flagellate *Dunaliella tertiolecta* under iron-limiting conditions. J. Mar. Biol. Assoc. UK 50, 65–86.

Davies, A.G., 1974. The growth kinetics of *Isochrysis galbana* in cultures containing sublethal concentrations of mercuric chloride. J. Mar. Biol. Assoc. UK 54, 157–169.

Davies, A.G., 1976. An assessment of the basis of mercury tolerance in *Dunaliella tertiolecta*. J. Mar. Biol. Assoc. UK 56, 39–57.

Delcourt, A., Mestre, J.C., 1978. The effects of phenylmercuric acetate on the growth of *Chlamydomonas variabilis* Dang. Bull. Environ. Contam. Toxicol. 20, 145–148.

Demayo, A., Taylor, M.C., Taylor, K.W., Hodson, P.V., 1982. Toxic effects of lead and lead compounds on human health, aquatic life, wildlife, plants, and livestock. CRC Crit. Rev. Environ. Control 12, 257–305.

Denton, G.R.W., Concepcion, L.R., Wood, H.R., Morrison, R.J., 2006. Trace metals in marine organisms from four harbours in Guam. Mar. Pollut. Bull. 52, 1784–1804.

Di Giulio, R.T., Scanlon, P.F., 1985. Heavy metals in aquatic plants, clams, and sediments from the Chesapeake Bay, U.S.A. Implications for waterfowl. Sci. Total Environ. 41, 259–274.

Donard, O.X.F., Short, F.T., Weber, J.H., 1987. Regulation of tin and methyltin compounds by the green alga *Enteromorpha* under simulated estuarine conditions. Can. J. Fish. Aquat. Sci. 44, 140–145.

Dongmann, G., Nurnberg, H.W., 1982. Observations with *Thalassiosira rotula* (Menier) on the toxicity and accumulation of cadmium and nickel. Ecotoxicol. Environ. Safety 6, 535–544.

Drbal, K., Elster, J., Komarek, J., 1992. Heavy metals in water, ice and biological material from Spitsbergen, Svalbard. Polar Res. 11, 99–101.

Dunstan, W.M., Windom, H.L., McIntire, G.L., 1975. The role of *Spartina alterniflora* in the flow of lead, cadmium and copper through the salt-marsh ecosystem. In: Howell, F.G., Gentry, J.B., Smith, M.H. (Eds.), Mineral Cycling in Southeastern Ecosystems. US Energy Res. Dev. Admin. CONF-740513, pp. 250–256.

Edgington, D.N., Gordon, S.A., Thommus, M.M., Almodovar, L.R., 1970. The concentration of radium, thorium, and uranium by tropical marine algae. Limnol. Oceanogr. 15, 945–955.

Edmonds, J.S., Francesconi, K.A., Stick, R.V., 1993. Arsenic compounds from marine organisms. Nat. Prod. Rep. 10, 421–428.

Eide, I., Jensen, A., 1979. Application of *in situ* cage cultures of phytoplankton for monitoring heavy metal pollution in two Norwegian fjords. J. Exp. Mar. Biol. Ecol. 37, 271–286.

Eisler, R., 1979. Copper accumulations in coastal and marine biota. In: Nriagu, J.O. (Ed.), Copper in the Environment, Part I: Ecological Cycling. John Wiley, New York, pp. 383–449.

Eisler, R., 1980. Accumulation of zinc by marine biota. In: Nriagu, J.O. (Ed.), Zinc in the Environment. Part II. Health Effects. John Wiley, New York, pp. 259–351.

Eisler, R., 1981. In: Trace Metal Concentrations in Marine Organisms. Pergamon Press, Elmsford, NY, 687 pp.

Eisler, R., 1987. Mercury hazards to fish, wildlife, and invertebrates: a synoptic review. US Fish Wildl. Serv. Biol. Rep. 85 (1.10), 1–90.

Eisler, R., 2000. Arsenic. In: Handbook of Chemical Risk Assessment. Health Hazards to Humans, Plants, and Animals, vol. 3. Lewis Publishers, Boca Raton, FL, pp. 1501–1566.

Eisler, R., 2000a. Zinc. In: Handbook of Chemical Risk Assessment. Health hazards to Humans, Plants, and Animals, vol. 1. Lewis Publishers, Boca Raton, FL, pp. 605–714.

Eisler, R., 2000b. Molybdenum. In: Handbook of Chemical Risk Assessment. Health Hazards to Humans, Plants, and Animals, vol. 3. Lewis Publishers, Boca Raton, FL, pp. 1613–1647.

Eisler, R., 2000c. Chromium. In: Handbook of Chemical Risk Assessment. Health Hazards to Humans, Plants, and Animals, vol. 1. Lewis Publishers, Boca Raton, FL, pp. 45–92.

Eisler, R., 2000d. Nickel. In: Handbook of Chemical Risk Assessment. Health Hazards to Humans, Plants, and Animals, vol. 1. Lewis Publishers, Boca Raton, FL, pp. 411–497.

Eisler, R., 2000e. Silver. In: Handbook of Chemical Risk Assessment. Health Hazards to Humans, Plants, and Animals, vol. 1. Lewis Publishers, Boca Raton, FL, pp. 499–550.

Eisler, R., 2000f. Tin. In: Handbook of Chemical Risk Assessment. Health Hazards to Humans, Plants, and Animals, vol. 1. Lewis Publishers, Boca Raton, FL, pp. 551–603.

Eisler, R., 2000g. Copper. In: Handbook of Chemical Risk Assessment. Health Hazards to Humans Plants and Animals, vol. 1. Lewis Publishers, Boca Raton, FL, pp. 93–200.

Eisler, R., 2004. Biogeochemical, Health, and Ecotoxicological Perspectives on Gold and Gold Mining. CRC Press, Boca Raton, FL, 355 pp.

Eisler, R., 2006. Mercury Hazards to Living Organisms. CRC Press, Boca Raton, FL, 312 pp.

El Ati-Hellal, M., Hedhili, A., Dachraoui, M., 2007. Contents of trace metals in water and macroalgae along the Mediterranean coast of Tunisia. Bull. Environ. Contam. Toxicol. 78, 33–37.

Erickson, S.J., 1972. Toxicity of copper to *Thalassiosira pseudonana* in unenriched inshore seawater. J. Phycol. 8, 318–323.

Erickson, S.J., Lackie, N., Maloney, T., 1970. A screening technique for estimating copper toxicity to estuarine phytoplankton. J. Water Pollut. Control Fed. 42 (8), Part 2, R270–R278.

Fang, S.C., 1973. Uptake and biotransformation of phenylmercuric acetate by aquatic organisms. Arch. Environ. Contam. Toxicol. 1, 18–26.

Faraday, W.E., Churchill, A.C., 1979. Uptake of cadmium by the eelgrass *Zostera marina*. Mar. Biol. 53, 293–298.

Farias, S., Smichowski, P., Velez, D., Montoro, R., Curtosi, A., Vodopivez, C., 2007. Total and inorganic arsenic in Antarctic macroalgae. Chemsphere 69, 1017–1024.

Favero, N., Cattalini, F., Bertaggia, D., Albergoni, V., 1996. Metal accumulation in a biological indicator (*Ulva rigida*) from the lagoon of Venice (Italy). Arch. Environ. Contam. Toxicol. 31, 9–18.

Fisher, N.S., Reinfelder, J.R., 1995. The trophic transfer of metals in marine systems. In: Tessier, A., Turner, D.R. (Eds.), Metal Speciation and Bioavailability in Aquatic Systems. John Wiley & Sons, New York, pp. 363–406.

Fisher, N.S., Bohe, M., Teyssie, J.L., 1984. Accumulation and toxicity of Cd, Zn, Ag, and Hg in four marine phytoplankters. Mar. Ecol. Prog. Ser. 18, 201–213.

Foster, P., 1976. Concentrations and concentration factors of heavy metals in brown algae. Environ. Pollut. 10, 45–53.

Fourqurean, J.W., Marba, N., Duarte, C.M., Diaz-Almela, E., Ruiz-Halpern, S., 2007. Spatial and temporal variation in the elemental and stable isotopic content of the seagrasses *Posidonia oceanica* and *Cymodocea nodosa* from the Illes Balears, Spain. Mar. Biol. 151, 219–232.

Freeman, M.C., 1985. The reduction of arsenate to arsenite by an *Anabaena* bacteria assemblage isolated from the Waikato River. NZ J. Mar. Freshw. Res. 19, 277–282.

Friberg, L., Boston, P., Nordberg, G., Piscator, M., Robert, K.H., 1975. Molybdenum: a toxicological appraisal. US Environ. Protect. Agen. Rep. 600/1-75-004, 142 pp.

Froelich, P.N., Kaul, L.W., Byrd, J.T., Andreas, M.O., Roe, K.K., 1985. Arsenic, barium, germanium, tin, dimethylsulfide and nutrient biogeochemistry in Charlotte Harbor, Florida, a phosphorus-enriched estuary. Estuar. Coast. Shelf Sci. 20, 239–264.

Fuhrman, J.A., Chisholm, S.W., Guillard, R.R.L., 1978. Marine alga *Platymonas* sp. accumulates silicon without apparent requirement. Nature 272, 244–246.

Fujita, T., 1971. Concentration of major chemical elements in marine plankton. Geochem. J. 4, 143–156.

Fujita, T., 1972. The zinc content in marine plankton. Rec. Oceangr. Works Jpn. 11 (2), 73–79.

Fukai, R., 1965. Analysis of trace amounts of chromium in marine organisms by the isotope dilution of Cr-51. In: Radiochemical Methods of Analysis. International Atomic Energy Agency, Vienna, pp. 335–351.

Fukai, R., Broquet, D., 1965. Distribution of chromium in marine organisms. Bull. Inst. Oceanogr. 65 (1336), 1–19.

Fukai, R., Meinke, W.W., 1959. Some activation analyses of six trace elements in marine biological ashes. Nature 184, 815–816.

Fukai, R., Meinke, W.W., 1962. Activation analyses of vanadium, arsenic, molybdenum, tungsten, rhenium, and gold in marine organisms. Limnol. Oceanogr. 7, 186–200.

Fukai, R., Suzuki, H., Watanake, K., 1962. Strontium-90 in marine organisms during the period 1857-1961. Bull. Inst. Oceanogr. Monaco 1251, 1–16.

Gambrell, R.P., Collard, V.R., Reddy, C.N., Patrick Jr., W.H., 1977. Trace and toxic metal uptake by marsh plants as affected by Eh, pH, and salinity. Dredged Mat. Res. Prog. Tech. Rep. D-77-40. U.S. Army Corps Engineer Waterways Experiment Station, Vicksburg, MS, pp. 1–124.

Gardner, W.S., Windom, H.L., Stephens, J.A., Taylor, F.E., Stickney, R.R., 1975. Concentrations of total mercury in fish and other coastal organisms: implications to mercury cycling. In: Howell, F.G., Gentry, J. B., Smith, M.H. (Eds.), Mineral Cycling in Southeastern Ecosystems. U.S. Energy Res. Dev. Admin. Available as CONF-740513 from NTIS. US Department of Commerce, Springfield, VA, pp. 268–278.

Giblin, A.E., Bourg, A., Valiela, I., Teal, J.M., 1980. Uptake and losses of metals in sewage sludge by a New England salt marsh. Am. J. Bot. 67, 1059–1068.

Glooschenko, W.A., 1969. Accumulation of ^{203}Hg by the marine diatom *Chaetoceros costatum*. J. Phycol. 5, 224–226.

Gnassi-Barelli, M., Romeo, M., Laumond, F., Pesando, D., 1978. Experimental studies on the relationship between natural copper complexes and their toxicity to phytoplankton. Mar. Biol. 47, 15–19.

Goering, J.J., Boisseau, D., Hattori, A., 1977. Effects of copper on silicic acid uptake by a marine phytoplankton population controlled ecosystem pollution experiment. Bull. Mar. Sci. 27, 58–65.

Gogate, S.S., Shah, S.M., Unni, C.K., 1975. Strontium, calcium and magnesium contents of some marine algae from the west coast of India. J. Mar. Biol. Assoc. UK 17, 28–33.

Goldberg, E.D., 1952. Iron assimilation by marine diatoms. Biol. Bull. 102, 243–248.

Greig, R.A., Wenzloff, D., Shelpuk, C., 1975. Mercury concentrations in fish, North Atlantic offshore waters—1971. Pestic. Monit. J. 9, 15–20.

Gryzhankova, L.N., Sayenko, G.N., Karyakin, A.V., Laktionova, N.V., 1973. Concentrations of some metals in the algae of the Sea of Japan. Oceanology 13, 206–210.

Gutknecht, J., 1965. Uptake and retention of cesium-137 and zinc-65 by seaweeds. Limnol. Oceanogr. 6, 426–431.

Hall Jr., L.W., Pinkney, A.E., 1985. Acute and sublethal effects of organotin compounds on aquatic biota: an interpretive literature evaluation. CRC Crit. Rev. Toxicol. 14, 159–209.

Hall, W.S., Pulliam, G.W., 1995. An assessment of metals in an estuarine wetlands ecosystem. Arch. Environ. Contam. Toxicol. 29, 164–173.

Hall, A., Fielding, A.H., Butler, M., 1979. Mechanisms of copper tolerance in the marine fouling alga *Ectocarpus siliculosis*-evidence for an exclusion mechanism. Mar. Biol. 54, 195–199.

Hallacher, L.E., Kho, E.B., Bernard, N.D., Orcutt, A.M., Dudley Jr., W.C., Hammond, T.M., 1985. Distribution of arsenic in the sediments and biota of Hilo Bay, Hawaii. Pac. Sci. 39, 266–273.

Hampson, M.A., 1967. Uptake of radioactivity by aquatic plants and location in the cells. I. The effect of pH on the Sr-90 and Y-90 uptake by the green alga *Ulva lactuca* and the effect of stable yttrium on Y-90 uptake. J. Exp. Bot. 18, 17–53.

Hannan, P.J., Wilkniss, P.E., Patouillet, C., Carr, R.A., 1973. Measurements of mercury sorption by algae. US Naval Res. Lab., Washington, DC. Rep. 7628, pp. 1–26.

Harris, J.E., Fabris, G.J., Statham, P.J., Tawfik, F., 1979. Biogeochemistry of selected heavy metals in Western Port, Victoria, and use of invertebrates as indicators with emphasis on *Mytilus edulis planulatus*. Aust. J. Mar. Freshw. Res. 30, 159–178.

Harrison, W.G., Renger, E.H., Eppley, R.W., 1978. Controlled ecosystem pollution experiment: effect of mercury on enclosed water columns. VII. Inhibition of nitrogen assimilation and ammonia regeneration by plankton in seawater samples. Mar. Sci. Comm. 4, 13–22.

Haug, A., Melsom, S., Omang, S., 1974. Estimation of heavy metal pollution on two Norwegian fjord areas by analysis of the brown alga *Ascophyllum nodosum*. Environ. Pollut. 7, 179–192.

Havens, K.E., 1994. An experimental comparison of the effects of two chemical stressors on a freshwater zooplankton assemblage. Environ. Pollut. 84, 245–251.

Hetherington, J.A., Jefferies, D.F., Mitchell, N.T., Pentreath, R.J., Woodhead, D.S., 1976. Environmental and public health consequences of the controlled disposal of transuranic elements to the marine environment. In: Transuranium Nuclides in the Environment. IAEA-SM-199/11. International Atomic Energy Agency, Vienna, pp. 139–154.

Hiyama, Y., Shimizu, M., 1964. On the concentration factors of radioactive Cs, Sr, Cd, Zn, and Ce in marine organisms. Rec. Oceangr. Works Jpn. 7(2), 43–77.

Hopkins, R., Kain, J.M., 1971. The effect of marine pollutants on *Laminaria hyperborea*. Mar. Pollut. Bull. 2, 75–77.

Horowitz, A., Presley, B.J., 1977. Trace metal concentrations and partitioning in zooplankton, neuston, and benthos from the south texas outer continental shelf. Arch. Environ. Contam. Toxicol. 5, 241–255.

Horvatic, J., Persic, V., 2007. The effect of Ni^{2+}, Co^{2+}, Zn^{2+}, Cd^{2+} and Hg^{2+} on the growth rate of marine diatom *Phaeodactylum tricornutum* Bohlin: microplate growth inhibition test. Bull. Environ. Contam. Toxicol. 79, 494–498.

Howarth, R.W., Cole, J.J., 1985. Molybdenum availability, nitrogen limitation, and phytoplankton growth in natural waters. Science 229, 653–655.

Hu, H., Zhen, H., Wan, Y., Gao, H., An, W., An, L., et al., 2006. Trophic magnification of triphenyltin in a marine food web of Bohai Bay, North China: comparison to tributyltin. Environ. Sci. Technol. 40, 3142–3147.

Ichikawa, R., 1961. On the concentration factors of some important radionuclides in marine food organisms. Bull. Jpn. Soc. Sci. Fish. 27, 66–74.

Ireland, M.P., 1973. Result of fluvial zinc pollution on the zinc content of littoral and sublittoral organisms in Cardigan Bay, Wales. Environ. Pollut. 4, 27–35.

Irgolic, K.J., Woolson, E.A., Stockton, R.A., Newman, R.D., Bottino, N.R., Zingaro, R.A., et al., 1977. Characterization of arsenic compounds formed by *Daphnia magna* and *Tetraselmis chuii* from inorganic arsenate. Environ. Health Perspect. 19, 61–66.

Ishibashi, M., Yamamoto, T., 1960. Inorganic constituents in seaweeds. Rec. Oceangr. Works Jpn. 5(2), 55–62.

Ishibashi, M., Yamamoto, T., Morii, F., 1962. Chemical studies on the ocean (part 85). Chemical studies on the seaweeds (11). Copper content in seaweeds. Rec. Oceangr. Works Jpn. 6, 157–162.

Ishibashi, M., Fujinaga, T., Morii, F., Kanchiku, Y., Kamiyama, F., 1964a. Chemical studies on the ocean (part 94). Chemical studies on the seaweeds (19). Determination of zinc, copper, lead, cadmium and nickel in seaweeds using dithizone extraction and polarographic method. Rec. Oceangr. Works Jpn. 7(2), 33–36.

Ishibashi, M., Yamamoto, T., Fujita, T., 1964b. Chemical studies on the ocean (part 93). Chemical studies on the seaweeds (18). Nickel content in seaweeds. Rec. Oceangr. Works Jpn. 7, 25–32.

Ishibashi, M., Yamamoto, T., Fujita, T., 1964c. Cobalt content in seaweeds. Rec. Oceangr. Works Jpn. 7(2), 17–24.

Ishibashi, M., Fujinaga, T., Yamamoto, T., Fujita, T., Watanabe, K., 1965. Zinc and iron in seaweeds. J. Chem. Soc. Jpn. 86, 728–733.

Ishii, T., Suzuki, H., Koyanagi, T., 1978. Determination of trace elements in marine organisms-I. Factors for variation of concentration of trace element. Bull. Jpn. Soc. Sci. Fish. 44, 155–162.

Jenkins, D.W., 1980a. Biological monitoring of toxic trace metals. Vol. 2. Toxic trace metals in plants and animals of the world. Part 1. US Environ. Protect. Agen. Rep. 600/3-80-090, pp. 30–138.

Jenkins, D.W., 1980b. Biological monitoring of toxic trace metals. Vol. 2. Toxic trace metals in plants and animals of the world. Part 3. US Environ. Protect. Agen. Rep. 600/3-80-092, p. 290.

Jensen, A., Rystad, B., Melsom, S., 1976. Heavy metal tolerance of marine phytoplankton. II. Copper tolerance of three species in dialysis and batch cultures. J. Exp. Mar. Biol. Ecol. 22, 249–256.

Johnson, D.L., Braman, R.S., 1975. The speciation of arsenic and the content of germanium and mercury in members of the pelagic *Sargassum* community. Deep Sea Res. 22, 503–507.

Johnson, D.L., Burke, R.M., 1978. Biological mediation of chemical speciation. II. Arsenate reduction during marine phytoplankton blooms. Chemosphere 8, 645–648.

Johnson, H.L., Stauber, J.L., Adams, M.S., Jolley, D.F., 2007. Copper and zinc tolerance of two tropical microalgae after copper acclimation. Environ. Toxicol. 22, 234–244.

Jones, R.F., 1960. The accumulation of nitrosyl ruthenium by fine particles and marine organisms. Limnol. Oceanogr. 5, 312–325.

Jones, A.M., Jones, Y., Stewart, W.D.P., 1972. Mercury in marine organisms of the Tay region. Nature 238, 164–165.

Kaldy, J.E., 2006. Carbon, nitrogen, phosphorus and heavy metal budgets: how large is the eelgrass (*Zostera marina* L.) sink in a temperate estuary? Mar. Pollut. Bull. 52, 342–353.

Kalesh, N.S., Nair, S.M., 2006. Spatial and temporal variability of copper, zinc, and cobalt in marine macroalgae from the southwest coast of India. Bull. Environ. Contam. Toxicol. 76, 293–300.

Kamala-Kannan, S., Batvari, B.P.D., Lee, K.J., Kannan, N., Krishnamoorthy, R., Shanthi, K., et al., 2008. Assessment of heavy metals (Cd, Cr and Pb) in water, sediment and seaweed (*Ulva lactuca*) in the Pulicat Lake, South East India. Chemosphere 71, 1233–1240.

Karadjova, I.B., Slaveykova, V.I., Tsalev, D.L., 2008. The biouptake and toxicity of arsenic species on the green microalga *Chlorella salina* in seawater. Aquat. Toxicol. 87, 264–271.

Kasprzak, K.S., 1987. Nickel. Adv. Mod. Environ. Toxicol. 11, 145–183.

Kayser, H., 1976. Waste-water assay with continuous algal culture: the effect of mercuric acetate on the growth of some marine dinoflagellates. Mar. Biol. 36, 61–72.

Khristoforova, N.R., Bogdanova, N.N., 1980. Mineral composition of seaweeds from coral islands of the Pacific Ocean as a function of environmental conditions. Mar. Ecol. Prog. Ser. 3, 25–29.

Khristoforova, N.K., Sin'kov, N.A., Saenko, G.N., Koryakova, M.D., 1976. Contents of the trace elements Fe, Mn, Ni, Cr, Cu and Zn in the proteins of marine algae. Sov. J. Mar. Biol. 2(2), 124–128.

Kim, C.Y., 1972. Studies on the contents of mercury, cadmium, lead and copper in edible seaweeds in Korea. Bull. Korean Fish. Soc. 5(3), 88–96.

Kim, C.Y., Won, J.H., 1974. Concentrations of mercury, cadmium, lead and copper in the surrounding seawater and in seaweeds, *Undaria pinnatifida* and *Sargassum fulvellum*, from Suyeong Bay in Busan. Bull. Korean Fish. Soc. 7, 169–178.

Knauer, G.A., Martin, J.H., 1973. Seasonal variations of cadmium, copper, manganese, lead, and zinc in water and phytoplankton in Monterey Bay, California. Limnol. Oceanogr. 18, 597–604.

Kniep, T.J., Re, G., Hernandez, T., 1974. Cadmium in an aquatic ecosystem: distribution and effects. In: Hemphill, D.D. (Ed.), Trace Substances in Environmental Health—VIII: Proceedings of University of Missouri's 8th Annual Conference on Trace Substances in Environmental Health. University of Missouri, Columbia, MO, pp. 173–177.

Koch, I., McPherson, K., Smith, P., Easton, L., Doe, K.G., Reimer, K.J., 2007. Arsenic bioaccessibility and speciation in clams and seaweed from a contaminated marine environment. Mar. Pollut. Bull. 54, 586–594.

Koczy, F.F., Titze, H., 1958. Radium content of carbonate shells. J. Mar. Res. 17, 302–311.

Kuyucak, N., Volesky, B., 1989. The mechanism of gold biosorption. Biorecovery 1, 219–235.

Laevastu, T., Thompson, T.G., 1956. The determination and occurrence of nickel in seawater, marine organisms, and sediments. J. du Conseil 21, 125–143.

Lafabrie, C., Pergent, G., Pergent-Martini, C., Capiomont, A., 2007a. *Posidonia oceanica*: a tracer of past mercury contamination. Environ. Pollut. 148, 688–692.

Lafabrie, C., Pergent, G., Kantin, R., Pergent-Martini, C., Gonzalez, J.L., 2007b. Trace metals assessment in water, sediment, mussel and seagrass species—validation of the use of *Posidonia oceanica* as a metal biomonitor. Chemosphere 68, 2033–2039.

Lafabrie, C., Pergent-Martini, C., Pergent, G., 2008a. Metal contamination of *Posidonia oceanica* meadows along the Corsican coastline (Mediterranean). Environ. Pollut. 151, 262–268.

Lafabrie, C., Pergent-Martini, C., Pergent, G., 2008b. First results on the study of metal contamination along the Corsican coastline using *Posidonia oceanica*. Mar. Pollut. Bull. 57, 155–159.

Lande, E., 1977. Heavy metal pollution in Trondheimsfjorden, Norway, and the recorded effects on the fauna and flora. Environ. Pollut. 12, 187–198.

Laughlin Jr., R.B., French, W., Guard, H.E., 1986. Accumulation of bis(tributyltin) oxide by the marine mussel *Mytilus edulis*. Environ. Sci. Technol. 20, 884–890.

Laumond, F., Neuburger, M., Donnier, B., Fourcy, A., Bittel, R., Aubert, M., 1973. Experimental investigations, at laboratory, on the transfer of mercury in marine trophic chains. In: 6th International Symposium on Medical Oceanography. Portoroz, Yugoslavia, September 26-30, 1973, pp. 47–53.

Leatherland, T.M., Burton, J.D., 1974. The occurrence of some trace metals in coastal organisms with particular reference to the Solent region. J. Mar. Biol. Assoc. UK 54, 457–468.

Levy, J.L., Stauber, J.L., Jolley, D.J., 2007. Sensitivity of marine microalgae to copper: the effect of biotic factors on copper adsorption and toxicity. Sci. Total Environ. 387, 141–154.

Levy, J.L., Angel, B.M., Stauber, J.L., Poon, W.L., Simpson, S.L., Cheng, S.H., et al., 2008. Uptake and internalisation of copper by three marine macroalgae: comparison of copper-sensitive and copper-tolerant species. Aquat. Toxicol. 89, 82–93.

Lewin, J., 1966. Silicon metabolism in diatoms. V. Germanium dioxide, a specific inhibitor of diatom growth. Phycologia 6, 1–12.

Lewin, J., Chen, C.H., 1971. Available iron: a limiting factor for marine phytoplankton. Limnol. Oceanogr. 16, 670–675.

Lewin, J., Chen, C.H., 1976. Effects of boron deficiency on the chemical composition of a marine diatom. J. Exp. Bot. 27, 916–921.

Lindberg, E., Harriss, C., 1974. Mercury enrichment in estuarine plant detritus. Mar. Pollut. Bull. 5, 93–95.

Lindsay, D.M., Sanders, J.G., 1990. Arsenic uptake and transfer in a simplified estuarine food chain. Environ. Toxicol. Chem. 9, 391–395.

Livingston, H.D., Bowen, V.T., 1976. Americium in the marine environment—relationships to plutonium. In: Miller, M.W., Stannard, J.N. (Eds.), Environmental Toxicity of Aquatic Radionuclides: Models and Mechanisms. Ann Arbor Science Publishers, Ann Arbor, MI, pp. 107–130.

Loring, D.H., Asmund, G., 1989. Heavy metal contamination of a Greenland fjord system by mine wastes. Environ. Geol. Water Sci. 14, 61–71.

Lowman, F.G., Phelps, D.K., Ting, R.Y., Escalera, R.M., 1966. In: Progress Summary Report No. 4, Marine Biology Program June 1965-June 1966. Puerto Rico Nucl. Cen., Rep. PRNC 85, pp. 1–57.

Lowman, F.G., Stevenson, R.A., Escalera, R.M., Ufret, S.L., 1967. The effects of river outflows upon the distribution patterns of fallout radioisotopes in marine organisms. In: Proceedings of an International Symposium on Radioecological Concentration Processes. Stockholm, 1966, pp. 735–752.

Lowman, F.G., Martin, J.H., Ting, R.Y., Barnes, S.S., Swift, D.J.P., Seiglie, G.A., et al., 1970. Bioenvironmental and radiological-safety feasibility studies, Atlantic-Pacific interoceanic canal. Estuar. Mar. Ecol. I-IV, Prepared for Battelle Memorial Institute, 505 King Avenue, Columbus, OH, contract AT (26-1)-171.

Lunde, G., 1970. Analysis of trace elements in seaweed. J. Sci. Food Agric. 21, 416–418.

Lunde, G., 1973. The synthesis of fat and water soluble arseno organic compounds in marine and limnetic algae. Acta Chem. Scand. 27, 1586–1594.

Lyngby, J.E., Brix, H., Schierup, H.H., 1982. Absorption and translocation of zinc in eelgrass (*Zostera marina* L.). J. Exp. Mar. Biol. Ecol. 58, 259–270.

Lytle, T.F., Lytle, J.S., Parker, P.L., 1973. A geochemical study of a marsh environment. Gulf Res. Rep. 4(2), 214–232.

MacFarlane, G.R., Pulkownik, A., Burchett, M.D., 2003. Accumulation and distribution of heavy metals in the grey mangrove, *Avicennia marina* (Forsk) Vierh.: biological indication potential. Environ. Pollut. 123, 139–151.

MacFarlane, G.R., Koller, C.E., Blomberg, S.P., 2007. Accumulation and partitioning of heavy metals in mangroves: a synthesis of field-based studies. Chemosphere 69, 1454–1564.

Maclean, F.I., Lucis, O.J., Shakh, Z.Z., Jansz, E.R., 1972. The uptake and subcellular distribution of Cd and Zn in microorganisms. Fed. Proc. Fed. Am. Soc. Exp. Biol. 31, 699.

Maddock, B.G., Taylor, D., 1980. The acute toxicity and bioaccumulation of some lead alkyl compounds in marine animals. In: Branica, M., Konrad, Z. (Eds.), Lead in the Marine Environment. Pergamon, Oxford, UK, pp. 233–261.

Maeda, S., Nakashima, S., Takeshita, T., Higashi, S., 1985. Bioaccumulation of arsenic by freshwater algae and the application to the removal of inorganic arsenic from the aqueous phase. II. By *Chlorella vulgaris* isolated from arsenic-polluted environment. Sep. Sci. Technol. 20, 153–161.

Maguire, R.J., Wong, P.T.S., Rhamey, J.S., 1984. Accumulation and metabolism of tri-*n*-butyltin cation by a green alga, *Ankistrodesmus falcatus*. Can. J. Fish. Aquat. Sci. 41, 537–540.

Maher, W.A., 1985. The presence of arsenobetaine in marine animals. Biochem. Physiol. 80C, 199–201.

Mahoney, J.B., 1982. The effects of trace metals on growth of a phytoflagellate *Olisthodiscus luteus*, which blooms in lower New York Bay. Bull. N. J. Acad. Sci. 27, 53–57.

Malea, P., Rijstenbil, J.W., Haritondis, S., 2006. Effects of cadmium, zinc and nitrogen on non-protein thiols in the macroalgae *Enteromorpha* spp. From the Scheldt Estuary (SW Netherlands, Belgium) and Thermaikos Gulf (N Aegean Sea, Greece). Mar. Environ. Res. 62, 45–60.

Mandelli, E.F., 1969. The inhibitory effects of copper on marine phytoplankton. Mar. Sci. 14, 47–57.

Marchetti, R., 1978. Acute toxicity of alkyl leads to some marine organisms. Mar. Pollut. Bull. 9, 206–207.

Marcias-Zamora, J.V., Sanchez-Osorio, J.L., Rios-Mendoza, L.M., Ramirez-Alvarez, N., Huerta-Diaz, M.A., Lopez-Sanchez, D., 2008. Trace metals in sediments and *Zostera marina* of San Ignacio and Ojo de Liebre Lagoons in the central Pacific coast of Baja California, Mexico. Arch. Environ. Contam. Toxicol. 55, 218–228.

Markham, J.W., Kremer, B.P., Sperling, K.R., 1980. Effects of cadmium on *Laminaria saccharina* in culture. Mar. Ecol. Prog. Ser. 3, 31–90.

Martin, J.H., Knauer, G.A., 1973. The elemental composition of plankton. Geochim. Cosmochim. Acta 37, 1639–1653.

Matida, Y., Kumada, H., 1969. Distribution of mercury in water, bottom mud and aquatic organisms of Minamata Bay, the River Agano and other water bodies in Japan. Bull. Freshw. Fish. Res. Lab. Tokyo 19(2), 73–93.

Matsuto, S., Kasuga, H., Okumoto, H., Takahashi, A., 1984. Accumulation of arsenic in blue-green alga, *Phormidium* sp. Comp. Biochem. Physiol. 78C, 377–382.

Mauchline, J., Templeton, W.L., 1966. Strontium, calcium and barium in marine organisms from the Irish Sea. J. Cons. Perm. Int. Explor. Mer. 30, 161–170.

Melhuus, A., Seip, K.L., Seip, M., Myklestad, S., 1978. A preliminary study on the use of benthic algae as biological indicators of heavy metal pollution in Sorfjorden, Norway. Environ. Pollut. 15, 101–107.

Melville, F., Pulkownik, A., 2007. Investigation of mangrove macroalgae as biomonitors of estuarine metal contamination. Sci. Total Environ. 387, 301–309.

Miao, A.J., Wang, W.X., 2007. Predicting copper toxicity with its intracellular or subcellular concentration and the thiol synthesis in a marine diatom. Environ. Sci. Technol. 41, 1777–1782.

Mills, A.L., Colwell, R.R., 1977. Microbiological effects of metal ions in Chesapeake Bay water and sediment. Bull. Environ. Contamin. Toxicol. 18, 99–103.

Miramand, P., Unsal, M., 1978. Acute toxicity of vanadium to some marine benthic and phytoplanktonic species. Chemosphere 10, 827–832.

Mitchell, N.T., 1977. Radioactivity in the surface and coastal waters of the British Isles, 1976. Part I. The Irish Sea and its environs. Tech. Rep. Fish. Radiobiol. Lab., MAFF Direct. Fish. Res., (FRL 13), Lowestoft, 15 pp.

Montgomery, J.R., Price, M., Thurston, J., de Castro, G.L., Cruz, L.L., Zimmerman, D.D., 1978. Biological availability of pollutants to marine organisms. US Environ. Protect. Agen. Narragansett, Rhode Island Rep. EPA-600/3-78-035, 134 pp.

Morel, N.M.L., Rueter, J.G., Morel, F.M.M., 1978. Copper toxicity to *Skeletonema costatum* (Bacillarophyceae). J. Phycol. 14, 43–48.

Morelli, E., Fantozzi, L., 2008. Phytochelatins in the diatom *Phaeodactylum tricornutum* Bohlin: an evaluation of their use as biomarkers of metal exposure in marine waters. Bull. Environ. Contam. Toxicol. 81, 236–241.

Morris, O.P., Russell, G., 1974. Inter-specific differences in responses to copper by natural populations of *Ectocarpus*. Br. Phycol. J. 269–272.

Morrison, L., Baumann, H.A., Stengel, D.B., 2008. An assessment of metal contamination along the Irish coast using the seaweed *Ascophyllum nodosum* (Fucales, Phaeophyceae). Environ. Pollut. 152, 293–303.

Munda, I.M., 1978. Trace metal concentrations in some Icelandic seaweeds. Botanica Marina 21, 261–263.

Myklestad, S., Eide, I., Melsom, S., 1978. Exchange of heavy metals in *Ascophyllum nodosum* (L.) Le Jod. *in situ* by means of transplanting experiments. Environ. Pollut. 16, 277–284.

Nakamura, R., Suzuki, Y., Ueda, T., 1977. Distribution of radionuclides among green alga, marine sediments and seawater. J. Radiat. Res. 18, 322–330.

Nakazawa, S., 1977. Development of *Fucus* eggs, as affected by iodine, lithium, and nitroprusside. Bull. Jpn. Soc. Phycol. 25 (Suppl.), 215–220.

National Academy of Sciences (NAS), 1975. Medical and Biological Effects of Environmental Pollutants. Nickel. National Research Council, NAS, Washington, DC, 277 pp.

NAS, 1977. Arsenic. National Academy of Sciences, Washington, DC, 332 pp.

National Research Council of Canada (NRCC), 1978. Effects of Arsenic in the Canadian Environment. Natl. Res. Coun. Canada Publ. No. NRCC 15391, 349 pp.

Nelson, D.M., Goering, J.J., 1978. Assimilation of silicic acid by phytoplankton in the Baja California and northwest Africa upwelling systems. Limnol. Oceanogr. 23, 508–517.

Nielsen, E.S., Kamp-Nielsen, L., 1970. Influence of deleterious concentrations of copper on the growth of *Chlorella pyrenoidosa*. Physiol. Plant. 23, 828–840.

Nielsen, E.S., Wium-Andersen, S., 1970. Copper ions as poison in the sea and in freshwater. Mar. Biol. 6, 93–97.

Nishimura, H., Kumagi, M., 1983. Mercury pollution of fishes in Minamata Bay and surrounding water: analysis of pathway of mercury. Water Air Soil Pollut. 20, 401–411.

Noda, K., Hirai, S., Sunayashiki, K., Danbara, H., 1979. Neutron activation analysis of selenium and mercury in marine products along the coast of Shikoku Island. Agric. Biol. Chem. 47, 1381–1386.

Norin, H., Vahter, M., Christakopoulos, A., Sandstrom, M., 1985. Concentration of inorganic and total arsenic in fish from industrially polluted water. Chemosphere 14, 325–334.

Noro, T., 1978. Effect of Mn on the growth of a marine green alga. *Dunaliella tertiolecta*. Jpn. J. Phycol. 26, 69–72.

Oliver, L.R., Perkins, W.T., Mudge, S.M., 2006. Detection of technetium-99 in *Ascophyllum nodosum* from around the Welsh coast. Chemosphere 65, 2297–2303.

Orduna-Rojas, J., Longoria-Espinoza, R.M., 2006. Metal content in *Ulva lactuca* (Linnaeus) from Navachiste Bay (Southeast Gulf of California) Sinaloa, Mexico. Bull. Environ. Contam. Toxicol. 77, 574–580.

Osterberg, C., Pearcy, W.G., Curl, H.J., 1964. Radioactivity and its relationship to oceanic food chains. J. Mar. Res. 22, 2–12.

Outridge, P.M., Scheuhammer, A.M., 1993. Bioaccumulation and toxicology of nickel: implications for wild mammals and birds. Environ. Rev. 1, 172–197.

Overnell, J., 1976. Inhibition of marine algal photosynthesis by heavy metals. Mar. Biol. 38, 335–342.

Paasche, E., 1973a. Silicon and the ecology of marine plankton diatoms. I. *Thalassiosira pseudonana* (*Cyclotella nana*) grown in a chemostat with silicate as a limiting nutrient. Mar. Biol. 19, 117–126.

Paasche, E., 1973b. Silicon and the ecology of marine plankton diatoms. II. Silicate-uptake kinetics in five diatom species. Mar. Biol. 19, 262–269.

Pace, F., Ferrara, R., Del Carratore, G., 1977. Effects of sublethal doses of copper sulfate and lead nitrate on growth and pigment composition of *Dunaliella salina* Teod. Bull. Environ. Contam. Toxicol. 17, 679–685.

Pak, C.K., Yang, K.R., Lee, I.K., 1977. Trace metals in several edible marine algae of Korea. J. Oceanol. Soc. Korea 12, 41–47.

Parchevskii, V.P., Polikarpov, G.G., Zabarunova, I.S., 1965. Certain regularities in the accumulation of yttrium and strontium by marine organisms. Dokl. Akad. Nauk. SSSR 164, 913–916.

Park, D., Yun, Y.S., Ahn, C.K., Park, J.M., 2007. Kinetics of the reduction of hexavalent chromium with the brown seaweed. *Ecklonia* biomass. Chemosphere 66, 939–946.

Parrish, K.M., Carr, R.A., 1976. Transport of mercury through a laboratory two-level marine food chain. Mar. Pollut. Bull. 7, 90–91.

Parry, G.D.R., Hayward., J., 1973. The uptake of ^{65}Zn by *Dunaliella tertiolecta* Butcher.. J. Mar. Biol. Assn. UK 53, 915–922.

Peng, L., Zheng, W., Li, Z., 1997. Distribution and accumulation of heavy metals in *Avicennia marina* community in Shenzhen, China. J. Environ. Sci. (China) 9, 427–429.

Penot, M., Videau, C., 1975. Absorption du ^{86}Rb et du ^{99}Mo par deux algues marines: le *Laminaria digitata* et le *Fucus serratus*. Z. Pflanzenphysiol. 76 (Suppl.), 285–293.

Penrose, W.R., Black, R., Hayward, M.J., 1975. Limited arsenic dispersion in seawater, sediments, and biota near a continuous source. J. Fish. Res. Board Can. 32, 1275–1281.

Perez, A.A., Farias, S.S., Strobl, A.M., Perez, L.B., Lopez, C.M., Pineiro, A., et al., 2007. Levels of essential and toxic elements in *Porphyra columbina* and *Ulva* sp. from San Jorge Gulf, Patagonia, Argentina. Sci. Total Environ. 376, 51–59.

Pershagen, G., Vahter, M., 1979. Arsenic—A Toxicological and Epidemiological Appraisal. Naturvaardsverket Rapp. SNV PM 1128. Liber Tryck, Stockholm, Sweden, 265 pp.

Phelps, D.K., Telek, G., Lapan Jr., R.L., 1975. Assessment of heavy metal distribution within the food web. In: Pearson, E.A., Frangipane, E.D. (Eds.), Marine Pollution and Marine Waste Disposal. Pergamon Press, Elmsford, NY, pp. 341–348.

Pillai, V.K., 1956. Chemical studies on Indian seaweeds. I. Mineral constituents. Proc. Indian Acad. Sci. 44(1), Sec. B, 3–29.

Pillai, K.C., Smith, R.C., Folsom, T.R., 1964. Plutonium in the marine environment. Nature 203, 568–571.

Polikarpov, G.G., 1961. The role of detritus formation in the migration of strontium-90, cesium-137, and cerium-144. Experiments with the seaweed *Cystoseira barbata*. CR Acad. Sci. URSS 136, 921–923 (Nucl. Sci. Abstr. 1961, 15, 1872).

Pouvreau, B., Amiard, J.C., 1974. Etude experimentale de l'accumulation de l'argent 110m chez divers organismes marins. Comm. Atom. France, 19 pp. Rep. CEA-R-4571.

Preda, M., Cox, M.E., 2002. Trace metal occurrence and distribution in sediments and mangroves, Pumicestone region, southeast Queensland, Australia. Environ. Int. 28, 433–439.

Preston, A., Jeffries, D.F., Dutton, J.W.R., Harvey, B.R., Steele, A.K., 1972. British isles coastal waters: the concentrations of selected heavy metals in sea water, suspended matter and biological indicators—a pilot survey. Environ. Pollut. 3, 69–82.

Pribil, S., Marvan, P., 1976. Accumulation of uranium by the chlorococcal alga *Scenedesmus quadricauda*. Arch. Hydrobiol. (Suppl. 49), 214–225.

Price, N.M., Morel, F.M.M., 1990. Cadmium and cobalt substitution for zinc in a marine diatom. Nature 344, 658–660.

Quan, W.M., Han, J.D., Shen, A.L., Ping, X.Y., Qian, P.L., Li, C.J., et al., 2007. Uptake and distribution of N, P and heavy metals in three dominant salt marsh macrophytes from Yangtze River estuary, China. Mar. Environ. Res. 64, 21–37.

Rao, D.V.S., 1981. Effect of boron on primary productivity of nanoplankton. Can. J. Fish. Aquat. Sci. 38, 52–58.

Rao, V.N.R., Govindarajan, S., 1992. Transfer of copper and zinc through a marine food chain. Acta Bot. Indica 20, 71–75.

Ratte, H.T., 1999. Bioaccumulation and toxicity of silver compounds: a review. Environ. Toxicol. Chem. 18, 89–108.

Reboreda, R., Cacador, I., 2007. Halophyte vegetation influences in salt marsh retention capacity for heavy metals. Environ. Pollut. 146, 147–154.

Rice, T.R., 1956. The accumulation and exchange of strontium by marine planktonic algae. Limnol. Oceanogr. 1, 123–138.

Rice, T.R., Willis, V.M., 1959. Uptake, accumulation, and loss of radioactive cerium-144 by marine planktonic algae. Limnol. Oceanogr. 4, 277–290.

Riekie, G.J., Williams, P.N., Raab, A., Meharg, A.A., 2006. The potential for kelp manufacture to lead to arsenic pollution of remote Scottish islands. Chemosphere 65, 332–342.

Riley, J.P., Roth, I., 1971. The distribution of trace elements in some species of phytoplankton grown in culture. J. Mar. Biol. Assoc. UK 51, 63–72.

Rivkin, R.B., 1979. Effects of lead on the growth of the marine diatom *Skeletonema costatum*. Mar. Biol. 50, 239–247.

Roberts, D.A., Poore, A.G.B., Johnston, E.L., 2006. Ecological consequences of copper contamination in macroalgae: effects on epifauna and associated herbivores. Environ. Toxicol. Chem. 25, 2470–2479.

Robinson, M.G., Brown, L.N., Hall, B.D., 1997. Effect of gold (III) on the fouling diatom *Amphora coffeaeformis*: uptake, toxicity and interactions with copper. Biofouling 11, 59–79.

Romeril, M.G., 1977. Heavy metal accumulation in the vicinity of a desalination plant. Mar. Pollut. Bull. 8, 84–87.

Rosemarin, A., Notine, M., Holmgren, K., 1985. The fate of arsenic in the Baltic Sea *Fucus vesiculosus* ecosystem. Ambio 14, 342–345.

Roth, I., Hornung, H., 1977. Heavy metal concentrations in water, sediments and fish from Mediterranean coastal area, Israel. Environ. Sci. Technol. 11, 265–269.

Ruelas-Inzunza, J., Paez-Osuna, P., 2006. Trace metal concentrations in different primary producers from Altata-Ensenada del Pabellon and Guaymas Bay (Gulf of California). Bull. Environ. Contam. Toxicol. 76, 327–333.

Ruelas-Inzunza, J., Paez-Osuna, F., 2008. Trophic distribution of Cd, Pb, and Zn in a food web from Alata-Ensenada del Pabellon subtropical lagoon, SE Gulf of California. Arch. Environ. Contam. Toxicol. 54, 584–596.

Russell, G., Morris, O.P., 1970. Copper tolerance in the marine fouling alga *Ectocarpus siliculosus*. Nature 228, 288–289.

Ryndina, D.D., 1976. Accumulation and fixation of radionuclides by algal polysaccharides. Hydrobiol. J. 9, 16–20.

Saboski, E.M., 1977. Effects of mercury and tin on frustular ultrastructure of the marine diatom *Nitzschia liebethrutti*. Water Air Soil Pollut. 8, 461–466.

Saenko, G.N., Koryakova, M.D., Makienko, V.F., Dobrosmyslova, I.G., 1976. Concentration of polyvalent metals by seaweeds in Vostok Bay, Sea of Japan. Mar. Biol. 34, 169–176.

Saifullah, S.M., 1978. Inhibitory effects of copper on marine dinoflagellates. Mar. Biol. 44, 299–308.

Sakaguchi, T., Nakajima, A., Horikoshi, T., 1981. Studies on the accumulation of heavy metal elements in biological systems. Accumulation of molybdenum by green microalgae. Eur. J. Appl. Microbiol. Biotechnol. 12, 84–89.

Sanders, J.G., 1978. Enrichment of estuarine phytoplankton by the addition of dissolved manganese. Mar. Environ. Res. 1, 59–66.

Sanders, J.G., 1985. Arsenic geochemistry in Chesapeake Bay: dependence upon anthropogenic inputs and phytoplankton species composition. Mar. Chem. 17, 329–340.

Sanders, J.G., 1986. Direct and indirect effects of arsenic on the survival and fecundity of estuarine zooplankton. Can. J. Fish. Aquat. Sci. 43, 694–699.

Sanders, J.G., Abbe, G.R., 1987. The role of suspended sediments and phytoplankton in the partitioning and transport of silver in estuaries. Continent. Shelf Res. 7, 1357–1361.

Sanders, J.G., Abbe, G.R., 1989. Silver transport and impact in estuarine and marine systems. In: Suter II, G.W., Lewis, M.A. (Eds.), Aquatic Toxicology and Environmental Fate: Eleventh Volume. Amer. Soc., Testing Mater., Spec. Tech. Publ. 1007, Philadelphia, PA, pp. 5–18.

Sanders, J.G., Cibak, S.J., 1988. Response of Chesapeake Bay phytoplankton communities to low levels of toxic substances. Mar. Pollut. Bull. 19, 439–444.

Sanders, J.G., Cibik, S.J., 1985. Adaptive behavior of euryhaline phytoplankton communities to arsenic stress. Mar. Ecol. Prog. Ser. 22, 195–205.

Sanders, J.G., Osman, R.W., 1985. Arsenic incorporation in a salt marsh ecosystem. Estuar. Coast. Shelf Sci. 20, 387–392.

Sanders, J.G., Abbe, G.F., Reidel, G.F., 1990. Silver uptake and subsequent effects on growth and species composition in an estuarine community. Sci. Total Environ. 97/98, 761–769.

Sanders, J.G., Riedel, G.F., Abbe, G.R., 1991. Factors controlling the spatial and temporal variability of trace metal concentrations in *Crassostrea virginica* (Gmelin). In: Elliott, M., Ducrotoy, J.P. (Eds.), Estuaries and Coasts: Spatial and Temporal Comparison. ECSA Sympos. 19 (Univ. Caen, France, 1989). Olsen and Olsen, Fredensborg, Denmark, pp. 335–339.

Santos, I.R., Silva-Filho, E.V., Schaefer, C., Sella, S.M., Silva, C.A., Gomes, V., et al., 2006. Baseline mercury and zinc concentrations in terrestrial and coastal organisms of Admiralty Bay, Antarctica. Environ. Pollut. 140, 304–311.

Saward, D., Stirling, A., Topping, G., 1975. Experimental studies on the effects of copper on a marine food chain. Mar. Biol. 29, 351–356.

Schell, W.R., Nevissi, A., 1977. Heavy metals from waste disposal in Central Puget Sound. Environ. Sci. Technol. 11, 887–893.

Schroeder, H.A., Balassa, J.J., Tipton, I.H., 1970. Essential trace metals in man: molybdenum. J. Chronic Dis. 23, 481–499.

Schulz-Baldes, M., Lewin, R.A., 1976. Lead uptake in two marine phytoplankton organisms. Biol. Bull. 150, 118–127.

Seeliger, U., Edwards, P., 1977. Correlation coefficients and concentration factors of copper and lead in seawater and benthic algae. Mar. Pollut. Bull. 8, 16–19.

Seeliger, U., Edwards, P., 1979. Fate of biologically accumulated copper in growing and decomposing thalli of two benthic red marine algae. J. Mar. Biol. Assoc. UK 59, 227–238.

Shiber, J., Washburn, E., 1978. Lead, mercury, and certain nutrient elements in *Ulva lactuca* (Linnaeus) from Ras Beirut, Lebanon. Hydrobiologia 61, 187–192.

Short, Z.F., Olson, P.R., Palumbo, R.F., Donaldson, J.R., Lowman, F.G., 1971. Uptake of molybdenum marked with ^{99}Mo, by the biota of Fern Lake, Washington, in a laboratory and field experiment. In: Nelson, D.J. (Ed.), Radionuclides in Ecosystems. Proceedings of the Third National Symposium on Radioecology, vol. 1, Oak Ridge, TN, May 10-12, 1971, pp. 474–485.

Sick, L.V., Windom, H.L., 1975. Effects of environmental levels of mercury and cadmium on rates of metal uptake and growth physiology of selected genera of marine phytoplankton. In: Howell, F.G., Gentry, J.B., Smith, M.H. (Eds.), Mineral Cycling in Southeastern Ecosystems. U.S. Energy Res. Dev. Admin., CONF-740513, Washington, DC, pp. 239–249.

Silva, C.A.R., Smith, B.D., Rainbow, P.S., 2006. Comparative biomonitors of coastal trace metal contamination in tropical South America (N. Brazil). Mar. Environ. Res. 61, 439–455.

Sivalingam, P.M., 1978a. Effects of high concentration stress of trace metals on their biodeposition modes in *Ulva reticulata* forskal. Jpn. J. Phycol. 26, 157–160.

Sivalingam, P.M., 1978b. Biodeposited trace metals and mineral content studies of some tropical marine algae. Botanica Marina 21, 327–330.

Sivalingam, P.M., 1980. Mercury contamination in tropical algal species of the Island of Penang, Malaysia. Mar. Pollut. Bull. 11, 106–107.

Skei, J.M., Saunders, M., Price, N.B., 1976. Mercury in plankton from a polluted Norwegian fjord. Mar. Pollut. Bull. 7, 34–35.

Skipnes, O., Roald, T., Haug, A., 1975. Uptake of zinc and strontium by brown algae. Physiol. Plant. 34, 314–320.

Slejkovec, Z., Kapolna, E., Ipolyi, I., van Elteren, J.T., 2006. Arsenosugars and other arsenic compound in littoral zone algae from the Adriatic Sea. Chemosphere 63, 1098–1105.

Slowey, J.F., Hayes, D., Dixon, B., Hood, D.W., 1965. Distribution of gamma-emitting radionuclides in the Gulf of Mexico. Occ. Publs. Narragansett Mar. Lab., Univ. Rhode Island, Kingston, Rhode Island, 3, pp. 109–129.

Smith, W.G., 1970. Spartina "die-back" on Louisiana marshlands. Coast. Stud. Bull. 5, 89–96.

Smyth, D.A., Dugger, W.M., 1980. Effects of boron deficiency on ^{86}rubidium uptake and photosynthesis in the diatom *Cylindrotheca fusiformis*. Plant Physiol. 51, 111–117.

Smyth, D.A., Dugger, W.M., 1981. Cellular changes during boron-deficient culture of the diatom. *Cylindrotheca fusiformis*. Physiol. Plant. 66, 692–695.

Sparling, D.W., Lowe, T.P., 1996. Environmental hazards of aluminum to plants, invertebrates, fish and wildlife. Rev. Environ. Contam. Toxicol. 145, 1–127.

Spear, P.A., 1981. Zinc in the aquatic environment: chemistry, distribution, and toxicology. Natl. Res. Coun. Can. Publ. NRCC 17589, 145 pp.

Steeg, P.F.T., Hanson, P.J., Paerl, H.W., 1986. Growth-limiting quantities and accumulation of molybdenum in *Anabaena oscillaroides* (Cyanobacteria). Hydrobiologia 140, 143–147.

Stenner, R.D., Nickless, G., 1974. Distribution of some heavy metals in organisms in Hardangerfjord and Skjerstadfjord, Norway. Water Air Soil Pollut. 3, 279–291.

Stenner, R.D., Nickless, G., 1975. Heavy metals in organisms of the Atlantic coast of S.W. Spain and Portugal. Mar. Pollut. Bull. 6, 89–92.

Stevenson, R.A., Ufret, S.L., 1966. Iron, manganese and nickel in skeletons and food of the sea urchins *Tripneustes esculentus* and *Echinometra lucunter*. Limnol. Oceanogr. 11, 11–17.

Stewart, J.G., 1977a. Relative sensitivity to lead of a naked green flagellate *Dunaliella tertiolecta*. Water Air Soil Pollut. 8, 243–247.

Stewart, J.G., 1977b. Effects of lead on the growth of four species of red algae. Phycologia 16, 31–36.

Stoddart, D.R., Bryan, G.W., Gibbs, P.E., 1973. Inland mangroves and water chemistry, Barbuda, West Indies. J. Nat. Hist. 7, 33–46.

Strohal, P., Huljev, D., Lulic, S., Picer, M., 1975. Antimony in the coastal marine environment, North Adriatic. Estuar. Coast. Mar. Sci. 3, 119–123.

Stromgren, T., 1979. The effect of zinc on the increase in length of five species of intertidal fucales. J. Exp. Mar. Biol. Ecol. 40, 95–102.

Stromgren, T., 1980. The effect of lead, cadmium, and mercury on the increase in length of five intertidal fucales. J. Exp. Mar. Biol. Ecol. 43, 107–119.

Sunda, W., Guillard, R.R., 1976. The relationship between cupric ion activity and the toxicity of copper to phytoplankton. J. Mar. Res. 34, 511–529.

Talarico, L., 2002. Fine structure and X-ray microanalysis of a red macrophyte cultured under cadmium stress. Environ. Pollut. 120, 813–821.

Tam, N.F.Y., Li, S.H., Lan, C.Y., Chen, G.Z., Li, M.S., Wong, Y.S., 1995. Nutrients and heavy metal contamination of plants and sediments in Futian mangrove forest. Hydrobiologia 295, 149–158.

Tam, N.R., Chong, A.M., Wong, Y.S., 2002. Removal of tributyltin (TBT) by live and dead microalgal cells. Mar. Pollut. Bull. 45, 362–371.

Tariq, J., Jaffar, M., Ashraf, M., Moazzam, M., 1993. Heavy metal concentrations in fish, shrimp, seaweed, sediment, and water from the Arabian Sea. Mar. Pollut. Bull. 26, 644–647.

Taylor, W.R., Odum, E.P., 1960. Uptake of iron-59 by marine benthic algae. Biol. Bull. 119, 343.

Thain, J.E., Waldock, M.J., 1986. The impact of tributyl tin (TBT) antifouling paints on molluscan fisheries. Water Sci. Technol. 18, 193–202.
Thomas, W.H., Dodson, A.N., 1974. Inhibition of diatom photosynthesis by germanic acid: separation of diatom productivity from total marine primary productivity. Mar. Biol. 27, 11–19.
Thomas, T.E., Robinson, M.G., 1986. The physiological effects of the leachates from a self-polishing organotin antifouling paint on marine diatoms. Mar. Environ. Res. 18, 215–229.
Thomas, T.E., Robinson, M.G., 1987. Initial characterization of the mechanisms responsible for the tolerance of *Amphora coffeaeformis* to copper and tributyltin. Bot. Mar. 30, 47–53.
Thomas, W.H., Hollibaugh, J.T., Siebert, D.L.R., Wallace Jr, G.T., 1980. Toxicity of a mixture of ten metals to phytoplankton. Mar. Ecol. Prog. Ser. 2, 213–220.
Thompson, J.A.J., Sheffer, M.G., Pierce, R.C., Chau, Y.K., Cooney, J.J., Cullen, W.R., et al., 1985. Organotin compounds in the aquatic environment: scientific criteria for assessing their effects on environmental quality. Natl. Res. Coun. Can. Publ., NRCC 22404, pp. 1–284.
Thursby, G.B., Steele, R.L., 1984. Toxicity of arsenite and arsenate to the marine macroalgae *Champia parvula* (Rhodophyta). Environ. Toxicol. Chem. 3, 391–397.
Tijoe, P.S., de Goeij, J.J.M., de Bruin, M., 1977. Determination of Trace Elements in Dried Sea-Plant Homogenate (SP-M-1) and in Dried Copepod Homogenate (MA-A-1) by Means of Neutron Activation Analysis. Interuniversity Reactor Institute, Rep., 133-77-05, Delft, Nederlands, 14 pp.
Trefry, J.H., Presley, B.J., 1976. Heavy metal transport from the Mississippi River to the Gulf of Mexico. In: Windom, H.L., Duce, R.A. (Eds.), Marine Pollutant Transfer. D.C. Heath, Lexington, MA, pp. 39–76.
Tripp, M., Harriss, R.C., 1976. Role of mangrove vegetation in mercury cycling in the Florida everglades. In: Nriagu, J.D. (Ed.), Environmental Biogeochemistry. Vol. 2. Metals Transfer and Ecological Mass Balances. Ann Arbor Science Publishers, Ann Arbor, MI, pp. 489–497.
Tsang, C.K., Lau, P.S., Tam, N.F.Y., Wong, Y.S., 1999. Biodegradation capacity of tributyltin by two *Chlorella* species. Environ. Pollut. 105, 289–297.
Ueda, T., Nakamura, R., Suzuki, Y., 1978. Comparison of influence of sediments and sea water on accumulation of radionuclides by marine organisms. J. Radiat. Res. 19, 93–99.
Ukeles, R., 1962. Growth of pure cultures of marine phytoplankton in the presence of toxicants. Appl. Microbiol. 10, 512–537.
U.S. Environmental Protection Agency (USEPA), 1980a. Ambient water quality criteria for copper. USEPA Rep. 440/5-80-036, 162 pp.
USEPA, 1980b. Ambient water quality criteria for nickel. USEPA Rep. 440/5-80-060, 206 pp.
USEPA, 1980c. Ambient water quality criteria for zinc. USEPA Rep. 440/5-80-079, 158 pp.
USEPA, 1980d. Ambient water quality criteria for chromium. USEPA Rep. 440/5-80-035 pp.
USEPA, 1980e. Ambient water quality criteria for nickel. USEPA Rep. 440/5-80-060, pp. 1–206.
USEPA, 1980f. Ambient water quality criteria for silver. USEPA, Rep. 440/5-80-071, pp. 1–212.
USEPA, 1985a. Ambient water quality criteria for arsenic—1984. USEPA Rep. 440/5-84-033, 66 pp.
USEPA, 1985b. Ambient water quality criteria for lead—1984. USEPA Rep. 440/5-84-027, 81 pp.
USEPA, 1986. Initiation of a special review of certain pesticide products containing tributyltins used as antifoulants; availability of support document. Fed. Regis. 51(5), 778–779 pp.
USEPA, 1987. Ambient water quality criteria for zinc—1987. USEPA Rep. 440/5-87-003, 207 pp.
USEPA, 1989. Ambient water quality criteria for antimony. USEPA Rep. 440/5-80-120, 103 pp.
USEPA, 1999. National recommended water quality criteria—correction. USEPA 822-Z-99-001. Office of Water, Washington, DC.
US Public Health Service (USPHS), 1990. Toxicological profile for silver. Agen. Toxic Subs. Dis. Regis. TP-90-124, 145 pp.
USPHS, 1993. Toxicological profile for nickel. USPHS, Agen. Toxic Subs. Dis. Regis., Atlanta, GA, pp. Rep. TP-92/14, 1–158 pp.
USPHS, 1995. Toxicological profile for nickel (update). USPHS, Agen. Toxic Subs. Dis. Regis., Atlanta, GA, pp. 1–244.

Vaishampayan, A., 1983. Mo-V interactions during N_2- and NO_3-metabolism in a N_2-fixing blue-green alga *Nostoc muscorum*. Experientia 39, 358–360.

Valega, M., Lillebo, A.I., Cacador, I., Pereira, M.E., Duarte, A.C., Pardal, M.A., 2008. Mercury mobility in a salt marsh colonised by *Halimione portulacoides*. Chemosphere 72, 1607–1613.

Van As, D., Fourie, H.O., Vlegaar, C.M., 1973. Accumulation of certain trace elements in the marine organisms from the sea around the Cape of Good Hope. In: Radioactive Contamination of the Marine Environment. International Atomic Energy Agency, Vienna, Austria, pp. 615–624.

Van As, D., Fourie, H.O., Vlegaar, C.M., 1975. Trace element concentrations in marine organisms for the Cape West Coast. S. Afr. J. Sci. 71, 151–154.

Vinogradova, Z.A., Koual'skiy, V.V., 1962. Elemental composition of the Black Sea plankton. Dokl. Acad. Sci. USSR Earth Sci. Sec. 147, 217–219.

Visser, F., Gerringa, J.A., Van der Gaast, S.J., de Baar, H.J.W., Timmermans, K.R., 2003. The role of the reactivity and content of iron of aerosol dust on growth rates of two Antarctic diatom species. J. Phycol. 39, 1085–1094.

Visviki, L., Rachlin, J.W., 1994a. Acute and chronic exposure of *Dunaliella salina* and *Chlamydomonas bullosa* to copper and cadmium: effects on growth. Arch. Environ. Contam. Toxicol. 26, 149–153.

Visviki, L., Rachlin, J.W., 1994b. Acute and chronic exposure of *Dunaliella salina* and *Chlamydomonas bullosa* to copper and cadmium: effects on ultrastructure. Arch. Environ. Contam. Toxicol. 26, 154–162.

Vosjan, J.H., 1969. Effect of chelation on the uptake and loss of yttrium-91 by *Porphyra*. Neth. J. Sea Res. 4, 310–316.

Walsh, G.E., McLaughlin, L.L., Lores, E.M., Louie, M.K., Deans, C.H., 1985. Effects of organotins on growth and survival of two marine diatoms, *Skeletonema costatum* and *Thalassiosira pseudonana*. Chemosphere 14, 383–392.

Wang, Z.S., Kong, H.N., Wu, D.Y., 2007. Reproductive toxicity of dietary copper to a saltwater cladoceran, *Moina monogolica* Daday. Environ. Toxicol. Chem. 26, 126–131.

Ward, T.J., Correll, R.L., Anderson, R.B., 1986. Distribution of cadmium, lead and zinc amongst the marine sediments, sea grasses and fauna, and the selection of sentinel accumulators, near a lead smelter in South Australia. Aust. J. Mar. Freshw. Res. 37, 567–585.

Weis, J.S., Weis, P., 1992. Transfer of contaminants for CCA-treated lumber to aquatic biota. J. Exp. Mar. Biol. Ecol. 161, 189–199.

Weiss, R.E., Blackwelder, P.L., Wilbur, K.M., 1976. Effects of calcium, strontium and magnesium on the coccolithophorid *Cricosphaera* (*Hymenomonas*) *carterae*. II. Cell division. Mar. Biol. 34, 17–22.

Williams, R.B., Murdock, M.B., 1969. The potential importance of *Spartina alterniflora* in conveying zinc, manganese, and iron into estuarine food chains. In: Proceedings of 2nd National Symposium on Radioecology. USAEC Conf. 670503, pp. 431–440.

Windom, H.L., 1973. Mercury distribution in estuarine-nearshore environment. J. Am. Soc. Civ. Eng. Waterways Harbors Coast. Eng. Div. 99 (WW2), 257–264.

Windom, H., 1975. Heavy metal fluxes through salt marsh estuaries. In: Cronin, L.E. (Ed.), Estuarine Research, Vol. 1. Chemistry, Biology, and the Estuarine System. Academic Press, New York, pp. 137–152.

Windom, H.L., Gardner, W., Dunstan, W.M., Paffenhofer, G.A., 1976. Cadmium and mercury transfer in a coastal marine ecosystem. In: Windom, H.L., Duce, R.A. (Eds.), Marine Pollutant Transfer. D.C. Heath, Lexington, MA, pp. 135–157.

Wolfe, D.A., 1974. The cycling of zinc in the Newport River estuary, North Carolina. In: Vernberg, F.J., Vernberg, W.J. (Eds.), Pollution and Physiology of Marine Organisms. Academic Press, New York, pp. 79–99.

Wong, K.M., 1971. Radiochemical determination of plutonium in sea water, sediments and marine organisms. Anal. Chim. Acta, 56, 355–364.

Wong, K.H., Chan, K.Y., Ng, S.L., 1979. Cadmium uptake by the unicellular green alga *Chlorella salina* CU-1 from culture media with high salinity. Chemosphere 11/12, 887–891.

Wong, P.T.S., Trevors, J.T., 1971. Chromium toxicity to algae and bacteria. In: Nriagu, J.O., Nieboer, E. (Eds.), Chromium in Natural and Human Environments. John Wiley, New York, pp. 305–315.

Woolery, M.L., Lewin, R.A., 1976. The effects of lead on algae. IV. Effects of Pb on respiration and photosynthesis of *Phaeodactylum tricornutum* (Bacillariophyceae). Water Air Soil Pollut. 6, 25–31.

Woolson, E.A. (Ed.), 1975. Arsenical Pesticides. Amer. Chem. Soc. Sympos. Ser. 7. 176 pp.

World Health Organization (WHO), 1991. Nickel. Environ. Health Crit. 108, 383.

Wort, D.J., 1955. The seasonal variation in chemical composition of *Macrocystis integrifolia* and *Nereocystis luetkeana* in British Columbia coastal waters. Can. J. Bot. 33, 323–340.

Wrench, J.J., 1978. Selenium metabolism in the marine phytoplankters *Tetraselmis tetrathele* and *Dunaliella minuta*. Mar. Biol. 49, 231–236.

Wrench, J., Fowler, S.W., Unlu, M.Y., 1979. Arsenic metabolism in a marine food chain. Mar. Pollut. Bull. 10, 18–20.

Yamamoto, Y., Fujita, T., Ishibashi, M., 1970. Chemical studies on the seaweeds (25). Vanadium and titanium contents in seaweeds. Rec. Oceangr. Works Jpn. 10(2), 125–135.

Yamamoto, T., Yamaoka, T., Fujita, T., Isoda, C., 1971. Chemical studies on the seaweeds (26). Boron content of seaweeds. Rec. Oceangr. Works Jpn. 11(1), 7–13.

Yamamoto, T., Otsuka, Y., Uemura, K., 1976. Gallium content in seaweeds. J. Ocean. Soc. Jpn. 32, 182–186.

Yamaoka, Y., Takimura, O., 1986. Marine algae resistant to inorganic arsenic. Agric. Biol. Chem. 50, 185–186.

Yeo, A.R., Flowers, T.J., 1977. Salt tolerance in the halophyte *Suaeda maritima* (L.) Dum.: interaction between aluminum and salinity. Ann. Bot. 41, 331–339.

Yoshimura, A., Tada, H., Sakai, M., Harada, T., Oishi, K., 1976. Distribution of inorganic constituents of kombu blade-III. Inorganic constituents of acceleratedly cultured ma-kombu at various growth stages. Bull. Jpn. Soc. Sci. Fish. 42, 661–664.

Young, M.L., 1975. The transfer of ^{65}Zn and ^{59}Fe along a *Fucus serratus* (L.)/*Littorina obtusa* food chain. J. Mar. Biol. Assoc. UK 55, 583–610.

Young, E.G., Langille, W.M., 1958. The occurrence of inorganic elements in marine algae of the Atlantic provinces of Canada. Can. J. Bot. 36, 301–310.

Zavodnik, N., 1977. Note on the effects of lead on oxygen production of several littoral seaweeds of the Adriatic Sea. Botanica Marina 20, 167–170.

Zbikowski, R., Szefer, P., Latala, A., 2006. Distribution and relationships between selected chemical elements in green alga *Enteromorpha* sp. from the southern Baltic. Environ. Pollut. 143, 435–448.

Zbikowski, R., Szefer, P., Latala, A., 2007. Comparison of green algae *Cladophora* sp. and *Enteromorpha* sp. as potential biomonitors of chemical elements in the southern Baltic. Sci. Total Environ. 387, 320–332.

Zhao, Y., Hao, H., Ramelow, G.J., 1994. Evaluation of treatment techniques for increasing the uptake of metal ions from solution by nonliving seaweed algal biomass. Environ. Monit. Assess. 33, 61–70.

Zheng, W., Lin, P., 1996. Accumulation and distribution of Cu, Pb, Zn and Cd in *Avicennia marina* mangrove community of Futian in Shenzhen. Oceanol. Limnol. Sinica 1996, 77.

Zheng, W.J., Chen, X.Y., Lin, P., 1997. Accumulation and biological cycling of heavy metal elements in *Rhizophora stylosa* mangroves in Yingluo Bay, China. Mar. Ecol. Prog. Ser. 159, 293–301.

Zingde, M.D., Singbal, S.Y.S., Moraes, C.F., Reddy, C.V.G., 1976. Arsenic, copper, zinc & manganese in the marine flora & fauna of coastal & estuarine waters around Goa. Indian J. Mar. Sci. 5, 212–217.

Zlobin, U.S., 1966. Accumulations of radioactive strontium by the brown seaweeds. Gig. Sanit. 31(12), 86–88 (Nucl. Sci. Abstr. 1967, 21, 2349).

CHAPTER 3
Protists

This group comprises all marine nonphotosynthetic unicellular organisms, such as protozoans, bacteria, yeasts, and fungi. Trace metal concentrations in protists are known initially from the results of three studies: Culkin and Riley (1958), who collected the foraminiferan *Ramulina* sp. from the Irish Sea; Krinsley (1960), who analyzed planktonic foraminiferan tests from Atlantic Ocean cores; and Martin and Knauer (1973), who sampled radiolarians from the Pacific Ocean near and between Hawaii, California, and Oregon.

3.1 Aluminum

Concentrations ranged from 137.0 to 3008.0 mg Al/kg DW (dry weight) (Table 3.1). Endosymbiotic bacteria (sulfur and methane oxidizers) from gills of a hydrothermal vent mussel, *Bathymodiolus azoricus*, collected in 2002 contained 209.0 mg Al/kg DW (Table 3.1). Authors aver that endosymbiotic bacteria sequester aluminum (and to a lesser extent manganese and mercury) from host gill and contribute to its removal (Kadar et al., 2006).

3.2 Arsenic

Organoarsenicals were less toxic to *Vibrio fischeri* marine bacteria than were inorganic arsenicals, based on luminescence tests (Fulladosa et al., 2007). In most cases, adverse effects of dimethylarsinic acid, monomethylarsonic acid, and 4-hydroxy-3-nitrobenzene arsonic acid could not be determined because of the low toxicity of these compounds. Arsenobetaine and monomethylarsonic acid, however, can stimulate bacterial growth (Fulladosa et al., 2007).

Bacterial cultures from the Sargasso Sea and from marine waters of Rhode Island were grown in media with added trivalent arsenic. During log growth phase, bacteria reduced all available pentavalent arsenic and used As^{3+}, presumably as an essential nutrient. The arsenate reduction rate was estimated at 75×10^{-11} mg As/cell/min (Johnson, 1972). Since As^{3+} is considered a threat to human health, while As^{5+} by all accounts remains comparatively innocuous, these findings have important toxicological implications.

Table 3.1: Trace Element Concentrations in Field Collections of Protists

Element and Organism	Concentration	Reference[a]
Aluminum		
Endosymbiotic bacteria from gills of a hydrothermal vent mussel, *Bathymodiolus azoricus*	209.0 DW	4
Foraminiferan tests	1269.0-3008.0 DW	1
Radiolarians	137.0 DW	2
Ramulina sp.	973.0 DW	3
Barium		
Radiolarians	17.0 DW	2
Cadmium		
Endosymbiotic bacteria from gills of mussel	0.3 DW	4
Radiolarians	6.4 DW	2
Cobalt		
Endosymbiotic bacteria from mussel gills	0.2 DW	4
Copper		
Foraminiferan tests	8.0-30.4	1
Radiolarians	6.5 (4.4-11.4) DW	2
Ramulina sp.	111.0 DW	3
Gallium		
Ramulina sp.	0.18 DW	3
Iron		
Radiolarians	315.0 DW	2
Ramulina sp.	1148.0 DW	3
Lead		
Radiolarians	2.1 DW	2
Manganese		
Endosymbiotic bacteria from mussel gills	1.4 DW	4
Foraminiferan tests	9.3-63.3 DW	1

(*Continues*)

Table 3.1: Cont'd

Element and Organism	Concentration	Reference[a]
Radiolarians	6.4 DW	2
Mercury		
Endosymbiotic bacteria from mussel gills	0.9 DW	4
Radiolarians	0.14 DW	2
Nickel		
Foraminiferan tests	15.4-23.0 DW	1
Radiolarians	3.7 DW	2
Strontium		
Foraminiferan tests	933.9-1782.9 DW	1
Radiolarians	163.0 DW	2
Titanium		
Foraminiferan tests	186.0-360.0 DW	1
Zinc		
Radiolarians	63.0-279.0 DW	2

Values are in mg element/kg dry weight (DW).
[a]1, Krinsley, 1960; 2, Martin and Knauer, 1973; 3, Culkin and Riley, 1958; 4, Kadar et al., 2006.

3.3 Barium

A single datum was available of 17.0 mg Ba/kg DW (Table 3.1).

3.4 Cadmium

Limited data show that cadmium ranged from 0.3 to 6.4 mg Cd/kg DW (Table 3.1). Cadmium uptake by a gram-positive bacterium, *Bacillus firmus*, isolated from Hong Kong sediments is best predicted by Cd^{2+} activity and passive diffusion through the plasma membrane (Keung et al., 2008). The trypanosomid flagellate *Crithidia fasciculate*, grown in culture media containing 1.79 mg Cd/L, removed 95% of the cadmium (Maclean et al., 1972). Under similar conditions, *Escherichia coli* bacteria from sediments of Corpus Christi Harbor, Texas, took up 70% of the metal (McLerran and Holmes, 1974). The use of protists as possible decontaminators of cadmium-impacted environments is suggested.

Extracts from cadmium-exposed clams, *Scapharca inaequivalvis*, inhibited the bacterial luminescence of *Vibrio logei*; the inhibition decreased as time of clam exposure to cadmium increased, suggesting a reduction of the biological component (Girotti et al., 2006).

3.5 Cesium

A foraminiferan, *Elphidium crispum*, showed a maximum concentration factor of 16 for radiocesium (Bryan, 1963). Accumulation rates were similar for dead and living *Elphidium*, suggesting that cesium metabolism by this species may be governed by the same mechanisms responsible for potassium balance (Bryan, 1963).

3.6 Chromium

A hexavalent chromium-resistant strain of *Enterobacter cloacae* can remove up to 70% of the chromium in solution and show enhanced growth and exopolysaccharide production at concentrations as high as 100.0 mg Cr^{6+}/L (Iyer et al., 2004). *V. fischeri* bacteria are comparatively resistant to hexavalent chromium owing to its capacity to reduce the toxic Cr^{6+} to the comparatively innocuous Cr^{3+}; the reducing capacity is attributed to unidentified thermostable nonproteic molecules (Fulladosa et al., 2006).

3.7 Cobalt

Bacteria from gills of mussels contained 0.2 mg Co/kg DW (Table 3.1). Cobalt uptake into cells of *E. coli* was inhibited by 0.24 mg magnesium/L and higher, and energy-dependent transport of Mg^{2+} into cells was inhibited by 0.59 mg Co^{2+}/L and higher (Nelson and Kennedy, 1971). Cobalt-magnesium interactions and energy transport mechanisms of protists merit additional research.

3.8 Copper

Concentrations in field collections ranged from 4.4 to 111.0 mg Cu/kg DW (Table 3.1). Extracts from copper-exposed clams inhibited the natural luminescence of *V. logei* bacteria (Girotti et al., 2006). A bacterial copper-resistant strain of *Vibrio* sp.—resistant to 50.0 mg Cu/L—was isolated from the scallop *Argopecten purpuratus* (Miranda and Rojas, 2006). On exposure to 15.0 mg Cu/L for 24 h, the *Vibrio* strain accumulated copper in cellular and loosely bound copper burdens of 201.1 and 493.2 mg/kg DW, respectively. Scallop larvae subjected to Cu-enriched *Vibrio* took up 20.4 mg Cu/kg DW after 12 h and 30.1 mg Cu/kg DW after 24 h, and suggest that bacterial copper accumulation could be active in copper transfer among marine food chains (Miranda and Rojas, 2006). Resistance mechanisms, such as accumulation of copper by resistant cells or reduced permeability to Cu^{2+} owing to the synthesis of new membrane proteins, could be active in aerobic aquatic environments (Cha and Cooksey, 1991; Harwood-Sears and Gordon, 1990, 1994). As a group, the bacteria are important agents in determining the form and distribution of copper in marine environments, playing a major role in the modification, activation, and detoxification of copper (Gordon et al., 2000).

The bacterial community of silty estuarine sediments withstands high concentrations of copper at the cost of reduced bacterial organic matter degradation and reduced bacterial production (Almeida et al., 2007). In one study, bacteria-containing silty estuarine sediments were spiked with 200.0 mg Cu/kg DW. After 21 days, bacterial abundance decreased by 98.6% when compared to controls, and leucine turnover rate decreased by 87%. The toxic effects caused by copper on protein and carbohydrate degradation were not rapidly repaired by erosion and oxygenation; however, bacterial biomass production and leucine turnover were rapidly reactivated (Almeida et al., 2007).

The marine protozoan *Euplotes vannus* subjected to various concentrations of copper for 6 days showed stimulated growth at low concentrations (0.1 mg/L), normal growth at 0.2 mg Cu/L, and adverse effects at 0.4 mg Cu/L, such as depressed growth, morphological cell alterations, and copper accumulations of 239.0 mg/kg DW (Coppellotti, 1998).

Copper-zinc mixtures are frequently more than additive in toxicity to marine biota; however, mixtures of copper (up to 0.09 mg/L) and zinc (up to 1.2 mg/L) are only additive in action to a marine bacterium (*Photobacterium phosphoreum*), decreasing its luminescence after exposure for 30 min (Parrott and Sprague, 1993). Phosphate interfered with the inhibitory effects of copper on marine bacteria and yeasts under conditions of manganese deficiency (Button and Dunker, 1971). It is probable that the complexed copper was not biologically available. High concentrations of copper had no apparent effect on *Cristigera*, a ciliate protozoan, but this may be due to the complexation or chelation of ionic copper with organic matter (Gray and Ventilla, 1971). Iron-reducing bacteria from copper-contaminated sediment were more tolerant to copper adsorbed on hydrous ferric oxide (HFO) than were pristine-sediment bacteria (Markwiese et al., 1998). Copper-tolerant bacteria were more efficient in reducing contaminated HFO, with greater potential for copper mobilization in aquatic sediments (Markwiese et al., 1998).

3.9 Gallium

A single datum was available of 0.18 mg Ga/kg DW (Table 3.1).

3.10 Iron

Iron concentrations in protists ranged from 315.0 to 1148.0 mg Fe/kg DW (Table 3.1).

3.11 Lead

A single datum was available of 2.1 mg Pb/kg DW in radiolarians (Table 3.1).

Reduced growth was documented in the protozoan *Cristigera* sp. after immersion for 12 h in seawater containing 0.15 mg inorganic Pb/L (USEPA, 1985a). Organolead compounds are, in general, more toxic than inorganic lead compounds to aquatic organisms. Ethyl derivatives were more toxic than methyl derivatives, and toxicity increased with increasing degree of alkylation, tetraethyllead being the most toxic (Babich and Borenfreund, 1990; Chau et al., 1980). Mixed species of coastal marine bacteria show marked reduction in oxygen consumption during exposure to tetramethyllead salts; however, tetraethyllead salts were more effective at doses only 10% that of tetramethyllead (Marchetti, 1978). It is reasonable to assume that lead is more readily accumulated by these species in the tetraethyl form than the tetramethyl form.

Interaction of lead with mercury and zinc are documented for *Cristigera* (Gray, 1974). At 0.3 mg Pb/L, about 12% growth reduction was observed; similar results occur at 0.005 mg Hg/L or 0.25 mg Zn/L. However, when mixtures containing 0.3 mg Pb/L plus 0.005 mg Hg/L plus 0.25 mg Zn/L were tested, the growth reduction was 67% rather than the expected 36%. It is possible that several physiologically active sites are involved.

3.12 Manganese

Concentrations in field collections ranged from 1.4 to 63.3 mg Mn/kg DW (Table 3.1).

Manganese is essential for normal metabolism of some marine bacteria. For example, bacteria isolated from ferromanganese nodules synthesized ATP at rates governed by available manganese in the ratio of 1 to 2 moles of ATP for each mole of Mn^{2+} oxidized (Ehrlich, 1976).

3.13 Mercury

Limited data show 0.14-0.9 mg Hg/kg DW, being highest in bacteria isolated from gills of mussels (Table 3.1).

Common features of foraminiferal assemblages in polluted environments include reduced foraminiferal population diversity and density, stunting of adult tests, frequent presence of deformed tests, and an increase in percentage of individuals belonging to a few opportunistic species (Di Leonardo et al., 2007). For example, sediment cores collected off Sicily in 2003-2004 near industrialized areas showed highly enriched mercury concentrations as well as polycyclic aromatic hydrocarbon burdens, and this was associated with a reduction in the abundance of benthic foraminifera, an increasing percentage of tests with morphological deformities, and the dominance of opportunistic species in the more recent sediments associated with human activities (Di Leonardo et al., 2007).

Mercury discharged into rivers, bays, or estuaries as metallic mercury, inorganic divalent mercury, or organomercurial can all be converted to water soluble methylmercury compounds of high toxicity (Jernelov, 1969). Mercury-resistant strains of bacteria that have been developed or discovered may have application in mercury mobilization or fixation from mercury-contaminated waters to the extent that polluted areas become innocuous (Colwell and Nelson, 1975; Eisler, 2006; Nelson et al., 1973; Vosjan and Van der Hoek, 1972). For example, a mercury-reducing strain of *Pseudomonas* reduced 0.025 mg of $HgCl_2$ to elemental mercury from solutions containing 6.0 mg $HgCl_2$/L (Colwell et al., 1976). Mercury-resistant strains of bacteria are common, and 364 strains resistant to $HgCl_2$ have been isolated from Chesapeake Bay alone; most were pseudomonads and almost all from seven genera (Colwell et al., 1976). The percentage of mercury-resistant strains in bay sediments was highest in the spring of the year (Colwell and Nelson, 1974). Some strains were not only mercury resistant, but were reported to also degrade petroleum (Walker and Colwell, 1974). Mercury-resistant and metabolizing strains of bacteria showed a wide range of metabolic rates depending on the chemical species of mercury, the bacterial biomass (Colwell and Nelson, 1975; Nelson and Colwell, 1975), and species competition in the case of *E. coli* and *Staphylococcus aureus* (Stutzenberger and Bennet, 1965). Other groups of bacteria that influence mercury fluxes include strains of *E. coli* that convert inorganic divalent mercury to metallic mercury (Summers and Silver, 1972), and strains of anaerobic methanogenic bacteria that enhance the transfer of methylcobalamin to Hg^{2+} in mild reducing conditions (Wood et al., 1968).

In the case of Minamata Bay, Japan, a site initially contaminated by inorganic mercury, the mercury was methylated chemically with methylcobalamin as the carbon methyl group donor (Baldi et al., 1993; Choi and Bartha, 1993). The methylcobalamin is synthesized by a range of bacterial species, especially *Desulfovibrio desulfuricans*, which is considered a major methylating source in anaerobic sediments (Choi and Bartha, 1993). At low (0.1 mg Hg^{2+}/kg) sediment mercury concentrations, up to 37% of the added mercury was methylated during fermentative growth of *D. desulfuricans*. But under conditions of sulfate reduction, mercury methylation was less efficient. The fermentative mercury methylating cultures were comparatively sensitive to Hg^{2+}, whereas the sulfate-reducing cultures that did not methylate at a high rate were more resistant to Hg^{2+} (Choi and Bartha, 1993). Some strains of *D. desulfuricans*, in addition to producing methylmercury, can also convert methylmercury to methane and Hg^{o}, as was true for aerobic species of bacteria isolated from Minamata Bay sediments (Baldi et al., 1993). In Minamata Bay, 72 strains of mercury-resistant *Pseudomonas* spp. bacteria were isolated from sediments near the drainage outlet to the bay (Nakamura et al., 1986). *Pseudomonas* spp. dominated the bacteria with the highest resistance to mercury, although *Bacillus* spp. strains were the most numerous among all bacterial species isolated from Minamata Bay sediments. Total bacterial concentrations in Minamata Bay were the same as a nearby reference site, but mercury-resistant bacteria were found in higher numbers in Minamata Bay. In 1984, additional bacteria were isolated. *Bacillus* strains dominated in sediments containing up to 23.0 mg total Hg/kg and *Pseudomonas* strains in sediments with higher (>52.0 mg total Hg/kg) mercury concentrations. The mercury-resistant *Pseudomonas* strains, when compared to *Bacillus* strains, were more resistant to inorganic mercury, methylmercury, and phenylmercury (Nakamura et al., 1986). Multiple organomercurial-volatilizing bacteria that can volatilize Hg^o from methylmercury chloride, ethylmercury chloride, phenylmercury acetate, and fluorescein mercuric acetate, and *p*-chloromercuric benzoate were found in the sediments of Minamata Bay (Nakamura, 1994; Nakamura and Silver, 1994). The bacteria detoxify mercury compounds by two separate enzymes—organomercurial lyase and mercuric reductase—acting sequentially. Organomercurial lyase cleaves the C-Hg bond of certain mercurials, and mercuric reductase then reduces Hg^{2+} to volatile Hg^o; bacteria remove mercury from the environment as mercury vapor (Nakamura, 1994). In 1988, a total of 4604 bacterial strains were isolated from Minamata Bay (1428 strains) and a reference site (3176 strains) and screened for their ability to volatilize mercuric chloride (Nakamura, 1994). Up to 38% of all bacterial strains isolated from Minamata Bay could grow on agar media containing 40.0 mg $MgCl_2$/L versus only 0.8% of bacteria from reference locations. A total of 67 *Pseudomonas* strains and 91 *Bacillus* strains from Minamata Bay were significantly more resistant to mercuric chloride than were *Bacillus* spp., both detoxify the mercury compounds in Minamata Bay by organomercurial lyase and mercuric reductase enzymes. The organomercurials-volatilizing bacteria strains were found only in sediments of Minamata Bay. A total of 78 strains of *Bacillus* that can degrade both inorganic and organic mercurials were collected from Minamata Bay and were identified as *Bacillus subtilis, B. firmus, B. lentus, B. bodius*, and two unidentified strains. The ability to detoxify mercury compounds is chromosomally encoded and similar to that of *Bacillus* spp. isolated from Boston Harbor, Massachusetts (Nakamura, 1994). Enzymatic detoxification was determined as the major resistance mechanism in all species of mercury-resistant bacteria (Furukawa and Tonomura, 1971; Komura and Izaki, 1971; Summers and Silver, 1972). For example, mercuric reductase was essential for volatilization of Hg^o from Hg^{2+}, and various organomercurial hydrolases were responsible for volatilization of methane (CH_4) from methylmercury, for ethane (C_2H_5) from ethylmercury, and for benzene from phenylmercury (Silver et al., 1994). Minamata Bay bacterial isolates can also volatilize Hg^o from added inorganic and organic mercurials (Nakamura, 1989; Nakamura et al., 1988, 1990). Genes that govern the chemistry of mercury detoxification were abundant in bacteria found in Minamata Bay and other mercury-polluted sites; these genetic strains of mercury-resistant bacteria show promise for bioremediation of mercury pollution (Misra, 1992; Nakamura, 1994; Silver and Waldrhaug, 1992; Silver et al., 1994).

Mercury was the most toxic metal tested to the ciliated marine protozoan *Uronema marinum*, with 50% survival in 24 h at 0.006 mg Hg/L (Parker, 1979). Similar values for lead and zinc were 60.0 and 400.0 mg/L, respectively. Interaction effects among the three were evident. Attempts to acclimatize the ciliates to trace metals over an 18-week period were unsuccessful (Parker, 1979). However, the marine ciliate *Uronema nigricans* did acquire tolerance to mercury after eating mercury-laden bacteria, although survival was lower than controls

(Berk et al., 1978). Acquired mercury tolerance was demonstrated within a single generation. Ciliates fed with mercury-free bacteria and subsequently exposed to increasing levels of mercury in solution show an elevated tolerance to concentrations that initially killed 83% of the ciliates (Berk et al., 1978).

3.14 Nickel

Concentrations in field collections ranged from 3.7 to 23.0 mg Ni/kg DW (Table 3.1).

Nickel is essential for the normal growth of many species of microorganisms and plants, and several species of mammals (USPHS, 1993, 1995; WHO, 1991). Nickel introduced into the environment from natural or human sources is circulated through the system by chemical and physical processes and though biological transport mechanisms of living organisms (NAS, 1975; WHO, 1991). In microorganisms, nickel binds mainly to the phosphate groups of the cell wall. From this site, an active transport mechanism designed for magnesium transport transports the nickel (Kasprzak, 1987).

Nickel salts gave no evidence of mutagenesis in tests with viruses (USPHS, 1977), and bacterial tests of nickel compounds have consistently yielded negative or inconclusive results (Sunderman, 1981; Sunderman et al., 1984; USPHS, 1977; WHO, 1991). However, nickel chloride and nickel sulfate were judged to be mutagenic or weakly mutagenic in certain bacterial eukaryotic test systems (USEPA, 1985b). Nickel subsulfide was positively mutagenic to the protozoan *Paramecium* sp. at 0.5 mg Ni/L (WHO, 1991). Ionic Ni^{2+} was mutagenic to *E. coli*; mutagenesis was enhanced by the addition of both hydrogen peroxide and tripeptide glycyl-L-histidine, suggesting that short-lived oxygen free radicals are generated (Tkeshelashvili et al., 1993). Nickel chloride hexahydrate induced respiratory deficiency in yeast cells, but this may be a cytotoxic effect rather than a gene mutation (USPHS, 1977; WHO, 1991).

Marine bacteria are comparatively tolerant of nickel, with growth inhibition evident at concentrations greater than 10.0 mg Ni/L (Babich and Stotzky, 1982).

3.15 Silver

Silver was lethal to sensitive protozoans at 0.009-0.015 mg/L (Nalecz-Jawecki et al., 1993). The ciliate protozoan *Fabrea salina* held in seawater solution containing radiosilver 110 m accumulated silver from the medium by factors that ranged from 7000 to 40,000 within 24 h (Fisher et al., 1995).

Marine bacteria isolated from tubes of deep-sea polychaete annelids were comparatively resistant to silver salts; 3.0 mg Ag/L was the lowest concentration tested that inhibited growth in 50% of strains tested during exposure for 10 days (Jeanthon and Prieur, 1990).

Silver-resistant strains (55% of all strains tested) survived at least 24 h during exposure to 20.0 mg Ag/L, and some strains survived exposure to 40.0 mg Ag/L (Jeanthon and Prieur, 1990).

3.16 Strontium

Concentrations of strontium were comparatively high in field collections, ranging from 163.0 to 1782.0 mg Sr/kg DW (Table 3.1). Calcium is an important variable mediating strontium accumulations in foraminiferans. Two studies show that the Sr/Ca ratios range from 0.0012-0.0015 (Bender et al., 1975) and 0.028-0.038 (Thompson and Chow, 1955), respectively.

3.17 Tin

Total tin accumulations in marine algae, molluscs, and crustaceans seldom exceeded 100.0 mg Sn/kg DW; however, certain species of tin-resistant bacteria contained a remarkable 3700.0-7700.0 mg Sn/kg DW (Eisler, 2000; Maguire, 1991; Maguire et al., 1984). Tributyltin (TBT)-resistant strains of *Pseudoalteromonas* sp. were unable to degrade the comparatively toxic TBT to the less toxic dibutyltin (DBT) and monobutyltin (MBT); resistance in this strain was attributed to the unique cell surface containing membrane-binding proteins that can pump TBT from the cytoplasm (Mimura et al., 2008).

Inorganic tin can be biomethylated by microorganisms in the aquatic environment and subsequently mobilized in the ecosystem (Tugrul et al., 1983; Yemenicioglu et al., 1987). The process is slow and usually does not proceed beyond the monomethyltin stage (Zuckerman et al., 1978), although dimethyltin formation by *Pseudomonas* bacteria is reported (Roy et al., 2004; Smith, 1978). The fungus *Cunninghamella elegans* can degrade TBTs to less toxic derivatives, that is, DBTs and MBTs; up to 89% of TBT is degraded within 3 days in oxygenated media containing glucose, ammonium chloride, potassium diphosphate, and magnesium sulfate (Bernat and Dlugonski, 2006). The ability of microorganisms to reduce various organotins to less toxic, easily excreted, metabolites seems to preclude food-chain biomagnification (Cardwell and Sheldon, 1986). Strains of *Pseudomonas* isolated from Chesapeake Bay were about 100 times more sensitive to divalent tin salts than to tetravalent tin salts (Iverson et al., 1973).

Organotins can be highly toxic to many prokaryotic and eukaryotic organisms and have been identified as immune system inhibitors and endocrine disruptors in humans (Dubey and Roy, 2003; Dubey et al., 2006). Mechanisms involved in TBT resistance in bacteria include transformation into less-toxic compounds, such as DBT and MBT, by abiotic and biotic processes (Clark et al., 1988; Pain and Cooney, 1998a). Marine bacteria that are resistant to TBT and can degrade TBT to less-toxic metabolites were isolated from water and sediments near Aveiro, Portugal (Cruz et al., 2007). Of the 50 TBT-resistant strains, an isolate of *Aeromonas veronii*

AV27 was highly resistant to TBT (about 330.0 mg TBT/L), uses TBT as a carbon source and degrades it to less-toxic compounds, including DBT and MBT (Cruz et al., 2007).

Antifouling paints containing organotin compounds are associated with reductions in natural bacterial assemblages (Prior and Riemann, 1998). Among the triorganotin compounds, tripropyltins (TPTs) were highly toxic to gram-negative bacteria, and TBTs to gram-positive bacteria (Eisler, 2000; Maguire, 1991; Mendo et al., 2003). Antifouling paints containing TPTs will not be a suitable substitute for TBTs in paints designed to inhibit microbial biofilms (Pain and Cooney, 1998b). Up to 98% of bacteria resistant to six organotins, including TBTs, isolated from estuarine sediments of Boston Harbor, Massachusetts, were also resistant to TPTs. All bacteria were resistant to at least six of eight metals tested, suggesting that resistance to metals—including nickel, cadmium, lead, mercury, copper, and zinc—may be associated with resistance to organotins (Pain and Cooney, 1998a,b).

3.18 Titanium

A single datum was available of 186.0-360.0 mg Ti/kg DW in tests of foraminiferans (Table 3.1).

3.19 Zinc

A single datum was available of 63.0-279.0 mg Zn/kg DW in radiolarians (Table 3.1).

Zinc uptake by a gram-positive bacterium, *B. firmus*, isolated from Hong Kong sediments, is best predicted by free Zn^{2+} ion activity and passive diffusion through the plasma membrane (Keung et al., 2008). Bacteria compete actively with sediments and higher trophic levels regarding zinc uptake in salt marsh ecosystems (Pomeroy et al., 1966). Bacteria and their metabolic byproducts also play an important role in removal of zinc and other metals from seawater and subsequent deposition in sediments. In one study, mixed cultures of bacteria removed up to 85% of ^{65}Zn from seawater in 120 h (McLerran and Holmes, 1974). Interaction effects of zinc with salts of other metals on protozoan growth are documented. *Cristigera*, a ciliate protozoan, had reduced growth—and presumably altered zinc uptake—when held in media containing salts of zinc, lead, and mercury (Gray, 1974). Growth reduction was observed in *Cristigera* during 5 h exposure in solutions containing 0.05-0.12 mg Zn/L (USEPA, 1980, 1987); other species of protists tested were more resistant to zinc (Spear, 1981).

3.20 Literature Cited

Almeida, A., Cunha, A., Fernandes, S., Sobral, P., Alcantara, F., 2007. Copper effects on bacterial activity of estuarine silty sediments. Estuar. Coastal Shelf Sci. 73, 743–752.

Babich, H., Borenfreund, E., 1990. In vitro cytotoxicities of inorganic lead and di- and trialkyl lead compounds to fish cells. Bull. Environ. Contam. Toxicol. 44, 456–460.

Babich, H., Stotzky, G., 1982. Nickel toxicity to microbes: Effect of pH and implications for acid rain. Environ. Res. 29, 335–350.

Baldi, F., Pepi, M., Filipelli, M., 1993. Methylmercury resistance in *Desulfovibrio desulfuricans* in relation to methylmercury degradation. Appl. Environ. Microbiol. 59, 2479–2485.

Bender, M.L., Lorens, R.B., Williams, D.F., 1975. Sodium, magnesium, and strontium in the tests of planktonic foraminifera. Micropaleontology 21, 448–459.

Berk, S.K., Mills, A.L., Henricks, D.L., Colwell, R.R., 1978. Effects of ingesting mercury containing bacteria on mercury tolerance and growth in ciliates. Microb. Ecol. 4, 319–330.

Bernat, P., Dlugonski, J., 2006. Acceleration of tributyltin chloride (TBT) degradation in liquid cultures of the filamentous fungus *Cunninghamella elegans*. Chemosphere 62, 3–8.

Bryan, G.W., 1963. The accumulation of radioactive caesium by marine invertebrates. J. Mar. Biol. Assoc. UK 43, 519–539.

Button, D.K., Dunker, S.S., 1971. Biological effects of copper and arsenic pollution. U.S. Environmental Protection Agency Contract 18050 DLW, Report R71-8. University of Alaska (College), Alaska, 59 pp.

Cardwell, R.D., Sheldon, A.W., 1986. A risk assessment concerning the fate and effects of tributyltins in the aquatic environment. In: Maton, G.L. (Ed.), Proceedings Oceans 86 Conference, September 23-25, 1986, vol. 4, Organotin Symposium. Available from Marine Technology Society, 2000 Florida Avenue NW, Washington, DC, pp. 1117–1129.

Cha, J.S., Cooksey, D.A., 1991. Copper resistance in *Pseudomonas syringae* mediated by periplasmic and outer mechanism proteins. Proc. Natl. Acad. Sci. USA 88, 8915–8919.

Chau, Y.K., Wong, P.T.S., Kramer, O., Bengert, G.A., Cruz, R.B., Kinrade, J.O., et al., 1980. Occurrence of tetraalkyllead compounds in the aquatic environment. Bull. Environ. Contam. Toxicol. 24, 265–269.

Choi, S.C., Bartha, R., 1993. Cobalamin-mediated mercury methylation by *Desulfovibrio desulfuricans* LS. Appl. Environ. Microbiol. 59, 290–295.

Clark, E.A., Sterritt, R.M., Lester, J.N., 1988. The fate of tributyltin in the aquatic environment. Environ. Sci. Technol. 22, 600–604.

Colwell, R.R., Nelson Jr., J.D., 1974. Bacterial mobilization of mercury in Chesapeake Bay. In: Proceedings of an International Conference on Transport of Persistent Chemicals in Aquatic Ecosystems, Ottawa, Canada, pp. 1–10.

Colwell, R.R., Nelson Jr., J.D., 1975. In: Metabolism of Mercury Compounds in Microorganisms. Report 600/3-75-077. U.S. Environmental Protection Agency, Narragansett, RI, pp. 1–84.

Colwell, R.R., Sayler, G.S., Nelson Jr., J.D., Justice, A., 1976. Microbial mobilization of mercury in the aquatic environment. In: Nriagu, J.O. (Ed.), Environmental Biogeochemistry, Vol. 2. Metals Transfer and Ecological Mass Balances. Ann Arbor Science Publication, Ann Arbor, MI, pp. 437–487.

Coppellotti, O., 1998. Sensitivity to copper in a ciliate as a possible component of biological monitoring in the Lagoon of Venice. Arch. Environ. Contam. Toxicol. 35, 417–425.

Cruz, A., Caetano, T., Suzuki, S., Mendo, S., 2007. *Aeromonas veronii*, a tributyltin (TBT)-degrading bacterium isolated from an estuarine environment, Rio de Aveiro in Portugal. Mar. Environ. Res. 64, 639–650.

Culkin, F., Riley, J.P., 1958. The occurrence of gallium in marine organisms. J. Mar. Biol. Assoc. UK 37, 607–615.

Di Leonardo, R., Bellanca, A., Capotondi, L., Cundy, A., Neri, R., 2007. Possible impacts of Hg and PAH contamination on benthic foraminiferal assemblages: An example from the Sicilian coast, central Mediterranean. Sci. Total Environ. 388, 168–183.

Dubey, S.K., Roy, U., 2003. Biodegradation of tributyltins (organotins) by marine bacteria. Appl. Organomet. Chem. 17, 3–8.

Dubey, S.K., Tokashiki, T., Suzuki, S., 2006. Microarray-mediated transcriptome analysis of the tributyltin (TBT)-resistant bacterium *Pseudomonas aeruginosa* 25W in the presence of TBT. J. Microbiol. 44, 200–205.

Ehrlich, H.L., 1976. Manganese as an energy source for bacteria. In: Nriagu, J.O. (Ed.), Environmental Biogeochemistry, Vol. 2. Metals Transfer and Ecological Mass Balances. Ann Arbor Science Publication, Ann Arbor, MI, pp. 633–644.

Eisler, R., 2000. Tin. In: Handbook of Chemical Risk Assessment. Health Hazards to Humans, Plants, and Animals. Vol. 1, Metals. Lewis Publishers, Boca Raton, FL, pp. 551–603.

Eisler, R., 2006. Mercury Hazards to Living Organisms. CRC Press, Boca Raton, FL, p. 312.

Fisher, N.S., Breslin, V.T., Levandowsky, M., 1995. Accumulation of silver and lead in estuarine microzooplankton. Mar. Ecol. Prog. Ser. 116, 207–215.

Fulladosa, E., Desjardin, V., Murat, J.C., Gourdon, R., Villaescusa, I., 2006. Cr(VI) reduction into Cr(III) as a mechanism to explain the low sensitivity of *Vibrio fischeri* bioassay to detect chromium pollution. Chemosphere 65, 644–650.

Fulladosa, E., Murat, J.C., Bollinger, J.C., Villaescusa, I., 2007. Adverse effects of organic arsenical compounds towards *Vibrio fischeri* bacteria. Sci. Total Environ. 377, 207–213.

Furukawa, A., Tonomura, K., 1971. Enzyme system involved in the decomposition of phenyl mercuric acetate by mercury-resistant *Pseudomonas*. Agric. Biol. Chem. 35, 604–610.

Girotti, S., Botelli, L., Fini, F., Monari, M., Andreani, G., Isani, G., et al., 2006. Trace metals in arcid clam *Scapharca inaequivalvis*: Effects of molluscan extracts on bioluminescent bacteria. Chemosphere 65, 627–633.

Gordon, A.S., Donat, J.R., Kango, R., Dyer, B., Stuart, L., 2000. Dissolved copper-complexing ligands in cultures of marine bacteria and estuarine water. Mar. Chem. 70, 149–160.

Gray, J.S., 1974. Synergistic effects of three heavy metals on growth rates of a marine ciliate protozoan. In: Vernberg, F.J., Vernberg, W.B. (Eds.), Pollution and Physiology of Marine Organisms. Academic Press, New York, pp. 465–485.

Gray, J.S., Ventilla, R.J., 1971. Pollution effects on micro- and meiofauna of sand. Mar. Pollut. Bull. 2, 39–43.

Harwood-Sears, V., Gordon, A.S., 1990. Copper-induced production of copper-binding supernatant proteins by the marine bacterium *Vibrio alginolyticus*. Appl. Environ. Microbiol. 56, 1327–1332.

Harwood-Sears, V., Gordon, A.S., 1994. Regulation of extracellular copper-binding proteins in copper-resistant and copper-sensitive mutants of *Vibrio alginolyticus*. Appl. Environ. Microbiol. 60, 1749–1753.

Iverson, W.P., Huey, C., Brinckman, F.E., Jewett, K.L., Blair, W., 1973. Biological and nonbiological transformations of mercury in aquatic systems. In: Krenkel, P.A. (Ed.), Heavy Metals in the Aquatic Environment. Pergamon, Elmsford, NY, pp. 193–195.

Iyer, A., Mody, K., Jha, B., 2004. Accumulation of hexavalent chromium by an exopolysaccharide producing marine *Enterobacter cloaceae*. Mar. Pollut. Bull. 49, 974–977.

Jeanthon, C., Prieur, D., 1990. Susceptibility to heavy metals and characterization of heterotrophic bacteria isolated from two hydrothermal vent polychaete annelids. Appl. Environ. Microbiol. 56, 3308–3314.

Jernelov, A., 1969. Conversion of mercury compounds. In: Miller, C.W., Berg, G.G. (Eds.), Chemical Fallout, Current Research on Persistent Pesticides. Chas. C. Thomas, Springfield, IL, pp. 68–74.

Johnson, D.L., 1972. Bacterial reduction of arsenate in seawater. Nature 240, 44–45.

Kadar, E., Santos, R.S., Powell, J.J., 2006. Biological factors influencing tissue compartmentalization of trace metals in the deep-sea hydrothermal vent bivalve *Bathymodiolus azoricus* at geochemically distinct vents sites of the mid-Atlantic Ridge. Environ. Res. 101, 221–229.

Kasprzak, K.S., 1987. Nickel. Adv. Mod. Environ. Toxicol. 11, 145–183.

Keung, C.F., Guo, F., Qian, P., Wang, W.X., 2008. Influences of metal-ligand complexes on the cadmium and zinc biokinetics in the marine bacterium, *Bacillus firmus*. Environ. Toxicol. Chem. 27, 131–137.

Komura, I., Izaki, K., 1971. Mechanism of mercuric chloride resistance in microorganisms. I. Vaporization of a mercury compound from mercuric chloride by multiple drug resistant strains of *Escherichia coli*. J. Biochem. (Tokyo) 70, 885–893.

Krinsley, D., 1960. Trace elements in the tests of planktonic foraminifera. Micropaleontology 6, 297–300.

Maclean, F.I., Lucis, O.J., Shakh, Z.Z., Jansz, E.R., 1972. The uptake and cellular distribution of Cd and Zn in microorganisms. Fed. Proc. FASEB 31, 699.

Maguire, R.J., 1991. Aquatic environmental aspects of non-pesticidal organotin compounds. Water Pollut. Res. J. Can. 26, 243–360.

Maguire, R.J., Wong, P.T.S., Rhamey, J.S., 1984. Accumulation and metabolism of tri-*n*-butyltin cation by a green alga, *Ankistrodesmus falcatus*. Can. J. Fish Aquat. Sci. 41, 537–540.

Marchetti, R., 1978. Acute toxicity of alkyl leads to some marine organisms. Mar. Pollut. Bull. 9, 206–207.

Markwiese, J.T., Meyer, J.S., Colberg, P.J.S., 1998. Copper tolerance in iron-reducing bacteria: Implications for copper mobilization in sediments. Environ. Toxicol. Chem. 17, 675–678.

Martin, J.H., Knauer, G.A., 1973. The elemental composition of plankton. Geochim. Cosmochim. Acta 37, 1639–1653.

McLerran, C.J., Holmes, C.W., 1974. Deposition of zinc and cadmium by marine bacteria in estuarine sediments. Limnol. Oceanol. 19, 988–1001.

Mendo, S.A., Nogueira, P.R., Ferreira, S.C.N., Silva, R.G., 2003. Tributyltin and triphenyltin toxicity on benthic estuarine bacteria. Fresenius Environ. Bull. 12, 1361–1367.

Mimura, H., Sato, R., Furyama, Y., Taniike, A., Yagi, M., Yoshida, K., 2008. Adsorption of tributyltin by tributyltin resistant marine *Pseudoalteromonas* sp. Cells. Mar. Pollut. Bull. 57, 877–882.

Miranda, C.D., Rojas, R., 2006. Copper accumulation by bacteria and transfer to scallop larvae. Mar. Pollut. Bull. 52, 293–300.

Misra, T.K., 1992. Bacterial resistance to inorganic mercury salts and organomercurials. Plasmid 27, 4–16.

Nakamura, K., 1989. Volatilization of fluorescein mercuric acetate by marine bacteria from Minamata Bay. Bull. Environ. Contam. Toxicol. 42, 785–790.

Nakamura, K., 1994. Mercury compounds-decomposing bacteria in Minamata Bay. In: Proceedings of the International Symposium on Assessment of Environmental Pollution and Health Effects from Methylmercury, Kumamoto, Japan, October 8-9, 1993. National Institute for Minamata Disease, Minamata City, Kumamoto, Japan, pp. 198–209.

Nakamura, K., Silver, S., 1994. Molecular analysis of mercury-resistant *Bacillus* isolates from sediment of Minamata Bay, Japan. Appl. Environ. Microbiol. 60, 4599–4956.

Nakamura, K., Fujisaki, T., Tamashiro, H., 1986. Characteristics of Hg-resistant bacteria isolated from Minamata Bay sediment. Environ. Res. 40, 58–67.

Nakamura, K., Fujisaki, T., Shibata, Y., 1988. Mercury-resistant bacteria in the sediment of Minamata Bay. Nippon Suisan Gakkaishi 54, 1359–1363.

Nakamura, K., Sakamoto, M., Uchiyama, H., Yagi, O., 1990. Organomercurial-volatilizing bacteria in the mercury-polluted sediment of Minamata Bay, Japan. Appl. Environ. Microbiol. 56, 304–305.

Nalecz-Jawecki, G., Demkowicz-Dobrzanski, K., Sawicki, J., 1993. Protozoan *Spirostomum ambiguum* as a highly sensitive bioindicator for rapid and easy determination of water quality. Sci. Total Environ. 2 (Suppl.), 1227–1234.

National Academy of Sciences (NAS), 1975. Medical and Biological Effects of Environmental Pollutants: Nickel. National Research Council, NAS, Washington, DC, 277 pp.

Nelson Jr., J.D., Colwell, R.R., 1975. The ecology of mercury-resistant bacteria in Chesapeake Bay. Microbiol. Ecol. 1, 191–218.

Nelson, D.L., Kennedy, E.P., 1971. Magnesium transport in *Escherichia coli*. J. Biol. Chem. 246, 3042–3049.

Nelson, J.D., Blair, W., Brinckman, F.E., Colwell, R.R., Iverson, W.P., 1973. Biodegradation of phenylmercuric acetate by mercury-resistant bacteria. Appl. Microbiol. 26, 321–326.

Pain, A., Cooney, J.J., 1998a. Characterization of organotin-resistant bacteria from Boston Harbor sediments. Arch. Environ. Contam. Toxicol. 35, 412–416.

Pain, A., Cooney, J.J., 1998b. Characterization of organotin-resistant bacteria from Boston Harbor sediments. Arch. Environ. Contam. Toxicol. 35, 412–416.

Parker, J.G., 1979. Toxic effects of heavy metals upon cultures of *Uronema marinum* (Ciliophora: uronematidae). Mar. Biol. 54, 17–24.

Parrott, J.L., Sprague, J.B., 1993. Patterns in toxicity of sublethal mixtures of metals and organic chemicals determined by ᵣmicrotox and by DNA, RNA, and protein content of fathead minnows (*Pimephales promelas*). Can. J. Fish. Aquat. Sci. 50, 2245–2253.

Pomeroy, L.R., Odum, E.P., Johannes, R.E., Roffman, B., 1966. Flux of phosphorus-32 and zinc-65 through a salt-marsh ecosystem. Proc. Symp. IAEA 177–188.

Prior, S., Riemann, B., 1998. Effects of tributyltin, linear alkylbenzenesulfonates, and nutrients (nitrogen and phosphorus) on nucleoid-containing bacteria. Environ. Toxicol. Chem. 17, 1473–1480.

Roy, U., Dubey, S.K., Bhosle, S., 2004. Tributyltin chloride-utilizing bacteria from marine ecosystem of west coast of India. Curr. Sci. 86(5), 702–705.

Silver, S., Waldrhaug, M., 1992. Gene regulation of plasmid and chromosome-determined inorganic ion transport in bacteria. Microbiol. Rev. 56, 195–228.

Silver, S., Endo, G., Nakamura, K., 1994. Mercury in the environment and in the laboratory. J. Jpn. Soc. Water Environ. 17, 235–243.

Smith, P.J., 1978. Structure/Activity Relationships for Di- and Triorganotin Compounds. I.T.R.I. Rep. 569. Available from International Tin Research Institute, Greenford, Middlesex, UK, 16 pp.

Spear, P.A., 1981. Zinc in the aquatic environment: Chemistry, distribution, and toxicology. Natl. Res. Coun. Can. Publ., NRCC 17589, 145.

Stutzenberger, F.J., Bennet, E.O., 1965. Sensitivity of mixed populations of *Staphylococcus aureus* and *Escherichia coli* to mercurials. Appl. Microbiol. 13, 570–574.

Summers, A.O., Silver, S., 1972. Mercury resistance in a plasmid-bearing strain of *Escherichia coli*. J. Bacteriol. 112, 1228–1236.

Sunderman Jr., F.W., 1981. Recent research on nickel carcinogenesis. Environ. Health Perspect. 40, 131–141.

Sunderman Jr., F.W., Aito, A., Berlin, A., Bishop, C., Buringh, E., Davis, W., et al. (Eds.), 1984. Nickel in the Human Environment. IARC Sci. Publ. No. 53. International Agency for Research on Cancer, Oxford University Press, Oxford, 530 pp.

Thompson, T.G., Chow, T.J., 1955. The strontium-calcium atom ratio in carbonate-secreting marine organisms. Deep Sea Res. 3 (Suppl.), 20–39.

Tkeshelashvili, L.K., Reid, T.M., McBride, T.J., Loeb, L.A., 1993. Nickel induces a signature mutation for oxygen free radical damage. Cancer Res. 53, 4172–4174.

Tugrul, S., Balkas, T.I., Goldberg, E.D., 1983. Methyltins in the marine environment. Mar. Pollut. Bull. 14, 297–303.

U.S. Environmental Protection Agency (USEPA), 1980. Ambient water quality criteria for zinc. USEPA Rep. 440/5-80-079, 158 pp.

USEPA, 1985a. Ambient water quality criteria for lead—1984. USEPA Rep. 440/5-84-027, 81 pp.

USEPA, 1985b. Drinking water criteria document for nickel. USEPA Rep. 600/X-84-193-1, pp. 1–64.

USEPA, 1987. Ambient water quality criteria for zinc—1987. USEPA Rep. 440/5-87-003, 207 pp.

U.S. Public Health Service (USPHS), 1977. Criteria for a recommended standard: occupational exposure to inorganic nickel. U.S. Dept. Health Educ. Welfare, PHS, Center Dis. Control. Nat. Inst. Occupat. Safety Health, DHEW (NIOSH), Publ. No. 77-164, pp. 1–282.

USPHS, 1993. In: Toxicological profile for nickel. Rep. TP-92/14. USPHS, Agen. Toxic Subs. Dis Regis., Atlanta, GA, pp. 1–158.

USPHS, 1995. Toxicological profile for nickel (update). USPHS, Agen. Toxic Subs. Dis. Regis., Atlanta, GA, pp. 1–244.

Vosjan, J.H., Van de Hoek, G.J., 1972. A continuous culture of *Desulfovibrio* on a medium containing mercury and copper ions. Neth. J. Sea Res. 5, 440–444.

Walker, J.D., Colwell, R.R., 1974. Mercury-resistant bacteria and petroleum degradation. Appl. Microbiol. 27, 285–287.

Wood, J.M., Kennedy, F.S., Rosen, C.G., 1968. Synthesis of methylmercury compounds by extracts of a methanogenic bacterium. Nature 220, 173–174.

World Health Organization (WHO), 1991. Nickel. Environ. Health Crit. 18, 1–383.

Yemenicioglu, D.R., Saydam, C., Salihoglu, I., 1987. Distribution of tin in the northeastern Mediterranean. Chemosphere 16, 429–443.

Zuckerman, J.J., Reisdorf, R.P., Ellis III, H.V., Wilkinson, R.R., 1978. Organotins in biology and the environment. In: Brinckman, F.E., Bellama, J.M. (Eds.), Organometals and Organometalloids, Occurrence and Fate in the Environment. American Chemical Society Symposium Series 82, Washington, DC, pp. 388–424.

CHAPTER 4
Sponges

There are approximately 5000 species in this group, most of them marine. The largest, the loggerhead sponge *Spheciospongia vesparum*, may reach 2 m in diameter. These multicellular organisms are sessile and aquatic, with a single body cavity lined in part or almost wholly by collared flagellate cells. Sponges possess numerous body pores in the body wall through which water enters, and one or more larger pores through which it exits. In general, sponges have a calcareous, horny, or siliceous skeleton. Sponges have been proposed as heavy metal biomonitors owing to their ability to accumulate metals (Cebrian et al., 2003; Hansen et al., 1995; Olesen and Weeks, 1994; Patel et al., 1985; Perez et al., 2005) and because they show biological and biochemical responses when submitted to metal insult (Agell et al., 2001; Berthet et al., 2005; Cebrian et al., 2003, 2006; Philp, 1999; Rao et al., 2006; Schroeder et al., 2000). For example, high accumulations of metals alter the composition of sugars, proteins, and lipids, especially in near-shore sponges which usually have higher accumulations than offshore conspecifics (Rao et al., 2006).

Despite their widespread geographic distribution, comparatively little is known of the trace metal composition of this group (Table 4.1).

4.1 Aluminum

Aluminum content ranges from 37.3 to 3700.0 mg Al/kg DW (dry weight) whole sponge (Table 4.1), being higher in polluted areas.

4.2 Antimony

A single datum is available of 0.089 mg Sb/kg DW whole sponge (Table 4.1).

4.3 Arsenic

Arsenic seems to concentrate in lipid extracts of sponges (6400.0 mg As/kg; Table 4.1); however, of all groups examined by Vaskovsky et al. (1972), the annelids, molluscs, and asteroid echinoderms had the highest arsenic contents, and sponges had the lowest.

Table 4.1: Trace Element Concentrations in Field Collections of Sponges

Element and Organism	Concentration	Reference[a]
Aluminum		
Halichondria panicea, whole	3700.0 DW	1
Petrosia testudinaria[b]	205.9 DW vs. 37.3 DW	10
Antimony		
Halichondria panicea, whole	0.089 DW	2
Arsenic		
Guam; 11 spp.; 4 locales; 1998-1999; whole	<0.11-47.7 DW	9
Halichondria panacea		
Lipid extract; whole	6400.0 FW	3
Whole	2.8 DW	2
Petrosia testudinaria[b]	2.3 DW vs. 1.3 DW	10
Barium		
Sponges; whole	0.22 FW	4
Cadmium		
Guam; 11 spp.; 4 locales; 1998-1999; whole	0.13-0.86 DW	9
Halichondria panicea; whole	0.85 DW	2
Sponges; whole	0.17-1.0 FW; 1.2-4.5 DW	5
Cobalt		
Petrosia testudinaria[b]	3.0 DW vs. 0.1 DW	10
Copper		
Sponges; whole	0.99-5.0 FW; 8.5-31.0 DW	6
Crambe crambe; Spain; contaminated site (97.0 mg Cu/kg DW sediment) vs. reference site (6.0 mg Cu/kg DW)	Max. 280.0 DW vs. <20.0 DW	12
Sponges; whole; 3 spp.	13.0-34.0 DW	8
Sponges; whole; 4 spp.; 3 sites; Catalonia, Spain; 2003-2004		
Chondrosia reniformis	8.2-11.3 DW	11
Crambe crambe	9.1-42.2 DW	11

(Continues)

Table 4.1: Cont'd

Element and Organism	Concentration	Reference[a]
Phorbas tenacior	34.0-91.0 DW	11
Dysidea avara	97.4-299.3 DW	11
Petrosia testudinaria[b]	2.5 DW vs. 2.0 DW	10
Gallium		
Halichondria panicea; whole	0.93 FW	1
Iron		
Halichondria panicea; whole	4040.0 DW	1
Petrosia testudinaria[b]	305.5 DW vs. 66.3 DW	10
Lead		
Catalonia, Spain; summer; 2003-2004; 3 sites; whole		
Chondrosia reniformis	1.5-2.1 DW	11
Crambe crambe	0.3-1.8 DW	11
Phorbas tenacior	0.5-0.8 DW	11
Dysidea avara	0.1-0.8 DW	11
Crambe crambe; Spain; contaminated site vs. reference site	Max. 130.0 DW vs. <60.0 DW	12
Guam; 11 spp.; 4 locales; 1998-1999; whole	(0.25-8.1) DW	9
Petrosia testudinaria[b]	0.3 DW vs. no data	10
Manganese		
Petrosia testudinaria[b]	91.0 DW vs. 2.0 DW	10
Mercury		
Halichondria panicea; whole	0.33 DW	2
Nickel		
Halichondria sp.; whole; Sweden	22.0 DW	8
Petrosia testudinaria[b]	5.0 DW vs. 4.0 DW	10
Selenium		
Petrosia testudinaria[b]	1.0 DW vs. 0.7 DW	10
Silver		
Guam; 11 spp; 4 locales; 1998-1999; whole	<0.11-0.47 DW	9

(Continues)

Table 4.1: Cont'd

Element and Organism	Concentration	Reference[a]
Strontium		
Sponges; whole	0.31 FW	4
Vanadium		
Crambe crambe; Spain; contaminated site vs. reference site	Max. 15.0 DW vs. <6.0 DW	12
Petrosia testudinaria[b]	0.6 DW vs. no data	10
Zinc		
Halichondria panicea; whole	89.0-152.0 DW	7
Sponges; whole	4.7-30.0 FW; 63.0-180.0 DW; 93.0-360.0 AW	6
Petrosia testudinaria[b]	0.6 DW vs. no data	10

Values are in mg element/kg fresh weight (FW), dry weight (DW), or ash weight (AW).
[a] 1, Culkin and Riley, 1958; 2, Leatherland and Burton, 1974; 3, Vaskovsky et al., 1972; 4, Mauchline and Templeton, 1966; 5, Bernhard and Zattera, 1975; 6, Lowman et al., 1966; 7, Ireland, 1973; 8, Jenkins, 1980; 9, Denton et al., 2006; 10, Rao et al., 2006; 11, Cebrian et al., 2007; 12, Cebrian et al., 2003.
[b] Gulf of Mannar; India; 2004; polluted area; whole less calcareous spicules; 0.5-1.0 km from shore vs. 5.0-7.0 km from shore.

4.4 Barium

A single datum is available of 0.22 mg Ba/kg FW (fresh weight) whole sponge (Table 4.1).

4.5 Cadmium

Maximum concentrations of 4.5 mg Cd/kg DW and 1.0 mg Cd/kg FW are reported in whole sponges (Table 4.1). Isolated cells of the sponge *Scopalina lyphropoda* subjected for 3 h to 0.005 or 0.010 mg Cd/L showed enhanced pseudopodia formation, cell motility, and cell aggregation at the concentrations tested (Cebrian and Uriz, 2007a). Larvae of two Mediterranean Sea sponges *Cranke cranke* and *Scopalina lophyropoda* exposed to 0.005 mg Cd/L for 1 week had normal settlement and juvenile survival (Cebrian and Uriz, 2007c).

Calcium is important in cadmium uptake (Cebrian and Uriz, 2007b). Isolated cells of *Scopalina lophyropoda* were held for 3 h and larvae for 7 days in calcium-free seawater (CFSW), seawater (SW), or SW with 0.005 mg Cd/L (CdCFSW). Sponge cells incubated in CdCFSW took up as much cadmium as SW controls, but little uptake in CFSW. Cadmium initially appeared to accelerate larval settlement, but after 5 days mortality increased and settlement decreased (Cebrian and Uriz, 2007b).

4.6 Cobalt

The maximum concentration of cobalt in sponges from polluted areas is 3.0 mg Co/kg DW (Table 4.1).

The introduction of lithogenic material into the sea from the rivers of the west coast of Puerto Rico, and fallout, had a direct effect on cobalt burdens in *Spheciospongia* sp. and *Iricinia* sp. (Lowman et al., 1967). Sponges distant from the outflows contained lower amounts of total ash and fallout radioisotopes of cobalt, cerium, zirconium, and ruthenium (Lowman et al., 1967). In general, organisms of a single species that is widely distributed geographically have lower trace metal concentrations with increasing distance from land masses.

4.7 Copper

Highest values recorded for copper in sponges are 5.0 mg/kg FW and 299.3 mg/kg DW (Table 4.1). High concentrations of copper in *Crambe crambe*, were associated with inhibited growth and decreased fecundity (Cebrian et al., 2003).

Isolated cells of *Scopalina lophyropoda* subjected to 0.03 and 0.1 mg Cu/L for 3 h, as was true for cadmium, showed enhanced pseudopodia formation, cell motility, and cell aggregation at the concentrations tested (Cebrian and Uriz, 2007a). Calcium seems to control copper metabolism in sponges. Isolated cells of *Scopalina lophyropoda* were subjected for 3 h and larvae for 7 days to CFSW, SW, or CFSW with copper (CuCFSW) at 0.03 mg Cu/L (Cebrian and Uriz, 2007b). Sponge cells incubated in CuCFSW took up Cu as SW controls, with little uptake in CFSW. In the short term, copper accelerated larval settlement but decreased larval settlement and increased mortality after 5 days (Cebrian and Uriz, 2007b). Larvae of two Mediterranean Sea sponges, *Crambe crambe* and *Scopalina lophyropoda* exposed to 0.03 mg Cu/L for 1 week had normal settlement and juvenile survival (Cebrian and Uriz, 2007c). However, copper (0.03 mg/L) in combination with 0.005 mg/L of a mixture of polycyclic aromatic hydrocarbons (benzo(*a*)pyrene, benzo(*e*)pyrene, benzo(*a*)anthracene), benzo(*b*)fluoranthene, benzo(*j*)fluoranthene, and benzo(*k*) fluoranthene) acted more than additively to inhibit the development of sponge larvae with resultant population declines expected (Cebrian and Uriz, 2007c).

4.8 Gallium

A single datum was available of 0.93 mg Ga/kg FW whole sponge (Table 4.1)

4.9 Iron

Iron content in whole sponges ranged from 66.3 to 4040.0 mg Fe/kg DW (Table 4.1).

4.10 Lead

The maximum lead concentration recorded of 130.0 mg Pb/kg DW was in *Crambe crambe* from a lead-contaminated site in Spain (Table 4.1). Concentrations in whole sponges collected from various locations in Guam ranged up to 8.1 mg Pb/kg DW; this animal was collected near a fuel station and commercial port (Denton et al., 2006).

4.11 Manganese

Manganese concentrations in whole sponge, *Petrosia testudinaria*, taken in 2004 from an industrialized area in the Gulf of Mannar, India, were significantly higher in specimens collected 0.5-1.0 km from shore than those collected 5.0-7.0 km from shore: 91.0 mg Mn/kg DW vs. 2.0 mg Mn/kg DW (Rao et al., 2006; Table 4.1).

4.12 Mercury

A single datum was available of 0.33 mg Hg/kg DW whole sponge (Table 4.1). Isolated cells of a Mediterranean Sea sponge, *Scopalina lyphropoda*, held in 0.001 or 0.005 mg Hg/L for 3 h, showed enhanced cell aggregation, a reduction in cell motility, and the absence of pseudopodia (Cebrian and Uriz, 2007a).

4.13 Nickel

Whole sponges, *Halichondria* sp., from Sweden contained an average of 22.0 m Ni/kg DW (Table 4.1).

4.14 Plutonium

Sponges reportedly concentrate ^{239}Pu from seawater by as much as 2300-fold (Noshkin et al., 1971).

4.15 Ruthenium

Soluble radioruthenium-106 was accumulated by sponges 5-10 times more rapidly than insoluble forms; similar observations were noted for shrimps, mussels, sea anemones, and algae (Ancellin and Bovard, 1971).

4.16 Selenium

Whole *Petrosia testudinaria* from the Gulf of Mannar, India, in 2004 contained up to 1.0 mg Se/kg DW (Rao et al., 2006; Table 4.1).

4.17 Silver

The maximum concentration recorded in whole sponges from Guam is 0.47 mg/kg DW whole animal (Denton et al., 2006; Table 4.1).

4.18 Strontium

A single datum was available of 0.31 mg Sr/kg FW whole sponge (Table 4.1).

4.19 Vanadium

Whole sponges contained up to 15.0 mg V/kg DW (Table 4.1; Cebrian et al., 2003).

4.20 Zinc

Limited data indicate that maximum zinc concentrations in whole sponges are 30.0 mg/kg fresh weight, 180.0 mg/kg dry weight, and 360.0 mg/kg ash weight (Table 4.1).

4.21 Literature Cited

Agell, G., Uriz, M.J., Cebrian, E., Marti, R., 2001. Does stress protein induction modify natural toxicity in sponges? Environ. Toxicol. Chem. 20, 2588–2593.

Ancellin, J., Bovard, P., 1971. Observations concernant les contaminations experimentales et les contaminations "in situ" d'especes marines par le ruthenium 106. Rev. Int. Ocean. Med. 21, 85–92.

Bernhard, M., Zattera, A., 1975. Major pollutants in the marine environment. In: Pearson, E.A., Frangipane, E.D. (Eds.), Marine Pollution and Marine Waste Disposal. Pergamon, Elmsford, New York, pp. 195–300.

Berthet, B., Mouneyrac, C., Perez, T., Amiard-Triquet, C., 2005. Metallothionein concentration in sponges (*Spongia officinalis*) as a biomarker of metal contamination. Comp. Biochem. Physiol. 141C, 306–313.

Cebrian, E., Uriz, M.J., 2007a. Contrasting effects of heavy metals on sponge cell behavior. Arch. Environ. Contam. Toxicol. 53, 552–558.

Cebrian, E., Uriz, M.J., 2007b. Do heavy metals play an active role in sponge cell behaviour in the absence of calcium? Consequences in larval settlement. J. Exp. Mar. Biol. Ecol. 346, 60–65.

Cebrian, E., Uriz, M.J., 2007c. Contrasting effects of heavy metals and hydrocarbons on larval settlement and juvenile survival in sponges. Aquat. Toxicol. 81, 137–143.

Cebrian, E., Marti, R., Uriz, M.J., Turon, X., 2003. Sublethal effects of contamination on the Mediterranean sponge *Crambe crambe*: metal accumulation and biological responses. Mar. Pollut. Bull. 46, 1273–1284.

Cebrian, E., Marti, R., Agell, G., Uriz, M.J., 2006. Response of the Mediterranean sponge *Chondrosia reniformis* Nardo to heavy metal pollution. Environ. Pollut. 141, 452–458.

Cebrian, E., Uriz, M.J., Turon, X., 2007. Sponges as biomonitors of heavy metals in spatial and temporal surveys in northwestern Mediterranean: multispecies comparison. Environ. Toxicol. Chem. 26, 2430–2439.

Culkin, F., Riley, J.P., 1958. The occurrence of gallium in marine organisms. J. Mar. Biol. Assoc. UK 37, 607–615.

Denton, G.R.W., Concepcion, L.P., Wood, H.R., Morrison, R.J., 2006. Trace metals in marine organisms from four harbours in Guam. Mar. Pollut. Bull. 52, 1784–1804.

Hansen, I.V., Weeks, J.M., Depledge, M.H., 1995. Accumulation of copper, zinc, cadmium and chromium by the marine sponge *Halichondria panicea* Pallas and the implications for biomonitoring. Mar. Pollut. Bull. 31, 133–138.

Ireland, M.P., 1973. Result of fluvial zinc pollution on the zinc content of littoral and sublittoral organisms in Cardigan Bay, Wales. Environ. Pollut. 4, 27–35.

Jenkins, D.W., 1980. Biological monitoring of toxic trace metals. Volume 2. Toxic trace metals in plants and animals of the world. Part II. U.S. EPA Report 600/3-80-091, pp. 505–618.

Leatherland, T.M., Burton, J.D., 1974. The occurrence of some trace metals in coastal organisms with particular reference to the Solent region. J. Mar. Biol. Assoc. UK 54, 457–468.

Lowman, F.G., Phelps, D.K., Ting, R.Y., Escalera, R.M., 1966. Progress Summary Report No. 4, Marine Biology Program June 1965-June 1966. Puerto Rico Nuclear Center Report PRNC 85, pp. 1–57.

Lowman, F.G., Stevenson, R.A., Escalera, R.M., Ufret, S.L., 1967. The effects of river outflows upon the distribution patterns of fallout radioisotopes in marine organisms. In: Proceedings of an International Symposium on Radioecological Concentration Processes, 1966. Stockholm, Int. Atom. Ener. Agency, pp. 735–752.

Mauchline, J., Templeton, W., 1966. Strontium, calcium, and barium in marine organisms from the Irish Sea. J. Cons. Perm. Int. Explor. Mer. 30, 161–170.

Noshkin, V.E., Bowen, V.T., Wong, K.M., Barke, J.C., 1971. Plutonium in North Atlantic ocean organisms: ecological relationships. In: Nelson, D.J. (Ed.), Radionuclides in Ecosystems. Proceedings of the Third National Symposium on Radioecology, May 10-12, 1971, Oak Ridge, Tennessee, Vol. 2, pp. 681–688. Available from NTIS Springfield, Virginia.

Olesen, T.M.E., Weeks, J.M., 1994. Accumulation of Cd by the marine sponge *Halichondria panicea* Pallas: effects upon filtration rate and its relevance for biomonitoring. Bull. Environ. Contam. Toxicol. 52, 722–728.

Patel, B., Balani, M.C., Patel, S., 1985. Sponge "sentinel" of heavy metals. Sci. Total. Environ. 41, 143–152.

Perez, T., Longet, D., Schembri, T., Rebouillon, P., Vacelet, J., 2005. Effects of 12 years' operation of a sewage treatment plant on trace metal occurrence within a Mediterranean commercial sponge (*Spongia officinalis*, Demospongiae). Mar. Pollut. Bull. 50, 301–309.

Philp, R.B., 1999. Cadmium content of the marine sponge *Microciona prolifera*, other sponges, water and sediment from the eastern Florida panhandle: possible effects on *Microciona* cell aggregation and potential roles of low pH and low salinity. Comp. Biochem. Physiol. 124C, 41–49.

Rao, J.V., Kavitha, P., Reddy, N.C., Rao, T.G., 2006. *Petrosia testudinaria* as a biomarker for metal contamination at Gulf of Mannar, southeast coast of India. Chemosphere 65, 634–638.

Schroeder, H.C., Shostak, K., Gamulin, V., Skorokhod, A., Kavsan, V., Muller, I.M., et al., 2000. Purification, cDNA cloning and expression of cadmium-inducible cysteine-rich metallothionein-like protein from the marine sponge *Suberites domuncula*. Mar. Ecol. Prog. Ser. 200, 149–157.

Vaskovsky, V.E., Korotchenko, O.D., Kosheleva, L.P., Levin, V.S., 1972. Arsenic in the lipid extracts of marine invertebrates. Comp. Biochem. Physiol. 41B(4), 777–784.

CHAPTER 5
Coelenterates

This phylum includes all diploblastic animals. Two distinct groups comprise this phylum: the Cnidarians, characterized by muscular movements, nematocyst presence, and occurrence as polyp or medusa; and the Ctenophores which retain the ciliary locomotion of the planula, are without nematocysts, and are not assigned either to the polyp or medusa type. There are about 9000 species of Cnidarians; the largest solitary Cnidarian is a sea anemone, *Stoichactis* sp., 0.5 m high and 1.5 m in diameter. Major subgroups within the Cnidaria include the Hydrozoa (about 2700 species of which 700 are free-living medusa, or jellyfish) and the Anthozoa. There are about 6000 species of Anthozoans, including anemones, corals, and sea pens; all are polyps, either solitary or colonial, and all are marine. There are about 80 species of Ctenophores, of which the largest is *Cestus veneris*, 1.5 m in length. Ctenophores, also known popularly as comb jellies or sea walnuts, have an apical statocyst, eight rows of combs, and usually two tentacles armed with numerous adhesive cells.

5.1 Aluminum

Highest values recorded for aluminum in different species of coelenterates are 453.0 mg/kg DW (dry weight), and 3600.0 mg/kg AW (ash weight) (Table 5.1). The significance of aluminum in coelenterates and other marine organisms is imperfectly understood.

5.2 Antimony

Limited data indicate that antimony concentrations range from 0.022 to 0.043 mg/kg DW (Table 5.1).

5.3 Arsenic

Maximum concentrations of arsenic recorded in coelenterates were 2200.0 mg/kg FW (fresh weight) in lipid extracts from a jellyfish and 72.0 mg/kg DW in whole anemones (Table 5.1).

Table 5.1: Aluminum, Antimony, and Arsenic Concentrations in Field Collections of Coelenterates

Element and Organism	Concentration	Reference[a]
Aluminum		
Coral, *Alcyonium digitatum*	90.0 DW	1
Coral, *Alcyonium digitatum*	453.0 DW	2
Jellyfish, *Aurelia* sp.	0.6 FW	3
Coral, *Montastrea annularis*, skeleton deposits	<125.0 DW	4
Anemone, *Tealia felina*	80.0 DW	1
Jellyfish, *Velella lata*	1600.0-3600.0 AW	5
Antimony		
Jellyfish, *Pelagia* sp.	0.043 DW	6
Anemone, *Tealia felina*	0.022 DW	7
Arsenic		
Anemone	6.6 DW	8
Jellyfish, *Cyanea arctica*, lipid extract of whole organism	2200.0 FW	9
Guam; 1998-1999; 4 locales; whole		
Soft coral, *Sinularia* sp.	0.01-2.33 DW	10
Hard coral, *Pocilopora damicornis*	<0.01-67.1 DW	10
Hard coral, *Fungia* sp.	0.19-0.25 DW	10
Hard coral, *Acropora formosa*	0.14 DW	10
Jellyfish, *Pelagia* sp.	11.0 DW	6
Anemone, *Tealia felina*	72.0 DW	7

Values are in mg element/kg fresh weight (FW), dry weight (DW), or ash weight (AW).

[a] 1, Riley and Segar, 1970; 2, Culkin and Riley, 1958; 3, Matsumoto et al., 1964; 4, Goreau, 1977; 5, Bieri and Krinsley, 1958; 6, Leatherland et al., 1973; 7, Leatherland and Burton, 1974; 8, Gorgy et al., 1948; 9, Vaskovsky et al., 1972; 10, Denton et al., 2006.

5.4 Barium

Corals, especially hermatypic inshore species, appear to selectively accumulate barium to a maximum of 85.0 mg Ba/kg DW (Table 5.2). Corals reportedly lose barium after death (Livingston and Thompson, 1971). As will be seen later, post-mortem changes in metal and metalloid content of many groups of organisms are well documented and contraindicate

Table 5.2: Barium, Bismuth, and Boron Concentrations in Field Collections of Coelenterates

Element and organism	Concentration	Reference[a]
Barium		
Coelenterates	0.03-0.55 FW	1
Corals, 34 spp.		
Deep open ocean	16.0-60.0 DW	2
Shallow open ocean	15.0-30.0 DW	2
Shallow coastal zone	9.0-85.0 DW	2
Ctenophore, *Pleurobrachia pileus*	60.0 AW	3
Jellyfish, *Velella lata*	<100.0 AW	4
Bismuth		
Corals, 34 spp.	<2.0 DW	2
Boron		
Ctenophore, *Beroe cucumis*	115.0 AW	5
Corals, 34 spp.		
Deep open ocean	50.0-85.0 DW	2
Shallow open ocean	65.0-100.0 DW	2
Shallow coastal zone	40.0-110.0 DW	2
Jellyfish, *Cyanea capillata*	100.0 AW	5

Values are in mg element/kg fresh weight (FW), dry weight (DW), or ash weight (AW).
[a] 1, Mauchline and Templeton, 1966; 2, Livingston and Thompson, 1971; 3, Vinogradova and Koual'skiy, 1962; 4, Bieri and Krinsley, 1958; 5, Nicholls et al., 1959.

the collection of dead organisms, including corals, for baseline residue purposes. Barium was initially incorporated into the coral skeleton along with calcium at the time of skeleton formation; this was also true for other elements including strontium, vanadium, copper, boron, lithium, and zinc (Livingston and Thompson, 1971).

5.5 Bismuth

All available data indicate that bismuth levels are less than 2.0 mg Bi/kg DW in all tissues analyzed (Table 5.2).

5.6 Boron

As was true for barium, corals accumulated boron from ambient seawater to a maximum of 110.0 mg B/kg DW (Table 5.2).

5.7 Cadmium

Cadmium concentrations in representative species of coelenterates ranged from 0.01 to 5.3 mg/kg DW whole organism (Table 5.3).

In laboratory studies, sea anemones, *Bundosoma cavernata*, held in seawater containing high sublethal concentrations of 7.0 mg Cd/L for up to 7 days had significant increases in glutamate and alanine (Kasschau et al., 1980); also, ambient concentrations as low as 0.04 mg Cd/L were associated with disrupted enzyme metabolism in the hydroid *Campanularia flexuosa* (Moore, 1977; Moore and Stebbing, 1976). The hydrozoan *Laomedea loveni* experienced 50% irreversible polyp retraction in 7 days at 0.003 mg Cd/L at 10 ppt salinity; organisms were comparatively resistant at elevated salinities and reduced temperatures (Theede et al., 1979).

Colonies of the tropical coral *Pocillopora damicornis* were exposed to 0, 0.005, or 0.05 mg Cd/L for up to 14 days (Mitchelmore et al., 2007). All corals exposed to 0.05 mg Cd/L

Table 5.3: Cadmium Concentrations in Field Collections of Coelenterates

Organism	Concentration	Reference[a]
Corals; 5 spp.; Arabian Sea		
Skeleton	0.3-2.6 DW	7
Tissues	0.04-2.2 DW	7
Coral, *Alcyonium digitatum*	4.1 DW	1
Jellyfish, *Aurelia* sp.	0.1 FW	2
Corals, 34 spp.	<2.0 DW	3
Guam; 1998-1999; 4 locales; whole		
Soft coral, *Sinularia* sp.	<0.05-0.16 DW	6
Hard coral, *Pocilopora damicornis*	<0.04-0.24 DW	6
Hard coral, *Fungia* sp.	0.01-0.08 DW	6
Hard coral, *Acropora formosa*	0.09 DW	6
Jellyfish, *Pelagia* sp.	5.3 DW	4
Anemone, *Tealia felina*	0.07 DW	5
Anemone, *Tealia felina*	0.7 DW	1

Values are in mg Cd/kg fresh weight (FW) or dry weight (DW).
[a]1, Riley and Segar, 1970; 2, Matsumoto et al., 1964; 3, Livingston and Thompson, 1971; 4, Leatherland et al., 1973; 5, Leatherland and Burton, 1974; 6, Denton et al., 2006; 7, Anu et al., 2007.

died by day 4 of exposure. After 14 days, cadmium concentrations, in mg Cd/kg DW, were 0.005 in controls and 0.0086 in the 0.005 mg Cd/L group (Mitchelmore et al., 2007). Exposure of temperate water sea anemones, *Anthopleura elegantissima* to 0.02 mg Cd/L for 14 days resulted in concentrations of 0.003 mg Cd/kg DW (Mitchelmore et al., 2003a), and this is similar to findings for the same species exposed to 0.10 mg Cd/L for 7 days (Mitchelmore et al., 2003b).

5.8 Cesium

Radiocesium accumulation in *Actinia equina, Tealia felina, Metridium senile*, and *Calliactes parasitica* was documented by Bryan (1963). In all cases, accumulation increased with increasing exposure time, suggesting that equilibrium may not be reached for months or years. Highest concentrations were observed in tentacles from all four species, with concentration values over the ambient medium ranging from 16.6 to 19.6. In general, the lowest concentration factors measured were in whole organism, with values between 4.2 and 11.6 (Bryan, 1963).

5.9 Chromium

Chromium concentrations in field collections of coelenterates ranged from <0.12 to 35.0 mg/kg DW (Table 5.4). Chromium, and also scandium and titanium, were largely associated with detrital phases in corals from chromium-rich areas (Livingston and Thompson, 1971).

5.10 Cobalt

Cobalt values were low among near-shore coral species from Hawaii, Tahiti, Samoa, and other areas, seldom exceeding 0.04 mg Co/kg DW (Table 5.4). Higher values were measured in medusa and especially among some species of deep-ocean ahermatypic corals (Table 5.4).

5.11 Copper

Copper concentrations in coelenterates were comparatively low except for anemones *T. felina* (52.0 mg/kg DW) and jellyfish *Cyanea capillata* (68.0 mg/kg DW) (Table 5.5). Highest accumulations of copper in marine invertebrate organisms were observed in molluscan tissues and soft parts, followed by crustaceans, macrophytes, annelids, tunicates, algae, echinoderms, and coelenterates, in that general sequence (Eisler, 1979). Lowest concentrations of copper were consistently found among the marine vertebrate groups examined (Eisler, 1979).

Table 5.4: Chromium and Cobalt Concentrations in Field Collections of Coelenterates

Element and Organism	Concentration	Reference[a]
Chromium		
Corals; 5 spp.; Arabian Sea		
Skeleton	0.82-5.44 DW	8
Tissues	2.8-15.9 DW	8
Coral, *Alcyonium digitatum*	<0.4 DW	1
Corals, 34 spp.		
Deep open ocean	0.8-3.0 DW	2
Shallow open ocean	2.0-35.0 DW	2
Shallow coastal zone	0.2-23.0 DW	2
Guam; 1998-1999; 4 locations; whole		
Soft coral, *Sinularia* sp.	<0.15-0.31 DW	7
Hard coral, *Pocilopora damicornis*	<0.12-0.33 DW	7
Hard coral, *Fungia* sp.	0.24-0.34 DW	7
Hard coral, *Acropora formosa*	0.27 DW	7
Ctenophore, *Pleurobrachia pileus*	10.0 AW	3
Anemone, *Tealia felina*	0.37 DW	1
Jellyfish, *Velella lata*	>100.0 AW	4
Cobalt		
Corals; 5 spp.; Arabian Sea		
Skeleton	0.79-9.82 DW	8
Tissues	0.2-8.9 DW	8
Coral, *Acropora* sp.; exoskeleton		
Samoa	0.004 DW	5
Tahiti	0.004 DW	5
Coral, *Alcyonium digitatum*	0.6 DW	1
Corals, 34 spp.		
Deep open ocean	0.4-105.0 DW	2
Shallow open ocean	0.2-3.0 DW	2
Shallow coastal zone	0.04-2.0 DW	2
Jellyfish, *Cyanea capillata*	3.0 AW	6
Coral, *Favia* sp.; exoskeleton	0.011 DW	5

(*Continues*)

Table 5.4: Cont'd

Element and Organism	Concentration	Reference[a]
Coral, *Fungia* sp.		
Septa	0.002 DW	5
Base	0.001 DW	5
Coral, *Leptastrea* spp.; exoskeleton	0.012 DW	5
Coral, *Leptoria* spp.; exoskeleton	0.002 DW	5
Coral, *Pocillopora* sp.; exoskeleton		
Hawaii	0.007 DW	5
Tahiti	0.001 DW	5
Anemone, *Tealia felina*	<0.2 DW	1

Values are in mg element/kg dry weight (DW) or ash weight (AW).
[a]1, Riley and Segar, 1970; 2, Livingston and Thompson, 1971; 3, Vinogradova and Koual'skiy, 1962; 4, Bieri and Krinsley, 1958; 5, Veek and Turekian, 1968; 6, Nicholls et al., 1959; 7, Denton et al., 2006; 8, Anu et al., 2007.

Laboratory studies demonstrate that marine hydroid coelenterates were quite sensitive to copper salts, and this may account, in part, for the low uptake of copper by coelenterates.

Colonies of the tropical coral *P. damicornis* were exposed to 0, 0.005, or 0.05 mg Cu/L for up to 9 days (Mitchelmore et al., 2007). All corals exposed to 0.05 mg Cu/L died by day 9. Copper concentrations at day 4, in mg Cu/kg DW were 0.004 in controls, 0.013 in the 0.005 mg Cu/L group, and 0.13 in the 0.05 mg Cu/L group (Mitchelmore et al., 2007). Serious adverse effects to hydroids were observed at nominal concentrations of 0.0012-0.05 mg Cu/L (Moore, 1977; Moore and Stebbing, 1976; Stebbing, 1976). The ctenophore, *Pleurobrachia pileus*, held in seawater containing 0.033 mg Cu/L for 24 h had a 50% death rate (USEPA, 1980). The coral, *Acropora formosa*, exposed to 0.01-0.04 mg Cu/L for 48 h lost most of their symbiotic algae (Jones, 1997). The sea anemone. *Anemonia viridis*, held in seawater containing 0.05-0.2 mg Cu/L for 5 days exhibited immediate tentacle retraction, copious production of mucus, progressive visible bleaching, and loss of zooxanthellae (Harland and Nganro, 1990). Zooxanthellae accumulate copper and their expulsion is considered a form of copper regulation (Harland and Nganro, 1990).

5.12 Gallium

A single value for gallium was available of 0.05 mg Ga/kg DW in a whole coral, *Alcyonium digitatum* (Culkin and Riley, 1958).

Table 5.5: Copper Concentrations in Field Collections of Coelenterates

Organism	Concentration	Reference[a]
Coral, *Alcyonium digitatum*	6.2 DW	1
Alcyonium digitatum	9.7 DW	2
Jellyfish, *Aurelia* sp.	0.7 FW	3
Ctenophore, *Beroe cucumis*	700.0 AW	4
Corals; 5 spp.; Arabian Sea		
Skeleton	0.49-1.88 DW	14
Tissues	0.18-6.48 DW	14
Corals, 34 spp.		
Deep open ocean	<2.0-14.0 DW	5
Shallow open ocean	2.0-10.0 DW	5
Shallow coastal zone	<2.0-10.0 DW	5
Jellyfish, *Cyanea capillata*; whole		
Pacific Ocean	13.0 AW	6
New England	8.2 DW	10
Sweden	68.0 DW	10
Guam; 1998-1999; 4 locales; whole		
Soft coral, *Sinularia* sp.	0.4-0.9 DW	13
Hard coral, *Pocilopora damicornis*	Max. 0.2 DW	13
Hard coral, *Fungia* sp.	0.4-1.0 DW	13
Hard coral, *Acropora formosa*	<0.1 DW	13
Coral, *Meandrina meandrites*	22.0 (14.0-33.0) FW	7
Octacorals; Venezuela; whole; 5 spp.	0.9-3.1 DW	11
Ctenophore, *Pleurobrachia pileus*	300.0 AW	8
Anemone, *Tealia felina*	57.0 DW	2
Jellyfish, *Velella lata*	71.0-220.0 AW	9
Wales; 1989; coastal area		
Anemones; whole	0.6 FW	12
Soft corals; whole	1.0 FW	12

Values are in mg Cu/kg fresh weight (FW), dry weight (DW), or ash weight (AW).

[a]1, Culkin and Riley, 1958; 2, Riley and Segar, 1970; 3, Matsumoto et al., 1964; 4, Nicholls et al., 1959; 5, Livingston and Thompson, 1971; 6, Nicholls et al., 1959; 7, Lowman et al., 1966; 8, Vinogradova and Koual'skiy, 1962; 9, Bieri and Krinsley, 1958; 10, Jenkins, 1980; 11, Jaffe et al., 1992; 12, Morris et al., 1989; 13, Denton et al., 2006; 14, Anu et al., 2007.

5.13 Iron

On a DW basis, the highest concentrations of iron recorded were among ahermatypic deep ocean corals (up to 6500.0 mg Fe/kg) and lowest values (<5.0 mg Fe/kg) among reef-building corals and jellyfish (Table 5.6).

5.14 Lead

In general, deep-sea ahermatypic corals contained the highest concentrations of lead recorded among coelenterates, while jellyfishes and coastal zone corals contained the lowest (Table 5.7).

Between 1954 and 1980—prior to the wide-spread use of unleaded gasoline products—there were significant increases in lead concentrations in marine corals and other biota as a direct result of increased global lead availability during that period (Dodge and Gilbert, 1984). The lead concentration in exoskeleton of *Porites* spp. corals from Ogasawara Island, Japan, gradually increased over the past 108 years with steep increases in the 1950s owing to

Table 5.6: Iron Concentrations in Field Collections of Coelenterates

Organism	Concentration	Reference[a]
Corals; 5 spp.; Arabian Sea		
Skeletons	0.53-12.8 DW	8
Tissues	1.2-62.9 DW	8
Coral, *Alcyonium digitatum*	438.0 DW	1
Alcyonium digitatum	250.0 DW	2
Jellyfish, *Aurelia* sp.	0.9 FW	3
Corals		
Deep open ocean	30.0-6500.0 DW	4
Shallow open ocean	14.0-780.0 DW	4
Shallow coastal zone	<5.0-880.0 DW	4
Coral, *Montastrea annularis*	363.0-506.0 DW	5
Ctenophore, *Pleurobrachia pileus*	5000.0 AW	6
Anemone, *Tealia felina*	730.0 DW	2
Jellyfish, *Velella lata*	>10.0 AW	7

Values are in mg Fe/kg fresh weight (FW), dry weight (DW), or ash weight (AW).
[a] 1, Culkin and Riley, 1958; 2, Riley and Segar, 1970; 3, Mauchline and Templeton, 1966; 4, Livingston and Thompson, 1971; 5, Goreau, 1977; 6, Vinogradova and Koual'skiy, 1962; 7, Bieri and Krinsley, 1958; 8, Anu et al., 2007.

Table 5.7: Lead and Lithium Concentrations in Field Collections of Coelenterates

Element and Organism	Concentration	Reference[a]
Lead		
Coral, *Alcyonium digitatum*	24.0 DW	1
Jellyfish, *Aurelia* sp.	0.8 FW	2
Ctenophore, *Beroe cucumis*	6.0 AW	3
Corals		
Deep open ocean	<2.0-42.0 DW	4
Shallow open ocean	<2.0-9.0 DW	4
Shallow coastal zone	<2.0-4.0 DW	4
Corals; 5 spp.; Arabian Sea		
Skeletons	0.7-23.8 DW	7
Tissues	0.3-20.7 DW	7
Jellyfish, *Cyanea capillata*	6.0 AW	3
Ctenophore, *Pleurobrachia pileus*	40.0 AW	6
Coral, *Porites* spp.; exoskeleton		
Japan; 1994-2000	0.400 DW	6
Hong Kong; 1990-1993	0.222 DW	6
Bermuda; 1980-1983	0.01-0.134 DW	7
St. Croix; 1954-1980	0.021-0.207 DW	8
Anemone, *Tealia felina*	2.6 DW	1
Jellyfish, *Velella lata*	>10.0 AW	5
Lithium		
Corals, 23 spp.		
Deep open ocean	1.0-3.0 DW	4
Shallow open ocean	1.0 DW	4
Shallow coastal zone	0.5-2.0 DW	4
Ctenophore, *Pleurobrachia pileus*	600.0 AW	6

Values are in mg element/kg fresh weight (FW), dry weight (DW), or ash weight (AW).
[a]1, Riley and Segar, 1970; 2, Matsumoto et al., 1964; 3, Nicholls et al., 1959; 4, Livingston and Thompson, 1971; 5, Bieri and Krinsley, 1958; 6, Vinogradova and Koual'skiy, 1962; 7, Anu et al., 2007; 8, Inoue et al., 2006; 7, Shen and Boyle, 1987; 8, Dodge and Gilbert, 1984.

lead-containing industrial discharges from Japan (Inoue et al., 2006). In *Porites* spp. exoskeletons from Hainan Island, China, there is a marked decline in lead content in recent years attributed to the introduction of unleaded gasoline in China and Indonesia (Inoue et al., 2006).

5.15 Lithium

Lithium was selectively accumulated by corals (Livingston and Thompson, 1971). Concentrations among corals on a DW basis show little variability with all values measured in the range 0.5-3.0 mg Li/kg (Table 5.7).

5.16 Manganese

Manganese concentrations were highest among the corals—especially non-reef-building species that frequented the deep open ocean—and lowest, in general, among polyps (Table 5.8).

Table 5.8: Manganese and Mercury Concentrations in Field Collections of Coelenterates

Element and Organism	Concentration	Reference[a]
Manganese		
Coral, *Alcyonium digitatum*	3.7 DW	1
Anemone, *Anemonia sulcata*	21.3 AW	2
Corals; 5 spp.; Arabian Sea		
Skeleton	0.3-4.7 DW	9
Tissue	0.2-2.5 DW	9
Corals, 34 spp.		
Deep open ocean	2.0-2100.0 DW	3
Shallow open ocean	<2.0-465.0 DW	3
Shallow coastal zone	2.0-130.0 DW	3
Ctenophore, *Pleurobrachia pileus*	20.0 AW	4
Anemone, *Tealia felina*	9.3 DW	1
Jellyfish, *Velella lata*	7.0-41.0 AW	5
Mercury		
Jellyfish, *Pelagia* sp.	0.07 DW	6
Anemone; Minamata Bay, Japan	41.0 DW	7
Anemone, *Tealia felina*	0.86 DW	8

Values are in mg element/kg dry weight (DW) or ash weight (AW).
[a] 1, Riley and Segar, 1970; 2, Chipman and Thommeret, 1970; 3, Livingston and Thompson, 1971; 4, Vinogradova and Koual'skiy, 1962; 5, Bieri and Krinsley, 1958; 6, Leatherland et al., 1973; 7, Matida and Kumada, 1969; 8, Leatherland and Burton, 1974; 9, Anu et al., 2007.

5.17 Mercury

Mercury concentrations were low among coelenterates except for species collected from the mercury-contaminated Minamata Bay, Japan (Table 5.8).

Mercury is among the most toxic metals known to marine organisms including coelenterates; this point will be developed in detail later in this volume. In the case of coelenterates, concentrations as low as 0.0017 mg Hg/L depressed the growth of the hydrozoan *C. flexuosa*; moreover, concentrations an order of magnitude lower adversely affected metabolism of various lysosomal enzymes in that species (Moore, 1977; Moore and Stebbing, 1976; Stebbing, 1976). Higher sublethal concentrations of 1.2 mg Hg/L were associated with increased concentrations of free amino acids in the anemone *B. cavernata* following exposure for up to 7 days (Kasschau et al., 1980).

5.18 Molybdenum

Concentrations of molybdenum in coelenterates examined were uniformly low and near the detection limits of the analytical procedures used (Table 5.9).

5.19 Nickel

Accumulations of nickel up to 23.0 mg Ni/kg DW are documented in corals, especially among nonhermatypic corals and selected medusa (Table 5.9).

5.20 Radium

Corals appear to reflect environmental levels of radium. The radium/calcium ratio in corals is reportedly similar to that of seawater (Moore et al., 1973).

5.21 Rubidium

Corals from different habitats and geographic areas all contained less than 10.0 mg Rb/kg DW whole organism (Livingston and Thompson, 1971).

5.22 Ruthenium

Laboratory studies with radioruthenium-106 and selected species of invertebrates, including coelenterates, demonstrated that all organisms examined accumulated ^{106}Ru over ambient seawater levels by 30- to 50-fold factors in 10 days (Ancellin et al., 1967).

Table 5.9: Molybdenum and Nickel Concentrations in Field Collections of Coelenterates

Element and Organism	Concentration	Reference[a]
Molybdenum		
Beroe cucumis	3.0 AW	1
Corals, 34 spp.	<2.0 DW	2
Jellyfish, *Cyanea capillata*	3.0 AW	1
Jellyfish, *Velella lata*	>10.0 AW	3
Nickel		
Coral, *Alcyonium digitatum*	17.0 DW	4
Corals; 5 spp.; Arabian Sea		
Skeleton	2.07-12.7 DW	6
Tissue	0.5-10.1 DW	6
Corals, 24 spp.		
Deep open ocean	<2.0-23.0 DW	2
Shallow open ocean	2.0-19.0 DW	2
Shallow coastal zone	<2.0-3.0 DW	2
Pleurobranchia pileus	50.0 AW	5
Anemone, *Tealia felina*	3.3 DW	4
Jellyfish, *Velella lata*	33.0-220.0 AW	3

Values are in mg element/kg dry weight (DW) or ash weight (AW).
[a] 1, Nicholls et al., 1959; 2, Livingston and Thompson, 1971; 3, Bieri and Krinsley, 1958; 4, Riley and Segar, 1970; 5, Vinogradova and Koual'skiy, 1962; 6, Anu et al., 2007.

5.23 Scandium

Scandium concentrations in corals never exceeded 0.25 mg Sc/kg DW (Table 5.10).

5.24 Silicon

Silicon was relatively abundant in nonhermatypic as well as reef-building corals, with concentrations up to 2000.0 mg Si/kg DW documented (Table 5.10).

5.25 Silver

All silver concentrations were less than 0.26 mg Ag/kg DW except for a single value of 2.7 mg/kg DW in *Sinularia* sp., a soft coral (Table 5.10). Measurable differences in silver accumulations were evident between tissues, and between geographic areas (Table 5.10).

Table 5.10: Scandium, Silicon, and Silver Concentrations in Field Collections of Coelenterates

Element and Organism	Concentration	Reference[a]
Scandium		
Corals, 34 spp.		
Deep open ocean	0.01-0.17 DW	1
Shallow open ocean	0.01-0.24 DW	1
Shallow coastal zone	0.005-0.25 DW	1
Silicon		
Corals, 34 spp.		
Deep open ocean	15.0-2000.0 DW	1
Shallow open ocean	45.0->2000.0 DW	1
Shallow coastal zone	20.0->2000.0 DW	1
Jellyfish, *Velella lata*	>10.0 AW	2
Silver		
Coral, *Acropora* spp.; exoskeleton		
Samoa	0.014 DW	3
Tahiti	0.022 DW	3
Coral, *Alcyonium digitatum*	<0.03 DW	4
Corals, 34 spp.	<1.0 DW	1
Coral, *Favia* spp.; exoskeleton	0.024 DW	3
Coral, *Fungia* spp.		
Septa	0.065 DW	3
Base	0.022 DW	3
Guam; 1998-1999; 4 locations; whole		
Soft coral, *Sinularia* sp.	<0.11-2.7 DW	6
Hard coral, *Pocilopora damicornis*	<0.10-0.26 DW	6
Hard coral, *Fungia* sp.	0.14-0.24 DW	6
Hard coral, *Acropora formosa*	<0.11 DW	6
Hard coral, *Leptastrea* spp.; exoskeleton	0.10 DW	3
Hard coral, *Leptoria* spp.; exoskeleton	0.057 DW	3
Pleurobranchia pileus	4.0 AW	5

(Continues)

Table 5.10: Cont'd

Element and Organism	Concentration	Reference[a]
Hard coral, *Pocilopora* spp.; exoskeleton		
Hawaii	0.034 DW	3
Tahiti	0.018 DW	3
Anemone, *Tealia felina*	0.05 DW	4
Jellyfish, *Velella lata*	>3.0 AW	2

Values are in mg element/kg dry weight (DW) or ash weight (AW).
[a] 1, Livingston and Thompson, 1971; 2, Bieri and Krinsley, 1958; 3, Veek and Turekian, 1968; 4, Riley and Segar, 1970; 5, Vinogradova and Koual'skiy, 1962; 6, Denton et al., 2006.

5.26 Strontium

Strontium was especially abundant among coelenterates with calcified exoskeletons and concentrations >6300.0-10,800.0 mg Sr/kg DW were not uncommon in that group; however, strontium concentrations were significantly lower among polyps and medusa (Table 5.11).

5.27 Thorium

Thorium concentrations were always less than 0.01 mg Th/kg DW (Table 5.11). This was attributed primarily to the very low thorium content of seawater (Moore et al., 1973).

5.28 Tin

Tin concentrations in coelenterates ranged between 4.0 and 20.0 mg Sn/kg AW (Table 5.11). Tributyltins are the most toxic of all tin compounds tested. In the case of the hydroid, *C. flexuosa*, complete growth inhibition occurred in 11 days during exposure to 0.001 mg Sn (as tributyltin)/L (Hall and Pinkney, 1985).

5.29 Titanium

Titanium residues in coelenterates ranged up to 330.0 mg Ti/kg on a DW basis and up to 1400.0 on an AW basis (Table 5.11). The significance of these very considerable titanium accumulations in coelenterates is unknown.

Table 5.11: Strontium, Thorium, Tin, and Titanium Concentrations in Field Collections of Coelenterates

Element and Organism	Concentration	Reference[a]
Strontium		
Coral, *Alcyonium digitatum*	220.0 DW	1
Coelenterates	0.02-1.4 FW	2
Corals, 14 spp.		
Deep open ocean	8200.0-10,000.0 DW	3
Shallow open ocean	8200.0-9320.0 DW	3
Shallow coastal zone	7070.0-8680.0 DW	3
Coral, *Montastrea annularis*	6300.0-7160.0 DW	4
Ctenophore, *Pleurobrachia pileus*	1000.0 AW	5
Coral, *Porites* sp., skeletal parts	2500.0-10,800.0 DW	6
Anemone, *Tealia felina*	9.5 DW	1
Jellyfish, *Velella lata*	>200.0 AW	7
Thorium		
Coelenterates	<0.01 DW	8
Tin		
Ctenophore, *Beroe cucumis*	7.0 AW	9
Corals, 34 spp.	<5.0 DW	3
Jellyfish, *Cyanea capillata*	4.0 AW	9
Ctenophore, *Pleurobrachia pileus*	20.0 AW	5
Titanium		
Corals, 34 spp.		
Deep open ocean	<50.0-330.0 DW	3
Shallow open ocean	<20.0-60.0 DW	3
Shallow coastal zone	<20.0-80.0 DW	3
Jellyfish, *Cyanea capillata*	3.0 AW	9
Ctenophore, *Pleurobrachia pileus*	30.0 AW	5
Jellyfish, *Velella lata*	460.0-1400.0 AW	7

Values are in mg element/kg fresh weight (FW), dry weight (DW), or ash weight (AW).
[a] 1, Riley and Segar, 1970; 2, Mauchline and Templeton, 1966; 3, Livingston and Thompson, 1971; 4, Goreau, 1977; 5, Vinogradova and Koual'skiy, 1962; 6, Polyakov and Krasnov, 1976; 7, Bieri and Krinsley, 1958; 8, Moore et al., 1973; 9, Nicholls et al., 1959.

5.30 Uranium

Uranium values of 1.5-4.8 mg/kg coral DW were not unusual (Table 5.12). Uranium values in corals reflected seawater values, with nonhermatypic corals of the open ocean containing more uranium than most near-shore hermatypes (Livingston and Thompson, 1971). Uranium content in corals also varied as a function of growth rate. For example, rapidly growing reef corals incorporated up to 60% more uranium and up to 9% more strontium than did ahermatypic corals (Moore et al., 1973).

Table 5.12: Uranium and Vanadium Concentrations in Field Collections of Coelenterates

Element and Organism	Concentration	Reference[a]
Uranium		
Coral, *Acropora* sp.; exoskeleton		
Samoa	2.82 DW	1
Tahiti	1.95 DW	1
Corals, 34 spp.		
Deep open ocean	Max. 4.8 DW	2
Shallow open ocean	Max. 4.7 DW	2
Shallow coastal zone	Max. 4.8 DW	2
Coral, *Favia* sp.; exoskeleton	1.52 DW	1
Coral, *Fungia* spp.		
Septa	2.06 DW	1
Base	2.19 DW	1
Coral, *Leptastrea* spp.; exoskeleton	2.83 DW	1
Coral, *Leptoria* spp.; exoskeleton	2.00 DW	1
Coral, *Pocilopora* spp.; exoskeleton		
Hawaii	2.11 DW	1
Tahiti	2.17 DW	1
Vanadium		
Ctenophore, *Beroe cucumis*	8.0 AW	3
Corals, 34 spp.	<10.0 DW	2
Jellyfish, *Cyanea capillata*	5.0 AW	3
Jellyfish, *Velella lata*	>70.0 AW	4

Values are in mg element/kg dry weight (DW) or ash weight (AW).
[a] 1, Veek and Turekian, 1968; 2, Livingston and Thompson, 1971; 3, Nicholls et al., 1959; 4, Bieri and Krinsley, 1958.

5.31 Vanadium

Vanadium concentrations in coelenterates were usually at or near the detection limits of the analytical methodologies used (Table 5.12).

5.32 Zinc

Zinc residues in coelenterates ranged widely (Table 5.13). The highest zinc residue reported of 603.0 mg Zn/kg DW in the anemone *Actina equina* was from specimens collected from a metals-contaminated area (Ireland, 1973); however, most values were substantially lower (Table 5.13).

Table 5.13: Zinc Concentrations in Field Collections of Coelenterates

Organism	Concentration	Reference[a]
Anemone, *Actina equina*	167.0-603.0 DW	1
Soft coral, *Alcyonia alcyonium*; whole	9.6 FW	9
Coral, *Alcyonium digitatum*	46.0 DW	2
Jellyfish, *Aurelia* sp.	3.4 FW	3
Corals; 5 spp.; Arabian Sea		
Skeleton	1.2-2.6 DW	11
Tissue	0.7-9.3 DW	11
Corals		
Deep open ocean, 7 spp.	2.0-9.0 DW	4
Shallow open ocean, 8 spp.	<2.0-70.0 DW	4
Shallow coastal zone, 8 spp.	2.0-7.0 DW	4
Coral, *Eusmilia fastigata*	74.0-80.0 FW; 75.0-80.0 DW; 77.0-83.0 AW	5
Coral, *Manicina aerolata*	59.0-63.0 FW; 60.0-68.0 DW; 65.0-69.0 AW	5
Coral, *Meandrina meandrites*	54.0-120.0 FW; 55.0-120.0 DW; 56.0-120.0 AW	5
Ctenophores; whole; Calcasieu River Estuary	(31.0-64.0) DW	10
Plumose anemone, *Metridium senile*; whole	18.0 FW	9
Jellyfish, *Pelagia* sp.	28.0 DW	6
Ctenophore, *Pleurobrachia pileus*	900.0 AW	7
Anemone, *Tealia felina*	200.0 DW	8
Anemone, *Tealia felina*	280.0 DW	2

Values are in mg Zn/kg fresh weight (FW), dry weight (DW), or ash weight (AW).
[a] 1, Ireland, 1973; 2, Riley and Segar, 1970; 3, Matsumoto et al., 1964; 4, Livingston and Thompson, 1971; 5, Lowman et al., 1966; 6, Leatherland et al., 1973; 7, Vinogradova and Koual'skiy, 1962; 8, Leatherland and Burton, 1974; 9, Morris et al., 1989; 10, Ramelow et al., 1989; 11, Anu et al., 2007.

5.33 Literature Cited

Ancellin, J., Bovard, P., Vilquin, A., 1967. New studies on experimental contamination of marine species by ruthenium-106. In: Actes du Congress Internationale Radioprotection Milieu. Soc. Franc. Radioprotect., Toulouse, France, January 14–16, 1967, pp. 213–234.

Anu, G., Kumar, N.C., Jayalakshmi, K.V., Nair, S.M., 2007. Monitoring of heavy metal partitioning in reef corals of Lakshadweep Archipelago, Indian Ocean. Environ. Monit. Assess. 128, 195–208.

Bieri, R., Krinsley, D.M., 1958. Trace elements in the pelagic coelenterate, *Velella lata*. J. Mar. Res. 16, 246–254.

Bryan, G.W., 1963. The accumulation of radioactive caesium by marine invertebrates. J. Mar. Biol. Assoc. UK 43, 519–539.

Chipman, W., Thommeret, J., 1970. Manganese content and the occurrence of fallout ^{54}Mn in some marine benthos of the Mediterranean. Bull. Inst. Oceanol. 69(1402), 1–15.

Culkin, F., Riley, J.P., 1958. The occurrence of gallium in some marine organisms. J. Mar. Biol. Assoc. UK 37, 607–615.

Denton, G.R.W., Concepcion, L.P., Wood, H.R., Morrison, R.J., 2006. Trace metals in marine organisms from four harbours in Guam. Mar. Pollut. Bull. 52, 1784–1804.

Dodge, R.E., Gilbert, T.R., 1984. Chronology of lead pollution contained in banded coral skeletons. Mar. Biol. 82, 9–13.

Eisler, R., 1979. Copper accumulations in coastal and marine biota. In: Nriagu, J.O. (Ed.), Copper in the Environment, Part 1: Ecological Cycling. John Wiley, New York, pp. 383–449.

Goreau, T.T., 1977. Coral skeletal chemistry: Physiological and environmental regulation of stable isotopes and trace metals in *Montastrea annularis*. Proc. R. Soc. Lond. B 196, 291–315.

Gorgy, S., Rakestraw, N.W., Fox, D.L., 1948. Arsenic in the sea. J. Mar. Res. 7, 22–32.

Hall Jr., L.W., Pinkney, A.E., 1985. Acute and sublethal effects of organotin compounds on aquatic biota: An interpretive literature evaluation. CRC Crit. Rev. Toxicol. 14, 159–209.

Harland, A.D., Nganro, N.R., 1990. Copper uptake by the sea anemone *Anemonia viridis* and the role of zooxanthellae in metal regulation. Mar. Biol. 104, 297–301.

Inoue, M., Hata, A., Suzuki, A., Nohara, M., Shikazono, N., Yim, W.W.S., et al., 2006. Distribution and temporal changes of lead in the surface seawater in the western Pacific and adjacent seas derived from coral skeletons. Environ. Pollut. 144, 1045–1052.

Ireland, M.P., 1973. Result of fluvial zinc pollution on the zinc content of littoral and sublittoral organisms in Cardigan Bay, Wales. Environ. Pollut. 4, 27–35.

Jaffe, R., Fernandez, C.A., Alvarado, J., 1992. Trace metal analyses in octacorals by microwave acid digestion and graphite furnace atomic-absorption spectrometry. Talanta 39, 113–117.

Jenkins, D.W., 1980. Biological Monitoring of Toxic Trace Metals. Volume 2. Toxic Trace Metals in Plants and Animals of the World. Part II. U.S. Environmental Protection Agency Report 600/3-80-091, pp. 505–618.

Jones, R.J., 1997. Zooxanthellae loss as a bioassay for assessing stress in corals. Mar. Ecol. Prog. Ser. 149, 163–171.

Kasschau, M.R., Skaggs, M.M., Chen, E.C.M., 1980. Accumulation of glutamate in sea anemones exposed to heavy metals and organic amines. Bull. Environ. Contam. Toxicol. 25, 873–878.

Leatherland, T.M., Burton, J.D., 1974. The occurrence of some trace metals in coastal organisms with particular reference to the Solent region. J. Mar. Biol. Assoc. UK 54, 457–468.

Leatherland, T.M., Burton, J.D., Culkin, F., McCartney, M.J., Morris, R.J., 1973. Concentrations of some trace metals in pelagic organisms and of mercury in Northeast Atlantic Ocean water. Deep Sea Res. 20, 679–685.

Livingston, H.D., Thompson, G., 1971. Trace element concentrations in some modern corals. Limnol. Oceanol. 16, 786–795.

Lowman, F.G., Phelps, D.K., Ting, R.Y., Escalera, R.M., 1966. Progress Summary Report No. 4, Marine Biology Program, June 1965-June 1966, Puerto Rico Nuclear Center Report PRNC 85, pp. 1–57.

Matida, Y., Kumada, H., 1969. Distribution of mercury in water, bottom mud, and aquatic organisms of Minamata Bay, the River Agano, and other water bodies in Japan. Bull. Freshw. Fish. Res. Lab. Tokyo 19(2), 73–93.

Matsumoto, T., Satake, M., Yamamoto, J., Haruna, S., 1964. On the micro constituent elements in marine invertebrates. J. Oceanol. Soc. Jpn. 20(3), 15–19.

Mauchline, J., Templeton, W.L., 1966. Strontium, calcium, and barium in marine organisms from the Irish Sea. J. Cons. 39, 161–170.

Mitchelmore, C.L., Ringwood, A.H., Weis, V.M., 2003a. Differential accumulation of cadmium and changes in glutathione levels as a function of symbiotic state in the sea anemone *Anthopleura elegantissima*. J. Exp. Mar. Biol. Ecol. 284, 71–85.

Mitchelmore, C.L., Verde, E.A., Ringwood, A.H., Weis, V.M., 2003b. Differential accumulation of heavy metals in the sea anemone *Anthopleura elegantissima* as a function of symbiotic state. Aquat. Toxicol. 64, 317–329.

Mitchelmore, C.L., Verde, E.A., Weis, V.M., 2007. Uptake and partitioning of copper and cadmium in the coral *Pocillopora damicornis*. Aquat. Toxicol. 85, 48–56.

Moore, N.N., 1977. Lysosomal responses to environmental chemicals in some marine invertebrates. In: Giam, C.S. (Ed.), Pollutant Effects on Marine Organisms. D.C. Heath, Lexington, MA, pp. 143–154.

Moore, M.N., Stebbing, A.R.D., 1976. The quantitative cytochemical effects of three metal ions on a lysosomal hydrolase of a hydroid. J. Mar. Biol. Assoc. UK 56, 995–1005.

Moore, W.S., Krishnaswami, S., Bhat, S.G., 1973. Radiometric determinations of coral growth rates. Bull. Mar. Sci. 23, 157–176.

Morris, R.J., Law, R.J., Allchin, C.R., Kelly, C.A., Fileman, C.F., 1989. Metals and organochlorines in dolphins and porpoises of Cardigan Bay, West Wales. Mar. Pollut. Bull. 20, 512–523.

Nicholls, G.D., Curl Jr., H., Bowen, V.T., 1959. Spectrographic analyses of marine plankton. Limnol. Oceanol. 4, 472–476.

Polyakov, D.M., Krasnov, E.V., 1976. Determination of the growth rate and age of corals *Porites* by the strontium and sodium content in their skeletons. Sov. J. Mar. Biol. 2(1), 391–396.

Ramelow, G.J., Webre, C.L., Mueller, C.S., Beck, J.N., Young, J.C., Langley, M.P., 1989. Variations of heavy metals and arsenic in fish and other organisms from the Calcasieu River and Lake, Louisiana. Arch. Environ. Contam. Toxicol. 18, 804–818.

Riley, J.P., Segar, D.A., 1970. The distribution of the major and minor elements in marine animals. I. Echinoderms and coelenterates. J. Mar. Biol. Assoc. UK 50, 721–730.

Shen, G.T., Boyle, E.A., 1987. Lead in corals: Reconstruction of historical industrial fluxes to the surface ocean. Earth Planet. Sci. Lett. 82, 289–304.

Stebbing, A.R.D., 1976. The effects of low metal levels on a colonial hydroid. J. Mar. Biol. Assoc. UK 56, 977–994.

Theede, H., Scholz, N., Fisher, H., 1979. Temperature and salinity effects on the acute toxicity of cadmium to *Laomedea loveni* (Hydrozoa). Mar. Ecol. Prog. Ser. 1, 13–19.

U.S. Environmental Protection Agency (USEPA), 1980. Ambient water quality criteria for copper. USEPA Rep. 440/5-80-036, pp. 1–162.

Vaskovsky, V.E., Korotchenko, O.D., Kosheleva, L.P., Levin, V.S., 1972. Arsenic in the lipid extracts of marine invertebrates. Comp. Biochem. Physiol. 41B(4), 777–784.

Veek, H.H., Turekian, K.K., 1968. Cobalt, silver, and uranium concentrations of reef-building corals in the Pacific Ocean. Limnol. Oceanol. 13, 304–308.

Vinogradova, Z.A., Koual'skiy, V.V., 1962. Elemental composition of the Black Sea plankton. Doklady Acad. Sci. USSR Earth Sci. Sec. 147, 217–219.

CHAPTER 6
Molluscs

Molluscs are unsegmented coelomate animals with a well-developed head, a ventral muscular foot, and a dorsal visceral hump; with soft skin, partly covering the visceral hump, the mantle, and often secreting a calcareous shell. Molluscs are characterized by a complex anatomy and highly diverse specializations. Many are of large size, especially the giant squids. There are about 128,000 species of molluscs. There are five major groups in this phylum: the pelycopods, with 20,000 recent species, including clams, oysters, mussels, cockles, and scallops; gastropods, comprising 105,000 species, such as the drills, conchs, cowries, whelks, nudibranchs, and mud snails; about 750 species of cephalopods, including octopus, squid, and cuttlefish; 305 species of scaphopods; and about 1000 species of amphineurans, including the chitons.

Bivalves exhibit several characteristics of ideal indicator organisms (Darracott and Watling, 1975; Eisler, 1981; Phillips, 1977a). This includes an ability to accumulate high concentrations of metals without dying, a sedentary life history, high numerical abundance, a lengthy life span that permits sampling of more than 1 year class throughout the monitoring period, a large size so that ample tissue is available for analysis, and good adaptation to laboratory conditions. Extensive monitoring programs have been implemented, mainly with mussels, to assess potentially toxic metals and other contaminants in marine environments (Goldberg et al., 1978; Phillips, 1976b). Although mussels are recommended as indicator organisms, additional biological variables need to be investigated (Eisler, 1981; Phillips, 1977a; Westernhagen et al., 1978), along with water and sediment analyses.

The model of Goldberg et al. (1978), for example, is based on the equation

$$dM/dt = KM_{sw} - TM$$

where M is metal concentration in bivalve tissue, t is age of organism, K is constant, T is biological half-life of metal, and M_{sw} is concentration of metal in seawater.

The age of the organism was correlated with shell thickness, and this relation appears valid. The variables T and M are easily measured. However, K, the uptake constant at time t may be dependent upon the age of the bivalve; further, the uptake constant assumes linearity, and this has not been established even today. The basic flaw in the equation is the assumption that M_{sw} is constant over time. It is probable that M_{sw} is variable, with pulses in M_{sw} due to variations in outfall discharge rates, as well as the changing composition of the waste effluent. It is likely, in many cases, that only during surges in metal concentrations above a given level will the sentinel organism bioconcentrate, and even this process is modified significantly by biological and abiotic variables. Accordingly, pending further refinement of this and similar models, predicted data on KM_{sw} derived from TM, and the reverse, should be accepted with caution.

In general, the highest concentrations of all metals and metalloids examined in marine molluscs were in gut and digestive gland, with moderate enrichment in gills, mantle, and gonads, and lowest residuals in muscle and shell (Eisler, 1981; Segar et al., 1971). Most of the initial literature on trace metal composition of molluscs was devoted to the filter-feeding bivalves, especially mussels and oysters. Bivalves have received—and continue to receive—extensive treatment in the literature owing to their reported ability to reflect environmental levels of trace metals in marine and estuarine ecosystems. It is now generally accepted that mussels from highly industrialized areas contain significantly higher levels in tissues of copper, lead, cobalt, chromium, nickel, iron, manganese, and other elements (Karouna-Renier et al., 2007; Szefer et al., 2006). Variability in metal concentrations in mussels from impacted environments is attributed mainly to variability in byssus metal concentrations: byssus, as was true for soft tissues, selectively and sensitively reflects variations of certain metals in ambient seawater and may prove useful as a reliable biomonitor of these metals in a variety of coastal and estuarine areas (Szefer et al., 2006). The presence of other contaminants—such as polychlorinated biphenyls (PCBs), dioxins, and dibenzofurans—in areas contaminated by metals, may present difficulty in the interpretation of tissue metal residues in molluscs (Karouna-Renier et al., 2007).

6.1 Aluminum

Aluminum concentrations in molluscan tissues were variable, especially among conspecifics collected from different portions of its range (Table 6.1). Byssus threads had aluminum concentrations more than an order of magnitude greater than soft tissues and appears to be an important detoxification route by sequestration of aluminum (and to a lesser degree for cadmium, cobalt, and manganese) within its proteinaceous fibers (Kadar et al., 2006; Table 6.1).

Highest concentrations reported on a dry weight (DW) basis were 12,475.0 mg Al/kg in soft parts, 470.0 in shell, and 860.0 in gut and digestive gland; however, most values were substantially lower (Table 6.1). Most of this variability is attributable to proximity to anthropogenic sources and some to differences in sample preparation techniques. In the case of *Pecten maximus*, for example, washing the sample prior to analysis lowers measured aluminum concentrations by more than half in mantle and gill samples, and by 10% in gonad (Segar et al., 1971).

Table 6.1: Aluminum Concentrations in Field Collections of Molluscs

Organism	Concentration	Reference[a]
Mussel, *Anodonta* sp.		
Shell	173.0 DW	1
Soft parts	107.0 DW	1
Mussel, *Bathymodiolus azoricus*; 2002; mid-Atlantic Ridge; 3 sites		
Gill	1.6-8.4 DW	9
Mantle	3.1-4.5 DW	9
Digestive gland	5.9-28.8 DW	9
Byssus threads	274.8-304.6 DW	9
Waved whelk, *Buccinum undatum*		
Shell	84.0 DW	1
Soft parts	300.0 DW	1
Cockle, *Cardium edule*		
Shell	84.0 DW	1
Soft parts	2700.0 DW	1
Scallop, *Chlamys opercularis*		
Soft parts	49.0 DW	2
Kidneys	58.0 DW	2
Digestive gland	43.0 DW	2
Shell	300.0-430.0 DW	2
American oyster, *Crassostrea virginica*		
Soft parts	20.0-540.0 DW	3
Soft parts; Savannah, Georgia; 9 sites; November 2000 and 2001	309.0-1460.0 DW	10, 12
Slipper limpet, *Crepidula fornicata*		
Shell	80.0 DW	1
Soft parts	150.0 DW	1
Ribbed mussel, *Geukensia demissa*; soft parts		
Southern Chesapeake Bay, Maryland	450.0 DW	7
San Francisco Bay, California	1435.0 DW	7
Periwinkle, *Littoraria scabra*; soft parts; Tanzania; winter 2005		
Polluted site	55.0 DW	11
Reference site	15.0 DW	11
Baltic clam, *Macoma nasuta*; soft parts		
Chesapeake Bay, Maryland; northern portion vs. southern portion	1994.0 DW vs. 3010.0 DW	7
San Francisco Bay, California	12,475.0 DW	7

(*Continues*)

Table 6.1: Cont'd

Organism	Concentration	Reference[a]
Hardshell clam, *Mercenaria mercenaria*		
Shell	71.0 DW	1
Soft parts	615.0 DW	1
Mussel, *Modiolus modiolus*		
Shell	150.0 DW	1
Soft parts	210.0-260.0 DW	1
Mantle and gills	210.0 DW	1
Muscle	95.0 DW	1
Gonad	230.0 DW	1
Gut and digestive gland	860.0 DW	1
Bivalves		
Shell	32.0-174.0 DW	4
Soft parts	186.0-346.0 DW	4
California mussel, *Mytilus californianus*		
Soft parts	6.0-440.0 DW	3
Common mussel, *Mytilus edulis*		
Shell	76.0 DW	1
Shell; Baltic Sea; 2005; 12 sites	119.0-214.0 DW	13
Soft parts	1230.0 DW	1
Soft parts	24.0-420.0 DW	3
Soft parts; Scheldt estuary; March 2000	925.0 DW	14
Dog whelk, *Nucella lapillus*		
Shell	170.0 DW	1
Soft parts	160.0 DW	1
Octopus, *Octopus* sp., whole	4.0 FW	5
Common limpet, *Patella vulgata*		
Shell	470.0 DW	1
Soft parts	160.0 DW	1
Scallop, *Pecten maximus*		
Soft parts	610.0 DW	1
Muscle	53.0 DW	1
Gut and digestive gland	340.0 DW	1
Mantle and gills		
Unwashed	200.0 DW	1
Washed	95.0 DW	1
Gonad		
Unwashed	130.0 DW	1

(*Continues*)

Table 6.1: Cont'd

Organism	Concentration	Reference[a]
Washed	95.0 DW	1
Shell	190.0-230.0 DW	1
Soft parts	55.0 DW	2
Kidneys	53.5 DW	2
Digestive gland	173.0 DW	2
Sea scallop, Placopecten magellanicus; soft parts	338.0 DW	8
Pteropods; shell	3.0-8.0 DW	6
Cuttlefish, Sepia sp.; whole	1.7 FW	5

Values are in mg Al/kg fresh weight (FW) or dry weight (DW).
[a] 1, Segar et al., 1971; 2, Bryan, 1973; 3, Goldberg et al., 1978; 4, Culkin and Riley, 1958; 5, Matsumoto et al., 1964; 6, Krinsley and Bieri, 1959; 7, Sparling and Lowe, 1996; 8, Reynolds, 1979; 9, Kadar et al., 2006; 10, Kumar et al., 2008; 11, De Wolf and Rashid, 2008; 12, Sajwan et al., 2008; 13, Protasowicki et al., 2008; 14, Wepener et al., 2008.

Scallops, *P. maximus*, subjected to aluminum loading via the ambient medium under controlled conditions show high bioconcentration factors (BCFs) in digestive gland (9000), kidneys (2100), and total soft parts (1400); however, in the case of *Chlamys opercularis*, another scallop, concentration factors were different: 2600 in kidney, 2300 in digestive gland, and 1100 in soft parts (Bryan, 1973). The softshell clam *Mya arenaria*, and the American oyster *Crassostrea virginica*, remove aluminum particles from the medium by filtering activity; however, there were no measurable increases in aluminum content of tissues when compared to controls (Hanks, 1965). High aluminum concentrations in shellfish may have some beneficial impact on human consumers. Thus, in butter clam, *Saxidomus giganteus*, the amount of PSP (paralytic shellfish poisoning) bound to clam melanin decreased by 68% in the presence of Al^{3+}, and 57% with Ba^{2+} (Price and Lee, 1972). These results confirmed an earlier conclusion by the same authors that the PSP-melanin interaction was reversible and electrostatic in nature.

6.2 Americium

Measurable levels of ^{241}Am were found in the rock jingle *Pododesmus macroschisma*, blue mussel *Mytilus trossulus*, and horse mussel *Modiolus modiolus* from Amchitka Island in the Aleutian chain during summer 2004; Amchitka was the site of nuclear tests between 1965 and 1971 (Burger et al., 2007b).

In laboratory studies, newly hatched juvenile cuttlefish (*Sepia officinalis*) were exposed for 29 days to ^{241}Am-spiked sediments (Bustamante et al., 2006a,b, 2008), and also uptake of ^{241}Am from seawater over 36 h by newly hatched juveniles and 8-h exposure by adults, both followed by 29 days in clean seawater. A final set of studies involved feeding newly hatched juveniles ^{241}Am-labeled mussels and brine shrimp for 1 h followed by 29 days in clean seawater. There was very little uptake from spiked sediments after 29 days, although digestive gland had 47% of the total radioactivity at that time. In seawater exposure, whole juveniles had a BCF of 6.2 after 36 h; 29 days later, 27% of the ^{241}Am was retained in digestive glands. Adults exposed to ^{241}Am via seawater for 8 h had a BCF of 42 in branchial hearts and 16 in branchial heart appendages; the biological half-time persistence of ^{241}Am in seawater-exposed adults was about 14 days. In the 1-h diet route study, 89% of the total radioactivity was in digestive gland after 24 h, and 98% after 29 days; half-time persistence of ^{241}Am via the diet was 28 days (Bustamante et al., 2006b, 2008). Gravid cuttlefish were fed crabs radiolabeled with ^{241}Am over a 2 week period to determine the rate of maternal transfer into eggs; there was no significant transfer of ^{241}Am (Lacoue-Labarthe et al., 2008b).

6.3 Antimony

Five species of bivalves from the Aegean Sea concentrated antimony over ambient seawater levels by factors that ranged from 13 in *Spondylus gaederopus* to 2600 for *Chama placentina* (Papadopoulu, 1973). In North America, most species of molluscs had antimony residues that ranged from 0.5 to 2.0 mg/kg fresh weight (FW) edible tissues (Hall et al., 1978; Table 6.2); among European collections, antimony residues were somewhat lower in similar tissues (Table 6.2).

Table 6.2: Antimony Concentrations in Field Collections of Molluscs

Organism	Concentration	Reference[a]
Mussel, *Anodonta* sp.; soft parts; United Kingdom		
Southampton area	0.029 DW	1
Chertsey area	0.054 DW	1
Cockle, *Cardium edule*		
Soft parts	0.034-0.063 DW	1
Mantle cavity fluid	0.18 DW	1
Mussel, *Choromytilus meridionalis*; soft parts	0.20 FW	2
Clam, *Donax serra*; soft parts	0.27 FW	2
Clam, *Ensis arcuatus*; soft parts	0.028 DW	3
Clam, *Ensis siliqua*; shell	0.014 DW	3

(Continues)

Table 6.2: Cont'd

Organism	Concentration	Reference[a]
Gastropods; 5 spp; soft parts	0.007-0.124 DW	1
Abalone, *Haliotis midea*; soft parts	0.083 FW	2
Hardshell clam or Quahaug, *Mercenaria mercenaria*; soft parts; age in years		
3	0.014 DW	1
4	0.011 DW	1
10	0.008 DW	1
15	0.006 DW	1
Molluscs; commercial species; edible tissues; 21 spp.		
3 spp.	<0.1 FW	4
3 spp.	0.08-0.20	5
3 spp.	0.5-0.7 FW	4
4 spp.	0.7-0.9 FW	4
5 spp.	0.0-1.0 FW	4
3 spp.	1.0-2.0 FW	4
Common mussel, *Mytilus edulis*		
Soft parts	0.042-0.047 DW	1
Shell	0.022 DW	6
Soft parts	0.11 DW	3
European oyster, *Ostrea edulis*; soft parts	0.010-0.015 DW	1
Pen shell, *Pinna nobilis*		
Soft parts	1.1 DW	7
Mantle and gills	0.9 DW	7
Muscle	1.0 DW	7
Nervous system	0.9 DW	7
Stomach and intestines	1.3 DW	7
Gonad	0.3 DW	7
Hepatopancreas	0.6 DW	7
Byssus	1.2 DW	7
Byssus gland	0.5 DW	7
Cuttlefish, *Sepia* sp.		
Gills	0.026 DW	1
Mantle	0.01 DW	1
Squid, whole	0.46 AW	8

Values are in mg Sb/kg fresh weight (FW), dry weight (DW), or ash weight (AW).
[a] 1, Leatherland and Burton, 1974; 2, Van As et al., 1973; 3, Bertine and Goldberg, 1972; 4, Hall et al., 1978; 5, Van As et al., 1975; 6, Bertine and Goldberg 1972; 7, Papadopoulu, 1973; 8, Robertson, 1967.

6.4 Arsenic

Early analytical interest in arsenic was prompted by a rash of arsenic homicides in the early 1900s. Subsequently, it was determined that seafoods, especially molluscs and crustaceans, were unusually rich in arsenic compounds, although these were present almost always in a nontoxic form, that is, as arsenobetaine or as an organic pentavalent compounds, with little risk to human consumers. The relation between arsenic transformation rates between nontoxic and toxic forms of arsenic, such as inorganic trivalent arsenites, is discussed later.

It is clear from Table 6.3 that arsenic concentrates in lipid extracts of representative molluscs, that wastes containing arsenic compounds raised biogenic arsenic levels in mussels and gastropods, that arsenic content in molluscs varied with the age of the organism as well as tissue analyzed, and that mean levels of arsenic in edible portions of commercially important North American molluscs do not exceed 20.0 mg As/kg FW. Unusually high levels of arsenic are reported in filter feeding bivalves from the vicinity of Athens, Greece (Papadopoulu, 1973). It was subsequently determined that *Pinna nobilis* concentrated arsenic from seawater by a factor of 46,000 and that concentration factors for other species of bivalves from the same Aegean area ranged from 1500 to 18,000 (Papadopoulu, 1973). Cultured Pacific oysters, *Crassostrea gigas*, from Taiwan contained up to 33.4 mg total arsenic/kg FW; however, less than 2% of the total arsenic was in the form of inorganic arsenicals (Table 6.3), and the remainder was in the form of arsenobetaine and other comparatively harmless organoarsenicals with little risk to health of human consumers (Liu et al., 2006).

Trivalent arsenic was lethal to embryos of the Pacific oyster, *C. gigas*, at 0.33 mg As^{3+}/L, with 50% dead in 96 h (USEPA, 1985a). And 2.0 mg As^{3+}/L depressed oxygen consumption of mud snails, *Nassarius obsoletus* within 72 h (USNAS, 1977). Bryant et al. (1985) showed that pentavalent As^{5+} was 15 times more toxic at 15 °C than at 5 °C to clams, *Macoma balthica* during exposure for 7 days.

Studies with radioarsenic and *Mytilus galloprovincialis* show that accumulation was modified at nominal arsenic concentrations in the medium, with tissue, with age of the mussel, and with temperature and salinity of the medium (Unlu and Fowler, 1979). Uptake of arsenate arsenic, increased with increasing arsenate concentration in the medium; however, the response was not linear, with accumulation suppressed at higher external arsenate concentrations. Smaller mussels accumulated more arsenic than larger mussels; in both size groups, highest levels were in byssus and digestive gland. In general, uptake and loss of arsenate was enhanced at higher temperatures, and lower salinities (19 vs. 38 ppt). Loss rate was the same at both salinities. Radioarsenic loss followed a biphasic pattern with biological half-times of 3 and 32 days for the fast and slow compartments, respectively. Byssal thread secretion played a key role in arsenic elimination from *M. galloprovincialis* (Unlu and Fowler, 1979).

Table 6.3: Arsenic Concentrations in Field Collections of Molluscs

Organism	Concentration	Reference[a]
Alaska; Cook Inlet; summer 1997; clams; soft parts; maximum values		
Total arsenic	5.0 FW	17
Dimethylarsinic acid	0.6 FW	17
As^{3+} plus As^{5+}	0.01 FW	17
Monomethylarsinic acid	<0.001 FW	17
Giant squid, *Architeuthis dux*; 2001-2005		
Gill; Bay of Biscay	1.0 DW	42
Hepatopancreas; Bay of Biscay vs. Mediterranean Sea	48.0 DW vs. 44.0 DW	42
Clam, *Asaphis violascens*; soft parts; American Samoa; 2001-2002; total arsenic vs. inorganic arsenic		
Industrialized sites	4.7-5.9 FW vs. 0.21-0.24 FW	38
Reference sites	1.2-1.5 FW vs. 0.07-0.10 FW	38
Bivalves; California; 1984-1985; soft parts		
Clam, *Corbicula* sp.	5.4-11.5 DW	18
Clam, *Macoma balthica*	7.6-12.1 DW	18
Bivalves; East China Sea; 5 spp.; soft parts; August 2002	0.50 (0.22-0.81) FW	31
Clam, *Anodonta cygnea*; soft parts	4.3 DW	2
Ivory shell, *Buccinum striatissimum*; total arsenic vs. arsenobetaine		
Muscle	38.0 FW vs. 24.2 FW	19
Midgut gland	18.0 FW vs. 10.8 FW	19
Channeled whelk, *Busycon canaliculatum*	9.0 FW	3
Cockle, *Cardium edule*		
Soft parts	5.1-6.3 DW	2
Mantle cavity fluids	8.4 DW	2
Scallop, *Comptopallium radula*; New Caledonia; October 2004; Santa Marie Bay		
Soft parts	44.7 (39.3-53.6) DW	36
Kidney	116.0 DW	36
Pacific oyster, *Crassostrea gigas*		
Taiwan; cultured; 4 sites; soft parts		
Total arsenic		

(Continues)

Table 6.3: Cont'd

Organism	Concentration	Reference[a]
All sites	8.4-9.5 (3.1-33.4) FW	32
April vs. February	22.9 FW vs. 4.2 FW	32
As^{5+}	0.038-0.087 FW	32
As^{3+}	0.071-0.125 FW	32
Monomethylarsonic acid	0.052-0.124 FW	32
Dimethylarsinic acid	0.3(0.06-0.47) FW	32
Taiwan; soft parts; 2002; 4 sites, 4 seasons		
Total arsenic	9.9 (8.5-10.7) DW	37
Arsenite	0.09 (0.06-0.12) DW	37
Arsenate	0.03 (0.02-0.05) DW	37
Dimethylarsinic acid	0.53 (0.47-0.62) DW	37
Monomethylarsonic acid	0.04 (0.03-0.05) DW	37
Arsenobetaine	3.9 DW	37
American oyster, *Crassostrea virginica*		
Soft parts	10.3 DW	20
Soft parts	2.9 FW	26
Soft parts; Pensacola, Florida: 2003-2004	Max. 2.8 FW	34
Soft parts; Savannah, Georgia; 2000-2001; 9 sites	0.2-3.0 DW	35, 40
Oyster, *Crassostrea* spp; soft parts	(0.3-3.4) FW; (1.3-10.0) DW	2,4,5,6
Gastropods, 5 spp.; soft parts	8.1-38.0 DW	2
Periwinkle, *Littoraria scabra*; soft parts		
Industrial zone; 1998 vs. 2005	6.5 DW vs. 1.4 DW	39
Reference site; 2005	1.1 DW	39
Spindle shell, *Hemifusus* spp.; Hong Kong; 1984; muscle; total arsenic vs. inorganic arsenic	Max. 500.0 FW vs. <0.5 FW	21
Limpet. *Littorina littorea*; soft parts		
Near arsenic source	11.5 DW	7
Offshore	4.0 DW	7
Squid, *Loligo vulgaris*; soft parts	(0.8-7.5) FW	26
Hardshell clam, *Mercenaria mercenaria*; soft parts; age in years		
3	3.8 DW	2
4	4.7 DW	2
10	9.3 DW	2
15	8.4 DW	2

(*Continues*)

Table 6.3: Cont'd

Organism	Concentration	Reference[a]
Clam, *Meretrix casta*; soft parts	10.9 DW	5
Molluscs; soft parts; lipid extracts; 7 spp.	4600.0-39,300.0 FW	1
Molluscs; commercial; edible tissues; North America		
6 spp.	2.0-3.0 FW	8
8 spp.	3.0-4.0 FW	8
3 spp.	4.0-5.0 FW	8
4 spp.	7.0-20.0 FW	8
Molluscs; commercial; edible portions; various locations; 21 spp.	7.1-83.8 (1.3-129.6) DW	9
Molluscs; edible tissues		
Hong Kong, 1976-1978		
Bivalves	3.2-39.6 FW	22
Gastropods	19.0-176.0 FW	22
Cephalopods	0.7-5.5 FW	22
Yugoslavia; northern Adriatic Sea; summer 1986	21.0-31.0 FW	23
Softshell clam, *Mya arenaria*; soft parts; arsenic-contaminated sites at Seal Harbour, Nova Scotia vs. reference sites		
Total arsenic	218.0-228.0 FW vs. 7.0-7.9 FW	28
Inorganic arsenic		
Trivalent arsenic	22.0 FW vs. 0.37 FW	28
Pentavalent arsenic	57.0 FW vs <0.25 FW	28
Bioaccessible fraction	89.2 FW vs. 6.0 FW	28
Common mussel, *Mytilus edulis*		
Soft parts	1.6-16.0 DW	2,7,10,11
Soft parts	2.5 (1.4-4.6) FW	24
Soft parts	Max. 20.0 DW	30
Soft parts; Scheldt estuary; March 2000	61.0 DW	41
Pacific blue mussel, *Mytilus edulis trossulus*; 2004; Adak Island, Alaska; soft parts; 5 sites	9.0 FW; max. 32.2 FW; 6.0-19.0 DW	33

(*Continues*)

Table 6.3: Cont'd

Organism	Concentration	Reference[a]
Mussel, *Mytilus galloprovincialis*; soft parts		
Western Mediterranean Sea; Balearic Islands; Port of Mahon; 1991-2005	5.4 (0.8-11.0) DW	29
Adriatic Sea; 2001-2005; winter vs. summer	11.8-17.0 DW vs. 10.8-37.6 DW	44
Mytilus spp.; soft parts	3.0 FW; (8.2-9.6) DW	5,12
North Sea; 1997-1998; soft parts		
Whelk, *Buccinum undatum*	Max. 65.0 FW	43
Giant scallop, *Pecten maximus*	1.0-3.6 FW	43
Octopus, *Octopus cyanea*; Guam; June 1998; tentacle vs. liver	96.4 DW vs. 44.3 DW	27
Octopus, *Octopus vulgaris*		
Portugal; 1999-2002; 3 sites		
Digestive gland	23.8-56.0 DW	45
Branchial heart	12.0-32.0 DW	45
Gill	16.0-59.0 DW	45
Muscle	33.0-56.0 DW	45
European flat oyster, *Ostrea edulis*		
Soft parts	2.8-6.2 DW	2
Soft parts; Balearic Islands; western Mediterranean Sea; Port of Mahon; 1991-2005	10.6 (0.8-19.0) DW	29
Scallop, *Pecten maximus*; soft parts	11.6 DW	13
Pen shell, *Pinna nobilis*		
Soft parts	670.0 DW	14
Mantle, gills	120.0 DW	14
Muscle	95.0 DW	14
Nervous system	180.0 DW	14
Stomach, intestine	1100.0 DW	14
Gonads	90.0 DW	14
Hepatopancreas	210.0 DW	14
Byssus	57.0 DW	14
Byssus gland	37.0 DW	14
Scallop, *Placopecten magellanicus*; soft parts	1.6 (1.3-2.4) FW	26
Cuttlefish, *Sepia officinalis*		
Gills	198.0 DW	2
Mantle	73.0 DW	2

(*Continues*)

Table 6.3: Cont'd

Organism	Concentration	Reference[a]
Surf clam, Spisula solidissima; muscle	1.3 FW	3
Taiwan; 1995-1996; soft parts Clam, Meretrix lusoria Pacific oyster, Crassostrea gigas	13.7 (9.5-20.1) DW 11.8 (7.2-16.3) DW	25 25
Clam, Tapes japonica; soft parts	<0.004 to <0.05 AW	15,16

Values are in mg As/kg fresh weight (FW), dry weight (DW), or ash weight (AW).
[a]1, Vaskovsky et al., 1972; 2, Leatherland and Burton, 1974; 3, Greig et al., 1977; 4, Mackay et al., 1975; 5, Zingde et al., 1976; 6, Sims and Presley, 1976; 7, Penrose et al., 1975; 8, Hall et al., 1978; 9, Costa and da Fonseca 1967; 10, Karbe et al., 1977; 11, Bohn, 1975; 12, Vecchio et al., 1962; 13, Lunde, 1970; 14, Papadopoulu, 1973; 15, Fukai and Meinke, 1959; 16, Fukai and Meinke, 1962; 17, Bigler and Crecelius, 1998; 18, Johns and Luoma, 1990, Shiomi et al., 1984; 20, Zaroogian and Hoffman, 1982; 21, Phillips and Depledge, 1986; 22, Phillips et al., 1982; 23, Ozretic et al., 1990; 24, Vos and Hovens, 1986; 25, Han et al., 1998; 26, Jenkins, 1980; 27, Denton et al., 2006; 28, Koch et al., 2007; 29, Deudero et al., 2007; 30, Mubiana and Blust, 2006; 31, Huang et al., 2007; 32, Liu et al., 2006; 33, Burger and Gochfeld, 2006; 34, Karouna-Renier et al., 2007; 35, Kumar et al., 2008; 36, Metian et al., 2008; 37, Liu et al., 2008; 38, Peshut et al., 2008; 39, De Wolf and Rashid, 2008; 40, Sajwan et al., 2008; 41, Wepener et al., 2008; 42, Bustamante et al., 2008; 43, De Gieter et al., 2002; 44, Fattorini et al., 2008; 45, Napoleao et al., 2005.

Experimental work with molluscs suggest only minor accumulations of stable arsenic, although arsenic seemed to be retained by gastropods and mussels (Rosemarin et al., 1985). American oysters, *C. virginica*, held in seawater containing 0.005 mg As^{5+}/L for 112 days concentrated arsenic by a factor of 350 (Zaroogian and Hoffman, 1982). There was no relation between oyster body burdens of arsenic and exposure concentrations; however, diet contributed more to arsenic uptake than did seawater concentrations (Zaroogian and Hoffman, 1982).

In another study, neither soft parts nor shell of *Mytilus edulis* retained more than 0.01 mg of arsenic during immersion in solutions containing 100,000.0 mg As/L. By contrast, the byssus accumulated up to 0.5 mg and the excreta contained up to 0.8 mg arsenic in an insoluble form (Sautet et al., 1964). Juveniles of bay scallops, *Argopecten irradians*, contained 29.2 mg As/kg FW soft parts after exposure to 20.0 mg As/L for 96 h, again suggesting little bioconcentration (Nelson et al., 1976).

6.5 Barium

Most of the information on barium in molluscs is restricted to residues in shell (Table 6.4). Factors known to modify incorporation of barium and other alkaline earths into shells include the ambient salinity of the medium (Pilkey and Goodell, 1963), direct relation to shell strontium (Pearce and Mann, 2006), and the chemical form of barium, with ionic species most readily incorporated (Turekian and Armstrong, 1960). Barium incorporation from seawater into soft tissues is reported: concentration factors for barium in four bivalve species from polluted waters of Greece ranged from 200 to 1100 (Papadopoulu, 1973).

Table 6.4: Barium, Bismuth, and Boron Concentrations in Field Collections of Molluscs

Element and Organism	Concentration	Reference[a]
Barium		
Gastropods; shell	4.0-15.0 DW	1
Molluscs		
Tissues and shell	0.02-0.1 FW	2
Shells, 7 spp.	3.0-48.0 DW	3
Common mussel, *Mytilus edulis*; soft parts	0.8-26.0 DW	4
Mussel, *Mytilus galloprovincialis*; soft parts; Adriatic Sea; 2001-2005; winter vs. summer	1.3-7.6 DW vs. 1.5-3.4 DW	11
Scallop, *Pecten* sp.; shell	6.0-12.0 DW	1
Pelycopods; shell	4.0-41.0 DW	1
Gastropod, *Pinna nobilis*		
Soft parts	120.0 AW	5
Mantle and gills	150.0 AW	5
Muscle	100.0 AW	5
Nervous system	140.0 AW	5
Stomach and intestine	140.0 AW	5
Gonads	79.0 AW	5
Hepatopancreas	66.0 AW	5
Byssus	63.0 AW	5
Byssus gland	120.0 AW	5
Bismuth		
Mussel, *Choromytilus meridionalis*; soft parts	5.0-6.0 DW	6
Pacific oyster, *Crassostrea gigas*; soft parts	4.0 DW	6
Boron		
Bivalve molluscs; soft parts; 11 spp.	2.0-4.5 FW	7
Bivalve molluscs; soft parts; 2 spp.	50.0-90.0 AW	8
American oyster, *Crassostrea virginica*; soft parts; Savannah, Georgia; 2000-2001	Max. 254.0 DW	9, 10
Octopus, *Polypus bimaculatus*; whole	1.3 FW	7
Squid, *Ommastrephes illicebrosa*; whole	420.0 AW	8

Values are in mg element/kg fresh weight (FW), dry weight (DW), or ash weight (AW).
[a] 1, Turekian and Armstrong, 1960; 2, Mauchline and Templeton, 1966; 3, Pilkey and Goodel 1963; 4, Karbe et al., 1977; 5, Papadopoulu, 1973; 6, Watling and Watling 1976a,b; 7, Thompson et al., 1976; 8, Nicholls et al., 1959; 9, Kumar et al., 2008; 10, Sajwan et al., 2008; 11, Fattorini et al., 2008.

Barium concentrations in shell of razor clam, *Ensis siliqua*, collected from the west coast of England tended to be elevated where sewage sludge dumping was operative during shell growth (Pearce and Mann, 2006).

6.6 Bismuth

The highest bismuth concentration recorded was 6.0 mg/kg DW in soft parts of a mussel (Table 6.4).

Bismuth concentrations in mussels, *Choromytilus meridionalis*, from South Africa were positively correlated with the age, sex, zinc content, and degree of metals contamination in the immediate environment. Bismuth accumulation rates were higher in larger mussels, in females, in mussels with elevated zinc content, and lower in mussels collected from waters of low metal content (Watling and Watling, 1976a).

6.7 Boron

The maximum boron concentration recorded of 254.0 mg B/kg DW was in soft parts of an American oyster, *C. virginica*, from Savannah, Georgia (Table 6.4). The significance of this high concentration on oyster health is unknown and merits additional research.

Boron concentrations in soft tissues of Pacific Northwest molluscs ranged from 1.6 to 4.5 mg B/kg FW (Table 6.4; Thompson et al., 1976). Oysters show little bioaccumulation potential or prolonged retention of boron after cessation of exposure to 10.0 mg B/L. Boron concentrations in industrial waste outfalls seldom exceed 1.0 mg B/L, and this concentration presents little risk to fish or oysters (Thompson et al., 1976).

6.8 Cadmium

Cadmium has no known biological function in molluscs and other living organisms. A criterion of 1.0 mg Cd/kg FW diet is proposed for human health protection (Burger and Gochfeld, 2006; Glynn et al., 2003; Julshamn et al., 2008; Sankar et al., 2006), although some localities suggest less than 2.0 mg Cd/kg FW in seafoods (Lekhi et al., 2008). Accordingly, the elevated concentrations of cadmium in some molluscan tissues (Table 6.5) is of concern to consumers, including humans. It is generally acknowledged that there is a high positive correlation between cadmium concentrations in molluscan tissues and proximity to heavily urbanized areas, especially areas featuring electroplating and mining operations (Bloom and Ayling, 1977; Eisler, 1981, 2000g; Eisler et al., 1978; Eustace, 1973a; Fowler and Oregioni, 1976; Graham, 1972; Maanan, 2007; Phillips, 1979; Talbot et al., 1976b; Thrower and Ratkowsky et al., 1974).

Table 6.5: Cadmium Concentrations in Field Collections of Molluscs

Organism	Concentration	Reference[a]
Abalone, Korea		
Muscle	Max. 0.1 FW	26
Viscera	1.3-2.0 FW	26
Cockle, *Anadara trapezia*; soft parts; April 2000; New South Wales, Australia	Max. 11.0 DW	75
Giant squid, *Architeuthis dux*; hepatopancreas; 2001-2005; Bay of Biscay vs. Mediterranean Sea	60.8 DW vs. 90.7 DW	102
Bay scallop, *Argopecten irradians*		
Soft parts	43.0 AW; 1.1 FW; 7.5 DW	14, 27
Adductor muscle	49.2 AW; 1.4 FW	14
Kidney concretion	19.7 DW	22
Hydrothermal vent mussel, *Bathymodiolus azoricus*; mid-Atlantic ridge		
2002; 3 sites		
Gills	8.2-28.6 DW	80
Mantle	0.01-2.1 DW	80
Digestive gland	13.3-28.8 DW	80
Byssus threads	11.4-89.8 DW	80
2001; shell; 4 sites	0.4-1.4 (0.2-2.4) DW	90
2001; 5 sites		
Gills	1.8-47.2 DW	91
Mantle	0.04-2.4 DW	91
Digestive gland	1.5-4.5 DW	91
Bivalves; 5 spp.; soft parts; East China Sea; August 2002	0.12 (0.01-0.33) FW	66
Mussel, *Brachidontes pharaonis*; Mersin coast, Turkey; 2002-2003; max. concentrations		
Muscle	2.2 DW	60
Gill	8.0 DW	60
Hepatopancreas	3.3 DW	60
Cephalopods; Brazil; June-September 2004; digestive gland		
Slender inshore squid, *Loligo plei*; adult	19.6 FW	78
Argentine short-finned squid, *Illex argentinus*		

(Continues)

Table 6.5: Cont'd

Organism	Concentration	Reference[a]
Immature	18.5 FW	78
Mature	1002.9 FW	78
Cockle, *Cerastoderma glaucum*; Gulf of Gabes, Tunisia; February 2001; max. concentrations		
Gill	3.5 DW	59
Digestive gland	0.5 DW	59
Remainder	0.4 DW	59
Soft parts, 7 sites	0.08-2.5 DW	59
Scallop, *Comptopallium radula*; Santa Marie Bay, New Caledonia; October 2004		
Soft parts	1.1 (0.5-1.9) DW	86
Kidney	57.8 DW	86
Oyster, *Crassostrea angulata*; soft parts	0.35-1.4 FW	28
Sydney rock oyster, *Crassostrea commercialis*		
Soft parts	0.4-18.6 FW	1
Soft parts	0.1-1.0 FW	2
Oyster, *Crassostrea corteziensis*; soft parts; Gulf of California; 1999-2000	7.2 DW	96
Pacific oyster, *Crassostrea gigas*		
Soft parts	0.2-2.1 FW	34
Soft parts	Max. 30.7 FW	1
Soft parts	3.7-9.0 DW	5
Soft parts; China; November 2003; impacted by industrial effluents; 3 sites	11.0-25.0 DW	61
Mangrove oyster, *Crassostrea rhizophorae* Sepetiba Bay, Brazil; soft parts; near zinc smelting plant closed in 1996; max. concentrations		
1978	10.9 DW	52
1980	1.9 DW	52
1983	20.5 DW	52
1989	4.6 DW	52
1996 (after a tailings spill following heavy rains)	29.0 DW	52
1997	4.0 DW	52

(Continues)

Table 6.5: Cont'd

Organism	Concentration	Reference[a]
1999	5.4 DW	52
2001	1.3 DW	52
2002	2.0 DW	52
Near Natal, Brazil; summer 2001; soft parts; contaminated estuary vs. reference site	1.3-2.4 DW; max. 3.1 DW vs. 1.5-2.5 DW; max. 3.1 DW	72
American oyster, *Crassostrea virginica*		
Soft parts; Pensacola, Florida; 2003-2004	Max. 0.61 FW	77
Soft parts; Savannah, Georgia; November 2000-2001; 9 sites	<1.5 to 2.9 DW	85
Mussel, *Crenomytilus grayanus*		
Soft parts; world wide	3.6-66.0 DW	74
Okhotsk Sea and Sea of Japan; summer 2001; upwelling region vs reference sites		
Kidney	1780.0 DW vs. 84.0-119.0 DW	74
Digestive gland	25.0 DW vs. 1.6-1.9 DW	74
Mussel, *Elliptio buckleyi*; soft parts; Marmara Sea, Turkey; 2004-2005	0.18-1.2 (0.04-1.40) FW	62
Razor clam, *Ensis siliqua*; shell; 23 stations; west coast of England		
18 stations	0.2-0.5 DW	53
2 stations	5.1-10.0 DW	53
1 station	25.7 DW	53
2 stations	127.0-156.0 DW	53
Greenland; 1975-1991; bivalves; 5 spp.; soft parts	0.4-3.3 FW	50
Abalone, *Haliotis corrugata*		
Digestive gland	Max. 82.2 FW	31
Muscle	0.3 FW	31
Red abalone, *Haliotis rufescens*		
Gill	4.0-10.0 DW	6
Mantle	2.8-12.8 DW	6
Digestive gland	183.0-1163.0 DW	6
Foot	0.2-0.5 DW	6
Periwinkle, *Littoraria scabra*; soft parts; Tanzania		
Industrial sites; 1998 vs. 2005	10.0 DW vs. 1.7 DW	88
Reference site; 2005	1.0 DW	88

(Continues)

Table 6.5: Cont'd

Organism	Concentration	Reference[a]
Periwinkle, *Littorina littorea*		
Soft parts	0.9-1.5 DW	7
Soft parts	0.0-0.5 FW	8
Soft parts	210.0 DW	9
Squid, *Loligo opalescens*; liver; California coast		
Central region	85.0 DW	10
Southern region	121.0 DW	10
Clam, *Macoma balthica*		
Whole	0.05 FW	32
Soft parts	0.2-0.5 FW; 0.21-0.85 DW	33, 34
Horse mussel, *Modiolus modiolus*; Norway; 2007		
Kidney	7.2 (2.5-12.0) FW	92
Digestive gland	0.6 (0.2-1.3) FW	92
Gills	0.9 (0.5-1.7) FW	92
Gonad	0.3 (0.1-0.6) FW	92
Muscle	0.15 (0.06-0.23) FW	92
Total soft parts	0.86 (0.34-1.52) FW	92
Mexico; Sinaloa; 2004-2005; soft parts; rainy season vs. dry season		
Oyster, *Crassostrea corteziensis*	7.3 DW vs. 6.0 DW	93
Mussel, *Mytella strigata*	7.1 DW vs. 5.1 DW	93
Clam, *Megapitaria squalida*	4.1 DW vs. 2.6 DW	93
Molluscs		
Soft parts		
12 spp.	0.04-2.0 FW	36
3 spp.	0.17-1.9 DW	37
5 spp.; Strait of Magellan, Chile; 2001; 17 stations	0.13-0.92 FW	76
Edible tissues		
11 spp.	<0.1 to 0.3 FW	38
3 spp.	0.3-1.0 FW	38
4 spp.	1.0-3.0 FW	38
2 spp.	0.1 FW	83
Morocco; 2001; clams; soft parts		
Scrobicularia plana	0.5 DW	95
Cerastoderma edule	0.7 DW	95

(*Continues*)

Table 6.5: Cont'd

Organism	Concentration	Reference[a]
Morocco; 2004-2005; soft parts		
Mussel, *Mytilus galloprovincialis*	7.2 (1.3-25.3) DW	87
Clam, *Venerupis decussatus*	2.2 (1.4-3.7) DW	87
Pacific oyster, *Crassostrea gigas*	4.5 (0.7-9.4) DW	87
Softshell clam, *Mya arenaria*		
Soft parts	0.3 (0.1-0.9) FW	3, 40
Soft parts	0.34-0.6 DW	41
Whole	0.1 FW	42
Common mussel, *Mytilus edulis*; soft parts		
United States West Coast	2.3-10.5 DW	12
United States East Coast	0.6-6.2 DW	12
Port Phillip Bay, Australia	0.2-1.3 FW	13
Western Port Bay, Australia	Max. 18.2 FW	13
Scottish waters	0.1-2.0 FW	14
Looe estuary, United Kingdom	0.8-2.6 DW	15
Tasmania	5.5 FW	16
Corio Bay, Australia	2.0-63.0 DW	17
The Netherlands; summer 2002	Max. 6.0 DW	67
Scheldt estuary; March 2000	4.0 DW	100
Common mussel, *Mytilus edulis*		
Shell	0.95-5.5 DW	19, 35, 43
Shell; Baltic Sea; 2005; 12 sites	0.009-0.11 DW	98
Hepatopancreas	Max. 18.0 DW	44
Mussel, *Mytilus edulis planulatus*; total dry weight soft parts		
0.09 g	0.6 DW	18
0.39 g	0.8 DW	18
0.48 g	1.1 DW	18
0.69 g	1.3 DW	18
Pacific blue mussel, *Mytilus edulis trossulus*; Adak Island, Alaska; 2004; soft parts; 5 sites	0.99 FW; max. 2.98 FW; 3.7-7.3 DW	70
Mussel, *Mytilus galloprovincialis*		
Soft parts	1.3 FW	45
Soft parts	1.9 DW	46
Soft parts; Marmara Sea, Turkey; 2005	1.1 FW	58
Soft parts; Izmir Bay, Turkey; 2004; near shore vs. mid-bay	0.03 DW vs. 0.01 DW	99

(*Continues*)

Table 6.5: Cont'd

Organism	Concentration	Reference[a]
Soft parts; Gulf of Gemlik, Turkey; 2004	2.4 DW	106
Soft parts; Adriatic Sea, Italy; 2 sites	0.26-0.35 FW	97
Soft parts; Sardinia; 2004 vs. 2006	2.8 DW vs. 1.1 DW	101
Soft parts; Adriatic Sea; 2001-2005; winter vs. summer	0.9-1.5 DW vs. 0.4-1.3 DW	104
Greece; 2004-2006; contaminated sites; max. mean concentrations; mantle vs. gills		
Summer	0.09 FW vs. 0.41 FW	97
Autumn	0.04 FW vs. 0.42 FW	97
Winter	0.07 FW vs 0.43 FW	97
Spring	0.05 FW vs. 0.36 FW	97
Soft parts; Adriatic Sea; 2001-2002; summer vs. winter		
Industrialized sites	0.54-0.59 DW vs. 0.44-0.63 DW	69
Reference sites	1.1 DW vs. 0.95 DW	69
Soft parts; Venice lagoon, Italy; 2005-2006; San Giuliano vs. Sacca Sessola		
April	6.6 DW vs. 2.5 DW	81
July	1.2 DW vs. 0.8 DW	81
October	2.0 DW vs. 1.2 DW	81
February	2.7 DW vs. 3.6 DW	81
Western Mediterranean Sea; Balearic Islands; Port of Mahon; 1991-2005	0.66 (0.25-1.7) DW	56
Aegean Sea; Turkey; 2002-2003	0.25 (0.04-0.52) FW	64
Soft parts; NW Mediterranean Sea; May 2003		
Italy; Genova and environs	0.27-0.55 DW	55
France; Marseille and vicinity	0.40-0.62 DW	55
Spain; Barcelona and vicinity	0.63-0.89 DW	55
Soft parts; Mediterranean Sea; summer 2004-2005; 3 sites	1.1-1.8 DW	84
Octopus, *Octopus cyanea*; Guam; June 1998; tentacle vs. liver	0.06 DW vs. 7.82 DW	57

Octopus, *Octopus vulgaris*
   Portugal; digestive gland
   Portugal; digestive gland; 2005-2006; northwest region vs. south coast
	19.6-761.5 DW	79

(*Continues*)

Table 6.5: Cont'd

Organism	Concentration	Reference[a]
Whole	150.0 (57.0-250.0) DW vs. 15.0 (10.0-30.0) DW	94
Nuclei	120.0 (28.0-500.0) DW vs. 22.0 (20.0-210.0) DW	94
Mitochondria	150.0 (36.0-360.0) DW vs. 21.0 (17.0-45.0) DW	94
Lysosomes	120.0 DW vs. 25.0 DW	94
Microsomes	230.0 (190.0-1080.0) DW vs. 21.0 (15.0-450.0) DW	94
Squid, *Ommastrephes bartrami*		
Liver	81.0-782.0 DW	10, 11
Muscle	0.7 DW	10, 11
Gonad	0.4 DW	10, 11
European flat oyster, *Ostrea edulis*; soft parts; 1991-2005; western Mediterranean Sea; Balearic Islands; Port of Mahon	1.3 (0.3-3.9) DW	56
Oyster, 5 species of *Ostrea*		
Soft parts	0-21.4 DW	51
Soft parts	0.2-2.3 FW	18
Soft parts	36.0-174.0 DW	17
Soft parts	Max. 10.7 FW	39
Limpet, *Patella candei gomesii*; soft parts; 1997; Azores		
Near hydrothermal vent	0.65-1.0 DW	89
Reference site	1.7-2.8 DW	89
Gastropod, *Patella piperata*; Canary Islands, Spain; soft parts; March-April 2003	0.36 (0.02-0.94) DW	65
Scallop, *Pecten maximus*		
Soft parts	13.0 DW	19
Muscle	1.9 DW	19
Gut and digestive gland	96.0 DW	19
Mantle and gills	3.2-17.0 DW	19
Gonad	2.5 DW	19
Shell	0.0 DW	19
Kidney	54.0-79.0 DW	20, 21
Kidney concretion	546.6 DW	22
Digestive gland	321.0 DW	20
Edible tissues	5.1-23.0 FW	14

(Continues)

Table 6.5: Cont'd

Organism	Concentration	Reference[a]
Giant scallop, *Pecten maximus*; Norway; 2006		
Digestive gland	52.0 (40.0-68.0) FW	92
Kidney	4.7 (2.3-8.8) FW	92
Muscle	0.15 (0.11-0.21) FW	92
Gonad	0.13 (0.07-0.20) FW	92
Total soft parts	5.4 (3.7-8.4) FW	92
Scallop, *Pecten novae-zelandiae*		
Stomach	137.0 FW	36
Gonad	1.5 FW	36
Adductor muscle	0.5 FW	36
Soft parts	0.2 FW	36
Soft parts	249.0 DW	49
Visceral mass	2000.0 DW	49
Other tissues	<20.0 DW	49
Green mussel, *Perna viridis*		
Hong Kong; juveniles vs. adults		
Reference sites		
Soft parts	3.1 DW vs. 5.4 DW	105
Byssus	1.5 DW vs 1.5 DW	105
Polluted sites		
Soft parts	1.3 DW vs. 1.1 DW	105
Byssus	0.8 DW vs. 1.3 DW	105
India; 28 sites; January 2002; soft parts	1.1-1.3 FW	73
False quahog, *Pitar morrhuana*; soft parts; near electroplating plant outfall vs. distant site	3.3 DW vs. <0.07 DW	29
Giant scallop, *Placopecten magellanicus*		
Muscle; March vs. rest of year	Max. 8.8 DW vs. <3.7 DW	23
Viscera; max. values		
March	104.1 DW	23
August	121.2 DW	23
February	161.8 DW	23
June	105.3 DW	23
Gonad	0.5-3.2 FW	24
Viscera	3.7-27.0 FW	24
Clam, *Protothaca staminea*		
Shell	2.9 DW	35
Soft parts	5.7 DW	35

(Continues)

Table 6.5: Cont'd

Organism	Concentration	Reference[a]
Manila clam, *Ruditapes philippinarum*; Korea; soft parts; 2002-2003		
Females vs. males	0.86 DW vs. 0.85 DW	54
Postspawn vs. prespawn	1.4 DW vs. 0.86 DW	54
Jiaozhou Bay, China; November 2004; sediments vs. soft parts	0.2 mg/kg DW vs. 0.51-0.67 DW	67
Clam, *Scrobicularia plana*; digestive gland		
Gannel estuary, United Kingdom	39.8 DW	25
Camel estuary, United Kingdom	1.7 DW	25
Transferred from Camel to Gannel estuary for 352 days	5.6 DW	25
Transferred from Gannel to Camel estuary for 352 days	21.0 DW	25
Common cuttlefish, *Sepia officinalis*		
Egg		
Eggshell	0.59 DW	71
Embryo	0.76 DW	71
Yolk	0.4 DW	71
Hatchlings		
Whole	0.5 DW	71
Cuttlebone	0.8 DW	71
Juveniles, age 1 month		
Whole	1.2 DW	71
Digestive gland	15.0 DW	71
Cuttlebone	0.06 DW	71
Immatures, age 12 months		
Whole	2.0 DW	71
Digestive gland	22.0 DW	71
Cuttlebone	0.22 DW	71
Adults, age 18 months		
Whole	1.1 DW	71
Digestive gland	25.0 DW	71
Cuttlebone	0.03 DW	71
Squid, *Sepia* spp.		
Gills	0.11 DW	7
Mantle	0.03 DW	7
Whole	0.7 FW	30
Whelk, *Thais lapillus*; soft parts	425.0 DW	9

(Continues)

Table 6.5: Cont'd

Organism	Concentration	Reference[a]
Japanese common squid, *Todarodes pacificus*; liver	336.0 DW	103
Clam, *Trivela mactroidea*; Venezuela; 2002; 14 sites; soft parts	<1.0 to 1.9 DW	82

Values are in mg Cd/kg fresh weight (FW), dry weight (DW), or ash weight (AW).

[a]1, Ratkowsky et al., 1974; 2, Mackay et al., 1975; 3, Pringle et al., 1968; 4, Kopfler and Mayer, 1967; 5, Watling and Watling 1976; 6, Anderlini, 1974; 7, Leatherland and Burton, 1974; 8, Topping, 1973; 9, Butterworth et al., 1972; 10, Martin and Flegal, 1975; 11, Hamanaka et al., 1977; 12, Goldberg et al., 1978; 13, Phillips 1976; 14, Eisler et al., 1972; 15, Bryan and Hummerstone 1973; 16, Eustace, 1974; 17, Talbot et al., 1976b; 18, Harris et al., 1979; 19, Segar et al., 1971; 20, Bryan, 1973; 21, George et al., 1980; 22, Carmichael et al., 1979; 23, Reynolds, 1979; 24, Greig et al., 1978; 25, Bryan and Hummerstone, 1978; 26, Won, 1973; 27, Pesch and Stewart, 1980; 28, Establier, 1977; 29, Eisler et al., 1978; 30, Matsumoto et al., 1964; 31, Vattuone et al., 1976; 32, White et al., 1979; 33, Bryan and Hummerstone, 1977; 34, Miettinen and Verta, 1978; 35, Graham, 1972; 36, Nielsen and Nathan, 1975; 37, Greig, 1979; 38, Hall et al., 1978; 39, Eustace, 1974; 40, Shuster and Pringle, 1968; 41, Eisler 1977b; 42, White et al., 1979; 43, Lande, 1977; 44, Young et al., 1979; 45, Uysal, 1978a; 46, Fowler and Oregioni, 1976; 47, Fukai et al., 1978; 48, Thomson, 1979; 49, Brooks and Rumsby, 1965; 50, Dietz et al., 1996; 51, Denton et al., 2006; 52, Lacerda and Molisani, 2006; 53, Pearce and Mann, 2006; 54, Ji et al., 2006; 55, Zorita et al., 2007; 56, Deudero et al., 2007; 57, Vlahogianni et al., 2007; 58, Keskin et al., 2007; 59, Machreki-Ajmi and Hamza-Chaffai, 2006; 60, Karayakar et al., 2007; 61, Liu and Deng, 2007; 62, Yarsan et al., 2007; 63, Mubiana and Blust, 2006; 64, Sunlu, 2006; 65, Bergasa et al., 2007; 66, Huang et al., 2007; 67, Li et al., 2006; 68, Liang et al., 2004a; 69, Scancar et al., 2007; 70, Burger and Gochfeld, 2006; 71, Miramand et al., 2006; 72, Silva et al., 2006; 73, Sasikumar et al., 2006; 74, Podgurskaya and Kavun, 2006; 75, Burt et al., 2007; 76, Espana et al., 2007; 77, Karouna-Renier et al., 2007; 78, Dorneles et al., 2007; 79, Raimundo et al., 2004; 80, Kadar et al., 2006; 81, Nesto et al., 2007; 82, LaBrecque et al., 2004; 83, Sankar et al., 2006; 84, Lafabrie et al., 2007; 85, Kumar et al., 2008; 86, Metian et al., 2008; 87, Maanan, 2008; 88, De Wolf and Rashid, 2008; 89, Cunha et al., 2008; 90, Cravo et al., 2008; 91, Cosson et al., 2008; 92, Julshamn et al., 2008; 93, Frias-Espericueta et al., 2008; 94, Raimundo et al., 2008; 95, Anajjar et al., 2008; 96, Ruelas-Inzunza and Paez-Osuna, 2008; 97, Conti et al., 2008; 98, Protasowicki et al., 2008; 99, Kucuksezgin et al., 2008; 100, Wepener et al., 2008; 101, Schintu et al., 2008; 102, Bustamante et al., 2008; 103, Arai et al., 2004; 104, Fattorini et al., 2008; 105, Nicholson and Szefer, 2003; 106, Unlu et al., 2008.

High BCFs of 10,000 and greater were recorded for most species of field-collected molluscs, with highest concentrations in digestive gland, kidney, and kidney concretions (Table 6.5; Belcheva et al., 2006; Bowen, 1958; Brooks and Rumsby, 1965; Bryan, 1973; Ketchum and Majori and Petronio, 1973; Podgurskaya and Kavun, 2006; Vattuone et al., 1976). Some of the observed variations in cadmium content from field collections of molluscs were attributable to the following: age and size of mollusc (Belcheva et al., 2006; Cossa et al., 1979; Nielsen, 1975; Noel-Lambot et al., 1980; Peden et al., 1973); moisture content (Phillips, 1976a); seasonal changes (Belcheva et al., 2006; Eisler, 1977b; Fowler and Oregioni, 1976; Phillips, 1976a); salinity of the medium (Huggett et al., 1973; Larsen, 1979; Phillips, 1976a, 1977); presence of other metals in solution (Huggett et al., 1973); chelating agents (George and Coombs, 1977); cadmium content of ambient seawater (Peden et al., 1973; Stenner and Nickless, 1974a); the diet (Peden et al., 1973; Phillips, 1977; Thorsson et al., 2008); the chemical form of cadmium (Chou et al., 1978); metabolic changes associated with increasing cadmium concentrations in tissues (Shore et al., 1975); calcium content of shell (Belcheva et al., 2006); and vertical concentration gradients (Nielsen, 1974; Podgurskaya and Kavun, 2006). Selected examples of biotic and abiotic modifiers of tissue cadmium concentrations follow.

6.8.1 Size of Organism

In the Manila clam, *Ruditapes philippinarum*, there is a positive correlation between cadmium concentration in soft tissues on a DW basis and increasing shell length (Ji et al., 2006). Cadmium in the oyster *Ostrea lutaria* increased with increasing body weight up to about 10 g; in larger oysters, concentrations were independent of body weight (Nielsen, 1975). In mussels, *M. edulis*, older specimens from the Gulf of St. Lawrence contained less cadmium per unit weight than younger stages; however, the correlations were variable with increasing age and may have been related to biochemical changes during the sexual cycle (Cossa et al., 1979). In mussels, *M. galloprovincialis*, cadmium increased in soft parts with increasing weight (Conti et al., 2008). Clams, *Gafrarium tumidum*, of shell width 3.5 cm held for 15 days in seawater containing ^{109}Cd, took up about 50 times more radiocadmium in soft parts on a weight/weight basis than in the medium; smaller clams of 2.5 cm took up more cadmium and larger clams of 5.0 cm took up the least radiocadmium (Hedouin et al., 2006). Limpets, *Patella vulgata*, from a metal-contaminated environment show a positive correlation between cadmium concentration and body weight (Noel-Lambot et al., 1980). Most of the cadmium present in limpets with high burdens was bound to thioneine. Because prolonged exposure of limpets to cadmium insult results in an increase in both tissue cadmium concentrations and in metallothioneins, authors conclude that the induced production of metallothioneins is responsible for the cumulative absorption of cadmium in limpets from the Bristol Channel, United Kingdom (Noel-Lambot et al., 1980); these findings have important implications in studies on ageing processes.

6.8.2 Tissue Specificity

Digestive gland of squid, *Illex argentinus*, contains up to 1002.9 mg Cd/kg FW (Dorneles et al., 2007). Authors suggest that upwelling, cannibalism, and cadmium pollution of the studied area are possible reasons for the comparatively elevated cadmium burden in digestive gland. Tissues from other species also contained elevated cadmium burdens: up to 2000.0 mg/kg DW in visceral mass of scallops; 1780.0 mg Cd/kg DW in mussel kidney; 1163.0 mg/kg in digestive gland of abalone; and 782.0 mg/kg DW in squid liver (Table 6.5).

Byssus threads from the hydrothermal vent mussel, *Bathymodiolus azoricus*, contained a maximum of 89.8 mg Cd/kg DW, or more than three times the level of any other tissue (Table 6.5); The role of byssus in cadmium metabolism of mussels is unknown.

6.8.3 Moisture Content

Variations in wet weight of mussels, *M. edulis*, have been recorded, with lowest weights in winter; however, the total cadmium content of individual mussels remains almost constant throughout the year, with weight fluctuations producing much of the observed seasonality of cadmium concentrations in mussels (Phillips, 1976a).

6.8.4 Seasonality

Significant differences between seasons in cadmium content of mussels from the Mersin coast, Turkey in 2002-2003(Karayakar et al., 2007), in three species of bivalves from Mexico in 2004-2005 (Frias-Espericueta et al., 2008) and in *M. galloprovincialis* from Morocco in 2004-2005 (Maanan, 2007) were documented (Table 6.5). Mean cadmium concentrations in mussels, *M. galloprovincialis*, from the NW Mediterranean Sea were similar to those measured in other species from different localities throughout the world (Fowler and Oregioni, 1976). Seasonal maxima in cadmium concentrations of mussels were observed, but this coincided with periods of high precipitation and runoff with a simultaneous increase in the suspended load in coastal waters. This increase could enhance the ambient concentrations of cadmium and other trace metals in both the soluble and particulate forms (Fowler and Oregioni, 1976). They concluded that the overall seasonal increase in metal content of mussels was more likely to be related to natural climatological conditions than to effects of local pollution.

6.8.5 Salinity

Salt content of the medium was inversely associated with cadmium body burdens. At low salinities, comparatively high cadmium concentrations were measured in *M. edulis* (Mubiana and Blust, 2006; Phillips, 1976a) and *Mercenaria* (Larsen, 1979). There is significant interaction between temperature, salinity, and cadmium burdens in *M. edulis*. Body burdens of cadmium were relatively constant at low temperatures and high salinities, elevated at low salinities and high temperatures, and lowest at low temperatures and low salinities (Phillips, 1976a). Cadmium concentrations in mussels from low salinity waters were, in general, considerably higher than conspecifics from relatively high salinity waters (Phillips, 1977). This was not attributed to cadmium concentrations in seawater or to availability of algal food, but to the greater accumulation of cadmium by algae in low salinity waters (Phillips, 1977). Clams, *Scrobicularia plana*, reportedly bioaccumulate cadmium more efficiently at higher salinities within the range 16-32 ppt (Garcia-Luque et al., 2007), but this needs verification.

Assimilation efficiency (AE) of ^{109}Cd-labeled phytoplankton by the Pacific blue mussel, *M. trossulus*, was significantly lower at higher salinities (Widmeyer and Bendell-Young, 2007). Differences in salinity, rather than ingestion rate or food quality—as percent organic carbon content—seemed to best define observed differences in ^{109}Cd AE by *M. trossulus* (Widmeyer and Bendell-Young, 2007).

6.8.6 Other Metals

The presence of other metals in solution, such as zinc, lead, and copper, had no effect on net uptake of cadmium from the medium by mussels (Phillips, 1976a) and this is consistent with the theory that no interaction occurred between these metals in molluscs during uptake. Also, unusually high levels of copper in whole body of American oysters, *C. virginica*, did not influence cadmium content of eggs. The two populations of oysters with different levels of cadmium and copper both contained less than 1.6 mg Cd/kg DW eggs (Greig et al., 1975).

In digestive gland of *M. galloprovincialis*, superoxide dismutase enzyme activity was significantly higher at inshore stations than offshore stations while the reverse was found for glutathione peroxidase activity; this may reflect significantly different concentrations of cadmium and other metals (lead, chromium, copper, zinc, and manganese) between the two sites (Kucuksezgin et al., 2008).

6.8.7 Chemical Form

Different chemical forms of cadmium exist in molluscan tissues, and these may be important in cadmium detoxification mechanisms. In scallop muscle that had 5.3 mg total Cd/kg FW, no cadmium was present in free form; however, soft parts of oysters with 0.8 mg C/kg FW contained up to 50% of the cadmium in free form (Chou et al., 1978).

6.8.8 Diet

Dietary cadmium toxicity in gastropods is low, even when accumulated cadmium body burdens are high; however, metal fractionation in prey can alter the subsequent cellular metal distribution in predators (Cheung et al., 2006). Thus, predatory gastropods, *Thais clavigera*, fed cadmium-contaminated rock oysters, *Saccostrea cucullata*, for up to 4 weeks, show increases in soft parts from 3.1 mg Cd/kg DW at start to 41.8 mg/kg DW after 4 weeks. Oysters and gastropods had been exposed to dissolved cadmium for 2 weeks (0.03-0.15 mg Cd/L) before oysters were fed to gastropods; oysters had up to 90.9 mg Cd/kg DW soft parts at that time. In gastropods, an increasing proportion of cadmium was distributed in metallothionein-like proteins and organelle fractions, but the metal-rich granule fraction decreased when gastropods were fed Cd-exposed prey; metallothionein content in these gastropods was significantly higher than controls (Cheung et al., 2006).

Dog whelks, *Nucella lapillus*, which feed extensively on limpets, *P. vulgata*, had more than twice as much cadmium in whole body than did whelks which do not feed on limpets or those which include them only occasionally in the diet (Peden et al., 1973). This was attributed to the observation that limpets rapidly reflected environmental cadmium levels. American oysters and hardshell clams show a continual increase in tissue cadmium concentrations during exposure to seawater and algae contaminated with cadmium. Algae were the main source of accumulation while direct absorption of soluble cadmium from seawater was another pathway of accumulation in molluscs (Kerfoot, 1979).

Cadmium uptake in fed mussels, *M. edulis*, was significantly higher from the medium than starved mussels. The elevated cadmium concentrations measured were not attributed to cadmium-contaminated food but to increased pumping rate when food is available. Highest cadmium concentrations and major body burdens were in the mid-gut gland (Janssen and Scholz, 1979). Transport via hemolymph, and selective discrimination at the basement lamina of mid-gut gland tubuli appears to be the main accumulation route. In the tubuli, cadmium is immobilized in membrane-bound vesicles and eventually excreted (Janssen and Scholz, 1979). Green mussels, *Perna viridis*, exposed to 0.005, 0.02, 0.05, or 0.2 mg Cd/L for 7 days, 16 h daily, all had higher metallothionein synthesis rates when compared to controls; dietary assimilation of cadmium increased by a factor of 2 at 0.02 mg Cd/L and higher (Ng et al., 2007).

Dissolved cadmium is the main source of cadmium to cultured Pacific oysters, *C. gigas*, where it concentrated mainly in gut tissues, with accumulation lower during summer; particulate matter was not a source of cadmium in oysters (Lekhi et al., 2008).

6.8.9 Ambient Cadmium Concentrations

Transplantation of molluscs from cadmium-free waters to areas heavily contaminated by cadmium and other metals wastes was reflected in elevated tissue concentrations of cadmium. For example, dog whelks transplanted from Beer, Dorset, United Kingdom, to Bristol Channel, United Kingdom, showed a rise in tissue levels from 36.0 mg Cd/kg DW at the start to 211.0 mg Cd/kg DW after 5 months; the rise was even more dramatic (from 11.0 to 220.0) for *Patella* sp. transplanted for 8 months (Stenner and Nickless, 1974a).

Cadmium sediment concentrations of 0.2 mg/kg DW in Jiaozhou Bay, China, were much lower than were cadmium concentrations in soft parts of clams, *R. philippinarum* (0.51-0.67 mg/kg DW); this was not the case for copper, zinc, lead, manganese, and chromium (Li et al., 2006). The reverse was true for cockles, *Anadara trapezia*, from a metals-contaminated estuary in New South Wales, Australia in April 2000: in that case, cadmium—as well as zinc and lead—concentrations in soft parts of the cockle were highly correlated with cadmium, zinc, and lead sediment burdens, respectively (Burt et al., 2007).

Sediment composition dramatically affects cadmium accumulation by softshell clams, *M. arenaria*, during the first 24 h (Phelps, 1979). Virtually all radiocadmium-109 sorbed to bentonite was taken up; for humic acid-^{109}Cd, this value was 60%; for mud sediment-Cd, 33%; and for albumin-Cd, only 12%. Shell sorbed twice as much cadmium than soft tissues during the first 230 min of exposure (Phelps, 1979). Availability of sediment-bound cadmium to cockles, *Cerastoderma edule*, is dependent on whether cadmium was bound to biogenic calcium carbonate (readily available) or precipitated calcium carbonate (low availability) (Cooke et al., 1979). Also, cadmium bound to iron oxide and manganese oxide was unavailable to cockles, at least in short-term exposure. In general, cadmium availability to cockles seems to be a function of the ability of cadmium to desorb from the sediment (Cooke et al., 1979).

6.8.10 Metabolism

Increasing body burdens of cadmium were associated with metabolic changes. In one case, as cadmium increased in digestive gland of limpets, *P. vulgata*, from 27.0 to 537.0 mg/kg DW, there was a significant decrease in glycolytic rate and corresponding increase in hemolymph glucose (Shore et al., 1975).

6.8.11 Depth

Vertical concentration gradients of heavy metals, including cadmium, in mussels, *Perna canaliculis*, are reported in some locations, but not others (Nielsen, 1974). Differences in mixing of the water column, which in turn causes variations in the type of food organisms with depth, or variations in the ratio of particulate/dissolved metal levels with depth, could account for this phenomenon (Nielsen, 1974). Similar findings were reported for *M. edulis* (Talbot et al., 1976b). Near freshwater inputs of trace metals, concentrations of cadmium in *M. edulis* varied with water depth of collection, being greater in shallower waters; however, this effect was absent in summer when freshwater runoff was reduced (Phillips, 1976a).

6.8.12 Laboratory Studies

Accumulation of cadmium by bivalve molluscs under controlled conditions is the subject of several investigations (Alquezar et al., 2007; Brooks and Rumsby, 1967; Chuang and Wang, 2006; Eisler et al., 1972; Hiyama and Shimizu, 1964; Mubiana and Blust, 2007; Nelson et al., 1976; Ng et al., 2007; Noel-Lambot, 1976; Pesch and Stewart, 1980; Pringle et al., 1968; Shuster and Pringle, 1969; Sturesson, 1978; Zaroogian, 1979; Zaroogian and Cheer, 1976).

Brooks and Rumsby (1967) report that oysters, *Ostrea sinuata*, held in high (50.0 mg Cd/L) concentrations of cadmium for 100 h accumulated this metal strongly in gills, visceral mass, and heart; high cadmium accumulations in oyster tissues had relatively little effect on concentrations of silver, manganese, lead, vanadium, or zinc in the same tissues, although some displacement may have occurred in visceral mass. Uptake was attributed to nonselective adsorption of cadmium, possibly by coordination to organic ligands (Brooks and Rumsby, 1967). American oysters, *C. virginica*, and scallops, *Aquipecten irradians*, that were immersed in flowing seawater containing 0.01 mg Cd/L for 21 days bioconcentrated cadmium by factors of 5.6 and 2.4 respectively in whole soft parts (Eisler et al., 1972). It has been stated that it is the *concentration* of cadmium in water and food and *not the absolute amount* that determines acute cadmium poisoning as indicated by vomiting (USPHS, 1962). On this basis, American oysters quickly accumulated emetic thresholds concentrations of cadmium, that is 13.0-15.0 mg Cd/kg FW, when cadmium levels in the medium exceeded 0.01 mg Cd/L. Emetic thresholds were surpassed in 5 and 2 weeks at cadmium water levels of 0.025 and 0.1 mg Cd/L, respectively (Shuster and Pringle, 1969). Similar results were documented in other species of oysters (Casterline and Yip, 1975) and in clams (Shuster and Pringle, 1968). Rates of accumulation in bivalve molluscs were not linearly related to ambient cadmium concentrations. For softshell clam, *M. arenaria*, daily accumulation rates, in mg Cd/kg FW, were 0.16 for clams held in solutions of 0.1 mg Cd/L, and 0.1 for those immersed in 0.05 mg Cd/L (Pringle et al., 1968). Concentrations as low as 0.005 mg Cd/L in flowing seawater were also accumulated to potentially hazardous levels (13.6 mg Cd/kg FW soft parts) by American oysters after an extended exposure of 40 weeks (Zaroogian and Cheer, 1976). Cadmium-loaded oysters retained virtually all accumulated cadmium (12.5 mg Cd/kg FW soft parts) during a period of 16 weeks immersion in clean seawater (Zaroogian, 1979), with attendant risk to human consumers.

Glycoproteins reportedly protect bivalve molluscs against cadmium (Ivanina and Sokolova, 2008). American oysters, *C. virginica*, exposed to 0.05 mg Cd/L for 30-40 days produced a 2.0- to 2.5-fold increase in glycoprotein content on cell membrane and mitochondria of gills and hepatopancreas, and a 3.5- to 7.0-fold increase in transport rate (Ivanina and Sokolova, 2008).

Scallops, *A. irradians*, held for 6 weeks in seawater solutions containing 0.06 mg Cd/L contained up to 1253.0 mg Cd/kg DW soft parts versus 7.6 in controls; higher residues were observed in scallops that survived 0.12, 0.25, and 0.5 mg Cd/L for a similar period (Pesch and Stewart, 1980).

Mussels, *M. edulis*, subjected to alternate immersion and emersion characteristic of tidal cycles accumulated significantly less cadmium that did continually immersed mussels (Coleman, 1980). It is noteworthy that mussels held 15 days in seawater solutions containing 0.035 mg Cd/L always accumulated more than 2.0 mg Cd/kg FW soft parts, which is the maximum safe value specified by Australian authorities for edible shell fish (Coleman, 1980); this value has since been modified for shellfish to <2.0 mg Cd/kg on a DW basis (Machreki-Ajmi and Hamza-Chaffai, 2006). Shells from *M. edulis* held in seawater solutions containing 0.2 mg Cd/L for 50 days had 75.0 mg Cd/kg DW in shell periostracum, 17.0 mg/kg in the nacreous layer, and 13.0 mg Cd/kg DW in newly formed calcite (Sturesson, 1978). In untreated *M. edulis*, cadmium was associated with molecular proteins of high weight, but after exposure of 90 days in 0.005 mg Cd/L, cadmium was bound to low-molecular-weight proteins, similar to vertebrate metallothioneins and apparently synthesized in a like manner as vertebrates (Noel-Lambot, 1976). Isolated gills of the common mussel, *M. edulis*, were held for 28 days in seawater containing 0.003 mg Cd/L at water temperatures between 6 and 26 °C (Mubiana and Blust, 2007). Increasing cadmium concentrations were measured at increasing water temperatures; whole mussels showed the same trend, but elimination was independent of water temperature (Mubiana and Blust, 2007). Uptake and elimination patterns for isolated gills and whole mussel were the same for lead exposures. Cobalt and copper uptake and elimination patterns in whole mussels were different from those of copper and lead, and it is not possible to generalize (Mubiana and Blust, 2007).

Mussels, *Mytilopsis sallei*, all survived immersion in 0.05 mg Cd/L for 96 h; however, 0.71 mg Cd/L was fatal to 50% in that same time period (Devi, 1996). Adaptation to environmental cadmium stress is reported for *M. sallei* exposed to sublethal concentrations >0.05 mg Cd/L for at least 96 h (Devi, 1996). The decrease in oxygen consumption and the increased metabolism of glycogen and carbohydrates during cadmium exposure suggest that *M. sallei* could shift to anaerobic metabolism to counter the environmental cadmium stress (Devi, 1996).

Uptake and retention of radiocadmium-109 was studied in the pygmy mussel, *Xenostrobus securis* (Alquezar et al., 2007). Mussels were held for 385 h in seawater spiked with ^{109}Cd followed by 189 h in ^{109}Cd-free media. During uptake, whole mussels concentrated radiocadmium by a factor of 81 over seawater; during the loss phase, 77% of the accumulated radiocadmium was excreted in a biphasic pattern. The biological half-time of the short-lived component was 11 h; for the long-lived component, it was 2140 h (about 89 days; Alquezar et al., 2007). Uptake of ligand-bound ^{109}Cd by the green mussel, *P. viridis*, indicates that metal-ligand complexes were available for uptake independent of free cadmium concentration (Chuang and Wang, 2006). Scallops, *Comptopallium radula*, exposed to ^{109}Cd for four days had a BCF of 62 in soft parts; however, uptake was greater in kidney with a BCF of 169 (Metian et al., 2008). Eggs of the cuttlefish, *S. officinalis*, exposed to ^{109}Cd during embryonic development and also stable cadmium at 0.001 mg Cd/L did not accumulate cadmium in yolk material; both stable cadmium and radiocadmium were mainly associated with eggshell, indicating that the eggshell acted as an efficient shield against penetration (Lacoue-Labarthe et al., 2008a,b).

Eggs of the European squid developed normally at 0.01-0.1 mg Cd/L; however, effects on survival and development were demonstrable at 1 mg Cd/L and higher (Sen and Sunlu, 2007). Feeding rate of clams, *Donax trunculus*, was inhibited 60% during immersion in seawater containing 0.1 mg Cd/L for 24 h (Neuberger-Cywiak et al., 2007).

6.9 Cerium

Concentrations of stable cerium in the tests of six species of pteropod molluscs ranged between 0.26 and 3.03 mg Ce/kg DW (Turekian et al., 1973). All species of molluscs examined accumulated cerium from ambient seawater by factors that ranged from 100 to 1200 (Avargues et al., 1968; LeGall and Ancellin, 1971). The addition of chelating agents to the medium reduced concentration factors to <1 for the clam *Venerupis philippinarum* (Hiyama and Shimizu, 1964). Radiocerium-144 concentrations in tissues of the clam *Anadara granosa* reflected sediment ^{144}Ce concentrations (Patel et al., 1978). On reaching maximum accumulations, the concentration of ^{144}Ce (and also cesium-137 and ruthenium-106) decreased exponentially over time, although ^{144}Ce was still available in sediments.

6.10 Cesium

Stable cesium concentrations in molluscan tissues never exceeded 0.13 mg Cs/kg DW, or 0.007 mg Cs/kg FW (Table 6.6).

Table 6.6: Cesium Concentrations in Field Collections of Molluscs

Organism	Concentration	Reference[a]
Mussel, *Choromytilus meridionalis*; soft parts	0.0070 FW	1
Clam, *Donax serra*; soft parts	0.0049 FW	1
Abalone, *Haliotis midea*; soft parts	0.0002 FW	1
Clam, *Meretrix lamarckii*; soft parts	0.022 DW	2
Molluscs, 3 spp; edible flesh	0.001-0.007 FW	3
Mussel, *Mytilus corscum*; soft parts	0.023 DW	2
Common mussel, *Mytilus edulis*; soft parts	0.0004-0.13 DW	4
Abalone, *Notohaliotis discus*; soft parts less viscera	0.029 DW	2
Limpet, *Patella candei gomesii*; soft parts; 2007; Azores		
Near hydrothermal vent	0.05-0.13 DW	5
Reference site	0.02 DW	5
Squid, *Sepia esculenta*; trunk	0.016 DW	2

Values are in mg Cd/kg fresh weight (FW) or dry weight (DW).
[a]1, Van As et al., 1973; 2, Ishii et al., 1978; 3, Van As et al., 1975; 4, Karbe et al., 1977; 5, Cunha et al., 2008.

As a result of the Chernobyl nuclear reactor plant accident on April 26, 1986 (Eisler, 1995b, 2003), radiocesium-137 levels (but not ^{134}Cs levels) in seven species of molluscs collected near a nuclear plant increased by a factor up to 6 (Lowe and Horrill, 1991). Mussels, *M. galloprovincialis*, accumulated ^{134}Cs and ^{137}Cs in soft parts following the Chernobyl nuclear reactor accident; however, between May 6 and August 14, 1986, mussels lost 98.4% of the accumulated ^{134}Cs and 94.4% of the ^{137}Cs (Whitehead et al., 1988). Using *M. galloprovincialis* as an indicator or ^{137}Cs activity in the Mediterranean Sea and Black Sea during 2004-2006, mussels with the highest radioactivity levels were confined to the Black Sea and North Aegean Sea and reflect fallout from Chernobyl (Thebault et al., 2008). In summer 2004, ^{137}Cs activity above minimum detectable levels was measured in Pacific octopus, *Enteroctopus dofleini*, from Amchitka Island in the Aleutian chain; Amchitka Island was the site of three nuclear tests between 1965 and 1971 (Burger et al., 2007). However, no ^{137}Cs activity was detected in summer 2004 samples of chiton, *Cryptochiton stelleri*, rock jingle, *P. macroschisma*, or plate limpet, *Tectura scutum* from Amchitka Island (Burger et al., 2007b).

Studies on radiocesium-137 accumulation and retention by marine molluscs under controlled environments (Argiero et al., 1966; Avargues et al., 1968; Bryan, 1963a,b; Hiyama and Shimizu, 1964; LeGall and Ancellin, 1971; Ueda et al., 1978; Van As et al, 1973) show similar results. In one of the more exhaustive studies, Bryan (1963b) established that six species of marine gastropods and bivalves reached equilibrium with ambient ^{137}Cs levels between 300 and 1400 h, that concentration factors for any tissue in any species never exceeded 15.3, that highest accumulations were usually in muscle and digestive gland, and the lowest concentration factors were in plasma, shell, and mucous. In one species, *Archidoris pseudoargus*, gonad was the second most active site of ^{137}Cs, next to digestive gland (Bryan, 1963b). Data seem needed on effect of cesium translocation to sex products on molluscan embryonic development. Accumulation of ^{137}Cs was measured in the cephalopods *Octopus vulgaris* and *Doryteuthis bleekeri* (Ueda et al., 1978). Concentration factors for *Octopus* after 14 days were highest for liver (12.8), ovotestis (10.9), kidney, funnel, branchial heart (8.0-8.3), heart, anus, tentacles, buccal bulb, ctenidia, gastric caecum (6.1-7.5) stomach, sucker, esophagus, salivary gland (4.7-5.7), and mantle (3.5), in that order. Most, that is, 90–95%, of the radioactivity was eliminated in 75 days. Concentration factors for *Doryteuthis* after 6 days ranged between 8 and 11 for the tissues examined (Ueda et al., 1978). Salinity of the medium affects cesium uptake mechanisms in the euryhaline clam *Rangia cuneata*. In one study, *Rangia* contained lower cesium concentrations in waters of higher salinities (Wolfe, 1970). Studies on the interaction of salinity with temperature on radiocesium-137 uptake by *Rangia* show that uptake in soft parts was lower with increasing salinity, as expected, but uptake increased with increasing temperature; in shell, however, the proportion of cesium decreased as both salinity or temperature was increased (Wolfe and Coburn, 1970).

Recent uptake studies (Bustamante et al., 2006) on ^{134}Cs uptake by different developmental stages of cuttlefish, *S. officinalis*, via the sediments, medium, and diet demonstrate that uptake from sediments was negligible compared to seawater and diet. In the seawater portion, for example, whole juveniles exposed to ^{134}Cs for 36 h had a BCF of 4.1; after 29 days in clean seawater; 61% of the radioactivity was retained in digestive gland. Adults held for 8 h in ^{134}Cs-spiked seawater, had a BCF <2 for all tissues, with half-time persistence of 6 days. Cuttlefish fed a radiolabeled ^{134}Cs diet of mussels and brine shrimp for 1 h then transferred to clean seawater for 29 days still retained 54% of the ^{134}Cs in digestive gland; the calculated half-time persistence of ^{134}Cs was 5 days (Bustamante et al., 2006). Gravid cuttlefish fed crabs radiolabeled with ^{134}Cs over a 2 week period to determine rate of maternal transfer into eggs showed no significant transfer (Lacoue-Labarthe et al., 2008b).

6.11 Chromium

Chromium concentrations were usually elevated in molluscs and other marine biota in the vicinity of electroplating plants, tanneries, ocean dump sites, oil drilling operations, sewage outfalls, and other sources of chromium-containing wastes (Eisler, 2000a; Maanan, 2007; Table 6.7).

Edible tissues of commercially important North American species of marine molluscs contained between 0.1 and 0.6 mg Cr/kg FW (Hall et al., 1978), and this is in general agreement with molluscan data from other geographic areas (Table 6.7). However, higher chromium values of 3.4 mg Cr/kg FW are documented in edible portions of American oysters, *C. virginica*, 5.8 mg Cr/kg FW in hardshell clam, *Mercenaria*, and 5.0 mg Cr/kg FW in softshell clam, *M. arenaria* (Shuster and Pringle, 1968).

The ability of marine molluscs to accumulate chromium far in excess of ambient seawater concentrations is documented in five species of bivalves from Greek waters; concentration factors ranged from 16,000 in *P. nobilis* to 260,000 in *Astralium rogosum* (Papadopoulu, 1973). No deleterious health effects are documented among consumers of molluscs with occasional high chromium content. Anthropogenic and natural chromium gradients in water column or sediments are reflected in the wide range of values reported for this element in field collections of *Mercenaria mercenaria* (Phelps et al., 1975), *M. edulis* (Karbe et al., 1977; Lande, 1977); *Pitar morrhuana* (Eisler et al., 1978), *Mytilus californianus* (Alexander and Young, 1976), and *M. galloprovincialis* (Fowler and Oregioni, 1976). Two variables linked to chromium accumulations in tissues of field-collected molluscs are the weight of the organism and the salinity of the medium. For example, decreasing concentrations of chromium in *R. cuneata* tissues occur with increasing body weight and increasing salinities (Olson and Harrel, 1973).

Table 6.7: Chromium Concentrations in Field Collections of Molluscs

Organism	Concentration	Reference[a]
Gastropod, *Acmaea digitalis*		
Shell	<5.7 DW	1
Soft parts	7.1-24.2 DW	1
Mussel, *Anodonta* sp.		
Shell	0.45 DW	2
Soft parts	0.84 DW	2
Giant squid, *Architeuthis dux*; 2001-2005; hepatopancreas		
Bay of Biscay	0.9 DW	67
Mediterranean Sea	0.5 DW	67
Ocean quahaug (clam), *Arctica islandica*; soft parts	0.3-2.5 FW	3
Bay scallop, *Argopecten irradians*; kidney concretion	49.3 DW	4
Cockle, *Austrovenus stutchburyi*; soft parts; 1993-1994; New Zealand	1.0-44.0 DW	50
Mussel, *Brachydontes pharaonis*; Mersin coast, Turkey; 2002-2003; max. concentrations		
Muscle	8.9 DW	47
Gill	19.0 DW	47
Hepatopancreas	6.5 DW	47
Whelk, *Busycon canaliculatum*		
Muscle	<0.8 FW	5
Digestive gland	<0.8 FW	5
Digestive diverticula	0.24 DW	6
Gills	4.2 DW	6
Cockle, *Cerastoderma edule*; soft parts	1.3-2.5 DW	7
Scallop, *Chlamys opercularia*		
Soft parts	2.2 DW	8
Kidneys	6.6 DW	8
Digestive gland	4.7 DW	8
Mussel, *Choromytilus meridionalis*		
Soft parts	1.4 DW	9
Soft parts	0.10-0.13 FW	10, 11

(Continues)

Table 6.7: Cont'd

Organism	Concentration	Reference[a]
Scallop, *Comptopallium radula*; Santa Marie Bay, New Caledonia; October 2004		
Soft parts	3.3 (2.4-5.0) DW	60
Digestive gland	16.8 DW	60
Pacific oyster, *Crassostrea gigas*; soft parts	0.1-0.3 FW	12
American oyster, *Crassostrea virginica*		
Soft parts	0.04-3.4 FW	13
Shell	1.8-7.3 FW	14
Soft parts; Pensacola, Florida; 2003-2004	Max. 0.71 FW	54
Soft parts; Savannah, Georgia; 2000-2001	<1.5 to 8.0 DW	59
Limpet, *Crepidula fornicata*; soft parts	2.0 DW	2
Clam, *Donax serra*; soft parts	0.19-0.24 FW	10, 11
Clam, *Ensis arcuatus*; shell	0.14 DW	15
Clam, *Ensis siliqua*; shell	0.13 DW	15
Abalone, *Haliotis midea*; soft parts	0.44-0.5 FW	10, 11
Red abalone, *Haliotis rufescens*		
Gill	0.6-4.0 DW	16
Mantle	0.0-12.6 DW	16
Digestive gland	2.0-13.2 DW	16
Foot	not detectable	16
Periwinkle, *Littoraria scabra*; soft parts; Tanzania		
Industrial sites; 1998 vs. 2005	7.0 DW vs. 8.5 DW	62
Reference site; 2005	4.0 DW	62
Limpet, *Littorina littorea*		
Soft parts plus operculum	0.1-1.0 DW	7
Soft parts; United Kingdom	<0.1 to 1.6 DW	40
Clam, *Macoma balthica*; soft parts	1.9-3.3 DW	7
Hardshell clam, *Mercenaria mercenaria*		
Soft parts	3.3-24.7 DW	17
Soft parts	0.2-5.8 FW	13
Soft parts	0.8 DW	2
Shell	0.37 DW	2

(*Continues*)

Table 6.7: Cont'd

Organism	Concentration	Reference[a]
Mussel, *Modiolus modiolus*		
Soft parts	0.14 DW	2
Mantle and Gills	0.15 DW	2
Muscle	0.2 DW	2
Gonad	0.2 DW	2
Gut and digestive gland	0.19 DW	2
Molluscs; commercially important; edible tissues		
3 spp.	0.1-0.2 FW	18
5 spp.	0.2-0.3 FW	18
8 spp.	0.3-0.4 FW	18
2 spp.	0.4-0.6 FW	18
6 spp.; Spain; April-May 1990; Mediterranean Sea	0.06-0.39 FW	39
3 spp.; India; 2003	0.3-0.9 FW	57
Morocco; 2001; clams; soft parts		
Scrobicularia plana	6.0 DW	63
Cerastoderma edule	1.0 DW	63
Morocco; 2004-2005; soft parts		
Mussel, *Mytilus galloprovincialis*	8.8 (4.0-20.6) DW	61
Clam, *Venerupis decussatus*	9.6 (4.3-15.3) DW	61
Pacific oyster, *Crassostrea gigas*	7.1 (1.2-15.3) DW	61
Gastropod, *Murex trunculus*; soft parts	1.3 DW	19
Softshell clam, *Mya arenaria*		
Soft parts	1.8-4.4 DW	20
Soft parts	0.1-0.5 FW	13
California mussel, *Mytilus californianus*		
Digestive gland	1.0-61.0 DW	21
Shell	<5.7 to 14.2 DW	1
Soft parts	<1.5 to 7.8 DW	1
Common mussel, *Mytilus edulis*		
Digestive gland	7.4 FW	22
Gonad	3.0 FW	22
Muscle	11.0 FW	22
Shell	1.0-2.0 DW	24
Shell	0.1 DW	15
Shell; Baltic Sea; 2005; 12 sites	1.1-1.4 DW	64
Hepatopancreas	3.5-15.0 DW	25

(Continues)

Table 6.7: Cont'd

Organism	Concentration	Reference[a]
Soft parts; Scheldt estuary; March 2000	5.8 DW	66
Soft parts	1.0-1.5 DW	25
Soft parts	0.9-2.7 DW	7
Soft parts	<1.0 to 7.6 DW	1
Soft parts	4.0-49.0 DW	24
Soft parts	Max. 3.0 DW	48
Gills; near leather tannery effluent vs. reference site; Ireland; River Calligan; 1992-1994	400.0-1000.0 DW vs. max. 6.0 DW	41
Mussel, *Mytilus edulis-aoteanus*		
Soft parts	16.0 DW	26
Gill	10.0 DW	26
Visceral mass	29.0 DW	26
Pacific blue mussel, *Mytilus edulis trossulus*; soft parts; 2004; 5 sites; Adak Island, Alaska	20.7 FW; max. 148.0 FW; 62.0-170.0 DW	52
Mussel, *Mytilus galloprovincialis*		
Soft parts; Venice lagoon, Italy; 2005-2006; San Giuliano vs. Sacca Sessola		
April	2.8 DW vs. 1.6 DW	55
July	0.2 DW vs. 0.7 DW	55
October	0.5 DW vs. 0.7 DW	55
February	0.8 DW vs. 2.0 DW	55
Soft parts	7.5 DW	27
Soft parts	4.4 FW	28
Soft parts; 1991-2005; western Mediterranean Sea; Balearic Islands; Port of Mahon	4.6 (0.7-14.8) DW	45
Soft parts; May 2003; NW Mediterranean Sea		
Italy; near Genova	1.7-5.1 DW	44
France; near Marseille	0.71-3.3 DW	44
Spain; near Barcelona	3.2-24.4 DW	44
Soft parts; summers 2004-2005; Mediterranean Sea; 3 sites	0.4-3.0 DW	58
Soft parts; Adriatic Sea; 2001-2005; winter vs. summer	1.1-2.3 DW vs. 0.4-0.8 DW	68
Mantle vs. gills; Saronikos Gulf of Greece; 2004-2006; max. mean concentrations; contaminated sites		

(*Continues*)

Table 6.7: Cont'd

Organism	Concentration	Reference[a]
Summer	2.3 FW vs. 4.2 FW	47
Autumn	0.39 FW vs. 0.57 FW	47
Winter	0.2 FW vs. 0.4 FW	47
Spring	1.4 FW vs. 0.55 FW	47
Soft parts; Adriatic Sea; 2001-2002; summer vs. winter		
Industrialized sites	3.7-4.7 DW vs. 4.4-10.3 DW	51
Reference site	3.4 DW vs. 2.7 DW	51
Soft parts; Izmir Bay, Turkey; 2004; near shore vs. mid-bay	0.17 DW vs. 0.12 DW	65
Soft parts; Gulf of Gemlik, Turkey; 2004	2.3 DW	70
Octopus, *Octopus cyanea*; Guam; June 1998; tentacle vs. liver	<0.16 DW vs. 1.87 DW	42
Oyster, *Ostrea edulis*; soft parts	0.1-2.1 DW	29
Oyster, *Ostrea sinuata*		
Soft parts	3.0 DW	26
Heart	9.0 DW	26
Scallop, *Pecten maximus*		
Soft parts	1.3-1.8 DW	2, 8
Kidneys	3.9 DW	8
Digestive gland	8.1 DW	8
Mantle and gills	0.76 DW	2
Gonad	0.45 DW	2
Scallop, *Pecten novae-zelandiae*		
Soft parts	10.0 DW	26
Gills	145.0 DW	26
Intestine	24.0 DW	26
Kidney	17.0 DW	26
Foot	8.0 DW	26
Pen shell, *Pinna nobilis*		
Soft parts	5.8 DW	30
Mantle and gills	16.0 DW	30
Muscle	3.9 DW	30
Nervous system	6.3 DW	30
Stomach and intestine	6.2 DW	30
Gonads	22.0 DW	30
Hepatopancreas	150.0 DW	30

(Continues)

Table 6.7: Cont'd

Organism	Concentration	Reference[a]
Byssus	32.0 DW	30
Byssus gland	12.0 DW	30
False quahaug, *Pitar morrhuana*; soft parts	3.1-14.2 DW	31
Pteropods; 9 spp.; tests	0.43-4.9 DW	32
Clam, *Rangia cuneata*; whole	0.32 FW	33
Manila clam, *Ruditapes philippinarum*; soft parts; Korea; 2002-2003		
Females vs. males	1.05 DW vs. 1.07 DW	43
Postspawn vs. prespawn	1.6 DW vs. 1.1 DW	43
Clam, *Spisula solidissima*; muscle	0.7-2.7 DW	6
Whelk, *Nucella lapillus*; soft parts plus operculum	0.4-5.6 DW	7
Limpet, *Patella vulgata*		
Soft parts	0.5-2.6 DW	7
Soft parts	7.0-17.0 DW	24
Shell	1.0-2.0 DW	24
European flat oyster, *Ostrea edulis*; soft parts; Balearic Islands; western Mediterranean Sea; 1991-2005; Port of Mahon	3.3 (0.8-11.0) DW	45
Green mussel, *Perna viridis*		
India; 28 sites; January 2002; soft parts		
Small	0.09 FW	53
Medium	0.07 FW	53
Large	0.01 FW	53
Hong Kong; juveniles vs. adults		
Reference site		
Soft parts	1.1 DW vs. 1.0 DW	69
Byssus	0.9 DW vs. 2.4 DW	69
Polluted site		
Soft parts	1.7 DW vs. 1.5 DW	69
Byssus	4.2 DW vs. 2.4 DW	69
Scallop, *Placopecten magellanicus*		
Soft parts	3.1 DW	34
Muscle	<0.3 to 0.6 FW	35

(*Continues*)

Table 6.7: Cont'd

Organism	Concentration	Reference[a]
Gonad	0.3-1.7 FW	35
Visceral mass	0.4-4.0 FW	35
Soft parts	0.2-2.4 FW	3
Clam, *Ruditapes philippinarum*		
Jiaozhou Bay, China; sediments vs. soft parts; November 2004	65.3 DW vs. <0.01 DW	49
Bohai Sea, China; soft parts	0.14-0.63 DW	49
Clam, *Scrobicularia plana*		
Looe estuary, England; soft parts	2.8 (1.1-3.9) DW	37
Tamar estuary, England		
Soft parts	2.1 (0.5-4.2) DW	36
Digestive gland	6.9 (3.9-12.8) DW	36
Gannel estuary, England		
Soft parts	1.5-2.2 DW	37
Digestive gland	3.6 DW	37
Mantle and siphons	2.1 DW	37
Foot and gonad	1.1 DW	37
Camel estuary, England		
Soft parts	1.2-1.3 DW	37
Digestive gland	2.9 DW	37
Mantle and siphons	1.4 DW	37
Foot and gonad	0.6 DW	37
Gills and palps	2.3 DW	37
Transferred from Camel estuary to Gannel estuary for 352 days		
Digestive gland	6.3 DW	37
Remaining tissues	1.2 DW	37
Transferred from Gannel estuary to Camel estuary for 352 days		
Digestive gland	3.7 DW	37
Remaining tissues	1.2 DW	37
Squid		
Muscle	5.4 DW	38
Muscle with skin	4.7 DW	38
Viscera	3.9 DW	38
Pen	3.1 DW	38
Clam, *Tapes semidecussata*		
Shell	10.3 DW	1
Soft parts	10.3 DW	1

(Continues)

Table 6.7: Cont'd

Organism	Concentration	Reference[a]
Gastropod, *Thais emarginata*		
Shell	<5.7 to 10.5 DW	1
Soft parts	<1.5 DW	1
Clam, *Trivela mactroidea*; Venezuela; 2002; 14 sites; soft parts	<1.0 to 6.2 DW	56

Values are in mg Cr/kg fresh weight (FW) or dry weight (DW).
[a]1, Graham, 1972; 2, Segar et al., 1971; 3, Palmer and Rand, 1977; 4, Carmichael et al., 1979; 5, Greig et al., 1977; 6, Greig, 1975; 7, Bryan and Hummerstone, 1977; 8, Bryan, 1973; 9, Watling and Watling, 1976a; 10, Van As et al., 1973; 11, Van As et al., 1975; 12, Pringle et al., 1968; 13, Shuster and Pringle, 1968; 14, Ferrell et al., 1973; 15, Bertine and Goldberg, 1972; 16, Anderlini, 1974; 17, Phelps et al., 1975; 18, Hall et al., 1978; 19, Fukai, 1965; 20, Eisler, 1977b; 21, Alexander and Young, 1976; 22, Young and McDermott, 1975; 23, Karbe et al., 1977; 24, Lande, 1977; 25, Young et al., 1979; 26, Brooks and Rumsby, 1965; 27, Fowler and Oregioni, 1976; 28, Uysal, 1978a; 29, Fukai et al., 1978; 30, Papadopoulu, 1973; 31, Eisler et al., 1978; 32, Turekian et al., 1973; 33, White et al., 1979; 34, Reynolds, 1979; 35, Greig et al., 1978; 36, Bryan and Uysal, 1978; 37, Bryan and Hummerstone, 1978; 38, Horowitz and Presley, 1977; 39, Schuhmacher et al., 1992; 40, Bryan et al., 1983; 41, Walsh and O'Halloran, 1998; 42, Denton et al., 2006; 43, Ji et al., 2006; 44, Zorita et al., 2007; 45, Deudero et al., 2007; 46, Vlahogianni et al., 2007; 47, Karayakar et al., 2007; 48, Mubiana and Blust, 2006; 49, Li et al., 2006; 50, Peake et al., 2006; 51, Scancar et al., 2007; 52, Burger and Gochfeld, 2006; 53, Sasikumar et al., 2006; 54, Karouna-Renier et al., 2007; 55, Nesto et al., 2007; 56, LaBrecque et al., 2004; 57, Sankar et al., 2006; 58, Lafabrie et al., 2007; 59, Kumar et al., 2008; 60, Metian et al., 2008; 61, Maanan, 2008; 62, De Wolf an Rashid, 2008; 63, Anajjar et al., 2008; 64, Protasowicki et al., 2008; 65, Kucuksezgin et al., 2008; 66, Wepener et al., 2008; 67, Bustamante et al., 2008; 68, Fattorini et al., 2008; 69, Nicholson and Szefer, 2003; 70, Unlu et al., 2008.

Hexavalent chromium accumulation rates by oysters, *C. virginica*, were studied over a 20 week period (Shuster and Pringle, 1968). Oysters were continuously subjected to seawater solutions containing 0.05 or 0.1 mg Cr^{6+}/L. After 5, 10, or 20 weeks in 0.05 mg Cr/L, maximum soft parts concentrations in mg total Cr/kg FW were 2.4, 3.7, and 6.3, respectively, up from control values of <0.12. Oysters held in 0.1 mg Cr/L contained 4.4 mg/kg FW after 5 weeks, 6.4 after 10 weeks, and 11.5 after 20 weeks (Shuster and Pringle, 1968). Preston (1971) concluded that *C. virginica*, under laboratory conditions, accumulated chromium more readily by direct absorption from the medium than from diet, using radiochromium-labeled algae, *Chlamydomonas* spp. However, in nature, chromium is likely to be greater in the food supply than in the water, and food supply would be the main source of chromium to oysters even though accumulation occurs more readily by direct absorption (Preston, 1971).

Studies with ^{51}Cr and clams, *G. tumidum*, of 3.5 cm shell width show that soft parts of clams after 15 days took up 10 to 40 times more radiochromium than in seawater; concentration factors were higher for clams of 2.5 cm shell width and lower at 5.0 cm shell width (Hedouin et al., 2006). Clams, oysters, and mussels take up chromium from seawater or from chromium-contaminated sediments at comparatively low concentrations. For example, American oysters subjected to 0.005 mg Cr^{6+}/L for 12 weeks contained 3.1 mg Cr/kg DW in soft parts and retained 52% of the accumulated chromium after they were transferred to chromium-free seawater for 28 weeks (Zaroogian and Johnson, 1983). Common mussels subjected to the same dose-time regimen contained 4.8 mg/kg but retained only 39% after 28 weeks of depuration. Both oysters and mussels contained higher residues after exposure to 0.01 mg Cr/L for 12 weeks—5.6 and 9.4 mg Cr/kg DW in soft parts, respectively—and both contained substantial (30-58%) residues after 28 weeks in a chromium-free environment (Zaroogian and Johnson, 1983). Chromium residues in cockle, *Austrovenus stutchburyi*, soft parts collected over the period 1993-1994 ranged between 1.0 and 44.0 mg Cr/kg DW; however, residues were highest in August (Peake et al., 2006).

It is emphasized that trivalent chromium (Cr^{3+}), probably because of its very low solubility in seawater, seems to have a much lower bioavailability to molluscs and other groups of marine animals than hexavalent chromium, which is more water soluble (Carr et al., 1982). The clam, *R. cuneata*, appears to be an exception: it accumulated up to 19.0 mg Cr/kg in soft parts, on a DW basis, during exposure for 16 days to chromium-contaminated muds, and retained most of it for an extended period; the estimated biological half-time was 11 days (Carr et al., 1982). Studies with the green mussel, *P. viridis*, and Cr^{3+} show that 1.0 mg Cr^{3+}/L is not fatal in 96 h, but oxygen consumption, filtration rate, and ATPase activity were all significantly decreased (Vijayavel et al., 2007).

Proposed national chromium criteria in seawater for marine life protection include <0.018 mg Cr^{6+}/L as 24 h average, not to exceed 1.26 mg/L at any time (USEPA, 1980a). For California, the following are proposed: <0.002 mg/L total chromium, 6 month median; <0.008 mg/L total chromium, daily maximum; <0.02 mg/L instantaneous mix (Ecological Analysts, 1981); and <0.005 mg total chromium/L for 50% of measurements, and <0.01 mg/L for 10% of measurements (Reish, 1977). To protect human consumers of molluscan shellfish, tissues should not exceed 13.0 mg Cr/kg FW (Sankar et al., 2006), although some locales now propose concentrations as low as 0.3-0.4 mg Cr/kg DW molluscan tissues (Kucuksezgin et al., 2008).

6.12 Cobalt

Maximum cobalt values recorded in molluscan tissues were 7.2 mg Co/kg on a FW basis and 121.0 mg Co/kg on a DW basis (Table 6.8), although most values were substantially lower. Cobalt content of soft tissues of clams increases with increasing shell length (Ji et al., 2006). Accumulation of cobalt and its isotopes from the medium is reported in many species of bivalve molluscs (Ichikawa, 1961; Papadopoulu, 1973; Shimizu et al., 1970, 1971; Van As et al., 1973; Weiss and Shipman, 1957). Among five species of bivalves collected from Greek waters, for example, cobalt concentration factors in tissues were 1000 to 9300 that of ambient seawater (Papadopoulu, 1973). Enhanced cobalt uptake occurs in mussels at lower salinities (Mubiana and Blust, 2006).

Table 6.8: Cobalt Concentrations in Field Collections of Molluscs

Organism	Concentration	Reference[a]
Mussel, *Anodonta* sp.		
Shell	0.2 DW	1
Soft parts	0.4 DW	1
Gastropod, *Aplysia benedicti*		
Ctenidium	3.0-4.0 DW	2
Gonad	5.0-8.0 DW	2
Parapodia	3.0-4.0 DW	2
Hepatopancreas	19.0-22.0 DW	2
Buccal mass	10.0-14.0 DW	2
Intestine	21.0 DW	2
Gut content	22.0 DW	2
Shell	4.0-11.0 DW	2
Giant squid, *Architeuthis dux*; hepatopancreas; 2001-2005; Bay of Biscay vs. Mediterranean Sea	3.3 DW vs. 4.8 DW	33
Mussel, *Bathymodiolus azoricus*; 2002; mid-Atlantic Ridge; 3 sites		
Gill	0.18-0.91 DW	28
Mantle	0.19-0.3 DW	28
Digestive gland	0.5-2.1 DW	28
Byssus threads	0.48-9.6 DW	28
Whelk, *Buccinum undatum*		
Shell	0.2 DW	1
Soft parts	0.8 DW	1
Cockle, *Cerastoderma edule*; soft parts	1.3-2.9 DW	3
Scallop, *Chlamys opercularis*		
Soft parts	0.33 DW	4
Body fluid	0.2 DW	4
Mantle	0.2 DW	4
Gills	0.45 DW	4
Digestive gland	1.03 DW	4
Striated muscle	0.03 DW	4
Gonad and foot	0.2 DW	4
Kidneys	15.1 DW	4
Shell	0.3 DW	4
Mussel, *Choromytilus meridionalis*		
Soft parts	0.037 FW	5
Soft parts	2.0-3.0 DW	6

(*Continues*)

Table 6.8: Cont'd

Organism	Concentration	Reference[a]
Scallop, *Comptopallium radula*; Santa Marie Bay, New Caledonia; October 2004		
Soft parts	1.8 (1.0-2.8) DW	30
Kidney	179.0 DW	30
Pacific oyster, *Crassostrea gigas*		
Soft parts	0.1-0.2 FW	7
Soft parts	1.0 DW	6
American oyster, *Crassostrea virginica*; soft parts	0.06-0.2 FW	7
Limpet, *Crepidula fornicata*		
Shell	0.2 DW	1
Soft parts	17.0 DW	1
Clam, *Donax serra*; soft parts	0.04 FW	5
Clam, *Ensis* sp.; shell	0.03 DW	8
Abalone, *Haliotis midea*; soft parts	0.023 FW	5
Limpet, *Littorina littorea*; soft parts plus operculum	0.8-3.0 DW	11
Clam, *Macoma balthica*; soft parts	3.7-6.8 DW	11
Hardshell clam, *Mercenaria mercenaria*		
Soft parts	0.1-0.2 FW	7
Soft parts	4.3 DW	1
Shell	1.2 DW	1
Clam, *Meretrix lamarcki*; soft parts	0.63 DW	9
Clam, *Meretrix meretrix*; soft parts	0.09-0.52 FW	10
Mussel, *Modiolus modiolus*		
Shell	0.3 DW	1
Soft parts	0.5-5.5 DW	1
Mantle and gills	2.9 DW	1
Muscle	0.3 DW	1
Gonad	0.7 DW	1
Gut and digestive gland	1.1 DW	1
Softshell clam, *Mya arenaria*; soft parts	0.1-0.2 FW	7
Mussel, *Mytilus corscum*; soft parts	0.64 DW	9
Common mussel, *Mytilus edulis*		
Shell	0.029 DW	8
Soft parts	1.7 DW	8

(*Continues*)

Table 6.8: Cont'd

Organism	Concentration	Reference[a]
Soft parts	0.1-2.4 DW	12
Shell	0.2 DW	1
Soft parts	1.6 DW	1
Soft parts	0.02-1.1 DW	11
Soft parts	Max. 3.0 DW	25
Soft parts; Scheldt estuary; March 2000	2.8 DW	32
Soft parts; Gulf of Gemlik, Turkey; 2004	2.0 DW	35
Mussel, *Mytilus galloprovincialis*		
Soft parts	7.2 FW	13
Soft parts	2.8 DW	14
Soft parts; Adriatic Sea; 2001-2002; summer vs. winter		
Industrialized sites	1.2-1.3 DW vs. 1.1-2.1 DW	26
Reference site	0.3 DW vs. 1.3 DW	26
Soft parts; summers 2004-2005; Mediterranean Sea; 3 sites	0.06-1.5 DW	29
Mussel, *Mytilus viridis*; soft parts	1.0-8.7 DW	15
Whelk, *Nucella lapillus*		
Soft parts plus operculum	0.2-1.2 DW	11
Shell	0.3 DW	1
Soft parts	0.3 DW	1
Squid, *Ommastrephes illicebrosa*; whole	3.0 AW	16
Oyster, *Ostrea edulis*; soft parts	0.26-0.54 DW	17
Limpet, *Patella candei gomesii*; soft parts; 2007; Azores		
Near hydrothermal vent	0.9-1.2 DW	31
Reference site	0.4-0.7 DW	31
Limpet, *Patella vulgata*		
Shell	0.8 DW	1
Soft parts	0.4 DW	1
Soft parts	0.2-1.6 DW	11
Scallop, *Pecten maximus*		
Soft parts	0.25 DW	4
Body fluid	0.07 DW	4
Mantle	0.14 DW	4
Gills	0.23 DW	4
Digestive gland	1.28 DW	4

(Continues)

Table 6.8: Cont'd

Organism	Concentration	Reference[a]
Striated muscle	0.009 DW	4
Gonad and foot	0.1 DW	4
Kidneys	9.05 DW	4
Soft parts	8.5 DW	1
Muscle	0.3 DW	1
Gut and digestive gland	0.7 DW	1
Mantle and gills		
Unwashed	0.6 DW	1
Washed	0.4 DW	1
Gonad		
Unwashed	2.7 DW	1
Washed	0.2 DW	1
Shell	0.3 DW	1
Green mussel, *Perna viridis*; Hong King; juveniles vs. adults		
Reference site		
Soft parts	1.9 DW vs. 0.6 DW	34
Byssus	0.7 DW vs. 3.9 DW	34
Polluted site		
Soft parts	1.5 DW vs. 2.8 DW	34
Byssus	26.2 DW vs. 7.3 DW	34
Pen shell, *Pinna nobilis*		
Soft parts	6.8 DW	18
Mantle and gills	11.0 DW	18
Muscle	10.0 DW	18
Nervous system	10.0 DW	18
Stomach and intestines	6.4 DW	18
Gonads	6.0 DW	18
Hepatopancreas	61.0 DW	18
Clam, *Pitar morrhuana*; soft parts	0.6-6.1 DW	19
Scallop, *Placopecten magellanicus*; soft parts	0.6 DW	20
Pteropods, 9 spp.; tests	0.064-1.01 DW	21
Manila clam, *Ruditapes philippinarum*; soft parts; Korea; 2002-2003		
Females vs. males	1.06 DW v. 1.05 DW	24
Postspawn vs. prespawn	3.7 DW vs. 1.1 DW	24
Clam, *Scrobicularia plana*		
Gannel estuary, England		
Soft parts	19.0-66.0 DW	22

(Continues)

Table 6.8: Cont'd

Organism	Concentration	Reference[a]
Digestive gland	156.0 DW	22
Mantle and siphons	6.3 DW	22
Foot and gonad	1.4 DW	22
Gills and palps	2.3 DW	22
Adductor muscle	0.5 DW	22
Kidney	1.1 DW	22
Camel estuary, England		
Soft parts	4.3-5.3 DW	22
Digestive gland	15.9 DW	22
Mantle and siphons	1.2 DW	22
Foot and gonad	0.8 DW	22
Gills and palps	not detectable	22
Adductor muscle	0.3 DW	22
Kidney	4.3 DW	22
Transferred from Camel estuary to Gannel estuary for 352 days		
Digestive gland	39.0 DW	22
Remaining tissues	1.9 DW	22
Transferred from Gannel estuary to Camel estuary for 352 days		
Digestive gland	66.0 DW	22
Remaining tissues	2.7 DW	22
Tamar estuary, England		
Soft parts	3.0-37.2 DW	23
Digestive gland	27.0-121.0 DW	23
Looe estuary, England; soft parts	2.4-17.7 DW	11
Squid, *Sepia esculata*; trunk	0.06 DW	9
Common cuttlefish, *Sepia officinalis*		
Egg shell	1.2 DW	27
Embryo	0.25 DW	27
Digestive gland		
Juveniles, age 1 month	2.6 DW	27
Immatures, age 12 months	6.8 DW	27
Adults, age 18 months	10.0 DW	27

Values are in mg Co/kg fresh weight (FW), dry weight (DW), or ash weight (AW).
[a]1, Segar et al., 1971; 2, Patel et al., 1973; 3, Bryan and Hummerstone, 1973; 4, Bryan, 1973; 5, Van As et al., 1973; 6, Watling and Watling, 1976a; 7, Pringle et al., 1968; 8, Bertine and Goldberg, 1972; 9, Ishii et al., 1978; 10, Bhatt et al., 1968; 11, Bryan and Hummerstone, 1977; 12, Karbe et al., 1977; 13, Uysal, 1978a; 14, Fowler and Oregioni, 1976; 15, Bhosle and Matondkar, 1978; 16, Nicholls et al., 1959; 17, Fukai et al., 1978; 18, Papadopoulu, 1973; 19, Eisler et al., 1978; 20, Reynolds, 1979; 21, Turekian et al., 1973; 22, Bryan and Hummerstone, 1978; 23, Bryan and Uysal, 1978; 24, Ji et al., 2006; 25, Mubiana and Blust, 2006; 26, Scancar et al., 2007; 27, Miramand et al., 2006; 28, Kadar et al., 2006; 29, Lafabrie et al., 2007; 30, Metian et al., 2008; 31, Cunha et al., 2008; 32, Wepener et al., 2008; 33, Bustamante et al., 2008; 34, Nicholson and Szefer, 2003; 35, Unlu et al., 2008.

Radiocobalt-60 was not detectable in mussels and chitons collected in summer 2004 from Amchitka Island in the Aleutian chain, the site of nuclear tests between 1965 and 1971 (Burger et al., 2007b). Laboratory studies with ^{57}Co and clams, *G. tumidum*, of various shell widths show that soft parts from 3.5 cm shell width clams contained 40-100 times more radiocobalt than seawater after 15 days; concentration factors were higher for clams of 2.5 cm shell width and lowest for those of 5.0 cm shell width (Hedouin et al., 2006). Scallops, *C. radula*, held in seawater containing ^{57}Co for four days had a BCF of 15 in total soft parts and 603 in kidney (Metian et al., 2008). Gravid cuttlefish, *S. officinalis*, fed crabs radiolabeled with ^{60}Co over a 2 week period to determine rate of maternal transfer into eggs showed no significant transfer (Lacoue-Labarthe et al., 2008b).

Results of laboratory studies in which American oysters were immersed for 20 weeks in 0.05 or 0.1 mg Co/L showed high tissue concentrations of 6.0-11.0 mg Co/kg FW soft parts (Shuster and Pringle, 1968); these values were significantly higher than those recorded in field populations. Cobalt most frequently accumulated in molluscan kidney or digestive gland tissues (Amiard-Triquet and Amiard, 1976; Bryan, 1973; Patel et al., 1973; Pentreath, 1973; Young and Folsom, 1967, 1973). In oceanic squid, *Stenoteuthis bartrami*, for example, 90 percent of all cobalt in the animal was found in digestive gland, which contributed less than 5 percent of the total body weight (Folsom and Young, 1965). Uptake of cobalt in molluscs is influenced by diet (Amiard-Triquet and Amiard, 1976; Pentreath, 1973), by the physicochemical form of cobalt (Lowman and Ting, 1973; Pentreath, 1973), by season of year (Bryan, 1973; Patel et al., 1973), by ambient sediment lithology in the case of burrowing bivalves (Luoma and Jenne, 1977), and by the number and types of protein binding materials (Shimizu et al., 1971). In general, molluscs accumulate cobalt mainly from the diet, with only minor accumulation from water; soluble forms of cobalt were preferred to particulates; and bivalves living on calcareous substrates accumulated cobalt more readily than those living on substrates of iron oxide, manganese oxide, or detrital organics (Eisler, 1981).

6.13 Copper

Molluscs tend to accumulate copper from seawater. For example, bivalve tissues from Greece had 10,000-15,000 times more copper than the ambient medium (Papadopoulu, 1973). In general, molluscs contain more copper in soft tissues than other marine groups examined, including plants, other invertebrate groups, and vertebrates (Eisler, 1979, 1981, 2000b). Among the molluscs, highest accumulations were generally observed in cephalopods and in ostreid and crassostreid oysters; blood, digestive gland, and kidney appeared to contain the highest copper concentrations (Table 6.9). Several species of molluscs other than cephalopods or oysters also had elevated copper burdens; in many cases, the high copper concentrations were associated with proximity to anthropogenic point sources of copper, and secondarily to various biological and abiotic variables that modify copper uptake and retention. Recognition of the importance of these variables and their interactions is essential to an understanding of copper kinetics in marine systems. Selected examples follow.

Table 6.9: Copper Concentrations in Field Collections of Molluscs

Organism	Concentration	Reference[a]
Abalones		
Muscle	1.8-3.4 FW	1
Viscera	4.2-17.2 FW	1
Gastropod, *Acmaea digitalis*		
Shell	7.5 DW	2
Soft parts	12.0 DW	2
Antarctic scallop, *Adamussium colbecki* Ross Sea; 1987-1988 vs. 1990		
Digestive gland	12.6 DW vs. 3.5 FW	106, 107
Gills	6.5 DW vs. 1.4 FW	106, 107
Gonads	2.7 DW vs. Not detectable (ND)	106
Kidneys	4.0 DW vs. ND	106
Mantle	3.5 DW vs. ND	106
Muscle	1.6 DW vs. ND	106
Blood clam, *Anadara granosa*; soft parts; Malaysia	0.7-0.8 FW; 6.3 4.5-8.0 DW	108, 109
Mussel, *Anodonta* sp.		
Shell	7.6 DW	4
Soft parts	3.0 DW	4
Gastropod, *Aplysia benedicti*		
Ctenidium	160.0-232.0 DW	5
Gonad	18.0-25.0 DW	5
Parapodia	12.0 DW	5
Hepatopancreas	7.0-9.0 DW	5
Buccal mass	11.0 DW	5
Intestine	23.0 DW	5
Gut content	16.0 DW	5
Shell	19.0-20.0 DW	5
Giant squid, *Architeuthis dux*; 2001-2005; hepatopancreas		
Bay of Biscay	108.0 DW	185
Mediterranean Sea	1218.0 DW	185
Ocean quahog, *Arctica islandica* Soft parts		
March	6.6-11.3 DW	6
August	4.5-13.2 DW	6
February	4.4-7.5 DW	6
June	4.5-9.3 DW	6

(Continues)

Table 6.9: Cont'd

Organism	Concentration	Reference[a]
Block Island Sound	10.0 DW	110
Chesapeake Bay	5.4 DW	110
Georges Bank	3.5-10.3 DW	110
New York Bight	11.3 DW	110
Western Baltic Sea; 1992-1993		
Adductor muscle	1.8-2.2 DW	110
Digestive gland	13.5 DW	110
Foot	3.1 DW	110
Gills	6.7 DW	110
Kidney	40.1 DW	110
Mantle	5.0 DW	110
Soft parts	14.3-15.3 DW	110
Scallop, *Argopecten gibbus*; kidney concretion	490.0 DW	7
Scallop, *Argopecten irradians*; kidney concretion	8.9 DW	7
Cockle, *Austrovenus stutchburyi*; soft parts; 1993-1994; Otago, New Zealand	3.0-60.0 DW	147
Hydrothermal vent mussel, *Bathymodiolus azoricus*; mid-Atlantic ridge; 2001		
5 sites		
Gills	52.0-130.0 DW	176
Mantle	1.0-81.0 DW	176
Digestive gland	11.0-172.0 DW	176
Various depths		
840 m deep		
Digestive gland	22.0 DW	169
Gills	47.0 DW	169
Mantle	10.0 DW	169
1700 m deep		
Digestive gland	5.2 DW	169
Gills	70.0 DW	169
Mantle	13.0 DW	169
2300 m deep		
Digestive gland	18.0 DW	169
Gills	67.0 DW	169
Mantle	4.0 DW	169
Shell	0.6-1.7 (0.1-2.7) DW	175
Summer 2002; start vs. 30 days in metals-free seawater		

(*Continues*)

Table 6.9: Cont'd

Organism	Concentration	Reference[a]
Gill	37.0 FW vs. 20.0 FW	149
Mantle	30.0 FW vs. 15.0 FW	149
Digestive gland	80.0 FW vs. 30.0 FW	149
Byssus threads	500.0 FW vs. 10.0 FW	149
Bivalves; 5 spp.; soft parts; East China Sea; August 2002	3.1 (1.3-9.1) FW	144
Mussel, *Brachydontes pharaonis*; Turkey; 2002-2003; max. concentrations		
Muscle	44.0 DW	138
Gill	245.0 DW	138
Hepatopancreas	62.0 DW	138
Whelk, *Buccinum undatum*		
Shell	9.1 DW	4
Soft parts	180.0 DW	4
Soft parts; Irish Sea vs. Scotland	180.0 DW vs. 78.0 DW	111
Channeled whelk, *Busycon canaliculatum*		
Soft parts	76.0 (58.0-116.0) FW	9
Digestive gland		
Summer months	400.0-900.0 FW	9
Winter months	62.0-120.0 FW	9
Blood		
Summer months	316.0 FW	9
Winter months	19.0 FW	9
Kidney	171.0 FW	9
Foot	20.0 FW	9
Osphradium	40.0 FW	9
Gill		
Summer months	70.0-80.0 FW	9
Winter months	15.0-30.0 FW	9
Gut	248.0 FW	10
Liver	146.0 FW	10
Kidney	76.0 FW	10
Pancreas	76.0 FW	10
Buccal mass	44.0 FW	10
Gill	40.0 FW	10
Gonad	21.0 FW	10
Heart	16.0 FW	10
Foot	8.0 FW	10
Blood	35.0-83.0 FW	10
Hemocyanin	67.0-137.0 FW	10

(Continues)

Table 6.9: Cont'd

Organism	Concentration	Reference[a]
Eggs	0.0023 mg per capsule FW	11
Muscle	11.9-21.0 FW	12
Digestive gland		
Virginia	32.4 FW	12
Long Island Sound, New York	1080.0-1135.0 FW	12
Cockle, *Cardium edule*		
Soft parts	5.0 FW	13
Shell	3.0 DW	4
Soft parts	11.0 DW	4
Cockle, *Cerastoderma edule*		
Soft parts	4.0-9.0 DW	14
Soft parts	9.8 (5.2-27.2) DW	15
Scallop, *Chlamys opercularis*		
Soft parts	15.4 DW	16
Body fluid	5.6 DW	16
Mantle	5.6 DW	16
Gills	16.7 DW	16
Digestive gland	36.7 DW	16
Striated muscle	1.3 DW	16
Unstriated muscle	3.1 DW	16
Gonad and foot	9.7 DW	16
Kidney	1285.0 DW	16
Shell	0.38-0.7 DW	4
Mussel, *Choromytilus meridionalis*; soft parts	7.0-14.0 DW	17
Scallop, *Comptopallium radula*; New Caledonia; October 2004		
Soft parts	3.7 (2.9-5.1) DW	171
Digestive gland	15.0 DW	171
Oyster, *Crassostrea angulata*; soft parts; Spain		
Sanlucar	264.7 FW	18
Sancti-Petri	28.0 FW	18
Barbate	35.0 FW	18
Rio Piedras	200.3 FW	18
Ayamonte	120.1 FW	18
Oyster, *Crassostrea commercialis*		
Soft parts	18.7 FW	19
Soft parts	3.0-48.0 FW	20

(Continues)

Table 6.9: Cont'd

Organism	Concentration	Reference[a]
Oyster, *Crassostrea cucullata*		
Soft parts	251.0-480.0 DW	21
Soft parts	20.6 FW	22
Pacific oyster, *Crassostrea gigas*		
Soft parts	9.4-192.2 FW	19
Soft parts	200.0-853.0 DW	14
Gill	513.0-1655.0 DW	14
Soft parts	6.6-37.9 FW	23
Soft parts	19.0-482.0 FW	24
Mantle	47.0-900.0 DW	24
Digestive diverticula	18.0-61.0 DW	24
Soft parts	110.0-643.0 DW	25
Soft parts	Up to 6480.0 DW in power station cooling water	26
Soft parts	19.0 (1.0-55.0) FW	27
Soft parts	33.0 DW	17
Soft parts	32.0 DW	28
Shell	1.6-2.9 DW	111
Soft parts		
England	340.0-6480.0 DW	111
South Africa	33.0 DW	111
Tasmania	9.4-84.4 DW	111
United States	7.8-38.0 FW	111
China; November 2003; impacted by industrial effluents; 3 sites	389.0-1411.0 DW	139
Hong Kong; 1989; various locations		
Gill	840.0 DW	112
Mantle	509.0 DW	112
Muscle	750.0 DW	112
Soft parts	344.0-422.0 DW; max. 1071.0 DW	112
Visceral mass	383.0 DW	112
Arcachon Bay, France; soft parts		
1979-1982	48.3-63.8 DW	113
1983-1987	67.7-116.2 DW	113
1988-1991	101.8-135.0 DW	113
Taiwan; soft parts		
1989; from discharge of copper recycling facility containing 0.005-0.024 mg Cu/L	4401.0 DW; green in color	114
1995-1996	909.0 (113.0-2805.0) DW	116
From Cu-contaminated environment	2225.0 DW	115

(*Continues*)

Table 6.9: Cont'd

Organism	Concentration	Reference[a]
As above; after 6 days in clean seawater	746.0 DW	115
As above; after 32 days in clean seawater	344.0 DW	115
Oyster, *Crassostrea gryphoides*; soft parts	175.0-210.0 DW	21
American oyster, *Crassostrea virginica* Soft parts		
Alabama	20.0 (4.0-78.0) FW	111
Atlantic Ocean; northwestern areas	46.0 (11.0-100.0) DW	111
Chesapeake Bay	5.0-240.0 FW	111
East coast, United States	91.5 (7.0-517.0) FW	29
East coast, United States	21.0-220.0 DW	39
Georgia	48.0-261.0 DW	111
Savannah, Georgia; 2000-2001; 9 sites	67.0-121.0 DW	170, 177
Gulf coast, United States	23.0-410.0 DW	39
Gulf coast, United States	16.0 (6.0-27.0) FW	111
Pensacola, Florida; 2003-2004	Max. 56.0 FW	165
West coast, United States	7.8-37.5 FW	29
Maryland	Max. 1120.0 DW	111
North Carolina; Newport River estuary	2.2-6.2 FW	36
Rhode Island	121.0 (92.0-140.0) FW	111
Texas	161.0 DW	111
Virginia, Rappahannock River estuary	3.0-29.0 FW	36
Virginia; 1972-1973; various salinities		
7.5 ppt	29.0 FW	36
9.5 ppt	16.0 FW	36
12 ppt	12.0 FW	36
13.5 ppt	3.0 FW	36
Purdy, Washington State, United States		
March-June 1965	20.2 FW	29
July-December 1965	20.3 FW	29
December-June 1966	14.2 FW	29
July-October 1966	19.2 FW	29
October-March 1967	12.5 FW	29
May-July 1967	14.3 FW	29

(Continues)

Table 6.9: Cont'd

Organism	Concentration	Reference[a]
Manchester, Washington State		
March-June 1965	24.0 FW	29
July-December 1965	22.4 FW	29
December-June 1966	20.6 FW	29
July-October 1966	25.8 FW	29
October-March 1967	17.4 FW	29
May-July 1967	18.3 FW	29
Shell	42.7-69.8 FW	30
Soft parts	290.0-913.0 DW	24
Mantle	968.0 DW	24
Soft parts	Max. 200.0 FW	31
Florida; soft parts		
From a canal lined with chromated-copper-arsenate wood vs. reference site	150.0-200.0 FW vs. 10.0 FW; elevated concentrations associated with greenish color and higher frequency of histopathology of digestive gland diverticula	118
Reference site oysters transplanted into above canal		
After 3 months	130.0 FW	118
After 4 months	220.0 FW; no increase in frequency of digestive gland lesions	118
North Carolina; soft parts		
Marina sites	36.7 FW	119
Open water sites	7.1 FW	119
At 12 ppt salinity	6.0 FW	36
At 33 ppt salinity	2.0 FW	36
Soft parts	55.0-261.0 DW	32
Soft parts	Mean 180.0 FW; Max. 584.0 FW	33
Soft parts		
Summer	250.0 DW	34
Autumn	50.0 DW	34
Shell	0.2 DW	34
Soft parts; metals-contaminated site vs. reference site	450.0 DW vs. 6.0 DW	35
Soft parts	1.0-702.0 FW	27, 37, 38
Soft parts	161.0 DW	40
Oyster, *Crassostrea rhizophorae*; soft parts; near Natal Brazil; summer 2001; contaminated estuary vs. reference site	18.0-39.0 DW vs. 16.9-28.2 DW	161

(*Continues*)

Table 6.9: Cont'd

Organism	Concentration	Reference[a]
Limpet, *Crepidula fornicata*		
Shell	1.8 DW	4
Soft parts	270.0 DW	4
Octopus, *Eledone cirrhosa*; English Channel; October, 1987		
Branchial hearts	335.0 DW	120
Digestive gland	448.0-463.0 DW	120
Genital tract	60.0-66.0 DW	120
Gill	268.0 DW	120
Kidney	594.0 DW	120
Mantle	102.0 DW	120
Muscle	17.0 DW	120
Whole	122.0 DW	120
Mussel, *Elliptio buckleyi*; soft parts; Marmara Sea, Turkey; 2004-2005	1.2-1.7 (0.51-2.99) FW	140
Red abalone, *Haliotis rufescens*		
Gills	18.0-123.0 DW	41
Mantle	8.0-20.0 DW	41
Digestive gland	11.0-78.0 DW	41
Foot	1.4-12.8 DW	41
Abalone, *Haliotis* sp.; soft parts	8.0 FW	13
Abalone, *Haliotis tuberculata*; United Kingdom locations		
Cobo Bay		
Soft parts	28.0 DW	42
Viscera	68.0 DW	42
Foot	15.4 DW	42
Digestive gland	20.7 DW	42
Fermain Bay		
Soft parts	39.0 DW	42
Viscera	77.0 DW	42
Foot	22.4 DW	42
Digestive gland	17.6 DW	42
All stations		
Blood	560.0 DW	42
Muscle	12.1 DW	42
Gills	103.0 DW	42
Left kidney	374.0 DW	42
Right kidney	24.2 DW	42

(*Continues*)

Table 6.9: Cont'd

Organism	Concentration	Reference[a]
Scallop, *Hinnites multirugosus*; power plant outfall vs. reference site		
Digestive gland	190.0 FW vs. 64.0 FW	43
Gonad	3.2 FW vs. 2.2 FW	43
Adductor muscle	0.4 FW vs. 0.2 FW	43
Mud snail, *Ilyanassa obsoleta*; soft parts; North Carolina		
Marina sites	402.2 FW	119
Open water sites	219.5 FW	119
Periwinkle, *Littoraria scabra*; soft parts; Tanzania		
Industrialized sites; 1998 vs. 2005	9.0 DW vs. 62.0 DW	173
Reference sites; 2005	8.0 DW	173
Limpet, *Littorina littorea*		
Soft parts	6.2-20.0 FW	44
Whole	23.8-36.1 FW	45
Soft parts plus operculum	124.0 (62.0-194.0) DW	15
Soft parts	42.7-248.8 DW	46
Digestive gland and gonad	91.4-92.5 DW	46
Shell	2.1-3.5 DW	46
Squid, *Loligo opalescens*; liver; California		
Central coast	5350.0 DW	47
Southern coast	8370.0 DW	47
Baltic clam, *Macoma balthica*		
Shell; Sweden		
West coast	41.0 FW	48
Baltic coast	117.0 FW	48
Soft parts	300.0 (96.0-615.0) DW	15
Soft parts	5.0-12.0 FW	49
Soft parts; 1996-1997; Gulf of Gdansk; age 3-4 years	25.0-150.0 DW	164
Soft parts; the Netherlands; 1990-1992; acid-soluble fraction vs. total copper	4.1 (2.7-6.7) DW vs. 13.8-22.6 DW	121
Hardshell clam, *Mercenaria mercenaria*		
Soft parts		
April		
3+ age group	42.0 DW	50
4+ age group	40.0 DW	50
5+ age group	35.0 DW	50

(Continues)

Table 6.9: Cont'd

Organism	Concentration	Reference[a]
August		
3+ age group	23.0 DW	50
4+ age group	28.0 DW	50
December; 5+ age group	27.0 DW	50
Shell	1.7 DW	4
Soft parts	25.0 DW	4
Soft parts	2.6 FW	29
Soft parts	1.0-16.5 FW	37
Soft parts	1.1-7.9 FW	51
Clam, *Meretrix casta*; soft parts	16.1 DW	21
Clam, *Meretrix lamarckii*; soft parts	15.0 DW	52
Mexico; Sinaloa; 2004-2005; soft parts; rainy season vs. dry season		
Oyster, *Crassostrea corteziensis*	73.2 DW vs. 70.3 DW	178
Mussel, *Mytella strigata*	27.8 DW vs. 20.5 DW	178
Clam, *Megapitaria squalida*	11.1 DW vs. 6.3 DW	178
Mussel, *Modiolus modiolus*		
Shell	1.9 DW	4
Soft parts	10.0-44.0 DW	4
Mantle and gills	42.0 DW	4
Muscle	14.0 DW	4
Gonads	19.0 DW	4
Gut	45.0 DW	4
Molluscs		
Soft parts		
West Norway coast	Max. 190.0 DW	53
Rio Tinto estuary, Spain	Max. 26.0 DW	54
5 spp; Chile, Strait of Magellan; 2001; 17 stations	0.7-7.5 FW	163
3 spp.; India; 2003	4.4-6.5 FW	168
Shell	1.6-7.9 DW	55
Soft parts	37.0-102.0 DW	55
Soft parts, 12 spp.	1.0-40.0 FW	57
Edible tissues; commercial species; 18 spp.		
3 spp.	0.1-0.4 FW	56
5 spp.	0.6-0.8 FW	56
3 spp.	1.0-3.0 FW	56
4 spp.	3.0-8.0 FW	56
3 spp.	10.0-40.0 FW	56

(*Continues*)

Table 6.9: Cont'd

Organism	Concentration	Reference[a]
Morocco; 2001; clams; soft parts		
Scrobicularia plana	50.0 DW	180
Cerastoderma edule	10.0 DW	180
Morocco; 2004-2005; soft parts		
Mussel, *Mytilus galloprovincialis*	26.8 (4.4-142.2) DW	172
Clam, *Venerupis decussatus*	11.1 (6.2-18.3) DW	172
Pacific oyster, *Crassostrea gigas*	25.9 (3.6-42.1) DW	172
Mussels; various species; soft parts; Wales; 1989; coastal area	1.2 FW	117
Softshell clam, *Mya arenaria*		
Soft parts	5.8 (1.2-90.0) FW	29, 37
Soft parts	9.8-14.6 DW	58
Lagoon mussel, *Mytella strigiata*; soft parts; Baja California; Mexico; 1989-1991	Max. 3.9 DW	122
Mussel, *Mytilus californianus*		
Shell	8.1-18.6 DW	2
Soft parts	9.0-30.3 DW	2
Digestive gland	14.0-69.0 DW	59
Soft parts	4.6-11.0 DW	39
Common mussel, *Mytilus edulis*		
Soft parts	3.7-65.4 DW	14
Whole	8.3-11.5 DW	60
Gill	6.9-10.8 DW	60
Foot	4.2-7.2 DW	60
Mantle	5.4-5.8 DW	60
Kidney	7.4-8.7 DW	60
Digestive gland	6.8-7.0 DW	60
Muscle	2.7-3.7 DW	60
Soft parts	5.0 FW	13
Shell	2.0 DW	4
Soft parts	9.6 DW	4
Shell	Max. 8.6 DW	2
Shell; Baltic Sea; 2005; 12 sites	2.2-4.2 DW	182
Soft parts	5.0-11.2 DW	2
Whole	1.9-2.3 FW	45
Soft parts	3.1 FW	61
Soft parts	0.6-0.9 FW	62

(Continues)

Table 6.9: Cont'd

Organism	Concentration	Reference[a]
Soft parts	Max. 1.4 FW	63
Soft parts	0.5-5.0 FW	44
Soft parts	9.0-17.0 DW	64
Digestive gland	Max. 77.0 FW	65
Gonad	Max. 23.0 FW	65
Muscle	Max. 53.0 FW	65
Hepatopancreas		
San Diego Harbor	39.0-73.0 DW	66
Offshore	17.0-22.0 DW	66
Protein fractions	0.1-0.2 FW	67
Soft parts	Max. 28.0 DW	141
Soft parts	5.0-88.0 DW	68
Shell	4.0-8.0 DW	68
Shell		
California	Max. 8.6 DW	111
England	9.6 DW	111
Japan	1.2-2.8 DW	111
New Zealand	3.0 DW	111
Soft parts		
United States West coast sites	3.5-8.6 DW	39
United States East coast sites	6.7-13.2 DW	39
Looe estuary, England	9.5 (3.9-13.6) DW	15
California	5.0-11.2 DW	111
Canada; Halifax	13.7-154.3 DW	123
England	7.0-11.0 DW	111
Long Island Sound, New York		
1983	1.0-2.3 FW	124
1986-1987	15.0 DW	125
Norway	3.0-130.0 DW	111
Portugal and Spain	6.5-14.0 DW	111
Mussel, *Mytilus edulis aoteanus*		
Soft parts	9.0 (5.0-11.0) DW	69
Mantle	11.0 DW	69
Gills	20.0 DW	69
Muscle	1.0 DW	69
Visceral mass	19.0 DW	69
Intestine	15.0 DW	69
Foot	13.0 DW	69
Gonads	6.0 DW	69
Shell	3.0 DW	69

(Continues)

Table 6.9: Cont'd

Organism	Concentration	Reference[a]
Mussel, *Mytilus edulis planulatus*		
Soft parts		
Dry weight 0.09 g	7.5 DW	3
Dry weight 0.39 g	6.9 DW	3
Dry weight 0.48 g	6.2 DW	3
Dry weight 0.69 g	6.0 DW	3
Soft parts		
June	19.6 FW	70
January	0.8 FW	70
Mussel, *Mytilus galloprovincialis*		
Protein extracts; Bulgarian Black Sea coast; polluted vs. reference site	23.2 DW vs. 12.9 DW	150
Soft parts; purged for 48 h in synthetic seawater vs. not purged	6.9 DW vs. 13.1 DW	126
Soft parts	18.0 DW	71
Soft parts	7.3 FW	72
Soft parts; May 2003; NW Mediterranean Sea		
Near Genova, Italy	5.7-40.0 DW	134
Near Marseille, France	5.5-8.5 DW	134
Near Barcelona, Spain	5.7-74.8 DW	134
Soft parts; 1991-2005; western Mediterranean Sea; Balearic Islands; Port of Mahon	21.2 (5.9-58.3) DW	135
Soft parts; Marmara Sea, Turkey; 2005	1.2 FW	137
Soft parts; Izmir Bay, Turkey; 2004; near shore vs. mid-bay	5.3 DW vs. 3.7 DW	183
Soft parts; Gulf of Gemlik, Turkey; 2004	5.5 DW	189
Soft parts; Adriatic Sea; 2001-2002; summer vs. winter		
Industrialized sites	6.5-11.7 DW vs. 8.3-49.0 DW	148
Reference site	7.6 DW vs. 9.7 DW	148
Soft parts; Adriatic Sea, Italy; 2 sites	0.74-0.82 FW	181
Soft parts; Adriatic Sea; 2001-2005; winter vs. summer	5.1-8.9 DW vs. 1.1-4.5 DW	186
Soft parts; Sardinia; 2004 vs. 2006	17.0 DW vs. 14.0 DW	184

(Continues)

Table 6.9: Cont'd

Organism	Concentration	Reference[a]
Soft parts; Aegean Sea, Turkey; 2002-2003	1.5 (0.95-1.8) FW	142
Mantle vs. gills; 2004-2006; Saronikos Gulf of Greece; max. mean concentrations; contaminated sites		
Summer	1.4 FW vs. 1.5 FW	136
Autumn	1.3 FW vs. 2.2 FW	136
Winter	0.0 FW vs. 2.2 FW	136
Spring	1.1 FW vs. 2.2 FW	136
Soft parts; Venice lagoon, Italy; 2005-2006; San Giuliano vs. Sacca Sessola		
April	9.3 DW vs. 10.8 DW	166
July	3.6 DW vs. 4.4 DW	166
October	4.7 DW vs. 6.3 DW	166
February	9.0 DW vs. 8.7 DW	166
Mussel, *Mytilus smarangdium*; soft parts; copper-contaminated environment vs. 6 days in clean seawater	20.2 DW vs. 1.8 DW	115
Mussels, *Mytilus* spp.; soft parts; United States; 1970s vs. 1980s		
Bodega Bay, California	6.9 DW vs. 7.7 DW	127
Narragansett Bay, Rhode Island	11.0 DW vs. 14.0 DW	127
Mussel, *Mytilus viridis*		
Soft parts	2.3 FW	73
Soft parts	3.0-16.0 DW	74
Soft parts	8.6-26.0 DW	21
Abalone, *Notohaliotis discus*; soft parts less viscera	17.0 DW	52
Whelk, *Nucella lapillus*		
Soft parts		
Reference site	70.0 DW	75
After transfer for 5 months to Cu-contaminated area	170.0 DW	75
Shell	0.29 DW	4
Soft parts	150.0 DW	4
Soft parts plus operculum	110.0 (51.0-141.0) DW	15

(Continues)

Table 6.9: Cont'd

Organism	Concentration	Reference[a]
Octopus, *Octopus* sp.		
Viscera	300.0 FW	1
Whole	7.0 FW	61
Whole	39.0 FW	76
Octopus, *Octopus cyanea*; Guam; June 1998		
Tentacle	12.1 DW	132
Liver	5680.0 DW	132
Octopus, *Octopus vulgaris*		
Hepatopancreas	2550.0 DW	77
Hepatopancreas	4880.0 DW	78
Branchial heart	93.0 DW	77
Gill	111.0 DW	77
Central heart	43.0 DW	77
Kidney	48.0 DW	77
Muscle	28.0 DW	77
Hemolymph	2450.0 DW	77
Portugal; 2005-2006; digestive gland; northwest region vs. south coast		
Whole	860.0 (64.0-1490.0) DW vs. 1390.0 (1120.0-1600.0) DW	179
Nuclei	1100.0 (90.0-1210.0) DW vs. 1160.0 (310.0-1670.0) DW	179
Mitochondria	740.0 (520.0-3260.0) DW vs. 890.0 (410.0-1280.0) DW	179
Lysosomes	620.0 (220.0-13,020.0) DW vs. 910.0 (440.0-2000.0) DW	179
Microsomes	950.0 (600.0-2890.0) DW vs. 1310.0 (720.0-1,570.0) DW	179
Portugal; 1999-2002; 3 sites		
Digestive gland	1411.0-2515.0 DW	187
Branchial heart	155.0-221.0 DW	187
Gill	100.0-151.0 DW	187
Muscle	26.0-36.0 DW	187
Squid, *Ommastrephes bartrami*; liver	195.0 DW	47
Squid, *Ommastrephes illicebrosa*; whole	2700.0 AW	79
Oyster, *Ostrea angasi*		
Whole	39.0-104.0 DW	3
Soft parts	4.0-80.5 FW	19

(*Continues*)

Table 6.9: Cont'd

Organism	Concentration	Reference[a]
Soft parts	57.9 FW	61
Soft parts		
June	20.6 FW	70
January	7.5 FW	70
Oyster, *Ostrea circumpicta*; soft parts; Japan; Nobeoka Bay		
South coast	356.0-686.8 FW	80
North coast	93.0 FW	80
European flat oyster, *Ostrea edulis*		
Soft parts; Balearic Islands; western Mediterranean Sea; 1991-2005; Port of Mahon	451.7 (0.8-1444.0) DW	135
Soft parts	86.0-597.0 DW	14
Gill	380.0-1420.0 DW	14
Soft parts	6.0 FW	13
Soft parts	38.0 DW	28
Soft parts	245.0-405.0 DW	82
Wales, United Kingdom vs. Cornwall, England		
Gills	253.0 FW vs. 2395.0 FW	81
Mantle	207.0 FW vs. 2175.0 FW	81
Kidney	124.0 FW vs. 500.0 FW	81
Viscera	77.0 FW vs. 1184.0 FW	81
Muscle	64.0 FW vs. 315.0 FW	81
Hemolymph		
Whole	2.0 FW vs. 21.0 FW	81
Cell-free	0.6 FW vs. 8.0 FW	81
Oyster, *Ostrea equestris*; soft parts; United States		
East coast	660.0 DW	39
Gulf coast	21.0-220.0 DW	39
Oyster, *Ostrea gigas*		
Soft parts; Japan; Nobeoka Bay		
South coast	320.2-434.6 FW	80
North coast	60.9-86.9 FW	80
Soft parts		
Reference site; seawater Cu levels 0.0005-0.0023 mg/L	25.0 FW	83

(Continues)

Table 6.9: Cont'd

Organism	Concentration	Reference[a]
Transplanted for 120 days to seawater Cu levels of 0.001-0.0064 mg/L		
Muscle	65.0 FW	84
Mantle	205.0 FW	84
Gill	255.0 FW	84
Labial palps	1237.0 FW	84
Transplants after depletion for 116 days in control waters		
Muscle	5.0 FW	85
Mantle	14.0 FW	85
Gill	15.0 FW	85
Labial palps	18.0 FW	85
Oyster, *Ostrea sinuata*		
Soft parts	41.0 (21.0-53.0) DW	69
Mantle	34.0 DW	69
Gills	100.0 DW	69
Muscle	23.0 DW	69
Striated muscle	6.0 DW	69
Visceral mass	49.0 DW	69
Kidney	22.0 DW	69
Heart	59.0 DW	69
Shell	<1.0 DW	69
Oyster, *Ostrea spinosa*; soft parts	40.0 FW	80
Oysters		
Soft parts	Max. 123.0 FW	86
Soft parts; from area of metal pollution vs. reference site	Max. 450.0 FW vs. Max. 100.0 FW	86
Clam, *Paphia undulata*; soft parts; Malaysia; 1993	0.9-1.1 FW	108
Limpet, *Patella caerulea*: soft parts	2.7-9.7 FW; 11.3-38.0 DW	8
Limpet, *Patella candei gomesii*; soft parts; 2007; Azores		
Near hydrothermal vent	6.4-7.2 DW	174
Reference site	5.2-5.7 DW	174
Gastropod, *Patella piperata*; Canary Islands, Spain; soft parts; March-April 2003	2.1 (0.6-5.0) DW	143

(*Continues*)

Table 6.9: Cont'd

Organism	Concentration	Reference[a]
Limpet, *Patella* sp.; soft parts		
Reference site	7.0 DW	75
Transplants to high copper areas for 8 months	30.0 DW	75
Limpet, *Patella vulgata*		
Bristol Channel, United Kingdom		
Foot	2.7 FW	87
Viscera	8.1 FW	87
Soft parts		
Near desalination plant outfall; May vs. August	25.4 vs. 282.0 DW	88
Reference site; May vs. August	8.7 DW vs. 6.6 DW	88
Soft parts	14.4 DW	89
Shell	42.0 DW	4
Soft parts	7.7 DW	4
Soft parts	12.0-30.0 DW	68
Shell	1.0-6.0 DW	68
Soft parts	19.0 (10.0-27.0) DW	15
Scallop, *Pecten jacobeus*, Adriatic Sea; June 1988		
Digestive gland	16.6 DW	106
Gills	6.3 DW	106
Gonad	10.3 DW	106
Kidney	17.5 DW	106
Mantle	3.3 DW	106
Muscle	1.1 DW	106
Scallop, *Pecten maximus*		
Soft parts	0.4-0.9 FW	44
Shell	1.1 DW	4
Soft parts	3.3 DW	4
Muscle	1.2 DW	4
Gut	25.0 DW	4
Mantle and gills	2.8 DW	4
Gonad	14.0 DW	4
Soft parts	8.9 DW	16
Body fluid	2.5 DW	16
mantle	2.7 DW	16
Gills	6.7 DW	16
Digestive gland	57.9 DW	16
Striated muscle	0.8 DW	16

(Continues)

Table 6.9: Cont'd

Organism	Concentration	Reference[a]
Unstriated muscle	1.6 DW	16
Gonad and foot	17.4 DW	16
Kidney	20.8 DW	16
Kidney	21.0 FW	90
Scallop, *Pecten novae-zelandiae*		
Soft parts	9.0 (2.0-14.0) DW	69
Mantle	15.0 DW	69
Gills	36.0 DW	69
Muscle	1.0 DW	69
Visceral mass	24.0 DW	69
Intestine	131.0 DW	69
Kidney	78.0 DW	69
Gonad	9.0 DW	69
Shell	2.0 DW	69
Soft parts	1.5 FW	57
Stomach	10.0 FW	57
Gonad	2.1 FW	57
Adductor muscle	6.6 FW	57
Green mussel, *Perna viridis*		
Hong Kong; soft parts; March 1986	Max. 35.1 DW	128
India; January 2002; soft parts; 28 sites		
Small	0.6 FW	162
Medium	0.6 FW	162
Large	0.23 FW	162
Hong Kong; juveniles vs. adults		
Reference site		
Soft parts	10.1 DW vs. 14.9 DW	188
Byssus	40.4 DW vs. 35.4 DW	188
Shell	11.8 DW vs. 12.3 DW	188
Polluted site		
Soft parts	17.5 DW vs. 16.4 DW	188
Byssus	60.2 DW vs. 96.3 DW	188
Shell	13.5 DW vs. 12.8 DW	188
Pearl oyster, *Pinctada fucata*		
Mantle	2.3 DW	190
Foot	3.7 DW	190
Pen shell, *Pinna nobilis*		
Soft parts	860.0 AW	91
Mantle and gills	700.0 AW	91

(*Continues*)

Table 6.9: Cont'd

Organism	Concentration	Reference[a]
Muscle	680.0 AW	91
Nervous system	370.0 AW	91
Stomach and intestines	610.0 AW	91
Gonads	460.0 AW	91
Hepatopancreas	800.0 AW	91
Byssus	960.0 AW	91
Byssus gland	700.0 AW	91
Clam, *Pitar morrhuana*; soft parts	2.3-29.6 DW	92
Scallop, *Placopecten magellanicus*		
Soft parts	7.3 DW	6
Muscle		
March	0.9-14.8 DW	6
August	0.8-4.4 DW	6
February	1.1-6.6 DW	6
June	0.9-2.6 DW	6
Viscera		
March	7.0-14.3 DW	6
August	8.0-20.1 DW	6
February	7.9-15.8 DW	6
June	7.2-9.6 DW	6
Muscle	0.27-1.1 FW	93
Gonad	1.1-10.6 FW	93
Visceral mass	1.3-5.6 FW	93
Clam, *Protothaca staminea*		
Shell	11.5 DW	2
Soft parts	7.5 DW	2
Pteropods; shells	Max. 20.0 DW	94
Clam, *Rangia cuneata*		
Soft parts	0.7-1.4 FW	95
Soft parts	25.0 DW	40
Whole	4.0 FW	96
Manila clam, *Ruditapes philippinarum*		
Soft parts; Korea; 2002-2003		
Females vs. males	8.6 DW vs. 8.1 DW	133
Postspawn vs. prespawn	8.9 DW vs. 8.4 DW	133
Jiaozhou Bay, China; sediments vs. soft parts; November 2004	25.8 DW vs. 6.4-19.8 DW	145
Bohai Sea; soft parts	1.3-4.4 DW	146

(*Continues*)

Table 6.9: Cont'd

Organism	Concentration	Reference[a]
Tropical rock oyster, *Saccostrea cucullata*; soft parts		
Australia; 1983-1984; near sewage discharge vs. reference site	285.0 DW vs. 34.0 DW	129
Hong Kong; March 1986	Max. 556.0 DW	128
Sydney rock oyster, *Saccostrea commercialis*; soft parts; Georges River, Australia; 1970s vs. 1980s	20.0-46.0 FW vs. 14.0-93.0 FW	130
Oyster, *Saccostrea glomerata*; soft parts	315.0-1146.0 DW	97
Clam, *Scrobicularia plana*		
Gannel estuary, England		
Soft parts	25.0-86.0 DW	98
Digestive gland	92.0 DW	98
Mantle and siphons	29.0 DW	98
Foot and gonads	48.0 DW	98
Gills and palps	36.0 DW	98
Adductor muscle	5.3 DW	98
Kidney	37.0 DW	98
Camel estuary, England		
Soft parts	25.0-77.0 DW	98
Digestive gland	51.0 DW	98
Mantle and siphons	30.0 DW	98
Foot and gonads	19.0 DW	98
Gills and palps	45.0 DW	98
Adductor muscle	8.3 DW	98
Kidney	28.0 DW	98
Transplants from Camel estuary to Gannel estuary for 352 days		
Digestive gland	170.0 DW	98
Remaining tissues	27.0 DW	98
Transplants from Gannel estuary to Camel estuary for 352 days		
Digestive gland	35.0 DW	98
Remaining tissues	16.0 DW	98
Tamar estuary, England		
Soft parts	13.0-52.0 DW	99
Digestive gland	26.0-102.0 DW	99
Looe estuary, England		
Soft parts	16.0-365.0 DW	15
Squid, *Sepia esculenta*; trunk	35.0 DW	52

(Continues)

Table 6.9: Cont'd

Organism	Concentration	Reference[a]
Common cuttlefish, *Sepia officinalis*		
Embryos	0.0038 mg/embryo FW	100
Mantle	1.5 FW	18
Dorsal shield	16.2 FW	18
Beak	2.9 FW	18
Gills	49.9 FW	18
Liver	1025.0 FW	18
Ink	4.6 FW	18
Blood		
Not dialyzed	175.0 DW	102
Dialyzed	73.0 DW	102
Liver		
Not dialyzed	165.0 DW	102
Dialyzed	73.0 DW	102
Gut		
Not dialyzed	42.0 DW	102
Dialyzed	33.0 DW	102
Branchial gland		
Not dialyzed	70.0 DW	102
Dialyzed	63.0 DW	102
Arterial heart		
Not dialyzed	30.0 DW	102
Dialyzed	22.0 DW	102
Ovary		
Not dialyzed	21.0 DW	102
Dialyzed	18.0 DW	102
Blood	157.0 DW	103
Hepatopancreas	100.0 FW	103
Branchial gland	22.0 FW	103
Branchial heart	20.0 FW	103
Branchial heart appendage	28.0 FW	103
Heart ventricle	23.0 FW	103
Pancreatic appendages	20.0 FW	103
English Channel; October 1987		
Branchial hearts	256.0 DW	131
Digestive gland	313.0–317.0 DW	131
Genital tract	55.0–56.0 DW	131
Gill	183.0 DW	131
Kidney	185.0 DW	131
Mantle	141.0 DW	131
Muscle	9.0 DW	131
Whole	59.0 DW	131
Egg		
Eggshell	14.0 DW	160

(Continues)

Table 6.9: Cont'd

Organism	Concentration	Reference[a]
Embryo	31.0 DW	160
Yolk	19.0 DW	160
Hatchlings		
Whole	60.0 DW	160
Cuttlebone	38.0 DW	160
Juveniles, age 1 month		
Whole	73.0 DW	160
Digestive gland	450.0 DW	160
Cuttlebone	12.0 DW	160
Immatures, age 12 months		
Whole	104.0 DW	160
Digestive gland	760.0 DW	160
Cuttlebone	8.0 DW	160
Adults, age 18 months		
Whole	70.0 DW	160
Digestive gland	600.0 DW	160
Cuttlebone	5.0 DW	160
Squid, *Sepia* sp.; whole	23.0 FW	76
Clam, *Spisula* sp.; whole	0.7-1.1 FW	104
Clam, *Spisula solidissima*; muscle	0.9-1.1 FW	12
Squid		
Flesh	15.0 DW	105
Flesh with skin	65.7 DW	105
Viscera	162.0 DW	105
Pen	65.0 DW	105
Gastropod, *Strombus gigas*		
Shell	5.3-39.0 FW; 11.0-40.0 DW	101
Foot	0.3-39.0 FW	101
Muscle	0.5-0.7 FW; 1.8-3.1 DW	101
Viscera	240.0-480.0 FW; 1500.0-2000.0 DW	101
Squid, *Symplectoteuthis oualaniensis*; liver	1720.0 DW	47
Clam, *Tapes decussatus*; soft parts	2.0-7.1 FW	18
Clam, *Tapes semidecussata*		
Shell	19.2 DW	2
Soft parts	19.2 DW	2
Gastropod, *Thais emarginata*		
Shell	5.9-8.2 DW	2
Soft parts	571.0 DW	2

(Continues)

Table 6.9: Cont'd

Organism	Concentration	Reference[a]
Gastropod, *Thais lapillus*		
Soft parts	167.6-458.1 DW	46
Digestive gland and gonad	506.7-554.0 DW	46
Shell	2.6-3.6 DW	46
Clam, *Trivela mactroidea*; Venezuela; 2002; 14 sites; soft parts	11.0-49.0 DW	167
Clam, *Venus mercenaria*; whole	0.91-19.2 FW	104

Values are in mg Cu/kg fresh weight (FW), dry weight (DW), or ash weight (AW).

[a]1, Won, 1973; 2, Graham, 1972; 3, Harris et al., 1979; 4, Segar et al., 1971; 5, Patel et al., 1973; 6, Reynolds, 1979; 7, Carmichael et al., 1979; 8, Shiber and Shatila, 1978; 9, Betzer and Pilson, 1974; 10, Townsley, 1954; 11, Betzer, 1972; 12, Greig et al., 1977; 13, George and Coombs, 1975; 14, Boyden, 1974; 15, Bryan and Hummerstone, 1977; 16, Bryan, 1973; 17, Watling and Watling, 1976b; 18, Establier, 1977; 19, Ratkowsky et al., 1974; 20, Mackay et al., 1975; 21, Zingde et al., 1976; 22, D'Silva and Quasim, 1979; 23, Jeng and Huang, 1973; 24, Ruddell and Rains, 1975; 25, Thornton et al., 1975; 26, Boyden and Romeril, 1974; 27, Kopfler and Mayer, 1967; 28, Watling and Watling, 1976a; 29, Pringle et al., 1968; 30, Ferrell et al., 1973; 31, Huggett et al., 1973; 32, Windom and Smith, 1972; 33, Drifmeyer, 1974; 34, Frazier, 1975; 35, Frazier, 1976; 36, Huggett et al., 1975; 37, Shuster and Pringle, 1968; 38, Greig and Wenzloff, 1978; 39, Goldberg et al., 1978; 40, Sims and Presley, 1976; 41, Anderlini, 1974; 42, Bryan et al., 1977; 43, Young and Jan, 1976; 44, Topping, 1973; 45, Wharfe and Van Den Broek, 1977; 46, Ireland and Wooton, 1977; 47, Martin and Flegal, 1975; 48, Sturesson and Reyment, 1971; 49, Miettinen and Verta, 1978; 50, Larsen, 1979; 52, Ishii et al., 1978; 53, Stenner and Nickless, 1974b; 54, Stenner and Nickless, 1975; 55, Culkin and Riley, 1958; 56, Hall et al., 1978; 57, Nielsen and Nathan, 1975; 58, Eisler, 1977b; 59, Alexander and Young, 1976; 60, Delhaye and Cornet, 1975; 61, Eustace, 1974; 62, Phillips, 1976a; 63, Phillips, 1976b; 64, Simpson, 1979; 65, Young and McDermott, 1975; 66, Young et al., 1979; 67, Talbot and Magee, 1978; 68, Lande, 1977; 69, Brooks and Rumsby, 1965; 70, Thomson, 1979; 71, Fowler and Oregioni, 1976; 72, Uysal, 1978a; 73, D'Silva and Kureishy, 1978; 74, Bhosle and Matondkar, 1978; 75, Stenner and Nickless, 1974a; 76, Matsumoto et al., 1964; 77, Ghiretti-Magaldi et al., 1958; 78, Rocca, 1969; 79, Nicholls et al., 1959; 80, Ikuta, 1967; 81, George et al., 1978; 82, Fukai et al., 1978; 83, Ikuta et al., 1968a; 84, Ikuta et al,. 1968b; 85, Ikuta et al., 1968c; 86, Thrower and Eustace, 1973a; 87, Noel-Lambot et al., 1980; 88, Romeril, 1977; 89, Preston et al., 1972; 90, George et al., 1980; 91, Papadopoulu, 1973; 92, Eisler et al., 1978; 93, Greig et al., 1978; 94, Krinsley and Bieri, 1959; 95, Wolfe and Jennings, 1973; 96, White et al., 1979; 97, Phillips, 1979; 98, Bryan and Hummerstone, 1978; 99, Bryan and Uysal, 1978; 100, Decleir et al., 1970; 101, Lowman et al., 1966; 102, Decleir et al., 1978; 103, Schipp and Hevert, 1978; 104, Craig, 1967; 105, Horowitz and Presley, 1977; 106, Mauri et al., 1990; 107, Viarengo et al., 1993; 108, Mat, 1994; 109, Mat et al., 1994;110, Swaileh and Adelung, 1994; 111, Jenkins, 1980; 112, Cheung and Wong, 1992; 113, Claisse and Alzieu, 1993; 114, Han and Hung, 1990; 115, Han et al., 1993; 116, Han et al., 1998; 117, Morris et al., 1989; 118, Weis et al., 1993a; 119, Byers, 1993; 120, Miramand and Bentley, 1992; 121, Bordin et al., 1994; 122, Paez-Osuna et al., 1994; 123, Ward, 1990; 124, Greig and Sennefelder, 1985; 125, Turgeon and O'Connor, 1991; 126, Fagioli et al., 1994; 127, Lauenstein et al., 1990; 128, Chu et al., 1990; 129, Talbot et al., 1985; 130, Brown and McPherson, 1992; 131, Miramand and Bentley, 1992; 132, Denton et al., 2006; 133, Ji et al., 2006; 134, Zorita et al., 2007; 135, Deudero et al., 2007; 136, Vlahogianni et al., 2007; 137, Keskin et al., 2007; 138, Karayakar et al., 2007; 139, Liu and Deng, 2007; 140, Yarsan et al., 2007; 141, Mubiana and Blust, 2006; 142, Sunlu, 2006; 143, Bergasa et al., 2007; 144, Huang et al., 2007; 145, Li et al., 2006; 146, Liang et al., 2004a; 147, Peake et al., 2006; 148, Scancar et al., 2007; 149, Kadar, 2007; 159, Gorinstein et al., 2006; 160, Miramand et al., 2006; 161, Silva et al., 2006; 162, Sasikumar et al., 2006; 163, Espana et al., 2007; 164, Sokolowski et al., 2007; 165, Karouna-Renier et al., 2007; 166, Nesto et al., 2007; 167, LaBrecque et al., 2004; 168, Sankar et al., 2006; 169, Colaco et al., 2006; 170, Kumar et al., 2008; 171, Metian et al., 2008; 172, Maanan, 2008; 173, De Wolf and Rashid, 2008; 174, Cunha et al., 2008; 175, Cravo et al., 2008; 176, Cosson et al., 2008; 177, Sajwan et al., 2008; 178, Frias-Espericueta et al., 2008; 179, Raimundo et al., 2008; 180, Anajjar et al., 2008; 181, Conti et al., 2008; 182, Protasowicki et al., 2008; 183, Kucuksezgin et al., 2008; 184, Schintu et al., 2008; 185, Bustamante et al., 2008; 186, Fattorini et al., 2008; 187, Napoleao et al., 2005; 188, Nicholson and Szefer, 2003; 189, Unlu et al., 2008; 190, Jing et al., 2007.

6.13.1 Age and Size

Most authorities agree that copper accumulations or rates of accumulation vary with the age of the organism, but even within the same species the pattern was not always predictable. In mussels there was an increase in copper per unit tissue weight with increasing size (Marks, 1938). In a cockle, *C. edule*, copper concentrations in tissues decreased with increasing age (Savari et al., 1991). In octopus, *Polypus*, however, copper levels were similar for all size groups (Marks, 1938). Decreasing copper concentrations with increasing body weight is reported for the clams *P. morrhuana* (Eisler et al., 1978) and *M. mercenaria* (Larsen, 1979), for the oyster *S. cucullata* (Mackay et al., 1975) and for the mussel *C. meridionalis* (Watling and Watling, 1976a). But copper was positively correlated with increasing body weight for other bivalve molluscs, including *M. edulis*, *M. mercenaria*, and *Venerupis decussata* (Boyden, 1974). Small oysters and other bivalve molluscs that have higher copper concentrations in soft tissues than larger bivalves reportedly take up copper at a greater rate and excrete it more slowly than larger conspecifics (Weis et al., 1993a).

Younger and smaller bivalve molluscs are more sensitive to copper insult. Juveniles of Asiatic clams (*Corbicula fluminea*) are more sensitive than adults to ionic copper (Belanger et al., 1990). And embryos of the common mussel are more sensitive to copper than older veliger larvae or postlarval spat (Hoare et al., 1995b).

6.13.2 Chelators

Chelators reduced the availability of potentially harmful levels of copper in seawater to molluscs and other marine groups. EDTA reduced toxic and behavioral effects of copper to the clam *V. decussata*, by making copper ions biologically unavailable (Stephenson and Taylor, 1975). Mussels, *M. edulis*, collected from the field contained almost as much copper as those subjected to high concentrations in the laboratory, suggesting that copper in field-collected mussels was in a relatively harmless organically bound form (Scott and Major, 1972). Reduced uptake of radiocopper-64 by Baltic clams, *M. balthica*, occurs at high concentrations of dissolved organic ligands (Absil et al., 1993).

6.13.3 Depth Effects

Mussels, *M. edulis*, collected near shore had nine times more copper in soft tissues than did conspecifics collected offshore in deeper waters (Young and McDermott, 1975). A similar pattern has been observed in other species with different metals and is attributed, in part, to salinity differences and food availability (Nielsen, 1974; Phillips, 1976a,b).

6.13.4 Inherent Species Differences

Copper concentrations in cephalopod molluscs are, in general, higher than those in bivalves; in cephalopods, 50-80% of the copper is localized in the digestive gland (Miramand and Bentley, 1992).

For crassostreid oysters, which strongly take up copper, green coloration is attributable to excessive accumulation of copper salts (Herdman and Boyce, 1899; Roosenburg, 1969). These oysters, which reportedly have a bitter taste, contain >1000.0 mg Cu/kg DW and have been observed to occur in areas close to steam electric stations (Roosenburg, 1969). Of three species of bivalves collected from Chesapeake Bay, Maryland, oysters contained up to 20 times more copper than clams and up to 200 times more copper than mussels (McFarren et al., 1962). In the mussel, *M. edulis*, copper was mainly associated with high-molecular weight proteins (Noel-Lambot, 1976); a similar pattern is reported for *O. vulgaris* hepatopancreas (Rocca, 1969). Loss rates of copper from molluscs varied widely. The gastropod, *Busycon canaliculatum*, depleted at the rate of 0.07 mg Cu/kg FW daily (Betzer and Pilson, 1975). For *Mercenaria*, this rate was about 0.05 mg Cu/kg FW daily (Pringle et al., 1968); and for *C. virginica* it was 3.95 mg Cu/kg FW daily (Mandelli, 1975). In all cases the copper depletion rates were independent of total body copper concentrations.

6.13.5 Interactions with Other Metals

Copper concentrations in soft parts of the green mussel, *P. viridis*, were significantly and positively correlated with concentrations of iron, nickel, lead, manganese, and zinc (Sasikumar et al., 2006). Copper uptake by mussels was erratic and seems to be influenced by available concentrations of zinc, cadmium, and lead salts (Phillips, 1976). Silver, at 0.0005-0.0155 mg Ag/L, enhances adverse effects of copper to embryos of the Pacific oyster when copper concentrations exceed 0.006 mg Cu/L (Coglianse and Martin, 1981).

Benthic communities near wooden bulkheads treated with chromated copper arsenate (CCA) had elevated concentrations of these elements, reduced species richness and diversity, and reduced numbers of total organisms when compared to reference sites (Weis and Weis, 1994; Weis et al., 1993b). Transfer of copper from CCA-treated wood occurs in estuarine algae, American oysters, mud snails (*N. obsoletus*), mussels, and crabs (Weis and Weis, 1992). American oysters (*C. virginica*) from a canal lined with CCA-treated wood had 150.0 mg Cu/kg FW soft parts versus 20.0 mg Cu/kg FW in conspecifics from a reference site (Weis and Weis, 1993). Copper is trophically transferred from CCA-exposed American oysters to predatory gastropods (*Thais* spp.), resulting in reduced gastropod feeding and growth (Weis and Weis, 1993).

6.13.6 Salinity

Copper concentrations in American oyster tissues are higher in low salinity waters than in high salinity (Roesijadi, 1994). Copper uptake by Baltic clams is higher at 10 ppt salinity than at 30 ppt (Absil et al., 1993). Clams, *R. cuneata*, were 20 times more sensitive to biocidal properties of ionic copper^{2+} at 1 ppt salinity than at 5 or 22 ppt (Olson and Harrel, 1973), and this was presumably reflected in tissue residues.

6.13.7 Proximity to Point Sources

Higher concentrations of copper are found in molluscan tissues near waste outfalls containing copper than in conspecifics from reference sites (Eisler, 1981). Mussels from the most polluted zones had altered enzyme chemistry and enlarged lysosomes (Zorita et al., 2007). Proximity to sewage discharge outfalls is associated with elevated copper burdens in mussels, *Mytilus* spp. (Lauenstein et al., 1990; Ward, 1990). Elevated copper concentrations in soft tissues of Baltic clams in San Francisco Bay are associated with anthropogenic inputs form nearby sources (Cain and Luoma, 1990). Copper concentrations in rock oysters, *Saccostrea glomerata*, collected from many locations in Hong Kong were highest in the vicinity of anthropogenic outfalls (Phillips, 1979). In the case of ocean-going and coastal vessels coated with several layers of antifouling paint—containing copper, and sometimes chromium, lead, mercury, tin, or zinc and designed for continual controlled release of toxicants—one result was elevated levels of copper and other paint components in harbor populations of *M. edulis* when compared to offshore populations (Young et al., 1979).

Pacific oysters near a copper recycling facility in Taiwan contained up to 4400.0 mg Cu/kg DW soft parts, a characteristic green color, and low survival after exposure to effluents for 3-7 months (Hung et al., 1990). Oysters in proximity to Australian urban and industrial discharge outfalls frequently exceed the recommended limit of 70.0 mg Cu/kg FW in shellfish edible tissues for protection of human health in than country (Brown and McPherson, 1992). Elevated copper concentrations in Pacific oysters (135.0 mg Cu/kg DW soft parts) near a marina in Arcachon Bay, France, are attributed to a ban on tributyltin (TBT) antifouling paints in 1982 and the subsequent growth in use of copper-based antifouling paints (Claisse and Alzieu, 1993).

6.13.8 Seasonal Variations

Copper burdens of bivalves in 2004-2005 from Sinaloa, Mexico, were higher in the rainy season than in the dry season (Frias-Espericueta et al., 2008). Copper accumulations in oysters, mussels, and scallops were highest in summer (Bryan, 1973; Hill and Helz, 1973; Karayakar et al., 2007; Roesijadi, 1994); similar results were observed in clams (Eisler, 1977b; Lee et al., 1975). Tissues of cockles, however, had lower concentrations in summer when compared to other seasons (Savari et al., 1991). Baltic clams, *M. balthica*, had highest concentrations in tissues in autumn and winter, and lowest in spring and summer (Sokolowski et al., 2007). These differences were not wholly explainable on the basis of thermal regimes. Possible causes of seasonal variation are attributed to reproductive cycle (Bryan, 1973), variations in moisture content of molluscs (Cain and Luoma, 1990; Phillips, 1976), food availability (Bryan, 1973), local pollution (Fowler and Oregioni, 1976), and drainage from land (Bryan, 1973; Fowler and Oregioni, 1976; Pentreath, 1973; Phillips, 1976).

6.13.9 Gender

Concentrations of copper that inhibit reproduction in the common mussel, *M. edulis*, from unpolluted or mildly contaminated sites did not inhibit embryonic development of mussels from copper-contaminated sites; cross-breeding of mussels from these sites suggest that copper tolerance in mussels in mostly maternally controlled (Hoare et al., 1995a).

Female *C. meridionalis* contain significantly more copper in whole soft parts than males, with copper concentrations positively correlated with zinc concentrations (Watling and Watling, 1976a). Copper does not necessarily translocate to sex products. In one case, Pacific oysters, *C. gigas*, from two geographic locations had significantly different tissue levels of copper; both groups were induced to spawn, and eggs from the two groups contained similar amount of copper: 28.0 and 29.0 mg Cu/kg DW (Greig et al., 1975). However, in the case of mussels, *M. californianus*, gonads of females contained up to twice as much copper as testes, with no-sex related differences in copper content of digestive gland or muscle (Alexander and Young, 1976).

6.13.10 Sediments

Copper concentrations in Baltic clam (*M. balthica*) and cockle (*C. edule*) tissues are positively correlated with copper concentrations in the surrounding sediments (Cain and Luoma, 1990; Savari et al., 1991; Sokolowski et al., 2007). However, copper burdens in tissues of another cockle (*Anadara trapezium*) are independent of sediment copper concentrations (Scanes, 1993). Larvae of Pacific oyster, *C. gigas*, accumulate copper and other metals from elutriates of metals-contaminated sediments (Geffard et al., 2007).

Proposed sediment copper criteria for protection of sensitive marine molluscs include <5.0 mg/kg DW to prevent avoidance by clams and <15.0 mg/kg DW to allow burrowing (Roper and Hickey, 1994). For more resistant species, less than 40.0 mg Cu/kg DW sediment is considered safe (Fagioli et al., 1994), but more than 200.0 mg Cu/kg DW may cause reduced species diversity and the disappearance of sensitive species (Bryan and Langston, 1992). Sediments containing more than 2000.0 mg Cu/kg DW are considered lethal to juvenile bivalve molluscs (Bryan and Langston, 1992).

6.13.11 Temperature

Groups of softshell clams, *M. arenaria*, had greatly reduced survival during immersion in given concentrations of copper salts at ambient summer temperatures (22 °C) that were nonlethal, and at autumn (17 °C) or winter (4 °C) thermal regimes (Eisler, 1977a). At 22 °C, 0.035 mg Cu/L was fatal to 50% of *M. arenaria* in 168 h; however, at 3.0 mg/L and 4 °C, none died in 336 h (Eisler, 1995a). Copper was also accumulated more rapidly by *Mya* at ambient summer temperatures than other seasonal thermal regimes (Eisler, 1977b).

Embryos of the giant clam, *Tridacna derasa*, were extremely sensitive to copper: at 27 °C and 0.0001 mg Cu/L, 50% died in 72 h (Soria-Dengg and Ochavillo, 1990). Copper is more toxic to embryos of the tropical giant clam than to bivalve embryos in temperate regions, possibly because many tropical species of shellfish live near their upper lethal thermal limits and are unable to withstand additional environmental stressors.

Isolated gills of the common mussel, *M. edulis*, took up copper from the medium (0.004 mg/L) over a 28 day period at temperatures between 6 and 26 °C; accumulations were directly proportional to increasing temperature (Mubiana and Blust, 2007). When the study was repeated with whole mussels, copper levels were inversely related to water temperature. In gills, temperature effects on uptake are due mainly to changes in solution chemistry and physical kinetics, which favor higher uptake at higher temperatures. At the whole organism level, however, complex physiological responses appear to mask the relationship, particularly for biologically essential metals such as copper (Mubiana and Blust, 2007).

6.13.12 Tissue Specificity

Copper concentrations in excess of 1000.0 mg/kg DW in molluscan tissues are documented, although tissues differ among groups (Table 6.9). For example, highest elevated copper burdens were in liver and hepatopancreas of octopods, squids, and gastropods; kidney in scallops; soft parts, gills, viscera, and mantle of oysters; and hemolymph of octopods (Table 6.9).

In cuttlefish, *S. officinalis*, copper tends to increase exponentially in digestive gland with increasing age and may contain up to 90% of the organisms total metal burden (Miramand et al., 2006). Highest accumulations of copper in the bivalve, *Meretrix casta*, were in gill tissues (Nambisan et al., 1977). In the gastropod, *B. canaliculatum*, copper concentrated in gill and osphradium, with histopathology observed at these sites (Betzer and Yevich, 1975). In the scallop, *P. maximus*, copper accumulated in digestive gland; however, copper was mainly in kidneys of the closely related *C. opercularis* (Bryan, 1973). In mussels, *M. edulis*, the digestive gland is the primary site of copper accumulation (Young and McDermott, 1975). Adverse effects of copper stress are not necessarily accompanied by measurable copper accumulations in tissues. Thus, copper-induced death was reported for the hardshell clam, *Mercenaria*, with no measured accumulations in tissues (Shuster and Pringle, 1968).

Copper accumulations, both natural and experimentally induced, in the gastropod *Littorina littorea*, shows copper in four types of cells, as determined by histochemical, spectrographic, and ion emission techniques (Martoja et al., 1980). Excretory cells of conjunctive tissues, in vascular conjunctive tissue and myocardium, contained precipitated CuS; older gastropods contained increasing concentrations of CuS and is interpreted as the result of hemocyanin degradation. Perivascular cells surrounding the aorta and arteries contain copper inside the spherocrystals; copper was associated with other elements, mainly magnesium and calcium, and intoxication by Cu^{2+} results in deposits at the periphery of these crystals. Hepatopancreas digestive cells are a third type; copper, in an unidentified unstable form, is found along the boundaries of these cells. Finally, the axial conjunctive cells of kidney lamellae contain copper overlying the sulfyrillated proteins, with highest concentrations in gastropods from copper-impacted areas; however, stress was not associated with environmental levels of copper. Progressive accumulation of copper sulfide (CuS), independent of environmental copper levels, makes *Littorina* an unreliable indicator of copper stress (Martoja et al., 1980).

6.13.13 Toxicity

Dissolved organic carbon, administered as humic acid, affects copper speciation and toxicity to embryos of the Pacific oyster, *C. gigas* (Brooks et al., 2007). Increasing humic acid concentration up to 1.0 mg/L reduces the toxicity of total copper by about 50%; however, the labile copper component—comprised of free ion and inorganic copper complexes—was unaffected. This finding suggests that the labile copper fraction—not total copper—is responsible for the observed toxicity to oyster embryos, and this is consistent with the free-ion activity model for copper toxicity (Brooks et al., 2007).

Copper was fatal at 0.005 mg Cu/L (50% dead in 119 days) to the bay scallop, *A. irradians*, at 0.025 mg Cu/L to the hardshell clam, *M. mercenaria* (53% dead in 77 days; USEPA, 1980b), and at 0.087 mg Cu/L (50% dead in 96 h) to the abalone, *Haliotis rubra* (Gorski and Nugegoda, 2006). Sublethal concentrations of copper to various shellfish species were associated with cytotoxicity (Zaroogian et al., 1992) and abnormal metabolism (Chelomin and Belcheva, 1992).

Adverse effects of copper to mussels include: inhibited mucous production at 0.01 mg Cu/L (Davies, 1992);reduced growth in 0.01-0.032 mg Cu/L (Calabrese et al., 1984; Sanders et al., 1991a, 1994); disrupted metabolism (Viarengo et al., 1981, 1990) and increased oxygen consumption (Mathew and Menon, 1993) at 0.015 mg Cu/L; a significant reduction in spawning frequency at 0.002-0.006 mg/L during immersion for 30 days (Stromgren and Nielsen, 1991); and death at 0.086 mg Cu/L (Krishnakumar et al., 1990). In mussels, copper impairs the structure and function of cellular membranes by stimulating the peroxidation of membrane lipids causing increased formation of lipofuscins (Viarengo et al., 1981, 1990). Copper-induced lysosomal lipofuscin accumulations, together with metallothioneins, control copper residues at the cellular levels and are responsible for the short half-time persistence (6-8 days) of copper in the digestive gland of mussels (Viarengo et al., 1990).

Juvenile clams, *Macoma liliana*, die when kept in sediments containing more than 30.0 mg Cu/kg DW; avoid sediments containing more than 25.0 mg Cu/kg DW (0.113 mg Cu/L sediment interstitial water); and show a reduction in ability to bury in sediments containing more than 15.0 mg Cu/kg DW (Roper and Hickey, 1994).

6.13.14 Accumulations by Laboratory Populations

Initial effects of copper on mussels (*Mytilus* spp.) include valve closure, a reduction in filtration rate, and cardiac inhibition; all of these responses serve to slow copper uptake through reduction in mussel contact with the ambient environment and a reduction in blood flow within the organism (Gainey and Kenyon, 1990). Eggs and larvae of the mussel, *M. galloprovincialis*, accumulate copper from the medium, but adverse effects on development occur only at seawater concentrations of 0.00946 mg Cu/L and higher, when whole embryos contain 47.8 mg Cu/kg DW, and when shell contains 26.8 mg Cu/kg DW or soft parts have more than 76.9 mg Cu/kg DW (Rosen et al., 2008).

BCFs for marine bivalve molluscs (ratio of mg Cu/kg FW soft parts to mg Cu/L medium) vary from 85 to 28,200. BCFs for copper are high for American oysters after exposure for 140 days (20,200-28,200) and lower (3100-3300) for bay scallops and softshell clams after exposure for 35-112 days (USEPA, 1980b).

Accumulations in Pacific oysters held on copper-loaded sediments are similar to those of oysters contaminated through ingestion of diatoms (*Haslea ostrearia*); however, accumulations are highest in oysters when exposed through the medium (Ettajani et al., 1992). In that study, a concentration of 0.03 mg Cu/L seawater for 21 days resulted in copper burdens of 137.0 mg/kg DW in whole diatoms and 1320.0 mg/kg DW in oyster soft tissues. Oysters fed contaminated diatoms in the study had 419.0 mg Cu/kg DW soft parts. Oysters held on sediments containing 108.0 mg Cu/kg DW had 401.0 mg Cu/kg DW (Ettajani et al., 1992). Han and Hung (1990) aver that diet is the major pathway by which greenish-colored Pacific oysters contain up to 4000.0 mg Cu/kg DW soft tissues. Initial daily accumulation rates up to 214.0 mg Cu/kg DW soft parts are reported (Han and Hung, 1990). Elimination of 50% of the copper from green Pacific oysters with elevated copper burdens takes only 11.6 days versus 25.1 days in reference oysters (Han et al., 1993).

Copper uptake kinetics in rock oysters, *Crassostrea cucullata*, was investigated (D'Silva and Quasim, 1979) and showed that exposure in media containing 0.01 mg Cu/L resulted in an increase in soft parts from 10.6 mg Cu/kg FW at start to 60.4 mg Cu/kg FW after 7 weeks. After 7 weeks in copper-free seawater, oysters lost 38% of the accumulated copper. Authors concluded that copper loss was closely correlated with total internal copper concentration rather than uptake rate as proposed by Pringle et al. (1968). However, copper uptake rate is considered an important variable in predicting copper toxicity to mussels, *M. edulis*, together with medium copper concentration (Martin, 1979).

High accumulations of copper in soft tissues were measured in several species of bivalves under various concentration/time regimes: *M. edulis* in 0.010-0.04 mg Cu/L for 2-30 days (Adema et al., 1972; Bordin et al., 1994); *Tellina tenuis* in 0.01 mg Cu/L for 100 days (Saward et al., 1975); *M. arenaria* in 0.01 mg Cu/L for 6 weeks at 0-10 °C temperatures and 7 days at 20-22 °C (Eisler, 1977a,b); *Haliotis* spp. in 0.01 mg Cu/L for 96 h (Martin et al., 1977); and *Mytilus viridis* in 0.01 mg Cu/L for 5 weeks (D'Silva and Kureishy, 1978). At higher copper concentrations and longer exposure periods, copper residues in molluscs were higher, survival lower, and damage effects more pronounced (Clarke, 1947; Eisler, 1977b; Jing et al., 2007; Martin et al., 1977; Pringle et al., 1968; Scott and Major, 1972). Most investigators now support the premise that copper in trace quantities is required by marine molluscs and that small increments above the required level are highly toxic.

6.14 Europium

The highest value known for stable europium in marine molluscs is 0.123 mg Eu/kg DW in pteropod tests (Table 6.10). No radioeuropium-152 activity was found in various species of molluscs collected in summer 2004 near Amchitka Island in the Aleutian chain; Amchitka Island was the site of three nuclear tests between 1965 and 1971 (Burger et al., 2007b).

6.15 Gallium

The highest value reported for gallium in marine molluscs is 0.16 mg Ga/kg DW in soft parts (Table 6.10).

6.16 Gold

The highest values known for gold concentrations in marine molluscs are 0.038 mg Au/kg DW and 0.079 mg Au/kg AW, both in soft parts (Table 6.10). High bioconcentration of gold was reported in three species of bivalves from the Aegean Sea. Although tissue gold concentrations were 770-2400 times higher than seawater levels (Papadopoulu, 1973), maximum values recorded in bivalves did not exceed 0.052 mg Au/kg AW soft parts (Table 6.10).

Table 6.10: Europium, Gallium, Gold, and Hafnium Concentrations in Field Collections of Molluscs

Metal and Organism	Concentration	Reference[a]
Europium		
Bivalve molluscs; 4 species; Aegean Sea		
Soft parts	0.020-0.044 DW	1
Muscle	0.009-0.03 DW	1
Mantle and gills	0.011-0.061 DW	1
Stomach and intestines	0.012-0.048 DW	1
Gonads	0.011-0.039 DW	1
Hepatopancreas	0.012-0.017 DW	1
Byssus	0.014-0.056 DW	
Common mussel, *Mytilus edulis*; soft parts	0.0005-0.11 DW	2
Pteropods; tests	0.007-0.123 DW	3
Gallium		
Molluscs		
Shell	0.008-0.036 DW	4
Soft parts	0.007-0.16 DW	4
Gold		
Molluscs; 3 species; soft parts	0.049-0.052 AW	5
Mytilus edulis; soft parts	0.002-0.038 DW	2
Clam, *Tapes japonica*		
Soft parts	0.079 AW	6
Soft parts	0.0057 DW	7
Hafnium		
Mytilus edulis; soft parts	0.0006-0.1000 DW	2

Values are in mg element/kg dry weight (DW) or ash weight (AW).
[a] 1, Papadopoulu and Hadzistelios, 1977; 2, Karbe et al., 1977; 3, Turekian et al., 1973; 4, Culkin and Riley, 1958; 5, Papadopoulu, 1973; 6, Fukai and Meinke, 1959; 7, Fukai and Meinke, 1962.

Common mussels, *M. edulis*, exposed for 24 h to gold-citrate nanoparticles had increased catalase activity in digestive gland and mantle, and other signs of oxidative stress (Tedesco et al., 2008). Smaller gold nanoparticles of about 1 nm in diameter are potentially toxic when compared to nanoparticles about 18 nm in diameter (Tedesco et al., 2008).

6.17 Hafnium

The only datum found on hafnium in molluscs is 0.0006-0.1 mg Hf/kg DW in mussel soft tissues (Table 6.10).

6.18 Iron

Iron concentrations in excess of 2000.0 mg/kg DW are documented for soft parts, mantle and digestive gland of clams; for shell, soft parts, gills, and visceral mass of mussels; for shell and soft parts of gastropods; and for shell, soft parts, gills, digestive gland, striated muscle, intestine, and kidney of scallops (Table 6.11).

Substantial variations in iron content between and among field-collected species are attributed, in part, to diet and pollution. For example, proximity to anthropogenic point sources such as electroplating plants, iron works, and waste outfalls was almost always associated with elevated iron concentrations in molluscs (Eisler et al., 1978; Fowler and Oregioni, 1976; Phillips, 1978, 1979; Sankaranarayanan et al., 1978). Diet was acknowledged as an important source of iron in marine molluscs. In the case of *Littorina obtusata*, food chain accounted for 42% of daily iron intake equivalent to 910.0 mg Fe/kg tissue/day to the total tissue concentration of 34,600.0 mg/kg DW (Young, 1975). Among predatory gastropods, such as *N. lapillus*, diet was the major source of iron, the input from food being about 100 times greater than seawater (Young, 1977). In *N. lapillus*, the fraction of iron assimilated from food was similar to rate of excretion, with relative iron concentrations reflecting the relative concentrations in prey organisms (Young, 1977). Mussels, *M. edulis*, may preferentially accumulate soluble forms of iron from seawater over particulates; however, accumulation of iron from the medium is relatively minor compared with that of dietary accumulation (Pentreath, 1973). Organism size and salinity of the medium also confound interpretation of residues: iron in soft tissues of clams is significantly correlated with increasing shell length (Ji et al., 2006), and with decreasing salinity of the medium (Mubiana and Blust, 2006). In the pearl oyster, *Pinctada fucata*, iron was most abundant in digestive gland, followed by mantle, and gill tissues in the ratio of 11:4:1 (Gonzalez et al., 2008).

M. edulis accumulates radiolabeled ferric hydroxide in linear proportion to concentrations in seawater in the range 0.007-0.5 mg Fe/L; iron uptake rates in whole soft parts were constant during exposure for 25 days although iron levels in some tissues show a tendency to equilibrate, presumably by transfer of iron from these tissues to a growing store elsewhere (George et al., 1976). Many mollusc cell types are rich in lysosomes and these have the ability to sequester—or otherwise bind—metals. Iron was demonstrated histochemically in cytoplasmic inclusions shown to be lysosomes in a number of cell types of *M. edulis* (Lowe and Moore, 1979). Exponential, rather than linear, uptake of radioiron from seawater by *Mytilus* was observed with equilibrium reached for individual tissues in 21-42 days (Pentreath, 1973).

Table 6.11: Iron Concentrations in Field Collections of Molluscs

Organism	Concentration	Reference[a]
Clam, *Andara trapezia*; whole	745.0-1285.0 DW	1
Mussel, *Anodonta* sp.		
Shell	2600.0 DW	2
Soft parts	5500.0 DW	2
Gastropod, *Aplysia benedicti*		
Ctenidium	320.0-750.0 DW	3
Gonad	194.0-325.0 DW	3
Parapodia	48.0-175.0 DW	3
Hepatopancreas	261.0-280.0 DW	3
Buccal mass	400.0-2300.0 DW	3
Intestine	456.0 DW	3
Gut Content	541.0 DW	3
Shell	647.0 DW	3
Giant squid, *Architeuthis dux*; 2001-2005; hepatopancreas; Bay of Biscay vs. Mediterranean Sea	497.0 DW vs. 158.0 DW	72
Scallop, *Argopecten irradians*; kidney concretion	608.3 DW	4
Hydrothermal vent mussel, *Bathymodiolus azoricus*; mid-Atlantic ridge		
Summer 2002; start vs. 30 days in metals-free seawater		
Gill	120.0 FW vs. 75.0 FW	55
Mantle	60.0 FW vs. 70.0 FW	55
Digestive gland	260.0 FW vs. 200.0 FW	55
Byssus thread	1400.0 FW vs. <10.0 FW	55
Shell; 2001	20.6-55.0 (15.6-62.1) DW	66
2001; 5 sites		
Gills	83.0-2066.0 DW	67
Mantle	28.0-229.0 DW	67
Digestive gland	198.0-1854.0 DW	67
Waved whelk, *Buccinum undatum*		
Shell	470.0 DW	2
Soft parts	110.0 DW	2
Cockle, *Cardium edule*		
Shell	1600.0 DW	2
Soft parts	590.0 DW	2

(*Continues*)

Table 6.11: Cont'd

Organism	Concentration	Reference[a]
Cockle, *Cerastoderma edule*		
Soft parts	390.0-1327.0 DW	5
Soft parts	406.0-991.0 DW	6
Scallop, *Chlamys opercularis*		
Shell	64.0-120.0 DW	2
Soft parts	113.0 DW	7
Body fluid	32.1 DW	7
Mantle	67.1 DW	7
Gills	134.0 DW	7
Digestive gland	853.0 DW	7
Striated muscle	10.7 DW	7
Unstriated muscle	16.3 DW	7
Gonad and foot	132.0 DW	7
Kidneys	330.0 DW	7
Mussel, *Choromytilus meridionalis*		
Soft parts	18.0 FW	8
Soft parts	60.0 DW	9
Scallop, *Comptopallium radula*; New Caledonia; October 2004		
Soft parts	221.0 (162.0-375.0) DW	63
Digestive gland	1195.0 DW	63
Pacific oyster, *Crassostrea gigas*		
Soft parts	15.3-91.4 FW	10
Soft parts	128.0 DW	11
Oyster, *Crassostrea margaritacea*; soft parts	57.0 DW	11
Oyster, *Crassostrea rhizophorae*; soft parts; summer 2001; Brazil; contaminated estuary vs. reference site	152.0-519.0 DW vs. 242.0-299.0 DW	57
American oyster, *Crassostrea virginica*		
Soft parts	31.0-238.0 FW	12
Soft parts; October vs. December	510.0 DW vs. 150.0 DW	13
Shell	19.0 DW	13
Soft parts	140.0-800.0 DW	14
Shell	15.0-79.0 DW	15
Soft parts	233.0-738.0 DW	15
Soft parts	140.0-690.0 DW	16
Soft parts; Savannah, Georgia; 2000-2001; 9 sites	232.0-1357.0 DW	62, 68

(*Continues*)

Table 6.11: Cont'd

Organism	Concentration	Reference[a]
Slipper limpet, *Crepidula fornicata*		
Shell	1600.0 DW	2
Soft parts	2000.0 DW	2
Clam, *Donax serra*; soft parts	47.0 FW	8
Clam, *Ensis arcuatus*; shell	66.0 DW	17
Clam, *Ensis siliqua*; shell	75.0 DW	17
Gastropods		
Flesh	13,000.0 AW	18
Shell	680.0 AW	18
Viscera	6100.0 AW	18
Operculum	50,000.0 AW	18
Abalone, *Haliotis midea*; soft parts	27.0 FW	8
Abalone, *Haliotis tuberculata*; soft parts	306.0-474.0 DW	19
Periwinkle, *Littoraria scabra*; soft parts; Tanzania		
Industrialized sites; 1998 vs. 2005	350.0 DW vs. 350.0 DW	64
Reference sites; 2005	110.0 DW	64
Limpet, *Littorina littorea*; soft parts plus operculum	227.0-784.0 DW	6
Squid, *Loligo opalescens*; liver; California		
Central coast	111.0 DW	20
Southern coast	87.0 DW	20
Clam, *Macoma balthica*		
Soft parts	502.0-1540.0 DW	6
Soft parts; 1996-1997; Gulf of Gdansk	650.0-1800.0 DW	60
Hardshell clam, *Mercenaria mercenaria*		
Soft parts	9.0-83.0 FW	12
Soft parts	18.0-81.0 FW	5
Shell	1600.0 DW	2
Soft parts	5400.0 DW	2
Soft parts	810.0 AW	21
Mantle	740.0 AW	21
Gills	1480.0 AW	21
Muscle	800.0 AW	21
Clam, *Meretrix lamarckii*; soft parts	130.0 DW	22

(*Continues*)

Table 6.11: Cont'd

Organism	Concentration	Reference[a]
Mussel, *Modiolus modiolus*		
Shell	15.0 DW	2
Soft parts	300.0-350.0 DW	2
Mantle and gills	220.0 DW	2
Muscle	51.0 DW	2
Gonad	130.0 DW	2
Gut and digestive gland	1100.0 DW	2
Molluscs		
Shell	41.0-239.0 DW	23
Soft parts	86.0-1415.0 DW	23
Soft parts; 5 spp.; Chile, Strait of Magellan; 2001; 17 stations	25.2-551.0 FW	59
Edible portions, 3 spp.	18.0-59.0 FW	24
Softshell clam, *Mya arenaria*		
Soft parts; winter vs. summer	4834.0 DW vs. 230.0-317.0 DW	25
Soft parts	49.7-1710.0 FW	12
Mussel, *Mytilus californianus*; soft parts	59.0-630.0 DW	26
Mussel, *Mytilus corscum*; soft parts	150.0 DW	22
Common mussel, *Mytilus edulis*		
Soft parts; Scheldt estuary; March 2000	3800.0 DW	71
Soft parts	14.0-1367.0 DW	27
Soft parts	112.0-1623.0 DW	28
Shell	17.0-55.0 DW	28
Shell; Baltic Sea; 2005; 12 sites	5.5-22.2 DW	69
Soft parts	81.0-710.0 DW	26
Soft parts	17.0-31.0 DW	29
Soft parts	152.0-401.0 DW	6
Shell	8.9 DW	17
Soft parts	64.0-589.0 DW	30
Soft parts	776.0 DW	17
Soft parts	41.0-707.0 DW	31
Soft parts	Max. 400.0 DW	54
Shell	290.0 DW	2
Soft parts	1700.0 DW	2
Mussel, *Mytilus edulis aoteanus*		
Soft parts	1960.0 DW	32
Mantle	180.0 DW	32
Gill	2940.0 DW	32

(Continues)

Table 6.11: Cont'd

Organism	Concentration	Reference[a]
Muscle	82.0 DW	32
Visceral mass	13,650.0 DW	32
Intestine	570.0 DW	32
Foot	<10.0 DW	32
Gonad	93.0 DW	32
Shell	<10.0 DW	32
Mussel, *Mytilus edulis planulatus*; soft parts; various dry weights		
0.09 g	1198.0 DW	1
0.39 g	671.0 DW	1
0.48 g	601.0 DW	1
0.69 g	616.0 DW	1
Mussel, *Mytilus galloprovincialis*		
Soft parts; Venice lagoon, Italy; 2005-2006; San Giuliano vs. Sacca Sessola		
April	207.9 DW vs. 234.9 DW	61
July	3.9 DW vs. 3.8 DW	61
October	131.6 DW vs. 131.0 DW	61
February	306.3 DW vs. 271.9 DW	61
Soft parts	443.0 DW	33
Soft parts	26.9 FW	34
Soft parts; Izmir Bay, Turkey; 2004	49.3 DW	70
Soft parts; Gulf of Gemlik, Turkey; 2004	205.4 DW	81
Soft parts; Adriatic Sea; 2001-2005; winter vs. summer	348.0-460.0 DW vs. 58.6-177.0 DW	73
Mantle vs. gills; 2004-2006; Greece; max. mean concentrations; contaminated sites		
Summer	98.4 FW vs. 151.0 FW	53
Autumn	90.0 FW vs. 130.0 FW	53
Winter	22.2 FW vs. 32.9 FW	53
Spring	30.0 FW vs. 43.8 FW	53
Mussel, *Mytilus viridis*; soft parts	549.0-2751.0 DW	35
Abalone, *Notohaliotis discus*; soft parts	110.0 DW	22
Gastropod, *Nucella conccina*; soft parts; Antarctica; 2004	2756.0 DW	52

(Continues)

Table 6.11: Cont'd

Organism	Concentration	Reference[a]
Gastropod, *Nucella lapillus*		
Soft parts with operculum	193.0-270.0 DW	6
Shell	4.5 DW	2
Soft parts	65.0 DW	2
Octopus, *Octopus* sp.; whole	37.0 FW	36
Octopus, *Octopus vulgaris*		
Hepatopancreas	1920.0 DW	37
Branchial heart	399.0 DW	37
Gill	188.0 DW	37
Central heart	160.0 DW	37
Kidney	112.0 DW	37
Body muscle	47.0 DW	37
Hemolymph	not detectable	37
Portugal; 1999-2002; 3 sites		
Digestive gland	636.0-844.0 DW	79
Branchial heart	464.0-690.0 DW	79
Gill	33.0-47.0 DW	79
Muscle	23.0-27.0 DW	79
Squid, *Ommastrephes bartrami*; liver	399.0 DW	20
Oyster, *Ostrea angasi*; whole	612.0-2585.0 DW	1
Oyster, *Ostrea edulis*		
Soft parts	232.0-352.0 DW	38
Soft parts	167.0 DW	11
Oyster, *Ostrea sinuata*		
Soft parts	682.0 DW	32
Mantle	840.0 DW	32
Gill	620.0 DW	32
Muscle, nonstriated	274.0 DW	32
Muscle, striated	2880.0 DW	32
Visceral mass	1156.0 DW	32
Kidney	384.0 DW	32
Heart	247.0 DW	32
Shell	570.0 DW	32
Limpet, *Patella coerulea*; soft parts	1535.6-5915.6 DW; 442.3-1283.7 FW	39
Limpet, *Patella candei gomesii*; soft parts; 2007; Azores		
Near hydrothermal vent	1635.0-2670.0 DW	65
Reference site	1024.0-1995.0 DW	65

(*Continues*)

Table 6.11: Cont'd

Organism	Concentration	Reference[a]
Limpet, *Patella vulgata*		
Soft parts	2454.0 DW	40
Soft parts	1289.0-2505.0 DW	28
Shell	39.0-64.0 DW	28
Shell	430.0 DW	2
Soft parts	150.0 DW	2
Soft parts	891.0-2330.0 DW	6
Scallop, *Pecten maximus*		
Shell	37.0-49.0 DW	2
Soft parts	170.0 DW	2
Muscle	30.0 DW	2
Gut and digestive gland	2900.0 DW	2
Mantle and gills	570.0 DW	2
Gonad	100.0 DW	2
Soft parts	196.0 DW	7
Body fluid	26.4 DW	7
Mantle	86.4 DW	7
Gills	137.0 DW	7
Digestive gland	1295.0 DW	7
Striated muscle	13.1 DW	7
Unstriated muscle	17.1 DW	7
Gonad and foot	143.0 DW	7
Kidney	149.0 DW	7
Kidney	42.0 FW	41
Scallop, *Pecten novae-zelandiae*		
Soft parts	2915.0 DW	32
Mantle	1540.0 DW	32
Gills	21,600.0 DW	32
Muscle	34.0 DW	32
Visceral mass	2200.0 DW	32
Intestine	6090.0 DW	32
Kidney	2470.0 DW	32
Foot	2380.0 DW	32
Shell	2000.0 DW	32
Gonad	228.0 DW	32
Pelycopods		
Soft parts	500.0 AW	18
Shell	142.0 AW	18
Green mussel, *Perna viridis*		
Soft parts; India; January 2002		
Small	81.0 FW	58

(Continues)

Table 6.11: Cont'd

Organism	Concentration	Reference[a]
Medium	64.9 FW	58
Large	40.3 FW	58
Hong Kong; juveniles vs. adults		
Reference site		
Soft parts	144.0 DW vs. 97.2 DW	80
Byssus	899.0 DW vs. 230.0 DW	80
Shell	4.9 DW vs. 5.2 DW	80
Polluted site		
Soft parts	345.0 DW vs. 354.0 DW	80
Byssus	527.0 DW vs. 1162.0 DW	80
Shell	12.4 DW vs. 8.9 DW	80
Clam, *Pitar morrhuana*; soft parts	250.0-799.0 DW	42
Scallop, *Placopecten magellanicus*; soft parts	803.0 DW	43
Pteropods; 9 spp.; tests	48.0-741.0 DW	44
Manila clam, *Ruditapes philippinarum*; soft parts; Korea; 2002-2003		
Females vs. males	573.0 DW vs. 571.0 DW	51
Postspawn vs. prespawn	740.0 DW vs. 572.0 DW	51
Oyster, *Saccostrea glomerata*; soft parts	202.0-731.0 DW	45
Clam, *Scrobicularia plana*		
Looe estuary, England; soft parts	1090.0 (559.0-3010.0) DW	6
Tamar estuary, England		
Soft parts	2370.0 (441.0-7280.0) DW	46
Digestive gland	1170.0 (676.0-2400.0) DW	46
Gannel estuary, England		
Soft parts	1120.0-1240.0 DW	47
Digestive gland	872.0 DW	47
Mantle and siphons	3480.0 DW	47
Foot and gonad	342.0 DW	47
Gills and palps	848.0 DW	47
Adductor muscle	236.0 DW	47
Kidney	819.0 DW	47
Camel estuary, England		
Soft parts	699.0-840.0 DW	47
Digestive gland	604.0 DW	47
Mantle and siphons	971.0 DW	47
Foot and gonad	298.0 DW	47
Gills and palps	680.0 DW	47

(*Continues*)

Table 6.11: Cont'd

Organism	Concentration	Reference[a]
Adductor muscle	177.0 DW	47
Kidney	1090.0 DW	47
Transferred from Camel estuary to Gannel estuary for 352 days		
Digestive gland	1940.0 DW	47
Other tissues	1020.0 DW	47
Transferred from Gannel estuary to Camel estuary for 352 days		
Digestive gland	728.0 DW	47
Other tissues	1290.0 DW	47
Squid, *Sepia esculenta*; trunk	16.0 DW	22
Cuttlefish, *Sepia officinalis*		
Blood	1.0 FW	48
Liver	37.0-42.0 DW	48
Gut	16.0-36.0 DW	48
Branchial gland	19.0-27.0 DW	48
Arterial heart	22.0-31.0 DW	48
Ovary	5.0-8.0 DW	48
Blood	26.0 FW	49
Hepatopancreas	152.0 FW	49
Branchial gland	22.0 FW	49
Branchial heart	46.0 FW	49
Branchial heart appendage	67.0 FW	49
Heart ventricle	37.0 FW	49
Pancreatic appendages	36.0 FW	49
Squid, *Sepia* sp.; whole	8.0 FW	36
Common cuttlefish, *Sepia officinalis*		
Egg		
Eggshell	398.0 DW	56
Embryo	14.0 DW	56
Yolk	1.1 DW	56
Hatchlings		
Whole	80.0 DW	56
Cuttlebone	21.0 DW	56
Juveniles, age 1 month		
Whole	38.0 DW	56
Digestive gland	210.0 DW	56
Cuttlebone	28.0 DW	56
Immature, age 12 months		
Whole	41.0 DW	56

(Continues)

Table 6.11: Cont'd

Organism	Concentration	Reference[a]
Digestive gland	340.0 DW	56
Cuttlebone	15.0 DW	56
Adults, age 18 months		
Whole	30.0 DW	56
Digestive gland	390.0 DW	56
Cuttlebone	2.5 DW	56
Squid		
Muscle	107.0 DW	50
Muscle with skin	19.3 DW	50
Viscera	60.6 DW	50
Pen	160.0 DW	50
Squid, *Symplectoteuthis oualaniensis*; liver	319.0 DW	20

Values are in mg Fe/kg fresh weight (FW), dry weight (DW), or ash weight (AW).

[a]1, Harris et al., 1979; 2, Segar et al., 1971; 3, Patel et al., 1973; 4, Carmichael et al., 1979; 5, Romeril, 1974; 6, Bryan and Hummerstone, 1977; 7, Bryan, 1973; 8, Van As et al., 1973; 9, Watling and Watling, 1976b; 10, Pringle et al., 1968; 11, Watling and Watling, 1976a; 12, Shuster and Pringle, 1968; 13, Frazier, 1975; 14, Galtsoff, 1953; 15, Windom and Smith, 1972; 16, Goldberg et al., 1978; 17, Bertine and Goldberg, 1972; 18, Lowman et al., 1970; 19, Bryan et al., 1977; 20, Martin and Flegal, 1975; 21, Eisler and Weinstein, 1967; 22, Ishii et al., 1978; 23, Culkin and Riley, 1958; 24, Van As et al., 1975; 25, Eisler, 1977b; 26, Goldberg et al, 1978; 27, Phillips, 1978; 28, Lande, 1977; 29, Hobden, 1967; 30, Simpson, 1979; 31, Karbe et al., 1977; 32, Brooks and Rumsby, 1965; 33, Fowler and Oregioni, 1976; 34, Uysal, 1978a,b; 35, Bhosle and Matondkar, 1978; 36, Matsumoto et al., 1964; 37, Ghiretti-Magaldi et al., 1958; 38, Fukai et al., 1978; 39, Shiber and Shatila, 1978; 40, Preston et al., 1972; 41, George et al., 1980; 42, Eisler et al., 1978; 43, Reynolds, 1979; 44, Turekian et al., 1973; 45, Phillips, 1979; 46, Bryan and Uysal, 1978; 47, Bryan and Hummerstone, 1978; 48, Decleir et al., 1978; 49, Schipp and Hevert, 1978; 50, Horowitz and Presley, 1977; 51, Ji et al., 2006; 52, Santos et al., 2006; 53, Vlahogianni et al., 2007; 54, Mubiana and Blust, 2006; 55, Kadar, 2007; 56, Miramand et al., 2006; 57, Silva et al., 2006; 58, Sasikumar et al., 2006; 59, Espana et al., 2007; 60, Sokolowski et al., 2007; 61, Nesto et al., 2007; 62, Kumar et al., 2008; 63, Metian et al., 2008; 64, De Wolf and Rashid, 2008; 65, Cunha et al., 2008; 66, Cravo et al., 2008; 67, Cosson et al., 2008; 68, Sajwan et al., 2008; 69, Protasowicki et al., 2008; 70, Kucukszegin et al., 2008; 71, Wepener et al., 2008; 72, Bustamante et al., 2008; 73, Fattorini et al., 2008; 79, Napoleao et al., 2005; 80, Nicholson and Szefer, 2003; 81, Unlu et al., 2008.

The concentration of iron in feces of *M. arenaria* and *Mercenaria* is increased by several orders of magnitude over that in seawater, that is, 15,400 for *Mercenaria*, 1000 for *Mya* and may represent an important factor in the transfer of iron to bottom sediments (Andrews and Warren, 1969).

Studies on the compartmentalization of iron in *M. edulis* (Hobden, 1967) showed that iron concentrations in whole mussels from Southampton, England, usually ranged between 20.0 and 40.0 mg Fe/kg DW, occasionally rising to 100.0 mg Fe/kg DW in the spring. Prolonged starvation in seawater of low iron content did not reduce average iron below 20.0 mg Fe/kg DW, as this represents a permanent store. Higher iron values were produced by a temporary store that was lost rapidly on starvation. Digestive gland was the major repository of the temporary store of iron, most of which was subjected to digestive processes. In a second study (Hobden, 1969), it was demonstrated that *Mytilus* absorbed soluble iron from citrated seawater either directly of by adsorption onto mucous used in feeding. Absorption slowed or ceased after 3-7 days at which time the total iron uptake was about 1% of the total body iron burden. Byssus threads had the highest iron concentration, followed by kidney, digestive gland, gut epithelia, gill, heart, gonad, and muscle, in that order. Accumulated iron persisted over a period of 10 days when the mussels were transferred to clean seawater (Hobden, 1969). Pteropods, unlike mussels, eliminated radioiron rapidly, at elimination rates of 5-6%/h (Kuenzler, 1969).

Iron-mercury interactions were observed in *M. mercenaria*. In the range 0.0-0.1 mg Hg/L in the medium there were significant decreases in iron content of mantle fringe tissues (Fowler et al., 1975). Other factors reported to modify iron accumulations in molluscs included gender, season of collection, and water turbidity, salinity, and depth. Thus, iron concentrations were higher in females of *C. meridionalis* than in males (Watling and Watling, 1976b); higher in *M. edulis* from waters of comparatively low salinity (Phillips, 1978); higher in oysters during winter than other seasons (Sankaranarayanan et al., 1978); higher in bivalves collected from relatively turbulent waters with high suspended solids content (Raymont, 1972); and higher in mussels *Perna canaliculus* held at increasing water depths (Nielsen, 1974).

6.19 Lanthanum

Limited data on lanthanum concentrations in molluscs indicate that tests of pteropods from the Red Sea contained lower concentrations than did pteropod tests from the South Atlantic Ocean, namely, 0.34-1.72 mg La/kg DW versus 5.28-6.09 mg La/kg DW (Turekian et al., 1973).

6.20 Lead

Maximum lead concentrations—in mg Pb/kg DW—in excess of 450.0 mg/kg DW among field collections of molluscs were 452.0 in soft parts of a mussel, *M. edulis planulatus*; 827.0 in kidneys of the scallop, *C. opercularis*; 820.0 in soft parts of the clam, *S. plana*; 931.0 in soft parts of the gastropod, *Acmaea digitalis*; 1130.0 in digestive gland of *S. plana*; and 3100.0 in soft parts of *M. edulis* (Table 6.12). The elevated levels reported in some species were almost always positively associated with degree of industrial or other anthropogenic lead discharges to the biosphere (Bloom and Ayling, 1977; Eisler, 2000c; Eisler et al., 1978; Fowler and Oregioni, 1976; Graham, 1972; Phillips, 1978; Schulz-Baldes, 1973). In one case, lead concentrations in the common mussel, *M. edulis*, exceeded the World Health Organization food standard of 10.0 mg Pb/kg DW in 19 of the 22 locations sampled in Port Phillip Bay, Australia; stormwater runoff and tidal movements were contributory causes to elevated lead

Table 6.12: Lead Concentrations in Field Collections of Molluscs

Organism	Concentration	Reference[a]
Abalone		
Muscle	0.33-0.39 FW	1
Viscera	0.6-2.5 FW	1
Gastropod, *Acmaea digitalis*		
Shell	<9.0 to 108.0 DW	2
Soft parts	7.5-931.0 DW	2
Cockle, *Anadara trapezia*; April 2000; soft parts; New South Wales, Australia	Max. 11.0 DW	87
Clam, *Anodonta* sp.		
Shell	0.6 DW	4
Soft parts	1.2 DW	4
Giant squid, *Architeuthis dux*; 2001-2005; hepatopancreas		
Bay of Biscay	0.4 DW	108
Mediterranean Sea	0.8 DW	108
Bivalves; 5 spp.; soft parts; East China Sea; August 2002	0.09 (0.003-0.24) FW	80
Mussel, *Brachydontes pharaonis*; Mersin coast, Turkey; 2002-2003; max. concentrations		
Muscle	2.0 DW	74
Gill	8.0 DW	74
Hepatopancreas	3.9 DW	74
Whelk, *Buccinum undatum*		
Shell	0.5 DW	4
Soft parts	5.4 DW	4
Channeled whelk, *Busycon canaliculatum*		
Muscle	<0.7 to 0.9 FW	6
Digestive gland	1.1-1.7 FW	6
Cockle, *Cardium edule*		
Shell	0.4 DW	4
Soft parts	0.8 DW	4
Cockle, *Cerastoderma edule*		
Soft parts	4.7-15.6 DW	7
Gulf of Gabes, Tunisia; February 2001		
Gill	Max. 14.0 DW	73
Digestive gland	Max. 2.0 DW	73

(Continues)

Table 6.12: Cont'd

Organism	Concentration	Reference[a]
Remainder	Max. 1.0 DW	73
Soft parts; 7 sites	0.13-1.1 DW	73
Chesapeake Bay; bivalves; 3 spp.; soft parts; 1979-1981	5.0 (0.6-27.0) DW	26
Scallop, *Chlamys opercularis*		
Soft parts	12.0 DW	8
Body fluid	1.1 DW	8
Mantle	1.0 DW	8
Gills	2.1 DW	8
Digestive gland	10.2 DW	8
Striated muscle	0.6 DW	8
Gonad and foot	2.5 DW	8
Kidneys	827.0 DW	8
Shell	0.6-1.7 DW	4
Mussel, *Choromytilus meridionalis*; soft parts	2.0-5.0 DW	9
Oyster, *Crassostrea commercialis*; soft parts	0.3-1.3 FW	10
Pacific oyster, *Crassostrea gigas*		
Soft parts	0.1-4.5 FW	11
Soft parts	1.0 DW	12
Soft parts; China; November 2003; contaminated sites	3.0-37.0 DW	75
Oyster, *Crassostrea corteziensis*; soft parts; Gulf of California; 1999-2000	3.4 DW	102
American oyster, *Crassostrea virginica*		
Soft parts		
San Antonio Bay, Texas	<0.8 DW	13
Galveston Bay, Texas; 1992-1993; summer vs. winter and spring	1.0-1.5 DW vs. 0.3-0.5 DW	63
United States East coast	<0.1 to 0.5 DW	14
United States Gulf coast	0.1-3.5 DW	14
Pensacola, Florida; 2003-2004	Max. 0.42 FW	90
Mississippi estuaries	0.63-1.6 FW	15
United States Atlantic Ocean coast	0.1-2.3 FW	16
Savannah, Georgia; 2000-2001	<1.5 to 4.0 DW	94
Shell	35.9-40.9 FW	17
Soft parts	<1.0 DW	18

(*Continues*)

Table 6.12: Cont'd

Organism	Concentration	Reference[a]
Limpet, *Crepidula fornicata*		
Shell	0.4 DW	4
Soft parts	3.9 DW	4
Mussel, *Elliptio buckleyi*; soft parts; Turkey	0.13-0.68 (0.08-0.96) FW	76
Razor clam, *Ensis siliqua*; shell; 23 stations; west coast of England		
5 stations	<10.0 DW	67
14 stations	10.1-20.0 DW	67
3 stations	21.0-30.0 DW	67
1 station	32.2 DW	67
Greenland; 5 spp.; soft parts; 1978-1993	0.07-0.09 FW	64
Gulf of Mexico; Coatzacoalcos estuary; 2005-2006; soft parts		
Asiatic clam, *Corbicula fluminea*	0.1 DW	89
Carolina marsh clam, *Polymesoda caroliniana*	0.6-1.5 DW	89
Abalone, *Haliotis rufescens*		
Digestive gland	9.0-40.0 DW	20
Remaining tissues	<0.1 DW	20
Periwinkle, *Littoraria scabra*; soft parts; Tanzania		
Industrialized sites; 1998 vs. 2005	8.2 DW vs. 1.6 DW	96
Reference site; 2005	0.8 DW	96
Limpet, *Littorina littorea*		
Soft parts	4.0-15.0 DW	21
Digestive gland and gonad	3.7-10.0 DW	21
Shell	28.8-41.5 DW	21
Soft parts	<0.1 to 2.8 FW	22
Soft parts	3.0 DW	23
Soft parts	19.0 (3.7-70.0) DW	7
Baltic clam, *Macoma balthica*		
Whole	0.56 FW	24
Soft parts	34.0 (15.0-61.0) DW	7
Soft parts	0.7-1.5 FW	25
Soft parts; 1996-1997; Gulf of Gdansk	0.2-8.5 DW	88

(Continues)

Table 6.12: Cont'd

Organism	Concentration	Reference[a]
Clam, *Macoma mitchelli*; whole	0.28 FW	24
Hardshell clam, *Mercenaria mercenaria*		
Soft parts	0.5 (0.1-7.5) FW	11
Soft parts	1.9 (<0.1 to 8.6) DW	27
Soft parts	18.0 DW	4
Shell	0.4 DW	4
Soft parts	3.5 DW	18
Bivalve, *Mesidotea entomon*; soft parts	0.3-1.5 FW	25
Mexico; Sinaloa; 2004-2005; soft parts; rainy season vs. dry season		
Oyster, *Crassostrea corteziensis*	7.7 DW vs. 8.8 DW	99
Mussel, *Mytella strigata*	4.5 DW vs. 6.3 DW	99
Clam, *Megapitaria squalida*	6.6 DW vs. 8.4 DW	99
Mussel, *Modiolus demissus*; soft parts	3.2 DW	18
Horse mussel, *Modiolus modiolus*		
Shell	1.9 DW	4
Soft parts	23.0-42.0 DW	4
Mantle and gills	18.0 DW	4
Muscle	7.5 DW	4
Gonad	25.0 DW	4
Gut and digestive gland	22.0 DW	4
Norway; 2007		
Kidney	44.5 (16.0-85.0) FW	98
Digestive gland	0.88 (0.16-2.2) FW	98
Gills	0.23 (0.05-0.49) FW	98
Gonad	0.15 (0.05-0.29) FW	98
Muscle	0.12 (0.04-0.32) FW	98
Total soft parts	3.3 (1.4-6.6) FW	98
Molluscs		
North America; 18 commercial species; edible tissues		
4 spp.	0.2-0.5 FW	28
12 spp.	0.5-0.7 FW	28
2 spp.	0.7-0.9 FW	28
Korea; edible tissues		
East coast	0.59 (0.25-1.2) FW	1
South coast	1.25 (0.32-3.2) FW	1
West coast	0.98 (0.11-3.8) FW	1
Soft parts; 12 spp.	0.04-1.8 FW	29

(*Continues*)

Table 6.12: Cont'd

Organism	Concentration	Reference[a]
Morocco; 2001; clams; soft parts		
Scrobicularia plana	11.0 DW	101
Cerastoderma plana	1.0 DW	101
Morocco; 2004-2005; soft parts		
Mussel, *Mytilus galloprovincialis*	9.6 (0.5-34.2) DW	95
Clam, *Venerupis decussatus*	4.1 (1.2-6.7) DW	95
Pacific oyster, *Crassostrea gigas*	4.2 (1.2-7.4) DW	95
Clam, *Mulinia lateralis*; whole	0.02 FW	24
Softshell clam, *Mya arenaria*		
Soft parts	0.1-10.2 FW	16
Soft parts	6.6-7.6 DW	30
Whole	0.3 FW	24
Mussel, *Mytilus californianus*		
Soft parts		
Near urban areas	8.0 DW	31
Near rural areas	0.3-0.8 DW	31
United States West coast	0.1-12.0 DW	14
Shell	<9.0 to 19.4 DW	2
Soft parts	<2.2 to 23.4 DW	2
Digestive gland	2.4-38.0 DW	36
Common mussel, *Mytilus edulis*		
Shell	<9.0 to 21.2 DW	2
Soft parts	<2.2 to 7.9 DW	2
Soft parts	1.9-6.4 DW	33
Shell	0.4 DW	4
Shell; Baltic Sea; 2005; 12 sites	0.01-0.43 DW	104
Soft parts	9.1 DW	4
Soft parts		
United States West coast	1.9-8.8 DW	14
United States East coast	<0.4 to 9.5 DW	14
Scandinavia	3.0-264.0 DW	34
Looe estuary, England	54.0 (30.0-105.0) DW	7
Scheldt estuary, Europe; March 2000	13.0 DW	106
Netherlands; summer 2002	Max. 9.0 DW	77
Puget Sound, Washington state	1.5-2.2 DW	35
Anaheim Bay, California	33.0 DW	31
Port Phillip Bay, Australia	up to 10.0 FW	36, 37
Scotland	<0.2 to 5.5 FW	22
New Zealand	12.0 (3.0-25.0) DW	39

(Continues)

Table 6.12: Cont'd

Organism	Concentration	Reference[a]
Florida	0.02-0.04 FW	38
Germany	2.0-6.0 DW	66
Norway	2.0-3100.0 DW	66
England	9.0 DW; 0.5-3.0 FW	66
Spain	2.0-15.0 DW	66
Greenland	2.0-21.0 FW	66
Mantle	<5.0 DW	39
Gills	36.0 DW	39
Muscle	<5.0 DW	39
Visceral mass	26.0 DW	39
Intestine	69.0 DW	39
Foot	<5.0 DW	39
Gonads	7.0 DW	39
Shell	<5.0 DW	39
Foot	Max. 8.2 DW	40
Gills	Max. 32.6 DW	40
Stomach	Max. 22.9 DW	40
Mantle	Max. 14.5 DW	40
Muscle	Max. 17.8 DW	40
Shell	Max. 81.0 DW	40
Soft parts	Max. 18.9 DW	40
Soft parts	30.0-138.0 DW	41
Hepatopancreas		
San Diego Harbor	6.9-22.0 DW	42
Offshore	<2.3 to 13.0 DW	42
Mussel, *Mytilus edulis aoteanus*		
Soft parts	12.0 DW	39
Gill	36.0 DW	39
Visceral mass	26.0 DW	39
Intestine	69.0 DW	39
Gonad	7.0 DW	39
Other tissues	<5.0 DW	39
Mussel, *Mytilus edulis planulatus*		
Soft parts	3.1-452.0 DW	19
Soft parts		
June	4.1 FW	43
January	0.5 FW	43
Soft parts	0.4-1.2 DW	3
Pacific blue mussel, *Mytilus edulis trossulus*; 2004; soft parts; Adak Island, Alaska; 5 sites	Max. 47.6 FW; 1.2-5.1 DW	84

(Continues)

Table 6.12: Cont'd

Organism	Concentration	Reference[a]
Mussel, *Mytilus galloprovincialis*		
Soft parts; Venice lagoon, Italy; 2005-2006; San Giuliano vs. Sacca Sessola		
April	4.3 DW vs. 3.2 DW	91
July	1.5 DW vs. 1.1 DW	91
October	2.3 DW vs. 1.9 DW	91
February	3.6 DW vs. 3.2 DW	91
Soft parts	13.4 FW	44
Soft parts	21.5 DW	45
Soft parts; Adriatic Sea, Italy; 2 sites	0.10-0.16 FW	103
Soft parts; Adriatic Sea; 2001-2005; winter vs. summer	1.2-2.0 DW vs. 0.2-1.0 DW	109
Soft parts; Marmara Sea, Turkey; 2005	0.82 FW	72
Soft parts; Gulf of Gemlik, Turkey; 2004	0.5 DW	112
Soft parts; Izmir Bay, Turkey; 2004; near shore vs. mid-bay	0.4 DW vs. 0.08 DW	105
Soft parts; 2002-2003; Aegean Sea, Turkey	0.98 (0.49-1.7) FW	78
Soft parts; 2001-2002; Adriatic Sea; summer vs. winter		
Industrialized sites	1.9-3.1 DW vs. 0.8-11.5 DW	83
Reference site	1.6 DW vs. 1.6 DW	83
Soft parts; May 2003; NW Mediterranean Sea		
Near Genova, Italy	1.9-29.8 DW	69
Near Marseille, France	2.2-4.0 DW	69
Near Barcelona, Spain	1.8-38.5 DW	69
Soft parts; Sardinia; 2004 vs. 2006	18.0 DW vs. 7.5 DW	107
Soft parts; 1991-2005; Balearic Islands; western Mediterranean Sea; Port of Mahon	10.0 (3.3-18.6) DW	70
Mantle vs. gills; 2004-2006; Saronikos Gulf of Greece; max. mean concentrations; contaminated sites		
Summer	10.5 FW vs. 21.0 FW	71
Autumn	5.4 FW vs. 8.7 FW	71
Winter	0.38 FW vs. 9.8 FW	71
Spring	0.34 FW vs. 0.95 FW	71
Soft parts; Mediterranean Sea; 3 sites; summers 2004-2005	1.0-1.4 DW	93
Mussel, *Mytilus viridis*; soft parts	1.1-12.7 DW	46

(*Continues*)

Table 6.12: Cont'd

Organism	Concentration	Reference[a]
Whelk, *Nucella lapillus*		
Shell	0.5 DW	4
Soft parts	4.9 DW	4
Soft parts	1.9-7.1 DW	7
Octopus, *Octopus* sp.; whole	1.0 FW	47
Octopus, *Octopus vulgaris*		
Portugal; 1999-2002; 3 sites		
Digestive gland	4.8-7.2 DW	110
Branchial heart	3.1-10.0 DW	110
Digestive gland; Portugal; 2005-2006; northwest region vs. south coast		
Whole	2.4 (1.5-4.1) DW vs. 4.8 (3.0-7.2) DW	100
Nuclei	5.4 (4.5-22.0) DW vs. 8.6 (2.7-16.0) DW	100
Mitochondria	6.4 (1.9-31.0) DW vs. 12.0 (4.7-21.0) DW	100
Lysosomes	3.2 (1.4-5.7) DW vs. 10.0 (2.4-26.0) DW	100
Microsomes	6.4 (1.5-25.0) DW vs. 7.0 (2.5-34.0) DW	100
Squid, *Ommastrephes illicebrosa*; whole	5.0 AW	48
Oyster, *Ostrea angasi*		
Soft parts	1.4-84.0 DW	19
Soft parts		
June	2.0 FW	43
January	0.6 FW	43
Whole	0.5-1.9 DW	3
European flat oyster, *Ostrea edulis*		
Soft parts	0.2-2.3 DW	49
Soft parts; 1991-2005; Balearic Islands; western Mediterranean Sea; Port of Mahon	7.5 (2.1-17.3) DW	70
Oyster, *Ostrea gigas*; soft parts		
United States East coast	1.4 DW	14
United States Gulf coast	0.1-0.3 DW	14
Oyster, *Ostrea sinuata*		
Soft parts	10.0 DW	39
Heart	15.0 DW	39
Other tissues	5.0 DW	39

(Continues)

Table 6.12: Cont'd

Organism	Concentration	Reference[a]
Limpet, *Patella candei gomesii*; soft parts; 2007; Azores		
Near hydrothermal vent	0.24-0.35 DW	97
Reference site	0.30-0.94 DW	97
Limpet, *Patella coerulea*; soft parts	<7.0 to 95.6 DW; <1.7 to 27.5 FW	5
Gastropod, *Patella piperata*; Canary Islands, Spain; soft parts; March-April 2003	1.6 (0.3-10.2) DW	79
Limpet, *Patella vulgata*		
Soft parts	5.1-38.0 DW	7
Soft parts	9.5 DW	23
Soft parts	7.9 DW	51
Shell	46.0 DW	4
Soft parts	32.0 DW	4
Giant scallop, *Pecten maximus*		
Soft parts	<0.1 to 1.0 FW	22
Soft parts	2.0 DW	8
Body fluid	0.9 DW	8
Mantle	0.7 DW	8
Gills	0.8 DW	8
Digestive gland	3.9 DW	8
Striated muscle	0.2 DW	8
Gonad and foot	1.7 DW	8
Kidney	159.0 DW	8
Kidney	44.0 FW	52
Soft parts	8.3 DW	4
Muscle	17.0 DW	4
Gut and digestive gland	1.7 DW	4
Mantle and gills	1.5-2.8 DW	4
Gonad		
Unwashed	31.0 DW	4
Washed	0.4 DW	4
Shell	0.6-2.4 DW	4
Norway; 2006		
Digestive gland	0.29 (0.17-0.50) FW	98
Kidney	8.2 (4.0-12.4) FW	98
Muscle	<0.01 (<0.01 to 0.02) FW	98
Gonad	0.07 (0.05-0.09) FW	98
Total soft parts	0.12 (0.09-0.14) FW	98

(Continues)

Table 6.12: Cont'd

Organism	Concentration	Reference[a]
Scallop, *Pecten novaezelandiae*		
Stomach	1.2 FW	29
Gonad	0.3 FW	29
Abductor muscle	0.5 FW	29
Soft parts	1.1 FW	29
Soft parts	16.0 DW	39
Gill	52.0 DW	39
Intestine	28.0 DW	39
Kidney	137.0 DW	39
Foot	14.0 DW	39
Gonad	78.0 DW	39
Green mussel, *Perna viridis*		
India; January 2002; 28 sites; soft parts		
Small	0.5 FW	86
Medium	0.25 FW	86
Large	0.09 FW	86
Hong Kong; juveniles vs. adults		
Reference site		
Soft parts	3.7 DW vs. 3.3 DW	111
Byssus	5.4 DW vs. 7.3 DW	111
Polluted site		
Soft parts	4.2 DW vs. 7.0 DW	111
Byssus	4.9 DW vs. 7.7 DW	111
Pearl oyster, *Pinctada fucata*		
Gills	3.2 DW	113
Mantle	1.4 DW	113
Digestive gland	2.1 DW	113
Foot	1.3 DW	113
Clam, *Pitar morrhuana*; soft parts	11.1-29.4 DW	53
Scallop, *Placopecten magellanicus*		
Soft parts	3.6 DW	54
Muscle	<0.4 to 1.7 FW	55
Gonad	<0.4 to 1.5 FW	55
Visceral mass	0.45-1.6 FW	55
Clam, *Protothaca staminea*		
Shell	<9.0 DW	2
Soft parts	5.2 DW	2
Clam, *Rangia cuneata*		
Soft parts	1.1 DW	56
Whole	0.37 FW	24

(Continues)

Table 6.12: Cont'd

Organism	Concentration	Reference[a]
Manila clam, *Ruditapes philippinarum*		
Soft parts; Korea; 2002-2003		
Females vs. males	0.78 DW vs. 0.73 DW	68
Postspawn vs. prespawn	1.1 DW vs. 0.76 DW	68
Jiaozhou Bay, China; sediments vs. soft parts; November 2004	34.8 DW vs. 0.31-1.0 DW	81
Bohai Sea; soft parts	0.13-0.33 DW	82
Clam, *Scrobicularia plana*		
Soft parts	8.0-327.0 DW	57
Soft parts	189.0 (62.0-473.0) DW	7
Camel estuary, England		
Soft parts	12.0-21.0 DW	58
Digestive gland	37.8 DW	58
Mantle and siphons	3.7 DW	58
Remaining tissues	not detectable	58
Gannel estuary, England		
Soft parts	234.0-828.0 DW	58
Digestive gland	1130.0 DW	58
Mantle and siphons	90.0 DW	58
Foot and gonads	8.0 DW	58
Gills and palps	29.0 DW	58
Adductor muscles	9.0 DW	58
Kidneys	74.0 DW	58
Transferred from Camel estuary to Gannel estuary for 352 days		
Digestive gland	402.0 DW	58
Remaining tissues	35.0 DW	58
Transferred from Gannel estuary to Camel estuary for 352 days		
Digestive gland	486.0 DW	58
Remaining tissues	20.0 DW	58
Tamar estuary, England		
Soft parts	94.0 (26.0-296.0) DW	59
Digestive gland	384.0 (235.0-983.0) DW	59
Common cuttlefish, *Sepia officinalis*		
Egg		
Eggshell	1.4 DW	85
Embryo	<0.1 DW	85
Yolk	<0.1 DW	85
Hatchlings		
Whole	<0.1 DW	85
Cuttlebone	0.6 DW	85

(*Continues*)

Table 6.12: Cont'd

Organism	Concentration	Reference[a]
Juveniles, age 1 month		
Whole	<0.1 DW	85
Digestive gland	1.6 DW	85
Cuttlebone	0.5 DW	85
Immatures, age 12 months		
Whole	0.65 DW	85
Digestive gland	2.4 DW	85
Cuttlebone	0.9 DW	85
Adults, age 18 months		
Whole	0.2 DW	85
Digestive gland	2.2 DW	85
Cuttlebone	0.3 DW	85
Squid, *Sepia* sp.; whole	0.4 FW	47
Spain; Tarragona coast; commercial species; edible portions	Max. 2.4 FW	65
Clam, *Spisula* sp.; soft parts; United States		
Delaware	7.4 FW	60
Massachusetts	1.2 DW	60
Squid		
Flesh	1.3 DW	61
Flesh with skin	2.0 DW	61
Viscera	2.7 DW	61
Pen	2.3 DW	61
Gastropod, *Strombus pugilis*		
Shell	88.0 (79.0-98.0) DW	62
Muscle	8.0 (7.3-8.6) DW	62
Soft parts	24.0 DW	62
Chiton, *Sypharochiton pellis-serpentis*; soft parts	1.9-152.0 DW	19
Clam, *Tapes semidecussata*		
Shell	<9.0 DW	2
Soft parts	<2.0 DW	2
Gastropod, *Thais lapillus*		
Soft parts	Max. 27.0 DW	23
Soft parts	3.9-19.6 DW	21
Digestive gland and gonad	4.2-11.0 DW	21
Shell	28.4-44.9 DW	21

(Continues)

Table 6.12: Cont'd

Organism	Concentration	Reference[a]
Clam, *Trivela mactroidea*; Venezuela; 2002; 14 sites; soft parts	<1.5 to 4.9 DW	92
Clam, *Venus mercenaria*; soft parts		
Delaware	1.1 FW	60
Massachusetts	2.6 FW	60

Values are in mg Pb/kg fresh weight (FW), dry weight (DW), or ash weight (AW).

[a]1, Won, 1973; 2, Graham, 1972; 3, Harris et al., 1979; 4, Segar et al., 1971; 5, Shiber and Shatila, 1978; 6, Greig et al., 1977; 7, Bryan and Hummerstone, 1977; 8, Bryan, 1973; 9, Watling and Watling, 1976b; 10, Mackay et al., 1975; 11, Pringle et al., 1968; 12, Watling and Watling, 1976a; 13, Sims and Presley, 1976; 14, Goldberg et al., 1978; 15, Harvey and Knight, 1978; 16, Shuster and Pringle, 1968; 17, Ferrell et al., 1973; 18, Valiela et al., 1974; 19, Bloom and Ayling, 1977; 20, Anderlini, 1974; 21, Ireland and Wooton, 1977; 22, Topping, 1973; 23, Butterworth et al., 1972; 24, White et al., 1979; 25, Miettinen and Verta, 1978; 26, Di Giulio and Scanlon, 1985; 27, Genest, 1979; 28, Hall et al., 1978; 29, Nielsen and Nathan, 1975; 30, Eisler, 1977b; 31, Chow et al., 1976; 32, Alexander and Young, 1976; 33, Schulz-Baldes, 1973; 34, Phillips, 1978; 35, Schell and Nevissi, 1977; 36, Phillips, 1976a; 37, Phillips, 1976b; 38, Horvath et al., 1972; 39, Brooks and Rumsby, 1965; 40, Talbot et al., 1976a; 41, Simpson, 1979; 42, Young et al., 1979; 43, Thomson, 1979; 44, Uysal, 1978a; 45, Fowler and Oregioni, 1976; 46, Bhosle and Matondkar, 1978; 47, Matsumoto et al., 1964; 48, Nicholls et al., 1959; 49, Fukai et al., 1978; 50, Hayashi, 1960; 51, Preston et al., 1972; 52, George et al., 1980; 53, Eisler et al., 1978; 54, Reynolds, 1979; 55, Greig et al., 1978; 56, Sims and Presley, 1976; 57, Luoma and Bryan, 1978; 58, Bryan and Hummerstone, 1978; 59, Bryan and Uysal, 1978; 60, Craig, 1967; 61, Horowitz and Presley, 1977; 62, Lowman et al., 1966; 63, Jiann and Presley, 1997; 64, Dietz et al., 1996; 65, Schuhmacher et al., 1990; 66, Jenkins, 1980; 67, Pearce and Mann, 2006; 68, Ji et al., 2006; 69, Zorita et al., 2007; 70, Deudero et al., 2007; 71, Vlahogianni et al., 2007; 72, Keskin et al., 2007; 73, Machreki-Ajmi and Hamza-Chaffai, 2006; 74, Karayakar et al., 2007; 75, Liu and Deng, 2007; 76, Yarsan et al., 2007; 77, Mubiana and Blust, 2007; 78, Sunlu, 2006; 79, Bergasa et al., 2007; 80, Huang et al., 2007; 81, Li et al., 2006; 82, Liang et al., 2004a; 83, Scancar et al., 2007; 84, Burger and Gochfeld, 2006; 85, Miramand et al., 2006; 86, Sasikumar et al., 2006; 87, Burt et al., 2007; 88, Sokolowski et al., 2007; 89, Ruelas-Inzunza et al., 2007; 90, Karouna-Renier et al., 2007; 91, Nesto et al., 2007; 92, LaBrecque et al., 2004; 93, Lafabrie et al., 2007; 94, Kumar et al., 2008; 95, Maanan, 2008; 96, De Wolf and Rashid, 2008; 97, Cunha et al., 2008; 98, Julshamn et al., 2008; 99, Frias-Espericueta et al., 2008; 100, Raimundo et al., 2008; 101, Anajjar et al., 2008; 102, Ruelas-Inzunza and Paez-Osuna, 2008; 103, Conti et al., 2008; 104, Protasowicki et al., 2008; 105, Kucuksezgin et al., 2008; 106, Wepener et al., 2008; 107, Schintu et al., 2008; 108, Bustamante et al., 2008; 109, Fattorini et al., 2008; 110, Napoleao et al., 2005; 111, Nicholson and Szefer, 2003; 112, Unlu et al., 2008; 113, Jing et al., 2007.

concentrations in mussels (Talbot et al., 1976a). Proposed lead criteria to protect human health include less than 5.0 mg Pb/kg DW in marine mollusks (Kucuksezgin et al., 2008), less than 1.5 mg Pb/kg FW in seafood products (Julshamn et al, 2008), less than 1.0 mg Pb/kg FW diet (Glynn et al., 2003; Ruelas-Inzunza et al., 2007), less than 1.5 mg Pb/kg DW diet (Unlu et al., 2008), and less than 0.5 mg Pb/kg FW in molluscan soft tissues (Burger and Gochfeld, 2006; Huang et al., 2007).

High lead concentrations were especially common in bivalves collected near industrialized and populated areas subject to continual atmospheric fallout of lead from automobile exhaust during the period preceding wide-scale use of lead-free gasoline. In general, shellfish collected near sewer outfalls, in regions near heavy vehicular traffic, and in proximity to bridges coated with lead-based paints, had especially high lead burdens (Eisler, 2000c;

Graham, 1972). *Mytilus* and *Acmaea*, two organisms that live on and feed from rock surfaces in the intertidal zone, had comparatively elevated lead content; incomplete rinsing during high tides may be a contributing factor in specimens living near point sources of lead contamination (Graham, 1972). Mussels from the most impacted zones showed enlarged lysosomes accompanied by reduced labilization period of lysosomal membranes, indicating disturbed health (Zorita et al., 2007).

In general, highest accumulations of lead in bivalve tissues were in kidney, gills, and digestive gland, and lowest concentrations in gonads, mantle and muscle (Table 6.12). Accumulation studies with *M. edulis* demonstrated that kidney contained 50-70% of the total lead while accounting for only 6-8% of the total biomass (Schulz-Baldes, 1973, 1974). Similar distributions were found in scallops (Brooks and Rumsby, 1965; Bryan, 1973), oysters (Pringle et al., 1968), and cephalopods (Heyraud and Cherry, 1979). Digestive gland was the major repository of lead in the abalone *Haliotis rufescens* (Stewart and Schulz-Baldes, 1976) and in the mussel *M. californianus* (Alexander and Young, 1976). Newly formed carbonate fractions of *M. edulis* shell contained 6-10 times more lead than older carbonate fractions; however, older periostracum contains more lead than newly formed periostracum (Sturesson, 1976).

Age and gender of the organism, and other variables influence lead content in tissues. For example, lead concentrations in soft parts of American oysters (*C. virginica*) were higher in smaller oysters than in larger ones, much lower than the lead burden of the surrounding sediments, and strongly correlated with iron content of the geochemical environment (Jiann and Presley, 1997). In the mussel, *M. galloprovincialis*, lead concentrations in soft parts decreased with increasing weight (Conti et al., 2008). Among various species of mussels, smaller and younger organisms had significantly higher lead uptake rates and lower loss rates than did larger, older conspecifics (Schulz-Baldes, 1973, 1974; Watling and Watling, 1976b); a similar pattern was observed among several species of bivalves from Greenland (Dietz et al., 1996). Lead concentrations were higher in soft tissues of male *C. meridionalis* than in females; lead was inversely related to iron concentrations in females and this may account for the depressed lead levels in females (Watling and Watling, 1976b). In *M. californianus*, lead concentrations were 1.7 times higher in testicular tissue than in ovarian tissue, although no sex-related differences were documented in lead content of digestive gland or muscle (Alexander and Young, 1976).

Temperature of ambient seawater in the 10-18 °C range made little difference in lead content in tissues of *M. edulis* (Phillips, 1976a). However, salinity of the medium was significant and lead residues were higher in mussels from lower salinity waters (Mubiana and Blust, 2006; Phillips, 1976a,b, 1978). The higher lead concentrations in mussels from low salinity waters are not correlated with water lead concentrations; differences are attributed to the biological availability of lead, that exists mainly in particulate form in seawater (Phillips, 1978).

Selectivity for lead and other metals among various molluscan species depends, in part, upon its availability in the environment, its chemical and physical properties, the kind and number of ligands available for chelation, its transportation and storage, and the stability of the complex formed. The relative toxicities of these compounds also play a prominent role (Pringle et al., 1968). For the clam *S. plana*, the biological availability of lead in the sediments is controlled mainly by the concentration of iron. The concentration of lead in *S. plana* may be predicted from the Pb/Fe ratio in acid extracts of surficial sediments (Luoma and Bryan, 1978). Mussels, *M. edulis*, show a linear relation between body burdens and lead in ambient seawater (Schulz-Baldes, 1973, 1974). However, Phillips (1976a,b) states that this relation is not valid for tetraethyllead salts owing to differences in uptake kinetics of this molecule and those of inorganic lead or other lead salts.

Diet, seasonality, depth of collection, and presence of other metals salts affect lead accumulations. Among abalones, *H. rufescens*, substantial uptake via diet occurred after ingestion of algae *Egretia laevigata* pretreated in solutions containing 1.0 mg Pb/L; abalones accumulated up to 21.0 mg Pb/kg FW soft parts, almost wholly in digestive gland (Stewart and Schulz-Baldes, 1976). Among mussels, however, there was no significant difference in lead uptake between lead offered via diet (algae) or by way of the medium (Schulz-Baldes, 1974). Seasonal variations in lead content of bivalves are attributed to seasonal variations in phytoplankton trace metals (Bryan, 1973) or to high precipitation and attendant runoff rather than to local pollution (Fowler and Oregioni, 1976). Phillips (1976a) maintains that the trace metal content of individual mussels remains constant throughout the year and that weight (moisture) fluctuations produced much of the seasonality in mussel lead concentrations. Differences in lead content of cultured mussels *P. canaliculus* with depth of collection is a function of food availability and to the ratio of particulate to dissolved lead with increasing vertical concentration gradients. Either or both of these conditions could influence the bioavailability of lead and other metals with increasing depth (Nielsen, 1974). The concentrations of zinc, lead, and cadmium attained in mussels are consistent with a theory that no interactions occurred between these metals during uptake (Phillips, 1977a,b); however, Phillips cautions that this relation does not hold for all metals, especially copper. In green mussel, *P. viridis*, lead content in soft parts is positively correlated with cadmium and chromium burdens, as well as proximity to industrial sites (Sasikumar et al., 2006).

Mussels are capable of acting as efficient, time-integrated indicators of lead and other metals over a wide range of environmental conditions, and are recommended as an alternative to the analysis of sediment and water (Phillips, 1976a,b,c 1977a,b,c, 1978). Fowler and Oregioni (1976) report that poor correlations between metals in mussel soft parts and metals in ambient seawater may arise from either of two conditions. First, filter feeders ingest particulates which could result in enhanced metal content during periods of high runoff; thus, metal concentrations in filtered seawater (used in some laboratory studies) may not be representative of total metal available to mussels in their immediate biosphere. Second, metal uptake and excretion kinetics in mussels may be slow relative to short-term fluctuations in metal concentrations occurring in the water; thus, metal concentrations in mussels at any given point in time will not reflect exactly those measured in seawater sampled at the same time. Clearly, care should be taken when using *Mytilus* and other filter-feeding bivalves to infer metal concentrations in ambient waters.

Many laboratory studies have been conducted to measure accumulation and loss of lead in various species of bivalves under controlled environmental conditions. At 0.01 mg Pb/L medium, there was no significant difference in lead uptake by *M. edulis* whether lead was offered via the diet or via the medium (Schulz-Baldes, 1974); the highest values recorded were in kidney (135.8 mg Pb/kg DW) after immersion for 63 days when compared to the start (10.0 mg Pb/kg DW kidney). For total mussel soft parts after 63 days immersed in seawater solutions containing 0.01 mg Pb/L, lead content was 18.3 mg Pb/kg DW versus 2.5 in controls. Schulz-Baldes (1974) showed that lead uptake was a linear function of lead concentrations in the medium for the first 40 days over the range of lead concentrations tested: 0.005, 0.01, 0.05, 0.1, 0.2, 0.5, 1.0, and 5.0 mg Pb/L. For all groups, lead burdens were similar at 0.26-0.31 mg Pb/kg daily, regardless of amount accumulated or route of administration (Schulz-Baldes, 1974). Mussels, *M. galloprovincialis*, exposed to 0.1 mg Pb/L for 60 days as $Pb(NO_3)_2$ had highest concentrations in gills, palps, and digestive gland, specifically in various lysosomal structures within these tissues (Dimitriadis et al., 2003). Remarkable lead accumulations of 12,840.0 mg Pb/kg DW soft parts were measured in *M. edulis* following exposure for 150 days in 0.5 mg Pb/L; controls contained 8.4 mg Pb/kg DW (Schulz-Baldes, 1972). Softshell clams, *M. arenaria*, in 0.014 mg Pb/L, accumulated up to 2.9 mg Pb/kg FW soft parts in 6 weeks under ambient winter thermal regimes versus 0.68 mg Pb/kg FW in controls. At 0.07 mg Pb/L, *Mya* took up 13.3 mg Pb/kg FW soft parts under similar conditions (Eisler, 1977b). At summer thermal regimes, these values in *Mya* were about 5.6 mg Pb/kg FW soft parts after 14 days in 0.014 mg Pb/L and 17.3 in 7 days in 0.07 mg Pb/L (Eisler, 1977b). Hardshell clams, *M. mercenaria*, contained 70.0 mg Pb/kg FW soft parts after exposure for 20 weeks in 0.2 mg Pb/L (Shuster and Pringle, 1968). American oysters, *C. virginica*, held in seawater solutions containing 0.025 mg Pb/L for 5 weeks contained up to 17.0 mg Pb/kg FW soft parts; after 10 weeks, oysters had 35.1 mg Pb/kg FW (Shuster and Pringle, 1968, 1969). Higher nominal lead concentrations and increasing periods of exposure in the above studies were always associated with higher lead body burdens in oysters. In another study with American oysters, Zaroogian et al. (1979) showed significant accumulations after long-term exposure to extremely low lead concentrations. At ambient levels of 0.001 mg Pb/L, oyster tissues contained 6.6 mg Pb/kg DW soft parts after 20 weeks; for 0.0033 mg Pb/L, this value was 11.4 mg Pb/kg DW. In all studies, the oysters and their progeny appeared normal. Zaroogian et al. (1979) maintain that 85-90% of the lead added to seawater was detected in the particulate fraction. The remainder is suggested to occur in a complex inorganic and a chelated form in the soluble fraction. The adsorptive behavior of lead is probably strongly pH dependent. Thus, when the pH is lowered, lead could be released from the particulates and a shift might occur in the equilibrium between the chemical species of lead. This change in equilibrium could favor ionic lead in the oyster digestive tract that has a pH range of 5.5-6.0.

Pearl oysters, *P. fucata*, exposed to 0.1 mg Pb/L for 72 h had 29.1 mg Pb/kg DW in gills (up from 3.2 mg Pb/kg DW at start), 11.1 mg Pb/kg DW in mantle (up from 1.4), 9.4 mg/kg DW in digestive gland (up from 2.1) and 9.8 mg Pb/kg DW in foot (up from 1.3 mg Pb/kg DW at start) (Jing et al., 2007). Mantle tissue was especially sensitive, with altered activities of antioxidant and immune enzymes (Jing et al., 2007).

Mechanisms to account for lead uptake and retention by molluscan soft tissues are proposed by Romeril (1971) and Bryan (1971). Uptake mechanisms, according to Romeril (1971), include adsorption of ions at membrane-water interfaces; absorption by active or passive diffusion of metal ions across semipermeable membranes into body fluids and subsequent distribution to other organs; and ingestion of ions with food or in combination with particulate matter or mucous, and absorption through the gut wall. Despite having no direct connection with the medium or intestine, the mussel kidney takes up lead most rapidly of all tissues and organs, presumably via the blood. It is likely that initial uptake is via gills and adductor muscle into the blood, with eventual deposition in the kidney. According to the last proposed mechanism of Romeril (1971), lead is taken up by stomach and digestive gland; however, gills accumulate lead rapidly, possibly owing to the high blood flow at that site. Bryan (1971) proposed three mechanisms for metals loss: excretion across the body surface or gills; excretion via the gut; and excretion via urine. It is probable that all three mechanisms are used by mussels in lead depuration.

In general, alkyl leads are more toxic than inorganic lead compounds, and tetraalkyl leads are more toxic than trialkyl lead compounds. For example, adverse effects of lead on the survival of the common mussel, *M. edulis*, include 50% dead in 96 h at 0.1 mg tetraethyllead/L, 0.27 mg tetramethyllead/L, 0.5 mg trimethyllead/L, 1.1 mg triethyllead/L (Maddock and Taylor, 1980), and 0.5 mg inorganic lead/L for mussel larvae (USEPA, 1985b).

Shell has been proposed as a suitable indicator of lead exposure in the Akoya pearl oyster, *Pinctada imbricata* (MacFarlane et al., 2006). Under controlled conditions, oysters were subjected to 0.18 mg Pb/L (as lead nitrate) for 9 weeks, or to two short term exposures of 0.18 mg Pb/L, namely, 3 weeks each, with an intervening depuration period of 3 weeks. Soft tissues after 9 week exposure had 539.0 mg Pb/kg DW; the 2×3 week exposure had 603.0 mg Pb/kg DW, and the controls 0.2 mg Pb/kg DW; for shell, these values were 249.0 mg Pb/kg DW, 207.0 mg/kg DW, and 6.0 mg Pb/kg DW, respectively. Lead accumulated in successively deposited nacreous layers of shell; patterns of deposition reflected the extent and frequency of lead insult (MacFarlane et al., 2006).

6.21 Lithium

Shells of the common mussel, *M. edulis*, collected from the Baltic Sea in 2005 at 12 sites contained 0.29-3.48 mg Li/kg DW (Protasowicki et al., 2008).

6.22 Manganese

Manganese concentrations in various molluscan tissues on a FW basis ranged from 0.5 to 110,000.0 mg Mn/kg; major sites of accumulation were in kidney concretions, kidney, shell, gill, and digestive gland (Table 6.13). In addition to tissue specific differences in ability to accumulate and retain manganese, other factors should be considered when interpreting manganese concentrations in field collections of molluscs. Elevated levels of manganese and other metals in flesh of mussels, for example, was linked to contamination from automobiles, small crafts, and to substrate composition, especially muddy substrates (Graham, 1972). It is generally acknowledged that mussels collected near industrialized port areas had higher manganese concentrations than did conspecifics collected some distance from these areas (Fowler and Oregioni, 1976; Sasikumar et al., 2006); however, manganese residues in mussels collected in Sweden showed little relation to proximity to known industrial areas (Phillips, 1978). Chemical availability of manganese to Swedish mussels is secondary to that of biological availability, and the latter is influenced by the presence of other metals in solution and water salinity (Phillips, 1978). Variations in manganese content of oyster shell and soft parts were attributed to fluctuations in water salinity; residues in shell (Rucker and Valentine, 1961) and soft tissues (Mubiana and Blust, 2006) being higher at lower salinities. Among common mussels, *M. edulis*, it seems that manganese is regulated in soft parts to some extent in waters of low salinity (Phillips, 1978). Seasonal variations in manganese content of molluscan tissues, as distinct from changes in thermal regimes, are documented (Bryan, 1973; Patel et al., 1973). In soft tissues of the green mussel, *P. viridis*, manganese concentrations were positively correlated with copper, iron, and nickel contents (Sasikumar et al., 2006).

Soft tissues of the Baltic clam, *M. balthica*, collected from the Gulf of Gdansk in 1996-1997 had highest concentrations in autumn and spring, and lowest in winter (Sokolowski et al., 2007). The main source of manganese was suspended and deposited particles. Uptake was inhibited by increasing concentrations of iron in the immediate biosphere. Uptake was higher at increasing depths, which was a function of extractable manganese and iron in the sediments (Sokolowski et al., 2007).

Diet is the major route of manganese accumulation in mussels; the role of seawater in manganese accumulation is minor when compared with food (Pentreath, 1973). Soluble chemical species of manganese were taken up by mussels more rapidly than particulate forms (Pentreath, 1973); the reverse was true for clams, *D. trunculus* (Orlando and Mauri, 1978).

Table 6.13: Manganese Concentrations in Field Collections of Molluscs

Organism	Concentration	Reference[a]
Gastropod, *Acmaea digitalis*		
Shell	19.3-30.7 DW	1
Soft parts	24.5-25.1 DW	1
Gastropod, *Aplysia benedicti*		
Ctenidium	13.0-15.0 DW	2
Gonad	14.0-21.0 DW	2
Parapodia	5.0-7.0 DW	2
Hepatopancreas	4.0-7.0 DW	2
Buccal mass	5.0-8.0 DW	2
Intestine	5.0 DW	2
Gut content	8.0 DW	2
Shell	101.0-106.0 DW	2
Giant squid, *Architeuthis dux*; Spain; 2001-2005		
Hepatopancreas	2.3-2.7 DW	67
Oviduct gland	27.0 DW	67
Skin	9.0 DW	67
Scallop, *Argopecten gibbus*; kidney concretion	24,000.0 DW	3
Scallop, *Argopecten irradians*; kidney concretion	23,643.0 DW	3
Cockle, *Austrovenus stutchburyi*; soft parts; New Zealand; 1993-1994	2.0-12.0 DW	45
Hydrothermal vent mussel, *Bathymodiolus azoricus*; mid-Atlantic Ridge		
2002; 3 sites		
Gill	4.0-24.6 DW	52
Mantle	3.1-21.5 DW	52
Digestive gland	5.9-17.7 DW	52
Byssus threads	12.4-208.5 DW	52
Shell; 2001; 4 sites	3.7-31.4 (0.1-62.1) DW	60
2001; 5 sites		
Gill	6.3-9.5 DW	61
Mantle	2.4-8.4 DW	61
Digestive gland	2.1-8.1 DW	61
Whelk, *Buccinum undatum*		
Shell	1.2 DW	4
Soft parts	1.7 DW	4

(*Continues*)

Table 6.13: Cont'd

Organism	Concentration	Reference[a]
Whelk, *Busycon canaliculatum*		
Muscle	2.4-3.5 FW	5
Digestive gland	3.5-5.7 FW	5
Cockle, *Cardium edule*		
Shell	2.0 DW	4
Soft parts	6.3 DW	4
Cockle, *Cerastoderma edule*; soft parts	6.2-44.6 DW	6
Scallop, *Chlamys opercularis*		
Shell	17.0-18.0 DW	4
Soft parts	158.0 DW	7
Body fluid	2.0 DW	7
Gills	105.0 DW	7
Mantle	18.6 DW	7
Digestive gland	29.7 DW	7
Striated muscle	4.0 DW	7
Unstriated muscle	4.4 DW	7
Gonad and foot	12.5 DW	7
Kidneys	17,300.0 DW	7
Mussel, *Choromytilus meridionalis*		
Soft parts	1.7 FW	8
Soft parts	9.0-11.0 DW	9
Scallop, *Comptopallium radula*; New Caledonia; October 2004		
Soft parts	6.1 (2.8-8.7) DW	56
Kidney	187.0 DW	56
Pacific oyster, *Crassostrea gigas*		
Soft parts	0.9-16.0 FW	10
Soft parts	12.0-16.0 DW	9, 11
Oyster, *Crassostrea gryphoides*; soft parts	4.9 (4.2-5.7) FW	12
Oyster, *Crassostrea margaritacea*; soft parts	2.0 DW	11
Oyster, *Crassostrea rhizophorae*; soft parts; summer 2001; Brazil; contaminated estuary vs. reference site	28.1-60.9 DW; max. 81.9 DW vs. 33.9-50.0 DW; max. 67.8 DW	48
American oyster, *Crassostrea virginica*		
Soft parts	4.3 (0.1-15.0) FW	10
Shell	33.0-121.0 DW	13
Soft parts	24.0-51.0 DW	13

(*Continues*)

Table 6.13: Cont'd

Organism	Concentration	Reference[a]
Soft parts; United States		
Gulf coast	5.9-41.9 DW	14
New England	9.0-60.0 DW	15
Chesapeake Bay		
October	30.0 DW	16
December	2.0 DW	16
Soft parts; Savannah, Georgia; 2000-2001	17.0-54.0 DW	55
Shell	21.0-227.0 DW	17
Shell	505.0 DW	16
Soft parts		
Winter	7.0-11.0 DW	18
Summer	38.6 DW	18
Ovaries, mature	51.0-59.6 DW	18
Testes, mature	4.6-7.3 DW	18
Gills		
Winter	17.1 DW	18
Summer	38.6 DW	19
Mantle		
Winter	8.7 DW	18
Summer	17.0 DW	18
Adductor muscle	3.6-9.2 DW	18
Limpet, *Crepidula fornicata*		
Shell	2.1 DW	4
Soft parts	17.0 DW	4
Clam, *Donax serra*; soft parts	0.84 FW	8
Gastropods		
Muscle	370.0 AW	19
Shell	140.0 AW	19
Viscera	23.0 AW	19
Operculum	17,000.0 AW	19
Abalone, *Haliotis midea*; soft parts	0.27 FW	8
Periwinkle, *Littoraria scabra*; soft parts; Tanzania		
Industrial sites; 1998 vs. 2005	300.0 DW vs. 50.0 DW	58
Reference site; 2005	15.0 DW	58
Limpet, *Littorina littorea*		
Soft parts	18.6-59.8 DW	20
Digestive gland and gonad	68.8-104.6 DW	20

(*Continues*)

Table 6.13: Cont'd

Organism	Concentration	Reference[a]
Shell	5.5-8.3 DW	20
Soft parts with operculum	18.0-133.0 DW	6
Clam, *Macoma balthica*		
Soft parts	19.0-24.0 DW	6
Soft parts; 1996-1997; Gulf of Gdansk; age 3-4 years	10.0-50.0 DW	51
Hardshell clam, *Mercenaria mercenaria*		
Shell	1.5 DW	4
Soft parts	5.8 (0.7-29.7) FW	10
Clam, *Meretrix lamarckii*; soft parts	5.4 DW	21
Clam, *Meretrix meretrix*; soft parts	11.5 (3.3-27.6) FW	12
Mussel, *Modiolus modiolus*		
Shell	20.0 DW	4
Soft parts	47.0-150.0 DW	4
Mantle and gills	160.0 DW	4
Muscle	71.0 DW	4
Gonad	49.0 DW	4
Gut and digestive gland	140.0 DW	4
Molluscs		
Edible tissues of 15 commercial species		
5 spp.	0.1-0.3 FW	22
2 spp.	0.3-0.5 FW	22
2 spp.	0.5-0.7 FW	22
3 spp.	0.8-2.0 FW	22
3 spp.	2.0-5.0 FW	22
3 spp.	6.0-8.0 FW	22
Edible tissues; 3 spp.; India; 2003	1.1-2.8 FW	54
Molluscs; 5 spp; soft parts; Chile, Strait of Magellan; September-December 2001; 17 stations	1.5-10.4 FW	50
Morocco; 2004-2005; soft parts		
Mussel, *Mytilus galloprovincialis*	20.8 (8.7-34.8) DW	57
Clam, *Venerupis decussatus*	18.8 (12.1-27.8) DW	57
Pacific oyster, *Crassostrea gigas*	26.4 (5.3-38.9) DW	57
Softshell clam, *Mya arenaria*		
Soft parts	29.0-70.0 DW	23
Soft parts	6.7 (0.1-29.9) FW	10, 24

(*Continues*)

Table 6.13: Cont'd

Organism	Concentration	Reference[a]
Mussel, *Mytilus californianus*		
Shell	8.4-14.2 DW	1
Soft parts	5.9-7.8 DW	1
Soft parts	2.5-17.0 DW	14
Mussel, *Mytilus corscum*; soft parts	7.7 DW	21
Common mussel, *Mytilus edulis*		
Shell	3.6 DW	4
Shell; Baltic Sea; 2005; 12 sites	23.4-74.4 DW	62
Soft parts	3.5 DW	4
Shell	9.3-45.8 DW	1
Soft parts	6.1-28.4 DW	1
Soft parts	21.0-22.0 DW	25
Soft parts		
United States West coast	8.2-25.0 DW	14
United States East coast	8.9-18.0 DW	14
Looe estuary, England	5.2-35.4 DW	6
Scandinavia	4.9-91.7 DW	26
Scheldt estuary; March 2000	130.0 DW	64
Tasmania	2.5 FW	27
Netherlands; summer 2002	Max. 32.0 DW	77
Mussel, *Mytilus edulis aoteanus*		
Soft parts	27.0 DW	28
Mantle	2.0 DW	28
Gill	28.0 DW	28
Muscle	<1.0 DW	28
Visceral mass	105.0 DW	28
Foot	<1.0 DW	28
Gonad	6.0 DW	28
Shell	<1.0 DW	28
Mussel, *Mytilus edulis planulatus*		
Whole	3.5-12.3 DW	29
Soft parts		
Dry weight 0.09 g	11.2 DW	29
Dry weight 0.39 g	8.9 DW	29
Dry weight 0.48 g	7.8 DW	29
Dry weight 0.69 g	7.1 DW	29
Pacific blue mussel, *Mytilus edulis trossulus*; soft parts; 2004; Adak Island, Alaska; 5 sites	3.6 FW; max. 47.6 FW; 3.0-36.0 DW	37

(Continues)

Table 6.13: Cont'd

Organism	Concentration	Reference[a]
Mussel, *Mytilus galloprovincialis*		
Soft parts; Venice lagoon, Italy; 2005-2006; San Giuliano vs. Sacca Sessola		
April	10.9 DW vs. 11.9 DW	53
July	7.4 DW vs. 5.3 DW	53
October	4.7 DW vs. 4.7 DW	53
February	9.2 DW vs. 8.7 DW	53
Soft parts	21.1 DW	30
Soft parts; Izmir Bay, Turkey; 2004; near shore vs. mid-bay	4.5 DW vs. 2.7 DW	63
Soft parts; Gulf of Gemlik, Turkey; 2004	5.8 DW	68
Soft parts; Adriatic Sea; 2001-2002; summer vs. winter		
Industrialized sites	9.6-10.9 DW vs. 14.0-29.8 DW	46
Reference site	12.9 DW vs. 14.0 DW	46
Soft parts; Adriatic Sea; 2002-2005; winter vs. summer	10.2-16.4 DW vs. 4.4-12.7 DW	66
Mussel, *Mytilus viridis*		
Soft parts	6.1 (4.2-8.1) FW	12
Soft parts	24.2-94.1 DW	31
Whelk, *Nucella lapillus*		
Shell	1.1 DW	4
Soft parts	12.0 DW	4
Soft parts and operculum	11.4-16.8 DW	6
Octopus, *Octopus* sp.; whole	0.6 FW	27
Octopus, *Octopus vulgaris*		
Portugal; 1999-2002; 3 sites		
Digestive gland	4.6-6.0 DW	67
Branchial heart	6.3-9.6 DW	67
Gill	5.0-7.1 DW	67
Muscle	1.6-1.9 DW	67
Oyster, *Ostrea angasi*		
Soft parts	2.5 FW	27
Whole	7.8-16.7 DW	29
Oyster, *Ostrea edulis*; soft parts		
Mediterranean Sea	6.1-85.0 DW	32
South Africa	6.0 DW	11

(*Continues*)

Table 6.13: Cont'd

Organism	Concentration	Reference[a]
Oyster, *Ostrea sinuata*		
Soft parts	8.0 DW	28
Mantle	4.0 DW	28
Gill	11.0 DW	28
Muscle	2.0 DW	28
Visceral mass	2.0 DW	28
Kidney	2.0 DW	28
Heart	2.0 DW	38
Shell	1.0 DW	28
Limpet, *Patella candei gomesii*; soft parts; 2007; Azores		
Near hydrothermal vent	77.2-204.5 DW	59
Reference site	6.8-20.9 DW	59
Limpet, *Patella vulgata*		
Shell	8.6 DW	4
Soft parts	13.0 DW	4
Soft parts	5.4-36.0 DW	6
Soft parts	42.0 DW	33
Scallop, *Pecten maximus*		
Soft parts	140.0 DW	4
Muscle	22.0 DW	4
Gut and digestive gland	410.0 DW	4
Mantle and gills		
Unwashed	200.0 DW	4
Washed	4.0 DW	4
Gonad		
Unwashed	31.0 DW	4
Washed	8.6 DW	4
Shell		
Upper valve	12.0 DW	4
Lower valve	4.9 DW	4
Kidney		
Soft parts	3206.0 FW	34
Excretory granules	110,000.0 FW	34
Kidney	15,300.0 DW	7
Soft parts	107.0 DW	7
Body fluid	1.0 DW	7
Mantle	8.8 DW	7
Gills	45.5 DW	7
Digestive gland	15.6 DW	7

(Continues)

Table 6.13: Cont'd

Organism	Concentration	Reference[a]
Striated muscle	1.8 DW	7
Unstriated muscle	0.8 DW	7
Gonad and foot	9.9 DW	7
Scallop, *Pecten novae-zelandiae*		
Soft parts	110.0 DW	28
Mantle	45.0 DW	28
Gill	353.0 DW	28
Muscle	2.0 DW	28
Visceral mass	24.0 DW	28
Kidney	2660.0 DW	28
Foot	27.0 DW	28
Gonad	5.0 DW	28
Shell	1.0 DW	28
Green mussel, *Perna viridis*		
Soft parts; India; January 2002; 28 sites	5.05–5.24 FW	49
Hong Kong; juveniles vs. adults		
Reference site		
Soft parts	26.6 DW vs. 27.3 DW	68
Byssus	13.2 DW vs. 24.2 DW	68
Shell	6.8 DW vs. 4.5 DW	68
Polluted site		
Soft parts	26.6 DW vs. 43.0 DW	68
Byssus	146.0 DW vs. 215.0 DW	68
Shell	9.0 DW vs. 8.2 DW	68
Pen shell (bivalve), *Pinna nobilis*		
Kidney	39,000.0 DW	35
Byssus	950.0 DW	35
Hepatopancreas	20.0 DW	35
Gonads	10.0 DW	35
Soft tissues	22,600.0 AW	36
Shell	34.0 AW	36
Mantle	4070.0 AW	36
Gill	730.0 AW	36
Byssus	820.0 AW	36
Hepatopancreas	32,000.0 AW	36
Muscle	1720.0 AW	36
Body liquids	1480.0 AW	36
Soft parts	5500.0 AW	37
Muscle	4900.0 AW	37

(Continues)

Table 6.13: Cont'd

Organism	Concentration	Reference[a]
Nervous system	2700.0 AW	37
Intestine	5200.0 AW	37
Gonads	500.0 AW	37
Hepatopancreas	3200.0 AW	37
Byssus	4700.0 AW	37
Byssus gland	4200.0 AW	37
Clam, *Pitar morrhuana*; soft parts	38.0-159.0 DW	38
Clam, *Protothaca staminea*		
Shell	16.8 DW	1
Soft parts	11.5 DW	1
Manila clam, *Ruditapes philippinarum*		
Soft parts; Korea; 2002-2003		
Females vs. males	37.7 DW vs. 32.7 DW	42
Postspawn vs. prespawn	82.3 DW vs. 35.5 DW	42
Jiaozhou Bay, China; sediments vs. soft parts; November 2004	433.0 DW vs. 27.5-67.6 DW	44
Clam, *Scrobicularia plana*		
Looe estuary, England; soft parts	43.0 (25.0-100.0) DW	6
Gannel estuary, England		
Soft parts	57.0-87.0 DW	39
Digestive gland	190.0 DW	39
Mantle and siphons	360.0 DW	39
Foot and gonads	15.0 DW	39
Camel estuary, England		
Soft parts	19.0-22.0 DW	39
Digestive gland	79.8 DW	39
Mantle and siphons	14.4 DW	39
Foot and gonads	4.0 DW	39
Transferred from Camel estuary to Gannel estuary for 352 days		
Digestive gland	271.0 DW	39
Remaining tissues	134.0 DW	39
Transferred from Gannel estuary to Camel estuary for 352 days		
Digestive gland	111.0 DW	39
Remaining tissues	36.0 DW	39
Tamar estuary, England		
Soft parts	48.0 (18.0-97.0) DW	40
Digestive gland	131.0 (72.0-183.0) DW	40

(*Continues*)

Table 6.13: Cont'd

Organism	Concentration	Reference[a]
Squid, *Sepia esculenta*; trunk	1.4 DW	21
Clam, *Spisula solidissima*; muscle	0.5-1.9 FW	5
Squid		
Flesh	6.5 DW	41
Flesh with skin	1.8 DW	41
Viscera	3.3 DW	41
Pen	2.1 DW	41
Clam, *Tapes decussata*		
Shell	3.0 AW	36
Mantle	77.0 AW	36
Gill	61.0 AW	36
Muscle	96.0 AW	36
Clam, *Tapes semidecussata*		
Shell	24.7 DW	1
Soft parts	24.7 DW	1
Gastropod, *Thais emarginata*		
Shell	5.3-6.0 DW	1
Soft parts	8.8 DW	1
Gastropod, *Thais lapillus*		
Soft parts	5.9-17.3 DW	20
Digestive gland and gonad	5.9-8.5 DW	20
Shell	2.8-4.2 DW	20

Values are in mg Mn/kg fresh weight (FW), dry weight (DW), or ash weight (AW).
[a]1, Graham, 1972; 2, Patel et al., 1973; 3, Carmichael et al., 1979; 4, Segar et al., 1971; 5, Greig et al., 1977; 6, Bryan and Hummerstone, 1977; 7, Bryan, 1973; 8, Van As et al., 1973; 9, Watling and Watling, 1976b; 10, Pringle et al., 1968; 11, Watling and Watling, 1976a; 12, Bhatt et al., 1968; 13, Windom and Smith, 1972; 14, Goldberg et al., 1978; 15, Galtsoff, 1953; 16, Frazier, 1975; 17, Rucker and Valentine, 1961; 18, Galtsoff, 1942; 19, Lowman et al., 1970; 20, Ireland and Wooton, 1977; 21, Ishii et al., 1978; 22, Hall et al., 1978; 23, Eisler, 1977b; 24, Shuster and Pringle, 1968; 25, Simpson, 1979; 26, Phillips, 1978; 27, Eustace, 1974; 28, Brooks and Rumsby, 1965; 29, Harris et al., 1979; 30, Fowler and Oregioni, 1976; 31, Bhosle and Matondkar, 1978; 32, Fukai et al., 1978; 33, Preston et al., 1972; 34, George et al., 1980; 35, Ghiretti et al., 1972; 36, Chipman and Thommeret, 1970; 37, Papadopoulu, 1973; 38, Eisler et al., 1978; 39, Bryan and Hummerstone, 1978; 40, Bryan and Uysal, 1978; 41, Horowitz and Presley, 1977; 42, Ji et al., 2006; 43, Mubiana and Blust, 2006; 44, Li et al., 2006; 45, Peake et al., 2006; 46, Scancar et al., 2007; 47, Burger and Gochfeld, 2006; 48, Silva et al., 2006; 49, Sasikumar et al., 2006; 50, Espana et al., 2007; 51, Sokolowski et al., 2007; 52, Kadar et al., 2006; 53, Nesto et al., 2007; 54, Sankar et al., 2006; 55, Kumar et al., 2008; 56, Metian et al., 2008; 57. Maanan, 2008; 58, De Wolf and Rashid, 2008; 59, Cunha et al., 2008; 60, Cravo et al., 2008; 61, Cosson et al., 2008; 62, Protasowicki et al., 2008; 63, Kucuksezgin et al., 2008; 64, Wepener et al., 2008; 65, Bustamante et al., 2008; 66, Fattorini et al., 2008; 67, Napoleao et al., 2005; 68, Nicholson and Szefer, 2003; 68, Unlu et al., 2008.

Manganese tends to accumulate in specific tissues, but this is not necessarily the same tissue among closely related species of molluscs. In mussels, for example, greatest manganese burdens were measured in stomach and digestive gland of *M. edulis*, and in musculature and digestive gland of *M. californianus* (Young and Folsom, 1967, 1973). In scallops, however, more than 90% of all manganese concentrated in kidney (Bryan, 1973); this was also the case in the clam *D. trunculus* (Orlando and Mauri, 1978).

Age, sex, and body weight all modify manganese uptake and retention in molluscs. Unlike several other metals, manganese concentrations in American oysters were not related to age or to body weight (Downes, 1957). In other species, however, increasing tissue concentrations of manganese were associated with increasing body weight in clams *P. morrhuana* (Eisler et al., 1978). Females of *C. meridionalis* contained significantly more manganese than males of similar size; these differences were especially pronounced for zinc to which manganese was positively correlated (Watling and Watling, 1976b).

Manganese uptake in molluscs under laboratory conditions is the subject of many studies. In one study, bivalves from the Baltic Sea did not accumulate manganese in tissues at ambient seawater concentrations of 5.0 mg Mn/L; at higher concentrations of 17.0-20.0 mg Mn/L, manganese accumulated in gonads, liver, and heart, accompanied by adverse metabolic effects, and finally death in about two months (Karpevich and Shurin, 1977). This contrasts sharply with known bioconcentration potential of field-collected molluscs. Thus, *P. nobilis* from Greek waters had manganese levels in soft tissues that were 180,000 times higher than seawater concentrations; lower concentration factors occurred in four other species of bivalves from the same area (Papadopoulu, 1973). In another study, Chipman and Thommeret (1970) found that high radioactivity from fallout ^{54}Mn was generally related to high manganese content in the organism; however, the specific activity of ^{54}Mn (the ratio of ^{54}Mn to stable manganese) varied widely. The manganese content in molluscs was more than twice that of tunicates, though the total ^{54}Mn radioactivity was the same in both groups. Scallops *C. radula* exposed to seawater containing ^{54}Mn for 96 h had BCFs of 15 for total soft parts, and about 30 for gill, 30 for kidney, and 30 for digestive gland tissues (Metian et al., 2008). Gravid cuttlefish fed crabs radiolabeled with ^{54}Mn over a 2 week period to determine rate of maternal transfer into eggs showed no significant transfer (Lacoue-Labarthe et al., 2008b).

6.23 Mercury

Toxicological aspects of mercury and mercury compounds in coastal and offshore environments as a result of anthropogenic or natural processes have been reviewed extensively (D'Itri, 1972; Eisler, 2000, 2006; Friberg and Vostal, 1972; Gavis and Ferguson, 1972; Harriss, 1971; Holden, 1973; Jernelov et al., 1975; Keckes and Miettinen, 1972). Most of these authorities agree on five points. First, forms of mercury with relatively low toxicity can be transformed into forms with very high toxicity through biological and abiotic processes. Second, uptake of mercury directly from seawater or through the diet returns mercury directly to upper level trophic consumers, including humans. Third, mercury uptake may result in genetic changes. Fourth, elevated levels of mercury in some marine fishes, such as tunas and swordfishes, emphasize the complexity of both natural mercury cycles and anthropogenic impacts on those cycles. Finally, human use of mercury should be curtailed because the difference between tolerable natural background levels of mercury and harmful effects in the environment is exceptionally small.

Any review of mercury hazards in the marine environment should include the Minamata Bay incident in southwestern Kyushu, Japan. This extensively documented case (Eisler, 2006; Fujiki, 1963; Irukayama, 1967; Irukayama et al., 1961, 1962a,b; Kiyoura, 1963; Kurland et al., 1960; Matida and Kumada, 1969; Matida et al., 1972; Takevich, 1972; Tsubaki et al., 1967) reveals effects on humans of chronic discharges of low-level methyl mercury wastes into coastal waters. The source of the mercury was waste discharged from an acetaldehyde plant into Minamata Bay beginning in 1952. Several years later, the mercury levels in sediments near the plant outfall were about 2010.0 mg/kg FW; this decreased sharply with increasing distance from the plant. Sediments in Bay contained between 0.4 and 3.4 mg Hg/kg FW. Concentrations of mercury in fish and shellfish decreased with increasing distance from the point of effluence and appeared to reflect sediment mercury levels. Late in 1953, a severe neurological disorder was recognized among inhabitants of the Minamata Bay region. By 1956, the outbreak had reached epidemic proportions: 111 cases of poisoning were diagnosed by the end of 1960, and 41 deaths by August 1965. The poisoning was caused by eating fish and shellfish from the Bay and environs, except for 19 congenital cases in children born of mothers who had eaten the same diet. In addition to humans, cats and waterfowl living near the Bay succumbed to the disease. Laboratory cats and rats fed fish and shellfish collected from the Bay developed the same signs as animals spontaneously affected. Signs included cerebellar ataxia, constriction of visual fields, dysarthria, and, in congenital cases, disturbance of physical and mental development. Pathological findings included regressive changes in the cerebellum and the cerebral complexes. The clinicopathological features resembled alkylmercury poisoning. Abnormal mercury content, namely more than 30.0 mg Hg/kg FW, was measured in fish, molluscs, and muds from the Bay and in organs of necropsied humans and cats that died of the disease. Years after the waste discharge situation was corrected, fish and shellfish still contained levels of mercury considered hazardous to human health.

Mercury levels in sediments vary widely and tend to be elevated near sewer outfalls (Klein and Goldberg, 1970), sludge disposal areas, and especially areas impacted by mercury wastes from industrial operations, such as Minamata Bay, Japan (Eisler, 2006; Irukayama, 1967). Vucetic et al. (1974) reported sediment mercury levels ranging from 0.13 to 1.5 mg/kg in the Adriatic Sea, and Williams and Weiss (1973) found 0.39 mg/kg in pelagic clays collected

430 km southeast of San Diego, California. It is reported that mercury in seawater exists almost entirely bound to suspended particles (Jernelov et al., 1972), that the surface area of sediment granules is instrumental in determining final mercury content (Renzoni et al., 1973), that conversion and transformation occur in the surface layer of the sediment or on suspended organic particles in the water (Dean, 1972; Fagerstrom and Jernelov, 1972; Jernelov et al., 1972), and that mercury-containing sediments require many decades to purge themselves naturally to background levels (Langley, 1973). Mercury-sediment-water interactions affect uptake by marine biota. It is generally acknowledged that marine organisms feeding in direct contact with sediment have higher overall mercury levels than those feeding above the sediment-water interface (Klemmer et al., 1976). Bivalves accumulate mercury directly from seawater; however, mercury concentrations were higher with increasing water turbidity (Raymont, 1972). Molluscs sampled before and immediately after their substrate was extensively dredged had significantly higher mercury concentrations postdredging; elevated concentrations persisted for at least 18 months (Rosenberg, 1977).

Some investigators report that mercury from point-source discharges, including sewer outfalls and chlor-alkali plants, is taken up by sediments; sediment mercury levels were eventually reflected by an increased mercury content in epibenthic fauna (Dehlinger et al., 1975; Hoggins and Brooks, 1973; Klein and Goldberg, 1970; Klemmer et al., 1973; Parsons et al., 1973; Takevich, 1972). Analysis of the effluent from the Hyperion sewer outfall in Los Angeles showed mercury content slightly below 0.001 mg/L (Klein and Goldberg, 1970). Concentrations of mercury in sediments near this outfall were as high as 0.82 mg/kg but decreased with increasing distance from the outfall; mercury levels in epibenthic fauna, including whelks and scallops, were also highest near the discharge and lowest at tens of kilometers distant. De Wolf (1975) found that mercury burdens in mussels along European coasts tend to reflect mercury levels in water and sediments to a greater degree than size of the mussel, season of collection, or position in the intertidal zone.

Rapid accumulation of mercury, especially organomercury compounds, by representative species of molluscs is documented (Cunningham and Tripp, 1975a,b; Fang, 1973; Fowler et al., 1975; Irukayama et al., 1962a; Laumond et al., 1973; Nelson et al., 1976). Mercury accumulation from seawater by molluscs is modified by many factors, including the chemical form of mercury administered (Cunningham and Tripp, 1975a; Irukayama et al., 1962a); water temperature (Cunningham and Tripp, 1975b; Fowler et al., 1978; Lee et al., 1975; Nelson et al., 1977); salinity of the medium (Dillon, 1977; Dillon and Neff, 1978; Olson and Harrel, 1973); presence of selenium (Fowler and Benayoun, 1976a; Glickstein, 1978); sexual condition (Cunningham and Tripp, 1975b; De Wolf, 1975; Norum et al., 2005); tissue specificity (Cunningham and Tripp, 1975a; Dimitriadis et al., 2003; Fowler et al., 1975, 1978; Wrench, 1978); previous acclimatization to mercury salts (Dillon, 1977); season of year (De Wolf, 1975); soluble protein content of organism (Wrench, 1978); and presence of mercury-resistant strains of bacteria (Nelson et al., 1971; Colwell and Nelson, 1975;

Colwell et al., 1976; Sayler et al., 1975). Although certain trends were evident among individual modifiers, the results were not applicable to all groups of molluscs examined, with two exceptions. First, there was general agreement that organomercury compounds were more toxic and taken up more rapidly than were inorganic mercury compounds. Second, that salinity stress, especially abnormally low salinities, reduced significantly the survival time of mercury-exposed molluscs. Jones (1973), suggests that species adapted to a fluctuating estuarine environment could be more vulnerable to the added stresses of heavy metal pollution, including mercury, than species inhabiting more uniformly stable environments.

Among cephalopods, total mercury concentrations are higher in the digestive gland than in the remaining tissues (Table 6.14). The percent of total body mercury burden of cephalopods that is organic is 67-93%; for digestive gland it is 3-16%, and for remaining tissues 84-98% (Bustamante et al., 2006). Marine molluscs, unlike certain species of marine mammals, teleosts, and elasmobranchs accumulate mercury only when the environment is contaminated with this element as a direct result of human activities. In every case reported wherein mercury levels in molluscan tissues exceed about 1.0 mg Hg/kg FW, this residue was associated with mercury pollution (Eisler, 1981, 2000, 2006; Table 6.14).

In general, salts of mercury and its organic compounds were more toxic to marine biota during short term bioassays than were salts of other trace metals tested (Berland et al., 1976; Connor, 1972; Eisler and Hennekey, 1977; Kobayashi, 1971; Reish et al., 1976; Schneider, 1972). To oyster embryos, for example, mercury salts were more toxic than were salts of silver, copper, zinc, nickel, lead, cadmium, arsenic, chromium, manganese, or aluminum (Calabrese et al., 1973); to clam embryos, mercury was the most toxic metal tested, followed by silver, zinc, nickel, and lead, in that order (Calabrese and Nelson, 1974). Mercury was extremely toxic to Pacific oyster embryos with 50% mortality reported at 0.0057 mg Hg/L; embryos were relatively insensitive to mercury 24 h postfertilization and survival was affected by a variety of factors including selenium concentrations in the ambient medium (Glickstein, 1978). Juveniles of bay scallops, *A. irradians*, subjected to 0.04 mg Hg/L for 96 h contained 48.9 mg Hg/kg FW, a concentration far in excess of the 0.5 mg Hg/kg FW limit considered hazardous to human health (Nelson et al., 1976). A concentration of 0.089 mg Hg/L was lethal to juvenile bay scallops in 96 h (Nelson et al., 1976). Immunotoxic effects of mercuric chloride were reported in adult common mussels, *M. edulis* after exposure for 7 days at 0.21 mg Hg/L with 29% dead in 21 days; no deaths or immunotoxic effects occurred at 0.02 mg Hg/L and lower (Duchemin et al., 2008).

Two studies, both on adult American ousters, *C. virginica*, are worth emphasizing. (Cunningham and Tripp, 1973; Kopfler, 1974). Cunningham and Tripp (1973) held oysters in seawater containing 0.01 mg Hg/L as mercuric acetate. After 45 days the total body mercury concentration of exposed oysters was 28.0 mg Hg/kg FW versus less than 0.02 mg Hg/kg FW in control oysters. Mercury burden in exposed oysters dropped to 18.0 mg/kg by

Table 6.14: Mercury Concentrations in Field Collections of Molluscs

Organism	Concentration	Reference[a]
Abalone		
Muscle	0.12 FW	1
Viscera	0.13 FW	1
Aleutian Islands; 2004-2005; edible portions; 6 spp.	0.010-0.038 FW; max. 0.096 FW	65
Clam, *Anadara tuberculosa*; soft parts	0.02-0.1 FW	2
Clam, *Anodonta cygnea*; soft parts; United Kingdom		
Southampton area	0.07 DW	3
Chertsey area	0.76 DW	3
Giant squid, *Architeuthis dux*; 2001-2005		
Hepatopancreas	0.47-1.6 DW	77
Branchial heart	2.2 DW	77
Muscle	2.4 DW	77
Mussel, *Bathymodiolus azoricus*		
Mid-Atlantic Ridge; hydrothermal vent fields; 2001		
840 m deep		
Digestive gland	4.6 DW	69
Gills	4.4 DW	69
Mantle	0.8 DW	69
Foot	0.8 DW	69
1700 m deep		
Digestive gland	2.4 DW	69
Gills	4.1 DW	69
Mantle	0.4 DW	69
Foot	0.2 DW	69
2300 m deep		
Digestive gland	1.3 DW	69
Gills	1.0 DW	69
Mantle	0.3 DW	69
Foot	0.2 DW	69
Mid-Atlantic Ridge; 3 sites; 2002		
Gill	2.4-8.2 DW	67
Mantle	0.27-0.44 DW	67
Digestive gland	2.2-12.5 DW	67
Byssus threads	1.6-5.1 DW	67
Bivalves and gastropods; 5 commercial spp.; edible portions	0.04 FW	4

(Continues)

Table 6.14: Cont'd

Organism	Concentration	Reference[a]
Bivalves; 5 spp.; soft parts; East China Sea; August 2002	0.012 (0.006-0.022) FW	61
Whelk, *Busycon canaliculatum*		
Muscle	0.06-0.14 FW	5
Digestive gland	0.16-0.52 FW	5
Cockle, *Cardium edule*		
Soft parts	0.8 DW	3
Soft parts	0.03-0.04 FW	6
Cephalopods; 3 spp.; edible portions	0.06 FW	4
Cephalopods; northeastern Atlantic Ocean between Bay of Biscay and Faroe Islands; 1996-2003		
Total mercury; 20 spp.		
Whole	0.06-0.62 (0.035-1.2) DW	63
Digestive gland	0.05-1.95 (0.01-3.6) DW	63
Remaining tissues	0.07-0.5 (0.04-1.0) DW	63
Organomercury compounds; 7 spp.		
Whole	0.07-0.8 DW	63
Digestive gland	0.04-0.71 DW	63
Remaining tissues	0.08-0.82 DW	63
China		
Coastal sites along Bohai Sea and Huangshi Sea; commercial species; soft parts; gastropods vs. bivalves	0.03 (0.002-0.09) FW vs. 0.01-0.08 FW	38
Zhejiang coastal area; May 1998		
Cephalopods; mantle; 2 spp.	0.016-0.024 (0.009-0.0047) FW	45
Bivalve molluscs; soft parts; 4 spp.	0.013-0.023 (0.009-0.044) FW	45
Spiny scallop, *Chlamys hastata*; British Columbia, Canada; July 1999-February 2000; males vs. females		
Gonad	0.4 DW vs. 0.2 DW	46
Gill	0.0 DW vs. 0.9 DW	46
Mantle	0.1 DW vs. 0.08 DW	46
Muscle	0.03 DW vs. 0.03 DW	46
Clams		
Soft parts	0.005-0.012 DW	8
Soft parts	0.05-0.3 FW	9

(*Continues*)

Table 6.14: Cont'd

Organism	Concentration	Reference[a]
Oyster, *Crassostrea amasa*; soft parts	0.07 FW	10
Oyster, *Crassostrea angulata*		
Soft parts	0.56 DW	3
Soft parts	0.03-0.05 FW	11
Oyster, *Crassostrea commercialis*		
Soft parts	0.001-0.1 FW	12
Soft parts	0.01-0.1 FW	13
Pacific oyster, *Crassostrea gigas*; Thailand; 1995-96; soft parts	0.2 (0.03-1.3) DW	47
Oyster, *Crassostrea glomerata*		
Soft parts	0.081 FW	14
Gills	0.165 FW	14
Mantle	0.137 FW	14
Muscle	0.08 FW	14
Shell	0.018 FW	14
Viscera	0.131 FW	14
American oyster, *Crassostrea virginica*		
Shell	0.089 FW	15
Soft parts		
Mexico	0.01-0.06 FW	2
Texas	0.05 DW	16
Mississippi	0.02-0.08 FW	17
Pensacola, Florida; 2003-2004	Max. 0.06 FW	66
Savannah, Georgia; 2000-2001	0.13-0.91 DW	71
Mussel, *Elliptio buckleyi*; soft parts; Marmara Sea, Turkey; 2004-2005	0.23-0.75 (0.07-0.90) FW	60
Clam, *Ensis arcuatus*; shell	0.11 DW	18
Clam, *Ensis siliqua*; shell	0.1 DW	18
Gastropods; 5 spp.; soft parts	0.07-0.8 DW	2
Greenland; 1983-1991; bivalves; 5 spp.; soft parts	0.01-0.02 FW	48
Abalone, *Haliotis ruber*; muscle	0.01-0.04 FW	13
Abalone, *Haliotis rufescens*		
Gills	0.08-0.27 DW	19
Mantle	0.02-0.33 DW	19
Digestive gland	0.12-4.64 DW	19
Foot	0.03-0.09 DW	19

(*Continues*)

Table 6.14: Cont'd

Organism	Concentration	Reference[a]
Italy; summer 1986-1987; soft parts		
Mussel, *Mytilus galloprovincialis*	0.01-0.07 FW	50
Gastropod, *Murex trunculus*	0.03-0.15 FW	50
Squid, *Illex illicebrosa*		
Whole	<0.05 to 0.06 FW	23
Soft parts	0.04 FW; 0.11 DW	25
Soft parts	0.06-0.16 DW	6
Limpet, *Littorina* sp.; soft parts	2.6 DW	26
Squid, *Loligo* sp.; muscle		
Total Hg	0.014 (0.006-0.027) FW	27
Organic Hg	0.008 (0.003-0.017) FW	27
Squid, *Loligo vulgaris*; edible portions	0.01-0.22 FW	28
Clam, *Macoma balthica*		
Soft parts	0.03-0.07 FW	6
Soft parts	0.04-0.1 FW	29
Marine mammal food items; whole; Atlantic Ocean vs. Mediterranean Sea		
Squid, *Loligo vulgaris*	0.056 (0.024-0.097) FW vs. 0.146 (0.068-0.250) FW	55
Squid, *Illex coindeti*	0.041 (0.014-0.069) FW vs. 0.277 (0.110-0.490) FW	55
Hardshell clam, *Mercenaria mercenaria*		
Soft parts	0.03-0.12 FW; 0.18-0.57 DW	30
Soft parts; various ages		
3 years	0.16 DW	3
4 years	0.2 DW	3
10 years	0.22 DW	3
15 years	0.22 DW	3
Minamata Bay, Japan; site of massive chronic mercury pollution easing in 1950s		
Scallop, *Chlamys ferrei nipponensis*; soft tissue	48.0 DW	7
Pacific oyster, *Crassostrea gigas*; soft tissues	10.0 DW	7
Clam, *Hormomya mutabilis*		
Soft tissues	18.0-48.0 DW	7
Soft parts		
December 1959	100.0 DW	20
January 1960	85.0 DW	21

(*Continues*)

Table 6.14: Cont'd

Organism	Concentration	Reference[a]
April 1960	50.0 DW	21
August 1960	31.0 DW	21
January 1961	56.0 DW	21
April 1961	30.0 DW	21
December 1961	9.0 DW	21
October 1963	12.0 DW	21
Ganglion	181.0 DW	22
Gills	87.0 DW	22
Ligament	62.0 DW	22
Mantle	63.0 DW	22
Muscle	25.0 DW	22
Byssus	20.0 DW	22
Digestive gland	73.0 DW	22
Genital gland	64.0 DW	22
Octopus, *Octopus vulgaris*		
Abdomen	62.0 DW	7
Tentacles	39.0 DW	7
Clam, *Pinctada martensii*; soft parts	5.2-32.0 DW	7
Bivalve, *Pinna attenuata*; soft parts	11.0-25.0 DW	7
Molluscs; commercial species; soft parts		
6 spp.	0.04-0.23 FW	11
18 spp.	<0.1 FW	31
Morocco; 2001; clams; soft parts		
Scrobicularia plana	0.4 DW	74
Cerastoderma edule	0.1 DW	74
Morocco; 2004-2005; soft parts		
Mussel, *Mytilus galloprovincialis*	0.6 (0.02-2.3) DW	72
Clam, *Venerupis decussatus*	0.3 (0.08-0.58) DW	72
Pacific oyster, *Crassostrea gigas*	0.4 (0.08-0.84) DW	72
Softshell clam, *Mya arenaria*; soft parts	0.04 FW	6
Mussel, *Mytilus californianus*		
Digestive gland	0.023-0.063 FW	32
Muscle	0.019-0.062 FW	32
Gonad	0.005-0.014 FW	32
Transferred to proximity of wastewater discharge for 24 weeks		
Digestive gland; start vs. end	0.025 FW vs. 0.040-0.063 FW	32
Muscle; start vs. end	0.027 FW vs. 0.032-0.046 FW	32
Gonads; start vs. end	0.005 FW vs. 0.009-0.018 FW	32

(*Continues*)

Table 6.14: Cont'd

Organism	Concentration	Reference[a]
Common mussel, *Mytilus edulis*		
Soft parts; Norway	Max. 0.04 FW	51
Soft parts	0.03-1.4 DW	33
Soft parts	0.42-1.9 DW	3
Soft parts	0.64 FW; 2.1 DW	25
Visceral mass	0.28 FW; 1.35 DW	25
Foot muscle	0.36 FW; 0.77 DW	25
Mantle	0.87 FW; 4.26 DW	25
Ctenidia	3.41 FW; 19.95 DW	25
Shell	0.49 DW	18
Shell; Baltic Sea; 2005; 12 sites	0.002-0.008 DW	75
Soft parts	0.97 DW	16
Soft parts	0.006-0.012 DW	8
Mussel, *Mytilus edulis planulatus*; soft parts	0.66-10.6 DW	34
Pacific blue mussel, *Mytilus edulis trossulus*; soft parts; 2004; Adak Island, Alaska; 5 sites	0.016 FW; max. 0.04 FW	62
Mussel, *Mytilus galloprovincialis*		
Cultured mussels (<0.15 mg/kg soft parts DW) transplanted to Kastela Bay, Croatia (Hg-contaminated 1950-1990), in 2000-2001 for 6 months		
Total mercury		
Digestive gland	0.44-1.35 DW	64
Soft tissues (all)	0.24-0.32 DW	64
Gills	0.24-0.38 DW	64
Methylmercury		
Digestive gland	0.16 DW	64
Soft tissues (all)	Max. 0.09 DW	64
Gills	Max. 0.04 DW	64
Soft parts	0.12 FW	35
Soft parts; 1991-2005; western Mediterranean Sea; Balearic Islands; Port of Mahon	0.78 (0.013-2.2) DW	58
Soft parts; Adriatic Sea; 2001-2005; winter vs. summer	0.04-0.19 DW vs. 0.02-0.03 DW	79
Soft parts; Izmir Bay, Turkey; 2004	0.18 DW	76
Soft parts; Marmara Sea, Turkey; 2005	1.75 FW	59

(*Continues*)

Table 6.14: Cont'd

Organism	Concentration	Reference[a]
Soft parts; May 2003; NW Mediterranean Sea		
Near Genova, Italy	0.08-0.31 DW	57
Near Marseille, France	0.11-0.26 DW	57
Near Barcelona, Spain	0.13-0.55 DW	57
Soft parts; summers 2004-2005; Mediterranean Sea; 3 sites	0.09-0.12 DW	70
Mussel, *Mytilus* spp.; soft parts	0.02-0.31 FW	36
Netted whelk, *Nassarius reticulatus*; Portugal; 2000; total mercury		
Soft parts; males vs. females	0.07-1.01 DW vs. 0.06-0.83 DW; 52% organomercury	54
Soft tissues	0.12-0.65 DW; 75-93% organomercury	54
Operculum	0.031-0.037 DW; 5-22% organomercury	54
Shell	0.008-0.013 DW; 1.9-2.5% organomercury	54
New Jersey (United States) supermarkets; July-October 2003; scallops; edible tissues	0.01 (0.007-0.02) FW	52
Gastropod, *Nucella conccina*; soft parts; Antarctica; 2004	0.026 DW	56
Whelk, *Nucella lapillus*; soft parts	0.48 FW; 1.47 DW	25
Octopus, *Octopus cyanea*; June 1998; Guam; 4 stations; tentacle vs. liver	0.1 DW vs. 0.1 DW	53
Octopus, *Octopus vulgaris*		
Tentacle muscle	0.75-2.32 FW	37
Digestive gland	15.5-202.0 FW	37
Kidney	4.0-7.5 FW	37
Gill	0.48-1.9 FW	37
Brain	0.99-1.24 FW	37
Gonad	0.40-1.03 FW	37
Edible portions	0.05-0.37 FW	28
Squid, *Ommastrephes sloani pacificus*		
Muscle	0.020-0.07 FW	39
Liver	0.03-0.13 FW	39
Oyster, *Ostrea angasi*; soft parts	1.5-8.2 DW	34

(Continues)

Table 6.14: Cont'd

Organism	Concentration	Reference[a]
European flat oyster, *Ostrea edulis*		
Soft parts	0.36-1.2 DW	3
Soft parts	0.08-0.28 DW	40
Soft parts; 1991-2005; Balearic Islands; western Mediterranean Sea; Port of Mahon	0.9 (0.15-2.2) DW	58
Oysters; soft parts	0.01-0.05 FW	9
Clam, *Paphies australe*		
Soft parts	0.023 FW	14
Foot	0.009 FW	14
Gill	0.02 FW	14
Mantle	0.013 FW	14
Muscle	0.014 FW	14
Shell	0.012 FW	14
Viscera	0.012 FW	14
Limpet, *Patella cerulea*		
Soft parts	0.10-2.72 FW	37
Shell	0.08-0.25 FW	37
Foot	0.13-0.62 FW	37
Visceral mass	2.04-5.92 FW	37
Limpet, *Patella candei gomesii*; soft parts; 2007; Azores		
Near hydrothermal vent	0.01 DW	73
Reference site	0.02-0.03 DW	73
Limpet, *Patella vulgata*; soft parts	0.19 FW; 0.54 DW	25
Mussel, *Perna canaliculus*		
Soft parts	0.017 FW	14
Foot	0.021 FW	14
Gills	0.032 FW	14
Mantle	0.028 FW	14
Shell	0.005 FW	14
Viscera	0.023 FW	14
Bivalve, *Pinna nobilis*		
Soft parts	6.0 DW	41
Mantle and gills	1.1 DW	41
Muscle	3.1 DW	41
Nerve tissue	1.4 DW	41
Stomach and intestines	14.0 DW	41
Gonads	3.0 DW	41

(Continues)

Table 6.14: Cont'd

Organism	Concentration	Reference[a]
Hepatopancreas	7.6 DW	41
Byssus	3.3 DW	41
Byssus gland	1.0 DW	41
Scallop, *Placopecten magellanicus*		
Muscle	<0.05 FW	25
Muscle	<0.08 to 0.18 FW	42
Gonad	<0.08 to 0.18 FW	42
Visceral mass	<0.08 to 0.18 FW	42
Cuttlefish, *Sepia officinalis*		
Gills	0.9 DW	3
Mantle	0.67 DW	3
Edible portions	0.10-0.13 FW	43
Edible portions	0.03-0.19 FW	35
Sepia sp.; India; 2003; edible portions	0.4 FW	68
Spain; Catalonia; November 1992- February 1993; edible tissues		
Cephalopods; 3 spp.	0.003-0.27 FW	49
Noncephalopods; 5 spp.	0.001-0.019 FW	49
Clam, *Spisula solidissima*; muscle	<0.05 FW	5
Clam, *Spisula* sp.; soft parts	0.8 FW	44
Clam, *Tapes decussata*; soft parts	0.03-0.06 FW	11
Japanese common squid, *Todarodes pacificus*; liver	0.19 DW	78

Values are in mg Hg/kg fresh weight (FW) or dry weight (DW).

[a]1, Won, 1973; 2, Reimer and Reimer, 1975; 3, Leatherland and Burton, 1974; 4, Kumagi and Saeki, 1978; 5, Greig et al., 1977; 6, Zauke, 1977; 7, Matida and Kumada, 1969; 8, Schell and Nevissi, 1977; 9, Hung and Lin, 1976; 10, Sorentino, 1979; 11, Establier, 1977; 12, Hussain and Bleiler, 1973; 13, Williams et al., 1976; 14, Hoggins and Brooks, 1973; 15, Ferrell et al., 1973; 16, Sims and Presley, 1976; 17, Harvey and Knight, 1978; 18, Bertine and Goldberg, 1972; 19, Anderlini, 1974; 20, Irukayama et al., 1961; 21, Irukayama, 1967; 22, Irukayama et al., 1962a; 23, Greig et al., 1975; 24, Windom et al., 1976; 25, Jones et al., 1972; 26, Windom et al., 1976; 27, Cheevaparanapivat and Menasveta, 1979; 28, Cumont et al., 1975; 29, Miettinen and Verta, 1978; 30, Burton and Leatherland, 1971; 31, Hall et al., 1978; 32, Eganhouse and Young, 1978; 33, Karbe et al., 1977; 34, Bloom and Ayling, 1977; 35, Uysal, 1978b; 36, Thibaud, 1973; 37, Renzoni et al, 1973; 38, Liang et al., 2004b; 39, Doi and Ui, 1975; 40, Fukai et al., 1978; 41, Papadopoulu, 1973; 42, Greig et al., 1978; 43, Yannai and Sachs, 1978; 44, Craig, 1967; 45, Fang et al., 2004; 46, Norum et al., 2005; 47, Han et al., 1998; 48, Dietz et al., 1996; 49, Schuhmacher et al., 1994; 50, Giordno et al., 1991; 51, Airas et al., 2004; 52, Burger et al., 2005; 53, Denton et al., 2006; 54, Coelho et al., 2006a,b; 55, Lahaye et al., 2006; 56, Santos et al., 2006; 57, Zorita et al., 2007; 58, Deudero et al., 2007; 59, Keskin et al., 2007; 60, Yarsan et al., 2007; 61, Huang et al., 2007; 62, Burger and Gochfeld, 2006; 63, Bustamante et al., 2006; 64, Kljakovic-Gaspic et al., 2006; 65, Burger et al., 2007a,b; 66, Karouna-Renier et al., 2007; 67. Kadar et al., 2006; 68, Sankar et al., 2006; 69, Colaco et al., 2006; 70, Lafabrie et al., 2007; 71, Kumar et al., 2008; 72 Maanan, 2008; 73, Cunha et al., 2008; 74, Anajjar et al., 2008; 75, Protasowicki et al., 2008; 76, Kucuksezgin et al., 2008; 77, Bustamante et al., 2008; 78, Arai et al., 2004; 79, Fattorini et al., 2008.

day 60, probably from loss owing to spawning. At day 60, oysters were transferred to mercury-free seawater for 160 days. During the first 18.0 days postexposure mercury levels declined to 15.0 mg/kg, but remained at that level for the next 142 days. Authors concluded that oysters concentrated 0.01 mg Hg/L by a factor of 2800 and that total self-purification was not achieved over a 6-month cleansing period. Kopfler (1974) found that continuous exposure to only 0.001 mg Hg/L in any of the three mercury compounds tested resulted in rapid accumulation in excess of 0.5 mg Hg/kg FW, that is, to concentrations potentially hazardous to humans ingesting these oysters, according to the action guideline established by the USFDA at that time. (Note: this action level is now 0.3 mg Hg/kg FW, as reviewed by Eisler, 2006). In China, the maximum allowable concentration of mercury in tissues of marine bivalves used as human food is 0.3 total mercury/kg FW (Huang et al., 2007).

Details of Kopfler's study (1974) follow. Accumulation of mercury compounds was determined in two experiments, each using three groups of 100 oysters. In the first experiment, conducted between 0 and 10 °C, mercury concentrations were maintained at 0.05 mg/L with flow rate adjusted to 1 L/oyster/h. In the second experiment, mercury levels were reduced to 0.001 mg/L, the water temperature varied between 25 and 35 °C, and flow rates maintained at 2 L/oyster/h because of the increased temperature. Controls were maintained for each study. In the first experiment, the administration of organic mercury compounds was terminated after 19 days because many of the oysters in groups receiving either methylmercury or phenylmercury salts were dead or moribund. Oysters classified as moribund had slow, incomplete valve closure when disturbed. When oysters in these groups (surviving 19 days exposure) were placed in mercury-free flowing seawater, about half in each group died within a week and all oysters exposed to either methylmercury or phenylmercury died within 14 days. However, oysters exposed to inorganic mercuric chloride had no apparent adverse effects over a 42-day exposure to 0.05 mg Hg/L; mean mercury levels in exposed oysters after exposure for 7 days were 15.0-17.0 mg/kg FW versus 0.02 mg/kg FW for controls. Copper and zinc metabolism was significantly altered in all exposure groups. In the second experiment, methylmercury and phenylmercury compounds were both significantly elevated in oyster tissues when compared to inorganic mercury; concentrations of organomercury compounds were about 20.0 mg Hg/kg FW after 20 days of exposure, and 30.0 mg Hg/kg FW after 60 days—a final concentration factor of about 30,000; for inorganic mercuric chloride, these values ware about 2.0 mg Hg/kg FW after exposure for 20 days and about 10.0 mg Hg/kg FW after 60 days (Kopfler, 1974).

Mercury is also accumulated from seawater by other marine molluscs. In radiotracer studies, clams *Tapes decussatus*, contained 10 times more ^{203}Hg per unit weight over the medium within 24 h (Unlu et al., 1972). Mercury can also be accumulated through food webs. Bacteria have a marked effect on mercury in food chains that include filter feeders. Thus, whole oyster accumulation of mercury more than doubled in the presence of mercury-resistant strains of *Pseudomonas* spp. (Colwell and Nelson, 1975; Colwell et al., 1976; Sayler et al., 1975). In clams, radiomercury seems to have a biological half-life of 10 days if accumulated via the food chain but only 5 days via the medium (Unlu et al., 1972). Retention times of radiomercury-203 in most marine species were comparatively lengthy, up to 1000 days for mussels, *M. galloprovincialis* (Miettinen et al., 1969, 1972). Miettinen and coworkers found that phylogenetically related species follow a similar pattern of methylmercury excretion, with biological half-life depending on water temperature and mode of entry into organisms; half-life is longer after intramuscular injection than after peroral administration.

Three points are with emphasizing at this juncture. first, mercury discharged into rivers, bays, or estuaries as metallic mercury, inorganic divalent mercury, phenylmercury, or alkoxyalkyl mercurials, can all be converted to methylmercurials by natural processes (Jernelov, 1969). Second, organomercury complexes are rapidly accumulated in tissues with high lipid content (Wood, 1973). Finally, mercury-resistant strains of bacteria which have been developed or discovered may have application in mercury mobilization of fixation from mercury-contaminated waters to the extent that polluted areas become innocuous (Colwell and Nelson, 1975; Nelson et al., 1983; Vosjan and Van der Hoek, 1972).

Establishment of mercury criteria for the protection of human health and sensitive marine resources includes abatement of existing discharges, development of technologies to decontaminate contaminated waterways, monitoring of bivalve molluscs and other sensitive indicators of mercury insult, and establishing predictive models of mercury stress in marine ecosystems (Eisler, 1978, 2000, 2006; Johnels and Westermark, 1969; USEPA, 1977; USNAS, 1973). Reduced mercury inputs to coastal area as a result of legislation and effective enforcement actions is reflected in mercury levels of common mussels, *M. edulis*, in Bergen Harbor, Norway. In 2002, mussels from Bergen Harbor contained a maximum of 0.04 mg Hg/kg FW soft parts; this was about 60% lower than mercury concentrations in mussels collected from the same area in 1993. The reduced mercury was attributed to reductions in mercury content to Bergen Harbor of municipal wastewater, urban runoff, and especially of mercury-containing dental wastes (Airas et al., 2004).

6.24 Molybdenum

Maximum concentrations of molybdenum recorded in molluscan tissues, in mg Mo/kg DW, are 240.0 in mantle and gills of *Arca noae* and 464.0 in branchial heart of *O. vulgaris*, although most values were significantly lower (Table 6.15). *A. noae* and *P. nobilis* collected from coastal waters of Greece exhibited unusually high bioconcentration up to 1300-fold over seawater molybdenum concentrations (Papadopoulu, 1973). Lower concentration factors were reported for mussels, scallops, and oysters; these groups contained

Table 6.15: Molybdenum Concentrations in Field Collections of Molluscs

Organism	Concentration	Reference[a]
Bivalve, *Arca noae*		
Soft parts	88.0 DW	1
Mantle and gills	240.0 DW	1
American oyster, *Crassostrea virginica*; soft parts; Savannah, Georgia; 2000-2001	Max. 0.64 DW	6
Molluscs; commercial species; edible tissues		
3 spp.	<0.1 to 0.2 FW	2
9 spp.	0.2-0.4 FW	2
3 spp.	0.4-0.6 FW	2
2 spp.	0.7-0.8 FW	2
1 spp.	3.0-4.0 FW	2
Common mussel, *Mytilus edulis*; soft parts; southeast Alaska; 1980-1982	<1.9 DW	5
Mussel, *Mytilus edulis aoteanus*		
Soft parts	0.6 DW	3
Gills	0.6 DW	3
Visceral mass	1.9 DW	3
Shell	11.0 DW	3
Octopus, *Octopus vulgaris*; Portugal; 1999-2002; 3 sites		
Digestive gland	2.7-19.0 DW	7
Branchial heart	233.0-464.0 DW	7
Squid, *Ommastrephes illicebrosa*; whole	4.0 AW	4
Oyster, *Ostrea sinuata*		
Soft parts	0.9 DW	3
Heart	1.3 DW	3
Visceral mass	0.5 DW	3
Gill	0.4 DW	3
Scallop, *Pecten novae-zelandiae*		
Soft parts	0.9 DW	3
Mantle	1.8 DW	3
Gill	3.1 DW	3
Intestine	3.6 DW	3
Kidney	3.4 DW	3
Foot	0.4 DW	3

(*Continues*)

Table 6.15: Cont'd

Organism	Concentration	Reference[a]
Bivalve, *Pinna nobilis*		
Mantle and gills	20.0 DW	1
Other tissues	<0.1 DW	1

Values are in mg Mo/kg fresh weight (FW), dry weight (DW), or ash weight (AW).
[a]1, Papadopoulu, 1973; 2, Hall et al., 1978; 3, Brooks and Rumsby, 1965; 4, Nicholls et al., 1959; 5, Franson et al., 1995; 6, Kumar et al., 2008; 7, Napoleao et al., 2005.

30-90 times the level of molybdenum normally encountered in ambient seawater (Brooks and Rumsby, 1965).

Molluscs are relatively resistant to molybdenum. Laboratory studies with larvae of the common mussel, *M. edulis*, show that 147.0 mg Mo/L was required for 50% inhibition of development in 48 h based on reduced survival and increased frequency of abnormalities (Morgan et al., 1986), and that 1375.0 mg Mo/L was associated with a 50% reduction in shell growth in 96 h of adult American oysters, *C. virginica* (Knothe and Van Riper, 1988). There is some evidence that pH of the medium directly affects molybdenum residues and biocidal properties in *Venerupis pallustra*, with higher residues and lower survival reported in the pH range 5.1-6.2 (Abbott, 1977).

6.25 Neptunium

During exposure of *M. galloprovincialis* for 58 days to ^{237}Np, about 90% of the accumulated isotope concentrated in the shell. The remainder was located in viscera, gill, mantle, and muscle, in descending order of accumulation (Guary and Fowler, 1978).

6.26 Nickel

Nickel concentrations in marine molluscs varied with season of collection, with geographic area, and among tissues examined (Table 6.16). Maximum nickel concentrations recorded in molluscan tissues, in mg Ni/kg DW, were 200.0 in byssus gland of *P. nobilis*, 342.0 in brachial heart of *O. vulgaris*, 850.0 in soft parts of *Crepidula fornicata*, and 876.0 in kidney of *C. radula* (Table 6.16), and were probably associated with industrial nickel wastes. BCFs for nickel of 800 in soft parts and 40 in shell of marine bivalve molluscs and 500 in brown algae suggest that some food chain biomagnification may occur (USNAS, 1975). For marine mussels and oysters, typical BCF values in soft tissues ranged between 279 and 414 (USEPA, 1980c). However, interpretation of nickel concentrations in molluscs is confounded by interspecies variations in uptake potential. For example, *Astralium rugosum* can bioconcentrate nickel up to 120,000 times that of ambient seawater levels (Papadopoulu,

Table 6.16: Nickel Concentrations in Field Collections of Molluscs

Organism	Concentration	Reference[a]
Mussel, *Anodonta* sp.		
Shell	1.2 DW	1
Soft parts	0.4 DW	1
Gastropod, *Aplysia benedicti*		
Ctenidium	8.0-10.0 DW	2
Gonad	6.0-10.0 DW	2
Parapodia	5.0 DW	2
Hepatopancreas	15.0-20.0 DW	2
Buccal mass	6.0-10.0 DW	2
Intestine	13.0 DW	2
Gut content	15.0 DW	2
Giant squid, *Architeuthis dux*; 2001-2005; Spain		
Hepatopancreas	0.6-1.4 DW	62
Branchial heart	1.4 DW	62
Ocean quahog, *Arctica islandica*; soft parts		
New England; offshore	4.2-13.4 DW	3
August	4.0-18.4 DW	3
February	5.4-29.3 DW	3
March	4.0-18.0 DW	3
June	3.1-22.7 DW	3
Long Island, New York; 1974-1975; offshore	1.1-7.0 FW	34
Cockle, *Austrovenus stutchburyi*; soft parts; 1993-1994; New Zealand	5.0-35.0 DW	46
Mussel, *Brachydontes variabilis*; soft parts	Max. 19.5 FW; max 98.4 DW	4
Whelk, *Buccinum undatum*		
Shell	0.1 DW	1
Soft parts	0.6 DW	1
Soft parts; near sludge dump site vs. reference site	8.5 DW vs. 0.6 DW	36
Cockle, *Cardium edule*		
Shell	0.1 DW	1
Soft parts	7.0 DW	1
Mussel, *Cerastoderma edule*; soft parts	34.0-62.0 DW	5

(Continues)

Table 6.16: Cont'd

Organism	Concentration	Reference[a]
Scallop, *Chlamys opercularis*		
Soft parts	1.6 DW	6
Body fluid	0.9 DW	6
Mantle	0.6 DW	6
Gills	1.6 DW	6
Digestive gland	4.3 DW	6
Striated muscle	0.2 DW	6
Gonad and foot	0.6 DW	6
Kidneys	78.2 DW	6
Shell	0.2-7.2 DW	1
Mussel, *Choromytilus meridionalis*; soft parts	2.0-3.0 DW	7
Scallop, *Comptopallium radula*; Santa Marie Bay (metals-contaminated), New Caledonia; October 2004		
Soft parts	12.8 (8.6-17.2) DW	56
Kidney	876.0 DW	56
Pacific oyster, *Crassostrea gigas*		
Soft parts	1.0-1.6 DW	7, 8
Soft parts	0.1-0.2 FW	9
Soft parts		
South Africa	Max. 2.0 DW	36
United Kingdom	1.0-10.0 DW	36
United States	Max. 0.2 DW	36
Oyster, *Crassostrea margaritacea*; soft parts	1.6 DW	8
Oyster, *Crassostrea rhizophorae*; soft parts; Brazil; summer 2001; contaminated estuary vs. reference site	0.5-2.1 DW; max. 9.0 DW vs. 1.2-1.8 DW; max. 4.3 DW	48
American oyster, *Crassostrea virginica*		
Soft parts	0.2 (0.08-1.8) FW	9
Soft parts	0.9-5.4 DW	10
Shell vs. soft parts	<1.0 DW vs. (0.9-5.4) DW; no data vs. 0.19 (0.08-1.8) FW	35, 36
Soft parts; Pensacola, Florida; 2003-2004	Max. 0.69 FW	52
Soft parts; Savannah, Georgia; 2000-2001	<1.5 to 2.5 DW	55

(*Continues*)

Table 6.16: Cont'd

Organism	Concentration	Reference[a]
Limpet, *Crepidula fornicata*		
Shell	1.6 DW	1
Soft parts	max. 850.0 DW	1
Abalone, *Haliotis rufescens*		
Gills	69.0-112.0 DW	11
Mantle	19.0-57.0 DW	11
Digestive gland	2.5-10.6 DW	11
Foot	0.2-1.6 DW	11
Abalone, *Haliotis tuberculata*; soft parts; England	13.6-15.9 DW	37
Clam, *Macoma balthica*		
Soft parts	6.9-7.9 DW	5
Soft parts; 1996-1997; Gulf of Gdansk	0.5-9.0 DW	51
Hardshell clam, *Mercenaria mercenaria*		
Shell	2.4 DW	1
Soft parts	max. 11.0 DW	1
Soft parts	0.2 (0.1-2.4) FW	9
Soft parts		
United Kingdom	2.2 FW; 6.5-19.2 DW	36
United States	1.2 (0.1-2.4) FW	36
Clam, *Meretrix lamarckii*; soft parts	3.1 DW	12
Mussel, *Modiolus modiolus*		
Shell	0.2 DW	1
Soft parts	9.4-133.0 DW	1
Mantle and gills	3.2 DW	1
Muscle	2.0 DW	1
Gonad	1.2 DW	1
Gut and digestive gland	0.6 DW	1
Molluscs; 21 commercial species; edible portions		
5 spp.	0.2-0.3 FW	13
4 spp.	0.3-0.5 FW	13
6 spp.	0.5-0.8 FW	13
3 spp.	1.0-3.0 FW	13
3 spp.	0.65-3.4 FW	14
Molluscs; 3 spp.; soft parts; Lebanon, near Ras Beirut	27.4-40.1 DW	4

(*Continues*)

Table 6.16: Cont'd

Organism	Concentration	Reference[a]
Molluscs; 5 spp; soft parts; Chile, Strait of Magellan; 2001; 17 stations	0.16-0.8 FW	50
Morocco; 2001; clams; soft parts		
Scrobicularia plana	6.0 DW	58
Cerastoderma edule	17.0 DW	58
Morocco; 2004-2005; soft parts		
Mussel, *Mytilus galloprovincialis*	32.8 (12.7-94.3) DW	57
Clam, *Venerupis decussatus*	22.4 (14.6-34.2) DW	57
Pacific oyster, *Crassostrea gigas*;	25.8 (6.0-37.8) DW	57
Softshell clam, *Mya arenaria*		
Soft parts	0.3 (0.1-2.3) FW	9
Soft parts	1.8-4.1 DW	15
Mussel, *Mytilus californianus*		
Digestive gland	3.0-20.0 DW	16
Soft parts	0.7-5.3 DW	10
Common mussel, *Mytilus edulis*		
Hepatopancreas	Max. 7.4 DW	17
Digestive gland	<6.0 FW	18
Gonads	<2.6 FW	
Muscle	<14.0 FW	18
Shell	2.1 DW	1
Shell; Baltic Sea; 2005; 12 sites	1.5-2.2 DW	59
Soft parts	3.7 DW	1
Soft parts	0.5-9.5 DW	19
Soft parts	0.0-3.5 DW	5
Soft parts	26.0-53.0 DW	20
Soft parts	0.4-6.3 DW	10
Soft parts		
France	0.5 FW; 2.4 DW	36
Norway	6.0-43.0 DW	36
United Kingdom	0.4 FW; max. 53.0 FW; 3.7 (5.0-12.0) DW	36
Scheldt estuary, Europe; March 2000	23.0 DW	60
The Netherlands		
1985-1990	0.33-0.52 FW	41
Summer, 2002	Max. 6.0 DW	45
Mussel, *Mytilus edulis aoteanus*		
Soft parts	7.0 DW	21
Gills	8.0 DW	21

(Continues)

Table 6.16: Cont'd

Organism	Concentration	Reference[a]
Visceral mass	32.0 DW	21
Gonads	5.0 DW	21
Intestine	42.0 DW	21
Mussel, *Mytilus galloprovincialis*		
Soft parts	4.3 DW	22
Soft parts; 1991-2005; western Mediterranean Sea	6.8 (0.7-35.2) DW	43
Soft parts; Sardinia; 2004 vs. 2006	26.0 DW vs. 7.0 DW	61
Soft parts; summers 2004-2005; Mediterranean Sea; 3 sites	1.1-3.7 DW	54
Soft parts; Gulf of Gemlik, Turkey; 2004	1.3 DW	66
Soft parts; Adriatic Sea; 2001-2002; summer vs. winter		
Industrialized sites	2.8-3.6 DW vs. 3.7-5.6 DW	47
Reference site	4.3 DW vs. 3.3 DW	47
Soft parts; Adriatic Sea; 2001-2005; winter vs. summer	2.6-6.5 DW vs. 1.8-2.6 DW	63
Mantle vs. gills; 2004-2006; Greece; max. mean concentrations; contaminated sites		
Summer	0.27 FW vs. 0.53 FW	44
Autumn	0.34 FW vs. 0.5 FW	44
Winter	0.3 FW vs. 0.6 FW	44
Spring	0.74 FW vs. 1.87 FW	44
Mussel, *Mytilus viridis*; soft parts	1.1-15.1 DW	23
Abalone, *Notohaliotis discus*; soft parts less viscera	2.4 DW	12
Whelk, *Nucella lapillus*		
Shell	0.3 DW	1
Soft parts	2.4 DW	1
Soft parts plus operculum	1.4-4.1 DW	5
Octopus, *Octopus vulgaris*; Portugal; 1999-2002; 3 sites		
Digestive gland	3.9-7.6 DW	64
Branchial heart	205.0-342.0 DW	64
Gill	1.5-1.6 DW	64
Muscle	1.5-2.2 DW	64
Squid, *Ommastrephes illicebrosa*; whole	<1.0 AW	24

(*Continues*)

Table 6.16: Cont'd

Organism	Concentration	Reference[a]
European flat oyster, *Ostrea edulis*		
Soft parts	1.7 DW	8
Soft parts; 1991-2005; Balearic Islands; Port of Mahon; western Mediterranean Sea	3.6 (0.8-7.5) DW	43
Oyster, *Ostrea equestris*; soft parts; United States		
East coast	<0.1 DW	10
West coast	1.4-7.3 DW	10
Oyster, *Ostrea sinuata*		
Soft parts	2.0 DW	21
Muscle	8.0 DW	21
Visceral mass	12.0 DW	21
Heart	2.0 DW	21
Limpet, *Patella coerulea*; soft parts	Max. 15.3 FW; max. 75.9 DW	4
Common limpet, *Patella vulgata*		
Shell	0.5 DW	1
Soft parts	2.5 DW	1
Shell	3.0-7.0 DW	26
Soft parts	1.7-3.7 DW	5
Soft parts		
Norway	4.0-11.0 DW	26, 36
United Kingdom	7.3 (2.5-24.0) DW	25, 36
Israel; near sewage discharge vs. reference site	12.0 DW vs. 5.0-9.0 DW	36
Scallop, *Pecten maximus*		
Soft parts	0.8 DW	6
Body fluid	0.7 DW	6
Mantle	0.4 DW	6
Gills	0.7 DW	6
Digestive gland	3.6 DW	6
Striated muscle	0.04 DW	6
Gonad and foot	0.3 DW	6
Kidneys	22.9 DW	6
Soft parts	49.0 DW	1
Muscle	1.7 DW	1
Gut and digestive gland	1.0 DW	1
Mantle and gills; washed vs. unwashed	0.3 DW vs. 0.8 DW	1
Gonad; washed vs. unwashed	0.4 DW vs. 1.5 DW	1
Shell; upper valve vs. lower valve	1.2 DW vs. 2.4 DW	1

(Continues)

Table 6.16: Cont'd

Organism	Concentration	Reference[a]
Scallop, *Pecten novae-zelandiae*		
Soft parts	6.0 DW	21
Gills	68.0 DW	21
Intestine	52.0 DW	21
Kidney	106.0 DW	21
Foot	22.0 DW	21
Gonads	<2.0 DW	21
Shell	<2.0 DW	21
Green mussel, *Perna viridis*		
Soft parts; India; January 2002; 28 sites	0.24-0.31 FW	49
Hong Kong; juveniles vs. adults		
Reference site		
Soft parts	5.4 DW vs. 2.8 DW	65
Byssus	10.0 DW vs. 10.4 DW	65
Polluted site		
Soft parts	6.5 DW vs. 13.4 DW	65
Byssus	12.6 DW vs. no data	65
Bivalve, *Pinna nobilis*		
Soft parts	21.0 DW	27
Nervous system	18.0 DW	27
Stomach and intestine	170.0 DW	27
Gonads	74.0 DW	27
Hepatopancreas	170.0 DW	27
Byssus gland	200.0 DW	27
Clam, *Pitar morrhuana*; soft parts	7.4-37.5 DW	28
Sea scallop, *Placopecten magellanicus*		
Soft parts	4.4 DW	3
Muscle		
March	0.0-22.3 DW	3
August	0.0-4.9 DW	3
February	0.0-0.3 DW	3
June	0.0 DW	3
Viscera		
March	1.5-13.1 DW	3
August	1.2-8.8 DW	3
February	1.5-7.6 DW	3
June	1.6-3.7 DW	3
Muscle	<0.28 to 0.68 FW	29
Gonad	0.23-2.5 FW	29

(Continues)

Table 6.16: Cont'd

Organism	Concentration	Reference[a]
Visceral mass	0.31-1.6 FW	29
Soft parts; Long Island, New York; 1974-1975	0.5-3.3 FW	34
North Atlantic Ocean coast; 42 stations		
Gonads	0.2-2.5 FW	38
Muscle	<0.3 to 0.7 FW	38
Viscera	0.3-1.6 FW	38
Soft parts; near ocean disposal sites	4.4 DW	39
Manila clam, *Ruditapes philippinarum*; soft parts; Korea; 2002-2003		
Females vs. males	4.9 DW vs. 4.9 DW	42
Postspawn vs. prespawn	21.0 DW vs. 4.9 DW	42
Clam, *Scrobicularia plana*		
Bidaron estuary, France-Spain border; April 1993; soft parts	4.1 (2.9-5.7) DW	40
Looe estuary, England; soft parts	5.3-13.9 DW	5
Camel estuary, England		
Soft parts	3.4-4.5 DW	30
Digestive gland	10.6 DW	30
Mantle and siphons	1.4 DW	30
Foot and gonad	4.0 DW	30
Gills and palps	10.5 DW	30
Gannel estuary, England		
Soft parts	9.8-11.9 DW	30
Digestive gland	43.1 DW	30
Mantle and siphons	3.2 DW	30
Foot and gonad	3.9 DW	30
Transferred from Camel Estuary to Gannel estuary for 352 days		
Digestive gland	32.0 DW	30
Remaining tissues	2.3 DW	30
Transferred from Gannel Estuary to Camel estuary for 352 days		
Digestive gland	29.0 DW	30
Remaining tissues	1.9 DW	30
Tamar estuary, England		
Soft parts	1.6-9.1 DW	31
Digestive gland	6.6-25.0 DW	31
Squid, *Sepia esculenta*; trunk	1.1 DW	12

(*Continues*)

Table 6.16: Cont'd

Organism	Concentration	Reference[a]
Squid; muscle; Texas outer continental shelf	2.5 DW	33
Clam, *Siliqua patula*; soft parts	0.74 DW	32
Squid		
Flesh	2.5 DW	33
Flesh with skin	2.5 DW	33
Viscera	1.7 DW	33
Pen	0.8 DW	33
Clam, *Trivela mactroidea*; Venezuela; 2002; 14 sites; soft parts	6.0–15.0 DW	53

Values are in mg Ni/kg fresh weight (FW), dry weight (DW), or ash weight (AW).

[a] 1, Segar et al., 1971; 2, Patel et al., 1973; 3, Reynolds, 1979; 4, Shiber and Shatila, 1978; 5, Bryan and Hummerstone, 1977; 6, Bryan, 1973; 7, Watling and Watling, 1976b; 8, Watling and Watling, 1976a; 9, Pringle et al., 1968; 10, Goldberg et al., 1978; 11, Anderlini, 1974; 12, Ishii et al., 1978; 13, Hall et al., 1978; 14, Ikebe and Tanaka, 1979; 15, Eisler, 1977b; 16, Alexander and Young, 1976; 17, Young et al., 1979; 18, Young and McDermott, 1975; 19, Karbe et al., 1977; 20, Simpson, 1979; 21, Brooks and Rumsby, 1965; 22, Fowler and Oregioni, 1976; 23, Bhosle and Matondkar, 1978; 24, Nicholls et al., 1959; 25, Preston et al., 1972; 26, Lande, 1977; 27, Papadopoulu, 1973; 28, Eisler et al., 1978; 29, Greig et al., 1978; 30, Bryan and Hummerstone, 1978; 31, Bryan and Uysal, 1978; 32, Laevastu and Thompson, 1956; 33, Horowitz and Presley, 1977; 34, Palmer and Rand, 1977; 35, Eisler, 1981; 36, Jenkins, 1980; 37, Bryan et al., 1977; 38, Greig et al., 1978; 39, Pesch et al., 1977; 40, Saiz-Salinas et al., 1996; 41, Stronkhorst, 1992; 42, Ji et al., 2006; 43, Deudero et al., 2007; 44, Vlahogianni et al., 2007; 45, Mubiana and Blust, 2006; 46, Peake et al., 2006; 47, Scancar et al., 2007; 48, Silva et al., 2006; 49, Sasikumar et al., 2006; 50, Espana et al., 2007; 51, Sokolowski et al., 2007; 52, Karouna-Renier et. al., 2007; 53, LaBrecque et al., 2004; 54, Lafabrie et al., 2007; 55, Kumar et al., 2008; 56, Metian et al., 2008; 57, Maanan, 2008; 58, Anajjar et al., 2008; 59, Protasowicki et al., 2008; 60, Wepener et al., 2008; 61, Schintu et al., 2008; 62, Bustamante et al., 2008; 63, Fattorini et al., 2008; 64, Napoleao et al., 2005; 65, Nicholson and Szefer, 2003; 66, Unlu et al., 2008.

1973), far greater than most species. Nickel concentrations in soft tissues of clams, *R. philippinarum*, increase significantly with increasing shell length (Ji et al., 2006). In cockles, *A. stutchburyi*, nickel burdens were lowest in soft parts during summer—unlike copper, chromium, and zinc (Peake et al., 2006). In soft parts of the green mussel, *P. viridis*, nickel content was positively correlated with chromium, copper, and manganese contents (Sasikumar et al., 2006).

Laboratory studies on nickel accumulation by the common mussel, *M. edulis*, were conducted by Freidrich and Felice (1976). At nominal seawater concentrations of 0.03 mg Ni/L and lower, they found that mussels did not accumulate nickel in soft parts after exposure for 28 days. At 0.056 mg Ni/L, however, mussels contained up to 33.0 mg Ni/kg DW soft parts after 28 days; at 0.107 mg Ni/L this was 41.0 mg Ni/kg DW after 28 days. At higher concentrations of 20.0, 40.0, and 80.0 mg Ni/L and exposure for 96 h, mussels contained

420.0, 450.0, and 820.0 mg Ni/kg DW, respectively. All mussels survived exposure for 96 h at all nickel concentrations tested, including 80.0 mg Ni/L (Freidrich and Felice, 1976). Nickel at 0.05 mg/L was accumulated from seawater by softshell clams, *M. arenaria*, more rapidly during summer at water temperatures of 16-22 °C than during winter at 0-10 °C; no accumulation occurred at 0.01 mg Ni/L in winter although clams took up twice as much nickel over controls in summer (Eisler, 1977a). At present, latent sublethal effects of short-term exposure to nickel salts have not been evaluated adequately in marine molluscs. Clam embryos were comparatively sensitive to ionic nickel. The LC-50 (48 h) value for nickel and embryos of the hardshell clam, *M. mercenaria*, was 0.31 mg Ni/L (Calabrese and Nelson, 1974). Embryos of the American oyster, *C. virginica*, were more resistant: the LC-50 (48 h) value for that life stage was 1.18 mg Ni/L (Calabrese and Nelson, 1974). No deaths occurred among adult softshell clams exposed to 100.0-500.0 mg Ni/L for 168 h (Eisler, 1977b; Eisler and Hennekey, 1977).

Three species of bivalves were exposed for 14 days to radionickel-63 via seawater or food (Hedouin et al., 2007). In the seawater portion of the study, oysters (*Isognomon isognomon, Malleus regula*) and a clam (*G. tumidum*) were subjected to stable nickel concentrations up to 0.0014 mg Ni/L; uptake was proportional to nickel concentration in the medium. Most of the accumulated nickel was lost within a few days; however, 7-47% of the ^{63}Ni was retained in tissues indefinitely. Feeding studies showed that ^{63}Ni ingested with phytoplankton was assimilated more efficiently in the clam (61%) than in oysters (17%) and was retained in tissues of all bivalves with an estimated biological half-life of more than 35 days (Hedouin et al., 2007).

To protect marine life, the USEPA has proposed a 24 h average of total recoverable nickel at less than 0.007 mg Ni/L and a maximum of 0.14 mg Ni/L that should not be exceeded at any time (USEPA, 1980c). However, 0.056 mg Ni/L is associated with nickel accumulations in mussels (Freidrich and Felice, 1976), suggesting reexamination of the 0.14 mg Ni/L proposed maximum criterion. To protect human consumers, edible portions of molluscs should not exceed 80.0 mg Ni/kg FW (Sankar et al., 2006).

6.27 Niobium

The single datum available on niobium in marine molluscs indicates that the common mussel, *M. edulis* contains less than 0.001 mg Nb/kg DW soft parts (Carlisle, 1958).

6.28 Plutonium

Plutonium was present in three species of molluscs as $^{239\,+\,240}$Pu—but not ^{238}Pu—in samples collected during summer 2004 near Amchitka Island in the Aleutians; Amchitka was the site of nuclear tests between 1965 and 1971 (Burger et al., 2007b). Elevated levels of plutonium were measured in gastropod and bivalve molluscs collected in the vicinity of French nuclear installations when compared to conspecifics collected 50 km distant (Guary and Frazier, 1977). Elevated plutonium levels were also recorded in the bivalve *Macoma* sp. and other benthic fauna from Bylot Sound, Greenland, following accidental contamination of that area with nuclear weapons in 1968; organisms lost about 90% of the accumulated $^{239+240}$Pu by 1970, but loss was negligible over the next few years (Aarkrog, 1971, 1977, 1990). Benthic bivalves reportedly can accumulate plutonium isotopes to a greater degree than all other marine biota (Aarkrog, 1971; Pillai et al., 1964).

Radioplutonium concentrations in the common mussel, *M. edulis*, seem to reflect global fallout levels (Crowley et al., 1990). Accumulation of transuranic elements, including $^{238+239+240}$Pu, and also americium-241, in *M. edulis* from the vicinity of a uranium processing plant is documented (Hamilton and Clifton, 1980). Uptake was probably from seawater rather than ingested sediments. Highest enrichment of plutonium and americium occurred in byssal threads and periostracum; lowest in shell with periostracum removed. Transuranics retained within the intestinal tract were excreted in feces, but those which enter the systemic circulation were widely diffused (Hamilton and Clifton, 1980).

6.29 Polonium

BCFs values for ^{210}Po from seawater to tissues of squids, *Todarodes pacificus*, from the East Japan Sea ranged from 6400 for muscle to 14 million for hepatopancreas; intermediate values were estimated for gill (2.4 million) and stomach (240,000) (Waska et al., 2008). Polonium-210 was measured in whole squid, muscle, and hepatopancreas (Heyraud and Cherry, 1979). The concentration factors from seawater to whole squid were about 1000, but were significantly lower in muscle, and higher in hepatopancreas. In general, ^{210}Po accumulations were highest in crustaceans and lowest in fish, cephalopods being intermediate (Heyraud and Cherry, 1979).

6.30 Protactinium

Lucu et al. (1969) showed that *M. galloprovincialis* held for 20 days in seawater solutions containing ^{233}Pa took up this isotope by factors of 2500 in byssus, 150 in shell, up to 200 in digestive tract, 60 in gill, 15 in gonad, and 10 in muscle.

6.31 Radium

Concentrations of radium in 18 species of molluscs collected from various locales parallel those of calcium; for all species, radium content ranged from 359.0×10^{-10} to 3580.0×10^{-10} mg/kg calcium (Koczy and Titze, 1958).

6.32 Rhenium

Clams, *Tapes japonicus*, contained 0.064 mg Re/kg ash weight (AW) soft parts; on a DW basis, this value was 0.0046 (Fukai and Meinke, 1959, 1962).

6.33 Rubidium

Rubidium concentrations in molluscan tissues ranged from 3.2 to 14.0 mg Rb/kg DW (Table 6.17).

Eversole (1978) studied rubidium uptake in hardshell clams, *M. mercenaria*. Clams held in seawater solutions containing 0.01 mg Rb/L and higher for 96 h had significantly more rubidium than controls. Clams held in 1.0 mg Rb/L for 96 h, showed measurable rubidium levels in tissues for periods up to 3 weeks postexposure when held in rubidium-free seawater. Clam embryos held for 96 h in vessels containing diatoms previously exposed to 1000.0 mg Rb/L took up more rubidium than embryos subjected to unlabeled diatoms, suggesting that diet is important as a source of rubidium for *Mercenaria*.

Table 6.17: Rubidium Concentrations in Field Collections of Molluscs

Organism	Concentration	Reference[a]
Clam, *Meretrix lamarckii*; soft parts	4.5 DW	1
Mussel, *Mytilus corscum*; soft parts	4.2 DW	1
Common mussel, *Mytilus edulis*; soft parts	3.4-14.0 DW	2
Abalone, *Notohaliotis discus*; soft parts less viscera	3.2 DW	1
Octopus, *Octopus vulgaris*; Portugal; 1999-2000; 3 sites		
Digestive gland	5.0-7.1 DW	4
Branchial heart	9.1-9.5 DW	4
Limpet, *Patella candei gomesii*; soft parts; 2007; Azores		
Near hydrothermal vent	8.2-8.4 DW	3
Reference site	5.8-5.9 DW	3
Squid, *Sepia esculenta*; trunk	3.4 DW	1

Values are in mg Rb/kg dry weight (DW).
[a] 1, 1, Ishii et al., 1978; 2, Karbe et al., 1977; 3, Cunha et al., 2008; 4, Seixas and Pierce, 2005.

6.34 Ruthenium

Ruthenium-106 concentrations in seven species of molluscs increased by a factor of at least 2.6 following release of ^{106}Ru and other isotopes from the Chernobyl nuclear reactor accident (Lowe and Horrill, 1991). Mussels, *M. galloprovincialis*, rapidly accumulated ^{103}Ru and ^{106}Ru after the Chernobyl incident in April 1986; however, loss was rapid, and between May 6 and August 14 the loss rate from mussel soft parts was 98% for ^{103}Ru and 91% for ^{106}Ru (Whitehead et al., 1988).

Molluscs accumulated radioruthenium from seawater by factors up to 200 (Ueda et al., 1978). However, uptake by clams was different for anionic, cationic, and neutral fractions of ruthenium; the cationic fraction was accumulated by an order of magnitude higher than the other fractions (Ishikawa et al., 1973, 1976). Similar results were observed in various species of mussels. In one case, *M. galloprovincialis* held for 5 days in seawater containing a ^{106}Ru-nitrate complex had ruthenium concentration factors of 50 in soft parts and 4 in shell; however, a ^{106}Ru-chloride complex had concentration factors of only 5 in soft parts and 1 in shell (Keckes et al., 1966). Uptake of ^{106}Ru-chloride fractions was faster than that of ^{106}Ru nitrosyl-nitrate fractions and uptake was greater in soft tissues than in shell (Keckes et al., 1966). On the other hand, several investigations on ruthenium distribution in mussels show that most of the ruthenium accumulated in shells (Jones, 1973; Zesenko and Polikarpov, 1965) with negligible uptake in edible meats (Avargues et al., 1968; LeGall and Ancellin, 1971).

6.35 Samarium

Samarium concentrations in tests of pteropods ranged from 0.04 to 0.72 mg Sm/kg DW (Turekian et al., 1973).

6.36 Scandium

The highest scandium concentration recorded in marine molluscs was 0.76 mg Sc/kg DW in mussel soft parts (Table 6.18).

6.37 Selenium

Maximum selenium concentrations in molluscan tissues were 11.1 mg Se/kg FW in mussel soft parts and 16.0 mg Se/kg DW in squid liver (Table 6.19). Selenium reportedly protects against mercury damage. Mercury-induced DNA damage in hemocytes of mussels, *M. edulis*, was produced in three days at 0.02 mg Hg^{2+}/L; however, effects were reduced with added selenium, as selenite, at 0.004 mg Se/L (Tran et al., 2007). Selenium pre-exposure also provided some protection against mercury-induced DNA damage. Glutathione peroxidase

Table 6.18: Scandium Concentrations in Field Collections of Molluscs

Organism	Concentration	Reference[a]
Clam, *Ensis arcuatus*; shell	0.005 DW	1
Clam, *Ensis siliqua*; shell	0.008 DW	1

Common mussel, *Mytilus edulis*
    Shell
    Soft parts
    Soft parts
	
	0.004 DW
	0.019 DW
	0.005-0.76 DW
		
		1
		1
		2

Pteropods; tests	0.028-0.247 DW	2
Squid, whole	0.01 AW	3

Values are in mg Sc/kg dry weight (DW) or ash weight (AW).
[a]1, Bertine and Goldberg, 1972; 2, Turekian et al., 1973; 3, Robertson, 1967.

Table 6.19: Selenium Concentrations in Field Collections of Molluscs

Organism	Concentration	Reference[a]
Cockle, *Anadara trapezia*; soft parts; April 2000; New South Wales, Australia	Max. 3.7 DW	11

Giant squid, *Architeuthis dux*; 2001-2005; Spain
    Hepatopancreas
    Branchial heart
	
	13.0 DW
	9.0 DW
		
		16
		16

Bivalves
    Muscle
    Viscera
	
	1.1-2.3 DW
	1.6-2.5 DW
		
		9
		9

Bivalves; edible portions; 3 spp.
    Total selenium
    Selenate
    Selenite plus selenide
	
	0.22 (0.16-0.31) FW
	0.05 FW
	0.17 FW
		
		6
		6
		6

American oyster, *Crassostrea virginica*; soft parts; Pensacola, Florida; 2003-2004	Max. 0.95 FW	13
Clam, *Ensis arcuatus*; shell	0.06 DW	1
Clam, *Ensis siliqua*; shell	0.03 DW	1
Greenland; bivalves; 1975-1991; 3 spp.; soft parts	0.2-0.9 FW	8

(*Continues*)

Table 6.19: Cont'd

Organism	Concentration	Reference[a]
Molluscs; commercial species; edible tissues		
2 spp.	0.1-0.3 FW	2
9 spp.	0.3-0.5 FW	2
7 spp.	0.5-0.9 FW	2
Molluscs; 5 spp.; soft parts; 2001; Chile, Strait of Magellan; 17 stations	0.15-0.29 FW	12
Common mussel, *Mytilus edulis*		
Shell	0.046 DW	1
Soft parts	4.5 DW	1
Soft parts	1.3-9.9 DW	3
Gills	2.0-16.0 DW	7
Viscera	Max. 5.0 DW	7
Pacific blue mussel, *Mytilus edulis trossulus*; 2004; soft parts; Adak Island, Alaska; 5 sites	0.9 FW; max. 11.1 FW; 2.2-8.8 DW	10
Mussel, *Mytilus galloprovincialis*; Adriatic Sea; 2001-2005; winter vs. summer	3.3-4.4 DW vs. 1.7-2.0 DW	18
Octopus, *Octopus vulgaris*; Portugal; 1999-2002; 3 sites		
Digestive gland	8.3-11.0 DW	19
Branchial heart	6.9-9.8 DW	19
Gill	1.9-2.2 DW	19
Muscle	1.3-1.7 DW	19
Oyster, *Ostrea edulis*		
Soft parts	1.7-2.9 DW	4
Soft parts	<0.5 FW	5
Limpet, *Patella candei gomesii*; soft parts; 2007; Azores		
Near hydrothermal vent	0.40-0.45 DW	14
Reference site	0.8 DW	14
Squid, *Todarodes pacificus*; Japan Sea; summer		
Muscle	1.6 DW	15
Gill	0.4 DW	15
Stomach	4.9 DW	15
Hepatopancreas	4.1 DW	15
Liver	16.0 DW	16

Values are in mg Se/kg fresh weight (FW) or dry weight (DW).
[a] 1, Bertine and Goldberg, 1972; 2, Hall et al., 1978; 3, Karbe et al., 1977; 4, Fukai et al., 1978; 5, Lande, 1977; Cappon and Smith, 1982; 7, Stump et al., 1979; 8, Dietz et al., 1996; 9, Maher, 1983; 10, Burger and Gochfeld, 2006; 11, Burt et al., 2007; 12, Espana et al., 2007; 13, Karouna-Renier et al., 2007; 14, Cunha et al., 2008; 15, Waska et al., 2008; 16, Bustamante et al., 2008; 17, Arai et al., 2004; 18, Fattorini et al., 2008; 19, Napoleao et al., 2005.

activity increased with added selenite, suggesting that selenium availability could offset the antioxidant status of mussels and levels of DNA damage (Tran et al., 2007).

Dietary uptake is the dominant pathway for selenium accumulation in clams; interspecies differences are attributed, in part, to different food ingestion rates (Lee et al., 2006). Radioselenium-75 uptake from diet or medium and subsequent retention was determined for the clams *C. fluminea* and *Potamocorbula amurensis* (Lee et al., 2006). *Corbicula* assimilated ^{75}Se from labeled diatoms more efficiently (66-87%) than selenium associated with sediments (20-37%); for *Potamocorbula*, there was no difference between food types (19-60%). Temperature and selenium of the medium had little effect on selenium assimilation from ingested food, but were significant in uptake from the medium. Influx of selenite increased by a factor of 3 when temperature increased from 5 to 21 °C for *Corbicula*, but salinity between 4 and 21 ppt had no effect; for *Potamocorbula*, uptake decreased with increasing salinity (Lee et al., 2006). Diet is also the major route of selenium intake by the common mussel (*M. edulis*). Studies with radioselenium-75 demonstrated that selenium efficiency was 28-34% from the diet and 0.03% from the medium (Wang et al., 1996). It was concluded that 96% of the selenium in mussels is obtained from ingested food under conditions typical of coastal waters (Wang et al., 1996).

Uptake from seawater by tissues of the squid, *T. pacificus*, captured in the East Japan Sea during summers—as judged by BCF values—ranged from 20,000 in muscle to 110,000 in hepatopancreas; intermediate values were estimated for stomach (62,000) and gill (84,000) (Waska et al., 2008). Uptake and retention of radioselenium-75 was studied in the pygmy mussel, *X. securis* (Alquezar et al., 2007). During the uptake phase of 388 h, whole mussels concentrated ^{75}Se from seawater by a factor of 162. During the loss phase of 189 h in ^{75}Se-free media, 71% was lost. The short-lived component had a biological half-time of 34 h; for the long-lived component, this was 555 h (Alquezar et al., 2007). Gravid cuttlefish, *S. officinalis*, were fed crabs radiolabeled with ^{75}Se for 2 weeks to determine rate of maternal transfer into eggs; during embryonic development ^{75}Se was progressively transferred to the embryo, and also to juveniles after hatching (Lacoue-Labarthe et al., 2008b).

Residue data for selenium in bivalve molluscs should be interpreted with caution. For example, Fowler and Benayoun (1976a,b) in studies on selenium uptake by *M. galloprovincialis* demonstrated the following: mussels selectively accumulate tetravalent selenium salts over hexavalent species; the smallest mussels (2.1 g FW) concentrate selenium from solution by a factor of 46 versus 29 for intermediate weight (13.2 g), and 13 for the largest (21.8 g) mussels; visceral mass had the highest selenium burdens, followed by gill, muscle, and mantle tissues, in that order; and mussels did not reach equilibrium with the medium during exposure for 9 weeks. Differences among molluscan species should also be considered in the interpretation of selenium residues. Adult oysters, as one case, accumulate radioselenium to a greater degree than other species of bivalve filter feeders examined, and this may prove useful in monitoring pollution due to selenium radionuclides (Patel and Ganguly, 1973). Incidentally, embryos of the Pacific oyster, *C. gigas*, were relatively insensitive to selenium salts; concentrations required to produce 50% abnormalities in 48 h being greater than 10.0 mg Se/L (Glickstein, 1978).

The accuracy and precision of analytical methodologies in trace metal and metalloid determination needs to be considered. In one interlaboratory analysis of selenium in homogenates of oyster, *Ostrea edulis*, from France, involving 87 laboratories, "acceptable" (viz., within two standard deviations of the mean) results ranged from 1.7 to 2.9 mg Se/kg DW (Fukai et al., 1978). On this basis, at least one laboratory would have reported a result near 1.1 or 3.3 mg Se/kg DW homogenate, and these values may have little worth in enforcement or abatement actions regarding water quality criteria for selenium. However, additional laboratory intercalibration studies are continually in progress using a variety of biological matrices, sample preparation techniques, and instrumentations.

6.38 Silicon

Soft parts of the American oyster, *C. virginica*, from Savannah, Georgia in November 2000 and November 2001 collected at nine sites contained 241.0-381.0 mg Si/kg DW (Kumar et al., 2008; Sajwan et al., 2008).

6.39 Silver

Of all trace metals tested to marine bivalve molluscs, silver is the most strongly accumulated (Luoma, 1994). Highest silver concentrations reported in tissues of marine molluscs, in mg Ag/kg DW, were 113.0 in soft parts of the mussel *P. canaliculus*, 129.0 in gills of the abalone *H. rufescens*, 133.0 in soft parts of the Baltic clam *M. balthica*, 166.0 in kidney of the scallop *C. radula*, 185.0 in soft parts of the clam *S. plana*, and 320.0 in digestive diverticula of the whelk *B. canaliculatum* (Table 6.20).

Silver content in molluscan tissues varies widely among closely related species as well as conspecifics collected from different locations (Table 6.20). In general, the highest silver concentrations were in samples collected nearest port cities and river discharges (Berrow, 1991; Fowler and Oregioni, 1976), electroplating plant outfalls (Eisler et al., 1978; Stephenson and Leonard, 1994), ocean dumpsites (Greig, 1979), urban point sources (Alexander and Young, 1976; Anderlini, 1992; Crecelius, 1993; Martin et al., 1988), and from calcareous sediments rather than detrital organic or iron oxide sediments (Luoma and Jenne, 1977). Seasonal (Fowler and Oregioni, 1976) and latitudinal (Anderlini, 1974) trends were also in evidence. Seasonal variations in silver concentrations of the Baltic clam, *M. balthica,* were associated with seasonal variations in moisture content of soft tissues, and frequently (66-87%) reflected the silver content in the surrounding sediments (Cain and

Table 6.20: Silver Concentrations in Field Collections of Molluscs

Organism	Concentration	Reference[a]
Gastropod, *Acmaea digitalis*		
Shell	6.3-11.1 DW	1
Soft parts	2.4-2.6 DW	1
Clam, *Anodonta* sp.		
Shell	0.15 DW	2
Soft parts	0.28 DW	2
Giant squid, *Architeuthis dux*; 2001-2005; Spain		
Gill	1.0 DW	58
Hepatopancreas; Bay of Biscay vs. Mediterranean Sea	1.9 DW vs. 14.0 DW	58
Hydrothermal vent mussel, *Bathymodiolus azoricus*; mid-Atlantic ridge; 2001; 5 sites		
Gills	1.8-5.2 DW	56
Mantle	0.1-0.7 DW	56
Digestive gland	0.5-0.9 DW	56
Whelk, *Buccinum undatum*		
Shell	0.11 DW	2
Soft parts	0.58 DW	2
Whelk, *Busycon canaliculatum*		
Muscle	0.1-0.18 FW	3
Digestive gland	6.4-19.7 FW	3
Gills	0.53 DW	4
Digestive diverticula	320.0 DW	4
Cockle, *Cardium edule*		
Shell	0.11 DW	2
Soft parts	0.04 DW	2
Cockle, *Cerastoderma edule*; soft parts	1.5 (0.11-6.5) DW	5
Scallop, *Chlamys opercularis*		
Soft parts	10.4 DW	6
Kidneys	35.0 DW	6
Digestive gland	77.0 DW	6
Mussel, *Choromytilus meridionalis*; soft parts	0.3 DW	7
Scallop, *Comptopallium radula*; Santa Marie Bay (metals-contaminated), New Caledonia; October 2004		
Soft parts	5.2 (3.9-6.5) DW	55
Kidney	160.0 DW	55

(Continues)

Table 6.20: Cont'd

Organism	Concentration	Reference[a]
Clam, *Corbicula* sp.; San Francisco Bay; 1983-1986; soft parts	0.07-0.2 DW	34
Pacific oyster, *Crassostrea gigas*; soft parts	1.9 DW	8
Oyster, *Crassostrea margaritacea*; soft parts	2.6 DW	8
American oyster, *Crassostrea virginica*		
Soft parts; United States		
East coast	0.3-5.0 DW	9, 45
Gulf coast	0.59-6.0 DW	9
Connecticut	6.1 FW	10, 50
Georgia	28.0-82.0 DW	11
Northeast	0.8-2.3 FW	12
Louisiana	1.0-5.0 DW	32, 39
Gulf coast	0.6-6.0 DW; max. 7.0 DW	9, 35, 36
Maryland, Chesapeake Bay; 1986-1988	2.0-6.0 DW	37
Gills	5.9 FW	10
Cephalopods; 7 spp.; digestive gland	3.0-46.0 DW	33
Cephalopods; coast of France on English Channel; October 1978		
Octopus, *Eledone cirrhosa*		
Digestive gland	2.0-4.4 DW	33
Digestive tract	0.6 DW	33
Other tissues	<0.3 DW	33
Whole	0.8 DW	33
Cuttlefish, *Sepia officinalis*		
Digestive gland	4.9-7.4 DW	33
Kidney	0.7 DW	33
Other tissues	<0.3 DW	33
Whole	0.7 DW	33
Limpet, *Crepidula fornicata*		
Shell	0.1 DW	2
Soft parts	0.97 DW	2
Clam, *Ensis arcuata*; shell	0.022 DW	13
Clam, *Ensis siliqua*; shell	0.017 DW	13
Abalone, *Haliotis rufescens*		
Gills	13.0-129.0 DW	14
Mantle	16.0-54.0 DW	14
Digestive gland	14.0-60.0 DW	14
Foot	1.1-44.4 DW	14

(Continues)

Table 6.20: Cont'd

Organism	Concentration	Reference[a]
Abalone, *Haliotis tuberculata*; soft part	2.9 DW	15
Limpet, *Littorina littorea*		
Soft parts	19.6 (3.2-73.0) DW	5
Soft parts; Looe estuary, United Kingdom, vs. reference site; 1988	10.7 (3.1-17.4) DW vs. 4.1 (3.4-5.0) DW	38
Squid, *Loligo opalescens*; liver; California		
Central coast	25.0 DW	16
Southern coast	45.0 DW	16
Baltic clam, *Macoma balthica*		
Soft parts	85.0 (19.0-128.0) DW	5
Soft parts; San Francisco Bay; near sewage site vs. reference site	32.0-133.0 DW vs. <1.0 DW	39, 40
Hardshell clam, *Mercenaria mercenaria*		
Body	0.4 FW	10
Gills	1.6 FW	10
Shell	0.21 DW	2
Soft parts	1.3 DW	2
Mussel, *Modiolus modiolus*		
Soft parts	0.26 DW	2
Mantle and gills	0.24 DW	2
Muscle	0.03 DW	2
Gonad	0.07 DW	2
Gut and digestive gland	0.81 DW	2
Molluscs; 21 commercial species; edible portions		
10 spp.	<0.1 FW	17
5 spp.	0.1-0.3 FW	17
3 spp.	0.3-0.7 FW	17
3 spp.	0.44-3.8 DW	18
Molluscs; South San Francisco Bay; 1982; soft parts; maximum values		
Mud snail, *Nassarius obsoletus*	320.0 DW	40
Clam, *Macoma nasuta*	5.1 DW	40
Softshell clam, *Mya arenaria*	34.0 DW	40
Clam, *Tapes japonica*	65.0 DW	40
Softshell clam, *Mya arenaria*; soft parts	0.3 FW	10

(Continues)

Table 6.20: Cont'd

Organism	Concentration	Reference[a]
California mussel, *Mytilus californianus*		
Soft parts		
California	<1.0-5.5 DW	1
United States West coast	0.1-2.0 DW	9
Digestive gland	0.7-33.0 DW	19
Shell	5.0-7.9 DW	1
Soft parts; California		
Bodega Bay; 1976-1978 vs. 1986-1988	0.15 DW vs. 0.1 DW	41
Coast; 1977-1981 vs. 1989-1991	Max. 10.0-70.0 DW vs. <2.0 DW	42
San Diego Bay; near municipal wastewater outfall vs. reference sites in northern California and Baja California, Mexico	59.0 DW vs. 0.08-0.22 DW	43
San Francisco Bay; 1987; North Bay vs. South Bay	0.04-0.16 DW vs. 0.7-2.9 DW	40
Common mussel, *Mytilus edulis*		
Soft parts		
North Sea	0.02-1.2 DW	20
United States West coast	0.3-2.3 DW	9
United States West coast; rural sites vs. urban areas	0.1 DW vs. max. 5.0 DW	45
United States East coast	0.04-0.7 DW	9
United States East coast; rural sites vs. urban areas	0.3 DW vs. max. 2.0 DW	45
Rhode Island; Narragansett Bay; 1976-1978 vs. 1986-1988	0.2 DW vs. 0.22 DW	41
Scheldt estuary, Europe; March 2000	1.5 DW	57
Belgium	0.15 DW	13
Connecticut, United States	0.2 FW	10
Looe estuary, United Kingdom	0.23 (0.10-0.55) DW	5
United Kingdom	0.03 DW	2
Norway	1.0-6.0 DW	21
California, United States	<1.0-1.3 DW	1
Ireland; February 1990; contaminated site (Cork Harbour) vs. reference sites (east coast of Ireland)	0.08-4.3 DW vs. <0.05-1.0 DW	44
Shell	0.006 DW	13
Shell	0.1 DW	2
Shell	2.0-4.0 DW	21
Shell	4.4-6.3 DW	1

(Continues)

Table 6.20: Cont'd

Organism	Concentration	Reference[a]
Mussel, *Mytilus edulis aoteanus*		
Soft parts	0.1 DW	22
Mantle	0.4 DW	22
Gill	1.0 DW	22
Visceral mass	0.2 DW	22
Gonad	0.1 DW	22
Soft parts; New Zealand; 1986-1987; various distances (meters) from sewage outfall		
50	5.3-7.7 DW	46
100	4.2 DW	46
200	3.9 DW	46
750	3.5-4.1 DW	46
1500	2.9 DW	46
3000	2.7-3.4 DW	46
Mussel, *Mytilus galloprovincialis*		
Soft parts	0.76 DW	23
Soft parts	0.9 (0.3-7.1) DW	53
Whelk, *Nucella lapillus*; soft parts	2.7 (1.3-4.2) DW	5
Squid, *Ommastrephes bartrami*; liver	12.0 DW	16
Oyster, *Ostrea edulis*		
Soft parts	6.4 DW	8
Homogenate	4.9-7.7 DW	24
Octopus, *Octopus cyanea*; June 1998; Guam; tentacle vs. liver	<0.12 DW vs. 4.4 DW	52
European flat oyster, *Ostrea edulis*; soft parts	2.9 (0.4-9.7) DW	53
Oyster, *Ostrea equestris*; soft parts; United States		
East coast	18.9 DW	9
West coast	0.7-1.6 DW	9
Oyster, *Ostrea sinuata*		
Soft parts	5.6 DW	22
Mantle	4.8 DW	22
Gill	4.6 DW	22
Muscle	1.8 DW	22
Visceral mass	68.0 DW	22
Kidney	1.0 DW	22

(Continues)

Table 6.20: Cont'd

Organism	Concentration	Reference[a]
Heart	48.0 DW	22
Shell	<0.1 DW	22
Limpet, *Patella vulgata*.		
Soft parts	<0.1-4.0 DW	21
Shell	4.0-5.0 DW	21
Soft parts	3.0 (1.5-6.0) DW	5
Soft parts	2.1 DW	25
Soft parts vs. shell; Israel; 1973		
Near sewage outfalls	6.7 DW vs. 5.7 DW	47
Reference site 80 km north of outfall	1.2 DW vs. 5.3 DW	47
Scallop, *Pecten maximus*		
Soft parts	2.7 DW	6
Kidneys	4.3 DW	6
Digestive gland	13.6 DW	6
Gut and digestive gland	8.9 DW	2
Mantle and gills	0.4 DW	2
Scallop, *Pecten novae-zelandiae*		
Soft parts	0.7 DW	22
Mantle	0.2 DW	22
Gills	1.0 DW	22
Muscle	<0.1 DW	22
Visceral mass	1.8 DW	22
Intestine	2.9 DW	22
Kidney	4.8 DW	22
Foot	1.1 DW	22
Gonad	0.2 DW	22
Shell	<0.1 DW	22
Mussel, *Perna canaliculus*; soft parts; New Zealand; 1986-1987; distance from sewage outfall, in meters		
200	35.0-113.0 DW	46
750	49.0-85.0 DW	46
1500-3000	8.0-13.0 DW	46
Clam, *Pitar morrhuana*; soft parts	1.2-4.6 DW	26
Giant scallop, *Placopecten magellanicus*		
Soft parts	1.8 DW	27, 28
Muscle	<0.08-0.24 FW	29
Gonads	0.13-0.69 FW	29
Visceral mass	0.22-1.9 FW	29

(*Continues*)

Table 6.20: Cont'd

Organism	Concentration	Reference[a]
Soft parts; ocean disposal site vs. reference site	Max. 9.1 DW vs. <0.1 DW	39
Clam, *Potamocorbula amurensis*; San Francisco Bay; 1991-1992	2.2 (0.3-7.0) DW; BCF of about 366,000	48
Clam, *Protothaca staminea*		
Soft parts	<1.0 DW	1
Shell	5.8 DW	1
Oysters, *Saccostrea* spp.; Australia; 1980-1983; soft parts	Max. 0.4 FW	49
Clam, *Scrobicularia plana*		
Digestive gland	0.8 DW	33
Kidney	0.4 DW	33
Soft parts; reference site vs. silver-contaminated estuary	0.1-1.5 (0.03-2.1) DW vs. 4.0-5.8 (1.1-185.0) DW	5, 30, 37, 38, 51
Soft parts	0.2 (0.03-0.38) DW	30
Digestive gland	0.38 (<0.1 to 1.1) DW	30
Soft parts	0.23-1.2 DW	31
Digestive gland	1.4 DW	31
Mantle and siphons	0.5 DW	31
Foot and gonad	0.67 DW	31
Gills and palps	1.58 DW	31
Adductor muscle	<0.1 DW	31
Kidney	0.24 DW	31
Camel estuary		
Digestive gland	0.55 DW	31
Remaining tissues	0.35 DW	31
Transferred from Camel estuary to Gannel estuary for 352 days		
Digestive gland	1.8 DW	31
Remaining tissues	0.25 DW	31
Gannel estuary		
Digestive gland	2.0 DW	31
Remaining tissues	0.28 DW	31
Transferred from Gannel estuary to Camel estuary for 352 days		
Digestive gland	0.43 DW	31
Remaining tissues	0.22 DW	31
Common cuttlefish, *Sepia officinalis*		
Egg		
Eggshell	0.07 DW	54

(Continues)

Table 6.20: Cont'd

Organism	Concentration	Reference[a]
Embryo	0.16 DW	54
Yolk	0.22 DW	54
Hatchlings		
Whole	1.8 DW	54
Cuttlebone	0.03 DW	54
Juveniles age 1 month		
Whole	1.4 DW	54
Digestive gland	14.0 DW	54
Cuttlebone	0.1 DW	54
Immatures, age 12 months		
Whole	1.8 DW	54
Digestive gland	19.0 DW	54
Cuttlebone	0.08 DW	54
Adults, age 18 months		
Whole	0.6 DW	54
Digestive gland	13.4 DW	54
Cuttlebone	0.08 DW	54
Clam, *Spisula solidissima*; muscle	0.39-1.3 DW	3, 4
Squid, *Symplectoteuthis oualaniensis*; liver	24.0 DW	16
Clam, *Tapes semidecussata*		
Shell	7.3 DW	1
Soft parts	7.3 DW	1
Gastropod, *Thais emarginata*		
Shell	6.4-11.1 DW	1
Soft parts	2.1 DW	1

Values are in mg Ag/kg fresh weight (FW) or dry weight (DW).
[a]1, Graham, 1972; 2, Segar et al., 1971; 3, Greig et al., 1977; 4, Greig, 1975; 5, Bryan and Hummerstone, 1977; 6, Bryan, 1973; 7, Watling and Watling, 1976b; 8, Watling and Watling, 1976a; 9, Goldberg et al., 1978; 10, Thurberg et al., 1974; 11, Windom and Smith, 1972; 12, Greig and Wenzloff, 1978; 13, Bertine and Goldberg, 1972; 14, Anderlini, 1974; 15, Bryan et al., 1977; 16, Martin and Flegal, 1975; 17, Hall et al., 1978; 18, Greig, 1979; 19, Alexander and Young, 1976; 20, Karbe et al., 1977; 21, Lande, 1977; 22, Brooks and Rumsby, 1965; 23, Fowler and Oregioni, 1976; 24, Fukai et al., 1978; 25, Preston et al., 1972; 26, Eisler, 1977b; 27, Pesch et al., 1977; 28, Reynolds, 1979; 29, Greig et al., 1978; 30, Bryan and Uysal, 1978; 31, Bryan and Hummerstone, 1978; 32, Ramelow et al., 1989; 33, Miramand and Bentley, 1992; 34, Luoma et al., 1990; 35, Morse et al., 1993; 36, Presley et al., 1990; 37, Sanders et al., 1991b; 38. Truchet et al., 1990; 39, USPHS, 1990; 40, Luoma and Phillips, 1988; 41, Lauenstein et al., 1990; 42, Stephenson and Leonard, 1994; 43, Martin et al., 1988; 44, Berrow, 1991; 45, Crecelius, 1993; 46, Anderlini, 1992; 47, Navrot et al., 1974; 48, Brown and Luoma, 1995; 49, Talbot, 1985; 50, Thurberg et al, 1974; 51, Bryan and Hummerstone, 1977; 52, Denton et al., 2006; 53, Deudero et al., 2007; 54, Miramand et al., 2006; 55, Metian et al., 2008; 56, Cosson et al., 2008; 57, Wepener et al., 2008; 58, Bustamante et al., 2008.

Luoma, 1990). Inherent differences among species in ability to accumulate silver from seawater are well-documented, with oysters >> scallops >> mussels (Brooks and Rumsby, 1965). Even closely related species of scallops collected from the same area differed in silver content of individual tissues by 3- to 9-fold (Bryan, 1973). Internal organs, especially digestive gland and kidney, usually contained the highest burdens of silver in all species of molluscs examined (Table 6.20). Studies with radiosilver-110 m suggest that the half-time persistence of silver is 27 days in mussels, 44-80 days in clams, and more than 180 days in oysters (Fisher et al., 1994; Preston et al., 1968). In oysters and other bivalve molluscs, the major pathway of silver accumulation was from dissolved silver. Uptake was negligible from silver adsorbed onto suspended sediments and algal cells, with elimination of adsorbed silver by oysters via the feces (Abbe and Sanders, 1990; Sanders et al., 1990). In oysters, silver associated with food was unavailable for incorporation, which may be due to the ability of silver to adsorb rapidly to cell surfaces and to remain tightly bound despite changes in pH or enzymatic activity (Connell et al., 1991). Studies with ^{110m}Ag and clams, *G. tumidum*, held for 15 days in seawater containing radiosilver had concentration factors of 4000-15,000 in soft parts; concentration factors were lower for smaller clams and higher for larger clams—which differs significantly from studies with the same species and isotopes of cadmium, cobalt, chromium, and zinc, wherein smaller clams had highest concentration factors (Hedouin et al., 2006).

Uptake of dissolved silver by oysters was higher at elevated temperatures in the 15-25 °C range tested (Abbe and Sanders, 1990). American oysters, *C. virginica*, maintained near a nuclear power plant in Maryland that discharged radionuclides on a daily basis into Chesapeake Bay accumulated ^{110m}Ag; accumulations were higher in summer and fall than in winter or spring (Rose et al., 1988). Oysters from the Gulf of Mexico vary considerably in whole-body content of silver and other trace metals. Variables known to modify silver concentrations in oyster tissues include the age, size, gender, reproductive stage, general health and metabolism of the oyster, water temperature and salinity, dissolved oxygen and turbidity of the medium, natural and anthropogenic inputs to the biosphere, and chemical species and interactions with other compounds (Presley et al., 1990). For example, silver concentrations in American oysters from the Chesapeake Bay were reduced in summer and at increasing water salinities, and were elevated near sites of human activity; chemical forms of silver taken up by oysters included the free ion (Ag^+) and the uncharged $AgCl^0$ (Daskalakis, 1995; Sanders et al., 1991a). Declines in tissue silver concentrations of the California mussel, *M. californianus*, were significant between 1977 and 1990; body burdens decreased from 10.0-70.0 mg Ag/kg DW to less than 2.0 mg Ag/kg DW and seem to be related to the termination of metal-plating facilities in 1974 and decreased mass emission rates by wastewater treatment facilities (Stephenson and Leonard, 1994).

Laboratory studies on silver accumulation and effects in marine gastropods, mussels, clams, scallops, and oysters is the subject of a growing number of studies. In one series, Thurberg et al. (1974) immersed American oysters, *C. virginica*, common mussels, *M. edulis*, and hardshell clams, *M. mercenaria*, in seawater solutions containing 1.0 mg Ag/L for 96 h. Whole-body silver burdens—in mg Ag/kg FW—rose during this period from 0.2 to 5.2 in mussels, from 0.4 to 1.0 in clams, and from 6.1 to 14.9 in oysters. Silver content in gill tissues during this interval rose from 5.9 mg/kg FW to 33.9 in oysters, and from 1.6 to 6.9 in clams. In a second series, Thurberg et al. (1975) exposed adult surf clams, *Spisula solidissima*, to 0.01 or 0.1 mg Ag/L for 96 h. Whole-body tissue concentrations in surf clams rose from 0.08 mg Ag/kg FW in controls to 1.0 in the low dose group and 2.0 in the high dose group; for gills, silver values were 5.7 mg/kg FW in the low dose group and 8.5 in the high dose group. Bivalve molluscs, especially oysters, took up radiosilver-110 from the medium by factors as high as 500- to 32,000-fold (Pouvreau and Amiard, 1974); these values were lower for gastropods.

Gastropod sea hares *Aplysia californica* were exposed to 0.001 or 0.01 mg Ag/L for 48 h (Bianchini et al., 2007). Afterwards, digestive gland contained 3.6 mg Ag/kg FW in the high dose group. Silver burdens in the high dose group were lowest in hemolymph, muscle, ganglia, and gills in that order, and highest in digestive gland (Bianchini et al., 2007). Slipper limpets (Gastropoda), *C. fornicata*, subjected to silver concentrations as low as 0.001 mg/L for up to 24 months showed histopathological damage and accumulations up to 34.0 mg Ag/kg FW soft parts; higher exposure concentrations of 0.005 and 0.01 mg Ag/L were associated with inhibited reproduction and whole-body burdens as high as 87.0 mg Ag/kg FW (Nelson et al., 1983). Maximum silver concentrations for the 0.005 mg Ag/L group after 6 months was 54.1 mg Ag/kg FW soft parts versus 2.8 in controls; after 24 months this was less than 8.0 mg/kg FW soft parts (Nelson et al., 1983). Respiration rate in gastropods was depressed by ionic silver; however, the reverse was observed in six species of bivalves tested (Gould and MacInnes, 1977). In marine gastropods, *L. littorea*, silver is stored in the basement membrane of the digestive system (Truchet et al., 1990).

Asiatic clams, *C. fluminea*, held in seawater containing 0.0045 mg Ag/L for 21 days showed growth inhibition; residues of 1.65 mg Ag/kg FW soft parts and higher were associated with reduced growth (Diamond et al., 1990). The LC50 (96 h) value for juveniles of the Asiatic clam followed by a 96 h observation period was 0.155 mg Ag/L (Diamond et al., 1990). Juveniles of the surf clam, *S. solidissima*, had increased oxygen consumption at 0.01 mg Ag/L after 96 h (Calabrese et al., 1977b) and abnormal embryo development at 0.014 mg/L following 1 h exposure after fertilization (Bryan and Langston, 1992); 0.1 mg/L was lethal to juveniles in 96 h (Calabrese et al., 1977a,b). In clams, *S. plana*, silver is stored in the basement membrane of the outer fold of the mantle edge and in the amoebocytes (Truchet et al., 1990).

Juvenile bay scallops, *A. irradians*, held in 0.01 mg Ag/L for 96 h contained 3.1 mg Ag/kg FW (Nelson et al., 1976). It is noteworthy that ionic silver is extremely toxic to molluscs, with concentrations lethal to 50% of bay scallops in 96 h as low as 0.033 mg Ag^+/L (Nelson et al., 1976); similar results were reported for clams and oysters (Calabrese et al., 1977b). Scallops, *Chlamys varia*, held in seawater containing 0.02 mg Ag/L for 14 days had 18.0 mg Ag/kg DW in soft tissues versus 1.7 in controls (Berthet et al., 1992). The LC50 (115 h) value for *C. varia* is 0.1 mg Ag/L (Berthet et al., 1992). In scallops and mussels, silver was stored in basement membranes and pericardial gland. In all species of bivalve molluscs, sequestered silver was in the form of silver sulfide (Berthet et al., 1992).

Trophic transfer of silver in mussels is dependent on the silver assimilation efficiency (AE) from ingested food particles, feeding rate, and silver efflux rate (Fisher and Wang, 1998). Silver AE is usually less than 30%, and is lower for sediment than for phytoplankton. Silver AE and distribution from ingested phytoplankton particles is modified by gut passage time, extracellular and intracellular digestion rates, and metal desorption at lowered pH. The kinetic model of Fisher and Wang (1998) for mussels predicts that either the solute or particulate pathway can dominate and is dependent on silver partition coefficients for suspended particles, and silver AE. Common mussels, *M. edulis*, exposed continuously for 21 months to 0.0 (controls), 0.001, 0.005, or 0.01 mg Ag/L from age 2.5 months, were observed for growth, accumulations, and histopathology (Calabrese et al., 1984). Growth was not affected. Silver burdens, in mg Ag/kg FW soft parts, for controls ranged from 0.2 to 0.7; maximum concentrations in the 0.001 mg/L group (9.1) occurred a 18 months; for the 0.005 mg/L group, residues were highest (11.9) at 18 months; for the 0.01 mg/L group, Concentrations were highest (15.3) at 12 months. All silver-exposed groups had histopathology of basement membranes and connective tissues; histopathology was typical of argyria in humans and other mammals that have absorbed organic or inorganic silver compounds (Calabrese et al., 1984). Adults of *M. galloprovincialis* had normal growth after 21-month exposure to 0.01 mg Ag/L; mussels held in 0.02 mg Ag/L for 28 days had 15.0 mg Ag/kg DW soft parts versus 0.08 in controls (Berthet et al., 1992). The LC50 (110 h) for *M. galloprovincialis* adults was 0.1 mg Ag/L (Berthet et al., 1992).

Eggs of the cuttlefish, *S. officinalis* were exposed to both ^{110m}Ag and stable silver at 0.002 mg Ag/L during embryonic development; both stable silver and radiosilver continued to increase in eggs beginning at day 30 of embryonic development, owing to eggshell permeability to silver at that time (Lacoue-Labarthe et al., 2008a,b). Silver concentrations in juvenile cuttlefish, *S. officinalis*, were much higher than those in embryos (Table 6.20), suggesting efficient incorporation from seawater (Miramand et al., 2006). Silver concentrations in digestive gland, unlike other metals analyzed wherein burdens increase exponentially with age, decreases as soon as adult cuttlefish migrate to the open sea; suggesting that silver is excreted from digestive gland on contact with less contaminated environments (Miramand et al., 2006). Embryos of the Pacific oyster, *C. gigas*, held in 0.002-0.01 mg Ag/L for 48 h had retarded shell growth, reduced size, and erratic swimming behavior in 5.8% of exposed groups versus 1% in controls (Coglianse and Martin, 1981).

Juvenile Pacific oysters held in 0.02 mg Ag/L for 14 days then transferred to uncontaminated seawater for 23 days had impaired glycogen storage capacity during exposure (Berthet et al., 1990). During depuration, silver concentrations decreased from 31.3 mg/kg DW soft parts to 12.8 versus <10.0 in controls. Most (69%-89%) of the insoluble accumulated silver was sequestered in amoebocytes and basement membranes as silver sulfide (Ag_2S), a stable mineral form that is not degradable, thereby limiting the risk of silver transfer through the food chain (Berthet et al., 1990). The LC50 (209 h) value for silver and adult Pacific oysters was 0.1 mg/L (Berthet et al., 1992). Adults of the American oyster, *C. virginica*, held in 0.002 mg Ag/L for 14 days in natural enclosures containing phytoplankton showed reduced growth; phytoplankton had 8.6 mg Ag/kg DW (vs. 0.03 in controls) and oyster soft parts 2.8 mg/kg DW (vs. 0.8 in controls). Oysters and phytoplankton held in 0.005 or 0.007 mg Ag/L under similar conditions had 24.0-44.0 mg Ag/kg DW phytoplankton and 4.8-6.8 mg Ag/kg DW soft part oysters (Abbe and Sanders, 1990; Sanders et al., 1990). The LC50 (48 h) value for American oyster embryos was 0.0058 mg Ag/L (Calabrese et al., 1977b; Ratte, 1999). American oysters excrete about 60% of their accumulated excess silver in soft tissues within 30 days of transfer to silver-free seawater; soluble forms were preferentially eliminated and insoluble forms retained (Berthet et al., 1992). Interspecies differences in ability to retain silver among bivalve molluscs are large, even among closely related species of crassostreid oysters. For example, the half-time persistence of silver was 149 days in American oysters but only 26 days in Pacific oysters (USPHS, 1990). Silver interactions with other metals and compounds in solution are not well defined. For example, mixtures of salts of silver and copper markedly increased the survival of oyster embryos, but only when copper concentrations were less than 0.006 mg/L and total silver less than 0.011 mg/L (Coglianse and Martin, 1981).

Silver interacts with salts of copper and cadmium in the green mussel, *P. viridis* in 2 week exposure studies (Ng et al., 2007). Mussels subjected to 0.005 mg Ag/L plus 0.02 mg Cd/L accumulated each metal in soft tissues, but independently. Interaction was observed at the subcellular level wherein 25% of the silver shifted from the insoluble fraction to the metallothionein-like protein in the presence of cadmium. Exposure to silver alone (0.005 mg Ag/L), for 2 weeks increased the dietary uptake of silver by 30%, although the effect was reduced in the presence of cadmium. Co-exposure of mussels to 0.005 mg Ag/L plus 0.03 mg Cu/L increased both silver and copper concentrations in soft parts by up to fivefold. Silver and copper subcellular distributions in co-exposed and singly exposed mussels were similar (Ng et al., 2007). Other studies with the green mussel show that 2.0 mg Ag/L is not fatal to adults in 96 h; however, there were significant decreases in filtration rate, oxygen consumption, and ATPase activity in gills, hepatopancreas, ovary, and muscle (Vijayavel et al., 2007).

Water quality criteria for silver for the protection of marine molluscs are proposed. Bryan and Langston (1992) state that tissue concentrations less than 1.0 mg total Ag/kg DW soft parts is normal and presumably safe for marine clams; however, more than 100.0 mg total Ag/kg DW is stressful and possibly fatal. This needs to be further refined. The proposed silver criterion of 0.0023 mg total Ag/L to protect marine molluscs needs to be reconsidered because some species of molluscs show extensive accumulations at 0.001-0.002 mg total Ag/L (Eisler, 2000d).

6.40 Strontium

Strontium concentrations in molluscs are highest in calcareous tissues, such as shell, and lowest in soft tissues (Table 6.21). Since the mean strontium to calcium ratios in pelycopods, amphineurans, scaphopods, and cephalopods all fall within the same narrow range of 0.00185-0.00806 (Thompson and Chow, 1955), it is probable that strontium residues closely parallel those of calcium concentrations in molluscan tissues.

Mussels, *M. galloprovincialis*, show increasing concentrations of strontium with increasing growth; moreover, this species reportedly can distinguish between stable strontium and radiostrontium-90, with selective accumulation of the stable form (Parchevskii et al., 1965). No ^{90}Sr activity was detectable in various species of molluscs collected near Amchitka Island in summer 2004; Amchitka Island was a nuclear test site between 1965 and 1971 (Burger et al., 2007b).

Some molluscs were inefficient accumulators of strontium. Among five species of bivalves taken from coastal waters of Greece, strontium concentration factors in soft tissues ranged from 1 to 15 (Papadopoulu, 1973). This is not the case with fecal materials of some species. Feces of bivalves contribute significantly to the transfer of strontium to bottom sediments. The concentration of strontium in feces of hardshell clams, as one example, is increased by a factor of 670 over that in seawater; for softshell clams, however, this value was only 14 (Andrews and Warren, 1969).

6.41 Tantalum

Soft tissues of the common mussel, *M. edulis*, contained between 0.004 and 0.33 mg Ta/kg DW (Karbe et al., 1977).

6.42 Technetium

No ^{99}Tc activity was detected in molluscs from the Aleutian Islands during summer 2004; Amchitka Island in the Aleutian chain was used for nuclear tests between 1965 and 1971 (Burger et al., 2007b).

Table 6.21: Strontium Concentrations in Field Collections of Molluscs

Organism	Concentration	Reference[a]
Mussel, *Anodonta* sp.; shell	150.0 DW	1
Gastropod, *Aplysia benedicti*		
Ctenidium	0.6-0.9 DW	2
Gonad	2.5-3.0 DW	2
Parapodia	0.3 DW	2
Hepatopancreas	5.0-7.0 DW	2
Buccal mass	1.9 DW	2
Gut content	0.8 DW	2
Shell	1300.0-1600.0 DW	2
Hydrothermal vent mussel, *Bathymodiolus azoricus*; mid-Atlantic ridge		
Shell; 2001; 4 sites	871.0-1063.0 (536.0-1261.0) DW	16
Whelk, *Buccinum undatum*		
Shell	750.0 DW	1
Soft parts	13.0 DW	1
Cockle, *Cardium edule*		
Shell	1300.0 DW	
Soft parts	17.0 DW	
Scallop, *Chlamys opercularis*; shell	270.0-410.0 DW	1
American oyster, *Crassostrea virginica*		
Shell	1142.0-1648.0 DW	3
Soft parts	13.0-28.0 DW	4
Limpet, *Crepidula fornicata*		
Shell	100.0 DW	1
Soft parts	20.0 DW	1
Gastropods		
Flesh	510.0 AW	5
Shell	2194.0 AW	5
Viscera	84.0 AW	5
Operculum	200.0 AW	5
Shell	860.0-2700.0 AW	6
Hardshell clam, *Mercenaria mercenaria*		
Shell	170.0 DW	1
Soft parts	13.0 DW	1
Mussel, *Modiolus modiolus*		
Shell	73.0 DW	1
Soft parts	30.0-47.0 DW	1

(Continues)

Table 6.21: Cont'd

Organism	Concentration	Reference[a]
Mantle and gills	41.0 DW	1
Muscle	23.0 DW	1
Gonad	33.0 DW	1
Gut and digestive gland	41.0 DW	1
Molluscs		
Shell	1000.0 to <5000.0 DW	6
Shell	1540.0-4097.0 DW	7
Soft parts	0.04-0.77 FW	8
Softshell clam, *Mya arenaria*; soft parts	63.0-163.0 DW	9
Mussel, *Mytilus californianus*; soft parts	16.0-48.0 DW	4
Common mussel, *Mytilus edulis*		
Soft parts	18.0-100.0 DW	10
Soft parts	19.0-56.0 DW	4
Soft parts	24.0 DW	1
Whelk, *Nucella lapillus*		
Shell	1200.0 DW	1
Soft parts	79.0 DW	1
Limpet, *Patella candei gomesii*; soft parts; 2007; Azores		
Near hydrothermal vent	63.3-84.7 DW	15
Reference site	111.3-154.3 DW	15
Limpet, *Patella vulgata*		
Shell	570.0 DW	1
Soft parts	23.0 DW	1
Scallop, *Pecten maximus*		
Soft parts	12.0 DW	1
Muscle	9.9 DW	1
Gut and digestive gland	13.0 DW	1
Mantle and gills	11.0-16.0 DW	1
Gonad	12.0-23.0 DW	1
Shell	270.0-660.0 DW	1
Scallop, *Pecten* sp.; shell	660.0-1200.0 DW	6
Bivalve, *Pinna nobilis*		
Soft parts	2300.0 AW	11
Mantle and gills	830.0 AW	11
Muscle	1400.0 AW	11
Nervous system	3300.0 AW	11

(Continues)

Table 6.21: Cont'd

Organism	Concentration	Reference[a]
Stomach and intestine	2900.0 AW	11
Gonads	1600.0 AW	11
Hepatopancreas	600.0 AW	11
Byssus	2200.0 AW	11
Byssus gland	4700.0 AW	
Pteropods		
Shell	220.0-620.0 DW	12
Shell	950.0-1400.0 DW	13
Clam, *Tapes japonica*		
Soft parts	110.0-360.0 AW	14
Shell	1700.0-1900.0 AW	14

Values are in mg Sr/kg fresh weight (FW), dry weight (DW), or ash weight (AW).
[a]1, Segar et al., 1971; 2, Patel et al., 1973; 3, Rucker and Valentine, 1961; 4, Goldberg et al., 1978; 5, Lowman et al., 1970; 6, Turekian and Armstrong, 1960; 7, Pilkey and Goodell, 1963; 8, Mauchline and Templeton, 1966; 9, Eisler, 1977b; 10, Karbe et al., 1977; 11, Papadopoulu, 1973; 12, Pyle and Tieh, 1970; 13, Krinsley and Bieri, 1959; 14, Fukai and Meinke, 1962; 15, Cunha et al., 2008; 16, Cravo et al., 2008.

6.43 Tellurium

Tellurium concentrations in tissues of the squid, *T. pacificus*, captured in the East Japan Sea near Korea during summer were highest in hepatopancreas (0.0025 mg Te/kg DW) and stomach (0.0024 mg/kg DW), and lowest in muscle (0.0006 mg Te/kg DW) and gill (0.0001 mg/kg DW) (Waska et al., 2008). BCF values from seawater are estimated at 5800 for muscle, 5900 for stomach, 16,000 for gill, and 22,000 for hepatopancreas (Waska et al., 2008).

6.44 Terbium

Terbium concentrations in soft tissues of the common mussel ranged from 0.006 to 0.03 mg Tb/kg DW (Karbe et al., 1977).

6.45 Thallium

Laboratory studies on thallium accumulation from the medium by the softshell clam and the common mussel were conducted by Zitko and Carson (1975). These studies demonstrated that clams accumulated 6.0 mg Tl/kg DW soft parts in 88 days from solutions containing 0.05 mg Tl/L; this value was 12.5 mg Tl/kg DW when environmental thallium levels were doubled to 0.1 mg/L. On transfer to thallium-free media for 30 days, clam soft parts contained less than 0.5 mg Tl/kg DW. Exposure of *M. edulis* for 40 days in seawater solutions containing 0.05 or 0.1 mg Tl/l produced thallium accumulations of 4.3 and 6.3 mg Tl/kg DW soft parts, respectively; thallium levels in mussels fell to less than 0.5 mg/kg DW within 7 days postexposure (Zitko and Carson, 1975).

6.46 Thorium

The maximum thorium concentration recorded in field collections of molluscs was 0.26 mg Th/kg DW in mussel soft parts (Table 6.22).

6.47 Tin

Inorganic tin concentrations up to 2.0 mg Sn/kg FW were recorded in edible tissues of commercial species of North American molluscs, and up to 35.0 mg Sn/kg FW in smoked canned oysters (Table 6.22). Organotin concentrations in edible tissues of 10 species of molluscs sold commercially in France in 2005 were usually less than 0.0076 mg total organotins/kg FW, but were somewhat higher in squid (0.0133 mg/kg FW) and scallop (0.0098 mg/kg FW); in all cases, TBTs accounted for the majority of all organotins (Table 6.22; Guerin et al., 2007). Larger bivalves (*D. trunculus*) collected from Alexandria, Egypt in 2004 have elevated concentrations of organotins and total tin when compared to smaller bivalves; intermediate sized groups had intermediate values (Table 6.22; Said et al., 2006).

Organic tin compounds are used extensively in antifouling marine paints and in molluscicides (Crowe, 1987). The use of antifoulants on ships was necessitated by the damage some organisms cause to wooden structures and by the reduced fuel efficiency and speed due to drag when vessels become heavily fouled (Laughlin et al., 1984). Organotins replaced copper in antifouling paints because copper has a short effective life and is comparatively expensive. Organic compounds of arsenic, mercury, and lead have also been used in antifouling paints, but present unacceptable toxicological risks during preparation and to the environment (Blunden et al., 1985; Hall et al., 1987). Organotin coatings were promoted because of their efficient antifouling action, long lifetime (up to 4 years) and lack of corrosion (Messiha and Ikladious, 1986). Use of organotin antifouling paints on recreational and commercial vessels has increased markedly. In Maryland, for example, at least 50% and up to 75% of the recreational boats used in Chesapeake Bay are covered with organotin paints (Hall et al., 1988). However, organotin coatings, especially TBTs, present potential environmental problems to nontarget aquatic biota due to their extreme toxicity (Eisler, 2000e; Maguire, 1991). The organotin biocide released by hydrolysis from the surface of the paint film into seawater provides the antifoulant action. In consequence, the depleted outer layer of paint film, containing hydrophilic carboxylate groups, is easily eroded by moving

Table 6.22: Thorium, Tin, Titanium, Tungsten, and Uranium Concentrations in Field Collections of Molluscs

Element and Organism	Concentration	Reference[a]
Thorium		
Common mussel, *Mytilus edulis*; soft parts	0.0003-0.26 DW	1
Pteropods; tests	0.023-0.147 DW	2
Tin		
Bohai Bay, North China; May-September, 2002; soft parts		
Bay scallop, *Argopecten irradians*		
Monobutyltins	0.009 FW	32
Dibutyltins	0.017 FW	32
Tributyltins	0.025 FW	32
Triphenyltins	0.004 FW	32
Veined rapa whelk, *Rapana venosa*		
Monobutyltins	0.004 FW	32
Dibutyltins	0.009 FW	32
Tributyltins	0.005 FW	32
Triphenyltins	0.005 FW	32
Short-necked clam, *Ruditapes philippinarum*		
Monobutyltins	0.004 FW	32
Dibutyltins	0.005 FW	32
Tributyltins	0.004 FW	32
Triphenyltins	0.003 FW	32
Pacific oyster, *Crassostrea gigas*		
Arcachon Bay, France; July; soft parts; total tin vs. organotins; max. values		
1982	7.0 DW vs. 1.6 DW	10
1983	4.0 DW vs. 0.8 DW	10
1985	0.9 DW vs. 0.4 DW	10
Soft parts; reared for 16 weeks in sea cages painted with tributyltin		
Total tin	1.4 FW	11
Tributyltin	0.9 FW	11
Total tin (unpainted)	0.1 FW	11
Tributyltin; soft parts; coastal United Kingdom estuaries		
1986	0.2-6.4 DW	12
1987	0.3-3.7 DW	12
1988	0.1-5.6 DW	12
1989	0.1-1.3 DW	12

(Continues)

Table 6.22: Cont'd

Element and Organism	Concentration	Reference[a]
Korea; January 1995; soft parts		
Tributyltins	0.1-0.9 DW; 28% of total body burden in gonads	13
Triphenyltins	0.16-0.7 DW; 19% of total body burden in gonads	13
American oyster, *Crassostrea virginica*		
Soft parts; Gulf of Mexico; 1989-1991; max. values		
Monobutyltins	0.15 DW	12
Dibutyltins	0.2 DW	12
Tributyltins	1.2 DW	12
Soft parts; Pensacola, Florida; 2003-2004	Max. 0.6 FW	28
Bivalve, *Donex trunculus*; Alexandria, Egypt; 2004; soft parts; shell length 1.5 cm vs. 2.5 cm		
Dibutyltins	0.011 DW vs. 0.028 DW	24
Tributyltins	0.16 DW vs. 0.25 DW	24
Diphenyltins	0.041 DW vs. 0.066 DW	24
Total tin	0.37 DW vs. 0.58 DW	24
France; January-April 2005; edible portions		
Total organotins; 8 spp.	<0.0076 FW	26
Squid		
Total organotins	0.0133 FW	26
Butyltins	0.0123 FW	26
Phenyltins	0.0006 FW	26
Octyltins	0.0003 FW	26
Scallop		
Total organotins	0.0098 FW	26
Butyltins	0.0086 FW	26
Phenyltins	0.0008 FW	26
Octyltins	0.0003 FW	26
Gastropod, *Hexaplex trunculus*; December 2002-2003; Venice lagoon, Italy (organotin antifouling paints banned in 1990) vs. Istrian region, Croatia (no restrictions on organotin antifouling paints)		
Visceral coil		
Tributyltin	0.1 DW vs. 0.41 DW	23

(*Continues*)

Table 6.22: Cont'd

Element and Organism	Concentration	Reference[a]
Dibutyltin	0.21 DW vs. 1.53 DW	23
Monobutyltin	0.12 DW vs. 2.4 DW	23
Monophenyltin	<0.0005 DW vs. 0.134 DW	23
Diphenyltin	0.004 DW vs. 0.064 DW	23
Triphenyltin	0.037 DW vs. 0.131 DW	23
Rest of soft parts		
Tributyltin	0.12 DW vs. 0.33 DW	23
Dibutyltin	0.12 DW vs. 0.73 DW	23
Monobutyltin	0.05 DW vs. 1.5 DW	23
Monophenyltin	<0.0005 DW vs. 0.026 DW	23
Diphenyltin	<0.0004 DW vs. 0.035 DW	23
Triphenyltin	0.03 DW vs. 0.029 DW	23
Japan; northern Kyushu; coastal areas; soft parts; maximum concentrations		
Common mussel, *Mytilus edulis*; 1998	0.135 FW	34
Manila clam, *Ruditapes philippinarum*; 1998 vs. 2001	0.125 FW vs. 0.033 FW	34
Pen shell, *Atrina pectinata*; 2001	0.095 FW	34
Molluscs; commercial species; edible portions; 18 species		
3 spp.	0.3-0.5 FW	3
7 spp.	0.5-0.7 FW	3
4 spp.	0.7-0.9 FW	3
4 spp.	0.9-2.0 FW	3
Softshell clam, *Mya arenaria*; soft parts; Bohai Sea, China; summer 2002; total butyltins	Max. 0.99 FW	31
Common mussel, *Mytilus edulis*		
Soft parts	1.3-7.1 DW	1
Soft parts; Lisbon, Portugal; max. values		
Monomethyltins	0.007 DW	14
Dimethyltins	0.01 DW	14
Trimethyltins	0.023 DW	14
Soft parts; Tokyo Bay; 1989; max. values		
Monobutyltins	0.12 FW	12
Dibutyltins	0.15 FW	12
Tributyltins	0.24 FW	12
Soft parts; Osaka Bay, Japan; 1986; tributyltins	0.02-0.39 DW	15

(*Continues*)

Table 6.22: Cont'd

Element and Organism	Concentration	Reference[a]
Soft parts; transplanted from reference site to River Tyne (United Kingdom), a TBT-impacted site for 6 months; March-September 2001		
Tributyltins; start vs. end	Max. 0.009 FW vs. 0.48 FW	27
Dibutyltins; start vs. end	Max. 0.011 FW vs. 0.49 FW	27
Soft parts; British Columbia; 1990; max. values		
Monobutyltins	0.05 DW	12
Dibutyltins	0.08 DW	12
Tributyltins	0.31 DW	12
Soft parts; tributyltins		
United States; 1986-1987	<0.005-1.4 FW	12
Maine; 1989	2.4 DW	12
Perth, Australia; 1991	<0.001-0.33 FW	12
San Diego Bay; 1988	0.86 FW	16
Hepatopancreas; tributyltins		
San Diego Harbor	1.9-3.5 DW	4
Offshore	<0.7-2.0 DW	4
Mussel, *Mytilus galloprovincialis*; soft parts Adriatic Sea; 2001-2002; summer vs. winter		
Monobutyltins		
Industrialized sites	0.4-0.6 DW vs. <0.01-0.27 DW	25
Reference site	0.03 DW vs. 0.4 DW	25
Dibutyltins		
Industrialized sites	1.2-2.2 DW vs. 0.09-1.1 DW	25
Reference sites	0.34 vs. 1.2 DW	25
Tributyltins		
Industrialized sites	1.2-1.7 DW vs. 1.5-7.9 DW	25
Reference site	0.34 vs. 1.1 DW	25
Monophenyltins		
Industrialized sites	0.04-0.07 DW vs. <0.01 DW	25
Reference site	<0.01 DW vs. 0.07 DW	25
Diphenyltins		
Industrialized sites	<0.005 DW vs. <0.005 DW	25
Reference site	<0.005 DW vs. <0.005 DW	25
Triphenyltins		
Industrialized sites	<0.005-0.028 DW vs. <0.005-0.007 DW	25
Reference site	<0.005 DW vs. 0.005 DW	25

(*Continues*)

Table 6.22: Cont'd

Element and Organism	Concentration	Reference[a]
Korea; 2004-2005; reference site vs. transplantation for 126 days near shipyard; 3 stations		
Monobutyltins	0.006-0.012 DW vs. 0.001-0.018 DW	33
Dibutyltins	0.013-0.017 DW vs. 0.008-0.07 DW	33
Tributyltins	0.006-0.017 DW vs. 0.016-0.155 DW	33
Gastropod, *Nassarius nitidus*		
Venice lagoon, Italy; 2004-2004; soft parts		
Tributyltins	Max. 0.244 DW	29
Dibutyltins	Max. 0.132 DW	29
Monobutyltins	Max. 0.082 DW	29
Southern Venice lagoon; 2004; soft parts		
Tributyltins	<0.004-0.976 DW	30
Dibutyltins	<0.004-1.43 DW	30
Common dogwhelk, *Nucellus lapillus*		
Soft parts; reference sites		
Total tin	Max. 0.3 FW	17
Total tin	0.1-0.2 DW	18
Tributyltins	0.1-0.2 DW	18
Dibutyltins	0.01-0.05 DW	18
Soft parts; transplanted to TBT-impacted site for 6 months (River Tyne estuary, United Kingdom)		
Tributyltins; start vs. end	Max. 0.006 FW vs. 0.056 FW	27
Dibutyltins; start vs. end	Max. 0.005 FW vs. 0.049 FW	27
Squid, *Ommastrephes illicebrosa*; whole	3.0 AW	5
European oyster, *Ostrea edulis*		
Reference site; all tissues	<0.1 FW	19
River Crouch, United Kingdom		
Soft parts	0.27-0.33 FW	19
Eggs	0.33 FW	19
Larvae	0.3 FW	19
Oysters; smoked; canned; soft parts; Canada	25.0-35.0 FW	6
Limpet, *Patella caerulea*; NE Mediterranean Sea; 1980; shell		

(Continues)

Table 6.22: Cont'd

Element and Organism	Concentration	Reference[a]
Total tin	0.013 DW	20
Methyltins	0.0004 DW	20
Dimethyltins	0.0002 DW	20
Trimethyltins	0.0009 DW	20
Scallop, *Pecten maximus*		
Reared for 3 weeks in sea pens with tributyltin-coated netting; total tin vs. tributyltin		
Gonad	0.6 FW vs. 0.4 FW	11
Adductor muscle	0.6 FW vs. 0.5 FW	11
Gills	0.6 FW vs. 0.6 FW	11
Digestive gland	1.0 FW vs. 0.5 FW	11
Green mussel, *Perna viridis*		
Soft parts; Thailand; 1994-1995		
Monobutyltins	0.014 (0.003-0.045) FW	12
Dibutyltins	0.013 (0.001-0.008) FW	12
Tributyltins	0.1 (0.004-0.8) FW	12
Soft parts; tributyltins		
Hong Kong; 1989	0.06-0.12 FW	12
Malaysia; 1992	0.01-0.02 FW	12
Portugal; Rio de Aveiro; May-June 2005 (2 years after European Union ban on TBT anti-fouling paints); whelk, *Nassarius reticulatus* vs. mussel, *Mytilus galloprovincialis*; soft parts; max. concentrations		
Monobutyltins	0.071 DW vs. 0.163 DW	22
Dibutyltins	0.05 DW vs. 0.094 DW	22
Tributyltins	0.073 DW vs. 0.5 DW	22
Monophenyltins	<0.0097 DW vs. 0.051 DW	22
Diphenyltins	<0.00016 DW vs. 0.00016 DW	22
Triphenyltins	0.0025 DW vs. 0.00054 DW	22
Monooctyltins	0.0118 DW vs. <0.0031 DW	22
Dioctyltins	0.0081 DW vs. <0.0036 DW	22
Trioctyltins	<0.00043 DW vs. <0.00043 DW	22
Titanium		
Squid, *Ommastrephes illicebrosa*; whole	6.0 AW	5
Scallop, *Pecten magellanicus*; soft parts	9.8 (1.1-26.2) DW	7

(Continues)

Table 6.22: Cont'd

Element and Organism	Concentration	Reference[a]
Tungsten		
Clam, *Tapes japonica*		
Soft parts	0.46 AW	8
Soft parts	0.03 DW	9
Uranium		
Razor clam, *Ensis siliqua*; shell; 23 stations; west coast of England		
17 stations	<0.5-1.0 DW	21
4 stations	1.1-2.0 DW	21
2 stations	3.1-4.7 DW	21

Values are in mg element/kg fresh weight (FW), dry weight (DW), or ash weight (AW).

[a]1, Karbe et al., 1977; 2, Turekian et al., 1973; 3, Hall et al., 1978; 4, Young et al., 1979; 5, Nicholls et al., 1959; 6, Dick and Pugsley, 1950; 7, Reynolds, 1979; 8, Fukai and Meinke, 1959; 9, Fukai and Meinke, 1962; 10, Alzieu et al., 1986; 11, Davies et al., 1986; 12, Kanatireklap et al., 1997; 13, Shim et al., 1998; 14, Hamasaki et al., 1995; 15, Harino et al., 1998; 16, Harrington, 1991; 17, Davies et al., 1987; 18, Bryan et al., 1986; 19, Thain and Waldock, 1986; 20, Tugrul et al., 1983; 21, Pearce and Mann, 2006; 22, Sousa et al., 2007; 23, Garaventa et al., 2007; 24, Said et al., 2006; 25, Scancar et al., 2007; 26, Guerin et al., 2007; 27, Smith et al., 2006; 28, Karouna-Renier et al., 2007; 29, Pavoni et al., 2007; 30, Berto et al., 2007; 31, Yang et al., 2006; 32, Hu et al., 2006; 33, Kim et al., 2008; 34, Inoue et al., 2006a.

seawater exposing a fresh surface layer or organotin acrylate polymer. In continuing tests by the United States Navy, ablative organotin fouling coatings have demonstrated more than 48 months of protection (Blunden et al., 1985).

Organotin compounds are more toxic than inorganic tin compounds; triorganotins are more toxic than mono-, di-, or tetraorgano forms; and TBT (TBT) compounds were the most toxic triorganotin compounds tested (Argese et al., 1998). TBTs cause adverse biological effects at levels far below those of any reported marine pollutant (Lawler and Aldrich, 1987; Matthiessen and Gibbs, 1998; van Slooten and Tarradellas, 1994), and, arguably, are the most toxic synthetic chemicals discharged into the marine environment (Eisler, 2000e). TBTs are highly toxic to aquatic plants and invertebrates, readily accumulate in molluscs from contaminated localities, and present in harbors where their release from antifouling paints—found usually on small boats and recreational craft—is the putative source (Chan et al., 2008; Horiguchi et al., 2006; Kim et al., 2008; Laughlin et al., 1986; USEPA, 1986; Walsh et al., 1985). TBT is a contributory factor and probably a major cause for the reproductive failure of the European flat oyster (*O. edulis*) in some locations (Thain and Waldock, 1986). Many organotins are slow-acting poisons; accordingly, short-term toxicity tests seriously underestimate the lethality of these compounds (Laughlin and Linden, 1985; Laughlin et al., 1988). Molluscs were the most sensitive group tested to TBT, especially gastropod and bivalve molluscs (Eisler, 2000e). Larvae of the Manila clam,

R. philippinarum, exposed to TBT for 13 days had abnormal development at all TBT concentrations tested (0.000033-0.0006 mg/L) when compared to controls (<0.000001 mg TBT/L; Inoue et al., 2007b). In a maternal exposure test, clams were exposed to 0.000061 or 0.00031 mg TBT/L for 3 weeks. No maternal effect on veliger larvae were observed, suggesting that declines in Manila clam populations attributed to TBT were due to inhibited development and reduced survival of veliger larvae (Inoue et al., 2007b). Adverse effects of TBT are recorded at 0.00001-0.00002 mg/L on growth of Pacific oysters (*C. gigas*) after 2-8 weeks (Lawler and Aldrich, 1987; Waldock and Thain, 1983) and concentrations of 0.001-0.0016 mg/L were lethal (Champ, 1986; Thompson et al., 1985). TBT was lethal to the common mussel (*M. edulis*) at 0.00005-0.00097 mg/L (Cardwell and Sheldon, 1986; Dixon and Prosser, 1986; Valkirs et al., 1987). Similar results are recorded in European oysters, American oysters, and mud snails (Eisler, 2000e). Energy metabolism of the pen shell, *Atrina pectinata japonica*, is disrupted by TBT subjected to 0.0001 mg TBT/L for 72 h under aerobic conditions; later, half the pen shells were subjected to anaerobic conditions for 12 h (Inoue et al., 2007a). Exposure to TBT under aerobic conditions significantly raised lactate, pyruvate, and fumarate concentrations over controls. Clams held under anaerobic conditions had slightly lower succinate concentration than controls (Inoue et al., 2007a).

In laboratory studies, crassostreid oysters, *C. cucullata* and *C. gryphoides*, reportedly accumulated radioisotopes of tin to a higher degree than other species of bivalve molluscs; this characteristic may be useful in the event of contamination from ^{113}Sn (Patel and Ganguly, 1973). Mussels exposed to organotins through a diet of algae showed slow accumulation when compared to exposure from the medium; the reverse was observed for crabs (Evans and Laughlin, 1984; Hall et al., 1987; Laughlin et al., 1986). Uptake of organotins by molluscs was highest when ambient tin concentrations were less than 0.001 mg/L, when exposure duration was comparatively lengthy, and when organism lipid content was elevated (Champ, 1986; Laughlin et al., 1986; Thain and Waldock, 1986; Thompson et al., 1985). Limpets (*Patella caerula*), for example, contain 35-75% of their total tin body burden as organotin (Tugrul et al., 1983). Studies on TBT uptake and depuration from seawater by crabs, oysters, and fish demonstrated that all species accumulate and metabolize TBT, at least partly, within 48 h to dibutyltins, monobutyltins, and more polar metabolites. However, American oysters, *C. virginica*, metabolize less TBT than other species tested (Lee, 1985). The accumulation of TBT compounds in different tissues correlated positively with lipid content and supports a partitioning mode of uptake (Laughlin et al., 1986). The mixed function oxygenase system from hepatic tissues was able to metabolize TBT by forming hydroxylated metabolites (Lee, 1985). TBT and its degradation products were measurable in tissues of American oysters from the Gulf of Mexico ten years after the use of TBT-based paints was regulated. The likely sources of the TBT compounds include sediments, TBT-based paints on vessels longer than 25 m, and shipyard wastes (Sericano et al., 1999). Softshell clams, *M. arenaria*, that were exposed to TBT concentrations of 0.00005-0.0008 mg Sn/L for 28 days showed high uptake, mostly in gills, and lengthy retention (Yang et al., 2006). Total maximum butyltin concentrations—mostly TBT —in soft parts at start was 0.19 mg/kg DW; after 28 days, maximum concentrations ranged between 4.6 and 12.4 mg total butyltins/kg DW, in a dose-dependent manner. The maximum total butyltin concentration value in clam soft parts after a 28 day-depuration period was 9.14 mg/kg DW, with a calculated half-time retention period of 94 days (Yang et al., 2006).

TBTs were not genotoxic to larvae of the common mussel as judged by results of sister chromatid exchange and analysis of chromosomal aberrations (Dixon and Prosser, 1986). However, genotoxicity was demonstrated in other species tested. Organotin-induced strand breakage in erythrocytes as an indicator of DNA damage was evaluated in the bivalve *Scapharca inaequivalvis* (Gabbianelli et al., 2006). Significant DNA damage was evident in 30 min of *in vitro* incubation with about 0.004 mg/L of various organotins. Tributyltin chloride was the most genotoxic compound followed by monobutyltin chloride and dibutyltin chloride. Molluscs exposed *in vivo* to 0.05 mg tributyltin chloride/kg FW for 11 days showed exacerbated damage effects (Gabbianelli et al., 2006).

Imposex—the superimposition of male characteristics onto a functionally normal female reproductive anatomy—is a phenomenon documented in populations of marine gastropods in the vicinity of yacht basins and marinas and is a sensitive indicator of TBT contamination. A female with imposex displays one or more male characteristics, such as a penis, a vas deferens, or convolution of the normally straight gonadal oviduct. Imposex is prevalent in mud snails (*N. obsoletus*) near estuarine marinas and has been induced experimentally in that species by exposure to TBT compounds at concentrations of 0.0045-0.0055 mg/L (Smith, 1981a,b). Imposex is documented extensively in declining populations of the common dogwhelk (*N. lapillus*), especially in southwestern England (Bryan et al., 1986, 1987b, Davies et al., 1987; Gibbs and Bryan, 1986; Gibbs et al., 1987; Smith et al., 2006). These authorities agree on six points: (1) dogwhelk populations near centers of boating and shipping show the highest frequency of imposex, coinciding with the introduction and increasing use of antifouling paints containing TBT compounds; (2) imposex is not correlated with tissue burdens of arsenic, cadmium, copper, lead silver, or zinc but is correlated with increasing concentrations of TBT and dibutyltin fractions; (3) transplantation of dogwhelks from a locality with little boating activity to a site near a heavily used marina caused a marked increase in frequency of imposex and in tissue accumulations of TBT; (4) imposex can be induced in female dogwhelks by exposure to 0.00002 mg Sn/L leached from a TBT antifouling paint; after exposure for 120 days, 41% of the females had male characteristics and whole-body burdens of 1.65 mg S/kg DW soft parts (vs. 0.1 in controls); concentrations as low as 0.0000015 mg TBT/L can initiate imposex in immature females; (5) declining dogwhelk populations were characterized by a moderate to high degree of imposex, relatively fewer functional females, few juveniles, and a general scarcity of deposited egg capsules. Many females in late imposex contained aborted capsules as a result of oviduct blockage, causing sterility and premature death; and (6) there was no evidence that loss of tin leads to any remission of imposex; in fact, all evidence indicates that gross morphological changes that occur in late imposex are irreversible. Additional studies with the common dogwhelk and TBT concentrations between 0.000001 and 0.00002 mg/L show that 13% of females develop male characteristics in 31 days, 27% in 91 days, and 41% in 120 days. BCFs from the medium were about 19,000 in 31 days, 36,000 in 91 days, and 78,000 in 120 days (Bryan et al., 1986). Others have shown that exposure of dogwhelks to TBT concentrations of 0.000002-0.000128 mg/L for 6-12 months resulted in a dose-related increase in frequency of imposex (Davies et al., 1997). There was no effect on feeding rate (predation on *M. edulis*), activity patterns, or male growth rate; however, female growth rate was inhibited. BCF values ranged from 7400 to 25,000 and decreased with increasing TBT exposure concentrations. Diet (mussels) accounted for about 40% of the accumulated TBT in whelks; the maximum concentration of TBT recorded in soft tissues of the 0.000128 mg/L group was 1.04 mg/kg FW after 12 months (Davies et al., 1997). Significant accumulation of butyltins—up to 0.196 mg/kg DW—in gastropod tissues along the western Iberian Peninsula plus high imposex frequency in the vicinity of shipping routes shows that organotin contamination is not confined to coastal areas (Gomez-Ariza et al., 2006; Rato et al., 2006). A similar situation is documented in areas of high shipping activity along the central west coast of India in which imposex frequency reached 100% in populations of the gastropod *Gyrineum natator* (Vishwakiran et al., 2006).

Signs of severe imposex documented in some populations of whelks (*Thais* spp., *Vasum turbinellus*) sampled in eastern Indonesia in 1993 are attributed to unrestricted use of paints containing TBT (Evans et al., 1995). Imposex frequency in soft parts of the gastropod, *Nassarius nitidus*, from the Lagoon of Venice in 2004-2005 increased with increasing concentrations of TBTs over the range 0.025-0.100 mg TBT/kg DW (Pavoni et al., 2007). Studies show that imposex in gastropods is caused by elevated testosterone levels that masculinized TBT-exposed females owing to inhibition of cytochrome P450-mediated aromatase (Matthiessen and Gibbs, 1998). TBT-induced masculinization in gastropods (imposex and intersex) is a convincing example of endocrine disruption described in invertebrates that is unequivocally linked to an environmental contaminant (Matthiessen and Gibbs, 1998). Imposex is also induced in the mud snail, *Ilyanassa obsoleta*, through injection or via the medium by tin compounds other than TBT (McClellan-Green et al., 2006). Mud snails held in seawater for 45 days containing 0.000005-0.00005 mg/L of either trioctyltin chloride, dioctyltin chloride, or tin tetrachloride show a dose-dependent increase in imposex frequency over controls (McClellan-Green et al., 2006). In the gastropod predatory muricid drill, *Stramonita haemastoma*, frequency of imposex and organotin concentrations in tissues were correlated (Limaverde et al., 2007). Body burden threshold of TBTs associated with

33–80% imposex frequency is 0.005-0.024 mg Sn/kg FW; for triphenyltins, 0.023 mg Sn/kg FW produced 50% imposex. Organotin uptake by drills via the medium or via a diet of contaminated mussels seem possible (Limaverde et al., 2007).

Once plentiful, no gastropods of the genus *Nucella* were found in locales near Vancouver, Canada in May-June 1999 owing to suspected TBT contamination from continuous use of TBT in antifouling paints and high (0.17 mg TBT/kg FW)concentrations of TBT in tissues of the mussel, *M. trossulus* (Horiguchi et al., 2003). During that same period, *Nucella* spp. from Victoria, and from Mission Point–where TBT contamination had decreased–seemed to be recovering from imposex; maximum TBT levels in tissues were 0.029 mg TBT/kg FW in *Nucella lima* and 0.044 in *N. lamellosa* (Horiguchi et al., 2003). Although TBT-antifouling paints are acknowledged causes of imposex in gastropod molluscs, this phenomenon has been recorded in more than 120 species of gastropods before the advent of TBT-antifouling paints (Garaventa et al., 2006). Factors other than TBT that can alter endocrine control of imposex development include exposure to heavy metals, changes in environmental conditions (Nias et al., 1993), infestation by parasites (Gorbushin, 1997), and the influence of other androgenic compounds (Cajaraville et al., 2000). More research is recommended on the imposex phenomenon in molluscs and its implications for vertebrates and other taxonomic groups (Eisler, 2000e).

Imposex frequency in *Thais* spp. gastropod populations in Hong Kong was positively correlated with butyltin body burdens (Leung et al., 2006). Risks associated with use of organotin compounds include imposex, growth inhibition, and impairment of immune function (Chan et al., 2008; Coelho et al., 2006a,b; Inoue et al., 2006b; Kim et al., 2008; Leung et al., 2006). TBT concentrations as low as 0.00005 mg/L were associated with abnormal shell growth and reduced weight gain rate of juvenile clams *Ruditapes decussatus* exposed for up to two years (Coelho et al., 2006a,b). Embryonic development of the Manila clam, *R. philippinarum*, was adversely affected when egg-bearing females were held for 3 weeks in seawater containing 0.00006 mg TBT/L (Inoue et al., 2006b). In a waterborne exposure test, 4-h-old zygotes experienced decreased developmental success at concentrations as low as 0.000062 mg TBT/L (Inoue et al., 2006b).

Antifouling paints containing TBT compounds are also used widely on netting panels of sea cages at fish and shellfish aquaculture units in order to minimize the obstruction of water change through the cages (Davies et al., 1987). Under these conditions, TBT paints were detrimental to the growth and survival of juvenile scallops and to calcium metabolism and growth of adult oysters (Paul and Davies, 1986). Scallops (*P. maximus*) reared in sea pens for 31 weeks on nets coated with TBT oxide contained 2.5 mg total Sn/kg FW soft parts (1.9 mg TBT/kg FW), but lost up to 40% during a 10 week depuration period (Davies et al., 1986). Scallop adductor muscle had 0.53 mg TBT/kg FW, suggesting that this tissue (the one consumed by humans) is a probable tin storage site (Davies et al., 1986). Pacific oysters (*C. gigas*) reared for 31 weeks on TBT-treated nets had up to 1.4 mg TBT/kg FW at week 16 (vs. 0.12 in controls), but lost 90% during a 10 week depuration period (Davies et al., 1986).

In 1979, due to the increasing use of the organotin compounds as a class, the Canadian government placed organotins on Canada's Category III Contaminant List, a designation requiring that more data be gathered on their occurrence, persistence, and toxicity in order to prepare informed environmental and human health risk assessments (Chau et al., 1984). In 1982, the use of TBT paints was curtailed in France, and in 1985 the United Kingdom introduced regulations to limit sales of TBT paints that released the biocide at high rates (Huggett et al., 1992). In 1985, tin and organotin concentrations in seawater and Pacific oysters in France were 5-10 times lower than 1982 (Alzieu et al., 1986). In Arcachon Bay, France, there was a decrease in the incidence and extent of anomalies in oyster calcification mechanisms that were correlated with decreases in tin contamination (Alzieu et al., 1986). In April 1987, the United Kingdom banned the retail sale of antifouling paints containing organotin concentrations greater than 4000.0 mg total tin/L (Side, 1987); continued monitoring of TBT was recommended, especially in areas of extensive boating activities (Simmonds, 1986). In 1990, Japan banned the use of TBT-containing paints (Horiguchi et al., 2006); however declining bivalve populations in coastal areas in northern Kyushu in 1998-2001 has been linked, in part to continued TBT contamination of bivalves (Inoue et al., 2006a). In 1992, field work on the River Crouch Estuary showed that sediments from areas most contaminated with TBT in 1987 contained 0.16 mg TBT/kg DW, but by 1992 this had declined to 0.02 mg TBT/kg DW. TBT declines were accompanied by increases in abundance and diversity of benthic fauna, especially bivalve molluscs (Rees et al., 1999; Waldock et al., 1999).

The United States Navy has implemented fleet wide use of organotin antifouling paints that contain TBT in order to reduce fuel consumption by up to 15%, and increase ship speed up to 40% as a direct result of reduced drag (USN, 1984). Since naval vessels seldom remain moored for extended periods in coastal areas, hazard effects to the environment should be minimal—despite the size of the vessel—when compared to boating practices at local marinas (USN, 1984). Accordingly, civilian use of marine paints containing TBT compounds has been severely restricted. The State of Virginia enacted legislation that prohibits the use of TBT paints on nonaluminum vessels under 25 m in length (Anonymous, 1987). Also in Virginia, TBT paints applied on large ships and aluminum craft should not exceed a daily leach rate of 0.000004 mg/cm^3 (Huggett et al., 1992); similar legislation is proposed in at least 12 other states. The most stringent criteria now proposed for tin and aquatic life protection are for triorganotins. These vary from 0.000002 to 0.000008 mg/L (Cardwell and Sheldon, 1986; Huggett et al., 1992; Pinkney et al., 1990; Side, 1987; USPHS, 1992). However, even these comparatively low concentrations will not protect certain species of gastropods from TBT impact, as discussed earlier.

6.48 Titanium

Up to 26.2 mg Ti/kg DW soft parts were measured in soft parts of scallops collected from an oceanic dump site rich in titanium (Table 6.22).

6.49 Tungsten

The maximum tungsten concentration recorded in field collections of molluscs was 0.03 mg W/kg DW in clam soft tissues (Table 6.22).

6.50 Uranium

Limited data indicate that the maximum concentration recorded in shell of the razor clam, *E. siliqua*, collected from the west coast of Britain was 4.7 mg U/kg DW (Table 6.22).

Uranium isotopes were detected in rock jingles *P. macroschisma*, blue mussels *M. trossulus*, and horse mussels *M. modiolus* collected near Amchitka Island during summer 2004; Amchitka Island in the Aleutian chain was the site of nuclear tests between 1965 and 1971 (Burger et al., 2007b). Naturally occurring levels of ^{234}U, ^{235}U, and ^{238}U were found in all three species, and anthropogenic ^{236}U only in the rock jingle (Burger et al., 2007b).

6.51 Vanadium

Vanadium concentrations in molluscan tissues vary substantially with season of year, with tissue specificity—shell being a prime repository—and between species (Table 6.23). For all species of molluscs collected, the highest concentration recorded was 290.0 mg V/kg DW in pteropod tests (Table 6.23). The ability to concentrate vanadium from ambient seawater by five species of bivalves from coastal waters of Greece ranged from 110 for the poorest accumulator to 940 for the best (Papadopoulu, 1973). The significance of vanadium in molluscs is imperfectly understood.

6.52 Ytterbium

The single datum available for ytterbium is 0.002-0.049 mg Yb/kg DW soft parts of the common mussel, *M. edulis* (Karbe et al., 1977).

6.53 Zinc

It is generally agreed that the highest concentrations of zinc in marine biota are found in tissues of filter-feeding molluscs, especially oysters (Eisler, 1980, 2000f). Crassostreid oysters, for example, frequently contain more than 4000.0 mg Zn/kg in soft tissues on a DW basis (Table 6.24). It is probable that zinc is not limiting to normal molluscan life processes

Table 6.23: Vanadium Concentrations in Field Collections of Molluscs

Organism	Concentration	Reference[a]
Giant squid, *Architeuthis dux*; 2001-2005; hepatopancreas; Bay of Biscay vs. Mediterranean Sea	2.2 DW vs. 1.7 DW	16
Clam, *Arctica islandica*; soft parts		
March	4.3-20.6 DW	1
August	1.6-4.3 DW	1
February	1.6-3.1 DW	1
June	1.3-2.9 DW	1
Molluscs		
Soft parts	0.18-0.82 FW	2
Edible tissues; 18 commercial species		
2 spp.	<0.1 to 0.2 FW	3
6 spp.	0.2-0.4 FW	3
7 spp.	0.4-0.6 FW	3
3 spp.	0.6-2.0 FW	3
Softshell clam, *Mya arenaria*; soft parts	5.8-8.4 DW	4
Common mussel, *Mytilus edulis*		
Soft parts	0.22 DW	5
Shell	1.0-1.4 DW	15
Mussel, *Mytilus edulis aoteanus*		
Soft parts	5.0 DW	6
Mantle	3.0 DW	6
Gill	5.0 DW	6
Muscle	2.0 DW	6
Visceral mass	23.0 DW	6
Foot	44.0 DW	6
Gonad	<2.0 DW	6
Shell	110.0 DW	6
Mussel, *Mytilus galloprovincialis*		
Soft parts; Adriatic Sea; 2001-2002; summer vs. winter		
Industrialized sites	1.6-8.5 DW vs. 2.9-3.9 DW	12
Reference site	5.1 DW vs. 4.6 DW	12
Soft parts; Adriatic Sea; 2001-2005; winter vs. summer	2.0-2.8 DW vs. 0.4-1.2 DW	17

(Continues)

Table 6.23: Cont'd

Organism	Concentration	Reference[a]
Octopus, *Octopus vulgaris*		
Portugal; 1999-2002; 3 sites		
Digestive gland	3.3-4.6 DW	18
Branchial heart	24.0-54.0 DW	18
Portugal; 1999-2000; 3 sites		
Digestive gland	5.0-7.2 DW	19
Branchial heart	27.5-53.0 DW	19
Squid, *Ommastrephes illicebrosa*; whole	4.0 AW	7
Oyster, *Ostrea sinuata*		
Soft parts	3.0 DW	6
Mantle	12.0 DW	6
Gills	3.0 DW	6
Muscle	<2.0 DW	6
Visceral mass	<2.0 DW	6
Kidney	4.0 DW	6
Heart	4.0 DW	6
Shell	49.0 DW	6
Scallop, *Pecten novae-zelandiae*		
Soft parts	9.0 DW	6
Mantle	7.0 DW	6
Gills	3.0 DW	6
Visceral mass	30.0 DW	6
Kidney	4.0 DW	6
Foot	4.0 DW	6
Gonad	32.0 DW	6
Shell	130.0 DW	6
Bivalve, *Pinna nobilis*		
Soft parts	28.0 AW	8
Mantle and gills	17.0 AW	8
Muscle	5.4 AW	8
Nervous system	23.0 AW	8
Stomach and intestine	53.0 AW	8
Gonads	74.0 AW	8
Hepatopancreas	4.1 AW	8
Byssus	2.5 AW	8
Byssus gland	6.5 AW	8
Scallop, *Placopecten magellanicus*		
Soft parts	28.5 (11.4-40.7) DW	1
Muscle		

(Continues)

Table 6.23: Cont'd

Organism	Concentration	Reference[a]
March	Max. 1.7 DW	1
August	<0.1 DW	1
February	Max. 0.8 DW	1
June	Max. 5.4 DW	1
Viscera		
February	7.8-21.3 DW	1
March	21.0-74.2 DW	1
June	4.6-13.2 DW	1
August	6.9-35.1 DW	1
Pteropods; shell	50.0-290.0 DW	9
Common cuttlefish, *Sepia officinalis*		
Egg		
Eggshell	3.4 DW	13
Embryo	<0.1 DW	13
Yolk	<0.1 DW	13
Hatchlings		
Whole	<0.1 DW	13
Cuttlebone	<0.5 DW	13
Juveniles, age 1 month		
Whole	<0.1 DW	13
Digestive gland	1.6 DW	13
Cuttlebone	<0.5 DW	13
Immatures, age 12 months		
Whole	0.47 DW	13
Digestive gland	2.9 DW	13
Cuttlebone	<0.5 DW	13
Adults, age 18 months		
Whole	0.5 DW	13
Digestive gland	3.3 DW	13
Cuttlebone	<0.5 DW	13
Clam, *Tapes japonica*		
Soft parts	15.0 AW	10
Soft parts	1.1 DW	11
Clam, *Trivela mactroidea*; Venezuela; 2002; 14 sites; soft parts	2.0-13.2 DW	14

Values are in mg V/kg fresh weight (FW), dry weight (DW), or ash weight (AW).
[a] 1, Reynolds, 1979; 2, Ikebe and Tanaka, 1979; 3, Hall et al., 1978; 4, Eisler, 1977b; 5, Carlisle, 1958; 6, Brooks and Rumsby, 1965; 7, Nicholls et al., 1959; 8, Papadopoulu, 1973; 9, Pyle and Tieh, 1970; 10, Fukai and Meinke, 1959; 11, Fukai and Meinke, 1962; 12, Scancar et al., 2007; 13, Miramand et al., 2006; 14, LaBrecque et al., 2004; 15, Protasowicki et al., 2008; 16, Bustamante et al., 2008; 17, Fattorini et al., 2008; 18, Napoleao et al., 2005; 19, Seixas and Pierce, 2005.

Table 6.24: Zinc Concentrations in Field Collections of Molluscs

Organism	Concentration	Reference[a]
Abalones; soft parts	55.0 (38.0-100.0) DW	88
Gastropod, *Acmaea digitalis*		
Shell	8.4-29.9 DW	1
Soft parts	420.0-763.0 DW	1
Cockle, *Anadara trapezia*; April 2000; soft parts; New South Wales, Australia	Max. 160.0 DW	149
Mussel, *Anodonta cygnea*; soft parts	141.0-165.0 DW	3
Mussel, *Anodonta* sp.		
Shell	6.4 DW	4
Soft parts	120.0 DW	4
Gastropod, *Aplysia benedicti*		
Ctenidium	55.0-60.0 DW	5
Gonad	33.0-40.0 DW	5
Parapodia	39.0 DW	5
Hepatopancreas	61.0-74.0 DW	5
Buccal mass	25.0 DW	5
Intestine	90.0 DW	5
Gut content	58.0 DW	5
Shell	15.0-28.0 DW	5
Giant squid, *Architeuthis dux*; 2001-2005; Spain; hepatopancreas		
Bay of Biscay	103.0 DW	172
Mediterranean Sea	219.0 DW	172
Clam, *Arctica islandica*; soft parts	2.4-25.8 FW	6
Scallop, *Argopecten gibbus*; kidney concretion	14,466.0 DW	7
Scallop, *Argopecten irradians*; kidney concretion	16,550.0 DW	7
Cockle, *Austrovenus stutchburyi*; soft parts; New Zealand; 1993-1994	40.0-118.0 DW	142
Hydrothermal vent mussel, *Bathymodiolus azoricus*; mid-Atlantic ridge		
2001		
840 m deep		
Digestive gland	40.0 DW	156
Gills	111.0 DW	156
Mantle	61.0 DW	156

(Continues)

Table 6.24: Cont'd

Organism	Concentration	Reference[a]
1700 m deep		
Digestive gland	201.0 DW	156
Gills	555.0 DW	156
Mantle	68.0 DW	156
2300 m deep		
Digestive gland	80.0 DW	156
Gills	106.0 DW	156
Mantle	46.0 DW	156
Summer 2002; start vs. 30 days in metals-free seawater		
Gill	150.0 FW vs. 130.0 FW	144
Mantle	140.0 FW vs. 60.0 FW	144
Digestive gland	152.0 FW vs. 80.0 FW	144
Byssus thread	2500.0 FW vs. <10.0 FW	144
Shell; 2001; 4 sites	9.4-13.0 (4.1-21.4) DW	162
2001; 5 sites		
Gills	106.0-1997.0 DW	163
Mantle	28.0-145.0 DW	163
Digestive gland	45.0-276.0 DW	163
Bivalves; 5 spp.; whole; Australia; lead smelter outfall		
2.5-5.2 km from outfall	4480.0 DW; max. 20,300.0 DW	100
18.0-18.8 km from outfall	2590.0 DW	100
Bivalves		
Soft parts	91.0-660.0 DW	107
Kidney granules	10,000.0-43,320.0 DW	106
Bivalves; 5 spp.; soft parts; East China Sea; August 2002	9.0 (3.3-17.9) FW	139
Mussel, *Brachidontes pharaonis*; Mersin cost, Turkey; 2002-2003; max. concentrations		
Muscle	105.0 DW	134
Gill	240.0 DW	134
Hepatopancreas	115.0 DW	134
Whelk, *Buccinum undatum*		
Shell	13.0 DW	4
Soft parts	620.0 DW	4
Whelk, *Busycon canaliculatum*		
Muscle	25.0 DW	8
Digestive diverticula	2600.0 DW	8
Muscle	15.3-29.5 FW	9
Digestive gland	405.0-2650.0 FW	9

(*Continues*)

Table 6.24: Cont'd

Organism	Concentration	Reference[a]
Calcasieu River Estuary; Louisiana; soft parts		
Hooked mussel, *Brachidontes exustus*	61.0 (39.0-86.0) DW	105
American oyster, *Crassostrea virginica*	3300.0 (1000.0-7794.0) DW	105
Cephalopods		
Soft parts	81.0-150.0 DW; max. 580.0 DW	88, 107
Whole	250.0 DW	118
Cockle, *Cardium edule*		
Soft parts	52.0-81.0 DW	3
Mantle fluids	95.0 DW	3
Shell	6.8 DW	4
Soft parts	130.0 DW	4
Mussel, *Cerastoderma edule*		
Soft parts	63.0-208.0 DW	10
Soft parts	55.0 (46.0-66.0) DW	11
Chitons; soft parts	290.0-700.0 DW	88
Scallop, *Chlamys opercularis*		
Soft parts	462.0 DW	12
Body fluids	45.1 DW	12
Mantle	152.0 DW	12
Gill	384.0 DW	12
Digestive gland	132.0 DW	12
Adductor muscle		
Striated	61.7 DW	12
Unstriated	111.0 DW	12
Gonad and foot	117.0 DW	12
Kidney	40,800.0 DW	12
Shell	7.3-11.0 DW	4
Mussel, *Choromytilus meridionalis*		
Soft parts	73.0-113.0 DW	13
Soft parts	16.0 FW	14
Clams		
Soft parts	81.0-115.0 DW; max. 510.0 DW	88
Soft parts	16.0-58.0 FW; 295.0-590.0 DW; 2825.0-5650.0 AW	15
Shell	8.7-28.0 FW; 9.1-44.1 DW; 43.0-88.0 AW	15

(Continues)

Table 6.24: Cont'd

Organism	Concentration	Reference[a]
Scallop, *Comptopallium radula*; Santa Marie Bay (metals-contaminated), New Caledonia; October 2004		
Soft parts	183.0 (109.0-308.0) EW	158
Digestive gland	1115.0 DW	158
Oyster, *Crassostrea angulata*; soft parts; Spain		
Sanlucar	2064.0 FW	16
Sancti-Petri	437.6 FW	16
Barbate	366.6 FW	16
Rio Pedras	1101.0 FW	16
Ayamonte	1027.0 FW	16
Sydney rock oyster, *Crassostrea commercialis*		
Soft parts	318.0 FW	17
Soft parts	80.0-665.0 FW	18
Soft parts; southeast Asia	800.0 (64.0-1920.0) DW	108
Oyster, *Crassostrea corteziensis*; soft parts; Gulf of California; 1999-2000	1420.0 DW	167
Oyster, *Crassostrea cucullata*		
Soft parts	254.5 FW	19
Soft parts	446.0-2800.0 DW	20
Pacific oyster, *Crassostrea gigas*		
Soft parts	86.0-344.0 FW	21
Soft parts	333.0-10,019.0 FW	17
Soft parts	99,220.0 DW	22
Soft parts	396.0 DW	23
Soft parts	424.0 DW	13
Soft parts	1272.0-3627.0 DW	135
Oyster, *Crassostrea gryphoides*; soft parts	323.0-550.0 DW	20
Oyster, *Crassostrea madrasensis*; soft parts	203.4 FW	19
Oyster, *Crassostrea margaritacea*; soft parts	886.0 DW	23
Mangrove oyster, *Crassostrea rhizophorae* Sepetiba Bay, Brazil; soft parts; near zinc smelting plant closed in 1996		
1978	2240.0 DW; max. 4150.0 DW	120
1980	1533.0 DW	121
1983	8073.0 DW; max. 16,130.0 DW	122
1989	4950.0 DW; max. 5016.0 DW	123, 124

(Continues)

Table 6.24: Cont'd

Organism	Concentration	Reference[a]
1996 tailings spill following heavy rains	80,724.0 DW	125
1997	9174.0 DW	126
1999	14,849.0 (8058.0-28,233.0) DW	125
2001	9770.0 DW	127
2002	11,984.0 (10,963.0-17,420.0) DW	128
Soft parts; near Natal Brazil; summer 2001; contaminated estuary vs. reference site	997.0-2082.0 DW; max. 4862.0 DW vs. 983.0-1491.0 DW; max. 1874.0 DW	147
American oyster, *Crassostrea virginica*		
Soft parts		
Lower Chesapeake Bay	310.0 FW	24
Near Baltimore	4000.0 FW	24
Chesapeake Bay	3975.0 (60.0-12,800.0) DW	108
Gulf of Mexico	2150.0 (485.0-10,000.0) DW	108
Galveston Bay, Texas; 1992-1993	2082.0 (602.0-4819.0) DW	109
Pensacola, Florida; 2003-2004	Max. 1000.0 FW	152
South Carolina	2410.0 (280.0-6305.0) DW	108
United States	1018.0-1641.0 (204.0-4000.0) DW	110
Savannah, Georgia; 2000-2001	978.0-2428.0 DW	157
Soft parts	230.0 (24.0-820.0) FW	25
Soft parts	1428.0 (180.0-4120.0) FW	21, 26, 27
Newport River, North Carolina		
April 1965		
Mantle	135.0 FW	28
Gill	182.0 FW	28
Labial palps	123.0 FW	28
Muscle	79.0 FW	28
Digestive gland	260.0 FW	28
February 1966		
Shell	24.7 FW	28
Soft parts	159.4 FW	28
January 1969		
Shell	245.1 FW	28
Soft parts	210.5 FW	28
North River, North Carolina, November 1966		
Mantle	450.6 FW	28
Gill	508.1 FW	28
Labial palps	501.2 FW	28
Muscle	126.1 FW	28

(Continues)

Table 6.24: Cont'd

Organism	Concentration	Reference[a]
Digestive gland	491.8 FW	28
Rectum	531.0 FW	28
Pericardial sac	223.9 FW	28
Gonad	435.9 FW	28
Shell	1.5-8.1 DW	29
Soft parts	1200.0-5700.0 DW	29
Soft parts	Max. 12,675.0 DW	30
Soft parts	mean 3915.0 DW; max. 10,000.0 DW	31
Soft parts		
North Carolina	125.0-325.0 FW	32
Virginia	275.0-425.0 FW	32
Soft parts		
August	5000.0 DW	33
November	2100.0 DW	33
Shell	2.5 DW	33
Soft parts; metals-contaminated area vs. reference site	4100.0 DW vs. 1700.0 DW	33
Soft parts	322.0 DW	34
Soft parts	1215.0-1980.0 FW	35
Soft parts; United States		
East coast	830.0-4060.0 DW	36
Gulf coast	370.0-7080.0 DW	36
Limpet, *Crepidula fornicata*		
Soft parts	116.0 DW	3
Shell	1.1 DW	4
Soft parts	940.0 DW	4
Clam, *Donax serra*; soft parts	16.0 FW	14
Clam, *Ensis arcuatus*; shell	1.4 DW	37
Razor clam, *Ensis siliqua*		
Shell	0.8 DW	37
Shell; 23 stations; West coast of England		
9 stations	23.5-66.7 DW	129
12 stations	70.1-125.0 DW	129
2 stations	184.0-295.0 DW	129
Gastropods		
Muscle	6500.0 AW	15
Shell	330.0 AW	15
Viscera	13.0 AW	15
Operculum	6000.0 AW	15
Soft parts	84.0-763.0 DW	107

(Continues)

Table 6.24: Cont'd

Organism	Concentration	Reference[a]
Gastropod drills; soft parts	536.0–3470.0 DW	88
Abalone, *Haliotis midea*; soft parts	12.0 FW	14
Abalone, *Haliotis rufescens*		
Gill	28.0–54.0 DW	38
Mantle	42.0–74.0 DW	38
Digestive gland	536.0–980.0 DW	38
Foot	38.0–46.0 DW	38
Abalone, *Haliotis tuberculata*		
Soft parts	98.0–103.0 DW	39
Foot	28.0–33.1 DW	39
Viscera	238.0–288.0 DW	39
Digestive gland	624.0–656.0 DW	39
Blood	19.9 DW	39
Muscle	38.0 DW	39
Gill	63.4 DW	39
Left kidney	124.0 DW	39
Right kidney	298.0 DW	39
Scallop, *Hinnites multirugosus*		
Kidney	37.2 FW	40
Digestive gland	68.8 FW	40
Periwinkle, *Littoraria scabra*; soft parts; Tanzania; 2005		
Industrial sites	750.0 DW	160
Reference site	95.0 DW	160
Limpet, *Littorina littoralis*; soft parts	312.0 DW	3
Limpet, *Littorina littorea*		
Soft parts	38.0 DW	3
Soft parts	6.6–20.2 FW	41
Soft parts	Max. 520.0 DW	42
Soft parts	28.0–274.0 DW	43
Soft parts	88.0 DW	39
Soft parts	15.5–24.3 FW	44
Soft parts	92.1–186.0 DW	45
Digestive gland and gonad	111.2–133.7 DW	45
Shell	2.6–3.0 DW	45
Soft parts and operculum	117.0 (45.0–284.0) DW	11
Limpets; soft parts		
18 species	112.0 (14.0–760.0) DW	88
7 species	196.0 (86.0–430.0) DW	88

(*Continues*)

Table 6.24: Cont'd

Organism	Concentration	Reference[a]
Squid, *Loligo opalescens*; liver	247.0-449.0 DW	46
Baltic clam, *Macoma balthica*		
Soft parts	804.0 (510.0-1160.0) DW	11
Soft parts	88.0-99.0 FW	47
Soft parts; adults; San Francisco Bay	200.0-600.0 DW	111
Soft parts; 1996-1997; Gulf of Gdansk; age 3-4 years	300.0-749.0 DW	151
Hardshell clam, *Mercenaria mercenaria*		
Soft parts	1680.0 AW	48
Mantle	810.0 AW	48
Gill	3070.0 AW	48
Muscle	1500.0 AW	48
Soft parts	20.6 (11.5-40.2) FW	21, 26
Shell	3.4 DW	4
Soft parts	94.0 DW	4
Soft parts	12.4-27.6 DW	10
Soft parts	Max. 298.0 DW	30
Soft parts, Maryland		
James River	11.8-35.5 FW	49
York River	7.5-14.6 FW	49
Chesapeake Bay	11.3-26.8 FW	49
Clam, *Meretrix casta*; soft parts	50.2 DW	20
Clam, *Meretrix lamarckii*; soft parts	63.0 DW	19
Mexico; Sinaloa; 2004-2005; sort parts; rainy season vs. dry season		
Oyster, *Crassostrea corteziensis*	1048.5 DW vs. 849.2 DW	164
Mussel, *Mytella strigata*	69.5 DW vs. 63.3 DW	164
Clam, *Megapitaria squalida*	98.3 DW vs. 94.2 DW	164
Mussel, *Modiolus demissus*; soft parts	Max. 52.0 DW	30
Mussel, *Modiolus modiolus*		
Shell	6.2 DW	4
Soft parts	320.0-530.0 DW	4
Mantle and gills	180.0 DW	4
Muscle	68.0 DW	4
Gonad	140.0 DW	4
Gut and digestive gland	190.0 DW	4
Morocco; 2001; clams; soft parts		
Scrobicularia plana	204.0 DW	166
Cerastoderma edule	80.0 DW	166

(Continues)

Table 6.24: Cont'd

Organism	Concentration	Reference[a]
Morocco; 2004-2005; soft parts		
Mussel, *Mytilus galloprovincialis*	292.0 (112.6-612.3) DW	159
Clam, *Venerupis decussatus*	103.1 (67.2-157.3) DW	159
Pacific oyster, *Crassostrea gigas*	481.7 (28.3-672.4) DW	159
Mussels; soft parts	109.0-267.0 DW; max. 7700.0 DW	88
Molluscs		
Soft parts		
West Norway coast	Max. 2800.0 DW	57
Rio Tinto estuary, Spain	Max. 370.0 DW	52
Chile, Strait of Magellan; 5 spp.; 2001; 17 stations	8.5-35.4 FW	150
Edible tissues; commercial species		
3 spp.	5.0-9.0 FW	53
13 spp.	10.0-20.0 FW	53
1 spp.	100.0-200.0 FW	53
1 spp.	300.0-400.0 FW	53
12 spp.	7.0-337.0 FW	54
3 spp.	12.0-18.0 FW	55
3 spp.; India; 2003	10.3-16.2 FW	155
Softshell clam, *Mya arenaria*		
Soft parts	17.0 (9.0-28.0) FW	21, 26
Soft parts	9.5-11.2 FW; 93.0-127.0 DW	56
Mussel, *Mytilus californianus*		
Shell	8.6-26.5 DW	1
Soft parts	164.0-310.0 DW	1
Digestive gland	46.0-110.0 DW	57
Soft parts	60.0-200.0 DW	36
Mussel, *Mytilus corscum*; soft parts	160.0 DW	50
Common mussel, *Mytilus edulis*		
Shell	0.04 DW	4
Shell; Baltic Sea; 2005; 12 sites	2.2-10.3 DW	169
Kidney; Newfoundland		
October 1984	144.0 (50.0-427.0) DW	113
April 1985	828.0 (94.0-3410.0) DW	113
Soft parts; weight 0.43 g		
Visceral mass	34.0-100.0 DW	112
Gills and palps	47.0-94.0 DW	112
Remainder	46.0-110.0 DW	112

(*Continues*)

Table 6.24: Cont'd

Organism	Concentration	Reference[a]
Soft parts; weight 0.22 g		
Visceral mass	28.0-112.0 DW	112
Gills and palps	38.0-158.0 DW	112
Remainder	40.0-130.0 DW	112
Soft parts	91.0 DW	4
Shell	0.6 DW	37
Soft parts	40.0 DW	37
Shell	6.9-14.2 DW	1
Soft parts	204.0-341.0 DW	1
Soft parts	253.0-779.0 DW	43
Soft parts	12.5-82.5 FW	41
Soft parts	45.8 FW	58
Soft parts	269.0 DW	3
Digestive gland	Max. 185.0 FW	59
Gonad	Max. 183.0 FW	59
Muscle	Max. 172.0 FW	59
Soft parts	19.3-60.1 FW	60
Soft parts	16.4-97.1 FW	61
Soft parts	29.0-41.8 FW	44
Soft parts	14.0-460.0 DW	62
Soft parts	38.0-575.0 DW	63
Soft parts	140.0-500.0 DW	64
Hepatopancreas		
San Diego harbor	120.0-210.0 DW	65
Offshore	120.0-140.0 DW	65
Soft parts	132.0 (57.0-199.0) DW	11
Soft parts		
Protein fraction I	0.03 FW	66
Protein fraction II	0.2 FW	66
Protein fraction III	0.2 FW	66
Soft parts U.S.		
West coast	90.0-260.0 DW	36
East coast	67.0-189.0 DW	36
Greenland; near lead zinc flotation mill outfall vs. reference site	502.0 (340.0-813.0) FW vs. 100.0 FW	119
The Netherlands; summer, 2002	Max. 170.0 DW	136
Scheldt estuary, Europe; March 2000	790.0 DW	171
Soft parts	85.0-359.0 DW	67
Shell	4.0-12.0 DW	67

(Continues)

Table 6.24: Cont'd

Organism	Concentration	Reference[a]
Mussel, *Mytilus edulis aoteanus*		
Soft parts	91.0 DW	68
Mantle	192.0 DW	68
Gill	336.0 DW	68
Visceral mass	525.0 DW	68
Gonad	260.0 DW	68
Mussel, *Mytilus edulis planulatus*		
Soft parts	301.0-702.0 DW	69
Soft parts		
June	22.1 FW	70
January	26.7 FW	70
Soft parts, dry weight		
0.09 g	142.0 DW	2
0.39 g	203.0 DW	2
0.48 g	207.0 DW	2
0.69 g	264.0 DW	2
Mussel, *Mytilus galloprovincialis*		
Protein extracts; Bulgarian Black Sea Coast; polluted vs. reference site	143.0 DW vs. 112.4 DW	145
Soft parts	209.0 DW	71
Soft parts	29.8 FW	72
Soft parts; Adriatic Sea, Italy; 2 sites	21.9-26.0 FW	168
Soft parts; Adriatic Sea; 2001-2002; summer vs. winter		
Industrialized sites	235.0-249.0 DW vs. 73.0-244.0 DW	143
Reference site	201.0 DW vs. 102.0 DW	143
Soft parts; Adriatic Sea; 2001-2005; winter vs. summer	66.2-119.0 DW vs. 23.2-105.0 DW	173
Soft parts; Aegean Sea, Turkey; 2002-2003	23.3 (16.1-37.1) FW	137
Soft parts; Izmir Bay, Turkey; 2004; near shore vs. mid-bay	27.7 DW vs. 17.9 DW	170
Soft parts; Gulf of Gemlik, Turkey; 2004	196.0 DW	176
Soft parts; May 2003; NW Mediterranean Sea		
Near Genova, Italy	74.0-240.0 DW	132
Near Marseille, France	135.0-153.0 DW	132
Near Barcelona, Spain	167.0-299.0 DW	132
Soft parts; Venice lagoon, Italy; 2005-2006; San Giuliano vs. Sacca Sessola		

(*Continues*)

Table 6.24: Cont'd

Organism	Concentration	Reference[a]
April	260.3 DW vs. 215.5 DW	153
July	186.2 DW vs. 164.0 DW	153
October	134.9 DW vs. 216.8 DW	153
February	144.1 DW vs. 400.2 DW	153
Soft parts	124.5 (48.9-316.7) DW	133
Mussel, *Mytilus viridis*		
Soft parts	11.0 FW	19
Soft parts	15.3-63.0 DW	20
Soft parts	14.1 FW	73
Soft parts	4.3-11.2 DW	74
Abalone, *Notohaliotis discus*; soft parts less viscera	33.0 DW	50
Gastropod, *Nucella conccina*; soft parts; Antarctica; 2004	64.4 DW	131
Whelk, *Nucella lapillus*		
Shell	0.59 DW	4
Soft parts	860.0 DW	4
Soft parts		
Beer, Dorset	345.0 DW	75
Transplants from Beer to Bristol Channel for 5 months	2530.0 DW	75
Soft parts plus operculum	416.0 (235.0-520.0) DW	11
Soft parts	415.0 DW	3
Oyster drill, *Ocenebra erinacea*; soft parts	1451.0-2160.0 DW	114
Octopus, *Octopus* sp.		
Whole	106.0 FW	76
Whole	18.5 FW	58
Octopus, *Octopus vulgaris*		
Portugal; 1999-2002; 3 sites		
Digestive gland	1247.0-1666.0 DW	174
Branchial heart	80.0-85.0 DW	174
Gill	67.0-77.0 DW	174
Muscle	Max. 76.0 DW	174
Portugal; 2005-2006; digestive gland; northwest region vs. south coast		
Whole	850.0 (410.0-2040.0) DW vs. 1840.0 (740.0-2870.0) DW	165
Nuclei	1070.0 (570.0-1660.0) DW vs. 3880.0 (1230.0-1660.0) DW	165

(*Continues*)

Table 6.24: Cont'd

Organism	Concentration	Reference[a]
Mitochondria	1300.0 (390.0-18,180.0) DW vs. 2600.0 (990.0-18,230.0) DW	165
Lysosomes	630.0 (220.0-10,100.0) DW vs. 1960.0 (630.0-6420.0) DW	165
Microsomes	1750.0 (1080.0-2250.0) DW vs. 1900.0 (1150.0-17,060.0) DW	165
Squid, *Ommastrephes bartrami*		
Liver	163.0 DW	46
Muscle	54.0 DW	77
Gonad	73.0 DW	77
Liver	115.0-124.0 DW	77
Oyster, *Ostrea angasi*		
Soft parts	371.0-8865.0 FW	17
Soft parts	5657.0 FW	58
Soft parts	3740.0-38,700.0 DW	69
Soft parts		
June	1217.0 FW	70
January	954.0 FW	70
European flat oyster, *Ostrea edulis*		
Soft parts	660.0 DW	23
Soft parts; 1991-2005; Balearic Islands; western Mediterranean Sea; Port of Mahon	1458.2 (204.5-3404.8) DW	133
Soft parts	2480.0-3280.0 DW	79
Soft parts		
Contaminated site	10,560.0 (4700.0-12,640.0) DW	115
Reference site	98.0 DW	115
Cornwall, England vs. Anglesey, North Wales		
Gills	14,746.0 FW vs. 10,027.0 FW	78
Mantle	12,770.0 FW vs. 6,189.0 FW	78
Kidney	6155.0 FW vs. 6050.0 FW	78
Viscera	5817.0 FW vs. 2988.0 FW	78
Muscle	2342.0 FW vs. 2683.0 FW	78
Hemolymph		
Whole	155.0 FW vs. 60.0 FW	78
cell-free	30.0 FW vs. 15.0 FW	78
Oyster, *Ostrea equestris*; soft parts; United States		
East coast	2800.0 DW	36
Gulf coast	310.0-3600.0 DW	36

(*Continues*)

Table 6.24: Cont'd

Organism	Concentration	Reference[a]
Oyster, *Ostrea gigas*		
Soft parts; transplants from control area to metal-contaminated area for 120 days		
Controls	200.0 FW	80
Transplants	580.0 FW	80
Transplants to metal-contaminated area for 120 days		
Adductor muscle	250.0 FW	81
Labial palps	800.0-900.0 FW	81
Mantle	800.0-900.0 FW	81
Gills	800.0-900.0 FW	81
Transplants from zinc-contaminated area vs. after 116 days in "clean" water		
Mantle	1072.0 FW vs. 95.0 FW	82
Gill	4270.0 FW vs. 171.0 FW	82
Labial palps	2178.0 FW vs. 112.0 FW	82
Muscle	221.0 FW vs. 17.0 FW	82
Remainder	1013.0 FW vs. 122.0 FW	82
Whole	953.0 FW vs. 108.0 FW	82
Oyster, *Ostrea sinuata*		
Soft parts	1103.0 DW	68
Mantle	4760.0 DW	68
Gill	3300.0 DW	68
Muscle	369.0-400.0 DW	68
Visceral mass	1122.0 DW	68
Kidney	1248.0 DW	68
Heart	2090.0 DW	68
Oysters; soft parts		
Metals-contaminated area	Max. 21,000.0 FW	83, 83
Reference site	Max. 7670.0 FW	83, 84
Oysters, various; soft parts	1960.0-7270.0 DW; max. 49,000.0 DW	88
Limpet, *Patella candei gomesii*; soft parts; 2007; Azores		
Near hydrothermal vent	44.4-48.4 DW	161
Reference site	40.2-40.4 DW	161
Limpet, *Patella* sp.		
Soft parts		
Beer, Dorset, United Kingdom	91.0 DW	75

(Continues)

Table 6.24: Cont'd

Organism	Concentration	Reference[a]
Transplants from Beer to Bristol Channel for 8 months	340.0 DW	75
Shell	51.0 DW	4
Soft parts	84.0 DW	4
Soft parts	Max. 580.0 DW	42
Soft parts	158.0 DW	85
Soft parts	95.0-229.0 DW	3
Soft parts	101.0 DW	39
Soft parts		
From desalinization plant outfall	255.0 DW	10
Reference site	175.0 DW	10
Gastropod, *Patella piperata*; Canary Islands, Spain; soft parts; March-April 2003	10.4 (1.6-24.1) DW	138
Limpet, *Patella vulgata*		
Soft parts	165.0 (83.0-224.0) DW	11
Soft parts	127.0-238.0 DW	67
Shell	3.0-18.0 DW	67
Scallop, *Pecten maximus*		
Upper valve	4.1 DW	4
Lower valve	6.1 DW	4
Soft parts	230.0 DW	4
Muscle	70.0 DW	4
Gut and digestive gland	1100.0 DW	4
Mantle and gills	160.0-420.0 DW	4
Gonad	180.0-360.0 DW	4
Soft parts	273.0 DW	12
Body fluid	30.1 DW	12
Mantle	59.4 DW	12
Gill	100.0 DW	12
Digestive gland	407.0 DW	12
Striated muscle	69.9 DW	12
Unstriated muscle	94.2 DW	12
Gonad and foot	196.0 DW	12
Kidney	19,300.0 DW	12
Edible tissues	19.1-45.5 FW	86
Kidney; soft parts vs. excretory granule	5898.0 FW vs. 140,000.0 FW	87
Scallop, *Pecten novae-zelandiae*		
Soft parts	283.0 DW	68
Muscle	108.0 DW	68

(*Continues*)

Table 6.24: Cont'd

Organism	Concentration	Reference[a]
Visceral mass	400.0 DW	68
Intestine	392.0 DW	68
Kidney	2630.0 DW	68
Foot	210.0 DW	68
Gonads	256.0 DW	68
Soft parts	20.8 FW	54
Stomach	24.0 FW	54
Gonad	27.0 FW	54
Adductor muscle	16.0 FW	54
Scallop, *Pecten* sp.		
Kidney	32,000.0 DW	88
Kidney granules	120,000.0 DW	88
Soft parts	200.0 DW	88
Pelycopods		
Soft parts	340.0 AW	15
Shell	2050.0 AW	15
Green mussel, *Perna viridis*		
Soft parts; 1986-1987	56.0-134.0 DW	116
Soft parts; 1986		
March	63.0-150.0 DW	117
May	77.0-94.0 DW	117
Soft parts; India; January 2002; 28 sites	9.5-10.5 FW	148
Hong Kong; juveniles vs. adults		
Reference site		
Soft parts	104.0 DW vs. 108.0 DW	175
Byssus	103.0 DW vs. 112.0 DW	175
Shell	0.25 DW vs. 0.39 DW	175
Polluted site		
Soft parts	116.0 DW vs. 152.0 DW	175
Byssus	314.0 DW vs. 297.0 DW	175
Shell	0.9 DW vs. 0.7 DW	175
Bivalve, *Pinna nobilis*		
Soft parts	4900.0 AW	89
Muscle	3200.0 AW	89
Nervous system	3400.0 AW	89
Gonads	4600.0 AW	89
Hepatopancreas	6900.0 AW	89
Byssus	3100.0 AW	89
Byssus gland	840.0 AW	89

(*Continues*)

Table 6.24: Cont'd

Organism	Concentration	Reference[a]
Clam, *Pitar morrhuana*; soft parts	145.0-276.0 DW	90
Scallop, *Placopecten magellanicus*		
Soft parts	105.0 DW	91
Soft parts	9.3-24.0 FW; max. 109.0 FW	92
Muscle	2.0-8.1 FW	93
Gonad	4.7-75.4 FW	93
Visceral mass	7.4-22.5 FW	93
Soft parts	105.0 DW	94
Clam, *Protothaca staminea*		
Shell	9.2 DW	1
Soft parts	67.7 DW	1
Pteropods; shell	36.0 DW; max. 312.0 DW	15
Clam, *Rangia cuneata*		
Soft parts	51.0 DW	34
Whole	6.8 DW	96
Manila clam, *Ruditapes philippinarum*		
Soft parts; Korea; 2002-2003		
Females vs. males	81.2 DW vs. 76.5 DW	130
Postspawn vs. prespawn	105.0 DW vs. 79.1 DW	130
Jiaozhou Bay, China; sediments vs. soft parts; November 2004	90.8 DW vs. 35.5-85.5 DW	140
Bohai Sea, China; soft parts	10.0-20.1 DW	141
Rock oyster, *Saccostrea cucullata*; soft parts; Hong Kong; 1986		
March	2082.0-3275.0 DW	117
May	2210.0-22,863.0 DW	117
Oyster, *Saccostrea glomerata*; soft parts	1908.0-5786.0 DW	97
Scallops; soft parts	105.0-212.0 DW; max. 462.0 DW	88
Clam, *Scrobicularia plana*		
Soft parts	974.0 DW	11
Gannel estuary, England		
Soft parts	1470.0-2940.0 DW	98
Digestive gland	9630.0 DW	98
Mantle and siphons	316.0 DW	98
Foot and gonads	235.0 DW	98
Gills and palps	406.0 DW	98
Adductor muscles	132.0 DW	98
Kidney	280.0 DW	98

(*Continues*)

Table 6.24: Cont'd

Organism	Concentration	Reference[a]
Camel estuary, England		
Soft parts	353.0-394.0 DW	98
Digestive gland	1104.0 DW	98
Mantle and siphons	170.0 DW	98
Foot and gonads	149.0 DW	98
Gills and palps	388.0 DW	98
Adductor muscles	87.0 DW	98
Kidney	315.0 DW	98
Transferred from Camel estuary to Gannel estuary for 352 days		
Digestive gland	3030.0 DW	98
Remaining tissues	199.0 DW	98
Transferred from Gannel estuary to Camel estuary for 352 days		
Digestive gland	4605.0 DW	98
Remaining tissues	249.0 DW	98
Tamar estuary; England; soft parts	1950.0 (501.0-3990.0) DW	99
Squid, *Sepia esculenta*; trunk	58.0 DW	50
Common cuttlefish, *Sepia officinalis*		
Gill	99.0 DW	3
Mantle	52.0 DW	3
Blood; dialyzed vs. not dialyzed	22.0 FW vs. 7.0 FW	101
Liver; dialyzed vs. not dialyzed	259.0 DW vs. 52.0 DW	101
Gut; dialyzed vs. not dialyzed	41.0 DW. vs. 12.0 DW	101
Branchial gland; dialyzed vs. not dialyzed	124.0 DW vs. 25.0 DW	101
Arterial heart; dialyzed vs. not dialyzed	122.0 DW vs. 21.0 DW	101
Ovary; dialyzed vs. not dialyzed	19.0 DW vs. 4.0 DW	101
Mantle	16.3 FW	16
Dorsal shield	12.6 FW	16
Beak	12.6 FW	16
Gills	24.7 FW	16
Liver	487.5 FW	16
Ink	1.4 FW	16
Egg		
Eggshell	69.0 DW	146
Embryo	68.0 DW	146
Yolk	50.0 DW	146
Hatchlings		
Whole	100.0 DW	146
Cuttlebone	30.0 DW	146

(*Continues*)

Table 6.24: Cont'd

Organism	Concentration	Reference[a]
Juveniles, age 1 month		
Whole	113.0 DW	146
Digestive gland	500.0 DW	146
Cuttlebone	75.0 DW	146
Immatures, age 12 months		
Whole	145.0 DW	146
Digestive gland	770.0 DW	146
Cuttlebone	116.0 DW	146
Adults, age 18 months		
Whole	156.0 DW	146
Digestive gland	1400.0 DW	146
Cuttlebone	150.0 DW	146
Squid, *Sepia* sp.; whole	50.0 FW	76
Clam, *Spisula solidissima*; muscle	18.4 FW	9
Squid		
Flesh	47.4 DW	102
Flesh with skin	144.0 DW	102
Viscera	94.3 DW	102
Pen	263.0 DW	102
Whole	710.0 AW	103
Gastropod, *Strombus pugilis*		
Shell	110.0-210.0 FW; 120.0-210.0 DW; 120.0-220.0 AW	104
Foot	12.0-25.0 FW; 55.0-110.0 DW; 510.0-1100.0 AW	104
Soft parts	39.0-250.0 FW; 250.0-1500.0 DW; 1500.0-3500.0 AW	104
Internal organs	23.0-93.0 FW; 96.0-560.0 DW; 280.0-2300.0 AW	104
Squid, *Symplectoteuthis oualeniensis*; liver	513.0 DW	46
Clam, *Tapes decussatus*; soft parts	16.1-25.5 FW	16
Clam, *Tapes semidecussata*		
Shell	11.2 DW	1
Soft parts	11.2 DW	1
Gastropod, *Thais emarginata*		
Shell	1.4-23.5 DW	1
Soft parts	1701.0 DW	1

(*Continues*)

Table 6.24: Cont'd

Organism	Concentration	Reference[a]
Gastropod, *Thais lapillus*		
Soft parts	Max. 3100.0 DW	42
Soft parts	916.0-1980.0 DW	43
Soft parts	492.1-2354.0 DW	45
Digestive gland and gonad	558.9-760.7 DW	45
Shell	2.7-3.1 DW	45
Clam, *Trivela mactroidea*; Venezuela; 2002; 14 sites; soft parts	5.0-166.0 DW	154
Whelks; soft parts	198.0 (13.0-650.0) DW	88

Values are in mg Zn/kg fresh weight (FW), dry weight (DW), or ash weight (AW).

[a]1, Graham, 1972; 2, Harris et al., 1979; 3, Leatherland and Burton, 1974; 4, Segar et al., 1971; 5, Patel et al., 1973; 6, Palmer and Rand, 1977; 7, Carmichael et al., 1979; 8, Greig, 1975; 9, Greig et al., 1977; 10, Romeril, 1974; 11, Bryan and Hummerstone, 1977; 12, Bryan, 1973; 13, Watling and Watling, 1976b; 14, Van As et al., 1973; 15, Lowman et al., 1970; 16, Establier, 1977; 17, Ratkowsky et al., 1974; 18, Mackay et al., 1975; 19, Sastry and Bhatt, 1965; 20, Zingde et al., 1976; 21, Pringle et al., 1968; 22, Boyden and Romeril, 1974; 23, Watling and Watling, 1976a; 24, McFarren et al., 1962; 25, Kopfler and Mayer, 1967; 26, Shuster and Pringle, 1968; 27, Shuster and Pringle, 1969; 28, Wolfe, 1970; 29, Windom and Smith, 1972; 30, Valiela et al., 1974; 31, Drifmeyer, 1974; 32, Huggett et al., 1975; 33, Frazier, 1975; 34, Sims and Presley, 1976; 35, Greig and Wenzloff, 1978; 36, Goldberg et al., 1978; 37, Bertine and Goldberg, 1972; 38, Anderlini, 1974; 39, Bryan et al., 1977; 40, Vattuone et al., 1976; 41, Topping, 1973; 42, Butterworth et al., 1972; 43, Ireland, 1973; 44, Wharfe and Van den Broek, 1977; 45, Ireland and Wooton, 1977; 46, Martin and Flegal, 1975; 47, Miettinen and Verta, 1978; 48, Eisler and Weinstein, 1967; 49, Larsen, 1979; 50, Ishii et al., 1978; 51, Stenner and Nickless, 1974b; 52, Stenner and Nickless, 1975; 53, Hall et al., 1978; 54, Nielsen and Nathan, 1975; 55, Van As et al., 1975; 56, Eisler, 1977b; 57, Alexander and Young, 1976; 58, Eustace, 1974; 59, Young and McDermott, 1975; 60, Phillips, 1976a; 61, Phillips, 1976b; 62, Phillips, 1977b; 63, Karbe et al., 1977; 64, Simpson, 1979; 65, Young et al., 1979; 66, Talbot and Magee, 1978; 67, Lande, 1977; 68, Brooks and Rumsby, 1965; 69, Bloom and Ayling, 1977; 70, Thomson, 1979; 71, Fowler and Oregioni, 1976; 72, Uysal, 1978a; 73, D'Silva and Kureishy, 1978; 74, Bhosle and Matondkar, 1978; 75, Stenner and Nickless, 1974a; 76, Matsumoto et al., 1964; 77, Hamanaka et al., 1977; 78, George et al., 1978; 79, Fukai et al., 1978; 80, Ikuta, 1968a; 81, Ikuta, 1968b; 82, Ikuta, 1968c; 83, Thrower and Eustace, 1973a; 84, Thrower and Eustace, 1973b; 85, Preston et al., 1972; 86, Topping, 1973; 87, George et al., 1980; 88, Sprague, 1986; 89, Papadopoulu, 1973; 90, Eisler et al., 1978; 91, Pesch et al., 1977; 92, Palmer and Rand, 1977; 93, Greig et al., 1978; 94, Reynolds, 1979; 95, Pyle and Tieh, 1970; 96, White et al., 1979; 97, Phillips, 1979; 98, Bryan and Hummerstone, 1978; 99, Bryan and Uysal, 1978; 100, Ward et al., 1986; 101, Decleir et al., 1978; 102, Horowitz and Presley, 1977; 103, Robertson, 1967; 104, Lowman et al., 1966; 105, Ramelow et al., 1989; 106, Sullivan et al., 1988; 107, White and Rainbow, 1985; 108, Presley et al., 1990; 109, Jiann and Presley, 1997; 110, USNAS, 1979; 111, Cain and Luoma, 1986; 112, Amiard et al., 1986; 113, Lobel, 1986; 114, Amiard-Triquet et al., 1988; 115, Bryan et al., 1987a; 116, Chan, 1988a; 117, Chu et al., 1990; 118, Young et al., 1980; 119, Loring and Asmund, 1989; 120, Lacerda, 1983; 121, Pfeiffer et al., 1985; 122, Lima et al., 1986; 123, Carvalho et al., 1991; 124, Carvalho et al., 1993; 125, Rebelo et al., 2003a; 126, Lacerda and Molisani, 2006; 127, Amaral et al., 2005; 128, Rebelo et al., 2003b; 129, Pearce and Mann, 2006; 130, Ji et al., 2006; 131, Santos et al., 2006; 132, Zorita et al., 2007; 133, Deudero et al., 2007; 134, Karayakar et al., 2007; 135, Liu and Deng, 2007; 136, Mubiana and Blust, 2006; 137, Sunlu, 2006; 138, Bergasa et al., 2007; 139, Huang et al., 2007; 140, Li et al., 2006; 141, Liang et al., 2004a; 142, Peake et al., 2006; 143, Scancar et al., 2007; 144, Kadar, 2007; 145, Gorinstein et al., 2006; 146, Miramand et al., 2006; 147, Silva et al., 2006; 148, Sasikumar et al., 2006; 149, Burt et al., 2007; 150, Espana et al., 2007; 151, Sokolowski et al., 2007; 152, Karouna-Renier et al., 2007; 153, Nesto et al., 2007; 154, LaBrecque et al, 2004; 155, Sankar et al., 2006; 156, Colaco et al., 2006; 157, Kumar et al., 2008; 158, Metian et al., 2008; 159, Maanan, 2008; 160, De Wolf and Rashid, 2008; 161, Cunha et al., 2008; 162, Cravo et al., 2008; 163, Cosson et al., 2008; 164, Frias-Espericueta et al., 2008; 165, Raimundo et al., 2008; 166, Anajjar et al., 2008; 167, Ruelas-Inzunza and Paez-Osuna, 2008; 168, Conti et al., 2008; 169, Protasowicki et al., 2008; 170, Kucuksezgin et al., 2008; 171, Wepener et al., 2008; 172, Bustamante et al., 2008; 173, Fattorini et al., 2008; 174, Napoleao et al., 2005; 175, Nicholson and Szefer, 2003; 176, Unlu et al., 2008.

in the marine environment and is accumulated in excess of the organism's immediate needs, at least on the basis of enzymatically bound zinc (Pequegnat et al., 1969). The information in Table 6.24 supports the observations that zinc is not limiting, and in several cases accumulated far in excess of biological requirements. It has also been stated (Robertson et al, 1972) that intraspecies zinc concentrations in a wide range of flora and fauna were relatively constant, suggesting definite physiological needs. However, much of the information in Table 6.24 contradicts the findings of Robertson and his coworkers. Marked variations in zinc content were evident both within and among closely related groups such as ostreid and crassostreid oysters, and among limpets of the genus *Patella*. In fact, numerous factors can account for the observed variations in zinc content of marine molluscs, of which proximity to anthropogenic sources of zinc was probably one of the most important. The influence of biotic and abiotic modifiers on zinc accumulations will be discussed later. At this time it is sufficient to note from Table 6.24 that zinc concentrations were markedly influenced by season of collection, geographic locale, and zinc-specific sites of accumulation, such as kidney and digestive gland (Eisler, 2000f).

Oysters and scallops are among the most efficient accumulators of zinc. Nonconcentration factors were highest in scallop kidney, with values 1.7-4.0 million times those of ambient seawater (Bryan, 1973). Concentration factors in excess of 100,000 were reported for soft parts of *O. sinuata* (Brooks and Rumsby, 1965) and *C. virginica* (Pringle et al., 1968). The lowest concentration factors reported among field collections of molluscs ranged from 1525 in surf clams, *S. solidissima* and 1700 in softshell clams, *M. arenaria*, to 2100 in hardshell clams, *M. mercenaria* (Pringle et al., 1968). Intermediate values were reported for many species, including oysters *C. gryphoides* (Bhatt et al., 1968), squid *Ommastrephes* spp. (Ketchum and Bowen, 1958), mussels *C. meridionalis* (Van As et al., 1973), and scallops *Pecten novae-zelandiae* (Brooks and Rumsby, 1965). Concentration factors for radiozinc-65 in Pacific oysters, *C. gigas*, collected from the outfall area of a nuclear-powered steam electric station generally followed changes in concentration of ^{65}Zn in the discharge waters (Salo and Leet, 1969); concentration factors for two distinct year classes of these oysters ranged from 5600 to 12,000. Salo and Leet (1969) concluded that the maximum concentration of ^{65}Zn that a human would derive from a protein diet of oysters raised in the canal would be well within the maximum permissible body burden established by United States regulatory agencies. However, concentration factors based on radiozinc accumulation data should be viewed with caution; the addition of stable zinc to the medium reduced radiozinc accumulation disproportionately in scallop tissues, suggesting an ability by this group to discriminate among chemical species of zinc (Bryan, 1973).

Ingestion of seafood products of commerce that contain extremely high concentrations of zinc probably do not represent a major threat to public health, although available evidence is not conclusive. In one case, high concentrations of zinc and other metals in oysters from Tasmania reportedly caused nausea and vomiting in some human consumers; these oysters exceeded the Australian food regulation at that time of 40.0 mg Zn/kg FW by a factor of about 500 (Thrower and Eustace, 1973a,b). The current zinc standard in Turkey is 50.0 mg Zn/kg DW for seafoods, and is sometimes exceeded in mussels from contaminated areas (Unlu et al., 2008).

Zinc accumulations in natural populations are influenced by many variables. Some of the more extensively documented modifiers, together with appropriate examples, follow.

6.53.1 Proximity to Anthropogenic Zinc Sources

When compared to more distant sites, elevated zinc concentrations were found in tissues of mussels and oysters from the vicinity of industrial operations in Australia (Bloom and Ayling, 1977), in clams *P. morrhuana* near an electroplating plant outfall in Narragansett Bay, Rhode Island, United States (Eisler, 1995a; Eisler et al., 1978), in mussels *M. edulis* near a lead-zinc flotation mill in Greenland (Loring and Asmund, 1989), and in various species of shellfish collected near heavily populated and industrialized areas of California (Graham, 1972). Similar observations are recorded in molluscs from widely separated geographic areas (Fowler and Oregioni, 1976; Maanan, 2007; Phillips, 1977b, 1979; Ratkowsky et al., 1974; Romeril, 1974; Thrower and Eustace, 1973a,b).

In the Fal estuary, England, long-term metal pollution over the past 125 years has resulted in zinc sediment burdens of 679.0-1780.0 mg/kg DW, producing benthic communities that favor zinc-tolerant organisms, such as oysters, and a general impoverishment of mussels, cockles, and gastropods (Bryan et al, 1987). It is not known if this tolerance is genetic or acquired.

6.53.2 Body Weight

Increasing concentrations of zinc per unit body weight are recorded with increasing body weight in various species of clams (Boyden, 1974; Eisler et al., 1978), oysters (Downes, 1957), mussels (Boyden, 1974), and squid (Bustamante et al., 2008). However, this pattern is reversed in other species of shellfish, with higher zinc concentrations recorded in smaller organisms (Boyden, 1974, 1977; Boyden et al., 1975; Watling and Watling, 1976a). Studies with ^{65}Zn and clams, *G. tumidum,* showed concentration factors in soft parts of 400-500 after 15 days; however, these values were higher for smaller clams and lower for larger clams (Hedouin et al., 2006).

6.53.3 Gender

Females of *C. meridionalis* contained zinc concentrations in total soft parts that were twice that of males (Watling and Watling, 1976b). However, testes of *M. californianus* contained 60% more zinc than ovarian tissue, with no sex-related differences evident for muscle or digestive gland (Alexander and Young, 1976).

Zinc was demonstrated histochemically in cytoplasmic inclusions shown to be lysosomes in a number of cell types in the common mussel, *M. edulis* (Lowe and Moore, 1979). Males had more zinc in kidneys than females, the latter appearing to use oocytes as an additional means of excretion. Lysosome-vascular systems are considered important detoxification mechanisms for zinc and other metals in molluscs and other organisms (Lowe and Moore, 1979).

In Pacific oysters, *C. virginica*, there is no evidence linking high body burdens of zinc with high concentrations in sex products. Unlike other trace metals, zinc residues in adults and eggs of *C. virginica* were essentially the same, although this was not the case for copper and cadmium (Greig et al., 1976).

6.53.4 Tissue Specificity

Zinc tends to accumulate in molluscan digestive gland and stomach as excretory granules, and in kidney as concretions (Eisler, 1981; Sprague, 1986; Sullivan et al., 1988; Table 6.24). Kidney is the preferred storage site in mussels and scallops (Sprague, 1986), and digestive gland in oysters (Sprague, 1986), and cuttlefish (Miramand et al., 2006). In oysters, granules may contain up to 60% of the total body zinc, and this may account, in part, how shellfish exist with grossly elevated body burdens of zinc (Sprague, 1986).

Different tissues are preferred as zinc storage sites in closely related species. For example, stomach and digestive gland were the major repositories for zinc in *M. edulis* (Pentreath, 1973); however, for *M. californianus*, it was digestive gland (Alexander and Young, 1976) and kidney (Young and Folsom, 1967). But in two species of scallops, highest accumulations of zinc were in kidney and digestive gland of both species (Bryan, 1973). Zinc persistence in selected organs shows considerable variability and may be significantly different from half-time persistence values seen in whole animal. For example, the half-time persistence of zinc in *M. edulis* kidney was estimated at 2-3 months (Lobel and Marshall, 1988) versus 4 days for whole mussel (USNAS, 1979). Gravid cuttlefish, *S. officinalis* were fed crabs radiolabeled with ^{65}Zn over a 2 week period to determine rate of maternal transfer into eggs; during embryonic development, ^{65}Zn was progressively transferred to the embryo, and also to juveniles after hatching (Lacoue-Labarthe et al., 2008b).

Concentrations of zinc in tissues of molluscs are usually far in excess of that required for normal metabolism; much of the excess zinc is bound to macromolecules or present as insoluble metal inclusions in tissues (Eisler, 2000f). In many organisms, zinc and other trace metals concentrate in membrane-limited granules (George et al., 1980). These could be considered as either a long-term store or a means of excretion or detoxification of toxic metals. Excretory granules, or concretions, in kidney of the scallop *P. maximus* are primarily inorganic in composition and contain calcium, manganese, zinc, and phosphorus, the total being equivalent to about 75% of the DW of the granule, with smaller amounts present of magnesium, copper, iron, cadmium, potassium, sulfur, and chlorine. It is postulated that the granules develop from lysomal membranes to mineralized membrane-limited vacuoles or residual bodies and eventually excreted via the urinary tract (George et al., 1980).

6.53.5 Protein Binding

Zinc was associated mainly with high molecular weight proteins in *M. edulis* (Noel-Lambot, 1976). In *C. virginica* nearly all zinc was bound either to soluble high-molecular-weight proteins or to structural cell components; dialysis of soluble tissue extracts removed up to 96% of total zinc without effect on alkaline phosphatase (Wolfe, 1970). Low-molecular-weight metal-binding proteins—not metallothionein—were induced in gastropods in metals-contaminated environments (Andersen et al., 1989). In rock oysters (*S. cucullata*) collected near an iron ore shipping terminal, some of the tissue zinc was bound to high-molecular-weight (around 550,000) iron-binding protein called ferritin (Webb et al., 1985). Ferritin accounts for about 40% of the protein-bound zinc in rock oysters, and probably in other bivalves containing elevated tissue levels of zinc (Webb et al., 1985).

6.53.6 Diet

The actual role of water in the accumulation of zinc by *M. edulis* is minor when compared to diet (Pentreath, 1973). Young (1977) agrees that the food chain is the major source of zinc in molluscan tissues. In *N. lapillus*, the input from food is about 100 times greater than that of seawater (Young, 1977). In *L. obtusata*, food accounts for about 60% of accumulated zinc with an estimated input of 1.1 mg Zn/kg tissue/day compared to the total tissue concentration of 46.7 mg Zn/kg tissue (Young, 1975).

6.53.7 Sediment Substrate

In general, zinc concentrations in both sediments and molluscan tissues are elevated in the vicinity of smelters and other point sources of zinc, and both decrease with increasing distance from the source (Ward et al., 1986). Extractable concentrations of sediment bound zinc were positively correlated with zinc concentrations on deposit-feeding clams (Luoma and Bryan, 1979; Romeril, 1974). Highest zinc concentrations in soft tissues of the clam *M. balthica* were observed in clams collected from calcareous substrates, such as crushed clam shells; lower accumulations were noted in *Macoma* collected from substrates of detrital organics, manganese oxides, and iron oxides, in that order (Luoma and Jenne, 1977).
In general, availability of sediment zinc to bivalve molluscs was higher at increased sediment concentrations of amorphous inorganic oxides or humic substances, and lower at increased concentrations of inorganic carbon and ammonium acetate-soluble manganese (Luoma and Bryan, 1979).

6.53.8 Inherent Species Differences

The half-time persistence of zinc in molluscs is variable and reported to range from 4 days in the common mussel (*M. edulis*) to 650 days in the duck mussel (*Anodonta nutalliana*); intermediate values were 23-40 days in limpet (*L. littorea*), 76 days in the California mussel (*M. californianus*), and 300 days in the Pacific oyster (USNAS, 1979).

Zinc accumulation kinetics in molluscs vary considerably between and within species (Chu et al., 1990) and are attributed to a variety of factors. Variations in zinc content of clam tissues, for example, were associated with seasonal changes in tissue weights (Cain and Luoma, 1986). Gastropods from the Mediterranean Sea near a ferronickel smelter had elevated zinc concentrations in hepatopancreas when compared to conspecifics from distant sites; however, there were no consistent seasonal variations (Nicolaidu and Nott, 1990). Fluctuations in zinc content in tissues of *M. edulis* related to mussel size or to season of collection were sufficient to conceal low chronic or short-term pollution (Amiard et al., 1986). Diet—the main route of zinc accumulation in most molluscs—had no significant effect on whole-body zinc content of certain predatory marine gastropods. For example, whole-body zinc concentrations in gastropod oyster drills (*Ocenebra erinacea*) ranged from 1451.0 to 2169.0 mg/kg DW, and remained unchanged after feeding for 6 weeks on Pacific oysters (*C. gigas*) containing 1577.0 mg Zn/kg DW or mussels (*M. edulis*) containing 63.0 mg Zn/kg DW (Amiard-Triquet et al., 1988).

Taxonomically close groups, such as crassostreid and ostreid oysters show marked differences in zinc contents. Adults of crassostreid oysters usually contain higher concentrations of zinc than ostreid adults, although tissue distributional patterns were similar in both groups (Romeril, 1971). Oysters, when compared to other groups of filter-feeding bivalves, show a greater capacity for zinc uptake. For example, in Chesapeake Bay, Maryland, oysters had 30-40 times more zinc than did clams and up to 90 times more zinc than mussels (McFarren et al., 1962).

6.53.9 Seasonality

Variations in zinc content of soft tissues from individual shellfish species are correlated with seasonal—as distinct from thermal—regimes (Bryan, 1973; Fowler and Oregioni, 1976; Frias-Espericueta et al., 2008; Karayakar et al., 2007; Maanan, 2007; Patel et al., 1973; Romeril, 1974). In one case, zinc content in soft parts of mussels, *M. galloprovincialis*, collected near shore coincided with a period of high precipitation and runoff with a concomitant increase in suspended particle loading of coastal waters (Fowler and Oregioni, 1976). Jiann and Presley (1997) demonstrate that zinc concentrations in American oyster soft parts were highest in summer and lowest in winter and spring.

6.53.10 Dredging

Zinc content of marine molluscs correlates positively with particulate matter produced as a result of dredging operations (Rosenberg, 1977). Zinc accumulated in this manner persists for lengthy periods. In the case of oysters, *O. edulis*, at least 31% of the zinc accumulated during dredging operations was retained for at least 18 months (Rosenberg, 1977). It is possible that some of the high body burdens reported in Table 6.24 may have been influenced by incomplete removal of zinc-contaminated sediments as well as food and other materials. For example, contamination of biological samples by ingested sediments is reported for several species of molluscs and crustaceans. In some locations, elemental zinc concentrations in organisms reflected elemental zinc in inorganic residues from gut contents (Flegal and Martin, 1977). In another study, corrosion of galvanized zinc suspension trays was the cause of grossly elevated zinc (99,000.0 mg Zn/kg DW) in soft parts of Pacific oysters (Boyden and Romeril, 1974).

6.53.11 Depth

Mussels, *P. canaliculus*, collected from different portions of the water column were analyzed for content of zinc, lead, iron, and cadmium (Nielsen, 1974). At one location, all metals except zinc increased in *Perna* with increasing depth. At another location, all metals were essentially the same from mussels collected from the surface to a depth of 9 m. Differences between stations were attributed to differences in mixing of the water column. The differences in mixing may cause variations in the type of food organisms available at the various depths, or variations in the particulate/dissolved ratio of metal levels with depth. Either or both of thee conditions could affect the biological availability of zinc and other metals at different depths (Nielsen, 1974).

6.53.12 Salinity

In general, bivalve molluscs residing in comparatively saline waters contained reduced zinc concentration in soft tissues when compared to conspecifics from lower salinity waters (Duke, 1967; Huggett et al., 1973, 1975; Larsen, 1979; Luoma and Bryan, 1979; Phillips, 1977a,b; Tabata, 1969). However, within the 1.5-3.5% salinity range (15-35 ppt), salinity of the medium was not an important factor governing uptake of zinc in the common mussel, *M. edulis* (Phillips, 1976a); in the case of the clam, *S. plana*, uptake was actually higher at elevated salinities (Garcia-Luque et al., 2007).

6.53.13 Temperature

Increasing water temperatures were associated with increasing concentrations of zinc in soft parts of clams (Duke, 1967; Eisler, 1977b), oysters, and scallops (Duke, 1967); however, common mussels, *M. edulis*, showed no trend within the 10-18 °C range (Phillips, 1976a).

6.53.14 pH

Duke (1967) reports that pH of the ambient medium significantly affected ^{65}Zn accumulation from seawater by a community of oysters, clams, and scallops; in all cases, uptake was higher at elevated pH levels.

6.53.15 Interactions

Zinc accumulations in molluscs are mediated by many factors, including interaction effects of zinc with salts of calcium, cobalt, iron, cadmium, copper, mercury, phosphates, various organic substances, and others. Zinc, for example, inhibited mercury accumulations in marine gastropods (Andersen et al., 1989). Radiozinc-65 was rapidly accumulated in southern quahog (*Mercenaria campechiensis*) over a 10-day period; accumulation in kidney was linear over time and was enhanced at elevated phosphate loadings in the medium (Miller et al., 1985).

In soft parts of *C. meridionalis*, concentrations of copper, lead, manganese, and bismuth were all positively correlated with zinc content, but not with each other (Watling and Watling, 1976b). In oysters, *O. edulis*, the total amount of zinc measured was far in excess of the amount of zinc contributed by zinc-dependent enzymes, such as carbonic anhydrase, alkaline phosphatase, carboxypeptidase A, malic dehydrogenase, and alpha-D-mannoside; however, the amount of nondialyzable zinc was of the same order of magnitude (Coombs, 1972). This apparent excess of dialyzable zinc was linked to the high levels of calcium in tissues, demonstrating a competition between calcium and zinc in uptake potential (Coombs, 1972). Mussels, *M. edulis*, during exposure to 0.2 mg cadmium/L for 20 days, showed only a slight increase in zinc content of soft parts (George and Coombs, 1977). On the other hand, zinc at 0.5 mg/L substantially decreased cadmium uptake in *M. edulis*, and clams, *Mulinia lateralis* (Jackim et al., 1977).

Also reported were interaction effects of zinc, lead, and cadmium salts on zinc accumulation in mussels (Phillips, 1976a); depressed zinc uptake in oysters by iron or cobalt (Romeril, 1971); increased zinc accumulation in mantle tissue of hardshell clams during exposure to organochlorine and organophosphorus insecticides (Eisler and Weinstein, 1967); and influence of dissolved carbonates (Tabata, 1969) and chelators (Keckes et al., 1968, 1969) on zinc uptake and retention in mussels. In the case of chelators, EDTA increased the loss rate of zinc from mussels at 50.0 mg EDTA/L, but uptake rates were depressed in the range of 0.01-5.0 mg EDTA/L (Keckes et al., 1968, 1969). To confound matters, the loss rate of zinc from some species of mussels was not constant, suggesting multicompartmental zinc metabolism, dependent, perhaps, on the solubility of various chemical species (Keckes et al., 1968; Tabata, 1969; Van Weers, 1973).

Mixtures of zinc and copper, and also mercury and zinc, were more than additive in toxicity to oyster larvae (Sprague, 1986). Pre-exposure of the common mussel, *M. edulis*, to 0.05 mg

Zn/L for 28 days conferred increased tolerance to 0.075 mg inorganic mercury/L (Roesijadi and Fellingham, 1987).

6.53.16 Laboratory Studies

Uptake, retention, and translocation of zinc by molluscs under controlled physicochemical conditions is a subject that has received extensive documentation (D'Silva and Kureishy, 1978; Eisler, 1977a, 2000f; Keckes et al., 1968, 1969; Mehran and Tremblay, 1965; Metian et al., 2008; Mishima and Odum, 1963; Pentreath, 1973; Phillips, 1977a; Pringle et al., 1968; Romeril, 1971; Shuster and Pringle, 1968, 1969; Young and Folsom, 1967).

M. edulis has been used extensively as a model for molluscan zinc kinetics. Results of selected studies follow. In the common mussel, zinc is taken up by the digestive gland, gills, and mantle and rapidly transported via hemolymph where it is stored as insoluble granules (Lobel and Marshall, 1988). There is a high degree of variability in zinc content of soft tissues of *M. edulis* due entirely to an unusually high degree of variability in kidney zinc of 97.0-7864.0 mg/kg DW (Lobel, 1987). This variability in kidney zinc content is attributed mainly to a low-molecular-weight zinc complex (MW 700-1300) that showed a high degree of variability and a positive correlation with kidney zinc concentration (Lobel and Marshall, 1988). However, at low ambient concentrations of 0.05 mg Zn/L, the most sensitive bioindicators of zinc exposure were gills and labial palps (Amiard-Triquet et al., 1988). Food composition had little effect on tissue distribution of radiozinc-65 in mussels, as judged by 5-day feeding studies of radiolabeled diatoms (*Thalassiosira pseudonana*), green alga (*Dunaliella tertiolecta*), glass beads, and egg albumin particles (Fisher and Teyssie, 1986). Soft part BCF values ranged between 12 and 35 times that of the medium and was probably de to a rapid desorption of radiozinc from the food particles into the acidic gut, followed by binding to specific ligands or molecules. The half-time persistence in mussel soft parts ranged from 42 to 80 days for all food items—including glass beads—and about 20 days in shell (Fisher and Teyssie, 1986; Wang et al., 1996). Zinc concentrations in mussels are proportionately related to zinc loadings in the water column and to the AE of ingested particles, which in *M. edulis* ranges from 32% to 41% (Wang et al., 1996). Elevated temperatures in the range 10-25 °C were associated with increased uptake rates of zinc from seawater by mussels (Watkins and Simkiss, 1988). If the temperature is oscillated through this range over a 6 h period, there is a further enhancement of zinc uptake. This effect parallels decreases in zinc content of cytosol fractions and increases in granular fractions (Watkins and Simkiss, 1988). Mussels were more sensitive to zinc than were other bivalves tested. The pumping rate of mussels completely stopped for up to 7 h on exposure to 0.47-0.86 mg Zn/L; other bivalves tested showed only a 50% reduction in filtration rates in the range 0.75-2.0 mg Zn/L (Redpath and Davenport, 1988). *M. edulis* accumulates zinc under natural conditions, but sometimes does not depurate (Luten et al., 1986). This conclusion was based on results of a study wherein mussels were transferred from a pristine environment in the Netherlands to a polluted estuary for 70 days, than back again for 77 days. At the start, zinc concentration was 106.0 mg/kg DW soft parts. By day 70, it had risen to 265.0 mg/kg DW, at a linear daily uptake rate of 0.47 mg/kg, but mussels contained 248.0 mg/kg DW on day 147, indicating that elimination was negligible (Luten et al., 1986). In another study, zinc depressed sperm motility through respiratory inhibition at 6.6 mg/L, a concentration much higher than those normally found environmentally (Earnshaw et al., 1986). In mussel spermatozoa, zinc caused reductions of bound calcium and phosphorus in both acrosomes and mitochondria, suggesting increased permeability of organelle membranes to both elements (Earnshaw et al., 1986).

Results of uptake studies of ligand-bound ^{65}Zn by the green mussel, *P. viridis*, show that radiozinc uptake could not be fully explained by free zinc concentrations in the presence of different ligands, indicating that metal-ligand complexes were available for uptake (Chuang and Wang, 2006). Scallops, *C. radula* held for 96 h in seawater containing ^{65}Zn had BCFs of 149 in total soft parts; however the BCF for kidney was 549 (Metian et al., 2008).

Marked interspecies variations were documented in ability to regulate zinc accumulation. At comparatively low ambient zinc concentrations of 0.1 or 0.2 mg/L, crassostreid oysters continued to accumulate zinc over a 20 week period (maximum value recorded of 3813.0 mg Zn/kg FW soft parts) with little evidence of regulation (Shuster and Pringle, 1968, 1969). But softshell clams, *M. arenaria*, at 0.5 mg Zn/L showed some regulatory ability at comparatively low water temperatures during exposure for 16 weeks, as evidenced by a drop in the concentration factor (Eisler, 1977b). This was not evident at higher temperatures during exposure for only 2 weeks, although accumulation was significantly more rapid (Eisler, 1977b). Mussels, *M. edulis*, however, appear to plateau or regulate zinc accumulations at concentrations above 0.4 mg Zn/L seawater (Phillips, 1977a). Maximum net daily accumulation rates recorded, in mg Zn/kg FW whole organism, were 7.7 for the common mussel, 19.8 for the American oyster, and 32.0 for the soft shell clam; in general, accumulation rates and total accumulations were higher at elevated water temperatures and at higher ambient zinc water concentrations (Eisler, 1980).

Loss rate of zinc from mussels, *M. galloprovincialis*, was greater with shorter contact time in zinc-contaminated seawater, and loss rate was greater from shell than soft tissues (Keckes et al., 1969). A similar pattern was observed in *L. obtusata* (Mehran and Tremblay, 1965). Stomach and digestive gland of mussels had the highest accumulations of zinc, but loss rate of zinc from these tissues was also highest (Pentreath, 1973). Mussels accumulate ^{65}Zn slowly from seawater and depurate it rapidly indicating a reduced risk of contamination of mussels following incidental release of ^{65}Zn in coastal waters (Van Weers, 1973). Loss rate of zinc in laboratory populations of bivalves may be lower than those in natural populations (Romeril, 1971), although this has not been resolved with certainty. For a marine gastropod, *Littorina irrorata*, zinc excretion seems governed by increasing respiration rate (linked to increasing water temperature) and increasing body size (Mishima and Odum, 1963).

Variability in daily zinc accumulation rates by the softshell clam, *M. arenaria*, is one example what may be typical for marine bivalve molluscs. During immersion in 0.5 mg Zn/L in flow-through system at elevated (16-22 °C) temperatures, *Mya* soft parts accumulated 2.0 mg Zn/kg FW daily during the first 24 h; between 24 h and 7 days this value was 7.7; and between 7 and 14 days, 3.3. Under the same conditions, except that ambient zinc levels were maintained at 2.5 mg/L, the accumulation rate during the first 24 h was 32.0 mg Zn/kg FW soft parts daily; between 24 h and 7 days, this dropped to 11.7 mg Zn/kg daily. At lower temperatures of 0-10 °C, and 0.5 mg Zn/L, *Mya* accumulated zinc at the rate of 0.9 mg/kg FW daily during the first 42 days; however, between days 42 and 112, *Mya* lost zinc at the rate of 0.24 mg/kg FW daily. Eisler (1977b) concluded that changes in accumulation rates of zinc by *Mya* reflected, in part, complex interaction effects between water temperature, ambient zinc concentrations, duration and season of exposure, and physiological saturation and detoxication mechanisms.

Deficiency and toxicity effects of zinc in molluscs are under investigation. Zinc deficiency effects have been produced experimentally in a wide variety of marine organisms, including molluscs, at nominal seawater concentrations between 0.00065 and 0.0065 mg Zn/L; however, zinc deficiency in natural waters has not yet been credibly documented for molluscs (Eisler, 2000f). Theoretically, whole molluscs should contain at least 34.5 mg total Zn/kg DW in order to avoid zinc deficiency (White and Rainbow, 1985), but this requires verification. Zinc was most toxic to representative molluscs at elevated temperatures and low salinities (Khangarot and Ray, 1987; Sprague, 1986), at earlier developmental stages (Munzinger and Guarducci, 1988), at low dissolved oxygen (Khangarot and Ray, 1987), and with increasing exposure to high zinc concentrations (Amiard-Triquet et al., 1988). The LC50(96 h) range for zinc and representative molluscs was 0.195 mg/L for embryos of the hardshell clam, *M. mercenaria*, to more than 320.0 mg/L for adults of the Baltic clam, *M. balthica* (Eisler, 2000f). Molluscan deaths and developmental abnormalities were common at 0.05 mg Zn/L in larvae of the red abalone, *H. rufescens* (Hunt and Anderson, 1989), at 0.12 mg Zn/L in larvae of the bay scallop (Yantian, 1989), at 0.195 mg/L in embryos of the hardshell clam (USEPA, 1987), at 0.23 mg Zn/L in embryos of the American oyster and larvae of the Pacific oyster (USEPA, 1987), at 1.75 mg Zn/L in adults of the common mussel in 48 h (Hunt and Anderson, 1989), and at 6.1 mg/L in adult green-lipped mussels *P. viridis* (Chan, 1988b).

Adverse sublethal effects of zinc to representative molluscs were most pronounced among embryos and larvae. Adverse effects documented at 0.01-0.02 mg Zn/L include reduced settlement of larvae of the Pacific oyster in 20 days (USEPA, 1987), and significant accumulations in 42 days in kidney and other tissues in adults of the gastropod *L. littorea*

(Mason, 1988). At 0.03-0.035 mg Zn/L, there was a reduction in larval settlement of the Pacific oyster in 6 days (USEPA, 1987). And at 0.05 mg Zn/L, there was a 22% reduction in growth rate in 9 days of larvae of the bay scallop, *A. irradians* (Yantian, 1989). Adults of the clam, *D. trunculus*, immersed for 24 h in seawater containing 0.1 mg Zn/L had feeding rate inhibited by 25% (Neuberger-Cywiak et al., 2007).

6.53.17 Proposed Criteria

Most species of molluscs can tolerate without harm seawater concentrations up to 0.054 mg total Zn/L (Sprague, 1986), although sensitive species—such as larvae of the Pacific oyster—are harmed at 0.01-0.02 mg Zn/L (USEPA, 1987). Proposed criteria to protect marine molluscs include a mean seawater concentration of less than 0.058 mg Zn/L, never to exceed 0.17 mg/L; for acid-soluble zinc, these values are less than 0.086 and 0.095 mg Zn/L, respectively (USEPA, 1980d, 1987). However, as shown earlier, sensitive species of molluscs are adversely affected at significantly lower concentrations, suggesting that these proposed criteria be further scrutinized before implementation.

6.54 Zirconium

The single datum available on zirconium and field collections of molluscs is that of Karbe et al. (1977) indicating that soft parts of the common mussel, *M. edulis* contained between 0.62 and 43.0 mg Zr/kg DW.

6.55 Literature Cited

Aarkrog, A., 1971. Radioecological investigations of plutonium in an arctic marine environment. Health Phys. 20, 31–47.
Aarkrog, A., 1977. Environmental behaviour of plutonium accidentally released at Thule, Greenland. Health Phys. 32, 271–284.
Aarkrog, A., 1990. Environmental radiation and radiation releases. Int. J. Radiat. Biol. 57, 619–631.
Abbe, G.R., Sanders, J.G., 1990. Pathways of silver uptake and accumulation by the American oyster (*Crassostrea virginica*) in Chesapeake Bay. Estuar. Coast. Shelf Sci. 31, 113–123.
Abbott, O.J., 1977. The toxicity of ammonium molybdate to marine invertebrates. Mar. Pollut. Bull. 8, 204–205.
Absil, M.C.P., Gerringa, L.S.A., Wolterbeek, B.T., 1993. The relation between salinity and copper complexing capacity of natural estuarine waters and the uptake of dissolved ^{64}Cu by *Macoma balthica*. Chem. Spec. Bioavail. 5, 119–128.
Adema, D.M.M., De Swaaf-Mooy, S.I., Bais, P., 1972. Laboratory investigations concerning the influence of copper on mussels (*Mytilus edulis*). TNO-Nieuws 27, 482–487.
Airas, S., Duinker, A., Julshamn, K., 2004. Copper, zinc, arsenic, cadmium, mercury, and lead in blue mussels (*Mytilus edulis*) in the Bergen Harbor area, western Norway. Bull. Environ. Contam. Toxicol. 73, 276–284.
Alexander, C.V., Young, D.R., 1976. Trace metals in southern California mussels. Mar. Pollut. Bull. 7, 7–9.
Alquezar, R., Markich, S.J., Twining, J.R., 2007. Uptake and loss of dissolved ^{109}Cd and ^{75}Se in estuarine macroinvertebrates. Chemosphere 67, 1201–1210.
Alzieu, C., Sanjuan, J., Deltreil, J.P., Borel, M., 1986. Tin contamination in Arcachon Bay: effects on oyster shell anomalies. Mar. Pollut. Bull. 17, 494–498.

Amaral, M.C.B., Rebelo, M.F., Torres, J.P.M., Pfeiffer, W.C., 2005. Bioaccumulation and depuration of Zn and Cd in mangrove oysters (*Crassostrea rhizophorae* Guilding, 1828) transplanted to and from a contaminated tropical lagoon. Mar. Environ. Res. 59, 277–285.

Amiard, J.C., Amiard-Triquet, C., Berthet, B., Metayer, C., 1986. Contribution to the ecotoxicological study of cadmium, lead, copper and zinc in the mussel *Mytilus edulis*. I. Field study. Mar. Biol. 90, 425–431.

Amiard-Triquet, C., Amiard, J.C., 1976. L'orgaotropisme du ^{60}Co chez *Scrobicularia plana* et *Carcinus maenas* en fonction du vecteur de contamination. Oikos 27, 122–126.

Amiard-Triquet, C., Amiard, J.C., Berthet, B., Metayer, C., 1988. Field and experimental study of the bioaccumulation of some trace metals in a coastal food chain: seston, oyster (*Crassostrea gigas*), drill (*Ocenebra erinacea*). Water Sci. Technol. 20, 13–21.

Anajjar, E.M., Chiffoleau, J.F., Bergayou, H., Moukrim, A., Burgeot, T., Cheggour, M., 2008. Monitoring of trace metal contamination in the Souss estuary (south Morocco) using the clams *Cerastoderma edule* and *Scrobicularia plana*. Bull. Environ. Contam. Toxicol. 80, 283–288.

Anderlini, V., 1974. The distribution of heavy metals in the red abalone, *Haliotis rufescens*, on the California coast. Arch. Environ. Contam. Toxicol. 2, 253–265.

Anderlini, V.C., 1992. The effect of sewage on trace metal concentrations on scope for growth in *Mytilus edulis aoteanus* and *Perna canaliculus* from Wellington Harbour, New Zealand. Sci. Total Environ. 125, 263–288.

Andersen, R.A., Eriksen, K.D.H., Bakke, T., 1989. Evidence of the presence of a low molecular weight, non-metallothionein-like metal-binding protein in the marine gastropod, *Nassarius reticulatus* L. Comp. Biochem. Physiol. 94B, 285–291.

Andrews, H.L., Warren, S., 1969. Ion scavenging by the eastern clam and quahaug. Health Phys. 17, 807–810.

Anonymous, 1987. Legislators consider plan to ban TBT paints on small craft only. Ecol. USA July 20, 1937, 124.

Arai, T., Ikemoto, T., Hokura, A., Terada, Y., Kunito, T., Tanabe, S., et al., 2004. Chemical forms of mercury and cadmium accumulated in marine mammals and seabirds as determined by XAFS analysis. Environ. Sci. Technol. 38, 6468–6474.

Argese, E., Bettiol, C., Ghirardini, A.V., Fasolo, M., Giurin, G., Ghetti, P.F., 1998. Comparison of in vitro submitochondrial particle and Microtox assays for determining the toxicity of organotin compounds. Environ. Toxicol. Chem. 17, 1005–1012.

Argiero, L., Manfredini, S., Palmas, G., 1966. Absorption de produits de fission par les organismes marins. Health Phys. 12, 1259–1265.

Avargues, M., Ancellin, J., Vilquin, A., 1968. Experimental investigation on the accumulation of radionuclides by marine organisms. In: Proceedings of the Third International Colloquium on Medical Oceanography, Fifth and Sixth Sessions, pp. 87–100.

Belanger, S.E., Farris, J.L., Cherry, D.S., Cairns Jr., J., 1990. Validation of *Corbicula fluminea* growth reductions induced by copper in artificial streams and river systems. Can. J. Fish. Aquat. Sci. 47, 904–914.

Belcheva, N.N., Zakhartsev, M., Silina, A.V., Slinko, E.N., Chelomin, V.P., 2006. Relationship between shell weight and cadmium content in whole digestive gland of the Japanese scallop *Patinopecten yessoensis* (Jay). Mar. Environ. Res. 61, 396–409.

Bergasa, O., Ramirez, R., Collado, C., Hernandez-Brito, J.J., Gelado-Caballero, M.D., Rodriguez-Somozas, M., et al., 2007. Study of metals concentration levels in *Patella piperata* throughout the Canary Islands, Spain. Environ. Monit. Assess. 127, 127–133.

Berland, B.R., Bonin, D.J., Kapkov, W.I., Maestrini, S.Y., Arlhac, D.P., 1976. Action toxique de quatre metaux lourds sur la croissance d'algues unicellulaires marines. C. R. Acad. Sci. Paris 282D, 633–636.

Berrow, S.D., 1991. Heavy metals in sediments and shellfish from Cork Harbour, Ireland. Mar. Pollut. Bull. 22, 467–469.

Berthet, B., Amiard-Triquet, C., Martoja, R., 1990. Effets chimiques et histologiques de la decontamination de l'huitre *Crassostrea gigas* Thurberg prealablement exposee a l'argent. Water Air Soil Pollut. 50, 355–369.

Berthet, B., Amiard, J.C., Amiard-Triquet, C., Martoja, M., Jeantet, A.Y., 1992. Bioaccumulation, toxicity and physico-chemical speciation of silver in bivalve molluscs: ecotoxicological and health consequences. Sci. Total Environ. 125, 97–122.

Bertine, K.K., Goldberg, E.D., 1972. Trace elements in clams, mussels, and shrimps. Limnol. Oceanogr. 17, 877–884.

Berto, D., Giani, M., Boscolo, R., Covelli, S., Giovanardi, O., Massironi, M., et al., 2007. Organotins (TBT and DBT) in water, sediments, and gastropods of the southern Venice lagoon (Italy). Mar. Pollut. Bull. 55, 425–435.

Betzer, S.B., 1972. Copper metabolism, copper toxicity, and a review of the function of hemocyanin in *Busycon canaliculatum* L. Ph.D. thesis, University of Rhode Island, Kingston, 133 pp.

Betzer, S.B., Pilson, M.E.Q., 1974. The seasonal cycle of copper concentration in *Busycon canaliculatum* L. Biol. Bull. 146, 165–175.

Betzer, S.B., Pilson, M.E.Q., 1975. Copper uptake and excretion by *Busycon canaliculatum*. L. Biol. Bull. 148, 1–15.

Betzer, S.B., Yevich, P.P., 1975. Copper toxicity in *Busycon canaliculatum*. L. Biol. Bull. 148, 16–25.

Bhatt, Y.M., Sastry, V.N., Shah, S.M., Krishnamoorthy, T.M., 1968. Zinc, manganese and cobalt contents of some marine bivalves from Bombay. Proc. Natl. Inst. Sci. India B Biol. Sec. 34(B6), 283–287.

Bhosle, N.B., Matondkar, S.G.P., 1978. Variation in trace metals in two populations of green mussel *Mytilus viridis* L. from Goa. Mahasagar Bull. Nat. Inst. Oceanogr. 11, 191–194.

Bianchini, A., Playle, R.C., Wood, C.M., Walsh, P.J., 2007. Short-term silver accumulation in tissues of three marine invertebrates: shrimp *Penaeus duorarum*, sea hare *Aplysia californica*, and sea urchin *Diadema antillarum*. Aquat. Toxicol. 84, 182–189.

Bigler, J., Crecelius, E., 1998. Methods for the analysis of arsenic speciation in seafood from Cook Inlet, Alaska. In: Society of Environmental Geochemistry and Health, 3rd International Conference on Arsenic Exposure and Health Effects, 9 pp.

Bloom, M., Ayling, G.M., 1977. Heavy metals in the Derwent estuary. Environ. Geol. 1, 3–22.

Blunden, S.J., Cusack, P.A., Hill, R., 1985. The Industrial Uses of Tin Chemicals. Royal Society of Chemistry, London, 337 pp.

Bohn, A., 1975. Arsenic in marine organisms from West Greenland. Mar. Pollut. Bull. 6, 87–89.

Bordin, G., McCourt, J., Rodriguez, A., 1994. Trace metals in the marine bivalve *Macoma balthica* in the Westerschelde estuary, the Netherlands. Part 2: intracellular partitioning of copper, cadmium, zinc and iron—variations of the cytoplasmic metal concentrations in natural and *in vitro* contaminated clams. Sci. Total Environ. 151, 113–124.

Boyden, C.R., 1974. Trace element content and body size in molluscs. Nature 251, 311–314.

Boyden, C.R., 1977. Effect of size upon metal content of shellfish. J. Mar. Biol. Assoc. UK 57, 675–714.

Boyden, C.R., Romeril, M.G., 1974. A trace metal problem in pond oyster culture. Mar. Pollut. Bull. 5, 74–78.

Boyden, C.R., Watling, H., Thornton, I., 1975. Effect of zinc on the settlement of the oyster *Crassostrea gigas*. Mar. Biol. 31, 227–234.

Brooks, R.R., Rumsby, M.G., 1965. The biogeochemistry of trace element uptake by some New Zealand bivalves. Limnol. Oceanogr. 10, 521–527.

Brooks, R.R., Rumsby, M.G., 1967. Studies on the uptake of cadmium by the oyster *Ostrea sinuata* (Lamarck). Aust. J. Mar. Freshw. Res. 15, 53–61.

Brooks, S.J., Bolam, T., Tolhurst, L., Bassett, J., La Roche, J., Waldock, M., et al., 2007. Effects of dissolved organic carbon on the toxicity of copper to the developing embryos of the Pacific oyster (*Crassostrea gigas*). Environ. Toxicol. Chem. 26, 1756–1763.

Brown, C.L., Luoma, S.N., 1995. Use of the euryhaline bivalve *Potamocorbula amurensis* as a biosentinel species to assess trace metal contamination in San Francisco Bay. Mar. Ecol. Prog. Ser. 124, 129–142.

Brown, K.R., McPherson, R.G., 1992. Concentrations of copper, zinc and lead in the Sydney rock oyster, *Saccostrea commercialis* (Iredale and Roughley) from the Georges River, New South Wales. Sci. Total Environ. 126, 27–33.

Bryan, G.W., 1963a. The accumulation of radioactive caesium by marine invertebrates. J. Mar. Biol. Assoc. UK 43, 519–539.

Bryan, G.W., 1963b. The accumulation of ^{137}Cs by brackish water invertebrates and its relation to the regulation of potassium and sodium. J. Mar. Biol. Assoc. UK 43, 541–565.

Bryan, G.W., 1971. The effects of heavy metals (other than mercury) on marine and estuarine organisms. Proc. R. Soc. Lond. 177B, 389–410.

Bryan, G.W., 1973. The occurrence and seasonal variation of trace metals in the scallops *Pecten maximus* (L.) and *Chlamys opercularis* (L.). J. Mar. Biol. Assoc. UK 53, 145–166.

Bryan, G.W., Hummerstone, L.G., 1973. Brown seaweed as an indicator of heavy metals in estuaries in south-west England. J. Mar. Biol. Assoc. UK 59, 89–108.

Bryan, G.W., Hummerstone, L.G., 1977. Indicators of heavy metal contamination in the Looe estuary (Cornwall) with particular regard to silver and lead. J. Mar. Biol. Assoc. UK 57, 75–92.

Bryan, G.W., Hummerstone, L.G., 1978. Heavy metals in the burrowing bivalve *Scrobicularia plana* from contaminated and uncontaminated estuaries. J. Mar. Biol. Assoc. UK 58, 401–419.

Bryan, G.W., Langston, W.J., 1992. Bioavailability, accumulation and effects of heavy metals in sediments with special reference to United Kingdom estuaries: a review. Environ. Pollut. 76, 89–131.

Bryan, G.W., Uysal, H., 1978. Heavy metals in the burrowing bivalve *Scrobicularia plana* from the Tamar estuary in relation to environmental levels. J. Mar. Biol. Assoc. UK 58, 89–108.

Bryan, G.W., Potts, G.W., Forster, G.R., 1977. Heavy metals in the gastropod mollusc *Haliotis tuberculata* (L.). J. Mar. Biol. Assoc. UK 57, 379–390.

Bryan, G.W., Langston, W.J., Hummerstone, L.S., Burt, G.R., Ho, Y.B., 1983. An assessment of the gastropod, *Littorina littorea*, as an indicator of heavy-metal contamination in United Kingdom estuaries. J. Mar. Biol. Assoc. UK 63, 327–345.

Bryan, G.W., Gibbs, P.E., Hummerstone, L.G., Burt, G.R., 1986. The decline of the gastropod *Nucella lapillus* around south-west England: evidence for the effect of tributyltin from antifouling paints. J. Mar. Biol. Assoc. UK 66, 611–640.

Bryan, G.W., Gibbs, P.E., Hummerstone, L.G., Burt, G.R., 1987a. Copper, zinc, and organotin as long-term factors in governing the distribution of organisms in the Fal estuary in southwest England. Estuaries 10, 208–219.

Bryan, G.W., Gibbs, P.E., Burt, G.R., Hummerstone, L.G., 1987b. The effects of tributyltin (TBT) accumulation on adult dog-whelks, *Nucella lapillus*: long-term field and laboratory experiments. J. Mar. Biol. Assoc. UK 67, 525–544.

Bryant, V., Newbery, D.M., McLusky, D.S., Campbell, R., 1985. Effect of temperature and salinity on the toxicity of arsenic to three estuarine invertebrates (*Corophium volutator*, *Macoma balthica*, *Tubifex costatus*). Mar. Ecol. Prog. Ser. 24, 129–137.

Burger, J., Gochfeld, M., 2006. Seasonal differences in heavy metals and metalloids in Pacific blue mussels *Mytilus [edulis] trossulus* from Adak Island in the Aleutian Chain, Alaska. Sci. Total Environ. 368, 937–950.

Burger, J., Stern, A.H., Gochfeld, M., 2005. Mercury in commercial fish: optimizing individual choices to reduce risk. Environ. Health Perspect. 113, 266–270.

Burger, J., Gochfeld, M., Jeitner, C., Burke, S., Stamm, T., Snigaroff, R., et al., 2007a. Mercury levels and potential risk from subsistence foods from the Aleutians. Sci. Total Environ. 384, 93–105.

Burger, J., Gochfeld, M., Jewett, S.C., 2007b. Radionuclide concentrations in benthic invertebrates from Amchitka and Kiska Islands in the Aleutian Chain, Alaska. Environ. Monit. Assess. 128, 329–341.

Burt, A., Maher, W., Roach, A., Krikowa, F., Honkoop, P., Bayne, B., 2007. The accumulation of Zn, Se, Cd, and Pb and physiological condition of *Anadara trapezia* transplanted to a contamination gradient in Lake Macquarie, New South Wales, Australia. Mar. Environ. Res. 64, 54–78.

Burton, J.D., Leatherland, T.M., 1971. Mercury in a coastal marine environment. Nature 231, 440–442.

Bustamante, P., Lahaye, V., Durnez, C., Churlaud, C., Caurant, F., 2006a. Total and organic Hg concentrations in cephalopods from the north eastern Atlantic waters: influence of geographical origin and feeding ecology. Sci. Total Environ. 368, 585–596.

Bustamante, P., Teyssie, J.L., Fowler, S.W., Warnau, M., 2006b. Assessment of the exposure pathway in the uptake and distribution of americium and cesium in cuttlefish (*Sepia officinalis*) at different stages of its life cycle. J. Exp. Mar. Biol. Ecol. 331, 198–207.

Bustamante, P., Gonzalez, A.F., Rocha, F., Miramand, P., Guerra, A., 2008. Metal and metalloid concentrations in the giant squid *Architeuthis dux* from Iberian waters. Mar. Environ. Res. 66, 278–287.

Butterworth, J., Lester, P., Nickless, G., 1972. Distribution of heavy metals in the Severn estuary. Mar. Pollut. Bull. 3, 72–74.

Byers, J.E., 1993. Variations in the bioaccumulation of zinc, copper, and lead in *Crassostrea virginica* and *Ilyanassa obsoleta* in marinas and open water environments. J. Elisha Mitchell Sci. Soc. 109, 163–170.

Cain, D.J., Luoma, S.N., 1986. Effect of seasonally changing tissue weight on trace metal concentrations in the bivalve *Macoma balthica*, in San Francisco Bay. Mar. Ecol. Prog. Ser. 28, 209–217.

Cain, D.J., Luoma, S.N., 1990. Influence of seasonal growth, age, and environmental exposure on Cu and Ag in a bivalve indicator, *Macoma balthica*, in San Francisco Bay. Mar. Ecol. Prog. Ser. 60, 45–55.

Cajaraville, M.P., Bebianno, M.J., Blasco, J., Porte, C., Sarasquete, C., Viarengo, A., 2000. The use of biomarker to assess the impact of pollution in coastal environments of Iberian Peninsula: a practical approach. Sci. Total Environ. 247, 295–311.

Calabrese, A., Nelson, D.A., 1974. Inhibition of embryonic development of the hard clam, *Mercenaria mercenaria*, by heavy metals. Bull. Environ. Contam. Toxicol. 11, 92–97.

Calabrese, A., Collier, R.S., Nelson, D.A., MacInnes, J.R., 1973. The toxicity of heavy metals to embryos of the American oyster *Crassostrea virginica*. Mar. Biol. 18, 162–166.

Calabrese, A., MacInnes, J.R., Nelson, D.E., Miller, J.E., 1977a. Survival and growth of bivalve larvae under heavy-metal stress. Mar. Biol. 41, 179–184.

Calabrese, A., Thurberg, F.P., Gould, E., 1977b. Effects of cadmium, mercury, and silver on marine animals. Mar. Fish. Rev. 39, 5–11.

Calabrese, A., MacInnes, J.R., Nelson, D.A., Greig, R.A., Yevich, P.P., 1984. Effects of long-term exposure to silver or copper on growth, bioaccumulation and histopathology in the blue mussel *Mytilus edulis*. Mar. Environ. Res. 11, 253–274.

Cappon, C.J., Smith, J.C., 1982. Chemical form and distribution of selenium in edible seafood. J. Anal. Toxicol. 6, 10–21.

Cardwell, R.D., Sheldon, A.W., 1986. A risk assessment concerning the fate and effects of tributyltins in the aquatic environment. In: Maton, G.L. (Ed.), Proceedings Oceans 86 Conference, Washington, DC, September 23-25, 1986, vol. 4. Organotin Symposium. Available from Marine Technology Society, 2000 Florida Avenue NW, Washington, DC, pp. 1117–1129.

Carlisle, D.B., 1958. Niobium in ascidians. Nature 181, 933.

Carmichael, N.G., Squibb, K.S., Fowler, B.A., 1979. Metals in the molluscan kidney: a comparison of two closely related bivalve species (*Argopecten*), using X-ray microanalysis and atomic absorption spectroscopy. J. Fish. Res. Board Can. 36, 1149–1155.

Carr, R.S., McCullough, W.L., Neff, J.M., 1982. Bioavailability of chromium from a used chrome lignosulphonate drilling mud to five species of marine invertebrates. Mar. Environ. Res. 6, 189–203.

Carvalho, C.E.V., Lacerda, L.D., Gomes, M.P., 1991. Heavy metal contamination of the marine biota along the Rio de Janeiro coast, SE Brazil. Water Air Soil Pollut. 57/58, 645–653.

Carvalho, C.E.V., Lacerda, L.D., Gomes, M.P., 1993. Metais pesados na biota bentica da Baia de Sepetiba e Angra dos Reis, RJ. Acta Limnol. Brasil 6, 222–229.

Casterline Jr., J.L., Yip, G., 1975. The distribution and binding of cadmium in oyster, soybean, and rat liver and kidney. Arch. Environ. Contam. Toxicol. 3, 319–329.

Champ, M.A., 1986. Organotin symposium: introduction and review. In: Maton, G.L. (Ed.), Proceedings Oceans 86 Conference, Washington DC, September 23-25, 1986, vol. 4. Organotin Symposium. Available from Marine Technology Society, 2000 Florida Avenue NW, Washington, DC, pp. 1093–1100.

Chan, H.M., 1988a. A survey of trace metals in *Perna viridis* (L.) (Bivalvia: mytilacea) from the coastal waters of Hong Kong. Asian Mar. Biol. 5, 89–102.

Chan, H.M., 1988b. Accumulation and tolerance to cadmium, copper, lead and zinc by the green mussel *Perna viridis*. Mar. Ecol. Prog. Ser. 48, 295–303.

Chan, K.M., Yeung, K.M.L., Cheung, K.C., Wong, M.H., Qiu, J.W., 2008. Seasonal changes in imposex and tissue burden of butyltin compounds in *Thais clavigera* populations along the coastal area of Mirs Bay, China. Mar. Pollut. Bull. 57, 645–651.

Chau, Y.K., Maguire, R.J., Wong, P.T.S., Glen, B.A., Bengert, G.A., Tkacz, R.J., 1984. Occurrence of methyltin and butyltin species in environmental samples in Ontario. Natl. Water Inst. Rep., 8401, pp. 1–25.

Cheevaparanapivat, V., Menasveta, P., 1979. Total and organic mercury in marine fish of the upper Gulf of Thailand. Bull. Environ. Contam. Toxicol. 23, 291–299.

Chelomin, V.P., Belcheva, N.N., 1992. The effect of heavy metals on processes of lipid peroxidation in microsomal membranes from the hepatopancreas of the bivalve mollusc *Mizuhopecten yessoensis*. Comp. Biochem. Physiol. 103C, 419–422.

Cheung, Y.H., Wong, M.H., 1992. Trace metal contents of the Pacific oyster (*Crassostrea gigas*) purchased from markets in Hong Kong. Environ. Manag. 16, 753–761.

Cheung, M.S., Fok, E.M.W., Ng, T.Y.T., Yen, Y.F., Wang, W.X., 2006. Subcellular cadmium distribution, accumulation, and toxicity in a predatory gastropod, *Thais clavigera*, fed different prey. Environ. Toxicol. Chem. 25, 174–181.

Chipman, W., Thommeret, J., 1970. Manganese content and the occurrence of fallout ^{54}Mn in some marine benthos of the Mediterranean. Bull. Inst. Oceanogr. 69(1402), 1–15.

Chou, C.L., Uthe, J.F., Zook, E.G., 1978. Polarographic studies on the nature of cadmium in scallop, oyster, and lobster. J. Fish. Res. Board Can. 35, 409–413.

Chow, T.J., Snyder, H.G., Snyder, C.B., 1976. Mussels (*Mytilus* sp.) as an indicator of lead pollution. Sci. Total Environ. 6, 55–63.

Chu, K.H., Cheung, W.M., Lau, S.K., 1990. Trace metals in bivalves and sediments from Tolo Harbour, Hong Kong. Environ. Int. 16, 31–36.

Chuang, C.Y., Wang, W.X., 2006. Co-transport of metal complexes by the green mussel *Perna viridis*. Environ. Sci. Technol. 40, 4523–4527.

Claisse, D., Alzieu, C., 1993. Copper contamination as a result of antifouling paint regulations? Mar. Pollut. Bull. 26, 395–397.

Clarke, G.L., 1947. Poisoning and recovery in barnacles and mussels. Biol. Bull. 92, 73–91.

Coelho, J.P., Pimenta, J., Gomes, R., Barroso, C.M., Pereira, M.E., Pardal, M.A., et al., 2006a. Can *Nassarius reticulatus* be used as a bioindicator for Hg contamination? Results for a longitudinal study of the Portuguese coastline. Mar. Pollut. Bull. 52, 674–680.

Coelho, M.R., Langston, W.J., Bebianno, M.J., 2006b. Effect of TBT on *Ruditapes decussatus* juveniles. Chemosphere 63, 1499–1505.

Coglianse, M.P., Martin, M., 1981. Individual and interactive effects of environmental stress on the embryonic development of the Pacific oyster, *Crassostrea gigas*. I. The toxicity of copper and silver. Mar. Environ. Res. 5, 13–27.

Colaco, A., Bustamante, P., Fouquet, Y., Sarradin, P.M., Serrao-Santos, R., 2006. Bioaccumulation of Hg, Cu, and Zn in the Azores triple junction hydrothermal vent fields food web. Chemosphere 65, 2260–2267.

Coleman, N., 1980. The effect of emersion on cadmium accumulation by *Mytilus edulis*. Mar. Pollut. Bull. 11, 359–362.

Colwell, R.R., Nelson Jr., J.D., 1975. Metabolism of mercury compounds in microorganisms. US Environ. Protect. Agen. Rep., 600/3-75-007, pp. 1–84.

Colwell, R.R., Sayler, G.S., Nelson Jr., J.D., Justice, A., 1976. Microbial utilization of mercury in the aquatic environment. In: Nriagu, J.O. (Ed.), Environmental Biogeochemistry, Volume 2. Metals Transfer and Ecological Mass Balances. Ann Arbor Science Publishers, Ann Arbor, MI, pp. 437–487.

Connell, D.B., Sanders, J.G., Riedel, G.F., Abbe, G.R., 1991. Pathways of silver uptake and trophic transfer in estuarine organisms. Environ. Sci. Technol. 25, 921–924.

Connor, P.M., 1972. Acute toxicity of heavy metals to some marine larvae. Mar. Pollut. Bull. 3, 190–192.

Conti, M.E., Iacobucci, M., Cecchetti, G., Alimonti, A., 2008. Influence of weight on the content of trace metals in tissues of *Mytilus galloprovincialis* (Lamarck, 1819): a forecast model. Environ. Monit. Assess. 141, 27–34.

Cooke, M., Nickless, G., Lawn, R.E., Roberts, D.J., 1979. Biological availability of sediment-bound cadmium to the edible cockle, *Cerastoderma edule*. Bull. Environ. Contam. Toxicol. 23, 381–386.

Coombs, T.L., 1972. The distribution of zinc in the oyster *Ostrea edulis* and its relation to enzymic activity and to other metals. Mar. Biol. 12, 170–178.

Cossa, D., Bourget, E., Piuze, J., 1979. Sexual maturation as a source of variation in the relationship between cadmium concentration and body weight of *Mytilus edulis* L. Mar. Pollut. Bull. 10, 174–176.

Cosson, R.P., Thiebaut, E., Company, R., Castrec-Rouelle, M., Colaco, A., Martins, I., et al., 2008. Spatial variation of metal bioaccumulation in the hydrothermal vent mussel *Bathymodiolus azoricus*. Mar. Environ. Res. 65, 405–415.

Costa, M.R.M., da-Fonseca, M.I.C., 1967. Teor de arsenico em mariscos. Rev. Portuguesa Farm. 17(1), 1–19.

Craig, S., 1967. Toxic ions in bivalves. J. Am. Osteopath. Assoc. 66, 1000–1002.

Cravo, A., Foster, P., Almeida, C., Bebianno, M.J., Company, R., 2008. Metal concentrations in the shell of *Bathymodiolus azoricua* from contrasting hydrothermal vent fields on the mid-Atlantic ridge. Mar. Environ. Res. 65, 338–348.

Crecelius, E.A., 1993. The concentration of silver in mussels and oysters from NOAA National Status and Trends Mussel Watch Sites. In: Andren, A.W., Bober, T.W., Crecelius, E.A., Kramer, J.R., Luoma, S.N., Rodgers, J.H., Sodergren, A. (Eds.), Proceedings of the First International Conference on Transport, Fate, and Effects of Silver in the Environment. Univ. Wisconsin Sea Grant Inst, Madison, WI, pp. 65–66.

Crowe, A.J., 1987. The chemotherapeutic properties of tin compounds. Drugs Future 12, 255–275.

Crowley, M., Mitchell, P.I., O'Grady, J., Vires, J., Sanchez-Cabeza, J.A., Vidal-Quadras, A., et al., 1990. Radiocaesium and plutonium concentrations in *Mytilus edulis* (L.) and potential dose implications for Irish critical groups. Ocean Shoreline Manag. 13, 149–161.

Culkin, F., Riley, J.P., 1958. The occurrence of gallium in marine organisms. J. Mar. Biol. Assoc. UK 37, 607–615.

Cumont, G., Gilles, G., Bernard, F., Briand, M.B., Stephan, G., Ramonda, G., et al., 1975. Bilan de la contamination des poissons de mer par le mercure a l'occasion d'un controle portant sur 3 annees. Ann. Hyg. L. Fr.—Med. et Nut. 11(1), 17–25.

Cunha, L., Amaral, A., Medeiros, V., Martins, G.M., Wallenstein, F.F.M.M., Couto, R.P., et al., 2008. Bioavailable metals and cellular effects in the digestive gland of marine limpets living close to shallow water hydrothermal vents. Chemosphere 71, 1356–1362.

Cunningham, P.A., Tripp, M.R., 1973. Accumulation and depuration of mercury in the American oyster *Crassostrea virginica*. Mar. Biol. 20, 14–19.

Cunningham, P.A., Tripp, M.R., 1975a. Factors affecting the accumulation and removal of mercury from tissues of the American oyster *Crassostrea virginica*. Mar. Biol. 31, 311–319.

Cunningham, P.A., Tripp, M.R., 1975b. Accumulation, tissue distribution and elimination of ^{203}HgCl$_2$ and CH$_3$^{203}HgCl in the tissues of the American oyster *Crassostrea virginica*. Mar. Biol. 31, 324–334.

Darracott, A., Watling, H., 1975. The use of molluscs to monitor cadmium levels in estuaries and coastal marine environments. Trans. R. Soc. S. Afr. 41(4), 325–338.

Daskalakis, K.D., 1995. Silver in oyster soft tissue: relations to site selection and sampling size. In: Andren, A.W., Bober, T.W. (Organizers), Transport, Fate and Effects of Silver in the Environment. 3rd International Conference, Washington, DC, August 6-9, 1995. Univ. Wisconsin Sea Grant Inst, Madison, WI.

Davies, M.S., 1992. Heavy metals in seawater: effects on limpet pedal mucus production. Water Res. 26, 1691–1693.

Davies, I.M., McKie, J.C., Paul, J.D., 1986. Accumulation of tin and tributyltin from anti-fouling paint by cultivated scallops (*Pecten maximum*) and Pacific oysters (*Crassostrea gigas*). Aquaculture 55, 103–114.

Davies, I.M., Bailey, S.K., Moore, D.C., 1987. Tributyltin in Scottish sea lochs, as indicated by degree of imposex in the dogwhelk, *Nucella lapillus* (L.). Mar. Pollut. Bull. 18, 404–407.

Davies, I.M., Harding, M.J.C., Bailey, S.K., Shanks, A.M., Lange, R., 1997. Sublethal effects of tributyltin oxide on the dogwhelk *Nucella lapillus*. Mar. Ecol. Prog. Ser. 158, 191–204.

Dean, R.B., 1972. The case against mercury. Nat. Tech. Infor. Serv. Springfield, VA, Doc. PB-213-6921–11.

Decleir, W., Lemaire, J., Richard, A., 1970. Determination of copper in embryos and very young specimens of *Sepia officinalis*. Mar. Biol. 5, 256–258.

Decleir, W., Vlaeminck, A., Geladi, P., Van Grieken, R., 1978. Determination of protein-bound copper and zinc in some organs of the cuttlefish, *Sepia officinalis*. L. Comp. Biochem. Physiol. 60B, 347–350.

De Gieter, M., Leermakers, M., Van Ryssen, R., Noyen, J., Goeyens, L., Baeyens, W., 2002. Total and toxic arsenic levels in North Sea fish. Arch. Environ. Contam. Toxicol. 43, 406–417.

Dehlinger, P., Fitzgerald, W.F., Feng, S.Y., Paskausky, D.F., Garvine, M.W., Bohlen, W.F., 1975. Determination of budgets of heavy metal wastes in Long Island Sound. Annual Rept. Parts I and II. Univ. Connecticut, Marine Sciences Inst, Groton, CT.

Delhaye, W., Cornet, D., 1975. Contribution to the study of the effect of copper on *Mytilus edulis* during reproductive period. Comp. Biochem. Physiol. 50A, 511–513.

Denton, G.R.W., Concepcion, L.P., Wood, H.R., Morrison, R.J., 2006. Trace metals in marine organisms from four harbours in Guam. Mar. Pollut. Bull. 52, 1784–1804.

Deudero, S., Box, A., March, D., Valencia, J.M., Grau, A.M., Tintora, J., et al., 2007. Temporal trends of metals in benthic invertebrate species from the Balearic Islands, Western Mediterranean. Mar. Pollut. Bull. 54, 1545–1558.

Devi, V.U., 1996. Bioaccumulation and metabolic effects of cadmium on marine fouling dressinid bivalve, *Mutilopsis sallei* (Recluz). Arch. Environ. Contam. Toxicol. 31, 47–53.

De Wolf, P., 1975. Mercury content of mussels from West European coasts. Mar. Pollut. Bull. 6, 61–63.

De Wolf, H., Rashid, R., 2008. Heavy metal accumulation in *Littoraria scabra* along polluted and pristine mangrove areas of Tanzania. Environ. Pollut. 152, 636–643.

Diamond, J.M., Mackler, D.G., Collins, M., Gruber, D., 1990. Deviation of a freshwater silver criteria for the New River, Virginia, using representative species. Environ. Toxicol. Chem. 9, 1425–1434.

Dick, J., Pugsley, L.I., 1950. The arsenic, lead, tin, copper, and iron content of canned clams, oysters, crabs, lobsters, and shrimps. Can. J. Res. 28F, 199–201.

Dietz, R., Riget, F., Johansen, P., 1996. Lead, cadmium, mercury and selenium in Greenland marine animals. Sci. Total Environ. 186, 67–93.

Di Giulio, R.T., Scanlon, P.F., 1985. Heavy metals in aquatic plants, clams, and sediments from the Chesapeake Bay U.S.A. Implications for waterfowl. Sci. Total Environ. 41, 259–274.

Dillon, T.M., 1977. Mercury and the estuarine marsh clam, *Rangia cuneata* Gray. I. Toxicity. Arch. Environ. Contam. Toxicol. 6, 249–255.

Dillon, T.M., Neff, J.M., 1978. Mercury and the estuarine marsh clam *Rangia cuneata* Gray. II. Uptake, tissue distribution and depuration. Mar. Environ. Res. 1, 67–77.

Dimitriadis, V.K., Domouhtsidou, G.P., Raftopoulou, E., 2003. Localization of Hg and Pb in the palps, digestive gland and the gills in *Mytilus galloprovincialis* (L.) using autometallography and X-ray microanalysis. Environ. Pollut. 125, 345–353.

D'Itri, F.M., 1972. Mercury in the aquatic ecosystem. Tech. Rep. Inst. Water Res. Michigan State Univ., Lansing, MI 23, 1–101.

Dixon, D.R., Prosser, H., 1986. An investigation of the genotoxic effects of an organotin antifouling compound (bis (tributyltin) oxide) in the chromosomes of the edible mussel, *Mytilus edulis*. Aquat. Toxicol. 8, 185–195.

Doi, R., Ui, J., 1975. The distribution of mercury in fish and its form of occurrence. In: Krenkel, P.A. (Ed.), Heavy Metals in the Aquatic Environment. Pergamon, New York, pp. 197–221.

Dorneles, P.R., Lailson-Brito, J., dos Santos, R.A., da Costa, P.A.S., Malm, O., Azevedo, A.F., et al., 2007. Cephalopods and cetaceans as indicators of offshore bioavailability of cadmium off Central South Brazil Bight. Environ. Pollut. 148, 352–359.

Downes, K.M., 1957. An investigation of manganese and zinc in the oyster *Crassostrea virginica* (Gmelin). Ph.D. thesis, Univ. Maryland, College Park, pp. 1–54.

Drifmeyer, N.E., 1974. Zn and Cu levels in the eastern oyster, *Crassostrea virginica*, from the lower James River. J. Wash. Acad. Sci. 64 (4), 292–294.

D'Silva, C., Kureishy, T.W., 1978. Experimental studies on the accumulation of copper and zinc in the green mussel. Mar. Pollut. Bull. 9, 187–190.

D'Silva, C., Quasim, S.Z., 1979. Bioaccumulation and elimination of copper in the rock oyster *Crassostrea cucullata*. Mar. Biol. 52, 343–346.

Duchemin, M.B., Auffret, M., Wessel, N., Fortier, M., Morin, Y., Pellerin, J., et al., 2008. Multiple experimental approaches of immunotoxic effects of mercury chloride in the blue mussel, *Mytilus edulis*, through in vivo, in tubo and in vitro exposures. Environ. Pollut. 153, 416–423.

Duke, T.W., 1967. Possible routes of zinc-65 from an experimental estuarine environment to man. J. Water Pollut. Control Fed. 39, 536–542.

Earnshaw, M.J., Wilson, S., Akberali, H.B., Butler, R.D., Marriott, K.R.M., 1986. The action of heavy metals on the gametes of the marine mussel, *Mytilus edulis* (L.)—III. The effect of applied copper and zinc on sperm motility in relation to ultrastructural damage and intracellular metal localisation. Mar. Environ. Res. 20, 261–278.

Ecological Analysts, Inc., 1981. In: The Sources, Chemistry, Fate, and Effects of Chromium in Aquatic Environments. Available from American Petroleum Institute, 2101 L Street NW, Washington, DC, 207 pp.

Eganhouse, R.P., Young, D.R., 1978. *In situ* uptake of mercury by the intertidal mussel, *Mytilus californianus*. Mar. Pollut. Bull. 9, 214–217.

Eisler, R., 1977a. Acute toxicities of selected heavy metals to the softshell clam, *Mya arenaria*. Bull. Environ. Contam. Toxicol. 17, 137–145.

Eisler, R., 1977b. Toxicity evaluation of a complex metal mixture to the softshell clam, *Mya arenaria*. Mar. Biol. 43, 265–276.

Eisler, R., 1978. Mercury contamination standards for marine environments. In: Thorp, J.H., Gibbons, J.W. (Eds.), Energy and Environmental Stress in Aquatic Systems. US Dept. Energy Sympos. Ser. 48. Available as CONF-771114 from NTIS, US Dept. Commerce, Springfield, VA, pp. 241–272.

Eisler, R., 1979. Copper accumulations in coastal and marine environments. In: Nriagu, J.O. (Ed.), Copper in the Environment, Part I: Ecological Cycling. Wiley, New York, pp. 383–449.

Eisler, R., 1980. Accumulation of zinc by marine biota. In: Nriagu, J.O. (Ed.), Zinc in the Environment, Part 2: Health Effects. Wiley, New York, pp. 259–351.

Eisler, R., 1981. Trace Metal Concentrations in Marine Organisms. Pergamon, New York, 687 pp.

Eisler, R., 1995a. Electroplating wastes in marine environments: a case history at Quonset Point, Rhode Island. In: Hoffman, D.J., Rattner, B.A., Burton Jr., G.A., Cairns Jr., A.J. (Eds.), Handbook of Ecotoxicology. Lewis Publishers, Boca Raton, FL, pp. 539–548.

Eisler, R., 1995b. Ecological and toxicological aspects of the partial meltdown of the Chernobyl nuclear plant reactor. In: Hoffman, D.J., Rattner, B.A., Burton Jr., G.A., Cairns Jr., A.J. (Eds.), Handbook of Ecotoxicology. Lewis Publishers, Boca Raton, FL, pp. 549–564.

Eisler, R., 2000. Mercury. In: Handbook of Chemical Risk Assessment: Health Hazards to Humans, Plants, and Animals. Vol. 1. Metals. Lewis Publishers, Boca Raton, FL, pp. 313–409.

Eisler, R., 2000a. Chromium. In: Handbook of Chemical Risk Assessment: Health Hazards to Humans, Plants, and Animals. Vol. 1. Metals. Lewis Publishers, Boca Raton, FL, pp. 45–92.

Eisler, R., 2000b. Copper. In: Handbook of Chemical risk Assessment: Health Hazards to Humans, Plants, and Animals. Vol. 1: Metals. Lewis Publishers, Boca Raton, FL, pp. 93–200.

Eisler, R., 2000c. Lead. In: Handbook of Chemical Risk Assessment: Health Hazards to Humans, Plants, and Animals. Vol. 1: Metals. Lewis Publishers, Boca Raton, FL, pp. 201–311.

Eisler, R., 2000d. Silver. In: Handbook of Chemical Risk Assessment: Health Hazards to Humans, Plants, and Animals. Vol. 1: Metals. Lewis Publishers, Boca Raton, FL, pp. 499–550.

Eisler, R., 2000e. Tin. In: Handbook of Chemical Risk Assessment: Health Hazards to Humans, Plants, and Animals. Vol. 1: Metals. Lewis Publishers, Boca Raton, FL, pp. 551–603.

Eisler, R., 2000f. Zinc. In: Handbook of Chemical Risk Assessment: Health Hazards to Humans, Plants, and Animals. Vol. 1: Metals. Lewis Publishers, Boca Raton, FL, pp. 605–714.

Eisler, R., 2000g. Cadmium. In: Handbook of Chemical Risk Assessment: Health Hazards to Humans, Plants, and Animals. Vol. 1: Metals. Lewis Publishers, Boca Raton, FL, pp. 1–43.

Eisler, R., 2003. The Chernobyl nuclear power plant reactor accident: ecotoxicological update. In: Hoffman, D.J., Rattner, B.A., Burton Jr., G.A., Cairns Jr., A.J. (Eds.), Handbook of Ecotoxicology, second ed. Lewis Publishers, Boca Raton, FL, pp. 703–736.

Eisler, R., 2006. Mercury Hazards to Living Organisms. CRC Press, Boca Raton, FL, 312 pp.

Eisler, R., Hennekey, R.J., 1977. Acute toxicities of Cd^{2+}, Cr^{+6}, Hg^{2+}, Ni^{2+}, and Zn^{2+} to estuarine macrofauna. Arch. Environ. Contam. Toxicol. 6, 315–323.

Eisler, R., Weinstein, M.P., 1967. Changes in metal composition of the quahaug clam, *Mercenaria mercenaria*, after exposure to insecticides. Chesapeake Sci. 8, 253–258.

Eisler, R., Zaroogian, G.E., Hennekey, R.J., 1972. Cadmium uptake by marine organisms. J. Fish. Res. Board Can. 29, 1367–1369.

Eisler, R., Barry, M.M., Lapan Jr., R.L., Telek, G., Davey, E.W., Soper, A.E., 1978. Metal survey of the marine clam *Pitar morrhuana* collected near a Rhode Island (USA) electroplating plant. Mar. Biol. 45, 311–317.

Espana, M.S.A., Rodriguez, E.M.R., Romero, C.D., 2007. Application of chemometric studies to metal concentrations in molluscs from the Strait of Magellan (Chile). Arch. Environ. Contam. Toxicol. 52, 519–524.

Establier, R., 1977. Estudio de la contaminacion marine por metales pesados y sus effectos biologicos. Inf. Tech. Inst. Invest. Pesq. 47, 1–36.

Ettajani, H., Amiard-Triquet, C., Amiard, J.C., 1992. Etude experimentale du transfert de deux elements traces (Ag, Cu) dans une chain trophique marine: eau—particles (sediment, natural, microalgue)—mollusques filtreurs (*Crassostrea gigas* Thunberg). Water Air Soil Pollut. 65, 215–236.

Eustace, I.J., 1974. Zinc, cadmium, copper and manganese in species of finfish and shellfish caught in the Derwent Estuary, Tasmania. Aust. J. Mar. Freshw. Res. 25, 209–220.

Evans, D.W., Laughlin Jr., R.B., 1984. Accumulation of bis(tributylin)oxide by the mud crab, *Rithropanopeus harrisii*. Chemosphere 13, 213–219.

Evans, S.M., Dawson, M., Frid, C.L.J., Gill, M.E., Pattisina, L.A., Porter, J., 1995. Domestic waste and TBT pollution in coastal areas of Ambon Island (eastern Indonesia). Mar. Pollut. Bull. 30, 109–115.

Eversole, A.G., 1978. Marking clams with rubidium. Proc. Nat. Shellfish. Assoc. 68, 78.

Fagerstrom, T., Jernelov, A., 1972. Some aspects of the quantitative ecology of mercury. Water Res. 6, 1193–1202.

Fagioli, F., Locatelli, C., Landi, S., 1994. Heavy metals in the Goro Bay: sea water, sediments and mussels. Ann. Chim. 84, 129–140.

Fang, S.C., 1973. Uptake and biotransformation of phenylmercuric acetate by aquatic organisms. Arch. Environ. Contam. Toxicol. 1, 18–26.

Fang, J., Wang, K.X., Tang, J.L., Wang, Y.M., Ren, S.J., Wu, H.Y., et al., 2004. Copper, lead, zinc, cadmium, mercury, and arsenic in marine products of commerce from Zhejiang coastal area, China, May, 1998. Bull. Environ. Contam. toxicol. 73, 583–590.

Fattorini, D., Notti, A., Di Mento, R., Cicero, A.M., Gabellini, M., Russo, A., et al., 2008. Seasonal, spatial and inter-annual variations of trace metals in mussels from the Adriatic Sea: a regional gradient for arsenic and implications for monitoring the impact of off-shore activities. Chemosphere 72, 1524–1533.

Ferrell, R.E., Carville, T.E., Martinez, J.D., 1973. Trace metals in oyster shells. Environ. Lett. 4(4), 311–316.

Fisher, N.S., Teyssie, J.L., 1986. Influence of food composition on the biokinetics and tissue distribution of zinc and americium in mussels. Mar. Ecol. Prog. Ser. 28, 37–43.

Fisher, N.S., Wang, W., 1998. Trophic transfer of silver to marine herbivores: a review of recent studies. Environ. Toxicol. Chem. 17, 562–571.

Fisher, N.S., Wang, W., Reinfelder, J.R., Luoma, S.N., 1994. Bioaccumulation of silver in marine bivalves. In: Andren, A.W., Bober, T.W., Kramer, J.R., Sodergren, A., Crecelius, E.A., Luoma, S.N., Rodgers, J.H. (Organizers), Transport, Fate and Effects of Silver in the Environment. 3rd International Conference. Proceedings of the 2nd International Conference. Univ. Wisconsin Sea Grant Inst, Madison, WI, pp. 139–140.

Flegal, A.R., Martin, J.H., 1977. Contamination of biological samples by ingested sediment. Mar. Pollut. Bull. 8, 90–92.

Folsom, T.R., Young, D.R., 1965. Silver-110m and cobalt-60 in oceanic and coastal organisms. Nature 206 (4986), 803–806.

Fowler, S.W., Benayoun, G., 1976a. Influence of environmental factors on selenium flux in two marine invertebrates. Mar. Biol. 37, 59–68.

Fowler, S.W., Benayoun, G., 1976b. Accumulation and distribution of selenium in mussel and shrimp tissues. Bull. Environ. Contam. Toxicol. 16, 339–346.

Fowler, S.W., Oregioni, B., 1976. Trace metals in mussels from the N.W. Mediterranean. Mar. Pollut. Bull. 7, 26–29.

Fowler, B.A., Wolfe, D.A., Hettler, W.F., 1975. Mercury and iron uptake by cytosomes in mantle epithelial cells of quahog clams (*Mercenaria mercenaria*) exposed to mercury. J. Fish. Res. Board Can. 32, 1767–1775.

Fowler, S.W., Heyraud, M., La Rosa, J., 1978. Factors affecting methyl and inorganic mercury dynamics in mussels and shrimp. Mar. Biol. 46, 267–276.

Franson, J.C., Koehl, P.S., Derksen, D.V., Ruth, T.C., Bunk, C.M., Moore, J.F., 1995. Heavy metals in seaducks and mussels from Misty Fjords National Monument in southeast Alaska. Environ. Monit. Assess. 36, 149–167.

Frazier, J.M., 1975. The dynamics of metals in the American oyster, *Crassostrea virginica*. I. Seasonal effects. Chesapeake Sci. 16, 162–171.

Frazier, J.M., 1976. The dynamics of metals in the American oyster, *Crassostrea virginica*. II. Environmental effects. Chesapeake Sci. 17, 188–197.

Freidrich, A.R., Felice, F.P., 1976. Uptake and accumulation of the nickel ion by *Mytilus edulis*. Bull. Environ. Contam. Toxicol. 16, 750–755.

Frias-Espericueta, M.G., Osuna-Lopez, J.I., Voltolina, D., Lopez-Lopez, G., Izaguirre-Fierro, G., Muy-Rangel, M.D., 2008. The metal content of bivalve molluscs of a coastal lagoon of NW Mexico. Bull. Environ. Contam. Toxicol. 80, 90–92.

Friberg, L., Vostal, J. (Eds.), 1972. Mercury in the Environment. CRC Press, Cleveland, OH, 215 pp.

Fujiki, M., 1963. Studies on the course that the causative agent of Minamata Disease was formed, especially on the accumulation of the mercury compound in the fish and shellfish of Minamata Bay. J. Kumamoto Med. Soc. 37, 494–521.

Fukai, R., 1965. Analysis of trace amounts of chromium in marine organisms by the isotope dilution of Cr-51. In: Radiochemical Methods of Analysis. Int. Atom. Ener. Agen, Vienna, Austria, pp. 335–351.

Fukai, R., Meinke, W.W., 1959. Some activation analyses of six trace elements in marine biological ashes. Nature 184, 815–816.

Fukai, R., Meinke, W.W., 1962. Activation analyses of vanadium, arsenic, molybdenum, tungsten, rhenium, and gold in marine organisms. Limnol. Oceanogr. 7, 186–200.

Fukai, R., Oregioni, B., Vas, D., 1978. Interlaboratory comparability of measurements of trace elements in marine organisms: results of intercalibration exercise on oyster homogenate. Oceanol. Acta 1, 391–396.

Gabbianelli, R., Moretti, M., Carpene, E., Falcioni, G., 2006. Effect of different organotins on DNA of mollusk (*Scapharca inaequivalvis*) erythrocytes assessed by the comet assay. Sci. Total Environ. 367, 163–169.

Gainey Jr., L.F., Kenyon, J.R., 1990. The effects of reserpine on copper induced cardiac inhibition in *Mytilus edulis*. Comp. Biochem. Physiol. 95C, 177–179.

Galtsoff, P.S., 1942. Accumulation of manganese and the sexual cycle in *Ostrea virginica*. Physiol. Zool. 15, 210–215.

Galtsoff, P., 1953. Accumulation of manganese, iron, copper, and zinc in the body of the American oyster, *Crassostrea* (*Ostrea*) *virginica*. Anatom. Rec. 117, 601–602.

Garaventa, F., Faimali, M., Terlizzi, A., 2006. Imposex in pre-pollution times. Is TBT to blame? Mar. Pollut. Bull. 52, 701–702.

Garaventa, F., Centanni, E., Pellizzato, F., Faimali, M., Terlizzi, A., Pavoni, B., 2007. Imposex and accumulation of organotin compounds in populations of *Hexaplex trunculus* (Gastropoda, Muricidae) from the Lagoon of Venice (Italy) and Istrian Coast (Croatia). Mar. Pollut. Bull. 54, 615–622.

Garcia-Luque, E., DelValls, A.T., Forja, J.M., Gomez-Parra, A., 2007. Biological adverse effects on bivalves associated with trace metals under estuarine conditions. Environ. Monit. Assess. 131, 27–35.

Gavis, J., Ferguson, J.F., 1972. The cycling of mercury through the environment. Water Res. 6, 989–1000.

Geffard, A., Geffard, O., Amiard, J.C., His, E., Amiard-Triquet, C., 2007. Bioaccumulation of metals in sediment elutriates and their effects on growth, condition index, and metallothionein contents in oyster larvae. Arch. Environ. Contam. Toxicol. 53, 57–65.

Genest, P.E., 1979. A study of some heavy metals. In: Mercenaria mercenaria, M.S. thesis, Southeastern Massachusetts Univ, 56 pp.

George, S.G., Coombs, T.L., 1975. A comparison of trace metal and metalloenzyme profiles in different molluscs and during development of the oyster. In: Barnes, H. (Ed.), Proceedings of the Nineth European Marine Biology Symposium, pp. 433–449.

George, S.G., Coombs, T.L., 1977. The effects of chelating agents on the uptake and accumulation of cadmium by *Mytilus edulis*. Mar. Biol. 39, 261–268.

George, S.G., Pirie, B.J.S., Coombs, T.L., 1976. The kinetics of accumulation and excretion of ferric hydroxide in *Mytilus edulis* (L.) and its distribution in the tissues. J. Exp. Mar. Biol. Ecol. 23, 71–84.

George, S.G., Pirie, B.J.S., Cheyne, A.R., Coombs, T.L., Grant, P.T., 1978. Detoxication of metals by marine bivalves: an ultrastructural study of the compartmentation of copper and zinc in the oyster *Ostrea edulis*. Mar. Biol. 45, 147–156.

George, S.G., Pirie, B.J.S., Coombs, T.L., 1980. Isolation and elemental analysis of metal-rich granules from the kidney of the scallop, *Pecten maximus* (L.). J. Exp. Mar. Biol. Ecol. 42, 143–156.

Ghiretti, F., Salvato, B., Carlucci, S., DePieri, R., 1972. Manganese in *Pinna nobilis*. Experientia 28, 232–233.

Ghiretti-Magaldi, A., Giuditta, A., Ghiretti, F., 1958. Pathways of terminal respiration in marine invertebrates. I. The respiratory system in cephalopods. J. Cell. Comp. Physiol. 52, 389–429.

Gibbs, P.E., Bryan, G.W., 1986. Reproductive failure in populations of the dog-whelk, *Nucella lapillus*, caused by imposex induced from tributyltin from antifouling paints. J. Mar. Biol. Assoc. UK 66, 767–777.

Gibbs, P.E., Bryan, G.W., Pascoe, P.L., Burt, G.R., 1987. The use of the dog-whelk, *Nucella lapillus*, as an indicator of tributyltin (TBT) contamination. J. Mar. Biol. Assoc. UK 67, 507–523.

Giordno, R., Arata, P., Ciaralli, L., Rinaldi, S., Giani, M., Cicero, A.M., et al., 1991. Heavy metals in mussels and fish from Italian coastal waters. Mar. Pollut. Bull. 22, 10–14.

Glickstein, N., 1978. Acute toxicity of mercury and selenium to *Crassostrea gigas* and *Cancer magister* larvae. Mar. Biol. 49, 113–117.

Glynn, D., Tyrrell, L., McHugh, B., Rowe, A., Monaghan, E., Costello, J., et al., 2003. Trace Metal and Chlorinated Hydrocarbon Concentrations in Shellfish from Irish Waters, 2001. Marine Institute, Marine Environment and Food Safety Services, Abbotstown, Dublin, pp. 1–15.

Goldberg, E.D., Bowen, V.T., Farrington, J.W., Harvey, G., Martin, J.H., Parker, P.L., et al., 1978. The mussel watch. Environ. Conserv. 5, 101–125.

Gomez-Ariza, J.L., Santos, M.M., Morales, E., Giraldez, I., Sanchez-Rodas, D., Vieira, N., et al., 2006. Organotin contamination in the Atlantic Ocean off the Iberian Peninsula in relation to shipping. Chemosphere 64, 1100–1108.

Gonzalez, P.M., Abele, D., Puntarulo, S., 2008. Iron and radical content in *Mya arenaria*. Possible sources of NO generation. Aquat. Toxicol. 89, 122–128.

Gorbushin, A.M., 1997. Field evidence of trematode-induced gigantism in *Hydrobia* spp. (Gastropoda: Prosobranchia). J. Mar. Biol. Assoc. UK 77, 785–800.

Gorinstein, S., Moncheva, S., Toledo, F., Avila, P.A., Trakhtenberg, S., Gorinstein, A., et al., 2006. Relationship between seawater pollution and qualitative changes in the extracted proteins from mussels *Mytilus galloprovincialis*. Sci. Total Environ. 364, 252–259.

Gorski, J., Nugegoda, D., 2006. Toxicity of trace metals to juvenile abalone, *Haliotis rubra* following short-term exposure. Bull. Environ. Contam. Toxicol. 77, 732–740.

Gould, E., MacInnes, J.R., 1977. Short term effects of two silver salts on tissue respiration and enzyme activity in the cunner (*Tautogolabrus adspersus*). Bull. Environ. Contam. Toxicol. 18, 401–408.

Graham, D.L., 1972. Trace metal levels in intertidal mollusks of California. Veliger 14, 365–372.
Greig, R.A., 1975. Comparison of atomic absorption and neutron activation analyses for determination of silver, chromium and zinc in various marine organisms. Anal. Chem. 47, 1682–1684.
Greig, R.A., 1979. Trace metal uptake by three species of mollusks. Bull. Environ. Contam. Toxicol. 22, 643–647.
Greig, R.A., Sennefelder, G., 1985. Metals and PCB concentrations in mussels from Long Island Sound. Bull. Environ. Contam. Toxicol. 35, 331–334.
Greig, R.A., Wenzloff, D.R., 1978. Metal accumulation and depuration by the American oyster, *Crassostrea virginica*. Bull. Environ. Contam. Toxicol. 20, 499–504.
Greig, R.A., Nelson, B.A., Nelson, D.A., 1975. Trace metal content in the American oyster. Mar. Pollut. Bull. 6, 72–73.
Greig, R.A., Wenzloff, D.R., Pearce, J.B., 1976. Distribution and abundance of heavy metals in finfish, invertebrates and sediments collected at a deepwater disposal site. Mar. Pollut. Bull. 7, 185–187.
Greig, R.A., Wenzloff, D.R., Adams, A., Nelson, B., Shelpuk, C., 1977. Trace metals in organisms from ocean disposal sites of the middle eastern United States. Arch. Environ. Contam. Toxicol. 6, 395–409.
Greig, R.A., Wenzloff, D.R., McKenzie Jr., C.L., Merrill, A.S., Zdanowicz, V.S., 1978. Trace metals in sea scallops, *Placopecten magellanicus* from eastern United States. Bull. Environ. Contam. Toxicol. 19, 326–334.
Guary, J.C., Frazier, A., 1977. Etude comparee des teneurs en plutonium chez divers mollusques de quelques sites littoraux fransais. Mar. Biol. 41, 263–267.
Guary, J.C., Fowler, S.W., 1978. Uptake from water and tissue distribution of neptunium-237 in crabs, shrimp, and mussels. Mar. Pollut. Bull. 9, 331–334.
Guerin, T., Sirot, V., Volatier, J.L., Leblanc, J.C., 2007. Organotin levels in seafood and its implications for health risk in high-seafood consumers. Sci. Total Environ. 388, 66–77.
Hall, R.A., Zook, E.G., Meaburn, G.M., 1978. National Marine Fisheries Service survey of trace elements in the fishery resources. In: US Dept. Commerce NOAA Tech. Rept., NMFS SSRF-721. pp. 1–313.
Hall Jr., L.W., Lenkevich, M.J., Hall, W.S., Pinkney, A.E., Bushong, S.J., 1987. Evaluation of butyltin compounds in Maryland waters of Chesapeake Bay. Mar. Pollut. Bull. 18, 78–83.
Hall Jr., L.W., Bushong, S.J., Hall, W.S., Johnson, W.E., 1988. Acute and chronic effects of tributyltin on a Chesapeake Bay copepod. Environ. Toxicol. Chem. 7, 41–46.
Hamanaka, T., Kato, H., Tsujita, T., 1977. Cadmium and zinc in ribbon seal, *Histriophoca fasciata*, in the Okhotsk Sea. Res. Inst. N. Pac. Fish. Hokkaido Univ. (Japan) Spec. Vol., 547–561.
Hamasaki, T., Nagase, H., Yoshioka, Y., Sato, T., 1995. Formation, distribution, and ecotoxicity of methyl-metals of tin, mercury, and arsenic in the environment. Crit. Rev. Environ. Sci. Technol. 25, 45–91.
Hamilton, E.I., Clifton, R.J., 1980. Concentration and distribution of the transuranium radionuclides $^{239+240}$Pu, ^{238}Pu and ^{241}Am in *Mytilus edulis*, *Fucus vesiculosus* and surface sediment of Esk estuary. Mar. Ecol. Prog. Ser. 3, 267–277.
Han, B.C., Hung, T.C., 1990. Green oysters caused by copper pollution on the Taiwan coast. Environ. Pollut. 65, 347–362.
Han, B.C., Jeng, W.L., Tsai, Y.N., Jeng, M.S., 1993. Depuration of copper and zinc by green oysters and blue mussels of Taiwan. Environ. Pollut. 82, 93–97.
Han, B.C., Jeng, W.L., Chen, R.Y., Fang, G.T., Hung, T.C., Tseng, R.J., 1998. Estimation of target hazard quotients and potential health risks for metals by consumption of seafood in Taiwan. Arch. Environ. Contam. Toxicol. 35, 711–720.
Hanks, R.W., 1965. Effect of metallic aluminum particles on oysters and clams. Chesapeake Sci. 6, 146–149.
Harino, H., Fukushima, M., Yamamoto, Y., Kuwai, S., Miyazaki, N., 1998. Organotin compounds in water, sediment, and biological samples from the port of Osaka, Japan. Arch. Environ. Contam. Toxicol. 35, 558–564.
Harrington, J.M., 1991. Tributyltin residues in Lake Tahoe and San Diego Bay, California, 1988. In: State of California Dept. Fish Game, Environ. Serv. Div., Admin. Rep. 91-1, pp. 1–29.
Harris, J.E., Fabris, G.J., Statham, P.J., Tawfik, F., 1979. Biogeochemistry of selected heavy metals in Western Port, Victoria, and use of invertebrates as indicators with emphasis on *Mytilus edulis planulatus*. Aust. J. Mar. Freshw. Res. 30, 159–178.

Harriss, R.C., 1971. Ecological indications of mercury pollution in aquatic systems. Biol. Conserv. 3, 279.

Harvey Sr., E.J., Knight Jr., L.A., 1978. Concentration of three toxic metals in oysters (*Crassostrea virginica*) of Biloxi and Pascagoula, Mississippi estuaries. Water Air Soil Pollut. 9, 255–261.

Hayashi, A., 1960. Biogeochemical studies on *Ostrea gigas*. IX. Lead content. Seikagaku 32, 871–873.

Hedouin, L., Metian, M., Teyssie, J.L., Fowler, S.W., Fichez, R., Warnau, M., 2006. Allometric relationships in the bioconcentration of heavy metals by the edible tropical clam *Gafrarium tumidum*. Sci. Total Environ. 366, 154–163.

Hedouin, L., Pringault, O., Metian, M., Bustamante, P., Warnau, M., 2007. Nickel bioaccumulation in bivalves from New Caledonia lagoon: seawater and food exposure. Chemosphere 66, 1449–1457.

Herdman, W.A., Boyce, R., 1899. Oysters and disease. An account of certain observations upon the normal and pathological histology and bacteriology of the oyster and other shellfish. Lancashire Sea Fish. Mem. 1, 1–60.

Heyraud, M., Cherry, R.D., 1979. Polonium-210 and lead-210 in marine food chains. Mar. Biol. 52, 227–236.

Hill, J.M., Helz, G.R., 1973. Copper and zinc in estuarine waters near a coal-fired electric power plant—correlation with oyster greening. Environ. Lett. 5(3), 165–174.

Hiyama, Y., Shimizu, M., 1964. On the concentration factors of radioactive Cs, Sr, Cd, Zn, and Ce in marine organisms. Rec. Oceanogr. Works Jpn. 7(2), 43–47.

Hoare, K., Beaumont, A.R., Davenport, J., 1995a. Variation among populations in the resistance of *Mytilus edulis* embryos to copper: adaptation to pollution? Mar. Ecol. Prog. Ser. 120, 155–161.

Hoare, K., Davenport, J., Beaumont, A.R., 1995b. Effects of exposure and previous exposure to copper on growth of veliger larvae and survivorship of *Mytilus edulis* embryos. Mar. Ecol. Prog. Ser. 120, 163–168.

Hobden, D.J., 1967. Iron metabolism in *Mytilus edulis*. I. Variation in total content and distribution. J. Mar. Biol. Assoc. UK 47, 597–606.

Hobden, D.J., 1969. Iron metabolism in *Mytilus edulis*. II. Uptake and distribution of radioactive iron. J. Mar. Biol. Assoc. UK 49, 661–668.

Hoggins, F.E., Brooks, R.R., 1973. Natural dispersion of mercury from Puhipuhi, Northland, New Zealand. N. Z. J. Mar. Freshw. Res. 7, 125–132.

Holden, A.V., 1973. Mercury in fish and shellfish, a review. J. Food Technol. 8, 1–25.

Horiguchi, T., Li, Z., Uno, S., Shimizu, M., Shiraishi, H., Morita, M., et al., 2003. Contamination of organotin compounds and imposex in molluscs from Vancouver, Canada. Mar. Environ. Res. 57, 75–88.

Horiguchi, F., Nakata, L., Ito, N., Okawa, K., 2006. Risk assessment of TBT in the Japanese short-neck clam (*Ruditapes philippinarum*) of Tokyo Bay using a chemical fate model. Estuar. Coast. Shelf Sci. 70, 589–598.

Horowitz, A., Presley, B.J., 1977. Trace metal concentrations and partitioning in zooplankton, neuston, and benthos from the south Texas Outer Continental Shelf. Arch. Environ. Contam. Toxicol. 5, 241–255.

Horvath, G.J., Harriss, R.C., Mattraw, H.C., 1972. Land development and heavy metal distribution in the Florida Everglades. Mar. Pollut. Bull. 3, 182–184.

Hu, J., Zhen, H., Wan, Y., Gao, J., An, W., An, L., et al., 2006. Trophic magnification of triphenyltin in a marine food web of Bohai Bay, North China: comparison to tributyltin. Environ. Sci. Technol. 40, 3142–3147.

Huang, H., Wu, J.Y., Wu, J.H., 2007. Heavy metal monitoring using bivalved shellfish from Zhejiang coastal waters, East China Sea. Environ. Monit. Assess. 129, 315–320.

Huggett, R.J., Bender, M.E., Slone, H.D., 1973. Utilizing metal concentration relationships in the eastern oyster (*Crassostrea virginica*) to detect heavy metal pollution. Water Res. 7, 451–460.

Huggett, R.J., Cross, F.A., Bender, M.E., 1975. Distribution of copper and zinc in oysters and sediments from three coastal plain estuaries. In: Howell, F.G., Gentry, J.B., Smith, M.H. (Eds.), Mineral Cycling in Southeastern Ecosystems, US Energy Res. Dev. Admin. CONF-740513, pp. 224–228.

Huggett, R.J., Unger, M.A., Seligman, P.F., Valkirs, A.O., 1992. The marine biocide tributyltin. Environ. Sci. Technol. 26, 232–237.

Hung, T.C., Lin, T.T., 1976. Study on mercury in the water, sediments, and benthonic organisms along Cahi—I. Coastal area. Acta Oceanogr. Taiwan. 6, 30–38.

Hung, T.C., Chuang, A., Wu, S.J., Tsai, C.C.H., 1990. Relationships among the species and forms of copper and biomass along the Erhjin Chi coastal water. Acta Oceanogr. Taiwan. 25, 65–76.

Hunt, J.W., Anderson, B.S., 1989. Sublethal effects of zinc and municipal effluents on larvae of the red abalone, *Haliotis rufescens*. Mar. Biol. 101, 545–552.

Hussain, M., Bleiler, E.L., 1973. Mercury in Australian oysters. Mar. Pollut. Bull. 4, 44.

Ichikawa, R., 1961. On the concentration factors of some important radionuclides in marine food organisms. Bull. Jpn. Soc. Sci. Fish. 27, 66–74.

Ikebe, K., Tanaka, R., 1979. Determination of vanadium and nickel in marine samples by flameless and flame atomic absorption spectrophotometry. Bull. Environ. Contam. Toxicol. 21, 526–532.

Ikuta, K., 1967. Studies on accumulation of heavy metals in aquatic organisms. I. On the copper content in oysters. Bull. Jpn. Soc. Sci. Fish. 33, 405–409.

Ikuta, K., 1968a. Studies on accumulation of heavy metals in aquatic organisms. II. On accumulation of copper and zinc in oysters. Bull. Jpn. Soc. Sci. Fish. 34, 112–116.

Ikuta, K., 1968b. Studies on accumulation of heavy metals in aquatic organisms. III. On accumulation of copper and zinc in the parts of oysters. Bull. Jpn. Soc. Sci. Fish. 34, 117–122.

Ikuta, K., 1968c. Studies on accumulation of heavy metals in aquatic organisms. IV. On disappearance of abnormally accumulated copper and zinc in oysters. Bull. Jpn. Soc. Sci. Fish. 34, 482–487.

Inoue, S., Abe, S.I., Oshima, Y., Kai, N., Honjo, T., 2006a. Tributyltin contamination of bivalves in coastal areas around northern Kyushu, Japan. Environ. Toxicol. 21, 244–249.

Inoue, S., Oshima, Y., Usuki, H., Hamaguchi, M., Hanamura, Y., Kai, N., et al., 2006b. Effects of tributyltin maternal and/or waterborne exposure on the embryonic development of the Manila clam, *Ruditapes philippinarum*. Chemosphere 63, 881–888.

Inoue, S., Oshima, Y., Abe, S.I., Wu, R.S.S., Kai, N., Honjo, T., 2007a. Effects of tributyltin on the energy metabolism of pen shell (*Atrina pectinata japonica*). Chemosphere 66, 1226–1229.

Inoue, A., Oshima, Y., Usuki, H., Hamaguchi, M., Hanamura, Y., Kai, N., et al., 2007b. Effect of tributyltin on veliger larvae of the Manila clam, *Ruditapes philippinarum*. Chemosphere 66, 1353–1357.

Ireland, M.P., 1973. Result of fluvial zinc pollution on the zinc content of littoral and sublittoral organisms in Cardigan Bay, Wales. Environ. Pollut. 4, 27–35.

Ireland, M.P., Wooton, R.J., 1977. Distribution of lead, zinc, copper and manganese in the marine gastropods, *Thais lapillus* and *Littorina littorea*, around the coast of Wales. Environ. Pollut. 12, 27–41.

Irukayama, K., 1967. In: The pollution of Minamata Bay and Minamata disease. In: Proceedings of the 3rd International Conference on Water Pollution Research, Munich, 1966, pp. 153–180.

Irukayama, K., Kondo, T., Kai, F., Fujiki, M., 1961. Studies on the origin of the causative agent of Minamata disease. I. Organic mercury compound in the fish and shellfish from Minamata Bay. Kumamoto Med. J. 14, 158–169.

Irukayama, K., Fujiki, M., Kai, F., Kondo, T., 1962a. Studies on the origin of the causative agent of Minamata disease. II. Comparison of the mercury compound in the shellfish from Minamata Bay with mercury compounds experimentally accumulated in the control shellfish. Kumamoto Med. J. 15, 1–12.

Irukayama, K., Kai, F., Fujiki, F., Kondo, T., 1962b. Studies on the causative agent of Minamata disease. III. Industrial wastes containing mercury compounds from Minamata factory. Kumamoto Med. J. 15, 57–68.

Ishii, T., Suzuki, H., Koyanagi, T., 1978. Determination of trace elements in marine organisms. I. Factors for variation of concentration of trace element. Bull. Jpn. Soc. Sci. Fish. 44, 155–162.

Ishikawa, M., Sumiya, M., Saiki, M., 1973. Chemical behaviour of ^{106}Ru in seawater and uptake by marine organisms. In: Radioactive Contamination of the Marine Environment. IAEA, Vienna, pp. 359–367.

Ishikawa, M., Koyanagi, T., Saiki, M., 1976. Studies on the chemical behaviour of ^{106}Ru in seawater and its uptake by marine organisms. I. Accumulation and excretion of ^{106}Ru by clam. Bull. Jpn. Soc. Sci. Fish. 42, 287–297.

Ivanina, A.V., Sokolova, I.M., 2008. Effects of cadmium exposure on expression and activity of P-glycoprotein in eastern oysters, *Crassostrea virginica* Gmelin. Aquat. Toxicol. 88, 19–28.

Jackim, E., Morrison, G., Steele, R., 1977. Effects of environmental factors on radiocadmium uptake by four species of marine bivalves. Mar. Biol. 40, 303–308.

Janssen, H., Scholz, N., 1979. Uptake and cellular distribution of cadmium in *Mytilus edulis*. Mar. Biol. 55, 133–141.

Jeng, S.S., Huang, Y.W., 1973. Heavy metals contents in Taiwan's cultured fish. Bull. Inst. Zool. Acad. Sin. 12, 79–85.

Jenkins, D.W., 1980. Biological Monitoring of Toxic Trace Metals. Vol. 2. Toxic Trace Metals in Plants and Animals of the World. Part I. US Environ. Protect. Agen. Rep. 600/3-80–090, pp. 30-138; Part II. US Environ. Protect Agen. Rep. 600/3-80-091, pp. 505–618.

Jernelov, A., 1969. Conversion of mercury compounds. In: Miller, M.W., Berg, G.G. (Eds.), Chemical Fallout, Current Research on Persistent Pesticides. Charles C Thomas Publishers, Springfield, IL, pp. 68–74.

Jernelov, A., Hartung, R., Trost, P.B., Bisque, R.E., 1972. Environmental dynamics of mercury. In: Hartung, R., Dinman, B.D. (Eds.), Environmental Mercury Contamination. Ann Arbor Science Publishers, Ann Arbor, MI, pp. 167–201.

Jernelov, A., Landner, L., Larsson, T., 1975. Swedish perspectives on mercury pollution. J. Water Control Pollut. Fed. 47, 810–822.

Ji, J., Choi, H.J., Ahn, I.Y., 2006. Evaluation of Manila clam *Ruditapes philippinarum* as a sentinel species for metal pollution monitoring in estuarine tidal flats of Korea: effects of size, sex, and spawning on baseline accumulation. Mar. Pollut. Bull. 52, 447–453.

Jiann, K.T., Presley, B.J., 1997. Variations in trace metal concentrations in American oysters (*Crassostrea virginica*) collected from Galveston Bay, Texas. Estuaries 20, 710–724.

Jing, G., Li, Y., Zhang, R., 2007. Different effects of Pb^{2+} and Cu^{2+} on immune and antioxidant enzyme activities in the mantle of *Pinctada fucata*. Environ. Toxicol. Pharmacol. 24, 122–128.

Johnels, A.G., Westermark, T., 1969. Mercury contamination of the environment in Sweden. In: Miller, M.W., Berg, G.G. (Eds.), Chemical Fallout, Current Research on Persistent Pesticides. Charles C Thomas Publishers, Springfield, IL, pp. 221–239.

Johns, C., Luoma, S.N., 1990. Arsenic in benthic bivalves of San Francisco Bay and the Sacramento/San Joaquin River delta. Sci. Total Environ. 97/98, 673–684.

Jones, M.B., 1973. Influence of salinity and temperature on the toxicity of mercury to marine and brackish water isopods (Crustacea). Estuar. Coast. Mar. Sci. 1, 425–431.

Jones, A.M., Jones, Y., Stewart, W.D.P., 1972. Mercury in marine organisms of the Tay region. Nature 238, 164–165.

Julshamn, K., Duinker, A., Frantzen, S., Torkildsen, L., Maage, A., 2008. Organ distribution and food safety aspects of cadmium and lead in great scallops, *Pecten maximus* L., and horse mussels, *Modiolus modiolus* L., from Norwegian waters. Bull. Environ. Contam. Toxicol. 80, 385–389.

Kadar, E., 2007. Postcapture depuration of essential metals in the deep sea hydrothermal mussel *Bathymodiolus azoricus*. Bull. Environ. Contam. Toxicol. 78, 99–106.

Kadar, E., Santos, R.S., Powell, J.L., 2006. Biological factors influencing tissue compartmentalization of trace metals in the deep-sea hydrothermal vent bivalve *Bathymodiolus azoricus* at geochemically distinct vent sites of the mid-Atlantic Ridge. Environ. Res. 101, 221–229.

Kanatireklap, S., Tanabe, S., Sanguansin, J., Tabucanon, M.S., Hungspreugs, M., 1997. Contamination by butyltin compounds and organochlorine residues in green mussel (*Perna viridis*, L.) from Thailand coastal waters. Environ. Pollut. 97, 78–89.

Karayakar, F., Erdem, C., Cicik, B., 2007. Seasonal variation in copper, zinc, chromium, lead and cadmium levels in hepatopancreas, gill and muscle tissues of the mussel *Brachidontes pharaonis* Fischer, collected along the Mersin coast, Turkey. Bull. Environ. Contam. Toxicol. 79, 350–355.

Karbe, L., Schnier, C., Siewers, H.O., 1977. Trace elements in mussels (*Mytilus edulis*) from coastal areas of the North Sea and the Baltic. Multielement analyses using instrumental neutron activation analysis. J. Radioanal. Chem. 37, 927–943.

Karouna-Renier, N.K., Snyder, R.A., Allison, J.G., Wagner, M.G., Rao, K.R., 2007. Accumulation of organic and inorganic contaminants in shellfish collected in estuarine waters near Pensacola, Florida: contamination profiles and risks to human consumers. Environ. Pollut. 145, 474–488.

Karpevich, A.F., Shurin, A.T., 1977. Manganese in the metabolic processes of mollusks of the Baltic Sea. Sov. J. Mar. Biol. 3, 437–442.

Keckes, S., Miettinen, J., 1972. Mercury as a marine pollutant. In: Ruivo, M. (Ed.), Marine Pollution and Sea Life. Fishing Trading News (Books), London, pp. 276–289.

Keckes, S., Pucar, Z., Marazovic, L., 1966. The influence of the physico-chemical form of ^{106}Ru on its uptake by mussels from sea water. In: Aberg, B., Hungate, F.P. (Eds.), Radioecological Concentration Processes. Proceedings of an International Symposium, Stockholm, 1966. Pergamon Press, New York, pp. 993–994.

Keckes, S., Ozretic, B., Krajnovic, M., 1968. Loss of Zn-65 in the mussel *Mytilus galloprovincialis*. Malacologia 7, 1–6.

Keckes, S., Ozretic, B., Krajnovic, M., 1969. Metabolism of zinc-65 in mussels (*Mytilus galloprovincialis* Lam). Uptake of zinc-65. Rapp. Comm. Int. Mer Medit. 19, 949–952.

Kerfoot, W.B., 1979. Artificial shellfish for monitoring ambient cadmium levels in seawater. In: Proceedings of the Workshop on Alternatives for Cadmium Electroplating in Metal Finishing, October 4-6, 1977. US Environ. Protect. Agen. Rep. 560/2-79-003, pp. 215–240.

Keskin, Y., Baskaya, R., Ozyaral, O., Yurdun, T., Luleci, N.E., Hayran, O., 2007. Cadmium, lead, mercury and copper in fish from the Marmara Sea, Turkey. Bull. Environ. Contam. Toxicol. 78, 258–261.

Ketchum, B.M., Bowen, V.T., 1958. Biological factors determining the distribution of radioisotopes on the sea. In: Proceeding of the Second Annual U.N. Conference on the Peaceful Uses of Atomic Energy, vol. 18. Waste Treat. Environ Aspect Atom. Ener., Geneva, pp. 429–433.

Khangarot, B.S., Ray, P.K., 1987. Zinc sensitivity of a freshwater snail, *Lymnaea luteola* L., in relation to seasonal variations in temperature. Bull. Environ. Contam. Toxicol. 39, 45–49.

Kim, N.S., Shim, W.J., Yim, U.H., Ha, S.Y., Park, P.S., 2008. Assessment of tributyltin contamination in a shipyard area using a mussel transplantation approach. Mar. Pollut. Bull. 57, 883–888.

Kiyoura, R., 1963. Water pollution and Minamata disease. Int. J. Air Water Pollut. 7, 459–470.

Klein, D.H., Goldberg, E.D., 1970. Mercury in the marine environment. Environ. Sci. Technol. 4, 765–768.

Klemmer, H., Luoma, S.N., Lau, L.S., 1973. Mercury levels in marine biota. Gov. Rep. Announ. 73(10), 76.

Klemmer, H.W., Unninayer, C.S., Ukubo, W.I., 1976. Mercury content of biota in coastal waters in Hawaii. Bull. Environ. Contam. Toxicol. 15, 454–457.

Kljakovic-Gaspic, Z., Odzak, N., Ujevic, I., Zvonaric, T., Horvat, M., Baric, A., 2006. Biomonitoring of mercury in polluted coastal area using transplanted mussels. Sci. Total Environ. 368, 199–209.

Knothe, D.W., Van Riper, G.G., 1988. Acute toxicity of sodium molybdate dehydrate (Molhibit 100) to selected saltwater organisms. Bull. Environ. Contam. Toxicol. 40, 785–790.

Kobayashi, N., 1971. Fertilized sea urchin eggs as an indicatory material for marine pollution bioassay, preliminary experiments. Publ. Seto Mar. Biol. Lab. 18, 379–406.

Koch, I., McPherson, K., Smith, P., Easton, L., Doe, K.G., Reimer, K.J., 2007. Arsenic bioavailability and speciation in clams and seaweed from a contaminated marine environment. Mar. Pollut. Bull. 54, 586–594.

Koczy, F.F., Titze, H., 1958. Radium content of carbonate shells. J. Mar. Res. 17, 302–311.

Kopfler, F.C., 1974. The accumulation of organic and inorganic mercury compounds by the eastern oyster (*Crassostrea virginica*). Bull. Environ. Contam. Toxicol. 11, 275–280.

Kopfler, F.C., Mayer, J., 1967. Studies of trace metals in shellfish. In: Proceedings of the Gulf and South Atlantic States Shellfish Sanitation Research Conference, pp. 67–80.

Krinsley, D., Bieri, R., 1959. Changes in the chemical composition of pteropod shells after deposition on the sea floor. J. Paleontol. 33, 682–684.

Krishnakumar, P.K., Asokan, P.K., Pillai, V.K., 1990. Physiological and cellular responses to copper and mercury in the green mussel *Perna viridis* (Linnaeus). Aquat. Toxicol. 18, 163–174.

Kucuksezgin, F., Kayatekin, B.M., Uluturhan, E., Oysal, N., Acikgoz, O., Gonenc, S., 2008. Preliminary investigation of sensitive biomarkers of metal pollution in mussel (*Mytilus galloprovincialis*) from Izmir Bay (Turkey). Environ. Monit. Assess. 141, 339–345.

Kuenzler, E.J., 1969. Elimination of iodine, cobalt, iron, and zinc by marine zooplankton. In: Proceedings of the 2nd National Symposium on Radioecology. US Atom. Ener. Comm. Conf. 67053, pp. 462–473.

Kumagi, H., Saeki, K., 1978. Contents of total mercury, alkyl mercury and methyl mercury in some coastal fish and shells. Bull. Jpn. Soc. Sci. Fish. 44, 807–811.

Kumar, K.S., Sajwan, K.S., Richardson, J.P., Kannan, K., 2008. Contamination profiles of heavy metals, organochlorine pesticides, polycyclic aromatic hydrocarbons and alkylphenols in sediment and oyster collected from marsh/estuarine Savannah GA, USA. Mar. Pollut. Bull. 56, 136–149.

Kurland, T., Faro, S.N., Siedler, H., 1960. Minamata disease. The outbreak of a neurologic disorder in Minamata, Japan, and its relationship to the ingestion of seafood contaminated by mercuric compounds. World Neurol. 1, 370–395.

LaBrecque, J.J., Benzo, Z., Alfonso, J.A., Cordoves, P.R., Quintal, M., Gomez, C.V., et al., 2004. The concentrations of selected trace elements in clams, *Trivela mactroidea* along the Venezuelan coast in the state of Miranda. Mar. Pollut. Bull. 49, 664–667.

Lacerda, L.D., 1983. Aplicacao da metodologia de abordagem peios parametros criticos no estudo da poluicao por metais pesados na Baia de Sepetiba, Rio de Janeiro. Ph.D. thesis, Inst. Biofisica, Univ. Fed. Rio de Janeiro, UFRJ, Rio de Janeiro, 134 pp.

Lacerda, L.D., Molisani, M.M., 2006. Three decades of Cd and Zn contamination in Sepetiba Bay, SE Brazil: evidence from the mangrove oyster *Crassostrea rhizophorae*. Mar. Pollut. Bull. 52, 974–977.

Lacoue-Labarthe, T., Warnau, M., Oberhansli, F., Teyssie, J.L., Koueta, N., Bustamante, P., 2008a. Differential bioaccumulation behaviour of Ag and Cd during the early development of the cuttlefish *Sepia officinalis*. Aquat. Toxicol. 86, 437–446.

Lacoue-Labarthe, T., Warnau, M., Oberhansli, F., Teyssie, J.L., Jeffree, R., Bustamante, P., 2008b. First experiments on the maternal transfer of metals in the cuttlefish *Sepia officinalis*. Mar. Pollut. Bull. 57, 826–831.

Laevastu, T., Thompson, T.G., 1956. The determination and occurrence of nickel in seawater, marine organisms, and sediments. J. Conseil 21, 125–143.

Lafabrie, C., Pergent, G., KAntin, R., Pergent-Martini, C., Gonzalez, J.L., 2007. Trace metals assessment in water, sediment, mussel and seagrass species—validation of the use of *Posidonia oceanica* as a metal biomonitor. Chemosphere 68, 2033–2039.

Lahaye, V., Bustamante, P., Dabin, W., Canney, O.V., Dhermain, F., Cesarini, C., et al., 2006. New insights from age determination on toxic element accumulation in striped and bottlenose dolphins from Atlantic and Mediterranean waters. Mar. Pollut. Bull. 52, 1219–1230.

Lande, E., 1977. Heavy metal pollution in Trondheimsfjorden, Norway, and the recorded effects on the fauna and flora. Environ. Pollut. 12, 187–198.

Langley, D.G., 1973. Mercury methylation in an aquatic environment. J. Water Pollut. Control Fed. 45, 44–51.

Larsen, P.F., 1979. The distribution of heavy metals in the hard clam, *Mercenaria mercenaria* in the lower Chesapeake Bay region. Estuaries 2, 1–8.

Lauenstein, G.G., Robertson, A., O'Connor, T.P., 1990. Comparison of trace metal data in mussels and oysters from a mussel watch programme of the 1970s with those from a 1980s programme. Mar. Pollut. Bull. 21, 440–447.

Laughlin, R., Linden, O., 1985. Sublethal responses of the tadpoles of the European frog *Rana temporaria* to two tributyltin compounds. Bull. Environ. Contam. Toxicol. 28, 494–499.

Laughlin, R., Nordlund, K., Linden, O., 1984. Long-term effects of tributyltin compounds on the Baltic amphipod, *Gammarus oceanicus*. Mar. Environ. Res. 12, 243–271.

Laughlin Jr., R.B., French, W., Guard, H.E., 1986. Accumulation of bis(tributyltin) oxide by the marine mussel *Mytilus edulis*. Environ. Sci. Technol. 20, 884–890.

Laughlin, R.B., Gustafson, J.R., Pendoley, P., 1988. Chronic embryo-larval toxicity of tributyltin (TBT) to the hard-shell clam *Mercenaria mercenaria*. Mar. Ecol. Prog. Ser. 50, 4829–4836.

Laumond, F., Neuberger, M., Donnier, B., Fourcy, A., Bittel, R., Aubert, M., 1973. Experimental investigations, at laboratory, on the transfer of mercury in marine trophic chains. In: Sixth International Symposium on Medical Oceanography, Portoroz, Yugoslavia, September 26-30, 1973, pp. 47–53.

Lawler, I.F., Aldrich, J.C., 1987. Sublethal effects of bis(tri-*n*-butyltin)oxide on *Crassostrea gigas* spat. Mar. Pollut. Bull. 18, 274–278.

Leatherland, T.M., Burton, J.D., 1974. The occurrence of some trace metals in coastal organisms with particular reference to the Solent region. J. Mar. Biol. Assoc. UK 54, 457–468.

Lee, R.F., 1985. Metabolism of tributyltin oxide by crabs, oysters and fish. Mar. Environ. Res. 17, 145–148.

Lee, E.H., Ryu, B.H., Yang, S.T., 1975. Suitability of shellfish for processing. Bull. Korean Fish. Soc. 8, 85–89.

Lee, B.G., Lee, J.S., Luoma, S.N., 2006. Comparison of selenium bioaccumulation in the clams *Corbicula fluminea* and *Potamocorbula amurensis*: a bioenergetic modeling approach. Environ. Toxicol. Chem. 25, 1933–1940.

LeGall, P., Ancellin, J., 1971. Concentration Factors for Cs 137, Ce 144, and Ru 106 in Several Species of Edible Marine Mollusks and Crustaceans. Available from NTIS, Springfield, VA, as CEA-N-1488, 17 pp.

Lekhi, P., Cassis, D., Pearce, C.M., Ebell, N., Maldonado, M.T., Orians, K.J., 2008. Role of dissolved and particulate cadmium in the accumulation of cadmium in cultured oysters (*Crassostrea gigas*). Sci. Total Environ. 393, 309–325.

Leung, K.M.Y., Kwong, R.P.Y., Ng, W.C., Horiguchi, T., Qiu, J.W., Yang, R., et al., 2006. Ecological risk assessments of endocrine disrupting organotin compounds using marine neogastropods in Hong Kong. Chemosphere 65, 922–938.

Li, Y., Yu, Z., Song, X., Mu, Q., 2006. Trace metal concentrations in suspended particles, sediments and clams (*Ruditapes philippinarum*) from Jiaozhou Bay of China. Environ. Monit. Assess. 121, 491–501.

Liang, L.N., He, B., Jiang, G.B., Chen, D.Y., Yao, Z.W., 2004a. Evaluation of mollusks as biomonitors to investigate heavy metal contaminations along the Chinese Bohai Sea. Sci. Total Environ. 324, 105–113.

Liang, L.N., Hu, J.T., Chen, D.Y., Zhou, Q.F., He, B., Jiang, G.B., 2004b. Primary investigation of heavy metal contamination status in molluscs collected from Chinese coastal sites. Bull. Environ. Contam. Toxicol. 72, 937–944.

Lima, N.R.W., Lacerda, L.D., Pfeiffer, W.C., Fiszman, M., 1986. Temporal and spatial variability in Zn, Cr, Cd and Fe concentrations in oyster tissues (*Crassostrea brasiliana* Lamarck, 1819) from Sepetiba Bay, Brazil. Environ. Technol. Lett. 7, 453–460.

Limaverde, A.M., Wagener, A.L.R., Fernandez, M.A., Scofield, A.L., Coutino, R., 2007. *Stramonita haemastoma* as a bioindicator for organotin contamination in coastal environments. Mar. Environ. Res. 64, 384–398.

Liu, W., Deng, P.Y., 2007. Accumulation of cadmium, copper, lead and zinc in the Pacific oyster, *Crassostrea gigas*, collected from the Pearl River estuary, southern China. Bull. Environ. Contam. Toxicol. 78, 535–538.

Liu, C.W., Liang, C.P., Huang, F.M., Hsueh, Y.M., 2006. Assessing the human health risks from exposure of inorganic arsenic through oyster (*Crassostrea gigas*) consumption in Taiwan. Sci. Total Environ. 361, 57–66.

Liu, C.W., Huang, Y.K., Hsueh, Y.M., Lin, K.H., Jang, C.S., Huang, L.P., 2008. Spatiotemporal distribution of arsineic species of oysters (*Crassostrea gigas*) in the coastal area of southwestern Taiwan. Environ. Monit. Assess. 138, 181–190.

Lobel, P.B., 1986. Role of the kidney in determining the whole soft tissue zinc concentration of individual mussels (*Mytilus edulis*). Mar. Biol. 92, 355–359.

Lobel, P.B., 1987. Short-term and long-term uptake of zinc by the mussel, *Mytilus edulis*: a study in individual variability. Arch. Environ. Contam. Toxicol. 16, 723–732.

Lobel, P.B., Marshall, H.D., 1988. A unique low molecular weight zinc-binding ligand in the kidney cytosol of the mussel *Mytilus edulis*, and its relationship to the inherent variability of zinc accumulation in this organism. Mar. Biol. 99, 101–105.

Loring, D.H., Asmund, G., 1989. Heavy metal contamination of a Greenland fjord system by mine wastes. Environ. Geol. Water Sci. 14, 61–71.

Lowe, V.P.W., Horrill, A.D., 1991. Caesium concentration factors in wild herbivores and the fox (*Vulpes vulpes* L.). Environ. Pollut. 70, 93–107.

Lowe, D.M., Moore, M.N., 1979. The cytochemical distributions of zinc (Zn II) and iron (Fe III) in the common mussel, *Mytilus edulis*, and their relationship with lysosomes. J. Mar. Biol. Assoc. UK 59, 851–858.

Lowman, F.G., Ting, R.Y., 1973. The state of cobalt in seawater and its uptake by marine organisms and sediments. In: Radioactive Contamination of the Marine Environment. IAEC, Vienna, Austria, pp. 369–384.

Lowman, F.G., Phelps, D.K., Ting, R.Y., Escalera, R.M., 1966. Progress Summary Report No.4, Marine Biology Program June 1965-June 1966. Puerto Rico Nucl. Cen. Rep.. PRNC 85, , pp. 1–57.

Lowman, F.G., Martin, J.H., Ting, R.Y., Barnes, S.S., Swift, D.J.P., Seiglie, G.A., et al., 1970. Bioenvironmental and radiological-safety feasibility studies, Atlantic-Pacific interoceanic canal. Estuar. Mar. Ecol.I-IV. Prepared for Battelle Memorial Institute, 505 King Avenue, Columbus, OH, Contract AT (26-1)-171.

Lucu, C., Jelisavcic, O., Lucic, S., Strohal, P., 1969. Interactions of ^{233}Pa with tissues of *Mytilus galloprovincialis* and *Carcinus mediterraneus*. Mar. Biol. 2, 103–104.

Lunde, G., 1970. Analysis of arsenic and selenium in marine raw materials. J. Sci. Food Agric. 21, 242–247.

Luoma, S.N., Bryan, G.W., 1978. Factors controlling the availability of sediment-bound lead to the estuarine bivalve *Scrobicularia plana*. J. Mar. Biol. Assoc. UK 58, 793–802.

Luoma, S.N., Bryan, G.W., 1979. Trace metal bioavailability: modeling chemical and biological interactions of sediment-bound zinc. In: Jenne, E.A. (Ed.), Chemical modeling in Aqueous Systems, Amer. Chem. Soc. Ser. 93, Washington, DC, 577–609.

Luoma, S.N., Jenne, E.A., 1977. The availability of sediment-bound cobalt, silver, and zinc to a deposit-feeding clam. In: Drucker, H., Wildung, R.E. (Eds.), Biological Implications of Metals in the Environment. ERDA Sympos. Ser. 42. Available as CONF-750929 from NTIS, Springfield, VA, pp. 213–230.

Luoma, S.N., Phillips, D.J.H., 1988. Distribution, variability, and impacts of trace elements in San Francisco Bay. Mar. Pollut. Bull. 19, 413–425.

Luoma, S.N., Dagovitz, R., Axtmann, E., 1990. Temporally intensive study of trace metals in sediments and bivalves from a large river-estuarine system: Suisan Bay/Delta in San Francisco Bay. Sci. Total Environ. 97/98, 685–712.

Luoma, S.N., 1994, Fate, Bioavailability and toxicity of silver in estuarine environments. In: Andren, A.W., Bober, T.W., Kramer, J.R., Sodergren, A., Crecelius, E.A., Luoma, S.N., Rodgers, J.H. (Organisers). Transport, Fate and Effects of Silver in the Environment, Proceedings of the 2nd International Conference, Univ. Wisconsin, Madison, 151-155.

Luten, J.B., Bouquet, W., Burggraaf, M.M., Rus, J., 1986. Accumulation, elimination, and speciation of cadmium and zinc in mussels, *Mytilus edulis*, in the natural environment. Bull. Environ. Contam. Toxicol. 35, 579–586.

Maanan, M., 2007. Biomonitoring of heavy metals using *Mytilus galloprovincialis* in Safi coastal waters, Morocco. Environ. Toxicol. 22, 525–531.

Maanan, M., 2008. Heavy metal concentrations in marine molluscs from the Moroccan coastal region. Environ. Pollut. 153, 176–183.

MacFarlane, G.R., Markich, S.J., Linz, K., Gifford, S., Dunstan, R.H., O'Connor, W., et al., 2006. Akoya pearl oyster shell as an archival monitor of lead exposure. Environ. Pollut. 143, 166–173.

Machreki-Ajmi, M., Hamza-Chaffai, A., 2006. Accumulation of cadmium and lead in *Cerastoderma glaucum* originating from the Gulf of Gabes, Tunisia. Bull. Environ. Contam. Toxicol. 76, 529–537.

Mackay, N.J., Williams, R.J., Kacprzac, J.L., Kazacos, M.N., Collins, A.J., Auty, E.H., 1975. Heavy metals in cultivated oysters (*Crassostrea commercialis* = *Saccostrea cucullata*) from the estuaries of New South Wales. Aust. J. Mar. Freshw. Res. 26, 31–46.

Maddock, B.G., Taylor, D., 1980. The acute toxicity and bioaccumulation of some lead alkyl compounds in marine animals. In: Branica, M., Konrad, Z. (Eds.), Lead in the Marine Environment. Pergamon Press, Oxford, UK, pp. 233–261.

Maguire, R.J., 1991. Aquatic environmental aspects of non-pesticidal organotin compounds. Water Pollut. Res. J. Can. 26, 243–360.

Maher, W.A., 1983. Selenium in marine organisms from St. Vincent's Gulf, South Australia. Mar. Pollut. Bull. 14, 35–36.

Majori, L., Petronio, F., 1973. Marine pollution by metals and their accumulation by biological indicators (accumulation factor). In: Sixth International Symposium on Medical Oceanography, Portoroz, Yugoslavia, September 26-30, 1973, pp. 55–90.

Mandelli, E.F., 1975. The effects of desalinization brines on *Crassostrea virginica* (Gmelin). Water Res. 9, 287–295.

Marks, G.W., 1938. The copper content and copper tolerance of some species of mollusks of the southern California coast. Biol. Bull. 75, 224–237.

Martin, J.L.M., 1979. Schema of lethal action of copper on mussels. Bull. Environ. Contam. Toxicol. 21, 808–814.

Martin, J.H., Flegal, A.R., 1975. High copper concentrations in squid livers in association with elevated levels of silver, cadmium, and zinc. Mar. Biol. 30, 51–55.

Martin, M., Stephenson, M.D., Martin, J.H., 1977. Copper toxicity experiments in relation to abalone deaths observed in a power plant's cooling waters. Calif. Fish Game 63, 95–100.

Martin, M., Stephenson, M.D., Smith, D.R., Gitierrez-Galindo, E.A., Munoz, G.F., 1988. Use of silver in mussels as a tracer of domestic wastewater discharge. Mar. Pollut. Bull. 19, 512–520.

Martoja, M., Tue, V.T., Elkaim, B., 1980. Bioaccumulation du cuivre chez *Littorina littorea* (L.) (gasteropode prosobranche): signification physiologique et ecologique. J. Exp. Mar. Biol. Ecol. 43, 251–270.

Mason, A.Z., 1988. The kinetics of zinc accumulation by the marine prosobranch gastropod *Littorina littorea*. Mar. Environ. Res. 24, 135–139.

Mat, I., 1994. Arsenic and trace metals in commercially important bivalves, *Anadara granosa* and *Paphia undulata*. Bull. Environ. Contam. Toxicol. 52, 833–839.

Mat, I., Maah, M.J., Johari, A., 1994. Trace metals in sediments and potential availability to *Anadara granosa*. Arch. Environ. Contam. Toxicol. 27, 54–59.

Mathew, P., Menon, N.R., 1993. Heavy metal stress induced variations in O:N ratio in two tropical bivalves *Perna indica* (Kuriakose & Nair) and *Donax incarnatus* Gmelin. Indian J. Exp. Biol. 31, 694–698.

Matida, Y., Kumada, H., 1969. Distribution of mercury in water, bottom mud and aquatic organisms of Minamata Bay, the River Agano and other water bodies in Japan. Bull. Freshw. Fish. Res. Lab. Tokyo 19(2), 73–79.

Matida, Y., Kumada, H., Kimura, S., Saiga, Y., Nose, T., Yokote, M., et al., 1972. Toxicity of mercury compounds to aquatic organisms and accumulation of the compounds by the organisms. Bull. Freshw. Fish. Res. Lab. Tokyo 21, 197–227.

Matsumoto, T., Satake, M., Yamamoto, J., Haruna, S., 1964. On the micro constituent elements in marine invertebrates. J. Oceanogr. Soc. Jpn 20(3), 15–19.

Matthiessen, P., Gibbs, P.E., 1998. Critical appraisal of the evidence for tributyltin-mediated endocrine disruption in mollusks. Environ. Toxicol. Chem. 17, 37–43.

Mauchline, J., Templeton, W.L., 1966. Strontium, calcium and barium in marine organisms from the Irish Sea. J. Conseil 30, 161–170.

Mauri, M., Orlando, E., Nigro, M., Regoli, F., 1990. Heavy metals in the Antarctic scallop *Adamussium colbecki*. Mar. Ecol. Prog. Ser. 67, 27–33.

McClellan-Green, P., Romano, J., Rittschof, D., 2006. Imposex induction in the mud snail, *Ilyanassa obsoleta* by three tin compounds. Bull. Environ. Contam. Toxicol. 76, 581–588.

McFarren, E.F., Campbell, J.E., Engle, J.B., 1962. The occurrence of copper and zinc in shellfish. In: Jensen, E.D. (Ed.), Proceedings of the 1961 Shellfish Sanitation Workshop. U.S. Dept. Health, Educa. Welfare, Publ. Health Serv, pp. 229–234.

Mehran, A.R., Tremblay, J.L., 1965. An aspect of zinc metabolism in *Littorina obtusa* L. and *Fucus edentatus*. Rev. Can. Biol. 24, 157–161.

Messiha, N.N., Ikladious, N.E., 1986. Antifouling performance of some new organotin polymers in the Mediterranean and Red Sea. J. Control. Release 3, 235–242.

Metian, M., Bustamante, P., Hedouin, L., Warnau, M., 2008. Accumulation of nine metals and one metalloid in the tropical scallop *Comptopallium radula* from coral reefs in New Caledonia. Environ. Pollut. 152, 542–552.

Miettinen, V., Verta, M., 1978. On the heavy metals and chlorinated hydrocarbons in the Gulf of Bothnia in Finland. Finn. Mar. Res. 244, 219–226.

Miettinen, J.K., Tillander, M., Risanen, K., Miettinen, V., Ohmomo, Y., 1969. Distribution and excretion rates of phenyl- and methylmercury nitrate in fish muscles, molluscs and crayfish. In: Proceedings of the Sixth Japan. Conf. Radioisotopes, Tokyo, pp. 474–478.

Miettinen, J.K., Heyraud, M., Keckes, S., 1972. Mercury as a hydrospheric pollutant. II. Biological half-time of methyl mercury in four Mediterranean species: a fish, a crab, and two molluscs. In: Ruivo, M. (Ed.), Marine Pollution and Sea Life. Fishing Trading News (Books), London, pp. 295–298.

Miller, W.L., Blake, N.J., Byrne, R.H., 1985. Uptake of Zn^{65} and Mn^{54} into body tissues and renal concretions by the southern quahog, *Mercenaria campechiensis* (Gmelin): effects of elevated phosphate and metal concentrations. Mar. Environ. Res. 17, 167–171.

Miramand, P., Bentley, D., 1992. Concentration and distribution of heavy metals in two cephalopods, *Eledone cirrhosa* and *Sepia officinalis*, from the French coast of the English Channel. Mar. Biol. 114, 407–414.

Miramand, P., Bustamante, P., Bentley, D., Koueta, N., 2006. Variation of heavy metal concentrations (Ag, Cd, Co, Cu, Fe, Pb, V, and Zn) during the life cycle of the common cuttlefish *Sepia officinalis*. Sci. Total Environ. 361, 132–143.

Mishima, J., Odum, E., 1963. Excretion rate of ^{65}Zn by *Littorina irrorata* in relation to temperature and body size. Limnol. Oceanogr. 8, 39–44.

Morgan, J.D., Mitchell, D.G., Chapman, P.M., 1986. Individual and combined toxicity of manganese and molybdenum to mussel, *Mytilus edulis*, larvae. Bull. Environ. Contam. Toxicol. 37, 303–307.

Morris, R.J., Law, R.J., Allchin, C.R., Kelly, C.A., Fileman, C.F., 1989. Metals and organochlorines in dolphins and porpoises of Cardigan Bay, West Wales. Mar. Pollut. Bull. 20, 512–523.

Morse, J.W., Presley, B.J., Taylor, R.J., Benoit, G., Santschi, P., 1993. Trace metal chemistry of Galveston Bay: water, sediments, and biota. Mar. Environ. Res. 36, 1–37.

Mubiana, V.K., Blust, R., 2006. Metal content of marine mussels from western Scheldt estuary and nearby protected marine bay, the Netherlands: impact of past and present contamination. Bull. Environ. Contam. Toxicol. 77, 203–210.

Mubiana, V.K., Blust, R., 2007. Effects of temperature on scope for growth and accumulation of Cd, Co, Cu and Pb by the marine bivalve *Mytilus edulis*. Mar. Environ. Res. 63, 219–235.

Munzinger, A., Guarducci, M.L., 1988. The effect of low zinc concentrations on some demographic parameters of *Biomphalaria glabrata* (Say), mollusca: gastropoda. Aquat. Toxicol. 12, 51–61.

Nambisan, P.N.K., Lakshmanan, P.T., Salih, K.Y., 1977. On the uptake of copper (II) by *Meretrix casta* (Cheminitz), an indicator species of pollution. Curr. Sci. 46, 437–440.

Napoleao, P., Pinheiro, T., Reis, C.S., 2005. Elemental characterization of tissues of *Octopus vulgaris* along the Portuguese coast. Sci. Total Environ. 345, 41–49.

Navrot, J., Amiel, A.J., Kronfeld, J., 1974. *Patella vulgata*: a biological monitor of coastal metal pollution—a preliminary study. Environ. Pollut. 7, 303–308.

Nelson, J.D., Blair, W., Brinckman, F.E., Colwell, R.R., Iverson, W.P., 1971. Biodegradation of phenylmercuric acetate by mercury-resistant bacteria. Appl. Microbiol. 26, 321–326.

Nelson, D.A., Calabrese, A., Nelson, B.A., MacInnes, J.R., Wenzloff, D.R., 1976. Biological effects of heavy metals on juvenile bay scallops, *Argopecten irradians*, in short-term exposures. Bull. Environ. Contam. Toxicol. 16, 275–282.

Nelson, D.A., Calabrese, A., MacInnes, J.R., 1977. Mercury stress on juvenile bay scallops, *Argopecten irradians*, under various salinity-temperature regimes. Mar. Biol. 43, 293–297.

Nelson, D.A., Calabrese, A., Greig, R.A., Yevich, P.P., Chang, S., 1983. Long-term silver effects on the marine gastropod *Crepidula fornicata*. Mar. Ecol. Prog. Ser. 12, 155–165.

Nesto, N., Romano, S., Moschino, V., Mauri, M., Da Ros, J., 2007. Bioaccumulation and biomarker responses of trace metals and micro-organic pollutants in mussels and fish from the Lagoon of Venice, Italy. Mar. Pollut. Bull. 53, 469–484.

Neuberger-Cywiak, L., Achituv, Y., Garcia, E.M., 2007. Effects of sublethal Zn^{++} and Cd^{++} concentrations on filtration rate, absorption efficiency and scope for growth in *Donax trunculus* (bivalvia; donacidae). Bull. Environ. Contam. Toxicol. 79, 622–627.

Ng, T.Y.T., Rainbow, P.S., Amiard-Triquet, C., Amiard, J.C., Wang, W.X., 2007. Metallothionein turnover, cytosolic distribution and the uptake of Cd by the green mussel *Perna viridis*. Aquat. Toxicol. 84, 153–161.

Nias, D.J., McKillup, S.C., Edyvane, K.S., 1993. Imposex in *Lepsiella vinosa* from southern Australia. Mar. Pollut. Bull. 26, 380–384.

Nicholls, G.D., Curl Jr., H., Bowen, V.T., 1959. Spectrographic analyses of marine plankton. Limnol. Oceanogr. 4, 472–476.

Nicholson, S., Szefer, P., 2003. Accumulation of metals in the soft tissues, byssus and shell of the mytilid mussel *Perna viridis* (Bivalvia: Mytilidae) from polluted and uncontaminated locations in Hong Kong coastal waters. Mar. Pollut. Bull. 46, 1039–1043.

Nicolaidu, A., Nott, J.A., 1990. Mediterranean pollution from a ferro-nickel smelter: differential uptake by some gastropods. Mar. Pollut. Bull. 21, 137–143.

Nielsen, S.A., 1974. Vertical concentration gradients of heavy metals in cultured mussels. N. Z. J. Mar. Freshw. Res. 8, 631–636.

Nielsen, S.A., 1975. Cadmium in New Zealand dredge oysters: geographic distribution. Int. J. Environ. Anal. Chem. 4, 1–7.

Nielsen, S.A., Nathan, A., 1975. Heavy metal levels in New Zealand molluscs. N. Z. J. Mar. Freshw. Res. 9, 467–481.

Noel-Lambot, F., 1976. Distribution of cadmium, zinc and copper in the mussel, *Mytilus edulis*. Existence of cadmium-binding proteins similar to metallothioneins. Experientia 32, 324–325.

Noel-Lambot, F., Bouquegneau, J.M., Frankenne, F., Disteche, A., 1980. Cadmium, zinc and copper accumulation in limpets (*Patella vulgata*) from the Bristol Channel with special reference to metallothioneins. Mar. Ecol. Prog. Ser. 2, 81–89.

Norum, U., Lai, V.W.M., Cullen, W.R., 2005. Trace element distribution during the reproductive cycle of female and male spiny and Pacific scallops, with implications for biomonitoring. Mar. Pollut. Bull. 50, 175–184.

Olson, K.R., Harrel, R.C., 1973. Effect of salinity on acute toxicity of mercury, copper, and chromium for *Rangia cuneata* (Pelycopoda, Mactridae). Contrib. Mar. Sci. 17, 9–13.

Orlando, E., Mauri, M., 1978. Accumulation of manganese in *Donax trunculus*. L. (Bivalvia). IVes J. Etud. Pollut. 297–299.

Ozretic, B., Krajinovic-Ozretic, M., Santin, J., Medjugorac, B., Kras, M., 1990. As, Cd, Pb, and Hg in benthic animals from the Kvarner-Rijeka Bay region, Yugoslavia. Mar. Pollut. Bull. 21, 595–597.

Paez-Osuna, F., Osuna-Lopez, J.I., Izaguirse-Fiero, G., Zazueta-Padilla, H.M., 1994. Trace metals in mussels from the Ensenada del Pallon Lagoon, Mexico. Mar. Pollut. Bull. 28, 124–126.

Palmer, J.B., Rand, G.M., 1977. Trace metal concentrations in two shellfish species of commercial importance. Bull. Environ. Contam. Toxicol. 18, 512–520.

Papadopoulu, C., 1973. In: The elementary composition of marine invertebrates as a contribution to the sea pollution investigation. Proceedings of MAMBO Meeting, Castellabate, Italy, June 18-22, 1973, pp. 1–18.

Papadopoulu, C., Hadzistelios, I., 1977. Radiochemical determination of europium and application of the method in marine samples. Rapp. Comm. Int. Mer Medit. 24, 89–93.

Parchevskii, V.P., Polikarpov, G.G., Zabarunova, I.S., 1965. Certain regularities in the accumulation of yttrium and strontium by marine organisms. Dokl. Akad. Nauk SSSR 164, 913–916.

Parsons, T.R., Bawden, C.A., Heath, W.A., 1973. Preliminary survey of mercury and other metals contained in animals from the Fraser River mudflats. J. Fish. Res. Board Can. 30, 1014–1016.

Patel, B., Ganguly, A.K., 1973. Occurrence of Se 75 and Sn 113 in oysters. Health Phys. 24, 559–562.

Patel, B., Valanju, P.G., Mulay, C.D., Balani, M.C., Patel, S., 1973. Radioecology of certain molluscs in Indian coastal waters. In: Radioactive Contamination of the Marine Environment. IAEA, Vienna, Austria, pp. 307–330.

Patel, B., Patel, S., Balani, M.C., Pawar, S., 1978. Flux of certain radionuclides in the blood-clam *Anadara granosa* Linnaeus under environmental conditions. J. Exp. Mar. Biol. Ecol. 35, 177–195.

Paul, J.D., Davies, I.M., 1986. Effects of copper- and tin-based anti-fouling compounds on the growth of scallops (*Pecten maximus*) and oysters (*Crassostrea gigas*). Aquaculture 54, 191–203.

Pavoni, B., Centanni, E., Valcanover, S., Fasolato, M., Ceccato, S., Tagliapietra, D., 2007. Imposex levels and concentrations of organotin compounds (TBT and its metabolites) in *Nassarius nitidus* from the Lagoon of Venice. Mar. Pollut. Bull. 55, 505–511.

Peake, B.M., Marsden, I.D., Bryan, A.M., 2006. Spatial and temporal variations in trace metal concentrations in the cockle, *Austrovenus stutchburyi* from Otago, New Zealand. Environ. Monit. Assess. 115, 119–144.

Pearce, N.J.G., Mann, V.L., 2006. Trace metal variations in the shells of *Ensis siliqua* record pollution and environmental conditions in the sea to the west of mainland Britain. Mar. Pollut. Bull. 52, 739–755.

Peden, J.D., Crothers, J.H., Waterfall, C.E., Beasley, J., 1973. Heavy metals in Somerset marine organisms. Mar. Pollut. Bull. 4, 7–10.

Penrose, W.R., Black, R., Hayward, M.J., 1975. Limited arsenic dispersion in seawater, sediments, and biota near a continuous source. J. Fish. Res. Board Can. 32, 1275–1281.

Pentreath, R.J., 1973. The accumulation from water of ^{65}Zn, ^{54}Mn, ^{58}Co and ^{59}Fe by the mussel, *Mytilus edulis*. J. Mar. Biol. Assoc. UK 53, 127–143.

Pequegnat, J.E., Fowler, S.W., Small, L.F., 1969. Estimates of the zinc requirements of marine organisms. J. Fish. Res. Board Can. 26, 145–150.

Pesch, G.G., Stewart, N.E., 1980. Cadmium toxicity to three species of invertebrates. Mar. Environ. Res. 3, 145–156.

Pesch, G., Reynolds, B., Rogerson, P., 1977. Trace metals in scallops from within and around two ocean disposal sites. Mar. Pollut. Bull. 8, 224–228.

Peshut, P.J., Morrison, R.J., Brooks, B.A., 2008. Arsenic speciation in marine fish and shellfish from American Samoa. Chemosphere 71, 484–492.

Pfeiffer, W.C., Lacerda, L.D., Fiszman, M., Lima, N.R.W., 1985. Metais pesados no pescado do Baia de Sepetiba, Estado de Rio de Janeiro. Ciencia Cultura 37, 297–302.

Phelps, H.L., 1979. Cadmium sorption in estuarine mud-type sediment and the accumulation of cadmium in the soft-shell clam, *Mya arenaria*. Estuaries 2, 40–44.

Phelps, D.K., Telek, G., Lapan Jr., R.L., 1975. Assessment of heavy metal distribution within the food web. In: Pearson, E.A., Frangipane, E.D. (Eds.), Marine Pollution and Marine Waste Disposal. Pergamon Press, Elmsford, NY, pp. 341–348.

Phillips, J.H., 1976a. The common mussel *Mytilus edulis* as an indicator of pollution by zinc, cadmium, lead and copper. I. Effects of environmental variables on uptake of metals. Mar. Biol. 38, 59–69.

Phillips, J.H., 1976b. The common mussel *Mytilus edulis* as an indicator of pollution by zinc, cadmium, lead and copper. II. Relationship of metals in the mussels to those discharged by industry. Mar. Biol. 38, 71–80.

Phillips, D.J.H., 1977a. The use of biological indicator organisms to monitor trace metal pollution in marine and estuarine environments—a review. Environ. Pollut. 13, 281–317.

Phillips, D.J.H., 1977b. Effects of salinity on the net uptake of zinc by the common mussel *Mytilus edulis*. Mar. Biol. 41, 79–88.

Phillips, D.J.H., 1977c. The common mussel *Mytilus edulis* as an indicator of trace metals in Scandinavian waters. I. Zinc and cadmium. Mar. Biol. 43, 283–291.

Phillips, D.J.H., 1978. The common mussel *Mytilus edulis* as an indicator of trace metals in Scandinavian waters. II. Lead, iron, and manganese. Mar. Biol. 46, 147–156.

Phillips, D.J.H., 1979. The rock oyster *Saccostrea glomerata* as an indicator of trace metals in Hong Kong. Mar. Biol. 53, 353–360.

Phillips, D.J.H., Depledge, M.H., 1986. Chemical forms of arsenic in marine organisms, with emphasis on *Hemifusus* species. Water Sci. Technol. 18, 213–222.

Phillips, D.J.H., Thompson, G.B., Gabuji, K.M., Ho, C.T., 1982. Trace metals of toxicological significance to man in Hong Kong seafood. Environ. Pollut. 3B, 27–45.

Pilkey, O.H., Goodell, H.G., 1963. Trace elements in recent mollusk shells. Limnol. Oceanogr. 8, 137–148.

Pillai, K.C., Smith, R.C., Folsom, T.R., 1964. Plutonium in the marine environment. Nature 203, 568–571.

Pinkney, A.E., Matteson, L.L., Wright, D.A., 1990. Effects of tributyltin on survival, growth, morphometry, and RNA-DNA ratio of larval striped bass, *Morone saxatilis*. Arch. Environ. Contam. Toxicol. 19, 235–240.

Podgurskaya, O.V., Kavun, V.Y., 2006. Cadmium concentration and subcellular distribution in organs of the mussel *Crenomytilus grayanus* from upwelling regions of Okhotsk Sea and Sea of Japan. Arch. Environ. Contam. Toxicol. 51, 567–572.

Pouvreau, B., Amiard, J.C., 1974. Etude experimentale de l'accumulation de l'argent 110 m chez divers organismes marins. Comm. Ener. Atom. France. Rept. CEA-R-4571, pp. 1–19.

Presley, B.J., Taylor, R.J., Boothe, P.N., 1990. Trace metals in Gulf of Mexico oysters. Sci. Total Environ. 97/98, 551–593.

Preston, E.M., 1971. The importance of ingestion in chromium-51 accumulation by *Crassostrea virginica* (Gmelin). J. Exp. Mar. Biol. Ecol. 6, 47–54.

Preston, A., Dutton, J.W.R., Harvey, B.R., 1968. Detection, estimation, and radiological significance of silver-110m in oysters in the Irish Sea and the Blackwater Estuary. Nature 218, 689–690.

Preston, A., Jeffries, D.F., Dutton, J.W.R., Harvey, B.R., Steele, A.K., 1972. British Isles coastal waters: the concentrations of selected heavy metals in sea water, suspended matter and biological indicators—a pilot survey. Environ. Pollut. 3, 69–82.

Price, R.J., Lee, J.S., 1972. Effects of cations on the interactions between paralytic shellfish poison and butter clam *Saxidomus giganteus* melanin. J. Fish. Res. Board Can. 29, 1659–1661.

Pringle, B.H., Hissong, D.E., Katz, E.L., Malawka, S.T., 1968. Trace metal accumulation by estuarine mollusks. J. Sanit. Eng. Div. 94(SA3), 455–475.

Protasowicki, M., Dural, M., Jaremek, J., 2008. Trace metals in the shells of blue mussels (*Mytilus edulis*) from the Poland coast of Baltic Sea. Environ. Monit. Assess. 141, 329–337.

Pyle, T.E., Tieh, T.T., 1970. Strontium, vanadium, and zinc in the shells of pteropods. Limnol. Oceanogr. 15, 153–154.

Raimundo, J., Caetano, M., Viale, C., 2004. Geographical variation and partition of metals in *Octopus vulgaris* along the Portuguese coast. Sci. Total Environ. 325, 71–81.

Raimundo, J., Vale, C., Duarte, R., Moura, J., 2008. Sub-cellular partitioning of Zn, Cu, Cd and Pb in the digestive gland of native *Octopus vulgaris* exposed to different metal concentrations (Portugal). Sci. Total Environ. 390, 410–416.

Ramelow, G.J., Webre, C.L., Mueller, C.S., Beck, J.N., Young, J.C., Langley, M.P., 1989. Variations of heavy metals and arsenic in fish and other organisms from the Calcasieu River and Lake, Louisiana. Arch. Environ. Contam. Toxicol. 18, 804–818.

Ratkowsky, D.A., Thrower, S.J., Eustace, I.J., Olley, J., 1974. A numerical study of the concentration of some heavy metals in Tasmanian oysters. J. Fish. Res. Board Can. 31, 1165–1171.

Rato, M., Sousa, A., Quinta, R., Langston, W., Barroso, C., 2006. Assessment of inshore/offshore tributyltin pollution gradients in the northwest Portugal continental shelf using *Nassarius reticulatus* as a bioindicator. Environ. Toxicol. Chem. 25, 3213–3220.

Ratte, H.T., 1999. Bioaccumulation and toxicity of silver compounds: a review. Environ. Toxicol. Chem. 18, 89–108.

Raymont, J.E.G., 1972. Some aspects of pollution in Southampton water. Proc. R. Soc. Lond. 180B (1061), 451–468.

Rebelo, M.F., Amaral, M.C.R., Pfeiffer, W.C., 2003a. High Zn and Cd accumulation in the oyster *Crassostrea rhizophorae* and its relevance as a sentinel species. Mar. Pollut. Bull. 46, 1341–1358.

Rebelo, M.F., Pfeiffer, W.C., Silva, H., Moraes, M.O., 2003b. Cloning and detection of metallothionein mRNA by PT-PCR in mangrove oysters (*Crassostrea rhizophorae*). Aquat. Toxicol. 64, 359–362.

Redpath, K.J., Davenport, J., 1988. The effect of copper, zinc and cadmium on the pumping rate of *Mytilus edulis*. L. Aquat. Toxicol. 13, 217–226.

Rees, H.I., Waldock, R., Matthiessen, P., Pendle, M.A., 1999. Surveys of the epibenthos of the Crouch Estuary (UK) in relation to TBT contamination. J. Mar. Biol. Assoc. UK 79, 209–223.

Reimer, A.A., Reimer, R.D., 1975. Total mercury in some fish and shellfish along the Mexican coast. Bull. Environ. Contam. Toxicol. 14, 105–111.

Reish, D.J., 1977. Effects of chromium on the life history of *Capitella capitata* (Annelida: Polychaeta). In: Vernberg, F.J., Calabrese, A., Thurberg, F.P., Vernberg, W.B. (Eds.), Physiological Responses of Marine Biota to Pollutants. Academic Press, New York, pp. 199–207.

Reish, D.J., Martin, J.M., Piltz, F.M., Word, J.Q., 1976. The effect of heavy metals on laboratory population of two polychaetes with comparison to the water quality conditions and standards in southern California marine waters. Water Res. 10, 299–302.

Renzoni, A., Bacci, E., Falciai, L., 1973. In: Mercury concentrations in the water, sediments and fauna of an area of the Tyrrhenian coast. Sixth International Symposium on Medical Oceanography, Portoroz, Yugoslavia, September 26-30, 1973, pp. 17–45.

Reynolds, B.H., 1979. Trace metals monitoring at two ocean disposal sites. US Environ. Protect. Agen., Rep. EPA 600/3-79-037, pp. 1–64.

Robertson, D.E., 1967. Trace elements in marine organisms. Rapp. Am. BNWL 481-2, pp. 56–59.

Robertson, D.E., Rancitelli, L.A., Langford, J.C., Perkins, R.W., 1972. In: Battelle-Northwest Contribution to the IDOE Base-Line Study. Battelle Pacific Northwest Laboratory, Richland, WA, pp. 1–46.

Rocca, E., 1969. Copper distribution in *Octopus vulgaris* Lam. hepatopancreas. Comp. Biochem. Physiol. 28, 67–82.

Roesijadi, G., 1994. Behavior of metallothionein-bound metals in a natural population of an estuarine mollusc. Mar. Environ. Res. 38, 147–168.

Roesijadi, G., Fellingham, G.W., 1987. Influence of Cu, Cd, and Zn preexposure on Hg toxicity in the mussel *Mytilus edulis*. Can. J. Fish. Aquat. Sci. 44, 680–684.

Romeril, M.G., 1971. The uptake and distribution of ^{65}Zn in oysters. Mar. Biol. 9, 347–354.

Romeril, M.G., 1974. Trace metals in sediments and bivalve mollusca in Southampton water and the Solent. Rev. Int. Oceanogr. Med. 33, 31–47.

Romeril, M.G., 1977. Heavy metal accumulation in the vicinity of a desalination plant. Mar. Pollut. Bull. 8, 84–87.

Roosenburg, W.H., 1969. Greening and copper accumulation in the American oyster, *Crassostrea virginica*, in the vicinity of a steam electric generating station. Chesapeake Sci. 10, 241–252.

Roper, D.S., Hickey, C.W., 1994. Behavioural responses of the marine bivalve *Macoma liliana* exposed to copper- and chlordane-exposed sediments. Mar. Biol. 118, 673–680.

Rose, K.A., Summers, J.K., McLean, R.I., Domotor, S.L., 1988. Radiosilver (Ag-110m) concentrations in Chesapeake Bay oysters maintained near a nuclear power plant: a statistical analysis. Environ. Monit. Assess. 10, 205–218.

Rosemarin, A., Notini, M., Holmgren, K., 1985. The fate of arsenic in the Baltic Sea *Fucus vesiculosus* ecosystem. Ambio 14, 342–345.

Rosen, G., Rivera-Duarte, I., Chadwick, D.B., Ryan, A., Santore, R.C., Paquin, P.R., 2008. Critical tissue copper residues for marine bivalve (*Mytilus galloprovincialis*) and echinoderm (*Strongylocentrotus purpuratus*) embryonic development: conceptual regulatory and environmental regulations. Mar. Environ. Res. 55, 327–336.

Rosenberg, R., 1977. Effects of dredging operations on estuarine benthic fauna. Mar. Pollut. Bull. 8, 102–104.

Rucker, J.B., Valentine, J.W., 1961. Salinity response of trace element concentration in *Crassostrea virginica*. Nature 190, 1099–1100.

Ruddell, C.L., Rains, D.W., 1975. The relationship between zinc, copper and the basophils of two crassostreid oysters, *C. gigas* and *C. virginica*. Comp. Biochem. Physiol. 51A, 591–595.

Ruelas-Inzunza, J., Paez-Osuna, F., 2008. Trophic distribution of Cd, Pb, and Zn in a food web from Alata-Ensenada del Pabellon subtropical lagoon, SE Gulf of Mexico. Arch. Environ. Contam. Toxicol. 54, 584–596.

Ruelas-Inzunza, J., Garate-Viera, Y., Paez-Osuna, F., 2007. Lead in clams and fish of dietary importance from Coatzacoalcos estuary (Gulf of Mexico), an industrialized tropical region. Bull. Environ. Contam. Toxicol. 79, 508–513.

Said, T.O., Farag, R.S., Younis, A.M., Shreadah, M.A., 2006. Organotin species in fish and bivalves samples collected from the Egyptian Mediterranean Coast of Alexandria, Egypt. Bull. Environ. Contam. Toxicol. 77, 451–458.

Saiz-Salinas, J.I., Ruiz, J.M., Frances-Zubillaga, G., 1996. Heavy metal levels in intertidal sediments and biota from the Bidasoa estuary. Mar. Pollut. Bull. 32, 69–71.

Sajwan, K.S., Kumar, K.S., Paramasivam, S., Compton, S.S., Richardson, J.P., 2008. Elemental status in sediment and American oyster collected from Savannah marsh/estuarine ecosystem: a preliminary assessment. Arch. Environ. Contam. Toxicol. 54, 245–258.

Salo, E.W., Leet, W.L., 1969. The concentration of zinc-65 by oysters maintained in the discharge canal of a nuclear power plant. In: Proceedings of the 2nd National Symposium on Radioecology, Ann Arbor, MI, 1967, pp. 363–371.

Sanders, J.G., Abbe, G.R., Riedel, G.F., 1990. Silver uptake and subsequent effects on growth and species composition in an estuarine community. Sci. Total Environ. 97/98, 761–769.

Sanders, B.M., Martin, L.S., Nelson, W.G., Phelps, D.K., Welch, W., 1991a. Relationships between accumulation of a 60 kDa stress protein and a scope-for-growth in *Mytilus edulis* exposed to a range of copper concentrations. Mar. Environ. Res. 31, 81–97.

Sanders, J.G., Riedel, G.F., Abbe, G.R., 1991b. Factors controlling the spatial and temporal variability of trace metal concentrations in *Crassostrea virginica* (Gmelin). In: Elliott, M., Ducrotoy, J.P. (Eds.), Estuaries and Coasts: Spatial and Temporal Comparisons. ECSA Sympos. 19 (Univ. Caen, France, 1989). Olsen & Olsen, Fredensborg, Denmark, pp. 335–339.

Sanders, B.M., Martin, L.S., Howe, S.R., Nelson, W.G., Hegre, E.S., Phelps, D.K., 1994. Tissue-specific differences in accumulation of stress proteins in *Mytilus edulis* exposed to a range of copper concentrations. Toxicol. Appl. Pharmacol. 125, 206–213.

Sankar, T.V., Zynudheen, A.A., Anandan, R., Nair, P.G.V., 2006. Distribution of organochlorine pesticides and heavy metal residues in fish and shellfish from Calicut region, Kerala, India. Chemosphere 65, 583–590.

Sankaranarayanan, V.N., Pusushan, K.S., Rao, T.S.S., 1978. Concentration of some of the heavy metals in the oyster, *Crassostrea madrasensis* (Preston) from the Cochin Region. Indian J. Mar. Sci. 7, 130–131.

Santos, I.R., Silva-Filho, E.V., Schaefer, C., Sella, S.M., Silva, C.A., Gomes, V., et al., 2006. Baseline mercury and zinc concentrations in terrestrial and coastal organisms of Admiralty Bay, Antarctica. Environ. Pollut. 140, 304–311.

Sasikumar, G., Krishnakumar, P.K., Bhat, G.S., 2006. Monitoring trace metal contaminants in green mussel, *Perna viridis* from the coastal waters of Karnataka, southwest coast of India. Arch. Environ. Contam. Toxicol. 51, 206–214.

Sastry, V.W., Bhatt, Y.M., 1965. Zinc content of some marine bivalves and barnacles from Bombay Shores. J. Indian Chem. Soc. 42, 121–122.

Sautet, J., Oliver, H., Quicke, J., 1964. Contribution to the study of the biological fixation and elimination of arsenic by *Mytilus edulis*. Second note. Ann. Med. Leg. Criminol. Police Sci. Toxicol. 44, 466–471.

Savari, A., Lockwood, A.P.M., Sheader, M., 1991. Effects of season and size (age) on heavy metal concentrations of the common cockle (*Cerastoderma edule* (L.)) from Southampton water. J. Molluscan Stud. 57, 45–57.

Saward, D., Stirling, A., Topping, G., 1975. Experimental studies on the effects of copper on a marine food chain. Mar. Biol. 29, 351–356.

Sayler, G.S., Nelson Jr., J.D., Colwell, R.R., 1975. Role of bacteria in bioaccumulation of mercury in the oyster *Crassostrea virginica*. Appl. Microbiol. 30, 91–96.

Scancar, J., Zuliani, T., Turk, T., Milacic, R., 2007. Organotin compounds and selected metals in the marine environment of northern Adriatic Sea. Environ. Monit. Assess. 127, 271–282.

Scanes, P., 1993. Trace metal uptake in cockles *Anadara trapezium* from Lake Macquarie, New South Wales. Mar. Ecol. Prog. Ser. 102, 135–142.

Schell, W.R., Nevissi, A., 1977. Heavy metals from waste disposal in Central Puget Sound. Environ. Sci. Technol. 11, 887–893.

Schintu, M., Durante, L., Maccioni, A., Meloni, P., Degetto, S., Contu, A., 2008. Measurement of environmental trace-metal levels in Mediterranean coastal areas with transplanted mussels and DGT techniques. Mar. Pollut. Bull. 57, 832–837.

Schipp, R., Hevert, F., 1978. Distribution of copper and iron in some central organs of *Sepia officinalis* (cephalopoda). A comparative study by flameless atomic absorption and electron microscopy. Mar. Biol. 47, 391–399.

Schneider, J., 1972. Lower fungi as test organisms of pollutants in sea and brackish water. The effects of heavy metal compounds and phenol on *Thraustochytrium striatum*. Mar. Biol. 16, 214–225.

Schuhmacher, M., Bosque, M.A., Domingo, J.L., Corbella, J., 1990. Lead and cadmium concentrations in marine organisms from the Tarragona coastal waters, Spain. Bull. Environ. Contam. Toxicol. 44, 784–789.

Schuhmacher, M., Domingo, J.L., Bosque, M.A., Corbella, J., 1992. Heavy metals in marine species from the Tarragona Coast, Spain. J. Environ. Sci. Health 27A, 1939–1948.

Schuhmacher, M., Batiste, J., Bosque, M.A., Domingo, J.L., Corbella, J., 1994. Mercury concentrations in marine species from the coastal area of Tarragona Province, Spain. Dietary intake of mercury through fish and seafood consumption. Sci. Total Environ. 156–273.

Schulz-Baldes, M., 1972. Toxicity and accumulation of lead in the common mussel *Mytilus edulis* in laboratory experiment. Mar. Biol. 16, 226–229.

Schulz-Baldes, M., 1973. Die Miesmuschel, *Mytilus edulis*, als Indikator fuer die Bleikonzentration in Weseraestuar un in der Deutschen Bucht. Mar. Biol. 21, 98–102.

Schulz-Baldes, M., 1974. Lead uptake from sea water and food, and lead loss in the common mussel, *Mytilus edulis*. Mar. Biol. 25, 177–193.

Scott, D.M., Major, C.W., 1972. The effect of copper (II) on survival, respiration and heart rate in the common blue mussel, *Mytilus edulis*. Biol. Bull. 143, 679–688.

Segar, D.A., Collins, J.D., Riley, J.P., 1971. The distribution of the major and some minor elements in marine animals. Part II. Molluscs. J. Mar. Biol. Assoc. UK 51, 131–136.

Seixas, S., Pierce, G.J., 2005. Vanadium, rubidium, and potassium in *Octopus vulgaris* (Mollusca: Cephalopoda). Sci. Mar. 69, 215–222.

Sen, H., Sunlu, U., 2007. Effects of cadmium ($CdCl_2$) on development and hatching of eggs in European squid (*Loligo vulgaris* Lamarck, 1798) (Cephalopoda: Loliginidae). Environ. Monit. Assess. 133, 371–378.

Sericano, J.L., Wade, T.L., Qian, Y., 1999. NOAA's national status and trends project: environmental significance of the uptake and depuration of the antifouling agent tributyltin by the American oyster *Crassostrea virginica*. In: 218th American Chemical Society National Meeting, New Orleans, August 22-26, 1999. Book of Abstracts, Part I, AGRO 59.

Shiber, J.G., Shatila, T.A., 1978. Lead, cadmium, copper, nickel and iron in limpets, mussels and snails from the coast of Ras Beirut, Lebanon. Mar. Environ. Res. 1, 125–134.

Shim, W.J., Oh, J.R., Kahng, S.H., Shim, J.H., Lee, S.H., 1998. Accumulation of tributyl-and triphenyltin compounds in Pacific oyster, *Crassostrea gigas,* from the Chinhae Bay system, Korea. Arch. Environ. Contam. Toxicol. 35, 41–47.

Shimizu, M., Kajihara, T., Hiyama, Y., 1970. Uptake of ^{60}Co by marine animals. Rec. Oceanogr. Works Jpn. 10, 137–145.

Shimizu, M., Kajihara, T., Suyama, I., Hiyama, Y., 1971. Uptake of ^{58}Co by mussel, *Mytilus edulis*. J. Rad. Res. 12, 17–28.

Shiomi, K., Shinigawa, A., Hirota, K., Yamanaka, H., Kikuchi, T., 1984. Identification of arsenobetaine as a major arsenic compound in the ivory shell *Buccinum striatissimum*. Agric. Biol. Chem. 48, 2863–2864.

Shore, R., Carney, G., Stygall, T., 1975. Cadmium levels and carbohydrate metabolism in limpets. Mar. Pollut. Bull. 6, 187–189.

Shuster Jr., C.N., Pringle, B.H., 1968. Effects of trace metals on estuarine molluscs. In: Proceedings of the First Mid-Atlantic Industrial Waste Conference, November 1967. Available from Dept. Civil Eng., Univ. Delaware, Newark, DE, pp. 285–304.

Shuster Jr., C.N., Pringle, B.H., 1969. Trace metal accumulation by the American oyster, *Crassostrea virginica*. 1968 Proc. Nat. Shellfish Assoc. 59, 91–103.

Side, J., 1987. Organotins—not so good relations. Mar. Pollut. Bull. 18, 205–206.

Silva, C.A.R., Smith, B.D., Rainbow, P.S., 2006. Comparative biomonitors of coastal trace metal contamination in tropical South America (N. Brazil). Mar. Environ. Res. 61, 439–445.

Simmonds, M., 1986. The case against tributyltin. Oryx 20, 217–220.

Simpson, R.D., 1979. Uptake and loss of zinc and lead by mussels (*Mytilus edulis*) and relationship with body weight and reproductive cycle. Mar. Pollut. Bull. 10, 74–78.

Sims Jr., R.R., Presley, B.J., 1976. Heavy metal concentrations in organisms from an actively dredged Texas bay. Bull. Environ. Contam. Toxicol. 16, 520–527.

Smith, B.S., 1981a. Male characteristics on female mud snails caused by antifouling bottom paints. J. Appl. Toxicol. 1, 22–25.

Smith, B.S., 1981b. Tributyltin compounds induce male characteristics on female mud snails *Nassarius obsoletus* = *Ilyanassa obsoleta*. J. Appl. Toxicol. 1, 141–144.

Smith, A.J., Thain, J.E., Barry, J., 2006. Exploring the use of caged *Nucella lapillus* to monitor changes to TBT hotspot areas: a trial in the River Tyne estuary (UK). Mar. Environ. Res. 62, 149–163.

Sokolowski, A., Wolowicz, M., Hummel, H., 2007. Metal sources to the Baltic clam *Macoma balthica* (mollusca: bivalvia) in the southern Baltic Sea (the Gulf of Gdansk). Mar. Environ. Res. 63, 236–256.

Sorentino, C., 1979. Mercury in marine and freshwater fish of Papua, New Guinea. Aust. J. Mar. Freshw. Res. 30, 617–623.

Soria-Dengg, S., Ochavillo, D., 1990. Comparative toxicities of trace metals on embryos of the giant clam, *Tridacna derasa*. Asian Mar. Biol. 7, 161–166.

Sousa, A., Matsudaira, C., Takahashi, S., Tanabe, S., Barroso, C., 2007. Integrative assessment of organotin contamination in a southern European estuarine system (Ria de Aveiro, NW Portugal): tracking temporal trends in order to evaluate the effectiveness of the EU ban. Mar. Pollut. Bull. 54, 1645–1653.

Sparling, D.W., Lowe, T.P., 1996. Environmental hazards of aluminum to plants, invertebrates, fish, and wildlife. Rev. Environ. Contam. Toxicol. 145, 1–127.

Sprague, J.B., 1986. Toxicity and Tissue Concentrations of Lead, Zinc, and Cadmium for Marine Molluscs and Crustaceans. Available from International Lead Zinc Research Organization, 2525 Meridian Parkway, Research Triangle Park, NC215.

Stenner, R.D., Nickless, G., 1974a. Absorption of cadmium, copper and zinc by dog whelks in the Bristol Channel. Nature 247, 198–199.

Stenner, R.D., Nickless, G., 1974b. Distribution of some heavy metals in organisms in Hardangerfjord and Skjerstadfjord, Norway. Water Air Soil Pollut. 3, 279–291.

Stenner, R.D., Nickless, G., 1975. Heavy metals in organisms of the Atlantic coast of S.W. Spain and Portugal. Mar. Pollut. Bull. 6, 89–92.

Stephenson, J.D., Leonard, G.H., 1994. Evidence for the decline of silver and lead and the increase of copper from 1977 to 1990 in the coastal marine waters of California. Mar. Pollut. Bull. 28, 148–153.

Stephenson, R.F., Taylor, D., 1975. The influence of EDTA on the mortality and burrowing activity of the clam (*Venerupis decussata*) exposed to sublethal concentrations of copper. Bull. Environ. Contam. Toxicol. 14, 304–308.

Stewart, J., Schulz-Baldes, M., 1976. Long term lead accumulation in abalone (*Haliotis* spp.) fed on lead-treated brown algae (*Egregia laevigata*). Mar. Biol. 36, 19–24.

Stromgren, T., Nielsen, M.V., 1991. Spawning frequency, growth and mortality of *Mytilus edulis* larvae exposed to copper and diesel oil. Aquat. Toxicol. 21, 171–180.

Stronkhorst, J., 1992. Trends in pollutants in blue mussel *Mytilus edulis* and flounder *Platichthys flesus* from two Dutch estuaries, 1985-1990. Mar. Pollut. Bull. 24, 250–258.

Stump, I.G., Kearney, J., D'Auria, J.M., Popham, J.D., 1979. Monitoring trace elements in the mussel, *Mytilus edulis*, using X-ray energy spectroscopy. Mar. Pollut. Bull. 10, 270–274.

Sturesson, U., 1976. Lead enrichment in shells of *Mytilus edulis*. Ambio 5, 235–236.

Sturesson, U., 1978. Cadmium enrichment in shells of *Mytilus edulis*. Ambio 7, 122–125.

Sturesson, U., Reyment, R.A., 1971. Some minor chemical constituents of the shell of *Macoma balthica*. Oikos 22, 414–416.

Sullivan, P.A., Robinson, W.E., Morse, M.P., 1988. Isolation and characterization of granules from the kidney of the bivalve *Mercenaria mercenaria*. Mar. Biol. 99, 359–368.

Sunlu, U., 2006. Trace metal levels in mussels (*Mytilus galloprovincialis* L. 1758) from Turkish Aegean sea coast. Environ. Monit. Assess. 114, 273–286.

Swaileh, K.M., Adelung, D., 1994. Levels of trace metals and effect of body size on metal content and concentration in *Arctica islandica* L. (Mollusca: Bivalvia) from Kiel Bay, western Baltic. Mar. Pollut. Bull. 28, 500–505.

Szefer, P., Fowler, S.W., Ikuta, K., Osuna, F.P., Ali, A.A., Kim, B.S., et al., 2006. A comparative assessment of heavy metal accumulation in soft parts and byssus of mussels from subarctic, temperate, and subtropical and tropical marine environments. Environ. Pollut. 139, 70–78.

Tabata, K., 1969. Studies on the toxicity of heavy metals to aquatic animals and the factors to decrease the toxicity—I. On the formation and the toxicity of precipitate of heavy metals. Bull. Tokai Fish. Res. Lab. Tokyo 58, 203–214.

Takevich, T., 1972. Distribution of mercury in the environment of Minamata Bay and inland Ariake Sea. In: Hartung, B.D., Dinman, B.D. (Eds.), Environmental Mercury Contamination. Ann Arbor Science Publications, Ann Arbor, MI, pp. 79–81.

Talbot, V., 1985. Heavy metal concentrations in the oysters *Saccostrea cucullata* and *Saccostrea* sp. From the Dampier Archipelago, Western Australia. Aust. J. Mar. Freshw. Res. 36, 169–175.

Talbot, V., Magee, R.J., 1978. Naturally occurring heavy metal binding proteins in invertebrates. Arch. Environ. Contam. Toxicol. 7, 73–81.

Talbot, V., Magee, R.J., Hussain, M., 1976a. Lead in Port Phillip Bay mussels. Mar. Pollut. Bull. 7, 234–237.

Talbot, V.W., Magee, R.J., Hussain, M., 1976b. Cadmium in Port Phillip Bay mussels. Mar. Pollut. Bull. 7, 84–86.

Talbot, V., Creagh, S., Schulz, R., 1985. The derivation of threshold mean concentrations of copper and zinc in seawater, to protect the edible tropical rock oyster *Saccostrea cucullata*, from exceeding the health (food) standards. Dept. Conserv. Environ. Perth, West. Aust. Bull. 12, 1–25.

Tedesco, S., Doyle, H., Redmond, G., Sheehan, D., 2008. Gold nanoparticles and oxidative stress in *Mytilus edulis*. Mar. Environ. Res. 66, 131–133.

Thain, J.E., Waldock, M.J., 1986. The impact of tributyl tin (TBT) antifouling paints on molluscan fisheries. Water Sci. Technol. 18, 193–202.

Thebault, H., Rodriguez y Baena, A.M., Andral, B., Barisic, D., Albaladejo, J.B., Bologa, A.S., et al., 2008. ^{137}Cs baseline levels in the Mediterranean and Black Sea: a cross-basin survey of the CIESM Mediterranean Mussel Watch programme. Mar. Pollut. Bull. 57, 801–806.

Thibaud, J., 1973. Teneur en mercure dans les moules du littoral francais. Sci. Peche Bull. Inst. Peches Marit. 221, 1–6.

Thompson, T.G., Chow, T.J., 1955. The strontium-calcium atom ratio in carbonate-secreting marine organisms. Deep Sea Res. 3 (Suppl.), 20–39.

Thompson, J.A.J., Davis, J.C., Drew, R.E., 1976. Toxicity, uptake and survey studies of boron in the marine environment. Water Res. 10, 869–875.

Thompson, J.A.J., Sheffer, M.G., Pierce, R.C., Chau, Y.K., Cooney, J.J., Cullen, W.R., et al., 1985. Organotin compounds in the aquatic environment: scientific criteria fo assessing their effects on environmental quality. Natl. Res. Coun. Can.. Publ. NRCC 22494, pp. 1–284.

Thomson, J.D., 1979. Heavy metals in the native oyster (*Ostrea angasi*) and mussel (*Mytilus edulis planulatus*) from Port Davey, south-western Tasmania. Aust. J. Mar. Freshw. Res. 30, 421–424.

Thornton, I., Watling, H., Darracott, A., 1975. Biochemical studies in several rivers and estuaries used for oyster rearing. Sci. Total Environ. 4, 325–345.

Thorsson, M.H., Hedman, J.E., Bradshaw, C., Gunnarsson, J.S., Gilek, M., 2008. Effects of settling organic matter on the bioaccumulation of cadmium and BDE-99 by Baltic Sea invertebrates. Mar. Environ. Res. 65, 264–281.

Thrower, S.J., Eustace, I.J., 1973a. Heavy metal accumulation in oysters grown in Tasmanian waters. Food Tech. Aust. 25, 546–553.

Thrower, S.J., Eustace, I.J., 1973b. Heavy metals in Tasmanian oysters in 1972. Aust. Fish. 32 (Oct. 1973), 7–10.

Thurberg, F.P., Calabrese, A., Dawson, M.A., 1974. Effects of silver on oxygen consumption of bivalves at various salinities. In: Vernberg, F.J., Vernberg, W.B. (Eds.), Pollution and Physiology of Marine Organisms. Academic Press, New York, pp. 67–78.

Thurberg, F.P., Cable, W.D., Dawson, M.A., MacInnes, J.R., Wenzloff, D.R., 1975. Respiratory response of larval, juvenile and adult surf clams, *Spisula solidissima*, to silver. In: Cech Jr., J.J., Bridges, D.W., Horton, D.B. (Eds.), Respiration of Marine Organisms. TRIGOM Publications, South Portland, ME, pp. 41–52.

Topping, G., 1973. Heavy metals in shellfish from Scottish waters. Aquaculture 1, 379–384.
Townsley, S.J., 1954. Studies on copper in mollusks, with particular reference to *Busycon canaliculatum* Linnaeus. Ph.D. thesis, Yale University, New Haven, CT, 126 pp.
Tran, D., Moody, A.J., Fisher, A.S., Foulkes, M.E., Jha, A.N., 2007. Protective effects of selenium on mercury-induced DNA damage in mussel haemocytes. Aquat. Toxicol. 84, 11–18.
Truchet, M., Martoja, R., Berthet, B., 1990. Consequences histologiques de la pollution metallique d'un estuaire sur deux mollusques, *Littorina littorea* L. et *Scrobicularia plana* da costa. C. R. Acad. Sci. Ser. III 311, 261–268.
Tsubaki, T., Sato, T., Kondo, K., Shirakawa, K., Kanbayashi, K., Hirota, K., et al., 1967. Outbreak of intoxication by organic mercury compound in Niigata Prefecture. An epidemiological and clinical study. Jpn. J. Med. 6, 132–133.
Tugrul, S., Balkas, T.I., Goldberg, E.D., 1983. Methyltins in the marine environment. Mar. Pollut. Bull. 14, 297–303.
Turgeon, D.D., O'Connor, T.P., 1991. Long Island Sound: distributions, trends, and effects of chemical contamination. Estuaries 14, 279–289.
Turekian, K.K., Armstrong, R.L., 1960. Magnesium, strontium, and barium concentrations and calcite-aragonite ratios of some recent molluscan shells. J. Mar. Res. 18, 133–151.
Turekian, K.K., Katz, A., Chan, L., 1973. Trace element trapping in pteropod tests. Limnol. Oceanogr. 18, 240–249.
Ueda, T., Nakamura, R., Suzuki, Y., 1978. Comparison of influences of sediments and sea water on accumulation of radionuclides by marine organisms. J. Radiat. Res. 19, 93–99.
Unlu, M.Y., Fowler, S.W., 1979. Factors affecting the flux of arsenic through the mussel *Mytilus galloprovincialis*. Mar. Biol. 51, 209–219.
Unlu, M.Y., Heyraud, M., Keckes, S., 1972. Mercury as a hydrospheric pollutant. I. Accumulation and excretion of $^{203}HgCl_2$ in *Tapes decussata* L. In: Ruivo, M. (Ed.), Marine Pollution and Sea Life. Fishing Trading News (Books), London, pp. 292–295.
Unlu, S., Topcuoglu, S., Alpar, B., Kurbasoglu, C., Yilmaz, Z.Y., 2008. Heavy metal pollution in surface sediment and mussel samples in the Gulf of Gemlik. Environ. Monit. Assess. 144, 169–178.
US Environmental Protection Agency (USEPA), 1977. Fed. Regist. 42(7), 2477.
USEPA, 1980a. Ambient water quality criteria for chromium. USEPA, Rep. 440/5-80-035, pp. 1–105.
USEPA, 1980b. Ambient water quality criteria for copper. USEPA, Rep. 440/5-80-036, pp. 1–162.
USEPA, 1980c. Ambient water quality criteria for nickel. USEPA, Rep. 440/5-80-060, pp. 1–206.
USEPA, 1980d. Ambient water quality criteria for zinc. USEPA, Rep. 440/5-80-079, pp. 1–158.
USEPA, 1985a. Ambient water quality criteria for arsenic—1984. USEPA, Rep. 440/5-84-033, pp. 1–66.
USEPA, 1985b. Ambient water quality criteria for lead—1984. USEPA, Rep. 440/5-84-027, pp. 1–81.
USEPA, 1986. Initiation of a special review of certain pesticide products containing tributyltins used as antifoulants; availability of support document. Fed. Regist. 51(5), 778–779.
USEPA, 1987. Ambient water quality criteria for zinc—1987. USEPA, Rep. 440/5-87-003, pp. 1–207.
US National Academy of Sciences (USNAS), 1973. Water Quality Criteria, 1972. US Environ. Protect. Agen. Rep. R3-73–033, 252 pp.
USNAS, 1975. Medical and Biological Effects of Environmental Pollutants. Nickel. Nat. Res. Coun., USNAS, Washington, DC, 277 pp.
USNAS, 1977. Arsenic. USNAS, Washington, DC, 332 pp.
USNAS, 1979. Zinc. USNAS, Natl. Res. Coun., Subcomm. Zinc., Univ. Park Press, Baltimore, MD, 471 pp.
US Navy (USN), 1984. Environmental Assessment of Fleetwide Use of Organotin Antifouling Paint. US Naval Sea Systems Command, Washington, DC, 184 pp.
US Public Health Service (USPHS), 1962. Drinking water standards. USPHS, Publ. 956, pp. 1–61.
USPHS, 1990. Toxicological profile for silver. Agen. Toxic Subs. Dis. Reg. TP-90-24, pp. 1–145.
USPHS, 1992. Toxicological profile for tin. Agen. Toxic Subs. Dis. Reg. TP-91/27, pp. 1–160.
Uysal, H., 1978a. The effects of some pollutants on *Mytilus galloprovincialis* Lam. and *Paracentrotus lividus* Lam. in the Bays of Izmir and Aliaga. IVes J. Etud. Pollut. 313–317.

Uysal, H., 1978b. Accumulation and distribution of heavy metals in some marine organisms in the bay of Izmir and on Aegean coasts. IV.^{es}J. Etud. Pollut. 213–217.

Valiela, I., Banus, M.D., Teal, J.M., 1974. Response of salt marsh bivalves to enrichment with metal-containing sewage sludge and retention of lead, zinc and cadmium by marsh sediments. Environ. Pollut. 7, 149–157.

Valkirs, A.O., Davidson, B.M., Seligman, P.F., 1987. Sublethal growth effects and mortality to marine bivalves from long-term exposure to tributyltin. Chemosphere 16, 201–220.

Van As, D., Fourie, H.O., Vleggaar, C.M., 1973. Accumulation of certain trace elements in the marine organisms from the sea around the Cape of Good Hope. Radioactive Contamination of the Marine Environment. IAEA, Vienna, Austria.

Van As, D., Fourie, H.O., Vleggaar, C.M., 1975. Trace element concentrations in marine organisms from the Cape West Coast. S. Afr. J. Sci. 71, 151–154.

van Slooten, K.B., Tarradellas, J., 1994. Accumulation, depuration and growth effects of tributyltin in freshwater bivalve *Dreissena polymorpha* under field conditions. Environ. Toxicol. Chem. 13, 755–762.

Van Weers, A.W., 1973. Uptake and loss of ^{65}Zn and ^{60}Co by the mussel *Mytilus edulis* L. In: Radioactive Contamination of the Marine Environment. IAEA, Vienna, Austria, pp. 401–853.

Vaskovsky, V.E., Korotchenko, O.D., Kosheleva, L.R., Levin, V.S., 1972. Arsenic in the lipid extracts of marine invertebrates. Comp. Biochem. Physiol. 41B(4), 777–784.

Vattuone, G.M., Griggs, K.S., McIntyre, D.R., Littlepage, J.L., Harrison, F.L., 1976. Cadmium concentrations in rock scallops in comparison with some other species. US Ener. Res. Dev. Admin., UCRL 52022, pp. 1–11.

Vecchio, P.V., Alasia, A.M., Gualdi, G., 1962. Determination of arsenic in molluscs (*Mytilus* Linn.). Ig. Sanita Pubbl. 18, 18–30.

Viarengo, A., Pertica, M., Mancinelli, G., Palermo, S., Zanicchi, G., Orunesu, M., 1981. Synthesis of Cu-binding proteins in different tissues of mussels exposed to the metal. Mar. Pollut. Bull. 12, 347–350.

Viarengo, A., Canesi, L., Pertica, M., Poli, G., Moore, M.N., Orunesu, M., 1990. Heavy metal effects on lipid peroxidation in the tissues of *Mytilus galloprovincialis* Lam. Comp. Biochem. Physiol. 97C, 37–42.

Viarengo, A., Canesi, L., Mazzucotelli, A., Ponzano, E., 1993. Cu, Zn and Cd content in different tissues of the Antarctic scallop *Adamussium colbecki*: role of metallothionein in heavy metal homeostasis and detoxication. Mar. Ecol. Prog. Ser. 95, 163–168.

Vijayavel, K., Gopalakrishnan, S., Balasubramanian, M.P., 2007. Sublethal effects of silver and chromium in the green mussel *Perna viridis* with reference to alterations in oxygen uptake, filtration rate and membrane bound ATPase system as biomarkers. Chemosphere 69, 979–986.

Vishwakiran, Y., Anil, A.C., Venkat, K., Sawant, S.S., 2006. *Gyrineum natator*: a potential indicator of imposex along the Indian coast. Chemosphere 62, 1718–1725.

Vlahogianni, T., Dassenakis, M., Scoullos, M.J., Valavanidis, A., 2007. Integrated use of biomarkers (superoxide dismutase, catalase, and lipid peroxidation) in mussels *Mytilus galloprovincialis* for assessing heavy metals' pollution in coastal areas from the Saronikos Gulf of Greece. Mar. Pollut. Bull. 54, 1361–1371.

Vos, G., Hovens, J.P.C., 1986. Chromium, nickel, copper, zinc, arsenic, selenium, cadmium, mercury and lead in Dutch fishery products 1977-1984. Sci. Total Environ. 52, 25–40.

Vosjan, J.H., Van der Hoek, G.J., 1972. A continuous culture of *Desulfovibrio* on a medium containing mercury and copper ions. Neth. J. Sea Res. 5, 440–444.

Vucetic, T., Vernberg, W.B., Anderson, G., 1974. Long-term annual fluctuations of mercury in zooplankton of the east central Adriatic. Rev. Int. Oceanogr. Med. 33, 75–81.

Waldock, M.J., Thain, J.E., 1983. Shell thickening in *Crassostrea gigas*: organotin antifouling or sediment induced? Mar. Pollut. Bull. 14, 411–415.

Waldock, R., Rees, H.I., Matthiessen, P., Pendle, M.A., 1999. Surveys of the benthic infauna of the Crouch Estuary (UK) in relation TBT contamination. J. Mar. Biol. Assoc. UK 79, 225–232.

Walsh, A.R., O'Halloran, J., 1998. Accumulation of chromium by a population of mussels (*Mytilus edulis* (L.)) exposed to leather tannery effluent. Environ. Toxicol. Chem. 17, 1429–1438.

Walsh, G.E., McLaughlin, L.L., Lores, E.M., Louie, M.K., Deans, C.H., 1985. Effects of organotins on growth and survival of two marine diatoms, *Skeletonema costatum* and *Thalassiosira pseudonana*. Chemosphere 14, 383–389.

Wang, W.X., Fisher, N.S., Luoma, S.N., 1996. Kinetic determinations of trace element bioaccumulation in the mussel *Mytilus edulis*. Mar. Ecol. Prog. Ser. 140, 91–113.

Ward, R.E., 1990. Metal concentrations and digestive gland lysomal stability in mussels from Halifax inlet, Canada. Mar. Pollut. Bull. 21, 237–240.

Ward, T.J., Correll, R.I., Anderson, R.B., 1986. Distribution of cadmium, lead and zinc amongst the marine sediments, sea grasses and fauna, and the selection of sentinel accumulators, near a lead smelter in South Australia. Aust. J. Mar. Freshw. Res. 37, 567–585.

Waska, H., Kim, S., Kim, G., Kang, M.R., Kim, G.B., 2008. Distribution patterns of chalcogens (S, Se, Te, and ^{210}Po) in various tissues of the squid, *Todarodes pacificus*. Sci. Total Environ. 392, 218–224.

Watkins, B., Simkiss, K., 1988. The effect of oscillating temperatures on the metal ion metabolism of *Mytilus edulis*. J. Mar. Biol. Assoc. UK 68, 93–100.

Watling, H.R., Watling, R.J., 1976a. Trace metals in oysters from Knysna Estuary. Mar. Pollut. Bull. 7, 45–48.

Watling, H.R., Watling, R.J., 1976b. Trace metals in *Choromytilus meridionalis*. Mar. Pollut. Bull. 7, 91–94.

Webb, J., Macey, D.J., Talbot, V., 1985. Identification of ferritin as a major high molecular weight zinc-binding protein in the tropical rock oyster, *Saccostrea cucullata*. Arch. Environ. Contam. Toxicol. 14, 403–407.

Weis, J.S., Weis, P., 1992. Transfer of contaminants from CCA-treated lumber to aquatic biota. J. Exp. Mar. Biol. Ecol. 168, 25–34.

Weis, J.S., Weis, P., 1993. Trophic transfer of contaminants form organisms living by chromated-copper-arsenate (CCA) wood to their predators. J. Exp. Mar. Biol. Ecol. 168, 25–34.

Weis, J.S., Weis, P., 1994. Effects of contaminants from chromated copper arsenate-treated lumber on benthos. Arch. Environ. Contam. Toxicol. 26, 103–109.

Weis, P., Weis, J.S., Couch, J., 1993a. Histopathology and bioaccumulation in oysters *Crassostrea virginica* living on wood preserved with chromated copper arsenate. Dis. Aquat. Org. 17, 41–46.

Weis, P., Weis, J.S., Lores, E., 1993b. Uptake of metals from chromated-copper-arsenate (CCA)-treated lumber by epibiota. Mar. Pollut. Bull. 26, 428–430.

Weiss, H.V., Shipman, W.H., 1957. Biological concentration by killer clams of cobalt-60 from radioactive fallout. Science 125, 695.

Wepener, V., Bervoets, L., Mubiana, V., Blust, R., 2008. Metal exposure and biological responses in resident and transplanted blue mussels (*Mytilus edulis*) from the Scheldt estuary. Mar. Pollut. Bull. 57, 624–631.

Westernhagen, H.V., Dethlefsen, V., Rosenthal, H., Furstenberg, G., Klinckmann, J., 1978. Fate and effects of cadmium in an experimental marine ecosystem. Helg. Wiss. Meer. 31, 471–484.

Wharfe, J.R., Van Den Broek, W.L.F., 1977. Heavy metals in macroinvertebrates and fish from the Lower Medway Estuary, Kent. Mar. Pollut. Bull. 8, 31–34.

White, S.L., Rainbow, P.S., 1985. On the metabolic requirements for copper and zinc in molluscs and crustaceans. Mar. Environ. Res. 16, 215–229.

White, D.H., Stendell, R.C., Mulhern, B.M., 1979. Relations of wintering canvasbacks to environmental pollutants—Chesapeake Bay, Maryland. Wilson Bull. 91, 279–287.

Whitehead, N.E., Ballestra, S., Holm, E., Huynh-Ngoc, L., 1988. Chernobyl radionuclides in shellfish. J. Environ. Radioact. 7, 107–121.

Widmeyer, J.R., Bendell-Young, L.I., 2007. Influence of food quality and salinity on dietary cadmium availability in *Mytilus trossulus*. Aquat. Toxicol. 81, 144–151.

Williams, P.M., Weiss, H.V., 1973. Mercury in the marine environment: concentration in sea water and in a pelagic food chain. J. Fish. Res. Board Can. 30, 293–295.

Williams, R.J., Mackay, N.J., Collett, L.C., Kacprzak, J.L., 1976. Total mercury concentration in some fish and shellfish from NSW estuaries. Food Technol. Aust. 18, 8–10.

Windom, H.L., Smith, R.G., 1972. Distribution of iron, magnesium, copper, zinc, and silver in oysters along the Georgia coast. J. Fish. Res. Board Can. 29, 450–452.

Windom, H.L., Gardner, W.S., Dunstan, W.M., Paffenhofer, G.A., 1976. Cadmium and mercury transfer in a coastal marine ecosystem. In: Windom, H.L., Duce, R.A. (Eds.), Marine Pollutant Transfer. D.C. Heath, Lexington, MA, pp. 135–137.

Wolfe, D.A., 1970. Fallout cesium-137 in clams (*Rangia cuneata*) from the Neuse River Estuary, North Carolina. Limnol. Oceanogr. 16, 797–805.

Wolfe, D.A., Coburn Jr., C.B., 1970. Influence of salinity and temperature on the accumulation of cesium-137 by an estuarine clam under laboratory conditions. Health Phys. 18, 499–505.

Wolfe, D.A., Jennings, C.D., 1973. Iron-55 and ruthenium-103 and −106 in the brackish-water clam *Rangia cuneata*. In: Nelson, D.J. (Ed.), Radionuclides in Ecosystems: Proceedings of the 3rd National Symposium on Radioecology, vol. 2. US Atom. Ener. Comm, pp. 783–790.

Won, J.H., 1973. The concentrations of mercury, cadmium, lead and copper in fish and shellfish of Korea. Bull. Korean Fish. Soc. 6, 1–19.

Wood, J.M., 1973. Metabolic cycles for toxic elements in aqueous systems. In: Sixth International Symposium, Med. Ocean., Portoroz, Yugoslavia, September 26-30, 1973, pp. 7–16.

Wrench, J.J., 1978. Biochemical correlates of dissolved mercury uptake by the oyster *Ostrea edulis*. Mar. Biol. 47, 79–86.

Yang, R., Zhou, Q., Jiang, G., 2006. Butyltin accumulation in the marine clam *Mya arenaria*: and evaluation of its suitability for monitoring butyltin pollution. Chemosphere 63, 1–8.

Yannai, S., Sachs, K., 1978. Mercury compounds in some eastern Mediterranean fishes, invertebrates, and their habitats. Environ. Res. 16, 408–418.

Yantian, L., 1989. Effect of zinc on the growth and development of larvae of the bay scallop *Argopecten irradians*. Chin. J. Oceanol. Limnol. 7, 318–326.

Yarsan, E., Baskaya, R., Yildiz, A., Altintas, L., Yesilot, S., 2007. Copper, lead, cadmium and mercury concentrations in the mussel *Elliptio*. Bull. Environ. Contam. Toxicol. 79, 218–220.

Young, M.L., 1975. The transfer of ^{65}Zn and ^{59}Fe along a *Fucus serratus* (L.)—*Littorina obtusata* (L.) food chain. J. Mar. Biol. Assoc. UK 55, 583–610.

Young, M.L., 1977. The roles of food and direct uptake from water in the accumulation of zinc and iron in the tissues of the dogwhelk *Nucella lapillus* (L.). J. Exp. Mar. Biol. Ecol. 30, 315–325.

Young, D.R., Folsom, T.R., 1967. Loss of zinc-65 from the California sea mussel. Mytilus californianus. Biol. Bull. 133, 438–447.

Young, D.R., Folsom, T.R., 1973. Mussels and barnacles as indicators of the variation of ^{54}Mn, ^{60}Co and ^{65}Zn in the marine environment. In: Radioactive Contamination of the Marine Environment. IAEA, Vienna, Austria, pp. 633–650.

Young, D.R., Jan, T.K., 1976. Metals in scallops. In: 1976 Annual Report of Southern California Coastal Water Research Projecct, El Segundo, CA, pp. 117–121.

Young, D.R., McDermott, D.J., 1975. Trace metals in harbor mussels. In: Annual Report of Southern California Coastal Water Research Project, June 30, El Segundo, CA, pp. 139–142.

Young, D.R., Alexander, G.V., McDermott-Ehrlich, D., 1979. Vessel-related contamination of southern California harbours by copper and other metals. Mar. Pollut. Bull. 10, 50–56.

Young, D.R., Jan, T.K., Hershelman, G.P., 1980. Cycling of zinc in the nearshore marine environment. In: Nriagu, J.O. (Ed.), Zinc in the Environment. Part 1: Ecological Cycling. Wiley, New York, pp. 297–355.

Zaroogian, G.E., 1979. Studies on the depuration of cadmium and copper by the American oyster, *Crassostrea virginica*. Bull. Environ. Contam. Toxicol. 23, 117–122.

Zaroogian, G.E., Cheer, S., 1976. Cadmium accumulation by the American oyster, *Crassostrea virginica*. Nature 261, 408–410.

Zaroogian, G.E., Hoffman, G.L., 1982. Arsenic uptake and loss in the American oyster, *Crassostrea virginica*. Environ. Monit. Assess. 1, 345–358.

Zaroogian, G.E., Johnson, M., 1983. Chromium uptake and loss in the bivalves *Crassostrea virginica* and *Mytilus edulis*. Mar. Ecol. Prog. Ser. 12, 167–173.

Zaroogian, G.E., Morrison, G., Heltshe, J.F., 1979. *Crassostrea virginica* as an indicator of lead pollution. Mar. Biol. 52, 189–196.

Zaroogian, G., Anderson, S., Voyer, R.A., 1992. Individual and combined cytotoxic effects of cadmium, copper, and nickel on brown cells of *Mercenaria mercenaria*. Ecotoxicol. Environ. Safety 24, 328–337.

Zauke, G.P., 1977. Mercury in benthic invertebrates of the Elbe estuary. Helg. Wiss. Meer. 29, 358–374.

Zesenko, A.Y., Polikarpov, G.G., 1965. Coefficients of accumulation and distribution of ruthenium-106 in organs and tissues of sea molluscs. Radiobiologiya 5, 320–322.

Zingde, M.D., Singbal, S.Y.S., Moraes, C.F., Reddy, C.V.G., 1976. Arsenic, copper, zinc & manganese in the marine flora & fauna of coastal & estuarine waters around Goa. Indian J. Mar. Sci. 5, 212–217.

Zitko, V., Carson, W.V., 1975. Accumulation of thallium in clams and mussels. Bull. Environ. Contam. Toxicol. 14, 530–533.

Zorita, I., Apraiz, I., Ortiz-Zarragoitia, M., Orbea, A., Cancio, I., Soto, M., et al., 2007. Assessment of biological effects of environmental pollution along the NW Mediterranean Sea using mussels as sentinel organisms. Environ. Pollut. 148, 236–250.

CHAPTER 7
Crustaceans

Crustaceans are aquatic arthropods which typically respire by gills and possess two pairs of antennae, three pairs of primary mouth parts, and numerous appendages on thorax and abdomen. This is an unusually diverse and widespread group that includes marine species of major economic importance—such as lobsters, shrimps, and crabs—as well as isopods, copepods, barnacles, and euphausiids. It is probable that all marine biological communities contain significant crustacean biomass and diversity, with representative species present as predators, parasites, and grazers, or as intermediary links in the food chain. Trace metal concentrations in crustaceans are comparatively well documented; however, the process of ecdysis, including the frequency of molting, is a major confounder of trace metal residue interpretation, especially among decapods.

7.1 Aluminum

Aluminum concentrations in whole crustaceans were variable and ranged—on a mg Al/kg dry weight (DW) basis—from 7.3 to 11,200.0, being highest in copepods from a Taiwanese sewage outfall (Table 7.1).

7.2 Antimony

Antimony values in crustacean tissues—on a fresh weight (FW) basis—did not exceed 2.0 mg Sb/kg (Table 7.2). Antimony is taken up from uncontaminated seawater: muscle of the lobster *Jasus lalandi* contained 170 times more antimony than the ambient medium (Van As et al., 1973). Antimony flux in the Mediterranean Sea due to fecal pellet deposition of euphausiids is substantial and equivalent to about 9.2 mg Sb/kg fecal pellets daily (Fowler, 1977). At present, there are no known health hazards associated with antimony accumulations and cycling rates via marine vectors.

7.3 Arsenic

Arsenic concentrations in tissues of marine biota—including crustaceans—show a wide range of values, being highest in lipids, liver, and muscle tissues, and varying with the age of the organisms, geographic locale, and proximity to anthropogenic activities (Eisler, 2000a;

Table 7.1: Aluminum Concentrations in Field Collections of Crustaceans

Organism	Concentration	Reference[a]
Copepods		
Whole	70.0 DW	2
Whole	55.0 DW	3
Whole; 3 spp.; Taiwan; near sewage outfall	2100.0-11,200.0 DW	6
Blue crab, *Callinectes sapidus*; muscle; Iskenderun Bay, Turkey; August 2005	7.3 (1.2-17.6) DW	5
Crustaceans; whole; Irish Sea	(166.0-1041.0) DW	4
Euphausiids; whole	31.0 DW	2
Crab, *Seserma cinerum*; whole; south Chesapeake Bay, Maryland	1847.0 DW	1
Crab, *Pachygrapsus* sp.; whole; San Francisco Bay, California	1965.0 DW	1

Values are in mg Al/kg dry weight (DW).
[a] 1, Sparling and Lowe, 1996; 2, Martin and Knauer, 1973; 3, Tijoe et al., 1977; 4, Culkin and Riley, 1958; 5, Turkmen et al., 2006; 6, Feng et al., 2006.

Table 7.3). Arsenic appears to be elevated in marine biota mainly owing to their ability to accumulate and retain arsenic from seawater or food sources, and not from localized pollution (Maher, 1985). The great majority of arsenic in marine organisms exists as water-soluble and lipid-soluble organoarsenicals that include arsenolipids, arsenosugars, arsenocholine, arsenobetaine [$(CH_3)AsCH_2COOH$], monomethylarsonate [$CH_3AsO(OH)_2$], and dimethylarsinate [$(CH_3)_2AsO(OH)$], as well as other forms (Edmonds et al., 1993). There is no convincing hypothesis to account for the existence of the various forms of organoarsenicals found in marine organisms. One suggested hypothesis is that each form involves a single anabolic/catabolic pathway concerned with the synthesis and turnover of phosphatidylcholine (Phillips and Depledge, 1986). Arsenosugars (arsenobetaine precursors) are the dominant arsenic species in brown kelp (*Ecklonia radiata*), giant clam (*Tridacna maxima*), shrimp (*Pandalus borealis*), and ivory shell (*Buccinum striatissimum*) (Francesconi et al., 1985; Matsuto et al., 1986; Phillips and Depledge, 1986; Shiomi et al., 1984a,b). For most marine species, however, there is general agreement that arsenic exists primarily as arsenobetaine, a water-soluble organoarsenical that has been identified in tissues of western rock lobster (*Panulirus cygnus*), American lobster (*Homarus americanus*), octopus (*Paraoctopus* sp.), sea cucumber (*Stichopus japonicus*), blue shark (*Prionace glauca*), sole (*Limanda* sp.), squid (*Sepioteuthis australis*), prawn (*Penaeus latisulcatus*), scallop (*Pecten alba*), and many other species, including teleosts, molluscs, tunicates, and crustaceans (Edmonds et al., 1993; Francesconi et al., 1985; Hanaoka and Tagawa, 1985a,b; Kaise and

Table 7.2: Antimony Concentrations in Field Collections of Crustaceans

Organism	Concentration	Reference[a]
Copepods; whole	0.027 DW	1
Crustaceans; 16 economically important species; edible tissues		
2 spp.	0.2-0.3 FW	2
1 spp.	0.7-0.8 FW	2
3 spp.	0.8-0.9 FW	2
7 spp.	0.9-1.0 FW	2
3 spp.	1.0-2.0 FW	2
Euphausiids; whole	1.9 AW	3
Lobster, *Jasus lalandi*; muscle	(0.12-0.14) FW	7,8
Euphausiid, *Meganyctiphanes norvegicus*		
Whole	<4.5 DW	4
Molts	0.8 DW	4
Whole	0.037 DW	5
Shrimp, *Palaemon elegans*; whole	0.016 DW	6
Amphipod, *Paramoera walkeri*; East Antarctica; 2003-2004; whole	(0.03-0.10) DW	10
Black tiger shrimp, *Penaeus monodon*; Vietnam; 2003-2005; max. values		
Muscle	0.02 DW	11
Exoskeleton	0.04 DW	11
Hepatopancreas	0.07 DW	11
Shrimps; coast of Belgium		
Proteinaceous molts	0.03 DW	9
Remainder	0.15 DW	9

Values are in mg Sb/kg fresh weight (FW), dry weight (DW), or ash weight (AW).
[a] 1, Tijoe et al., 1977; 2, Hall et al., 1978; 3, Robertson, 1967; 4, Fowler, 1977; 5, Leatherland et al., 1973; 6, Leatherland and Burton, 1974; 7, Van As et al., 1973; 8, Van As et al., 1975; 9, Bertine and Goldberg, 1972; 10, Palmer et al., 2006; 11, Tu et al., 2008.

Fukui, 1992; Maher, 1985; Matsuto et al., 1986; Norin et al., 1985; Ozretic et al., 1990; Phillips, 1990; Shiomi et al., 1984b). The potential risks associated with consumption of seafoods containing arsenobetaine seems to be minor (Eisler, 2000a).

Decapod crustaceans contained relatively high concentrations of arsenic; concentrations varied widely, but usually ranged from 1.0 to 100.0 mg kg on a DW basis (Fowler and Unlu, 1978). Edible tissues of crustaceans from coastal waters of the United States usually contained 3.0-10.0 mg As/kg FW (Table 7.3); these values were elevated when compared to

Table 7.3: Arsenic Concentrations in Field Collections of Crustaceans

Organism	Concentration	Reference[a]
Amphipods; whole	7.9 DW	1
Blue crab, *Callinectes sapidus*		
Whole	0.6 DW	2
Whole	0.36 (0.20-0.88) FW; 1.8 (1.0-4.4) DW	3
Whole; Florida	7.7 FW	28
Soft parts; Maryland	(0.5-1.8) FW	28
Pensacola, Florida; 2003-2004		
Hepatopancreas	Max. 9.6 FW	31
Muscle	Max. 8.3 FW	31
Crab, *Cancer irroratus*		
Muscle	1.9 FW	4
Gills	(0.57-0.87) FW	4
Dungeness, crab, *Cancer magister*		5
Soft parts	37.8 FW	5
Muscle	4.0 FW	22
Alaskan snow crab, *Chinocetes bairdi*; muscle	7.4 FW	22
Copepods		
Whole	0.88 (0.42-1.25) FW; 4.2 (2.0-6.0) DW	3
Whole	1.4 FW	6
Whole	5.6 DW	1
Whole	6.0 DW	7
Whole	8.2 DW	8
Crabs		
Muscle; total arsenic vs. inorganic arsenic	3.7 FW vs. < 0.5 FW	9
Muscle	(6.1-6.4) FW	10
Sand shrimp, *Crangon crangon*; Netherlands; 1977-1984; muscle	3.0 (2.0-6.8) FW	23
Crustaceans; whole; 4 species; lipid extracts	7300.0-24,400.0 FW	11
Crustaceans; economically important species; edible tissues; North America		
6 spp.	3.0-5.0 FW	12
3 spp.	5.0-10.0 FW	12
4 spp.	10.0-20.0 FW	12

(Continues)

Table 7.3: Cont'd

Organism	Concentration	Reference[a]
2 spp.	20.0-30.0 FW	12
1 spp.	40.0-50.0 FW	12
9 spp.	5.6-220.2 DW	13
Crustaceans; edible tissues; Hong Kong; 1976-1978		
Crabs	(5.4-19.1) FW	24
Lobsters	(26.7-52.8) FW	24
Prawns and shrimps	(1.2-44.0) FW	24
Euphausiids; whole	1.8 FW	6
Gammarids; whole	6.6 FW	6
Mantis shrimp, *Gonodactylus* sp; Guam; June 1998; muscle vs. gonad	5.06 DW vs. 4.58 DW	29
Lobster, *Homarus americanus*		
Muscle	(3.8-7.6) DW	28
Hepatopancreas	22.5 FW	28
Whole	(3.8-7.5) DW	14
Whole	(0.95-3.00) FW; (5.0-16.0) DW	3
Lobster; total arsenic vs. inorganic arsenic		
Tail	40.5 FW vs. < 0.5 FW	9
Hepatopancreas	22.5 FW vs. 0.9 FW	9
Euphausiid, *Meganyctiphanes norvegica*		
Whole	42.0 DW	16
Whole	(1.9-5.5) FW	3
Stone crab, *Menippe mercenaria*; whole	(9.0-11.8) DW	28
Shrimp, *Metapenaeus affinis*; edible portions	13.6 DW	17
North Sea; 1997-1998; muscle		
Shrimp, *Crangon crangon*	5.2 FW	34
Crab, *Cancer pagurus*	Max. 40.0 FW	34
Shrimp, *Palaemon elegans*; whole	16.0 DW	18
Deep sea prawn, *Pandalus borealis*		
Whole	(7.3-11.5) FW	6
Muscle	80.0 DW	7
Head and shell	68.3 DW	28
Muscle	61.6 DW	28

(Continues)

Table 7.3: Cont'd

Organism	Concentration	Reference[a]
Oil	42.0 DW; 10.1 FW	28
Egg	(3.7-14.0) FW	28
Shrimp, *Pandalus montagui*; whole	(7.4-10.8) FW	6
Shrimp, *Pandalus* sp.		
Whole	0.83 AW	19
Soft parts	0.05 DW	20
Lobster, *Panulirus argus*; muscle	3.8 FW; 14.0 DW	3
Reef lobster, *Panulirus* spp.; American Samoa; 2001-2002; soft parts		
Total arsenic	(19.8-98.2) FW	32
Inorganic arsenic	(<0.009-0.083) FW	32
Alaskan king crab, *Paralithodes camtschatica*; muscle	8.6 FW	22
Marsh crab, *Parasesarma erythodactyla*; soft parts; 2002; New South Wales, Australia; metals-contaminated estuary		
Males	Max. 7.2 DW	30
Females	Max. 32.6 DW	30
Brown shrimp, *Peneus aztecus*		
Whole	0.6 DW	2
Muscle	(3.1-5.2) FW	28
Black tiger shrimp, *Penaeus monodon*		
Edible portions	(9.3-11.2) DW	17
Vietnam; 2003-2005; juveniles; max. values		
Muscle	18.0 DW	33
Exoskeleton	7.7 DW	33
Hepatopancreas	20.0 DW	33
White shrimp, *Peneus setiferus*		
Whole	1.3 (0.8-2.2) FW	3
Whole	3.8 (2.2-6.3) DW	3
Muscle		
Mississippi	(1.7-4.4) FW	28
Florida	(2.8-7.7) FW	28
Crab, *Portunus pelagicus*; edible portions	(11.3-25.2) DW	17
Crab, *Portunus* sp.; edible portions	(5.5-6.5) FW	15

(*Continues*)

Table 7.3: Cont'd

Organism	Concentration	Reference[a]
Crab, *Pugettia producta*; feces	(190.0–340.0) AW	21
Shrimp, *Sergestes lucens*; muscle; total arsenic vs. arsenobetaine	5.5 FW vs. 4.5 FW	25
Shrimps; whole	12.7 FW	15
Taiwan; 1995–1996; shrimps; muscle	Max. 5.1 DW	27
South Texas, Lower Laguna Madre; 1986–1987; whole		
Grass shrimp, *Palaemonetes* sp.	26.9 (9.7–55.0) DW	26
Brown shrimp, *Penaeus aztecus*	17.9 (8.1–50.0) DW	26
Blue crab, *Callinectes sapidus*	18.4 (2.7–50.0) DW	26

Values are in mg As/kg fresh weight (FW), dry weight (DW), or ash weight (AW).

[a] 1, Bohn and McElroy, 1976; 2, Sims and Presley, 1976; 3, Bernhard and Zattera, 1975; 4, Greig et al., 1977a; 5, LeBlanc and Jackson, 1973; 6, Kennedy, 1976; 7, Bohn, 1975; 8, Tijoe et al., 1977; 9, Reinke et al., 1975; 10, Hoover et al., 1974; 11, Vaskofsky et al., 1972; 12, Hall et al., 1978; 13, Costa and da Fonseca, 1967; 14, Penrose et al., 1975; 15, Johnson and Braman, 1975; 16, Leatherland et al., 1973; 17, Zingde et al., 1976; 18, Leatherland and Burton, 1974; 19, Fukai and Meinke, 1959; 20, Fukai and Meinke, 1962; 21, Boothe and Knauer, 1972; 22, Francesconi et al., 1985; 23, Vos and Hovens, 1986; 24, Phillips et al., 1982; 25, Shiomi et al., 1984b; 26, Custer and Mitchell, 1993; 27, Han et al., 1998; 28, Jenkins, 1980; 29, Denton et al., 2006; 30, MacFarlane et al., 2006; 31, Karouna-Renier et al., 2007; 32, Peshut et al., 2008; 33, Tu et al., 2008; 34, De Gieter et al., 2002.

finfish and molluscan tissues (Hall et al., 1978). Arsenic in crustaceans and other marine species is concentrated in lipid fractions primarily as an organoarsenic compound of negligible toxicity, and this is believed to account for the lack of adverse effects to consumers of crustaceans with relatively high arsenic body burdens. The carnivorous shrimp *Lysmata seticaudata* derived organoarsenicals indirectly from primary producers, which efficiently transfer these compounds along the food chain (Wrench et al., 1979). No organic arsenic could be formed by shrimp; in the case of *Lysmata*, arsenate taken up from seawater was converted mostly to arsenite (Wrench et al., 1979). Factors known to modify rates of arsenic accumulation and retention in *L. seticaudata* include water temperature and salinity, arsenic concentration in diet and ambient medium, age of shrimp, and especially the frequency of molting (Fowler and Unlu, 1978). There is only slight uptake of inorganic arsenicals by crustaceans from the medium or from food (Rosemarin et al., 1985). In a simplified estuarine food chain, there was no significant increase in arsenic content of grass shrimp, *Palaemonetes pugio*, exposed to arsenate-contaminated food or to elevated water concentrations of arsenate (Lindsay and Sanders, 1990).

Sensitive species of crustaceans died at 0.23–0.51 mg As^{3+}/L, with 50% mortality recorded in 96 h for juvenile copepods, *Acartia clausi*, *Eurytemora affinis*, and zoeae of dungeness crab, *Cancer magister* (Sanders, 1986; USEPA, 1985a). Arsenic toxicity increases with increasing water temperature between 5 °C and 15 °C to the marine amphipod *Corophium volutator* (Bryant et al., 1985). Studies with the mysid, *Mysidopsis bahia*, showed that lifetime exposure (about 28 days) to As^{3+} produced a no-effect level at 0.63 mg/L and an observed effect at 1.27 mg/L; short-term tests with As^{5+} produced an LC50, 96 h concentration of 2.3 mg/L (USEPA, 1985a). Copepods, *Tigriopus japonicus*, exposed for two generations to As^{3+} or As^{5+} had reduced fecundity at 0.0001 mg/L and higher, and an increase in the proportion of males and reduced survival at 0.01 mg/L and higher (Lee et al., 2008).

7.4 Barium

Whole copepods contained up to 2000.0 mg Ba/kg on an ash weight (AW) basis, and whole hermit crabs, *Eupagurus bernhardus*, contained up to 84.0 mg Ba/kg on a FW basis (Table 7.4).

Table 7.4: Barium, Beryllium, and Boron Concentrations in Field Collections of Crustaceans

Element and Organism	Concentration	Reference[a]
Barium		
Copepods		
Whole	17.0 DW	1
Whole	2000.0 AW	2
Shrimp, *Crangon* sp.; whole	0.3 DW	6
Crustaceans; whole; 12 spp.	11.0-21.0 FW	3
Crab, *Eupagurus bernhardus*; whole	84.0 FW	3
Euphausiids; whole	24.0 DW	1
Black tiger shrimp, *Penaeus monodon*; juveniles; Vietnam; 2003-2005; max. values		
Muscle	0.18 DW	7
Exoskeleton	76.0 DW	7
Hepatopancreas	2.6 DW	7
Beryllium		
Copepods; whole	3.0 AW	2
Boron		
Copepod, *Calanus finmarchicus*; whole	760.0 AW	4
Crab, *Cancer magister*; whole	1.8 FW	5
Euphausiid, *Euphausia krohni*; whole	440.0 AW	4

Values are in mg element/kg fresh weight (FW), dry weight (DW), or ash weight (AW).
[a]1, Martin and Knauer, 1973; 2, Vinogradova and Koual'skiy, 1962; 3, Mauchline and Templeton, 1966; 4, Nicholls et al., 1959; 5, Thompson et al., 1976; 6, Zumholz et al., 2006; 7, Tu et al., 2008.

7.5 Beryllium

The single datum available for beryllium is 3.0 mg Be/kg AW in whole copepods (Table 7.4).

7.6 Bismuth

Maximum bismuth concentrations in tissues of juvenile tiger shrimp, *Penaeus monodon*, from Vietnam during 2003-2005 were (in mg Bi/kg DW) 0.02 in muscle, 0.006 in exoskeleton, and 0.06 in hepatopancreas (Tu et al., 2008). Plankton, mostly crustaceans, collected at Eniwetok Atoll in 1964 had the highest concentration of ^{207}Bi of all biological samples analyzed (Welander, 1969).

7.7 Boron

Whole copepods contained up to 760.0 mg B/kg on an AW basis and the euphausiid *Euphausia krohnii* had 440.0 mg B/kg AW (Table 7.4). Boron is a common ingredient in many dishwashing detergents and cleansers, and this should be considered as one source of the element into coastal waters of the United States (Eisler, 2000b).

7.8 Cadmium

The highest cadmium concentrations were usually found in digestive gland, hepatopancreas, or kidney, and the lowest concentrations in edible muscle (Table 7.5). Some of these data suggest that cadmium levels in edible tissues may approach or exceed 13.0-15.0 mg Cd/kg FW, a concentration that will induce vomiting in human consumers. However, edible tissues in those instances included viscera, wherein cadmium burdens were greatest. There is no demonstrable hazard to human health associated with ingestion of edible muscle tissues of marine crustaceans (Eisler, 1981). The highest concentration of cadmium in whole crustaceans reported in a nationwide survey of 736 stations in estuaries of the United States during 2000-2001 was 0.70 mg Cd/kg FW in the shrimp, *Farfantepenaeus aztecus*; about 29% of all samples exceeded the putative guideline of 0.4 mg Cd/kg FW (Harvey et al., 2008). Amphipods, *Paramoera walkeri*, from the Australian Antarctic Territory in 1999-2000 had whole-body bioconcentration factors that ranged between 150 and 630 when seawater contained between 0.00006 and 0.0001 mg Cd/L; BCFs decreased with increasing seawater concentration (Clason et al., 2003). Amphipods showed increased sensitivity when seawater contained 0.0009-0.003 mg Cd/L (Clason et al., 2003).

Cadmium concentrations in gill and hepatopancreas of a prawn, *Penaeus indicus*, were mostly in the insoluble fraction, wherein cadmium was associated with phosphorus, calcium, magnesium, and silicon (Nunez-Nogueira et al., 2006a). Exposure to 0.1 mg Cd/L for 10 days caused a 50-fold increase in cadmium concentration in hepatopancreas and a 60-fold increase

Table 7.5: Cadmium Concentrations in Field Collections of Crustaceans

Organism	Concentration	Reference[a]
Amphipods; whole	7.0 DW	1
Copepod, *Anomalocera patersoni*; whole; males vs. females	1.4 DW vs. 1.6 DW	2
Barents Sea and environs; 1991-2000; whole		
Copepods; 4 spp.	1.2-6.3 DW	48
Euphausid, *Meganyctiphanes norvegica*	0.2 DW	48
Amphipod, *Themisto abyssorum*	10.5 DW	48
Shrimp, *Pandalus borealis*	1.6 DW	48
Barnacles; soft parts	12.1 DW	3
Barnacles; summer 2001; near Natal, Brazil; soft parts		
Fistulobalanus citerosum	(2.7-4.2) DW	40
Balanus amphitrite	5.1 DW	40
Blue crab, *Callinectes sapidus*		
Whole	0.1 DW	4
Muscle; Iskenderun Bay, Turkey; August 2005	1.8 (0.09-5.6) DW	38
Pensacola, Florida; 2003-2004		
Hepatopancreas	Max. 4.6 FW	42
Muscle	Max. 0.08 FW	42
Crab, *Cancer irroratus*		
Muscle	(0.08-1.0) FW	5
Digestive gland	(1.1-4.8) FW	5
Gills	(0.7-2.7) FW	5
Crab, *Cancer pagurus*		
Edible tissues	(3.6-13.0) FW	6
Brown meat	(0.03-3.4) FW	7
Whole	5.0 FW	8
Crab, *Carcinus maenas*; whole	(14.3-33.1) FW	8
Copepods		
Whole	5.0 DW	1
Whole	4.1 DW	9
Whole	0.72 DW	10
Whole; 3 spp.; Taiwan; near sewage outfall	0.23-1.8 DW	39

(*Continues*)

Table 7.5: Cont'd

Organism	Concentration	Reference[a]
Crabs		
Muscle	0.19 FW	11
Shell	0.14 FW	11
Crustaceans		
Whole	(0.98-2.8) FW	12
Whole	8.8 DW	13
Soft parts	(0.01-0.40) DW	14
Commercial species; edible tissues		
3 spp.	<0.1 FW	15
9 spp.	0.1-0.2 FW	15
4 spp.	0.2-0.4 FW	15
Euphausiid, *Euphausia pacifica*; whole	(0.05-0.36) FW; (0.33-2.8) DW	7
Euphausiid, *Euphausia superba*; East Antarctica; whole	3.3 DW	35
Euphausiids		
Whole	2.8 DW	9
Whole	0.14 FW	17
Greenland; 1975-1991; 6 spp.; whole	0.2-4.6 FW	33
Mantis shrimp, *Gonodactylus* sp.; Guam; June 1998; muscle vs. gonad	0.36 DW vs. 9.1 DW	34
Lobster, *Homarus americanus*		
Muscle	0.03 FW	18
Whole	0.51 FW; 5.3 AW	19
Muscle	0.20 FW; 10.0 AW	19
Exoskeleton	0.59 FW; 4.1 AW	19
Gill	0.49 FW; 17.2 AW	19
Viscera	1.21 FW; 33.8 AW	19
Lobster, *Homarus vulgaris*; muscle	(<0.03-0.09) FW	6
Blue shrimp, *Litopenaeus stylirostris*; Gulf of California; 2004-2005; muscle	0.66 (0.38-1.1) DW	36
Shrimp, *Lysmata seticaudata*		
Whole	0.41 FW; 1.7 DW	7
Exoskeleton	0.76 FW; 3.1 DW	7
Viscera	0.91 FW; 2.8 DW	7
Muscle	0.09 FW; 0.39 DW	7
Eyes	0.42 FW; 1.3 DW	7
Molts	1.3 FW; 5.5 DW	7

(Continues)

Table 7.5: Cont'd

Organism	Concentration	Reference[a]
Euphausiid, *Megancytiphanes norvegica*		
Whole	0.7 DW	20
Molts	2.1 DW	20
Feces	9.6 DW	20
Eggs	0.3 DW	20
Stomach contents	2.1 DW	20
Whole	<4.5 DW	21
Feces	<80.0 DW	21
Whole	0.25 DW	16
Peneid shrimp, *Melicertus kerathurus*; Lesinalagoon, Italy; October 2004; 3 sites		
Muscle	0.10-0.23 DW	47
Exoskeleton	0.91-2.9 DW	47
Mexico; Gulf of California; 1999-2000		
Barnacle, *Balanus eburneus*; soft parts	1.1 DW	45
Shrimps; muscle		
Leptopenaeus stylirostris	0.5 DW	45
Leptopenaeus vanamei	3.1 DW	45
Lobster, *Nephrops norvegicus*; edible tissues	(<0.03-0.1) FW	6
Shrimp, *Palaemon elegans*; whole	0.31 DW	22
Shrimp, *Palaemonetes pugio*; whole	3.3 (1.4-6.2) DW	23
Amphipod, *Paramoera walkeri*; East Antarctica; 200-2004; whole	2.3-7.0 DW	35
Shrimp, *Pandalus jordani*; whole	0.16 (0.06-0.30) FW; 0.49 (0.18-0.96) DW	7
Shrimp, *Pandalus montagui*		
Tail	0.03 DW	24
Egg	0.12 DW	24
Carcass	0.34 DW	24
Hepatopancreas	6.4 DW	24
Whole	0.47 DW	24
Lobster, *Panulirus interruptus*		
Muscle	(0.28-0.31) FW	25
Hepatopancreas	(5.6-29.3) FW	25

(*Continues*)

Table 7.5: Cont'd

Organism	Concentration	Reference[a]
Marsh crab, *Parasesarma erythodactyla*; soft parts; 2002; New South Wales, Australia; metals-contaminated estuary		
Males	Max. 25.9 DW	41
Females	Max. 22.5 DW	41
Prawn, *Penaeus indicus*; juveniles		
Hepatopancreas		
Whole	0.8 DW	43
Soluble fraction	0.2 DW	43
Insoluble fraction	0.6 DW	43
Gills		
Whole	0.6 DW	43
Soluble fraction	0.1 DW	43
Insoluble fraction	0.5 DW	43
Black tiger shrimp, *Penaeus monodon*; juveniles; Vietnam; 2003-2005; max. values		
Muscle	0.04 DW	46
Exoskeleton	0.03 DW	46
Hepatopancreas	8.4 DW	46
Shrimp, *Peneus aztecus*		
Muscle	0.16 DW	26
Exoskeleton	0.50 DW	26
Viscera	2.6 DW	26
Whole	<0.4 DW	4
Shrimp, *Peneus duorarum*; whole	0.3 FW	28
White shrimp, *Peneus setiferus*; muscle; Gulf of Mexico	6.1 DW	37
Crab, *Portunus holsatus*; whole	0.14 FW	29
Crab, *Pugettia producta*; feces	(3.1-3.2) AW	30
Shrimps		
Muscle	(<0.02-0.35) FW	11
Shell	(0.28-0.35) FW	11
Muscle	0.25 DW	26
Exoskeleton	0.49 DW	26

(*Continues*)

Table 7.5: Cont'd

Organism	Concentration	Reference[a]
Zooplankton		
Surface vs. > 99 m depth	16.0 AW vs. 15.0 AW	31
Whole	2.5 (0.3-10.9) DW	32
Baltic Sea; May 1999; cladocerans vs. copepods; whole	2.0 DW vs. 0.7 DW	44

Values are in mg Cd/kg fresh weight (FW), dry weight (DW), or ash weight (AW).
[a]1, Bohn and McElroy, 1976; 2, Polikarpov et al., 1979; 3, Stenner and Nickless, 1975; 4, Sims and Presley, 1976; 5, Greig et al., 1977b; 6, Topping, 1973; 7, Bernhard and Zattera, 1975; 8, Peden et al., 1973; 9, Martin and Knauer, 1973; 10, Tijoe et al., 1977; 11, Won, 1973; 12, Wright, 1976; 13, Stenner and Nickless, 1974b; 14, Stickney et al., 1975; 15, Hall et al., 1978; 16, Leatherland et al., 1973; 17, Greig and Wenzloff, 1977; 18, Chou et al., 1978; 19, Eisler et al., 1972; 20, Benayoun et al., 1975; 21, Fowler, 1977; 22, Leatherland and Burton, 1974; 23, Pesch and Stewart, 1980; 24, Ray et al., 1980; 25, Vattuone et al., 1976; 26, Horowitz and Presley, 1977; 28, Nimmo et al., 1977; 29, DeClerck et al., 1979; 30, Boothe and Knauer, 1972; 31, Martin, 1970; 32, Hardstedt-Romeo and Laumond, 1980; 33, Dietz et al., 1996; 34, Denton et al., 2006; 35, Palmer et al., 2006; 36, Frias-Espericueta et al., 2007; 37, Vazquez et al., 2001; 38, Turkmen et al., 2006; 39, Feng et al., 2006; 40, Silva et al., 2006; 41, MacFarlane et al., 2006; 42, Karouna-Renier et al., 2007; 43, Nunez-Nogueira et al., 2006a; 44, Pempkowiak et al., 2006; 45, Ruelas-Inzunza and Paez-Osuna, 2008; 46, Tu et al., 2008; 47, D'Adamo et al., 2008; 48, Zauke and Schmalenbach, 2006.

in gill; more than 80% of the cadmium was in the insoluble fraction of the Cd-exposed prawns (Nunez-Nogueira et al., 2006a). Juvenile mysids, *Neomysis integer*, exposed to 0.045 mg Cd/L for 7 days at different salinities were analyzed for metallothionein induction and cytosolic metal concentrations (Erk et al., 2008). Metallothionein induction occurred at all salinities and cytosolic cadmium increased 29% from the lowest salinity tested of 5 ppt to the highest salinity tested of 25 ppt; however, cytosolic copper decreased 44% within this salinity range, although there was a significant positive correlation between metallothionein and cytosolic copper concentration (Erk et al., 2008). Metal assimilation from food plays an important role in cadmium accumulation, especially in estuarine stages of the life cycle (Nunez-Nogueira et al., 2006b). Juvenile prawns, *P. indicus*, fed with radiolabeled ^{109}Cd squid muscle (*Loligo vulgaris*) or filaments of the green alga (*Cladophora* sp.) for 60 min, were then transferred to clean seawater. The assimilation efficiency (AE) was calculated from the time at which no radioactivity was detected in feces, or about 6 h after ingestion of algae and 10 h after ingestion of squid muscle. The AE of ^{109}Cd was 42% from algae at 6 h and 75% (64-83%) from squid muscle at 10 h. During the depuration period of 49 h, assimilated cadmium was lost at a constant rate from the diet. Prawns fed ^{109}Cd-labeled algae had most of the assimilated ^{109}Cd in hepatopancreas (71%) after 49 h depuration, 12% in exoskeleton, 5% in foregut, 4% in external organs, 3% in muscle, and 2% in gills. Prawns fed ^{109}Cd-labeled squid muscle showed (after 134 h) 53% of the radioactivity in hepatopancreas, 12% in exoskeleton, and 8% in foregut (Nunez-Nogueira et al., 2006b). Thorsson et al. (2008) aver that settling organic matter (OM) is the major food source for heterotrophic benthic fauna. OM attached to ^{109}Cd-labeled microalgae *Tetraselmis* spp. fed to the amphipod *Monoporeia affinis* for 34 days was biomagnified by a factor of 3.0 (Thorsson et al., 2008).

Crustaceans are comparatively sensitive to cadmium insult. A crab, *Pontoporia affinis*, held in 0.0065 mg Cd/L for 265 days had a reduced F1 life span (Sundlein, 1983). An LC-50 range of 0.015-0.019 mg Cd/L was reported for two species of mysid shrimp during lifetime exposure (i.e., 23-27 days) to cadmium salts (Gentile et al., 1982). Crustaceans immersed in ambient cadmium concentrations between 0.005 and 0.019 mg/L had decreased growth, abnormal respiration, delayed molt, shortened life span, altered enzyme levels, and abnormal muscular contractions (as quoted in Eisler, 2000d). Cadmium was fatal to 50% of nauplii of copepods (*E. affinis*) at 0.051 mg/L and 5 ppt salinity, and 0.083-0.213 mg Cd/L at higher salinities (Hall et al., 1995). In general, resistance to cadmium is higher in marine than in fresh water organisms, and survival is usually higher at lower temperatures and higher salinities tested for any given level of cadmium in the medium (Eisler, 2000d). Studies with *Leptocheirus plumulosus*, an estuarine amphipod, showed that gravid females were more resistant than males or mature females to the biocidal properties of Cd^{2+}; juveniles were more sensitive than adults; sensitivity was greatest among starved animals and immediately after molting; field-collected amphipods were more sensitive to dissolved cadmium than were laboratory strains, regardless of the season of collection (McGee et al., 1998). Acute toxicity tests of 96 h duration with adults of *T. japonicus*—a species proposed as a benchmark for routine ecotoxicity testing in the western Pacific Ocean region—showed no effect on survival at 10.0 mg Cd/L, 10% mortality at 13.9 mg/L, and 50% mortality at 25.2 mg Cd/L (Lee et al., 2007). Laboratory studies on cadmium accumulation in crustaceans and on the factors modifying cadmium uptake rates show that adult fiddler crabs, *Uca pugilator*, as one example, accumulated maximum concentrations of cadmium under conditions of comparatively high water temperatures and low salinities; concentrations were greatest in green gland, gill, hepatopancreas, and muscle, in that order (O'Hara, 1973a,b). However, *Leander adspersus* took up significantly more cadmium at 6°C than at 15°C (Bengtsson, 1977). Most authorities concluded that cadmium-stressed crabs and shrimps showed greatest uptake at lower salinities (Hutcheson, 1974; Vernberg and DeCoursey, 1977; Vernberg et al., 1977; Wright, 1977a,b). Cadmium was more toxic to juveniles of the euryhaline crab *Chasmagnathus granulatus* at 5 ppt salinity than at 25 ppt salinity, as judged by LC50 (96 h) values of 2.2 mg Cd/L at 5 ppt and 14.4 mg Cd/L at 25 ppt salinity (Beltrame et al., 2008).

Viscera were acknowledged as the main site of cadmium accumulation. Cadmium-loaded shrimp contained highest residues in hepatopancreas, followed by gill, exoskeleton muscle, and serum (Nimmo et al., 1977); similar patterns were demonstrated for shrimp, *Leander* sp. (Hiyama and Shimizu, 1964), blue crab, *Callinectes sapidus* (Hutcheson, 1974), green crab, *Carcinus maenas* (Wright, 1977c), the euphausiid, *Megancytiphanes norvegica* (Benayoun et al., 1975), and the prawn, *Pandalus montagui* (Ray et al., 1980). The effects of cadmium on fiddler crabs depend upon stage of the life cycle, thermal history, and salinity of the medium; larvae were more sensitive than adults, with survival lowest at elevated temperatures and depressed salinities (Vernberg and DeCoursey, 1977).

Transfer of cadmium to shrimp was less efficient via the diet than transfer directly from seawater. In order to produce equivalent whole-body cadmium residues in shrimp, about 15,000 times more cadmium must be introduced in food than could be obtained from the medium (Nimmo et al., 1977). A similar conclusion was reached for the euphausiid *M. norvegica* (Benayoun et al., 1975). Amphipods (*Gammarus lawrencianus*) labeled with ^{109}Cd via the medium were fed to grass shrimp, *P. pugio*; about 50% of the radiocadmium was retained by the shrimp after 7 days (Seebaugh et al., 2006). Cadmium accumulation by brine shrimp, *Artemia salina*, directly from solution and from diet was investigated at seawater cadmium concentrations of 0.1, 1.0, and 10.0 mg Cd/L (Jennings and Rainbow, 1979b). At lower cadmium concentrations, diet was the major route for accumulation. At higher cadmium concentrations, cadmium-laden food items displaced cadmium-rich water from the gut with less cadmium accumulated by *Artemia*. It seems that food chain is the major source of cadmium to *Artemia* provided that lower trophic levels biomagnify cadmium to the extent that it is more available to *Artemia* than direct uptake from seawater (Jennings and Rainbow, 1979b).

Depuration of cadmium from whole green crab, *C. maenas*, is reported: crabs containing 19.0 mg Cd/kg FW after immersion in cadmium-containing solutions for 37 days lost about half during a postexposure period of 11 days (Wright, 1977a). Two heavy metal-binding proteins occur naturally in the midgut glands of green crabs; these proteins of approximately 27.000 molecular weight (MW) and 11,500 MW together bound about 32% of the soluble cadmium in the midgut gland and may represent a possible defense mechanism against atypically high cadmium insults (Rainbow and Scott, 1979). Cadmium accumulation by green crabs was measured under controlled conditions (Jennings and Rainbow, 1979a; Jennings et al., 1979). When cadmium was offered via the diet, most of the cadmium accumulated was adsorbed onto the exoskeleton surface, with midgut gland containing about 10% of the total body burden. When cadmium was introduced via the diet, midgut gland contained about 17% of the total absorbed cadmium; levels in exoskeleton were dramatically lower (Jennings et al., 1979). Dependent on tissue and MW of cadmium-binding proteins, which ranged between <6000 and >50,000, differential tissue uptake was evident. In the midgut gland, for example, which had the highest concentrations of cadmium, at least four cadmium binding proteins were identified (Jennings et al., 1979). Duration of exposure and nominal cadmium concentrations in ambient seawater affected final residue values in crustaceans. In general, uptake was continuous over time, with little evidence of equilibrium or plateauing (Vernberg et al., 1977); also, lower concentration factors were evident at higher initial cadmium levels in the medium (Nimmo et al., 1977). At relatively low cadmium concentrations of 0.003-0.010 mg Cd/L, the uptake patterns were different for some species. Thus, adult lobsters, *H. americanus*, held for 21 days in flowing seawater solutions containing 0.010 mg Cd/L had up to 41% more cadmium than controls on a FW basis (Eisler et al., 1972). Not all tissues accumulated at the same rate: gills showed the largest increase of 78% and viscera the least, with no increase (Eisler et al., 1972). Shrimp, *Crangon crangon*, held for 8 days in media containing 0.005 mg Cd/L had a bioconcentration factor of 860 for the whole animal (Jung and Zauke, 2008). At 0.003 and 0.006 mg Cd/L, lobsters held for 30 or 60 days showed little accumulation in digestive glands, but muscle cadmium levels almost doubled when compared to controls (Thurberg et al., 1977). Grass shrimp, *P. pugio*, after exposure for 6 weeks in seawater solutions containing 0.06 mg Cd/L, had up to 20.4 mg Cd/kg DW whole body versus 6.2 in controls (Pesch and Stewart, 1980). At higher experimental cadmium concentrations of 0.12, 0.25, 0.5, and 1.0 mg Cd/L, whole-body residues in shrimp ranged up to 31.0, 58.2, 63.9, and 106.0 mg Cd/kg DW, respectively (Pesch and Stewart, 1980). High mortality was documented in grass shrimp at seawater concentrations of more than 0.25 mg Cd/L. Hermit crabs, *Pagurus longicarpus*, were even more sensitive than grass shrimp, with many deaths noted at the lowest test concentration of 0.06 mg Cd/L, commencing at week 6 of exposure and continuing until week 10, when the study was terminated (Pesch and Stewart, 1980). Cadmium uptake rates in various tissues of a cold water pandalid shrimp, *P. montagui*, held in seawater solutions containing 0.037 mg Cd/L, ranged from 0.002 to 0.06 mg Cd/kg dry tissue per hour (Ray et al., 1980). After 14 days, hepatopancreas had 25.6 mg Cd/kg DW; values for carcass, egg, and tail were 8.2, 4.3, and 0.8, respectively, in a similar period. During depuration for 57 days, cadmium concentrations in most tissues decreased, but continued to rise in hepatopancreas, indicating cadmium redistribution (Ray et al., 1980).

Radiocadmium uptake studies with *C. maenas* show that most of the cadmium entering hemolymph was freely labile and quickly displaced by stable cadmium taken up subsequently (Wright and Brewer, 1979). Some hemolymph cadmium was probably translocated to hepatopancreas; however, a significant portion of whole-body cadmium was adsorbed onto gills and exoskeleton. The rapid efflux of radiocadmium from *Carcinus* over a period of 7 days was due mainly to cadmium loss from the outer body surface (Wright and Brewer, 1979). Radiocadmium-109 uptake during exposure of 385 h and loss during 189 h in ^{109}Cd-free seawater was measured in semaphore crab, *Heloecius cordiformis*, soldier crab, *Mictryris platycheles*, and ghost shrimp, *Trypnaea australiensis* (Alquezar et al., 2007). Cadmium concentration factors over the medium during uptake were 5.7 for semaphore crab, 9.7 for soldier crab, and 68.0 for ghost shrimp; loss during immersion in ^{109}Cd-free media was 59% for semaphore crab, 66% for soldier crab, and 69% for ghost shrimp. Excretion followed a biphasic pattern. The biological half-time for short-lived and long-lived components ranged from 5 to 12 h, and from 1370 to 2590 h, respectively (Alquezar et al., 2007).

Grass shrimp, *P. pugio*, were exposed for 8 months to Cd^{2+} concentrations of 0.0015 or 0.0025 mg/L for an entire life cycle from larva to reproductive adult, and through the production of second-generation larvae (Manyin and Rowe, 2008). Cadmium concentrations in whole adult shrimp, in mg Cd/kg DW, after 8 months were 0.64 in controls, 59.6 in the low-dose group, and 80.4 in the high-dose group. Adult survival was significantly reduced in the high-dose group, but had no effect on survival or stage duration of embryos, larvae, or juveniles. Brood size was reduced by 27% in the low-dose group and by 36% in the high-dose group. Both populations showed a dose-dependent decrease in population growth, up to 12% for the high-dose group. The ability of populations to withstand predation pressure is unknown, but would probably be compromised (Manyin and Rowe, 2008).

The chemical form of cadmium, as well as interaction with other substances in solution, may affect the bioavailability of ionic cadmium. Nimmo et al. (1977) aver that at certain given cadmium concentrations, nitrate salts were accumulated to higher levels in whole shrimp than were sulfate, chloride, or acetate salts, in that order. Interactive toxicity among metal-polycyclic aromatic hydrocarbons mixtures are common among benthic copepods (Fleeger et al., 2007). For example, cadmium-phenanthrene mixtures were more than additive in toxicity by way of sediments or medium to the copepod *Schizopera knabeni*; however, cadmium-mercury-lead mixtures were less than additive in toxicity. In the copepod *Amphiascoides atopus*, mixtures of cadmium-phenanthrene and cadmium-fluoranthrene were more than additive in toxicity (Fleeger et al., 2007). Chelating agents, such as humates and alginates, significantly reduced cadmium accumulation in barnacles, *Semibalanus balanoides*, during the first 7 days of exposure, but not afterwards (Rainbow et al., 1980). Authors concluded that the amount of cadmium accumulated by barnacles is a function of the free Cd^{2+} ions available in solution and that uptake of cadmium does not require external binding of ionic cadmium to released metabolites prior to uptake (Rainbow et al., 1980). In studies with Aesop shrimp, *P. montagui*, and mixtures of cadmium and zinc (0.04 mg Cd/L plus 0.07 mg Zn/L), cadmium concentrations in *Pandalus* hepatopancreas doubled when compared to concentrations accumulated during exposure to cadmium alone (Ray et al., 1980). Increased amounts of zinc (0.41 mg Zn/L) resulted in a depression of hepatopancreatic cadmium, although the cadmium burden was still greater than in the case of exposure to cadmium alone; there was no significant accumulations of zinc in all studies at the concentrations studied (Ray et al., 1980). Weis (1978) reports that limb regeneration and ecdysis in three species of fiddler crabs were severely retarded by low concentrations of cadmium in the ambient medium; a similar effect was observed with methylmercury. However, mixtures of cadmium and methylmercury, or cadmium and calcium, ameliorated the severe effects of cadmium, indicating antagonism or site-specific competition of cadmium-mercury and cadmium-calcium mixtures.

At present, seawater concentrations in excess of 0.0045 mg/L of total cadmium at any time should be considered potentially hazardous to marine life until additional data prove otherwise (Eisler, 2000d).

7.9 Cerium

The high concentrations of cerium in fecal pellets of the euphausiid *M. norvegica* (Table 7.6) indicated that diet was the most important transfer mechanism for this element in crustaceans. Cerium flux attributable to euphausiid fecal pellets in the Mediterranean Sea was small, but measurable (Fowler, 1977). Following the Chernobyl nuclear reactor accident in April 1986 (Eisler, 1995, 2003), radiocerium-141 and -144 concentrated in fecal pellets of copepods (Fowler et al., 1987). Fecal pellets, when compared to whole copepods, contained 45 times more ^{141}Ce and 25 times more ^{144}Ce (Fowler et al., 1987).

Laboratory studies with radiocerium and the prawn, *Leander* sp., showed that viscera contained 200 times more cerium than seawater (Hiyama and Shimizu, 1964); for muscle and exoskeleton, these values were 25 and 2, respectively. The addition of chelating agents to the medium, such as ethylenediaminetetraacetic acid (EDTA), decreased the bioavailability of cerium, with the result that all tissue cerium values were lowered by an order of magnitude (Hiyama and Shimizu, 1964). The euphausiid, *M. norvegica* accumulated radiocerium from seawater by a factor of about 65; after 12 h in cerium-free seawater, about 10% of the radiocerium was lost (Antonini-Kane et al., 1972). In another study with *M. norvegica*,

Table 7.6: Cerium and Cesium Concentrations in Field Collections of Crustaceans

Element and Organism	Concentration	Reference[a]
Cerium		
Amphipod, *Paramoera walkeri*; East Antarctica; 2003-2004; whole	(0.11-0.25) DW	5
Euphausiid, *Meganyctiphanes norvegica*		
Whole	<4.5 DW	1
Fecal pellets	200.0 DW	1
Cesium		
Lobster, *Jasus lalandi*; muscle	0.01 FW	2.3
Black tiger shrimp, *Penaeus monodon*; juveniles; Vietnam; 2003-2005; max. values		
Muscle	0.04 DW	6
Exoskeleton	0.03 DW	6
Hepatopancreas	0.09 DW	6
Shrimp, *Peneus japonica*; muscle	0.046 DW	4
Stomatopod, *Squilla oratoria*; muscle	0.028 DW	4

Values are in mg element/kg fresh weight (FW), or dry weight (DW).
[a]1, Fowler, 1977; 2, Van As et al., 1973; 3, Van As et al., 1975; 4, Ishii et al., 1978; 5, Palmer et al., 2006; 6, Tu et al., 2008.

molting was responsible for up to 99% loss of total body burden of radiocerium at first molt, and about 45% of the remaining radiocerium at the second molt (Fowler et al., 1973). Fecal pellets did not contain measurable ^{144}Ce activity when the euphausiids accumulated the isotope from the medium, suggesting that surface adsorption was the key accumulating process from the medium (Fowler et al., 1973).

7.10 Cesium

The maximum concentration of stable cesium recorded in crustacean tissues from field collections was 0.09 mg Cs/kg DW in hepatopancreas of shrimp (Table 7.6).

Cesium is taken up from seawater to various degrees in different species of crustaceans. For whole copepods, *Centropages* sp. and *Calanus* sp., the concentration factors ranged from 0.1 to 1.0 times seawater levels (Ketchum and Bowen, 1958). In muscle tissues of the lobster, *J. lalandi*, cesium was concentrated by a factor of 32 over ambient seawater (Van As et al., 1973). Soft tissues and exoskeleton of a prawn, *Leander* sp., both took up radiocesium over nominal cesium levels by a factor of 15; time to reach equilibrium was 4 days for soft tissues, and 2 days for exoskeleton (Hiyama and Shimizu, 1964). Whole isopods, *Sphaeroma serratum*, contained 6-9 times more cesium than the medium (Bryan, 1963). Plasma of the stomatopod, *Squilla desmaresti* show low potential for cesium accumulation with a concentration factor of 1.4 times seawater; for muscle and digestive gland, these values ranged from 19.4 to 22.0 (Bryan, 1963). Cesium concentrations increased with increasing weight in amphipods (Palmer et al., 2006), and this should be considered when interpreting cesium residues.

Uptake of ^{134}Cs by *C. maenas* was not affected by a nonradioactive diet; muscle was the main limiting factor in the attainment of equilibrium by whole crabs (Bryan, 1961). Radiocesium accumulation occurs primarily across the body surface in starved animals; however, uptake of ^{134}Cs from the diet is rapid and complete and would probably enhance attainment of equilibrium in a natural environment. The squat lobster, *Galathea squamifera*, shows evidence of regulating radiocesium at blood plasma levels 1.1 times those of seawater; maximum radiocesium accumulation in *Galathea* was in hepatopancreas and excretory organs at levels 23-29 times that of seawater (Bryan, 1965). Cesium recycling via fecal pellets is documented in copepods after radiocesium contamination from the Chernobyl nuclear accident. Three Mediterranean Sea species contained 155 times more ^{134}Cs in fecal pellets than did whole organism; for ^{137}Cs, this value was 185 (Fowler et al., 1987).

7.11 Chromium

Whole crustaceans from chromium-contaminated environments in Egypt contained up to 483.0 mg Cr/kg FW, and those from Taiwan and Texas 195.0-463.0 mg Cr/kg on a DW basis (Table 7.7). Chromium concentrations, however, seldom exceed 0.3 mg/kg FW in edible

Table 7.7: Chromium Concentrations in Field Collections of Crustaceans

Organism	Concentration	Reference[a]
Copepod, *Acartia clausi*; whole	1.52 DW	1
Copepod, *Anomalocera patersoni*; whole; males vs. females	2.08 DW vs. 2.16 DW	2
Barnacle, *Balanus amphitrite*; soft parts	(2.1-3.9) DW	3
Blue crab, *Callinectes sapidus*		
Muscle; Iskenderun Bay, Turkey; August 2005	4.5 (0.6-11.6) DW	22
New Jersey; Hackensack River wetlands vs. reference site		
Hepatopancreas	5.2 (1.8-7.3) DW vs. 1.4 (0.8-1.8) DW	17
Muscle	0.45 DW vs. 0.54 DW	17
Pensacola, Florida; 2003-2004		
Hepatopancreas	Max. 1.1 FW	25
Muscle	Max. 0.98 FW	25
Crab, *Cancer irroratus*		
Muscle	(<0.3-0.6) FW	4
Digestive gland	(<0.5-1.2) FW	4
Gills	(0.8-2.5) FW	4
Copepod, *Centropages* spp.; whole	260.0 AW	5
Copepods		
Whole	2.1 DW	6
Whole	0.23 DW	7
Whole; 3 spp.; Taiwan; near sewage outfall	16.5-195.0 DW	23
Crustaceans; commercial species; edible tissues		
7 spp.	0.1-0.2 FW	8
9 spp.	0.2-0.3 FW	8
5 spp; Spain; April-May 1990; Mediterranean Sea	0.10-0.38 FW	19
Crustaceans; whole; 2 spp.; Egypt; chromium-contaminated Bay	212.0-483.0 FW	18
Crab, *Eriphia verrucosa*		
Carapace	0.4 DW	9
Soft parts	0.7 DW	9
Muscle	0.48 DW	10

(*Continues*)

Table 7.7: Cont'd

Organism	Concentration	Reference[a]
Euphausiids; whole	0.46 FW	11
Mantis shrimp, *Gonodactylus* sp.; Guam; June 998; muscle vs. gonad	0.57 DW vs. 0.91 DW	21
Lobster, *Jasus lalandi*; muscle	(0.08-0.11) FW	12, 13
Peneid shrimp, *Melicertus kerathurus*; Lesina lagoon, Italy; October 2004; 3 sites		
Muscle	1.9-4.3 DW	28
Exoskeleton	3.1-5.0 DW	28
Crab, *Paralithodes kamtschatica*; muscle	0.09 DW	6
Shrimp, *Parapenaeus longirostris*		
Carapace	0.05 DW	9
Soft parts	1.1 DW	10
Marsh crab, *Parasesarma erythodactyla*; soft parts; 2002; New South Wales, Australia; metals-contaminated estuary		
Males	Max. 4.5 DW	24
Females	Max. 3.5 DW	24
Shrimp, *Penaeus aztecus*		
Muscle	2.1 DW	16
Exoskeleton	10.6 DW	16
Viscera	4.0 DW	16
Black tiger shrimp, *Penaeus monodon*; juveniles; Vietnam; 2003-2005; max. values		
Muscle	0.35 DW	27
Exoskeleton	0.94 DW	27
Hepatopancreas	0.79 DW	27
Crab, *Portunus holsatus*; whole	0.33 FW	14
Crab, *Pugettia producta*; feces	(1.1-7.6) AW	15
Shrimp		
Muscle	2.8 DW	16
Exoskeleton	1.7 DW	16
Texas; Laguna Madre; 1986-1987		
Grass shrimp, *Palaemonetes* sp.; whole	68.0 (15.0-463.0) DW	20
Brown shrimp, *Penaeus aztecus*; whole	60.0 (8.0-130.0) DW	20
Blue crab, *Callinectes sapidus*; whole minus legs, carapace and abdomen	1.4 (0.5-5.7) DW	20

(Continues)

Table 7.7: Cont'd

Organism	Concentration	Reference[a]
Crab, *Ucides cordatus*; adult males		
Gills	4.8 DW	29
Hepatopancreas	1.7 DW	29
Muscle	2.5 DW	29
Zooplankton; Baltic Sea; May 1999; cladocerans vs. copepods; whole	1.5 DW vs. 16.5 DW	26

Values are in mg Cr/kg fresh weight (FW), dry weight (DW), or ash weight (AW).
[a]1, Fukai, 1965; 2, Polikarpov et al., 1979; 3, Barbaro et al., 1978; 4, Greig et al., 1977b; 5, Nicholls et al., 1959; 6, Fukai and Broquet, 1965; 7, Tijoe et al., 1977; 8, Hall et al., 1978; 9, Bernhard and Zattera, 1975; 10, Fukai, 1965; 11, Greig and Wenzloff, 1977; 12, Van As et al., 1973; 13, Van As et al., 1975; 14, DeClerck et al., 1979; 15, Boothe and Knauer, 1972; 16, Horowitz and Presley, 1977; 17, Hall and Pulliam, 1995; 18, Dahab et al., 1990; 19, Schuhmacher et al., 1992; 20, Custer and Mitchell, 1993; 21, Denton et al., 2006; 22, Turkmen et al., 2006; 23, Feng et al., 2006; 24, MacFarlane et al., 2006; 25, Karouna-Renier et al., 2007; 26, Pempkowiak et al., 2006; 27, Tu et al., 2008; 28, D'Adamo et al., 2008; 29, Correa et al., 2005.

tissues of commercial species of crustaceans from North America (Table 7.7). A value of 0.6 mg Cr/kg FW in edible muscle of the crab *Cancer irroratus* was from specimens collected near an ocean dump site receiving large quantities of metals; digestive gland and gills from these crabs also contained elevated chromium residues (Table 7.7). Chromium concentrations in tissues of amphipods from east Antarctica increased with increasing weight of the organism (Palmer et al., 2006). Chromium burdens in soft tissues of the marsh crab, *Parasesarma erythodactyla*, were significantly correlated with sediment chromium concentrations; this pattern held for both males and females (MacFarlane et al., 2006).

Of the 65,000 tons of chromium compounds used annually in exploratory oil drilling, a significant portion enters the marine environment through the discharge of used drilling muds. It is estimated that more than 225 tons of drilling mud may be used in a single 3000 m well (Carr et al., 1982). One of the most frequently used muds in offshore drilling operations is a chrome lignosulfonate mud containing barium sulfonate, bentonite clay, and ferrochrome or chrome lignosulfonates (Carr et al., 1982). The bioavailability of chromium to shrimp from used chrome lignosulfonate drilling muds is most pronounced at the mud aqueous layers. At chromium concentrations of 0.248 mg/L in the mud aqueous layer, grass shrimp took up 23.7 mg Cr/kg DW whole body after 7 days (Carr et al., 1982). Concentrations of drilling mud of 1% or greater in seawater were toxic to sensitive species of crustaceans (Neff et al., 1981). Uptake of 4.0-5.0 mg/kg was reported in grass shrimp exposed to sediments containing 188.0 mg Cr/kg (Neff et al., 1978). The toxicity of chromium-contaminated drilling muds to grass shrimp may sometimes be attributable to large residuals of petroleum hydrocarbons in the sediments (Conklin et al., 1983).

Sediment chromium concentrations of 3200.0 mg/kg in the New Bedford (Massachusetts) Acushnet estuary and 100.0 mg/kg in New York Bight have been recorded (Doughti et al., 1983). Massive cuticular lesions suggestive of shell disease characterized up to 30% of the lobsters, crabs, and shrimp collected from New York Bight, and these lesions could also be induced in crustaceans exposed to New York Bight sediments in the laboratory. This shell disease syndrome has been induced in 41% of grass shrimp (*P. pugio*) during exposure to 0.5 mg Cr^{6+}/L for 28 days (Doughti et al., 1983). It is proposed that chromium interferes with the normal function of subcuticular epithelium, particularly cuticle formation, and subsequently causes structural weaknesses or perforations to develop in the cuticle of newly molted shrimp. As a result of these chromium-induced exoskeletal deficiencies, a viaduct for pathogenic bacteria and direct chromium influx is formed, which perpetuates the development of the lesion.

Adult male mangrove crabs, *Ucides cordatus*, exposed for 96 h to a very high sublethal concentration of approximately 100.0 mg Cr^{3+}/L, contained grossly elevated concentrations in tissues—when compared to controls—of 17,396.4 mg Cr/kg DW in gills versus 4.8 in controls, 316.5 in hepatopancreas versus 1.7 in controls, and 47.3 in muscle versus 2.5 mg/kg DW in controls (Correa et al., 2005). Trivalent chromium reportedly precipitates on the gill surface colocalized with epiphyte bacteria (Correa et al., 2005).

Barnacles (*Balanus* sp.) incorporated hexavalent chromium (Cr^{6+}) in soft tissues up to 1000 times over ambient seawater concentrations, reaching equilibrium in 7 days (biological half-time for some components was 120 days); however, trivalent chromium (Cr^{3+}), which precipitates in seawater, was quickly removed by filtering activity, was not concentrated in soft tissues, and was rapidly excreted by way of the digestive system (Van Weerelt et al., 1984). Marine crustaceans were among the most sensitive organisms tested to hexavalent chromium; for example, LC50 values for early life stages of the blue crab, *C. sapidus*, were 0.93 mg/L at 96 h and 0.32 mg/L at 40 days (as quoted in Eisler, 2000e). Chelators, such as EDTA, reduce the toxicity of trivalent and hexavalent chromium compounds to adult crabs, *Petrolisthes laevigatus*, sometimes by as much as 41-48% (Urrutia et al., 2008). In other laboratory studies, Sather (1967) observed that uptake and loss of radiochromium by the crab *Podophthalmus vigil* was independent of gender and eyestalk hormone influence, Most of the radiochromium in that study accumulated in gills. Gills and muscle reached equilibrium in 2-3 days, but midgut and hemolymph required 4-5 days. Iron interfered with chromium uptake and retention in *Podophthalmus* (Sather, 1967). In another study, Tennant and Forster (1969) demonstrated that chromium concentrated in setae, gills, and hepatopancreas of the dungeness crab *C. magister*, and concluded that surface adsorption and physiological processes were both instrumental in chromium accumulation.

To protect human consumers of crustaceans, it is proposed that edible tissues should not exceed concentration of 12.0 mg Cr/kg FW (Sankar et al., 2006).

7.12 Cobalt

Concentrations of cobalt in field collections, in mg Co/kg, never exceeded 0.004 on a FW basis, 5.6 on a DW basis, or 44.0 on an AW basis (Table 7.8). Crustaceans concentrated cobalt from the surrounding medium, but this ability varied considerably among species. Concentration factors of 200-600 are reported for copepods (Ketchum and Bowen, 1958), 800 for euphausiids (Ketchum and Bowen, 1958), 4000 for whole spiny lobster (Ichikawa, 1961), and 140 for muscle of the lobster *J. lalandi* (Van As et al., 1973).

Table 7.8: Cobalt Concentrations in Field Collections of Crustaceans

Organism	Concentration	Reference[a]
Copepod, *Calanus finmarchicus*; whole	3.0 AW	1
Blue crab, *Callinectes sapidus*; muscle; August 2006; Iskenderun Bay, Turkey	2.95 (0.65-5.6) DW	11
Copepod, *Centropages* spp; whole	4.0 AW	1

Copepods
  Whole
  Whole
	
	15.0 AW
	0.12 DW
		
		2
		3

Euphausiid, *Euphausia krohni*; whole	5.0 AW	1
Euphausiids; whole	0.63 AW	4
Lobster, *Jasus lalandi*; muscle	0.004 FW	5, 6
Shrimp, *Penaeus japonica*; muscle	0.05 DW	7

Black tiger shrimp, *Penaeus monodon*; Vietnam; juveniles; 2003-2005; max. values
  Muscle
  Exoskeleton
  Hepatopancreas
	
	0.08 DW
	0.38 DW
	2.9 DW
		
		13
		13
		13

Crab, *Pugettia producta*; feces	(0.6-0.9) AW	8
Shrimps; proteinaceous molts vs. muscle	0.47 DW vs. 0.46 DW	9
Stomatopod; *Squilla oratoria*; muscle	0.16 DW	7

Zooplankton
  Surface vs. > 99 m depth
  Baltic Sea; May 1999; cladocerans vs. copepods; whole
	
	44.0 AW vs. 37.0 AW
	0.4 DW vs. 2.4 DW
		
		10
		12

Values are in mg Co/kg fresh weight (FW), dry weight (DW), or ash weight (AW).
[a]1, Nicholls et al., 1959; 2, Vinogradova and Koual'shiy, 1962; 3, Tijoe et al., 1977; 4, Robertson, 1967; 5, Van As et al., 1973; 6, Van As et al., 1975; 7, Ishii et al., 1978; 8, Boothe and Knauer, 1972; 9, Bertine and Goldberg, 1972; 10, Martin, 1970; 11, Turkmen et al., 2006; 12, Pempkowiak et al., 2006; 13, Tu et al., 2008.

In laboratory studies, concentration factors for cobalt in shrimp, *Leander pacificus*, over a 30-day period were 5 for muscle and 18 for exoskeleton (Shimizu et al., 1970); hermit crab, *Clibanarius virescens*, had a whole-body concentration factor of 520 in a similar period (Shimizu et al., 1970). Elimination rates of accumulated radiocobalt were estimated at 4-6%/h for copepods and 1%/h for euphausiids (Kuenzler, 1969). Chitin from ground lobster exoskeletons can adsorb 40-60% of trace metals and metalloids—including cobalt—in 183 h from distilled water (Yoshinari and Subramanian, 1976); this technique appears to hold considerable potential, together with biomonitoring in assessing marine water quality.

Transfer of cobalt—an essential element for normal growth and metabolism of crustaceans—through the food chain is documented. Crabs, *C. maenas*, feeding on annelid worms, *Arenicola marina*, containing radiocobalt show high assimilation rates with most of the radiocobalt accumulating in hepatopancreas (Amiard-Triquet and Amiard, 1974). A similar pattern was observed when radiocobalt-containing clams, *Scrobicularia plana*, were fed to *C. maenas* (Amiard-Triquet and Amiard, 1975). It was concluded that the surrounding medium was the main source of external cobalt contamination in *C. maenas*, that diet was the primary source of internal cobalt contamination, and that accumulation was largely in digestive glands (Amiard-Triquet and Amiard, 1976).

7.13 Copper

Among the highest copper concentrations documented in field collections of marine crustaceans, in mg Cu/kg DW, were 603.0 in hepatopancreas of lobster, 1120.0 in hepatopancreas of shrimp, 1000.0-2050.0 in various tissues of deep sea shrimps, and up to 22,900.0 in pylorus of deep sea shrimps (Table 7.9). In marine ecosystems, the high copper levels encountered in heavily contaminated coastal areas sometimes approach the incipient lethal concentrations for some organisms (Neff and Anderson, 1977). Elevated copper concentrations in saline environments may result from atmospheric deposition, industrial and municipal wastes, urban runoff, rivers, shoreline erosion (Eisler, 2000f), and proximity to deep sea hydrothermal vents (Table 7.9; Kadar et al., 2006). For example, Chesapeake Bay, Maryland, receives more than 1800 kg of copper daily from these sources (Hall et al., 1988a,b). In the Elizabeth River estuary of Chesapeake Bay, anthropogenic copper and other chelatable metals are present at concentrations sufficient to adversely affect growth and survival of the copepod *Acartia tonsa* (Sunda et al., 1990).

Blood and hepatopancreatic tissues usually contain the greatest accumulations of copper and muscle the least (Table 7.9). The presence of copper-containing respiratory pigments in crustaceans and some other invertebrate groups could account for the elevated levels found in blood. For example, copper accounted for 93% of the weight of hemocyanin in blood of the green crab *C. maenas* (Martin et al., 1977). Hepatopancreas is a major storage and regulatory organ for copper in crustaceans, with the result that copper residues are almost

Table 7.9: Copper Concentrations in Field Collections of Crustaceans

Organism	Concentration	Reference[a]
Copepod, *Acartia clausi*; whole	55.3 (34.0-107.0) DW	1
Amphipods; whole	26.0 DW	2
Copepod, *Anomalocera patersoni*; whole; males vs. females	42.4 DW vs. 41.3 DW	3
Antarctic Ocean; 1985-1988; 17 spp.; whole; various crustaceans		
2 spp.	5.5-7.7 DW	42
3 spp.	37.0-42.0 DW	42
6 spp.	53.0-68.0 DW	42
4 spp.	81.0-107.0 DW	42
2 spp.	123.0-149.0 DW	42
Barents Sea and environs; 1991-2000; whole		
Copepods; 4 spp.	6.0-9.0 DW	59
Euphausid, *Meganyctiphanes norvegica*	47.0 DW	59
Amphipod, *Themisto abyssorum*	9.0 DW	59
Shrimp, *Pandalus borealis*	61.0 DW	59
Barnacles; Natal, Brazil; summer 2001; soft parts		
Fistulobalanus citerosum	(26.4-62.8) DW	52
Balanus amphitrite	22.8 DW	52
Barnacle, *Balanus amphitrite*; soft parts	(44.0-109.0) DW	4
Barnacle, *Balanus balanoides*; soft parts; March vs. August	(0.3-0.7) DW vs. (0.1-0.2) DW	5
Barnacles; whole	Max. 600.0 DW	8
Copepod, *Calanus finmarchicus*; whole	1350.0 AW	6
Blue crab, *Callinectes sapidus*		
Whole	54.0 DW	32
Muscle; May vs. June	19.0 (5.0-97.0) FW vs. 11.0 (5.0-18.0) FW	7
Muscle; August 2005; Turkey	7.0 (3.9-9.4) DW	47
Pensacola, Florida; 2003-2004		
Hepatopancreas	Max. 99.0 FW	54
Muscle	Max. 14.0 FW	54
Crab, *Cancer irroratus*		
Muscle	(13.0-25.4) FW	9
Digestive gland	(48.0-161.0) FW	9
Gill	(26.5-46.4) FW	9

(*Continues*)

Table 7.9: Cont'd

Organism	Concentration	Reference[a]
Crab, *Cancer pagurus*		
Blood	76.0 FW	10
Muscle	7.5 FW	10
Stomach fluid	7.1 FW	10
Hepatopancreas	137.0 FW	10
Urine	0.4 FW	10
Excretory organs	17.0 FW	10
Gills	32.0 FW	10
Vas deferens	7.6 FW	10
Shell	0.5 FW	10
Ovary	9.0 FW	10
External eggs	39.0 FW	10
Muscle	(8.0-125.0) FW	11
Muscle	14.3 (4.6-32.8) FW	12
Crab, *Carcinus maenas*		
Whole	(15.4-31.4) FW	13
Whole	22.0 DW	10
Blood	46.0 FW	10
Muscle	5.7 FW	10
Stomach fluid	3.6 FW	10
Hepatopancreas	42.0 FW	10
Urine	0.5 FW	10
Excretory organs	16.0 FW	10
Gills	18.0 FW	10
Shell	0.6 FW	10
Vas deferens	4.0 FW	10
Hemolymph	44.7 FW	14
Fibrin	2.1 FW	14
Blood cells, lipids, carotenoids	0.4 FW	14
Copepod, *Centropages typicus*; whole	600.0 AW	6
Crab, *Clibanarius strigimanus*; whole	15.8 FW	35
Copepods		
Whole	3.7 DW	2
Whole; surface collection vs. > 99 m depth collection	115.0 (50.0-215.0) AW vs. 132.0 (50.0-620.0) AW	15
Whole	10.5 (9.0-22.6) DW	16
Whole	15,000.0 AW	17
Whole	7.1 DW	18
Whole; 3 spp.; Taiwan; near sewage outfall	14.0-160.0 DW	51

(Continues)

Table 7.9: Cont'd

Organism	Concentration	Reference[a]
Crabs; muscle vs. shell	10.9 (5.8-24.4) FW vs. 6.4 (4.1-9.2) FW	19
Crabs; hepatopancreas; Wales 1989; coastal area	58.0 FW	41
Shrimp, *Crangon vulgaris*		
Whole	18.5 FW	13
Whole	32.0 DW	10
Blood	68.0 DW	10
Muscle	4.0 FW	10
Hepatopancreas	520.0 FW	10
Crustaceans		
Various tissues; 12 spp.	13.0-74.0 DW	20
Muscle	Max. 94.0 DW	21
Whole	(6.8-10.9) FW	23
Commercial species; North America; edible tissues; 16 spp.		
4 spp.	1.0-3.0 FW	22
4. spp.	3.0-5.0 FW	22
4 spp.	5.0-7.0 FW	22
4 spp.	7.0-20.0 FW	22
Deep sea shrimps; Rainbow hydrothermal vent site; mid-Atlantic Ridge; summer 2002; adults; 2 spp.		
Rimicaris exoculata		
Whole body	800.0 DW	48
Cuticle	300.0 DW	48
Gill	1060.0 DW	48
Antennae	not detectable	48
Muscle	300.0 DW	48
Digestive gland	700.0 DW	48
Pylorus	22,900.0 DW	48
Microcaris fortunata		
Whole body	1000.0 DW	48
Cuticle	80.0 DW	48
Gill	1700.0 DW	48
Antennae	not detectable	48
Muscle	200.0 DW	48
Digestive gland	400.0 DW	48
Pylorus	26,300.0 DW	48

(Continues)

Table 7.9: Cont'd

Organism	Concentration	Reference[a]
Benthic crab, *Dorippe granulata*; Hong Kong; contaminated harbor		
Exoskeleton	7.7 DW	36
Gills	123.9 DW	36
Hemolymph	53.2 FW	36
Midgut gland	114.9 DW	36
Muscle	36.6 DW	36
Crab, *Eupagurus bernhardus*		
Whole	25.0 DW	10
Blood	89.0 FW	10
Leg muscle	8.4 FW	10
Abdominal muscle	5.4 FW	10
Stomach fluid	2.0 FW	10
Hepatopancreas	69.0 FW	10
Excretory organs	16.0 FW	10
Gills	31.0 FW	10
Shell	4.9 FW	10
Vas deferens	6.9 FW	10
Ovary	11.0 FW	10
Euphausiids		
Whole	10.1 FW	24
Whole	15.6 (7.5-21.3) DW	16
Antarctic Ocean and Atlantic Ocean; whole; 1985-1986		
Euphausia superba	55.0 (30.0-86.0) DW	37
Meganyctiphanes norvegica	58.0 (40.0-83.0) DW	37
Euphausiid, *Euphausia krohni*; whole	600.0 AW	6
Euphausiid, *Euphausia superba*; Antarctica; whole	103.0 DW	44
Hydrothermal vent field; mid-Atlantic Ridge; 2001		
Shrimp, *Chlorocaris chacei*; 1700 m deep; digestive gland vs. muscle	264.0 DW vs. 44.0 DW	56
Hydrothermal crab, *Segonzacia mesatlantica*; digestive gland vs. muscle		
840 m deep	150.0 DW vs. 47.0 DW	56
1700 m deep	2840.0 DW vs. 125.0 DW	56
2300 m deep	2850.0 DW vs. 125.0 DW	56

(*Continues*)

Table 7.9: Cont'd

Organism	Concentration	Reference[a]
Lobster, *Galathea squamifera*		
Whole	29.0 DW	10
Blood	77.0 FW	10
Muscle	7.2 FW	10
Hepatopancreas	603.0 FW	10
Gills	26.0 FW	10
Mantis shrimp, *Gonodactylus* sp.; Guam; June 1998; muscle vs. gonad	11.0 DW vs. 3195.0 DW	43
Lobster, *Homarus vulgaris*		
Whole	33.0 DW	10
Blood	82.0 FW	10
Muscle	6.0 FW	10
Stomach fluid	5.0 FW	10
Hepatopancreas	438.0 FW	10
Urine	1.3 FW	10
Excretory organs	17.0 FW	10
Gills	44.0 FW	10
Vas deferens	8.0 FW	10
Ovary	15.0 FW	10
Muscle	(6.2-13.2) FW	11
Muscle	10.3 (6.0-18.0) FW	12
Blood	32.0 FW	38
Exoskeleton	3.0 FW	38
Gill	26.0 FW	38
Liver	335.0 FW	38
Muscle	4.0 FW	38
Ovaries	50.0 FW	38
Stomach fluid	10.0 FW	38
Testes	1.0 FW	38
Urine	2.0 FW	38
Whole	17.0 FW	38
Blue shrimp, *Litopenaeus stylirostris*; Gulf of California, Mexico; 2004-2005; muscle	25.4 DW; max. 38.3 DW	45
Crab, *Maia squinado*		
Whole	25.0 DW	10
Blood	43.0 FW	10
Muscle	3.2 FW	10
Stomach fluid	1.2 FW	10
Hepatopancreas	65.0 FW	10

(Continues)

Table 7.9: Cont'd

Organism	Concentration	Reference[a]
Urine	0.2 FW	10
Excretory organs	5.0 FW	10
Gills	25.0 FW	10
Shell	1.4 FW	10
Vas deferens	3.0 FW	10
Ovary	11.0 FW	10
Euphausiid, *Meganyctiphanes norvegica*		
Whole	48.0 DW	25
Fecal pellets	226.0 FW	25
Peneid shrimp, *Melicertus kerathurus*; Lesina lagoon, Italy; October 2004; 3 sites		
Muscle	6.6-26.0 DW	58
Exoskeleton	39.2-57.9 DW	58
Shrimp, *Metapenaeus affinis*; muscle	28.9 DW	26
Lobster, *Nephrops norvegicus*; muscle	(2.1-11.3) FW	11
Beach hopper (amphipod), *Orchestia gammarellus*; whole; North Sea; 1989-1990; reference site vs. contaminated site	Usually < 70.0 DW vs. > 145.0 DW; max. 340.0 DW	39
Amphipod, *Orchestoidea corniculata*; whole	(10.0-125.0) DW	27
Shrimp, *Palaemon serratus*		
Whole	30.0 DW	10
Blood	97.0 FW	10
Muscle	3.5 FW	10
Hepatopancreas	185.0 FW	10
Gills	55.0 FW	10
External eggs	50.0 FW	10
Shrimp, *Palaemon squilla*; whole	31.0 DW	10
Shrimp, *Palaemonetes varians*		
Whole	32.0 DW	10
Blood	180.0 FW	10
Muscle	7.9 FW	10
Hepatopancreas	137.0 FW	10
Lobster, *Palinurus vulgaris*		
Blood	103.0 FW	10
Muscle	(9.0-11.0) FW	10

(*Continues*)

Table 7.9: Cont'd

Organism	Concentration	Reference[a]
Stomach fluid	10.0 FW	10
Hepatopancreas	163.0 FW	10
Urine	6.0 FW	10
Excretory organs	20.0 FW	10
Gills	57.0 FW	10
Shell	3.0 FW	10
Ovary	17.0 FW	10
External eggs	5.0 FW	10
Amphipod, *Paramoera walkeri*; East Antarctica; 2003-2004; whole	(7.2-42.0) DW	44
Shrimp, *Pandalus jordani*; muscle	(1.1-8.2) FW; (6.7-40.0) DW	12
Marsh crab, *Parasesarma erythodactyla*; soft parts; 2002; New South Wales, Australia; metals-contaminated estuary		
Males	Max. 86.8 DW	53
Females	Max. 239.0 DW	53
Brown Shrimp, *Penaeus aztecus*		
Whole	34.0 DW	32, 38
Muscle	24.2 (18.0-29.0) DW	33, 38
Exoskeleton	32.4 DW	33, 38
Viscera	173.0 (65.0-260.0) DW	33, 38
Shrimp, *Penaeus japonica*; muscle	14.0 DW	28
Black tiger shrimp, *Penaeus monodon*		
Muscle	(21.5-48.3) DW	26
Vietnam; 2003-2005; juveniles; max. values		
Exoskeleton	42.0 DW	57
Hepatopancreas	1120.0 DW	57
Muscle	28.2 DW	57
White shrimp, *Peneus setiferus*; muscle; Gulf of Mexico	17.3 DW	46
Shrimp; *Penaeus* spp.; whole	47.0 DW	29
Isopod, *Porcellana platycheles*; whole	27.0 DW	10
Crab, *Portunus depurator*		
Whole	18.0 DW	10
Blood	87.0 FW	10
Muscle	7.1 FW	10

(*Continues*)

Table 7.9: Cont'd

Organism	Concentration	Reference[a]
Stomach fluid	8.1 FW	10
Hepatopancreas	103.0 FW	10
Urine	0.2 FW	10
Excretory organs	11.0 FW	10
Gills	35.0 FW	10
Shell	1.1 FW	10
Vas deferens	14.0 FW	10
Crab, *Portunus holsatus*; whole	12.2 FW	30
Crab, *Portunus pelagicus*; muscle	(29.0-94.4) DW	26
Crab, *Portunus puber*		
Whole	21.0 DW	10
Blood	113.0 FW	10
Muscle	8.4 FW	10
Stomach fluid	36.0 FW	10
Hepatopancreas	134.0 FW	10
Urine	0.9 FW	10
Excretory organs	28.0 FW	10
Gills	36.0 FW	10
Shell	9.5 FW	10
Vas deferens	17.0 FW	10
Ovary	18.0 FW	10
Crab, *Pugettia producta*; whole	(15.0-60.0) AW	31
Shrimp		
Muscle	31.1 DW	33
Muscle; 2 spp.; Taiwan; 1995-1996	Max. 27.5 DW	40
Exoskeleton	16.3 DW	33
Muscle	(2.4-4.5) FW	19
Shell	(4.9-10.7) FW	19
Amphipod, *Themisto* spp.; whole; Antarctic Ocean and Atlantic Ocean; 1985-1986	28.0-31.0 (13.0-79.0) DW	37
Crab, *Xantho incisus*; whole	20.0 DW	10
Zooplankton; mostly copepods; NW Mediterranean Sea; whole	32.6 (5.9-121.1) DW	34
Fiddler crab, *Uca pugnax*; New Jersey Intermolt; contaminated site		
Soft tissue	639.7 DW	50
Carapace	152.7 DW	50

(Continues)

Table 7.9: Cont'd

Organism	Concentration	Reference[a]
Intermolt; reference site		
Soft tissue	518.9 DW	50
Carapace	136.6 DW	50
Immediate postmolt; contaminated site		
Soft tissue	306.5 DW	50
Exoskeleton	75.3 DW	50
Immediate postmolt; reference site		
Soft tissue	194.4 DW	50
Exoskeleton	16.7 DW	50
Zooplankton; Baltic Sea; May 1999; cladocerans vs. copepods; whole	2.0 DW vs. 25.1 DW	55

Values are in mg Cu/kg fresh weight (FW), dry weight (DW), or ash weight (AW).
[a]1, Zafiropoulos and Grimanis, 1977; 2, Bohn and McElroy, 1976; 3, Polikarpov et al., 1979; 4, Barbaro et al., 1978; 5, Ireland, 1974; 6, Nicholls et al., 1959; 7, Boon, 1973; 8, Stenner and Nickless, 1975; 9, Greig et al., 1977b; 10, Bryan, 1968; 11, Topping, 1973; 12, Bernhard and Zattera, 1975; 13, Wharfe and Van Den Broek, 1977; 14, Martin et al., 1977; 15, Martin, 1970; 16, Martin and knauer, 1973; 17, Vinogradova and Koual'skiy, 1962; 18, Tijoe et al., 1977; 19, Won, 1973; 20, Stickney et al., 1975; 21, Stenner and Nickless, 1974a; 22, Hall et al., 1978; 23, Wright, 1976; 24, Greig and Wenzloff, 1977; 25, Fowler, 1977; 26, Zingde et al., 1976; 27, Bender, 1975; 28, Ishii et al., 1978; 29, Knauer, 1970; 30, DeClerck et al., 1979; 31, Boothe and Knauer, 1972; 32, Sims and Presley, 1976; 33, Horowitz and Presley, 1977; 34, Hardstedt-Romeo and Laumond, 1980; 35, Eustace, 1974; 36, Depledge et al., 1993; 37, Rainbow, 1989; 38, Jenkins, 1980; 39, Moore et al., 1991; 40, Han et al., 1998; 41, Morris et al., 1989; 42, Petri and Zauke, 1993; 43, Denton et al., 2006; 44, Palmer et al., 2006; 45, Frias-Espericueta et al., 2007; 46, Vazquez et al., 2001; 47, Turkmen et al., 2006; 48, Kadar et al., 2006; 50, Bergey and Weis, 2007; 51, Feng et al., 2006; 52, Silva et al., 2006; 53, MacFarlane et al., 2006; 54, Karouna-Renier et al., 2007; 55, Pempkowiak et al., 2006; 56, Colaco et al., 2006; 57, Tu et al., 2008; 58, D'Adamo et al., 2008; 59, Zauke and Schmalenbach, 2006.

always highest at that site. However, sites of copper accumulation, storage, and action vary widely among crustaceans. In *C. maenas* as one case, heavy metal-binding proteins in midgut gland protect midgut against high environmental levels of copper; the two proteins, of about 27,000 and 11,500 MW are associated with 0.9% and 34.4%, respectively, of the total midgut gland copper content (Rainbow and Scott, 1979).

Biological and abiotic variables need to be considered in the interpretation of copper concentrations in crustaceans, including developmental stage, prior exposure to copper, chemical form of copper, temperature and salinity of the medium, chelators, season of collection, tissue specificity, pH, and many others. Selected examples follow for various crustacean groups.

Molting is a feasible mechanism of copper depuration in the fiddler crab, *Uca pugnax* (Bergey and Weis, 2007). Average percent of total body copper burden eliminated during molting from a relatively clean site was 3% versus 12% from a metal-contaminated site. Prior to molting, crabs shifted copper from the carapace into soft tissues (Bergey and Weis, 2007;

Table 7.9). Soft tissues of male, but not female, marsh crabs, *P. erythodactyla*, from a metal-contaminated estuary in Australia in 2002 reflected sediment copper concentrations; at higher salinities, copper content increased in males (MacFarlane et al., 2006).

In the amphipod, *Orchestia gammarellus*, copper concentrations vary seasonally due to variable copper loadings, are higher in organisms from contaminated sites, and higher in females with juveniles in the brood pouch than females without juveniles (Moore et al., 1991). Marine amphipods readily accumulate dissolved copper from seawater in a dose-dependent manner (Weeks and Rainbow, 1991). But some species of talitrid amphipods are unable to meet their copper requirements from seawater alone and depend on dietary sources of copper (Weeks and Rainbow, 1993). The tolerance of talitrid amphipods to high concentrations of ambient copper is attributable, in part, to the formation of intracellular copper granules within the cells of the ventral caeca (Weeks, 1992). The existence of copper-rich granules is common to all invertebrate phyla; these granules are usually found in the digestive gland or its evolutionary equivalent, and their formation is related to high concentrations of copper in the immediate environment (Weeks, 1992). Gender-related differences in uptake, retention, and accumulation of copper and other metals are reported in crustaceans (Burger et al., 2007a; Rabitsch, 1995; Zodle and Whitman, 2003), but results are variable. Adult amphipods, *Peramphithoe parmerong*, were less likely to select copper-contaminated habitat or diet (Roberts et al., 2006). In a 30-day study, juvenile amphipods fed copper-contaminated macroalgae had 75% mortality, although no effect was observed on growth of survivors (Roberts et al., 2006).

Among barnacles, certain developmental stages were extremely sensitive to Cu^{2+}, resulting in impaired settling and growth (Pyefinch and Mott, 1948); it is probable that copper accumulation rates were higher in these stages. Barnacles, *S. balanoides*, held in seawater containing 0.02-0.09 mg Cu/L for 100 days show dose-dependent increases in copper loadings in body and egg masses (Powell and White, 1990). In barnacles, *Balanus amphitrite nireus*, the respiratory surface in the thorax is a primary copper absorption site with excretion usually through hindgut epithelia (Bernard and Lane, 1961, 1963). Inorganic granules found in parenchyma cells surrounding the midgut of *Balanus balanoides* contained 1200.0 mg Cu/kg, and this could account for the high copper burdens found in that species (Walker et al., 1975a,b; Walker, 1977). Nauplii of the barnacle *Balanus amphitrite albicostatus*, when compared to oyster eggs, sea urchin eggs, and adult rotifers, were 5-10 times more resistant to biocidal properties of copper sulfate during a 24-h period; for barnacle nauplii, the highest concentration tested without significant adverse effect was 1.0 mg Cu/L (Kobayashi, 1979). Interaction effects with other metals are documented and need to be considered in residue interpretation. For example, barnacles tend to accumulate cadmium—a potentially toxic metal—in the presence of copper (Powell and White, 1990). Barnacles attacked by flatworm predators can be treated successfully with copper. For example, barnacles, *Balanus variegatus*, infested with free-living flatworms, *Stylochus pygmaeus*, treated with 0.025 mg Cu/L for 10 days resulted in reduced egg deposition and hatching success of flatworms by 80%, although barnacle feeding rate was inhibited during copper treatment (Lee and Johnston, 2007).

Whole-body copper accumulation in the euryhaline copepod *A. tonsa* is dependent on salinity, decreasing as salinity increased (Pinho et al., 2007). In the salinity range 5-30 ppt, copper-exposed copepods had higher whole-body sodium concentrations at higher salinities and lower sodium concentrations at lower salinities. Physiological effects induced by waterborne copper exposure to *A. tonsa* acclimatized to salinities of 15-30 ppt are due to a combined effect of food restriction and copper exposure and not due to differences in whole-body copper burden (Pinho et al., 2007). Based on 96-h toxicity tests, copepods, *T. japonicus*, were more sensitive to copper in summer than in winter, and more sensitive to copper acetate than to copper chloride or copper sulfate (Bao et al., 2008). Adverse effects of copper on survival of copepods are reduced or eliminated by the presence of clay minerals, diatoms, ascorbic acid, sewage effluents, water extracts of humic acid, and certain soil types (Lewis et al., 1972). Chelators, such as EDTA, and more alkaline pH increase the survival and larval developmental rates of copepods challenged with copper through increased complexation of cupric ions (Sunda et al., 1990). Copper uptake by brine shrimp, *Artemia franciscana*, increases with decreasing pH and decreasing carbonate complexation (Blust et al., 1991). Copper is toxic to the copepod, *A. tonsa*, when fed a copper-contaminated phytoplankton diet (Bielmyer et al., 2006). In that study, diatoms, *Thalassiosira pseudonana*, were held in seawater containing 0.0012 mg Cu/L for 7 days, reaching a copper concentration of 22.3 mg Cu/kg DW diatom; this, in turn, was fed to *Acartia* for 7 days, producing a 20% reduction in copepod reproduction. Authors note that the current water quality criterion for marine life protection is 0.003 mg Cu/L, far in excess of the 0.0012 mg Cu/L used in this study to produce a toxic diet (Bielmyer et al., 2006). Copper is increasingly toxic to the copepod *E. affinis* with increasing salinity in the range 5-25 ppt, and increasingly less toxic with increasing dissolved organic carbon in the range 2.0-8.0 mg/L (Hall et al., 2008). Acute toxicity bioassays of 96-h duration with adults of *T. japonicus* copepods—a species recommended for routine ecotoxicity testing in the western Pacific Ocean region—showed that all survived 1.0 mg Cu/L, 10% died at 2.2 mg/L, and 30% died at 3.9 mg Cu/L (Lee et al., 2007).

Soldier crabs, *Mictyris longicarpus*, take up copper mainly from sediments rather than the water column (Weimin et al., 1994). The fine particles of sediment trapped as food contain bioavailable fractions of copper and other metals, and these significantly correlate with metal concentrations in crab tissues. However, copper accumulated from sediments by soldier crabs occurs only at an artificially high level (1900.0 mg Cu/kg DW sediment), which also had toxic effects. Soldier crabs seem unable to regulate copper within the body (Weimin et al., 1994); however, many marine decapod crustaceans reportedly regulate tissue copper concentrations within the range of 25.0-35.0 mg kg DW tissue (Neff and Anderson, 1977). Transfer of copper from wood treated with chromated copper arsenate is documented in fiddler crabs, *Uca* spp. (Weis and Weis, 1992). In the crab *Carcinus mediterraneus*, tissue copper concentrations are lower in winter than in summer and correlate positively with total protein and hemolymph copper contents (Devescovi and Lucu, 1995). Elevated copper burdens in crab hemolymph probably reflect the incorporation of copper atoms in the structure of hemocyanin, the major hemolymph protein (Depledge et al., 1993). Starved shore crabs show a reduction in carapace copper content; starvation in combination with copper exposure (0.5 mg Cu/L) results in an increase in carapace copper and a decrease in carapace calcium (Scott-Fordsmand and Depledge, 1993). In *C. maenas*, ionic copper displaces ionic magnesium in gills, leading to inhibition of phosphoryl transfer (Hansen et al., 1992b).

The question of physiological adaptation to copper is unresolved. In one case, progeny of brine shrimp, *A. salina*, showed high tolerance to 1.0 mg Cu/L, when compared to controls, after exposure for two or three generations to 0.025-0.1 mg Cu/L; however, this tolerance diminished in successive generations (Saliba and Ahsanullah, 1973). At any given copper concentration, growth inhibition of *A. salina* previously acclimatized to 0.100, 0.050, or 0.025 mg Cu/L was least in cupric sulfate, and increased progressively in chloride, acetate, and carbonate salts (Saliba and Krzyz, 1976). But in survival tests with unacclimatized *Artemia* adults and larvae, cupric chloride was least toxic, followed progressively by sulfate, chloride, and acetate salts (Saliba and Krzyz, 1976).

Temperature of the medium affects the biocidal properties of copper salts, and presumably copper accumulation rates. In one study the shrimp, *Pandalus danae*, was more sensitive to copper at 20°C than at 10°C (Gibson et al., 1976), and lobsters, *H. americanus*, stressed by copper survived longer at 5°C than at 13°C (McLeese, 1974). The season of collection was linked to copper burdens in crab muscle (Boon, 1973) and whole copepods (Hardstedt-Romeo and Laumond, 1980); these differences were not wholly explainable on the basis of water temperature. Decreasing salinity of the medium was associated with decreasing survival of isopods at tested copper concentrations (Jones, 1975), and this appears to corroborate trends reported by others for different groups of marine biota. Copper absorption was coincident with the presence of other substances, especially calcium in barnacles (Bernard and Lane, 1961), mercury in copepods (Barnes and Stanbury, 1948; Reeve et al., 1977), and chelating agents in copepods (Lewis et al., 1972, 1973). In isopods, copper seems to move freely in an easily dissociable state between storage cells and other cells of the hepatopancreas (Wieser, 1967). Granular chitin from lobster shells absorbed 40-60% of copper in seawater in 183 h, with accumulation independent of initial copper concentration in the range 1.0-16.0 mg Cu/L (Yoshinari and Subramanian, 1976). This last observation may have application in environmental programs to control or monitor metals wastes in saline waters.

No documented report of fatal copper deficiency is available for any species of aquatic organism, and no correlation is evident in aquatic biota for the presumed nutritional copper requirements and its sensitivity to dissolved copper (Neff and Anderson, 1977). However, extremely low copper concentrations in whole bodies of 2 of 17 species of crustaceans (5.5 and 6.7 mg/kg DW) from the Antarctic Ocean support the hypothesis that certain Antarctic crustaceans may show copper deficiency or reduced metal requirements (Petri and Zauke, 1993). Amphipods collected in 2003-2004 from Antarctica show increasing concentrations of whole-body copper with increasing age (Palmer et al., 2006), suggesting a need for copper in reproduction. Amphipods, *P. walkeri*, from the Australian Antarctic Territory in 1999-2000 had whole-body bioconcentration factors of 1700-3800 when the seawater medium contained 0.0001-0.0002 mg Cu/L; however, BCFs decreased with increasing seawater concentration (Clason et al., 2003). *Paramoera* showed increased sensitivity when seawater concentrations were between 0.002 and 0.003 mg Cu/L (Clason et al., 2003).

Copper is lethal to sensitive crustaceans in the range of 0.01-0.16 mg/L. Some deaths were reported in amphipods, *Allorchestes compressa*, during immersion in 0.01 mg Cu/L for 4 weeks (Ahsanullah and Williams, 1991), in *T. japonicus* copepods at 0.01 mg/L for one generation (Lee et al., 2008), in *Melita* spp. amphipods at 0.04 mg/L in 96 h (King et al., 2006a,b), in copepods *Acartia* spp., in 96 h at 0.031-0.052 mg Cu/L (USEPA, 1980b), and in a marine cladoceran, *Moina monogolica* in 48 h at 0.099-0.112 mg Cu/L (Wang et al., 2007a). Norway lobsters, *Nephrops norvegicus*, all died in 14 days when held in seawater containing 0.1 mg Cu/L (Canli and Furness, 1993). Half the larvae of a peneid shrimp, *Metapeneus ensis*, held in 0.16 mg Cu/L for 48 h died (Wong et al., 1993). Among amphipods, juveniles were more sensitive to copper than were adults, and epibenthic amphipods were more sensitive than infaunal tube-dwelling amphipods (King et al., 2006a,b). Tissue hypoxia through gill destruction is probably the major cause of copper-induced death in crustaceans (Nonnotte et al., 1993). The horseshoe crab, *Limulus polyphemus*, is relatively resistant to copper, although populations from New Jersey and Delaware show differential embryo sensitivity: LC-50 (72 h) values for embryos were 2.0 mg Cu/L for the New Jersey population versus 171.0 mg/L for the Delaware population (Botton et al., 1998). Other species were remarkably resistant, with significant mortality at more than 2.0 mg Cu/L for *C. maenas*, and more than 10.0 mg Cu/L for others (Scott-Fordsmand and Depledge, 1993).

Sublethal effects include adverse effects on growth of a copepod, *Tisbe furcata*, at 0.057 mg Cu/L (Bechmann, 1994); increased copper uptake and lowered egg production in the amphipod *C. volutator* at 0.1 mg Cu/L, with effects exacerbated at low dissolved oxygen (Ericksson and Weeks, 1994); and reduced growth and high uptake in amphipods surviving 0.01 mg Cu/L for 4 weeks (Ahsanullah and Williams, 1991). Other sublethal effects of copper intoxication include abnormal variation in enzyme activities. In *A. salina*, for example, amylase activity was disrupted within 24 h and trypsin activity within 72 h by 2.0 mg Cu/L (Alayse-Danet et al., 1979). At high sublethal concentrations of waterborne copper for several days, shore crabs experienced extensive damage to the gill epithelium (Nonnotte et al., 1993), and altered enzyme and ATPase activity (Hansen et al., 1992a,b). Chinese mitten crabs, *Eriocheir sinensis*, exposed to graded concentrations of Cu^{2+} for 10 days showed gill and hepatopancreas histopathology and metallothionein induction (Yang et al., 2007, 2008). No damage effects were evident at 0.01 mg Cu/L; however, metallothionein induction was maximal at 0.1 mg Cu/L and histopathology evident at 0.1-5.0 mg Cu/L. The current Chinese National Water Quality Standard for Fisheries is 0.01 mg Cu/L, but this is routinely exceeded in many regions to at least 0.07 mg Cu/L from copper sulfate, used extensively as an algicide in culture of the Chinese mitten crab and other crustaceans (Yang et al., 2007). The maximum acceptable toxicant concentration (MATC) for the cladoceran *Moina mongolica*, wherein the lower value indicates no measurable effect (here the test period is 21 days), and the higher value has some observable effect (here reduced brood size) is in the range 0.0051-0.0089 mg Cu/L (Wang et al., 2007a). Dietary toxicity of copper is also reported for *M. mongolica* fed green alga, *Chlorella pyrenoidosa*, containing various concentrations of copper for 21 days (Wang et al., 2007b). Adverse effects were documented when *Moina* ate algae exposed to 0.044 mg Cu/L and higher for 96 h; effects included reduction in brood size and net reproductive rate (Wang et al., 2007b).

The formation of copper granules as a cellular response to copper detoxification along with normal storage of copper during the molt cycle is a proposed strategy to cope with copper insult by crustaceans (Correia et al., 2002). In one case, amphipods, *Gammarus lacustris*, that were subjected to sublethal concentrations of copper via seawater or copper-spiked sediments, accumulated copper as Cu-S granules stored in the hepatopancreas. The abundance of the granules increased with increasing whole-body copper content. The presence of sulfur within the granule represents an organic detoxification mechanism for copper (Correia et al., 2002). In another case, *T. japonicus* copepods were exposed for 7 days to high sublethal concentrations of copper (Barka, 2007). Copper localized in digestive epithelium cells and cuticular integument. Copper-containing granules were found in copper-exposed copepods and in controls within lysosomes, intracellular calcospherites, and extracellular granules. Similar results were reported for *T. japonicus* and other metals tested, suggesting a generalized model for crustaceans coping with metal stress (Barka, 2007).

Proposed copper criteria to protect most marine resources include less than 0.004-0.005 mg Cu/L and less than 5.0-15.0 mg Cu/kg DW sediments; for human health protection, this value is less than 30.0 mg Cu/kg DW in seafood (Eisler, 2000f).

7.14 Dysprosium

Whole amphipods from east Antarctica in 2003-2004 contained a maximum of 0.014 mg Dy/kg DW (Palmer et al., 2006).

7.15 Erbium

Whole amphipods, *P. walkeri*, from east Antarctica contained a maximum of 0.007 mg Er/kg DW (Palmer et al., 2006).

7.16 Europium

In 1977, whole euphausiids contained less than 4.5 mg Eu/kg DW, the detection limit at that time (Table 7.10). In 2004, however, whole amphipods contained up to 0.005 mg Eu/kg DW, the improved detection limit (Palmer et al., 2006).

7.17 Gadolinium

Whole amphipods, *P. walkeri*, from east Antarctica in 2003-2004 contained 0.007-0.016 mg Gd/kg DW (Palmer et al., 2006).

7.18 Gallium

Maximum gallium concentrations in field collections—based on limited data—were 40.0 mg Ga/kg on an AW basis and 0.36 mg Ga/kg on a DW basis (Table 7.10).

Table 7.10: Europium, Gallium, Germanium, and Gold Concentrations in Field Collections of Crustaceans

Element/Organism	Concentration	Reference[a]
Europium		
Amphipod, *Paramoera walkeri*; East Antarctica; 2003-2004; whole	Max. 0.005 DW	7
Euphausiid, *Meganyctiphanes norvegica*; whole	<4.5 DW	1
Gallium		
Copepods; whole	40.0 AW	2
Crustaceans; whole	(0.03-0.36) DW	3
Germanium		
Barnacle, *Lepas* sp.; whole	0.006 FW	4
Crab, *Portunus* sp.; whole	(0.00-0.01) FW	4
Shrimp; whole	(0.002-0.025) FW	4
Gold		
Shrimp, *Pandalus* sp.		
Whole	0.0046 AW	5
Soft parts	0.00028 DW	6

Values are in mg element/kg fresh weight (FW), dry weight (DW), or ash weight (AW).
[a] 1, Fowler, 1977; 2, Vinogradova and Koual'skiy, 1962; 3, Culkin and Riley, 1958; 4, Johnson and Braman, 1975; 5, Fukai and Meinke, 1959; 6, Fukai and Meinke, 1962; 7, Palmer et al., 2006.

7.19 Germanium

Germanium concentrations in field collections—on a FW basis—ranged from nondetectable to 0.025 mg Ge/kg (Table 7.10).

7.20 Gold

Data on gold concentrations in field collections of crustaceans are scarce and concentrations—based on a single species—were 0.0046 mg Au/kg AW and 0.00028 mg Au/kg on a DW basis (Table 7.10).

A laboratory study was conducted with radiogold and blue crab, *C. sapidus*. Crabs given an oral dose of radiogold were measured for radioactivity 4 days post exposure; almost all of the retained radiogold was in digestive gland and gut (Duke et al., 1966).

7.21 Holmium

A single datum is available indicating that whole amphipods, *P. walkeri*, from east Antarctica in 2003-2004 contained less than 0.005 mg Ho/kg DW, the detection limit at that time (Palmer et al., 2006).

7.22 Iron

It is clear that iron concentrations vary widely between and among species with different tissues, depth of collection, and geographic locale (Table 7.11). Maximum iron concentrations, in mg Fe/kg DW, are documented for whole copepods collected near a Taiwan sewage outfall (7255.0), whole crustaceans from Antarctica (1080.0), all tissues from deep sea shrimps (920.0-179,720.0), and euphausiid fecal pellets (24,000.0) (Table 7.11).

The ability of various crustacean groups to accumulate iron from the medium, as judged by bioconcentration factors, ranged from 250,000 for whole copepods (Ketchum and Bowen, 1958) to 4000 for whole noncopepod crustaceans (Ichikawa, 1961), to 1800 for lobster muscle (Van As et al., 1973). During intermolt stages, blue crabs, *C. sapidus*, accumulated greatest iron concentrations in gill tissue: up to 500.0 mg Fe/kg gill on a FW basis. A concentration ratio of only 30 was reported for the euphausiid, *M. norvegica*, with about a third of the accumulated iron lost on transfer to iron-free water for 12 h (Antonini-Kane et al., 1972); however, the uptake portion of that study was of relatively short duration.

High levels of iron occur in barnacles, *B. balanoides*, from Menai Straits, United Kingdom (Walker et al., 1975a,b). This species contains inorganic granules or concretions in parenchyma cells surrounding the midgut; concentrations of iron in these granules are estimated at 1100.0 mg/kg FW, and this might account for the elevated iron levels sometimes

Table 7.11: Iron Concentrations in Field Collections of Crustaceans

Organism	Concentration	Reference[a]
Copepod, *Acartia clausi*; whole	5200.0 FW	1
Amphipods; whole	87.0 DW	2
Barnacles; Natal, Brazil; summer 2001; soft parts		
Fistulobalanus citerosum	(371.0-527.0) DW	24
Balanus amphitrite	466.0 DW	24
Blue crab, *Callinectes sapidus*		
Whole	143.0 FW	4
Muscle; August 2005; Turkey	14.3 (2.1-45.3) DW	21
Crab, *Cancer irroratus*; gills	500.0 FW	5
Copepods		
Whole	78.0 DW	2
Whole	197.0 DW	6
Whole	59.0 DW	7
Whole	40,000.0 AW	8
Whole; surface vs. > 99 m collection depth	2900.0 AW vs. 4200.0 AW	9
Whole; 3 spp.; Taiwan; near sewage outfall	256.0-7255.0 DW	23
Crustaceans; whole	(146.0-1045.0) DW	10
Crustaceans; whole; 4 spp.; 2004; Antarctica	72.0-1108.0 DW	20
Deep sea shrimps; 2 spp.; adults; Rainbow hydrothermal vent site; mid-Atlantic Ridge; summer 2002		
Rimicaris exoculata		
Whole body	35,600.0 DW	22
Cuticle	3200.0 DW	22
Gill	90,460.0 DW	22
Antennae	920.0 DW	22
Muscle	610.0 DW	22
Digestive gland	6640.0 DW	22
Pylorus	138,990.0 DW	22
Microcaris fortunata		
Whole body	6600.0 DW	22
Cuticle	6960.0 DW	22
Gill	122,970.0 DW	22
Antennae	12,360.0 DW	22

(*Continues*)

Table 7.11: Cont'd

Organism	Concentration	Reference[a]
Muscle	940.0 DW	22
Digestive gland	3010.0 DW	22
Pylorus	179,720.0 DW	22
Euphausiid, *Euphausia superba*; East Antarctica; 2003-2004; whole	15.3 DW	19
Euphausiids		
Whole	92.0 DW	6
Whole	100.0 FW	1
Lobster, *Jasus lalandi*		
Muscle	17.0 FW	11
Muscle	2.9 FW	12
Lobster		
Muscle	34.0 AW	13
Carapace	100.0 AW	13
Head	130.0 AW	13
Euphausiid, *Meganyctiphanes norvegica*		
Whole	64.0 DW	14
Fecal pellets	24,000.0 DW	14
Molts	232.0 DW	14
Eggs	330.0 DW	14
Amphipod, *Orchestoidea corniculata*; whole	(100.0-225.0) DW	15
Amphipod, *Paramoera walkeri*; East Antarctica; 2003-2004; whole	(98.0-164.0) DW	19
Shrimp, *Penaeus aztecus*		
Muscle	14.2 DW	3
Exoskeleton	45.1 DW	3
Viscera	338.0 DW	3
Shrimp, *Penaeus japonica*; muscle	76.0 DW	16
Shrimp, *Penaeus* sp.		
Whole	76.0 FW	4
Whole	50.0 DW	17
Crab, *Pugettia producta*; feces	(750.0-1650.0) AW	18
Shrimps		
Muscle	40.2 DW	3
Exoskeleton	235.0 DW	3

(*Continues*)

Table 7.11: Cont'd

Organism	Concentration	Reference[a]
Muscle	920.0 AW	13
Carapace	8300.0 AW	13
Whole	640.0 AW	13
Stomatopod, *Squilla oratoria*; muscle	29.0 DW	16

Values are in mg Fe/kg fresh weight (FW), dry weight (DW), or ash weight (AW).
[a] 1, Fujita et al., 1969; 2, Bohn and McElroy, 1976; 3, Horowitz and Presley, 1977; 4, Wolfe et al., 1973; 5, Martin, 1973; 6, Martin and Knauer, 1973; 7, Tijoe et al., 1977; 8, Vinogradova and Koual'skiy, 1962; 9, Martin, 1970; 10, Culkin and Riley, 1958; 11, Van As et al., 1975; 12, Van As et al., 1973; 13, Lowman et al., 1970; 14, Fowler, 1977; 15, Bender, 1975; 16, Ishii et al., 1978; 17, Knauer, 1970; 18, Boothe and Knauer, 1972; 19, Palmer et al., 2006; 20, Santos et al., 2006; 21, Turkmen et al., 2006; 22, Kadar et al., 2006; 23, Feng et al., 2006; 24, Silva et al., 2006.

observed in *Balanus* spp. (Walker et al., 1975a,b). Iron was associated with soluble proteins in the hepatopancreas and hemolymph of *Cancer pagurus*. In hepatopancreas, iron was generally associated with high molecular weight proteins of about 450,000, which might be crustacean ferritin, an iron-storing protein (Guary and Negrel, 1980). In hemolymph, a common carrier protein for iron and plutonium of MW near 250,000 was demonstrated (Guary and Negrel, 1980).

Iron content of fecal pellets from the crab *Pugettia producta* fed with brown alga *Macrocystis pyrifera* was up to 14 times higher than algal iron content, suggesting that fecal material is significant in iron cycling in the marine environment (Boothe and Knauer, 1972). Fecal deposition of euphausiids, *M. norvegica*, contributed substantially to iron fluxes in the Mediterranean Sea: a daily rate of 910.0 mg Fe/kg DW pellets was estimated for that species (Fowler, 1977). A similar pattern was observed, though not as pronounced, for arsenic, cobalt, manganese, lead, and zinc (Boothe and Knauer, 1972). Frequency of molting is also a factor in iron transfer through marine ecosystems owing to the observations that some species contained more than 28% of total organism iron in molts (Fowler, 1977; Horowitz and Presley, 1977; Lowman et al., 1970).

7.23 Lanthanum

Whole amphipods, *P. walkeri*, from east Antarctica had 0.10-0.17 mg La/kg DW (Palmer et al., 2006).

7.24 Lead

Most of the lead in crustaceans is localized in the exoskeleton, with comparatively low residues in other tissues (Table 7.12).

Table 7.12: Lead Concentrations in Field Collections of Crustaceans

Organism	Concentration	Reference[a]
Copepod, *Anomalocera patersoni*; whole; males vs. females	6.0 DW vs. 3.5 DW	1
Barnacle, *Balanus amphitrite*; soft parts	(7.1-11.7) DW	2
Copepod, *Calanus finmarchicus*; whole	575.0 AW	3
Blue crab, *Callinectes sapidus*		
Pensacola, Florida; 2003-2004; hepatopancreas vs. muscle	Max. 0.36 FW vs. max. 0.39 FW	31
Whole	<0.2 DW	6
Muscle; August 2005; Iskenderun Bay, Turkey	3.5 (0.43-6.9) DW	27
Crab, *Cancer irroratus*		
Muscle	(<0.5-3.4) FW	4
Digestive gland	(<0.6-1.3) FW	4
Gills	(0.9-2.9) FW	4
Crab, *Cancer pagurus*; whole	<0.8 FW	5
Copepod, *Centropages* spp.; whole	1300.0 AW	3
Copepods		
Whole	150.0 AW	7
Whole	3.3 DW	8
Whole; 3 spp.; Taiwan; near sewage outfall	2.6-56.2 DW	28
Crabs		
Muscle	0.50 FW	9
Shell	0.66 FW	9
Crustaceans		
Whole	Max. 31.0 DW	10
Whole; 12 spp.	0.03-0.94 DW	11
16 commercial species; North America; edible tissues		
5 spp.	0.4-0.6 FW	12
9 spp.	0.6-0.7 FW	12
2 spp.	0.7-0.9 FW	12
Euphausiid, *Euphausia krohni*; whole	20.0 AW	3
Euphausiid, *Euphausia superba*; Antarctica; 2003-2004; whole	0.01 DW	24

(*Continues*)

Table 7.12: Cont'd

Organism	Concentration	Reference[a]
Euphausiids; whole	2.1 DW	8
Greenland; 1978-1993; 6 spp.		
Exoskeleton	0.08-0.18 FW	20
Muscle	0.01-0.05 FW	20
Lobster, *Homarus vulgaris*; muscle	<0.4 FW	5
Blue shrimp, *Litopenaeus stylirostris*; muscle; Gulf of California, Mexico; 2004-2005	5.3 (3.2-9.6) DW	25
Lobster; digestive gland; total lead vs. tetraalkyllead	0.20 FW vs. 0.16 FW	13
Euphausiid, *Meganyctiphanes norvegica*; molts	22.0 DW	14
Peneid shrimp, *Melicertus kerathurus*; Lesina lagoon, Italy; October 2004; 3 sites		
Muscle	0.47-0.68 DW	35
Exoskeleton	1.8-5.7 DW	35
Mexico; Gulf of California; 1999-2000		
Barnacle, *Balanus eburneus*; soft parts	2.1 DW	38
Shrimps; muscle		
Litopenaeus stylirostris	0.9 DW	38
Litopenaeus vanamei	0.5 DW	38
Lobster, *Nephrops norvegicus*; muscle	(<0.2-0.5) FW	5
Grass shrimp, *Palaemonetes pugio*; whole; Virginia		
Natural marsh area	0.2 DW	22
Spoil disposal area	11.0 DW	22
Amphipod, *Paramoera walkeri*; East Antarctica; 2003-2004; whole	(0.4-1.7) DW	24
Shrimp, *Pandalus montagui*; soft parts		
Sewage dump site	31.0 DW	23
Reference site	24.0 DW	23
Marsh crab, *Parasesarma erythodactyla*; soft parts; 2002; New South Wales, Australia; metals-contaminated estuary		
Males	Max. 132.0 DW	30
Females	Max. 193.0 DW	30

(Continues)

Table 7.12: Cont'd

Organism	Concentration	Reference[a]
Shrimp, *Penaeus aztecus*		
Muscle	1.1 DW	15
Exoskeleton	17.1 DW	15
Viscera	2.8 DW	15
Whole	20.2 DW	6
Black tiger shrimp, *Penaeus monodon*; Vietnam; 2003-2005; juveniles; max. values		
Muscle	0.02 DW	34
Exoskeleton	0.33 DW	34
Hepatopancreas	0.34 DW	34
White shrimp, *Peneus setiferus*; muscle; Gulf of Mexico	7.7 DW	26
Crab, *Portunus holsatus*; whole	1.3 FW	16
Crab, *Pugettia producta*; feces	(27.0-40.0) AW	17
Shrimps		
Muscle	1.6 DW	15
Exoskeleton	15.4 DW	15
Muscle	(0.24-1.2) FW	9
Shell	(0.44-1.2) FW	9
Spain; Tarragona coast; commercial species; edible parts	Max. 1.6 FW	21
Fiddler crab, *Uca pugnax*; New Jersey		
Intermolt; contaminated site		
Soft tissue	41.4 DW	28
Carapace	41.2 DW	28
Intermolt; reference site		
Soft tissue	20.4 DW	28
Carapace	27.0 DW	28
Immediate postmolt; contaminated site		
Soft tissue	18.3 DW	28
Exoskeleton	129.8 DW	28
Immediate postmolt; reference site		
Soft tissue	11.5 DW	28
Exoskeleton	33.1 DW	28
Shrimp, *Xiphopenaeus krogeri*		
Whole	3.7 (2.0-7.0) FW	18
Whole	13.4 (3.0-26.0) DW	18

(*Continues*)

Table 7.12: Cont'd

Organism	Concentration	Reference[a]
Zooplankton		
Surface vs. > 99 m collection	117.0 AW vs. 183.0 AW	190
Baltic Sea; May 1999; cladocerans vs. copepods; whole	1.0 DW vs. 13.3 DW	32

Values are in mg Pb/kg fresh weight (FW), dry weight (DW), or ash weight (AW).

[a]1, Polikarpov et al., 1979; 2, Barbaro et al., 1978; 3, Nicholls et al., 1959; 4, Greig et al., 1977b; 5, Topping, 1973; 6, Sims and Presley, 1976; 7, Vinogradova and Koual'skiy, 1962; 8, Martin and Knauer, 1973; 9, Won, 1973; 10, Stenner and Nickless, 1974a,b; 11, Stickney et al., 1975; 12, Hall et al., 1978; 13, Sirota and Uthe, 1977; 14, Fowler, 1977; 15, Horowitz and Presley, 1977; 16, DeClerck et al., 1979; 17, Boothe and Knauer, 1972; 18, Bernhard and Zattera, 1975; 19, Martin, 1970; 20, Dietz et al., 1996; 21, Schuhmacher et al., 1990; 22, Drifmeyer and Odum, 1975; 23, Mackay et al., 1972; 24, Palmer et al., 2006; 25, Frias-Espericueta et al., 2007; 26, Vazquez et al., 2001; 27, Turkmen et al., 2006; 28, Bergey and Weis, 2007; 29, Feng et al., 2006; 30, MacFarlane et al., 2006; 31, Karouna-Renier et al., 2007; 32, Pempkowiak et al., 2006; 33, Ruelas-Inzunza znd Paez-Osuna, 2008; 34, Tu et al., 2008; 35, D'Adamo et al., 2008.

Lead tissue burdens in the red fingered marsh crab, *P. erythodactyla*, collected in 2002 from an Australian estuary contaminated with lead and other metals, reflected lead sediment levels; this pattern held for both males and females (MacFarlane et al., 2006; Table 7.12). Glutathione peroxidase activity was elevated in crabs with elevated metal burdens, and, in females only, an increase in lipid peroxidation products (MacFarlane et al., 2006). Amphipods, *Elasmopus laevis*, held on sediments containing 58.0-424.0 mg Pb/kg DW as lead acetate for 60 days and two generations had reduced fecundity and increased body lead burdens proportional to lead sediment concentrations (Ringenary et al., 2007). Amphipods held on sediments containing 118.0 mg Pb/kg DW and higher had reduced population size when compared to control sediments containing 30.0 mg Pb/kg DW. On the basis of this study, authors propose a sediment lead concentration guideline of less than 118.0 mg Pb/kg DW (Ringenary et al., 2007).

Amphipods, *P. walkeri*, from the Australian Antarctic Territory in 1999-2000 had whole-body bioconcentration factors of 1600-7000 when seawater contained 0.00001-0.0001 mg Pb/L; BCFs decreased at higher seawater concentrations (Clason et al., 2003). *Paramoera* showed increased sensitivity when seawater concentrations ranged from 0.00012 to 0.00025 mg Pb/L (Clason et al., 2003).

Molting is an important pathway for lead depuration in the fiddler crab, *U. pugnax* (Bergey and Weis, 2007). Average percent of total body lead eliminated by *Uca* during molting from a relatively clean site was 56%, and for a metal-contaminated site it was 76%. Prior to molting, crabs shifted lead from soft tissues to exoskeleton in contaminated areas, and the reverse for the relatively clean site (Table 7.12; Bergey and Weis, 2007). In the euphausiid,

M. norvegica, molts represent most of the lead body burden and are probably more significant than fecal pellets in lead cycling processes (Fowler, 1977). The influence of deposited lead on surface sand meiofauna communities is not known with certainty, although harpacticoid copepods were among the most sensitive groups in that niche when subjected to lead contamination (Roberts and Maguire, 1976). Radiolead-210 concentrations in whole crustaceans were approximately 100 times higher than seawater; the highest ^{210}Pb concentration factors were found in shrimp, *Sergestes* spp., ranging from a low of 10 in muscle up to 1,000,000 in hepatopancreas (Heyraud and Cherry, 1979). Whole shrimp, *C. crangon*, held in seawater containing 0.018 mg Pb/L for 8 days had a bioconcentration factor of 750 (Jung and Zauke, 2008). Many factors influence lead accumulation rates in crustaceans. For example, lead concentrations in barnacles, *B. balanoides*, vary according to the time of the year when samples were taken, site of collection, reproductive state of the barnacles, tidal flows, and fluvial lead concentrations (Ireland, 1974). Significant differences were evident in lead content of male and female copepods, *Anomalocera patersoni*, from the Mediterranean Sea (Polikarpov et al., 1979). These differences were demonstrated for other metals examined, including chromium, copper, and cadmium. Data on lead accumulations by crustaceans under controlled conditions show that *U. pugilator* fiddler crabs contain 2.04 mg Pb/kg whole fresh organism after exposure for 14 days in 0.1 mg Pb/L (Weis, 1978). In another study, zinc-lead mixtures prolong hatching times of *Rithropanopeus harrisii* megalops (Benijts-Claus and Benijts, 1975); zinc may have influenced lead uptake rates in that study. Decreasing salinity and increasing temperature were associated with elevated body burdens in lead-stressed marine isopods (Jones, 1975).

Lethal properties of lead compounds to crustaceans are dependent on the formulation. For the shrimp, *C. crangon,* LC50 (96 h) values—in mg Pb/L—are 375.0 for inorganic ionic lead, 8.8 for trimethyllead, 5.8 for triethyllead, 0.110 for tetramethyllead, and 0.02 for tetraethyllead (Maddock and Taylor, 1980). American lobsters, *H. americanus*, held for 30 days in 0.05 mg inorganic Pb/L all survived, but show reduced *Delta* aminolevulinic acid dehydratase activity, biochemical alterations in antennal gland, and bioconcentration factors of 2760 for antennal gland and 58 for gill (Gould and Greig, 1983; USEPA, 1985b).

7.25 Lithium

Planktonic copepods from the Black Sea reportedly contain 400.0 mg Li/kg whole organism on an AW basis (Vinogradova and Koual'skiy, 1962).

7.26 Lutetium

Amphipods collected in east Antarctica in 2003-2004 had <0.006 mg Lu/kg DW whole organism or below the current detection limit (Palmer et al., 2006).

7.27 Manganese

In general, manganese concentrations in field collections of crustaceans were highest in calcified tissues and lowest in muscle (Table 7.13). In the dungeness crab, *C. magister*, manganese concentrations were highest in setae and calcareous exoskeleton, presumably because of surface-adsorbed MnO_2 (Tennant and Forster, 1969). Manganese values in barnacle shells range from 80.0 to 3800.0 mg/kg on a DW basis; these values were directly related to the concentration of manganese in the ambient seawater, which in turn was directly related to salinity (Gordon et al., 1970).

About 98% of the manganese in the body of lobsters, *Homarus vulgaris*, is in the calcified exoskeleton (Bryan and Ward, 1965). Manganese is lost from lobster blood during starvation and is absorbed so slowly from seawater that most manganese is absorbed from the diet. Primary losses of manganese from Mn-loaded lobsters occur via urinary excretion (20-40%), across the body surface (40-80%), and via the feces (up to 20%). It is possible that some absorption and release of manganese by exoskeleton occurs and this may assist in maintaining blood and tissue manganese concentrations (Bryan and Ward, 1965). Manganese

Table 7.13: Manganese Concentrations in Field Collections of Crustaceans

Organism	Concentration	Reference[a]
Copepod, *Acartia clausi*; whole	188.0 FW	1
Barnacles; summer 2001; Natal, Brazil; soft parts		
Fistulobalanus citerosum	(9.4-28.1) DW	19
Balanus amphitrite	9.6 DW	19
Blue crab, *Callinectes sapidus*		
whole	25.0 FW	2
Muscle; August 2005; Turkey	4.8 (0.1-15.4) DW	17
Crab, *Cancer irroratus*		
Muscle		
New York Bight	1.1 FW	3
New York Bight dumpsite	0.8 FW	3
Long Island Sound	28.7 FW	3
Delaware dumpsite	1.4 FW	3
Chincoteague inlet, Virginia	0.8 FW	3
Digestive gland	(<0.6-3.0) FW	3
Gills		
Long Island Sound	22.1 FW	3
Chincoteague inlet	5.7 FW	3

(*Continues*)

Table 7.13: Cont'd

Organism	Concentration	Reference[a]
Copepods		
Whole	500.0 AW	4
Whole	4.4 DW	5
Whole; 3 spp.; Taiwan; near sewage outfall	5.5-80.8 DW	18
Shrimp, *Crangon* sp.; whole	1.2 DW	20
Crustaceans; North America; 16 commercial species; edible tissues		
3 spp.	0.1-0.3 FW	6
7 spp.	0.3-0.5 FW	6
3 spp.	0.5-0.7 FW	6
3 spp.	0.7-2.0 FW	6
Euphausiid, *Euphausia superba*; Antarctica; whole	3.2 DW	16
Euphausiids		
Whole	3.6 DW	5
Whole	2.0 FW	1
Lobster, *Homarus vulgaris*		
Blood	(0.5-2.4) FW	7
Muscle	(0.3-0.8) FW	7
Hepatopancreas	(3.2-4.8) FW	7
Gills	(8.6-23.3) FW	7
Exoskeleton	(218.0-252.0) FW	7
Gastric mill teeth	150.0 FW	7
Stomach fluid	(0.2-2.8) FW	7
Hindgut and rectum	(1.0-10.5) FW	7
Excretory organs	(3.9-4.6) FW	7
Ovary	1.6 FW	7
Stomach contents	77.0 FW	7
Carapace		
Semi-transparent layer of shell	0.0 FW	7
Shell with inner surface removed	(245.0-267.0) FW	7
Complete calcified shell	(223.0-241.0) FW	7
Shell with outer surface removed	(189.0-218.0) FW	7
Lobster, *Jasus lalandi*; muscle	0.22-0.27 FW	89
Lobster		
Muscle	11.0 AW	10
Carapace	21.0 AW	10
Head	26.0 AW	10

(Continues)

Table 7.13: Cont'd

Organism	Concentration	Reference[a]
Euphausiid, *Meganyctiphanes norvegica*; fecal pellets	243.0 DW	11
Shrimp, *Penaeus aztecus*		
Muscle	1.5 DW	12
Exoskeleton	17.8 DW	12
Viscera	14.2 DW	12
Peneid shrimp, *Melicertus kerathurus*; Lesina lagoon, Italy; October 2004; 3 sites		
Muscle	1.0-5.0 DW	22
Exoskeleton	12.3-13.8 DW	22
Shrimp, *Penaeus monodon*		
Muscle	(14.9-21.3) DW	13
Vietnam; 2003-2005; max. values		
Muscle	2.3 DW	21
Exoskeleton	94.5 DW	21
Hepatopancreas	20.0 DW	21
Shrimp, *Penaeus japonica*; muscle	1.0 DW	14
Shrimp, *Penaeus* spp.		
Whole	6.1 DW	15
Whole	2.4 FW	2
Shrimp		
Muscle	8.0 DW	12
Exoskeleton	89.8 DW	12
Muscle	24.0 AW	10
Carapace	100.0 AW	10
Whole	90.0 AW	10
Stomatopod, *Squilla oratoria*; muscle	1.9 DW	14
Crab, *Ucides cordatus*; adult males		
Gill	5.4 DW	23
Hepatopancreas	10.1 DW	23
Muscle	1.2 DW	23

Values are in mg Mn/kg fresh weight (FW), dry weight (DW), or ash weight (AW).
[a]1, Fujita et al., 1969; 2, Wolfe et al., 1973; 3, Greig et al., 1977b; 4, Vinogradova and Koual'skiy, 1962; 5, Martin and Knauer, 1973; 6, Hall et al., 1978; 7, Bryan and Ward, 1965; 8, Van As et al., 1973; 9, Van As et al., 1975; 10, Lowman et al., 1966; 11, Fowler, 1977; 12, Horowitz and Presley, 1977; 13, Zingde et al., 1976; 14, Ishii et al., 1978; 15, Knauer, 1970; 16, Palmer et al., 2006; 17, Turkmen et al., 2006; 18, Feng et al., 2006; 19, Silva et al., 2006; 20, Zumholz et al., 2006; 21, Tu et al., 2008; 22, D'Adamo et al., 2008; 23, Correa et al., 2005.

is a naturally abundant metal in marine sediments where it occurs mainly as manganese dioxide (MnO_2); however, during hypoxic conditions it is converted into the bioavailable Mn^{2+} (up to 15.0 mg Mn/L) and, when accumulated, can adversely affect immune competent cells of the lobster, *N. norvegicus* (Oweson et al., 2006).

Adult male crabs, *U. cordatus*, subjected to a high sublethal concentration of 50.0 mg Mn^{2+}/L for 96 h had grossly elevated manganese burdens in gills (317.3 mg/kg DW versus 5.4 in controls), hepatopancreas (101.6 versus 10.1 in controls), and muscle (7.7 versus 1.2) (Correa et al., 2005).

7.28 Mercury

Most tissue mercury concentrations in field collections of crustaceans worldwide were below the United States Food and Drug Administration action guideline of 0.5 mg Hg/kg FW, except for samples collected in Minamata Bay, Japan, and other areas heavily impacted by mercury-containing anthropogenic wastes (Eisler, 1981, 2006; Schuhmacher et al., 1994; Srinivasin and Mahajan, 1989; Table 7.14).

Organomercury concentrations in the green crab, *C. maenas* comprised 70-95% of the total mercury burdens in muscle and hepatopancreas (Pereira et al., 2006). Green crabs continue to accumulate mercury throughout their life, although rates differ depending on degree of environmental contamination (Coelho et al., 2008). For example, under conditions of low mercury contamination, where diet is the major pathway, mercury tends to accumulate in muscle and hepatopancreas with 80-90% of the muscle mercury as organomercury. Under these conditions, mercury concentrations in tissues increase with increasing age. Under conditions of high mercury contamination where seawater is the major uptake pathway, mercury preferentially accumulates in gills; organomercury content is usually less than 40% of the total mercury burden, and organism age was not a significant factor (Coelho et al., 2008). Methylmercury concentrations in hepatopancreas of Chinese mitten crabs, *E. sinensis*, declined with increasing crab size, possibly through molting, suggesting a mechanism for mercury excretion (Hui et al., 2005), with important implications for crab predators that select larger crabs. Uptake rates of total mercury in field collections of *C. maenas* were about the same regardless of size: larger crabs took up 0.011-0.015 mg/kg body weight weekly versus 0.008-0.015 mg/kg BW weekly for smaller crabs. However, weekly excretion rates were different, with larger crabs excreting up to 0.034 mg/kg BW and smaller crabs up to 0.024 mg/kg BW (Pereira et al., 2006). Differences in reported mercury concentrations can significantly affect mercury intake estimates of seafood by human consumers of seafoods in the United States (Sunderland, 2007). National exposure estimates are most influenced by reported methylmercury concentrations in imported shrimp (0.04 mg/kg FW) and in Atlantic Ocean crabs (0.026 mg/kg FW; Sunderland, 2007).

Table 7.14: Mercury Concentrations in Field Collections of Crustaceans

Organism	Concentration	Reference[a]
Aleutian Islands, Alaska; 2004-2005; edible portions		
Golden king crab, *Lithodes aquispina*	0.043 FW; max. 0.070 FW	50
Red king crab, *Paralithodes camtschaticus*	0.10 FW; max 0.21 FW	50
Amphipod, *Asellus aquaticus*; whole	(0.22-0.56) FW	1
Barnacle, *Balanus amphitrite*; soft parts	(0.06-1.35) DW	2
Barnacle, *Balanus* spp.; soft parts	(0.09-0.22) FW	3
Blue crab, *Callinectes sapidus*		
Whole	0.26 (0.02-1.54) FW	4
Whole	1.3 (0.1-7.7) DW	4
Muscle; Southeast United States	0.45	5
Pensacola, Florida; 2003-2004		
Hepatopancreas	Max. 1.1 FW	51
Muscle	Max. 0.24 FW	51
Crab, *Cancer irroratus*		
Muscle	(0.15-0.19) FW	6
Digestive gland	(0.07-1.9) FW	6
Gills	0.03 FW	6
Shore crab, *Carcinus maenas*; Portugal; 1999-2000; high mercury site (max. 5.4 mg Hg/kg DW sediment) vs. low mercury site (max. 0.23 mg Hg/kg DW sediment)		
Larger crabs		
Muscle	(0.03-0.63) FW vs. max. 0.1 FW	47
Hepatopancreas	(0.02-0.34) FW vs. max < 0.1 FW	47
Smaller crabs		
Muscle	0.45 FW vs max. 0.2 FW	47
Hepatopancreas	Max. 0.3 FW vs. max. 0.1 FW	47
Copepods		
Whole	0.11 DW	7
Whole	0.27 DW	8
Amphipod, *Corophium volutator*; whole	(0.04-0.13) FW	1
Crabs		
Muscle	(0.09-0.53) FW	9
Shell	(0.06-0.23) FW	9

(*Continues*)

Table 7.14: Cont'd

Organism	Concentration	Reference[a]
Shrimp, *Crangon crangon*		
Muscle	0.19 FW	10
Whole	(0.2-1.7) DW	4
Whole	(0.03-0.12) FW	1
Crustaceans		
Edible portions		
Japan, 4 spp.	0.078 FW	11
Japan, Minamata Bay; 2 spp.; 1960s	41.0-100.0 DW	12
North America; 16 spp.		
6 spp.	<0.1 FW	13
9 spp.	0.1-0.2 FW	13
1 spp.	0.2-0.3 FW	13
Spain; Catalonia; November 1992 to February 1993; 4 spp.	0.006-0.72 FW	44
Whole		
12 spp.	0.06-1.57 DW	14
5 spp.	0.00-0.10 FW	15
6 spp.; Greenland; 1983-1991	Max. 0.33 DW	41
4 spp.; Antarctica; 2004	0.034-0.037 DW	48
Deep sea shrimps; 2 spp.; Rainbow hydrothermal vent site; mid-Atlantic Ridge; adults; summer 2003		
Rimicaris exoculata		
Whole body	0.07 DW	49
Cuticle	0.03 DW	49
Gill	0.2 DW	49
Antennae	0.9 DW	49
Muscle	0.01 DW	49
Digestive gland	0.4 DW	49
Pylorus	2.8 DW	49
Microcaris fortunata		
Whole body	0.05 DW	49
Cuticle	0.07 DW	49
Gill	0.5 DW	49
Antennae	0.6 DW	49
Muscle	0.02 DW	49
Digestive gland	0.09 DW	49
Pylorus	2.2 DW	49

(*Continues*)

Table 7.14: Cont'd

Organism	Concentration	Reference[a]
Chinese mitten crab, *Eriocheir sinensis*; San Francisco Bay, California; July-August 2002; total mercury vs. methylmercury		
Hepatopancreas	0.25 (0.04-1.03) DW vs. 0.036 (0.006-0.069) DW	40
Other tissues	0.15 (0.04-0.69) DW vs. 0.04 (0.007-0.095) DW	40
Crab, *Eupagurus bernhardus*; whole	0.67 FW	16
Euphausiid, *Euphausia pacifica*		
Whole	(0.008-0.058) FW	4
Whole	(0.25-0.52) DW	4
Euphausiids		
Whole	0.08 DW	7
Whole	<0.09 FW	17
Amphipod, *Gammarus* sp.: whole	(0.01-0.15) FW	1
Lobster, *Homarus americanus*		
Muscle	0.31 FW	18
Liver	0.60 FW	18
Lobster, *Homarus vulgaris*; muscle	(0.21-0.54) FW	19
Hydrothermal vent field; mid-Atlantic Ridge; 2001		
Shrimp, *Chlorocaris chacei*; 1700 m deep; digestive gland vs. muscle	1.4 DW vs. 0.06 DW	52
Hydrothermal crab, *Segonzacia mesatlantica* digestive gland vs. muscle		
840 m deep	1.4 DW vs. 0.7 DW	52
1700 m deep	0.4 DW vs. 0.3 DW	52
2300 m deep	0.2 DW vs. 0.2 DW	52
Shrimp, *Mirocaris fortunata*; digestive gland vs. muscle		
840 m deep	1.3 DW vs. 0.5 DW	52
1700 m deep	0.7 DW vs. 0.08 DW	52
2300 m deep	no data vs. 1.0 DW	52
Barnacle, *Lepas* sp.; whole	0.08 FW	20
Euphausiid, *Meganyctiphanes norvegica*; whole	0.26 DW	21
Mysidaceans; whole	(0.06-0.09) FW	22

(Continues)

Table 7.14: Cont'd

Organism	Concentration	Reference[a]
Spiny lobster, *Nephrops norvegicus*		
Muscle	(0.10-0.22) FW	23
Edible portions; Tyrrhenian Sea; 1981	2.9 FW	42
Crab, *Pachygrapsus marmoratus*; whole	(0.50-4.5) FW	16
Shrimp, *Palaemon debilis*; whole	(0.015-0.30) FW	24
Shrimp, *Palaemon elegans*; whole	0.98 DW	25
Shrimp, *Pandalus jordani*; muscle	(0.06-0.18) FW	3
Shrimp, *Penaeus aztecus*; muscle; Tampico, Mexico	0.06 (0.01-0.67) FW	26
Shrimp, *Penaeus californiensis*; muscle	0.12 (0.02-0.46) FW	26
Shrimp, *Penaeus kerathurus*		
Muscle	(0.09-0.26) FW	27
Muscle	(0.01-0.04) FW	28
Black tiger shrimp, *Penaeus monodon*; Vietnam; 2003-2005; juveniles; max. values		
Muscle	0.07 DW	53
Exoskeleton	<0.05 DW	53
Hepatopancreas	0.09 DW	53
Shrimp, *Penaeus orientalis*; muscle	Max. 0.04 FW	29
Shrimp, *Penaeus setiferus*; muscle; Mexico		
Vera Cruz	0.04 (0.02-0.06) FW	26
Ciudad del Carmen	0.03 (0.01-0.16) FW	26
Shrimp, *Penaeus* sp.; muscle; total mercury vs. organic mercury	0.011 (0.002-0.016) FW vs. 0.006 (0.001-0.010) FW	30
Shrimp, *Penaeus stylirostris*; muscle; Mexico		
Vera Cruz	0.09 (0.01-0.20) FW	26
Ciudad del Carmen	0.05 (0.01-0.15) FW	26
Crab, *Portunus holsatus*; whole	0.10 FW	31
Crab, *Portunus* sp.; whole	(<0.01-0.03) FW	20
Prawns		
Soft parts	<0.03 FW	32
Muscle	0.02 FW	33

(Continues)

Table 7.14: Cont'd

Organism	Concentration	Reference[a]
Puerto Rico; estuaries; 1988		
Blue crab, *Callinectes sapidus*; all tissues	Not detectable	43
Shrimps, *Palaemonetes* sp.; all tissues	Not detectable	43
Shrimp		
Whole		
Gulf of Mexico	(0.03-0.09) FW	20
Persian Gulf	0.24 (0.08-0.88) FW	34
North Sea	(0.04-0.18) FW	32
Persian Gulf	(0.005-0.012) FW	35
Texas	<0.02 DW	26
Muscle		
Korea	(0.08-0.17) FW	9
Southeastern United States; 4 spp.	0.22 DW	5
New Jersey supermarkets; July-October 2003	0.01 (0.002-0.02) FW	46
Belgium	1.3 DW	36
China; Zhejiang coastal area; May 1998; 3 spp.	0.01-0.015 (0.004-0.028) FW	39
Taiwan; 1995-1996; 2 spp.	2.2-2.4 (0.7-5.4) DW	45
Shell	(0.02-0.05) FW	9
Molts	1.3 DW	36
Stomatopod, *Squilla mantis*; muscle	0.12 FW	27
Crab, *Thalamita crenata*		
Muscle; chela vs. body	(0.020-0.058) FW vs. (0.027-0.061) FW	37
Viscera	(0.021-0.044) FW	37
Gills	(0.033-0.119) FW	37
Crab, *Uca* sp.; whole	0.04 DW	38

Values are in mg Hg/kg fresh weight (FW) or dry weight (DW).

[a]1, Zauke, 1977; 2, Barbaro et al., 1978; 3, Yannai and Sachs, 1978; 4, Bernhard and Zattera, 1975; 5, Gardner et al., 1975; 6, Greig et al., 1977a,b, Martin and Knauer, 1973; 8, Tijoe et al., 1977; 9, Won, 1973; 10, DeClerck et al., 1974; 11, Kumagai and Saeki, 1978; 12, Matida and Kumada, 1969; 13, Hall et al., 1978; 14, Stickney et al., 1975; 15, Ramos et al., 1979; 16, Renzoni et al., 1973; 17, Greig and Wenzloff, 1977; 18, Greig et al., 1975; 19, Holden and Topping, 1972; 20, Johnson and Braman, 1975; 21, Leatherland et al., 1973; 22, Nuorteva and Hasanen, 1975; 23, Cumont et al., 1975; 24, Luoma, 1977; 25, Leatherland and Burton, 1974; 26, Reimer and Reimer, 1975, Establier, 1977; 28, Tuncel et al., 1980; 29, Doi and Ui, 1975; 30, Cheevaparanapivat and Menasveta, 1979; 31, DeClerck et al., 1979; 32, Anonymous, 1978; 33, Sorentino, 1979; 34; Parveneh, 1977; 35, Eftekhari, 1975; 36, Bertine and Goldberg, 1972; 37, Luoma, 1976; 38, Windom et al., 1976; 39, Fang et al., 2004; 40, Hui et al., 2005; 41, Dietz et al., 1996; 42, Schreiber, 1983; 43, Burger et al., 1992; 44, Schuhmacher et al., 1994; 45, Han et al., 1998; 46, Burger et al., 2005; 47, Pereira et al., 2006; 48, Santos et al., 2006; 49, Kadar et al., 2006; 50, Burger et al., 2007b; 51, Karouna-Renier et al., 2007; 52, Colaco et al., 2006; 53, Tu et al., 2008.

Laboratory studies demonstrate that many biological and abiotic variables modify mercury accumulation, retention, and effects among crustaceans. Biological modifiers include tissue specificity, diet, feeding niche, gender, life stage, and health of the organism. Abiotic modifiers include the temperature and salinity of the medium, duration of exposure to mercury, chemical form of mercury, and presence of other compounds. Selected studies illustrating these points follow.

American lobsters, *H. americanus*, held in seawater containing up to 0.006 mg Hg/L for 30 days, show significant accumulations of total mercury in various tissues: 15.2 mg Hg/kg FW digestive gland, up from 0.12 at the start; 85.3 in gills, up from 0.14; and 1.0 mg Hg/kg FW muscle, up from 0.23 (Thurberg et al., 1977). With increasing exposure of 60 days, all tissues contained increasing burdens of mercury: gill, for example, contained 119.5 mg Hg/kg FW (Thurberg et al., 1977). Altered enzyme activity levels were documented in the green crab, *C. maenas*, at sublethal levels of 0.09 mg Hg^{2+}/L (Elumalai et al., 2007).

Effects of high sublethal concentrations of mercury (0.18 mg/L) on fiddler crabs, *U. pugilator*, is modified by life stage, gender, and thermosaline regime. Adults were significantly more resistant than larvae, adult males more sensitive than females, and resistance lowest at low temperatures and low salinities (Vernberg and DeCoursey, 1977). Exposure for 3 h in 0.18 mg Hg/L, as $HgCl_2$, show that gill tissues of fiddler crabs are the major site of mercury concentration; lesser amounts accumulate in hepatopancreas and green gland. Crabs are more susceptible to biocidal properties of mercury at low temperatures (5°C) than at elevated (33°C) temperatures. The inability to transfer mercury from gill to hepatopancreas at low temperatures is a factor in the higher toxicity of mercury to fiddler crabs at low temperatures (Vernberg and O'Hara, 1972; Vernberg and Vernberg, 1972).

Corner and Rigler (1958) found that larvae of a barnacle, *Elminius modestus*, accumulate up to 920.0 mg Hg/kg DW during exposure for 3 h in 0.2 mg Hg/L as mercuric chloride. Higher accumulations (700.0 mg Hg/kg) were observed with amylmercuric chloride (n-$C_5H_{11}HgCl$) during exposure for 3 h in only 0.01 mg Hg/L solutions. Similar trends were observed for brine shrimp, *A. salina* (Corner and Rigler, 1958). Studies with adult prawns, *Leander serratus*, held in toxic solutions of mercury (50.0 mg Hg/L) as $HgCl_2$; prawns dying during 3 h exposure to 1.0 mg Hg/L as n-$C_5H_{11}HgCl$ contained 320.0-460.0 mg Hg/kg DW in antennary gland (Corner and Rigler, 1958).

Corner (1959) reports that crabs, *Maia squinado*, held in seawater containing inorganic mercuric chloride show major accumulations in gills and other sites. Eventually, the mercury concentrations in blood surpass that of the external medium, the concentration in antennary gland above that in blood, and the crabs excrete small, but increasing, amounts of mercury in the urine. Most (95%) of the mercury in blood is attached to protein, with mercury concentrations in blood remaining constant for several weeks following mercury exposure. When *Maia* was poisoned with *n*-methylmercuric chloride, mercury again concentrates in gills and various internal organs; however, mercury burdens in blood were low and no mercury was found in urine (Corner, 1959).

Mercury content among crustaceans feeding in contact with sediment is highest among carnivores and lowest in herbivores; a similar pattern exists for crustaceans feeding above the sediment/water interface (Klemmer et al., 1976). Although crustaceans biomagnify mercury in tissues via the diet, some studies found that mercury accumulations from solution are more rapid and more pronounced than food sources (Luoma, 1976, 1977). In an algae to copepod link, however, copepods show no impairment of egg laying or egg development, and no retention of mercury in tissues, eggs, or feces (Parrish and Carr, 1976); in that study, the algae *Croomonas salina*, had 1400.0 mg Hg/kg DW.

Fowler et al. (1978) report that 45% of all mercury in the shrimp *L. seticaudata* localizes in viscera, 39% in muscle, 15% in exoskeleton, and 1.8% in molts. It is probable that most of the mercury was in methylated form as has been demonstrated for tissues of blue crabs (Gardner et al., 1975), fiddler crabs (Windom et al., 1976), and other decapods (Eganhouse and Young, 1978). It is generally agreed that organomercury compounds are accumulated by crustaceans more rapidly and to a higher level than inorganic mercury compounds (D'Agostino and Finney, 1974; Fowler et al., 1978). Some evidence exists that iodide salts of mercury have greater uptake potential than chloride salts in copepods and brine shrimp (Boney et al., 1959).

Larvae of dungeness crab, *C. magister*, were extremely sensitive to mercury with 50% dead within 48 h at 0.0066 mg/L (Glickstein, 1978). High levels of selenium (>5.0 mg Se/L) increased mercury toxicity but moderate levels of 0.010-1.0 mg Se/L decreased mercury toxicity (Glickstein, 1978) and accumulations (Fowler and Benayoun, 1976a,b,c).

Larvae and newly molted crustaceans were more sensitive to mercury effects than were adults of the same species (Shealy and Sandifer, 1975; Vernberg et al., 1974; Wilson and Connor, 1971); also, male adult fiddler crabs were more sensitive than females (Vernberg et al., 1974). Starved larvae of grass shrimp, *Palaemonetes vulgaris* were more sensitive to mercury insult than were fed larvae (Shealy and Sandifer, 1975).

Biocidal properties of mercury salts, and presumably uptake rates, were more pronounced for crab *Petrolisthes armatus* at lower salinities within the range 7-35 ppt (Roesijadi et al., 1974). A similar pattern was recorded for the fiddler crab *U. pugilator* (Vernberg et al., 1974). However, salinities as low as 6 ppt did not affect uptake patterns in the shrimp *Palaemon debilis* (Luoma, 1977).

Negligible uptake or effects were demonstrated in *U. pugilator* during exposure for 2 weeks in seawater containing 0.1 mg Hg/L (Weis, 1978). Similarly, pre-exposure of shrimp, *Penaeus setiferus*, for 57 days to 0.001 mg Hg/L had no effect on measured parameters during subsequent mercury stress experiments with that species (Green et al., 1976).

7.29 Molybdenum

Highest molybdenum values recorded in field collections of crustaceans were 0.4 mg/kg on a FW basis, 2.6 on a DW basis, and 50.0 mg/kg on an AW basis (Table 7.15). Molybdenum is relatively nontoxic to the crabs *E. bernhardus* and *C. maenas* at alkaline pH typical of seawater, but death and presumably accumulation occur at pH values near 5.0 (Abbott, 1977).

The concentration of molybdenum fatal to 50% of the euphausiid, *Euphausia pacifica*, in 8 days is 560.0 mg/L (Anderson and Mackas, 1986); to 50% of the amphipod *A. compressa* in 96 h it is 247.0 mg/L (Ahsanullah, 1982); to 50% of the copepod, *Calanus marshallae*, in 19 days it is 560.0 mg/L (Anderson and Mackas, 1986); and to 50% in 96 h of pink shrimp, *Penaeus duorarum* it is 1909.0 mg/L and to mysid shrimp, *M. bahia* it is 1205.0 mg/L (Knothe and Van Riper, 1988).

Table 7.15: Molybdenum Concentrations in Field Collections of Crustaceans

Organism	Concentration	Reference[a]
Copepod, *Calanus finmarchicus*; whole	12.0 AW	1
Copepod, *Centropages* spp.; whole	10.0 AW	1

Copepods
  Whole
  Whole
	
	50.0 AW
	1.3 DW
		
		2
		3

Crustaceans; North America; 16 commercial species; edible tissues
  11 spp.
  3 spp.
  2 spp.
	
	0.1-0.2 FW
	0.2-0.3 FW
	0.3-0.4 FW
		
		4
		4
		4

Euphausiid, *Euphausia krohni*; whole	17.0 AW	1

Tiger shrimp, *Penaeus monodon*; Vietnam; 2003-2005; juveniles; max. values
  Muscle
  Exoskeleton
  Hepatopancreas
	
	0.05 DW
	0.09 DW
	2.6 DW
		
		6
		6
		6

Zooplankton; Baltic Sea	2.0 DW	5

Values are in mg Mo/kg fresh weight (FW), dry weight (DW), or ash weight (AW).
[a] 1, Nicholls et al., 1959; 2, Vinogradova and Koual'skiy, 1962; 3, Tijoe et al., 1977; 4, Hall et al., 1978; 5, Prange and Kremling, 1985; 6, Tu et al., 2008.

7.30 Neodymium

Whole amphipods collected from East Antarctica in 2003-2004 contained a maximum of 0.11 mg Nd/kg DW (Palmer et al., 2006).

7.31 Neptunium

The uptake of ^{237}Np was followed in tissues of a crab, *C. pagurus*, and shrimp, *L. seticaudata*, for 48 days. Accumulation was observed in all tissues measured with 92-98% of the organism's total ^{237}Np content associated with exoskeleton; most of the remainder was in hepatopancreas and gut (Guary and Fowler, 1978). This artificial element will assume increasing environmental importance with further expansion of the nuclear fuel reprocessing industry.

7.32 Nickel

Large variations in nickel content of field collections of crustaceans were evident between species, tissues, depths of collection, and other variables (Table 7.16). Maximum nickel concentrations in crustacean tissues were 27.0 mg/kg FW in whole spiny lobster, *Panulirus argus*, 92.0 mg/kg DW in soft parts of sand shrimp, *Crangon allmani*, from a waste dump site, and 150.0 mg/kg AW in zooplankton collected at depths greater than 99 m (Table 7.16).

Most investigators showed that nickel concentrations were lowest in muscle and highest in exoskeleton. Molts of the euphausiid *M. norvegica*, for example, contained 78% of the total nickel body burden of that species (Fowler, 1977). Yoshinari and Subramanian (1976) demonstrated that granular chitin from lobster shell has high nickel-absorbing capability from distilled water; as was true for other metals tested, absorption decreased with increasing salinity.

Mysid shrimp were comparatively sensitive to ionic nickel, with 50% mortality documented in 96 h at 0.15 mg Ni/L for *Mysidopsis formosa* and 0.51-0.64 mg Ni/L for *M. bahia* (USEPA, 1980c). Nickel is toxic to the copepod, *A. tonsa*, when fed a nickel-contaminated algal diet (Bielmyer et al., 2006). In that study, diatoms *T. pseudonana*, were held in seawater containing 0.0024 mg Ni/L for 7 days, reaching a concentration of 15.3 mg Ni/kg DW diatom; this, in turn, was fed to *Acartia* for 7 days, producing a measurable reduction in copepod reproduction. Authors note that the proposed water quality criterion for marine life protection is 0.008 mg Ni/L, which is far in excess of the 0.0024 mg Ni/L used in this study to produce a toxic diet (Bielmyer et al., 2006).

To protect human consumers of crustaceans, it is proposed that edible seafood tissues contain less than 70.0 mg Ni/kg FW (Sankar et al., 2006).

Table 7.16: Nickel Concentrations in Field Collections of Crustaceans

Organism	Concentration	Reference[a]
Amphipods; Whole; Antarctica	2.2 DW	12
Barnacles; summer 2001; soft parts; Natal, Brazil		
Fistulobalanus citerosum	(3.7-12.0) DW	18
Balanus amphitrite	9.1 DW	18
Copepod, *Calanus finmarchicus*; whole	165.0 AW	1
Blue crab, *Callinectes sapidus*		
Muscle; August 2005; Turkey	2.8 (0.3-15.4) DW	17
Pensacola, Florida; 2003-2004		
Hepatopancreas	Max. 0.56 FW	19
Muscle	Max. 0.29 FW	19
Green crab, *Carcinus maenas*; all tissues	(6.2-12.3) FW	13
Copepod, *Centropages* spp.; whole	55.0 AW	1
Copepods		
Whole	2.0 DW	2
Whole	1.5 DW	3
Sand shrimp, *Crangon allmani*; Scotland; soft parts; reference site vs. waste dump site	15.0 DW vs. 92.0 DW	13
Crustaceans		
Whole	(6.5-9.8) FW	4
Edible tissues; 16 commercial species; North America		
4 spp.	0.2-0.3 FW	5
8 spp.	0.3-0.4 FW	5
4 spp.	0.4-0.9 FW	5
Euphausiid, *Euphausia krohni*; whole	6.0 AW	1
Euphausiid, *Euphausia superba*; Antarctica; whole	4.4 DW	16
Euphausiids; whole	3.8 DW	2
American lobster, *Homarus americanus*		
Serum	0.012 (0.008-0.020) FW	14, 15
Euphausiid, *Meganyctiphanes norvegica*; molts	6.7 DW	6

(*Continues*)

Table 7.16: Cont'd

Organism	Concentration	Reference[a]
Aesop shrimp, *Pandalus montagui*; soft parts; Scotland; reference site vs. waste dump site	25.0 DW vs. 70.0 DW	13
Caribbean spiny lobster, *Panulirus argus* Soft parts; Puerto Rico, Anasco Bay vs. West coast	1.3 (1.0-2.0) FW; 4.5 DW vs. 4.6 (1.4-5.0) FW; 36.0 (22.0-60.0) DW	13
Whole	(1.3-27.0 FW); (4.5-50.0) DW	7
Brown shrimp, *Penaeus aztecus*; Texas Muscle	1.4 DW; max. 1.9 DW	8
Exoskeleton	6.2 DW; max. 17.9 DW	8
Viscera	5.7 DW; max. 5.8 DW	8
Shrimp; 2 spp.; Texas; outer continental shelf Muscle	1.6 DW	8
Exoskeleton	5.8 DW	8
Whole	1.4-1.6 DW	8
Stomatopod, *Squilla oratoria*; muscle	1.2 DW	9
Barnacle, *Tetraclota* sp.; whole	0.43 FW	10
Shrimp, *Xiphopenaeus kroyeri*; whole	5.1 (2.0-7.0) FW	7
Zooplankton Surface vs. > 99 m depth collections	100.0 AW vs. 150.0 AW	11
Baltic Sea; May 1999; cladocerans vs. copepods; whole	2.6 DW vs. 10.7 DW	20

Values are in mg Ni/kg fresh weight (FW), dry weight (DW), or ash weight (AW).
[a]1, Nicholls et al., 1959; 2, Martin and Knauer, 1973; 3, Tijoe et al., 1977; 4, Wright, 1976; 5, Hall et al., 1978; 6, Fowler, 1977; 7, Bernhard and Zattera, 1975; 8, Horowitz and Presley, 1977; 9, Ishii et al., 1978; 10, Ikebe and Tanaka, 1979; 11, Martin, 1970; 12, Szefer et al., 1993; 13, Jenkins, 1980; 14, USEPA, 1980b; 15, Mushak, 1980; 16, Palmer et al., 2006; 17, Turkmen et al., 2006; 18, Silva et al., 2006; 19, Karouna-Renier et al., 2007; 20, Pempkowiak et al., 2006.

7.33 Niobium

Whole shrimp, *Crangon* sp., contained 0.00086 mg Nb/kg DW (Zumholz et al., 2006).

7.34 Plutonium

Crabs, *C. pagurus* collected near the La Haque, France, nuclear fuel reprocessing plant were examined for plutonium isotope activity. Gills contained the highest concentration of radioactivity, followed by hepatopancreas, gut, exoskeleton, gonad, and hemolymph, in that order (Guary et al., 1976). In hepatopancreas and hemolymph, plutonium was associated with low MW fractions of soluble proteins (Guary and Negrel, 1980). The significance of these results in terms of human health or effects on marine food chains is unclear.

Accidental contamination of crustaceans with plutonium resulted when a military aircraft carrying nuclear weapons crashed near Greenland. Benthic crustaceans in the crash area contained up to 1000 times more plutonium radioactivity than background; however, pelagic zooplankton did not contain significant accumulations (Aarkrog, 1971, 1990).

7.35 Polonium

Zooplankton, mainly calanoid and cyclopoid copepods, from a Caribbean Sea location accumulated ^{210}Po from ambient seawater by a factor of 970 (Kharkar et al., 1976). Polonium-210 concentrations in whole Mediterranean Sea crustaceans were up to 10,000 higher than ambient seawater ^{210}Po concentrations (Heyraud and Cherry, 1979). Concentration factors were highest in *Sergestes* spp., especially in hepatopancreas which contained up to 90% of all ^{210}Po (Heyraud and Cherry, 1979). Laboratory feeding studies with euphausiids, *M. norvegica*, on food low in ^{210}Po resulted in an approximate biological half-life of only 6.5 days in hepatopancreas (Heyraud and Cherry, 1979).

7.36 Praseodymium

Whole amphipods, *P. walkeri*, collected in East Antarctica in 2003-2004 had 0.013-0.027 mg Pr/kg DW (Palmer et al., 2006).

7.37 Protactinium

Lucu et al (1969) in uptake studies of 20-day duration with *C. mediterraneus* and ^{233}Pa showed that gills had the greatest ability to accumulate this isotope (1500 times over medium). For reproductive system and muscle, these values ranged between 5 and 20; no uptake was observed in hemolymph (Lucu et al., 1969).

7.38 Radium

The ability of mixed zooplankton cultures, mainly copepods, to accumulate ^{226}Ra from seawater is about five times over ambient levels (Kharkar et al., 1976).

7.39 Rhenium

Pandalus sp. from the coast of Japan have low body burdens of rhenium, estimated at <0.005 mg/kg on an AW basis (Fukai, 1965) and <0.0003 mg/kg on a DW basis (Fukai and Meinke, 1962).

7.40 Rubidium

Edible tissues of shrimp, *Penaeus japonica*, and stomatopod, *Squilla oratoria* contained 4.1 and 3.2 mg Rb/kg DW, respectively (Ishii et al., 1978). Rubidium concentrations in juvenile black tiger shrimp, *P. monodon*, from Vietnam in 2003-2005 were as high as 12.2 mg Rb/kg DW in muscle, 4.8 in exoskeleton, and 8.4 in hepatopancreas (Tu et al., 2008). In a controlled study, crabs, *Panopeus herbstii*, were fed clams preloaded with rubidium; rubidium accumulated in the hepatopancreas but most was lost—mainly via fecal excretion—4 to 7 days post feeding (Eversole, 1978).

7.41 Ruthenium

Accumulation of ^{106}Ru by the euphausiid, *M. norvegica*, was influenced by the concentration in solution and the chemical form of the radioisotope, ruthenium chloride fractions being accumulated at a faster rate than nitrosyl-nitrate complexes (Keckes et al., 1972). Molting was the primary loss route for accumulated ^{106}Ru, with first molts accounting for up to 80% of the loss. When euphausiids accumulated ^{106}Ru from the diet, initial loss was rapid due to fecal excretion. Molts from these individuals did not contain ^{106}Ru, thus loss from euphausiids obtaining the isotope through feeding is mainly due to fecal pellet deposition and other excretion or exchange processes (Keckes et al., 1972). Similar findings were reported for this species (Antonini-Kane et al., 1972) and other species of crustaceans (Ancellin et al., 1967).

Radioruthenium recycling via crustacean feces is an important pathway. On May 6, 1986—about 2 weeks after the Chernobyl nuclear power reactor accident (Eisler, 1995, 2003) — fecal pellets of copepods contained 57 and 82 times more ^{103}Ru and ^{106}Ru, respectively, than did whole organism (Fowler et al., 1987).

7.42 Samarium

Whole amphipods, *P. walkeri*, from East Antarctica in 2003-2004 contained 0.009-0.017 mg Sm/kg DW (Palmer et al., 2006).

7.43 Scandium

The single datum available for scandium in field collections of crustaceans is 0.016 mg Sc/kg AW whole euphausiid (Table 7.17). Fowler (1977) reports that scandium levels in *M. norvegica* were below analytical detection limits of 4.5 mg/kg DW for whole organisms, and below 80.0 mg/kg DW for fecal pellets; scandium flux was <0.8 mg/kg/day due to fecal deposition.

Table 7.17: Scandium, Selenium, and Silver Concentrations in Field Collections of Crustaceans

Element and Organism	Concentration	Reference[a]
Scandium		
Euphausiids; whole	0.016 AW	1
Selenium		
Blue crab, *Callinectes sapidus* Pensacola, Florida; 2003-2004; hepatopancreas vs. muscle	Max. 2.0 FW vs. max 1.0 FW	21
Copepods; whole	3.4 DW	2
Crustaceans; digestive system; 3 spp.	3.0-3.5 DW	11
Crustaceans; 16 commercial species; North America; edible tissues 5 spp. 6 spp. 5 spp. 2 spp. Total selenium Selenate Selenite plus selenide	0.2-0.4 FW 0.4-0.7 FW 0.7-2.0 FW 0.21 FW 0.05 FW 0.16 FW	3 3 3 12 12 12
Greenland; 1975-1991; 6 spp.; whole	0.2-3.2 FW	13
Euphausiid, *Meganyctiphanes norvegica* Viscera Eyes Muscle Exoskeleton	11.7 DW 7.8 DW 1.8 DW 0.84 DW	4 4 4 4
Marsh crab, *Parasesarma erythodactyla*; soft parts; 2002; New South Wales, Australia; metals-contaminated estuary Males Females	Max. 12.4 DW Max. 20.2 DW	20 20
Black tiger shrimp, *Penaeus monodon*; Vietnam; 2003-2005; max. values Muscle Exoskeleton Hepatopancreas	1.5 DW 0.63 DW 7.9 DW	22 22 22
Shrimps Muscle Molts	6.1 DW 5.0 DW	5 5

(Continues)

Table 7.17: Cont'd

Element and Organism	Concentration	Reference[a]
Texas; Lower Laguna Madre; shoalgrass community; 1986-1987; whole		
Grass shrimp, *Palaemonetes* sp.	1.2 (0.7-1.8) DW	14
Brown shrimp, *Penaeus aztecus*	1.8 (0.6-3.2) DW	14
Silver		
Amphipods; whole; Antarctica; February-March 1989	1.2 (0.7-1.4) DW	17
Arabian Sea; shrimps; edible tissues; 2 spp.	0.40-0.76 FW	15
Barnacle, *Elminius modestus*; pyrophosphate granules	10.5 (9.7-11.3) DW	18
Calcasieu River estuary, Louisiana		
Blue crab, *Callinectes sapidus*; muscle	0.1 DW	16
Shrimps; 2 spp.; muscle	0.04 DW	16
Crab, *Cancer irroratus*		
Muscle	(0.24-0.79) FW	6
Digestive gland	(2.1-3.4) FW	6
Muscle	0.21 DW	7
Digestive gland	6.3 DW	7
Copepods; whole	30.0 AW	8
Crustaceans; 16 commercial species; North America; edible tissues		
8 spp.	<0.1 FW	3
5 spp.	0.1-0.2 FW	3
3 spp.	0.3-0.5 FW	3
Euphausiids; whole	0.15 FW	9
Mantis shrimp, *Gonodactylus* sp.; Guam; June 1988; muscle vs. gonad	0.27 DW vs. 1.4 DW	19
Lobster, *Homarus americanus*; muscle	(0.37-0.54) DW	10
Black tiger shrimp, *Penaeus monodon*; Vietnam; 2003-2005; juveniles; max. values		
Muscle	0.02 DW	22
Exoskeleton	0.02 DW	22
Hepatopancreas	8.4 DW	22

(*Continues*)

Table 7.17: Cont'd

Element and Organism	Concentration	Reference[a]
Shrimps		
Muscle	0.24 DW	5
Molts	1.1 DW	5

Values are in mg element/kg fresh weight (FW), dry weight (DW), or ash weight (AW).
[a]1, Robertson, 1967; 2, Tijoe et al., 1977; 3, Hall et al., 1978; 4, Fowler and Benayoun, 1976a; 5, Bertine and Goldberg, 1972; 6, Greig et al., 1977a; 7, Greig et al., 1977b; 8, Vinogradova and Koual'skiy, 1962; 9, Greig and Wenzloff, 1977; 10, Greig, 1975; 11, Maher, 1983; 12, Cappon and Smith, 1982; 13, Dietz et al., 1996; 14, Custer and Mitchell, 1993; 15, Tariq et al., 1993; 16, Ramelow et al., 1989; 17, Szefer et al., 1993; 18, Pullen and Rainbow, 1991; 19, Denton et al., 2006; 20, MacFarlane et al., 2006; 21, Karouna-Renier et al., 2007; 22, Tu et al., 2008.

7.44 Selenium

The highest selenium concentrations recorded in crustacean tissues from field collections were 3.2 mg Se/kg FW and 20.2 mg Se/kg DW (Table 7.17). Selenium levels in soft tissues of the marsh crab, *P. erythodactyla*, increased with increasing concentrations of selenium in the ambient sediments, and increased with decreasing salinities; this pattern held for both males and females (MacFarlane et al., 2006).

Selenium accumulates in internal tissues of *M. norvegica*, with about 67% of all selenium in the viscera (Fowler and Benayoun, 1976a). With shrimp, *L. seticaudata*, highest concentrations were in exoskeleton when selenium was administered via the medium and in viscera and exoskeleton when administered via the diet (Fowler and Benayoun, 1976c). About half the artificially accumulated selenium in *Lysmata* was eliminated in 58-60 days (Fowler and Benayoun, 1976b). Marine crabs exposed to seawater concentration of 0.25 mg Se/L accumulated selenium over water concentration level by a factor of 25 for carapace, and 3.8 for gill; accumulations in muscle and hepatopancreas were negligible (Bjerragaard, 1982). Cadmium in solution enhanced selenium uptake (Bjerragaard, 1982).

Uptake of radioselenium-75 by semaphore crab, *H. cordiformis*, soldier crab, *M. platycheles*, and ghost shrimp, *T. australiensis* during immersion for 385 h, and loss in ^{75}Se-free seawater over the next 189 h was determined (Alquezar et al., 2007). Uptake factors (loss) were 39 (28%) for semaphore crab, 17 (42%) for soldier crab, and 54 (91%) for ghost shrimp. Excretion followed a biphasic pattern with short-lived and long-lived components. The biological half-time for short-lived components (long-lived) components were 23 h (161 h) for semaphore crab, 22 h (330 h) for soldier crab, and 16 h (1500 h) for ghost shrimp (Alquezar et al., 2007).

LC50 (96 h) values—in mg Se/L—for copepods, *Acartia* spp., ranged from 0.8-1.74; for blue crab, *C. sapidus*, it was 4.6; for mysid shrimp, *M. bahia*, adults and juveniles, it was 1.5 and 0.6, respectively; and for brown shrimp, *Penaeus aztecus*, it was 1.2 (USEPA, 1980a;

Ward et al., 1981). Results of a 28-day chronic exposure life cycle test with *M. bahia*, showed a NOAEL (no observed adverse effect level) of 0.027 mg Se/L and an LOAEL (lowest observed adverse effect level) of 0.143 mg Se/L (USEPA, 1980a). The concentration of total acid-soluble selenium currently recommended for marine life protection is less than 0.071 mg/L (Eisler, 2000c; USEPA, 1987a).

7.45 Silicon

The copepod *A. tonsa* can extract sufficient silicon necessary to form siliceous teeth from media so low in silicon that diatom growth was negligible (Miller et al., 1980). This suggests that low concentrations of silicates are not important limiting factors for *A. tonsa* or other species of pelagic copepods, unlike many species of marine alga.

7.46 Silver

Highest values recorded for silver in field collections of crustaceans were 3.4 mg Ag/kg FW in crab digestive gland, 11.3 mg/kg DW in barnacle pyrophosphate granules, and 30.0 mg Ag/kg AW in whole copepods (Table 7.17). Among arthropods, pyrophosphate granules isolated from barnacles can bind and effectively detoxify silver and other metals under natural conditions (Pullen and Rainbow, 1991).

Hepatopancreas of the pink shrimp, *Penaeus duorarum* readily accumulates ionic silver from the medium (Bianchini et al., 2007). In one study, pink shrimp that were exposed to 0.001 or 0.01 mg Ag/L for 48 h had silver burdens in hepatopancreas, in mg Ag/kg FW, of 123.7 in the low-dose group and 265.1 in the high-dose group; other tissues (hemolymph, gill muscle, eyestalk) were always <1.5 mg Ag/kg FW (Bianchini et al., 2007). Large variations were observed in ability of whole decapod crustaceans to accumulate ^{110}Ag from seawater, with concentration factors ranging between 70 and 4000 (Pouvreau and Amiard, 1974). The reasons for this variability are unknown, but is generally agreed that hepatopancreas or digestive gland are the major repositories of silver in decapods (Beasley and Held, 1971; Greig, 1975; Greig et al., 1977a,b). Silver is accumulated both from the medium and from the diet by grass shrimp, *P. pugio* (Connell et al., 1991). Silver accumulations from the medium is documented by *P. pugio* immersed in 0.002, 0.005, or 0.010 mg Ag/L for 2 weeks: 0.36 mg Ag/kg DW whole shrimp in controls, 0.5 in the 0.002 mg/L group, 3.7 in the 0.005 mg/L group, and 4.5 mg/kg DW whole shrimp in the 0.010 mg/L group (Connell et al., 1991). Silver accumulations from the diet is reported for grass shrimp fed *Artemia* nauplii containing 0.72 mg Ag/kg DW or bryozoans (*Victorella* sp.) containing 38.0 or 180.0 mg Ag/kg DW, or control bryozoans (11.5 mg Ag/kg DW). Silver concentrations (mg/kg DW whole body) in shrimp on *Artemia* diet were 0.19 versus 0.09 in silver-free *Artemia* diet; 0.26-0.62 in the high-silver bryozoan diet and 0.36 in the control bryozoan diet (Connell et al., 1991). Authors concluded that grass shrimp rapidly incorporate silver dissolved in brackish water in proportion to its concentration, but not from planktonic or detrital food sources containing elevated silver burdens (Connell et al., 1991).

Ionic silver is lethal to copepods, *A. tonsa*, at 0.036 mg/L (USEPA, 1980d), to amphipods, *Ampelisca abdita* at 0.03 mg/L (Berry et al., 1999), and to mysid shrimp, *M. bahia*, at 0.25 mg/L (USEPA, 1980d). Sediment chemistry affects silver toxicity to *A. abdita* exposed for 10 days on silver-loaded sediments (Berry et al., 1999). In general, sediments with an excess of acid volatile sulfide (AVS) relative to simultaneously extracted metal (SEM) were toxic to marine amphipods. Sediments with an excess of SEM relative to AVS, and no measurable AVS, were generally toxic. Sediments with measurable AVS were not toxic (Berry et al., 1999). In acute toxicity studies with the euryhaline *A. tonsa*, silver was about six times more toxic to starved copepods than fed animals, and more sensitive with decreasing salinity in the range 5-30 ppt (Pedroso et al., 2007a). Na^+, K^+, ATPase molecules seem to be a key site of acute silver toxicity in *A. tonsa* (Pedroso et al., 2007b). The mysid, *Americamysis bahia*, held in seawater for 28 days at 10, 20, or 30 ppt and dissolved organic carbon up to 6.0 mg/L were more sensitive to silver at 10 ppt (0.0039 mg dissolved Ag/L for reduced survival) than at 30 ppt salinity (0.060 mg/L); dissolved organic carbon had no effect on silver toxicity (Ward et al., 2006).

Silver is also toxic to the copepod, *A. tonsa,* when fed a silver-contaminated algal diet (Bielmyer et al., 2006). In that study, diatoms (*T. pseudonana*) were held in seawater containing 0.00064 mg Ag/L (as traced by radiosilver-110 m) for 7 days, reaching a silver burden of 5.4 mg Ag/kg DW diatom; this, in turn, was fed to *Acartia* for 7 days, causing a significant reduction in copepod fertility. Authors note that the proposed silver criterion for marine life protection is 0.003 mg Ag/L, and is in excess of 0.00064 mg Ag/L used in this study to produce a toxic diet (Bielmyer et al., 2006).

7.47 Strontium

All crustacean tissues and organs contain significant concentrations of strontium, especially exoskeleton (Table 7.18).

Strontium concentrations tend to decrease with increasing size in the amphipod, *P. walkeri* (Palmer et al., 2006), and this may confound the interpretation of strontium residues in this and other species. The ability of whole crustaceans to accumulate strontium from seawater is low, ranging from 0.3 to 1.0 times the concentration in seawater (Ichikawa, 1961; Ketchum and Bowen, 1958). Some organs show greater strontium accumulation potential than others. *Leander* sp., for example, shows strontium concentration factors of 0.2 in muscle, 1.3 in viscera, and 15.0 in exoskeleton (Hiyama and Shimizu, 1964). Molts of *M. norvegica* account for 23% of the strontium body burden in that species and molts

Table 7.18: Strontium Concentrations in Field Collections of Crustaceans

Organism	Concentration	Reference[a]
Copepods		
Whole	125.0 DW	1
Whole	1500.0 AW	2
Shrimp, *Crangon* sp.; whole	42.1 DW	9
Crustaceans		
Whole	(0.97-17.0) FW	3
Exoskeleton; 15 spp.	2210.0-6970.0 DW	4
Euphausiid, *Euphausia pacifica*; whole	(440.0-1000.0) AW	5
Euphausiids; whole	163.0 DW	1
Lobster		
Muscle	59.0 AW	6
Carapace	4000.0 AW	6
Head	6800.0 AW	6
Euphausiid, *Meganyctiphanes norvegica*		
Molts	350.0 DW	7
Whole	117.0 DW	7
Mysidacean, *Neomysis japonica*; whole	360.0 AW	5
Black tiger shrimp, *Penaeus monodon*; juveniles; Vietnam; 2003-2005; max. values		
Muscle	9.3 DW	10
Exoskeleton	2060.0 DW	10
Hepatopancreas	81.7 DW	10
Shrimp		
Muscle	510.0 AW	6
Carapace	600.0 AW	6
Whole	1500.0 AW	6
Zooplankton; surface vs. > 99 m depth collections	890.0 AW vs. 1140.0 AW	8

Values are in mg Sr/kg fresh weight (FW), dry weight (DW), or ash weight (AW).
[a] 1, Martin and Knauer, 1973; 2, Vinogradova and Koual'skiy, 1962; 3, Mauchline and Templeton, 1966; 4, Gibbs and Bryan, 1972; 5, Fukai et al., 1962; 6, Lowman et al., 1970; 7, Fowler, 1977; 8, Martin, 1970; 9, Zumholz et al., 2006; 10, Tu et al., 2008.

contribute significantly (3.2 mg Sr/kg euphausiid molt per day) to strontium cycling in the Mediterranean Sea (Fowler, 1977). Strontium can also accumulate in viscera of the crabs *C. maenas* (Martin, 1972) and *Birgus latro* (Chakravarti and Eisler, 1961). In *Birgus*, the fat content of liver accounted for an average of 47% of the wet weight, but less than 0.5% of the total strontium content; variability in fat content of crustacean liver should be considered in evaluating mineral composition of this organ, preferably by expressing results on a fat-free basis (Chakravarti and Eisler, 1961).

7.48 Terbium

Amphipods from East Antarctica contained <0.006 mg Tb/kg DW, or below the current detection limit (Palmer et al., 2006).

7.49 Thallium

Maximum thallium concentrations in tissues of juvenile tiger shrimp, *P. monodon*, from Vietnam in 2003-2005, in mg Tl/kg DW, were 0.005 in muscle, 0.01 in exoskeleton, and 0.005 in hepatopancreas (Tu et al., 2008).

7.50 Thorium

A concentration factor of 20 from seawater is reported for ^{228}Th and zooplankton samples comprised mainly of copepods (Kharkar et al., 1976).

7.51 Thulium

Whole amphipods collected from East Antarctica in 2003-2004 had <0.005 mg Tm/kg DW, or below the current detection limit (Palmer et al., 2006).

7.52 Tin

Edible tissues of crustaceans from North American waters contained a maximum of 2.0 mg total Sn/kg fresh tissue (Table 7.19). Total organotin concentrations in edible tissues of six species of crustaceans sold commercially in France in 2005 contained a maximum of 0.014 mg total organotins/kg FW, mostly tributyltins (Guerin et al., 2007; Table 7.19). However, triphenyltin concentrations in muscle of crustaceans from Bohai Bay, China in 2002 were higher than tributyltin concentrations, suggesting trophic magnification of triphenyltins through the algae-mollusc-crustacean food chain (Table 7.19; Hu et al., 2006).

In aquatic systems, triorganotins were the most toxic group of organotins tested, followed in decreasing order of toxicity by diorganotins, tetraorganotins, and monoorganotins (Argese et al., 1998). Within each series, butyltin and phenyltin compounds seemed most toxic (Argese et al., 1998). The toxicity of triorganotin compounds is attributed, in part, to their ability to bind to proteins and to inhibit oxidative phosphorylation (Blunden and Chapman, 1986; Davies and Smith, 1982; Duncan, 1980; Smith, 1978; WHO, 1980). The derangement of triorganotin-induced energy-coupled processes, which occur at the membrane level, appears

Table 7.19: Tin, Titanium, and Tungsten Concentrations in Field Collections of Crustaceans

Element and Organism	Concentration	Reference[a]
Tin		
Bohai Bay, North China; 2002; soft tissues		
Crab, *Portunus trituberculatus*		
Monobutyltins	0.003 FW	11
Dibutyltins	0.005 FW	11
Tributyltins	0.005 FW	11
Triphenyltins	0.017 FW	11
Burrowing shrimp, *Upogebia* sp.		
Monobutyltins	0.015 FW	11
Dibutyltins	0.005 FW	11
Tributyltins	0.003 FW	11
Triphenyltins	0.017 FW	11
Copepod, *Calanus finmarchicus*; whole	<1.0 AW	1
Blue crab, *Callinectes sapidus*		
Pensacola, Florida; 2003-2004; hepatopancreas vs. muscle	Max. 0.75 FW vs. max. 0.55 FW	10
Copepod, *Centropages* spp.; whole	50.0 AW	1
Copepods; whole	70.0 AW	2
Crustaceans; 16 commercial species; North America; edible tissues		
5 spp.	0.6-0.7 FW	3
8 spp.	0.7-0.9 FW	3
3 spp.	0.9-2.0 FW	3
France; January-April 2005; edible portions; sold commercially; organotins		
Total organotins; 5 spp.	0.0016-0.0087 FW	9
Crab, unidentified		
Total organotins	0.0140 FW	9
Butyltins	0.0115 FW	9
Phenyltins	0.0022 FW	9
Octyltins	0.0003 FW	9
American lobster, *Homarus americanus*; muscle	0.6 FW	6
Amphipod, *Paramoera walkeri*; East Antarctica; 2003-2004; whole	(0.05-0.12) DW	8

(Continues)

Table 7.19: Cont'd

Element and Organism	Concentration	Reference[a]
Black tiger shrimp, *Penaeus monodon*; Vietnam; 2003-2005; juveniles; max. values		
Muscle	0.17 DW	12
Exoskeleton	0.18 DW	12
Hepatopancreas	0.28 DW	12
Zooplankton; Osaka Bay, Japan; 1990; whole; tributyltins	(0.3-4.2) FW	7
Titanium		
Copepod, *Calanus finmarchicus*; whole	10.0 AW	1
Copepod, *Centropages* spp.; whole	3.0 AW	1
Copepods; whole	700.0 AW	2
Euphausiid, *Euphausia krohni*; whole	9.0 AW	1
Tungsten		
Shrimp, *Pandalus* spp; soft parts	<0.005 AW; < 0.0003 DW	4, 5

Values are in mg element/kg fresh weight (FW), dry weight (DW), or ash weight (AW).
[a]1, Nicholls et al., 1959; 2, Vinogradova and Koual'skiy, 1962; 3, Hall et al., 1978; 4, Fukai and Meinke, 1959; 5, Fukai and Meinke, 1962; 6, Jenkins, 1980; 7, Harino et al., 1998; 8, Palmer et al., 2006; 9, Guerin et al., 2007; 10, Karouna-Renier et al., 2007; 11, Hu et al., 2006; 12, Tu et al., 2008.

to be the most probable mode of action in aquatic organisms (Argese et al., 1998). Triorganotins also interfere with phagocytosis and other pathways where sulfhydryl groups play a pivotal role (Elferink et al., 1986) and inhibit uptake of gamma-aminobutyric acid and Na^+-K^+-ATPase in brain (Costa, 1985). Impairment of phagocytosis and related activities of polymorphonuclear eukaryotes may enhance susceptibility for infection (Elferink et al., 1986).

Adverse effects of TBT compounds to sensitive species of crustaceans tested are documented at 0.0000011-0.001 mg/L. Exposure of juvenile amphipods, *Caprella danilevskii*, for 49 days to tributyltin concentrations near or below ambient levels in seawater adversely affected adult reproduction; the total number of juveniles per female decreased from 39.5 in controls to 24.5 in those held in 0.0000011 mg TBT/L, and 17.5 in 0.0000107 mg TBT/L (Aono and Takeuchi, 2008). Female copepods, *Pseudodiaptomus marinus*, exposed to 0.000006 mg TBT/L for 72 h show a significant delay in egg sac production; at 0.00002 mg/L, fecundity was increased and frequency of males increased (Huang et al., 2006). Other effects include reduced survival in 96 h of *Pseudodiaptomus* at 0.00057 mg/L (Huang et al., 2006); death in 7 days of the copepod *E. affinis* at 0.0001 mg/L (Hall et al., 1988a,b); reduced survival in

5 weeks of the Baltic amphipod *Gammarus oceanicus* at 0.0003 mg/L (Laughlin et al., 1984); 50% immobilization in 6 days of the copepod *A. tonsa* at 0.0004 mg/L (U'ren, 1983); a decrease in the proportion of females of *T. japonicus* copepods after long-term exposure to 0.0001 mg/L (Lee et al., 2008); abnormal limb regeneration of the fiddler crab *U. pugilator* at 0.0005 mg/L (Weis and Kim, 1988; Weis et al., 1987); and death in 96 h of juvenile mysidaceans *Metamysidopsis elongatus* at 0.0005-0.001 mg/L (Hall and Pinkney, 1985). Acute toxicity bioassays of 96-h duration with adult *T. japonicus* copepods, a species recommended for routine ecotoxicity testing in the western Pacific Ocean region, showed that all survived 0.02 mg tributyltin/L, 10% died at 0.03, and 50% died at 0.05 mg TBT/L (Lee et al., 2007). Several factors are known to affect sensitivity, and accumulations of TBT and other organotins in crustaceans. For example, early developmental stages of mysid shrimps were more sensitive to organotins than later developmental stages (Thompson et al., 1985). Mortality of zoeae of fiddler crab, *U. pugilator*, to trimethyltins was greatest at comparatively elevated temperatures and reduced salinities (Thompson et al., 1985). Survival of grass shrimp, *P. pugio*, was decreased by several orders of magnitude when exposed simultaneously to organotins through contaminated water and contaminated sediments (Clark et al., 1987). Uptake of organotins from the medium by crustaceans was variable and greatest at low ambient tin concentrations (<0.001 mg/L), lengthy exposures, and high organism lipid content (Champ, 1986; Laughlin et al., 1986; Thain and Waldock, 1986; Thompson et al., 1985); however, diet remains the major pathway of tin uptake in crustaceans (Evans and Laughlin, 1984; Hall and Pinkney, 1985; Laughlin et al., 1986).

Structure-activity relations seem to have high predictive capacity in hazard assessment, and those for organotins are particularly promising (Vighi and Calamari, 1985). For example, studies on the biocidal properties of structurally distinct diorganotins (R_2SnX_2) and triorganotins (R_3SnX) to zoeae of the mud crab, *R. harrisii*, show, within a homologous series, that diorganotins are less toxic than the corresponding triorganotins, and the toxicity of organotins to crab zoeae are a function of the hydrophobic characteristics conferred by the number and structure of the organic ligands (Laughlin et al., 1985). For example, at least 50% mortality occurs in diorganotins (triorganotins), in mg Sn/L, at 10.8 dimethyltin (0.067 trimethyltin), at 2.4 diethyltin (0.053), at 2.2 dipropyltin (0.03), at 0.78 dibutyltin (0.011), at 0.27 diphenyltin (0.007), and at 0.082 dicyclohexyltin (0.0023) (Laughlin et al., 1985).

Abiotic and biological degradation of organotins generally occurs through sequential dealkylation and dearylation (Blunden et al., 1985; Chau et al., 1984; Eisler, 2000g; Smith, 1981; WHO, 1980; Zuckerman et al., 1978). Organotin compounds undergo successive cleavage of tin-carbon bonds to ultimately produce inorganic tin as follows: R_4Sn (via k_4) to R_3SnX (k_3) to R_2SnX_2 (k_2) to $RSnX_3$ (k_1) to $SnX4$. The reaction rate, k, usually proceeds as $k_4 > k_3 > k_2 = k_1$. The breaking of a Sn-C bond can occur via ultraviolet radiation, biological cleavage, gamma irradiation, and thermal cleavage (Blunden and Chapman, 1982, 1986; Blunden et al., 1985; Thompson et al., 1985; WHO, 1980). In general, UV and biological cleavage are the most important processes. The main abiotic factors that limit organotin persistence include elevated temperatures, increased sunlight, and aerobic conditions (Eisler, 2000g). The biological effects of the tetraorganotin compounds, R_4Sn, seem to be caused entirely by the R_3SnX derivative that is produced by their rapid *in vivo* dealkylation (Blunden and Chapman, 1986; Davies and Smith, 1982; Duncan, 1980).

Marine sediment concentrations less than 0.161 mg TBT/kg are considered safe to mysid shrimp, although proposed sediment criteria range from <0.001 to <0.007 mg/kg (Cardwell and Sheldon, 1986). For seawater, proposed TBT concentrations considered safe to aquatic life range from less than 0.000002 mg/L (Huggett et al., 1992) to the less stringent <0.00001 to <0.000002 in the United States (Harrington, 1991; Pinkney et al., 1990) and <0.00002 mg/L in the United Kingdom (Side, 1987). The banning of TBT-containing antifouling paints in the United Kingdom in 1987 was accompanied by increases in abundance and diversity of benthic fauna, especially amphipod crustaceans in River Crouch Estuary (Rees et al., 1999; Waldock et al., 1999).

7.53 Titanium

The maximum titanium value recorded in field collections of crustaceans was 700.0 mg Ti/kg AW from Black Sea copepods (Table 7.19).

7.54 Tungsten

The single datum available for tungsten indicated that this element was below analytical detection limits in a pandalid shrimp (Table 7.19).

7.55 Uranium

Whole shrimp, *Crangon* sp., had 0.00076 mg U/kg DW (Zumholz et al., 2006). Uranium-238 in zooplankton samples consisting mainly of copepods was only 0.17 times ^{238}U levels in ambient seawater (Kharkar et al., 1976).

7.56 Vanadium

Vanadium concentrations in any crustacean tissue did not exceed 0.5 mg V/kg FW, 3.3 mg V/kg DW, or 70.0 mg V/kg AW (Table 7.20). The significance of vanadium in crustaceans in not known with certainty.

7.57 Ytterbium

A maximum of <0.007 mg Yb/kg DW, or below the current detection limit, was found in amphipods from east Antarctica (Palmer et al., 2006).

Table 7.20: Vanadium Concentrations in Field Collections of Crustaceans

Organism	Concentration	Reference[a]
Copepod, *Calanus finmarchicus*; whole	21.0 AW	1
Copepod, *Centropages* spp.; whole	16.0 AW	1

Copepods
  Whole
  Whole
	
	70.0 AW
	3.3 DW
		
		2
		3

Crustaceans; 16 commercial species; North America; edible tissues
  3 spp.
  5 spp.
  8 spp.
	
	0.1-0.3 FW
	0.3-0.4 FW
	0.4-0.5 FW
		
		4
		4
		4

Euphausiid, *Euphausia krohni*; whole	45.0 AW	1

Shrimp, *Pandalus* sp.
  Whole
  Soft parts
	
	1.1 AW
	0.07 AW
		
		5
		6

Tiger shrimp, *Penaeus monodon*; Vietnam; 2003-2005; juveniles; max. values
  Muscle
  Exoskeleton
  Hepatopancreas
	
	0.063 DW
	0.80 DW
	0.80 DW
		
		8
		8
		8

Barnacle, *Tetraclita* sp.; whole	0.36 FW	7

Values are in mg V/kg fresh weight (FW), dry weight (DW), or ash weight (AW).
[a]1, Nicholls et al., 1959; 2, Vinogradova and Koual'skiy, 1962; 3, Tijoe et al., 1977; 4, Hall et al., 1978; 5, Fukai and Meinke, 1959; 6, Fukai and Meinke, 1962; 7, Ikebe and Tanaka, 1979; 8, Tu et al., 2008.

7.58 Yttrium

Whole shrimp, *Crangon* sp., contained 0.0033 mg Y/kg DW (Zumholz et al., 2006).

7.59 Zinc

There is general agreement that zinc is present in all tissues of all crustaceans examined; concentrations are elevated in organisms near anthropogenic point sources of zinc contamination; concentrations are normally grossly elevated in some species of barnacles; zinc-specific sites of accumulation in crustaceans include the hepatopancreas; interspecies variations in zinc content are considerable, even among closely related species; intraspecies differences in zinc content are modified by a host of biologic and abiotic variables; and many species regulate zinc within a threshold range of concentrations (as quoted in Eisler, 2000h).

In the case of deep sea shrimps collected from hydrothermal vent sites, extremely high concentrations of zinc (>100,000.0 mg/kg DW) were measured in gut or adsorbed onto tissues (Kadar et al., 2006; Table 7.21). Concentrations of total zinc in organs and tissues of representative species of marine, estuarine, and oceanic crustaceans from numerous geographic locations are listed (Table 7.21). According to Pequegnat et al. (1969), zinc is not limiting to normal life processes in the marine environment and is accumulated in excess of the organism's immediate needs, at least on the basis of enzymatically bound zinc. The information in Table 7.21 tends to support the observations that zinc is not limiting, and in several cases probably accumulated far in excess of biological requirements.

Zinc concentrations in crustacean tissues are usually less than 75.0 mg/kg FW or <100.0 mg/kg DW; exceptions include hepatopancreas, molts, eggs, fecal pellets, barnacles, and proximity to hydrothermal vents (Table 7.21). In crustaceans, zinc is slightly elevated in hepatopancreas, but most tissues are only 2-3 times higher than muscle (Sprague, 1986). The high concentration recorded in crustacean muscle of 57.0 mg Zn/kg FW in king crab, *Paralithodes camtschaticus* (USNAS, 1989), was associated with two metal-binding proteins of MW 11,500 and 27,000 (Eisler, 1981). In crustacean tissues, zinc levels were usually highest in summer, at low salinities, and in young animals (Eisler, 1981), although young amphipods had higher zinc residues than older stages (Rainbow, 1989). Seasonal accumulations of whole-body zinc in the shrimp *Palaemon elegans* during spring and summer and loss in winter seem to reflect water zinc concentration in the range 0.000-0.009 mg/L (Alliot and Frenet-Piron, 1990). Zinc is present in crustacean serum at concentrations of more than 1000 times higher than ambient seawater; in serum, it serves mainly as a cofactor of carbonic anhydrase, the principal enzyme involved in calcification. Serum zinc concentrations in crustaceans seem to be independent of season, water temperature, or salinity (Sprague, 1986).

In the prawn, *P. indicus*, most of the zinc in hepatopancreas and gill is in the insoluble fraction (Table 7.21) and is presumably bound by phosphorous, calcium, magnesium, and silicon (Nunez-Nogueira et al., 2006a). Exposure to 0.1 mg Zn/L for 10 days resulted in negligible uptake or change in soluble/insoluble ratios of gill and hepatopancreas (Nunez-Nogueira et al., 2006a). In another study, Nunez-Nogueira et al. (2006b) demonstrated the importance of diet in zinc assimilation; they fed juvenile *P. indicus* radiolabeled ^{65}Zn squid muscle (*L. vulgaris*) or filaments of a green alga (*Cladophora* sp.) for 60 min then transferred to clean seawater. The AE was calculated from the point at which no radioactivity was detected in feces or about 6 h after ingestion of algae and 10 h after ingestion of squid muscle. The AE of zinc ranged between 50% and 69% from algae and between 60% and 70% from squid muscle. Prawns fed algae and dissected after 49 h depuration had 53% of the ^{65}Zn in hepatopancreas, 24% in exoskeleton, and 13% in muscle. Prawns fed squid muscle and dissected 134 h after initial feeding had 29% of the ^{65}Zn in abdominal muscle and 21% in hepatopancreas (Nunez-Nogueira et al., 2006b).

Table 7.21: Zinc Concentrations in Field Collections of Crustaceans

Organism	Concentration	Reference[a]
Copepod, *Acartia clausi*; whole	1270.0 DW	1
Amphipods; whole	43.0 DW	2
Amphipods; whole; western Britain; coastal waters; 4 spp.	104.0-506.0 DW	47
Barnacle, *Balanus amphitrite*; soft parts	Max. 1937.0 DW	49
Barnacle, *Balanus amphitrite communis*; soft parts	71.5 FW	4
Barents Sea and environs; 1991-2000; whole		
Copepods; 4 spp.	108.0-509.0 DW	76
Euphausid, *Meganyctiphanes norvegica*	73.0 DW	76
Amphipod, *Themisto abyssorum*	110.0 DW	76
Shrimp, *Pandalus borealis*	79.0 DW	76
Barnacle, *Balanus balanoides*; soft parts	(4500.0-23,100.0) DW	5
Barnacles; summer 2001; soft parts; Natal, Brazil		
Fistulobalanus citerosum	(1624.0-1719.0) DW	67
Balanus amphitrite	1185.0 DW	67
Barnacles		
Soft parts	Max. 3300.0 DW	6
Whole, 3 spp.	138.0-3438.0 FW	7, 8
Soft parts, 3 spp.	110.0-2980.0 FW	7, 8
Gut and parenchyma, 3 spp.	Max. 2220.0 FW	7, 8
Calcasieu River Estuary, Louisiana		
Blue crab, *Callinectes sapidus*; muscle	112.0 (106.0-213.0) DW	50
Brown shrimp, *Penaeus aztecus*; whole	(46.0-61.0) DW	50
White shrimp, *Penaeus setiferus*; whole	(44.0-62.0) DW	50
Blue crab, *Callinectes sapidus*		
Whole	14.0 DW	9
Whole	17.0 FW	11
Muscle; August 2005; Iskenderun Bay, Turkey	9.1 (3.6-18.3) DW	63
Pensacola, Florida; 2003-2004		
Hepatopancreas	Max. 48.0 FW	69
Muscle	Max. 54.0 FW	69

(Continues)

Table 7.21: Cont'd

Organism	Concentration	Reference[a]
Crab, *Cancer irroratus*		
Digestive diverticula	(35.0-43.0) DW	12
Muscle	(32.1-64.6) FW	13, 14
Digestive gland	(18.1-31.5) FW	13, 14
Crab, *Cancer pagurus*		
Whole	27.0 FW	3
Muscle	64.0 FW	3
Stomach fluid	15.0 FW	3
Hepatopancreas	45.0 FW	3
Urine	0.6 FW	3
Excretory organ	29.0 FW	3
Gill	42.0 FW	3
Shell	3.0 FW	3
Vas deferens	16.0 FW	3
External eggs	94.0 FW	3
Blood	49.0 FW	3
Whole	(21.3-34.5) FW	15
Claw muscle	(51.0-61.0) FW	15
Abdominal muscle	(29.2-66.0) FW	15
Gills	(10.0-30.0) FW	15
Crab, *Carcinus maenas*		
Whole	24.5 FW	16
Hepatopancreas	56.0 FW	16
Leg muscle	44.0 FW	16
Whole	22.0 FW	3
Blood	36.0 FW	3
Muscle	44.0 FW	3
Stomach fluid	18.0 FW	3
Hepatopancreas	56.0 FW	3
Urine	0.4 FW	3
Excretory organ	19.0 FW	3
Gill	26.0 FW	3
Shell	3.0 FW	3
Vas deferens	23.0 FW	3
Whole	(25.1-38.8) FW	17
Hemolymph	32.2 FW	18
Copepods		
Whole	60.0 DW	2
Whole	(62.0-170.0) DW	11
Whole	4000.0 AW	20

(*Continues*)

Table 7.21: Cont'd

Organism	Concentration	Reference[a]
Whole	(160.0-165.0) DW	21
Whole; 3 spp.; Taiwan; near sewage outfall	132.0-3891.0 DW	66
Crabs		
Muscle	(24.0-51.0) FW; (120.0-250.0) DW; (750.0-1900.0) AW	22
Carapace	(7.0-46.0) FW; (15.0-120.0) DW; (23.0-110.0) AW	22
Shrimp, *Crangon* sp.; whole	2.8 DW	71
Shrimp, *Crangon vulgaris*		
Whole	34.0 FW	3
Blood	23.0 FW	3
Muscle	14.0 FW	3
Hepatopancreas	78.0 FW	3
Whole	26.4 FW	17
Crustaceans		
Whole	(23.7-28.9) FW	23
Soft parts, 12 spp.	55.0-290.0 DW	24
Whole	Max. 275.0 DW	25
Whole; 4 spp.; Antarctica; 2004	50.2-84.1 DW	60
Edible tissues, 16 spp.		
8 spp.	10.0-20.0 FW	26
5 spp.	20.0-40.0 FW	26
3 spp.	40.0-60.0 FW	26
Northeast Atlantic Ocean; July 1985; whole		
Decapods	(35.0-57.0) DW	51
Euphausiids	(44.0-96.0) DW	51
Mysids	(24.0-44.0) DW	51
Soft parts		
Amphipods	(73.0-109.0) DW	52
Barnacles	(690.0-27,837.0) DW	52
Barnacles	1050.0-5144.0 DW; max. 113,000.0 DW	53
Copepods	(60.0-170.0) DW	52
Copepods	164.0-177.0 DW; max. 1300.0 DW	53
Crabs	68.00-102.0 DW; max. 340.0 DW	53
Euphausiids	(53.0-83.0) DW	52
Isopods	94.0 DW	52
Shrimps	14.0-69.0 DW; max. 150.0 DW	53

(*Continues*)

Table 7.21: Cont'd

Organism	Concentration	Reference[a]
Near marine outfall of lead smelter; South Australia; 5 spp.; whole		
2.5-5.2 km from outfall	148.0 DW; max. 767.0 DW	54
18-18.8 km from outfall	68.0 DW	54
Deep sea shrimps; Rainbow hydrothermal vent site; mid-Atlantic Ridge; adults; summer 2002		
Rimicaris exoculata; adults		
Whole body	1800.0 DW	64
Cuticle	500.0 DW	64
Gill	2900.0 DW	64
Antennae	200.0 DW	64
Muscle	300.0 DW	64
Digestive gland	800.0 DW	64
Pylorus	25,900.0 DW	64
Microcaris fortunata; adults		
Whole body	2500.0 DW	64
Cuticle	300.0 DW	64
Gill	4300.0 DW	64
Antennae	900.0 DW	64
Muscle	200.0 DW	64
Digestive gland	600.0 DW	64
Pylorus	40,900.0 DW	64
Crab, *Eupagurus bernhardus*		
Whole	28.0 FW	3
Blood	11.6 FW	3
Leg muscle	36.0 FW	3
Abdominal muscle	24.0 FW	3
Stomach fluid	26.0 FW	3
Hepatopancreas	69.0 FW	3
Excretory organ	23.0 FW	3
Gill	69.0 FW	3
Vas deferens	27.0 FW	3
Ovary	46.0 FW	3
Euphausiid, *Euphausia pacifica*		
Whole	13.0 FW	27
Whole	(50.0-110.0) DW	28
Euphausiid, *Euphausia superba*		
Whole	68.0 (42.0-75.0) DW	48
Whole; Antarctica	41.3 DW	59

(*Continues*)

Table 7.21: Cont'd

Organism	Concentration	Reference[a]
Euphausiids		
Whole	(53.0-83.0) DW	19
Whole	430.0 AW	29
Whole	17.1 FW	30
Lobster, *Galathea squamifera*		
Whole	18.0 FW	3
Blood	2.5 FW	3
Muscle	10.0 FW	3
Stomach fluid	47.0 FW	3
Hepatopancreas	49.0 FW	3
Gill	27.0 FW	3
Shell	9.0 FW	3
American lobster, *Homarus americanus*		
Gills	(102.0-126.0) DW	55
Green gland	(114.0-148.0) DW	55
Hepatopancreas	(70.0-135.0) DW	55
Pincer muscle	(100.0-127.0) DW	55
Tail muscle	80.0 DW	55
Lobster, *Homarus vulgaris*		
Muscle		
Slow-contracting	(93.0-105.0) DW	31
Fast contracting	(13.0-20.0) DW	31
Whole	(13.8-16.7) FW	15
Whole	23.0 FW	3
Blood	7.4 FW	3
Leg muscle	60.0 FW	3
Abdominal muscle	15.0 FW	3
Stomach fluid	0.7 FW	3
Hepatopancreas	34.0 FW	3, 16
Urine	2.2 FW	3
Excretory organ	20.0 FW	3
Gill	15.0 FW	3
Shell	5.0 FW	3
Vas deferens	13.0 FW	3
Ovary	50.0 FW	3
Whole	22.0 FW	16
Leg muscle	64.0 FW	16
Blood	(4.7-11.8) FW	32
Urine	(0.1-6.1) FW	32
Excretory organs	(17.3-21.9) FW	32
Abdominal muscle	(14.3-17.7) FW	32

(*Continues*)

Table 7.21: Cont'd

Organism	Concentration	Reference[a]
Hepatopancreas	(23.3-59.5) FW	32
Stomach fluid	(0.5-1.3) FW	32
Gill	(12.8-19.9) FW	32
Shell	(0.9-8.9) FW	32
Vas deferens	13.3 FW	32
Ovary	50.2 FW	32
Hydrothermal vent field; mid-Atlantic Ridge; 2001		
Shrimp, *Chlorocaris chacei*; 2001; 1700 m deep; digestive gland vs. muscle	2330.0 DW vs. 50.0 DW	72
Hydrothermal crab, *Segonzacia mesatlantica*; digestive gland vs. muscle		
840 m deep	439.0 DW vs. 111.0 DW	72
1700 m deep	133.0 DW vs. 268.0 DW	72
2300 m deep	108.0 DW vs. 271.0 DW	72
Lobster, *Jasus lalandi*; muscle	17.0 FW	33, 34
Lobster		
Carapace	88.0 AW	22
Head	140.0 AW	22
Muscle	600.0 AW	22
Blue shrimp, *Litopenaeus stylirostris*; Gulf of California, Mexico; 2004-2005; muscle	57.8 DW; max. 80.4 DW	61
Crab, *Maia squinado*		
Whole	21.0 FW	3
Blood	2.4 FW	3
Muscle	63.0 FW	3
Stomach fluid	31.0 FW	3
Hepatopancreas	71.0 FW	3
Urine	0.3 FW	3
Excretory organ	15.0 FW	3
Gill	10.0 FW	3
Shell	5.0 FW	3
Vas deferens	16.0 FW	3
Ovary	45.0 FW	3
Euphausiid, *Meganyctiphanes norvegica*		
Whole	104.0 DW	35
Whole	62.0 DW	36

(Continues)

Table 7.21: Cont'd

Organism	Concentration	Reference[a]
Molts	146.0 DW	36
Eggs	318.0 DW	36
Fecal pellets	950.0 DW	36
Whole		
Firth of Clyde	43.0 (27.0-62.0) DW	48
Northeast Atlantic Ocean	102.0 (40.0-281.0) DW	48
Peneid shrimp, *Melicertus kerathurus*; Lesina lagoon, Italy; October 2004; 3 sites		
Muscle	41.8-60.4 DW	75
Exoskeleton	46.3-59.3 DW	75
Shrimp, *Metapenaeus affinis*; muscle	58.3 DW	37
Mexico; Gulf of California; 1999-2000		
Barnacle, *Balanus eburneus*; soft parts	1210.0 DW	73
Shrimps; muscle		
Litopenaeus stylirostris	61.0 DW	73
Litopenaeus vanamei	53.0 DW	73
Lobster, *Nephrops norvegicus*; muscle	8.5-12.2 FW	15
Amphipod, *Orchestoidea corniculata*; whole	150.0-300.0 DW	38
Crab, *Paguristes sericeus*; whole	(29.0-37.0) FW; (92.0-120.0) DW; (230.0-360.0) AW	39
Shrimp, *Palaemon serratus*		
Whole	21.0 FW	3
Blood	38.0 FW	3
Muscle	10.0 FW	3
Hepatopancreas	64.0 FW	3
Gill	35.0 FW	3
External eggs	24.0 FW	3
Shrimp, *Palaemon squilla*; whole	30.0 FW	3
Grass shrimp, *Palaemonetes pugio*		
From sediments containing 627 mg Zn/kg DW		
Exoskeleton	58.0 FW	56
Muscle	55.0 FW	56
From sediments containing 8 mg Zn/kg DW		
Exoskeleton	18.0 FW	56
Muscle	30.0 FW	56

(*Continues*)

Table 7.21: Cont'd

Organism	Concentration	Reference[a]
Shrimp, *Palaemonetes varians*		
Whole	20.0 FW	3
Blood	87.0 FW	3
Muscle	14.0 FW	3
Hepatopancreas	65.0 FW	3
Lobster, *Palinurus vulgaris*		
Blood	3.1 FW	3
Leg muscle	66.0 FW	3
Abdominal muscle	20.0 FW	3
Stomach fluid	48.0 FW	3
Hepatopancreas	97.0 FW	3
Urine	0.6 FW	3
Excretory organ	20.0 FW	3
Gill	20.0 FW	3
Shell	16.0 FW	3
Ovary	82.0 FW	3
External eggs	107.0 FW	3
Amphipod, *Paramoera walkeri*; East Antarctica; 2003-2004; whole	(48.0-82.0) DW	59
Shrimp, *Pandalus jordani*		
Whole	58.0 FW	27
Muscle	11.0 FW	27
Prawn, *Pandalus montagui*		
Tail	44.0 DW	40
Whole	70.0 DW	40
Carcass	78.0 DW	40
Egg	100.0 DW	40
Hepatopancreas	118.0 DW	40
Cuticle	57.0 DW	57
Eye	70.0 DW	57
Gill	106.0 DW	57
Hepatopancreas	30.0 DW	57
Muscle	57.0 DW	57
Whole	58.0 DW	57
Lobster, *Panulirus argus*; muscle	32.8 FW; 97.0 DW	22
Marsh crab, *Parasesarma erythodactyla*; soft parts; 2002; New South Wales, Australia; metals-contaminated estuary		

(*Continues*)

Table 7.21: Cont'd

Organism	Concentration	Reference[a]
Males	Max. 403.0 DW	68
Females	Max. 292.0 DW	68
Shrimp, *Penaeus aztecus*		
Whole	14.0 DW	9
Muscle	47.7 DW	10
Exoskeleton	28.3 DW	10
Viscera	105.0 DW	10
Shrimp, *Penaeus brasiliensis*; adults; whole	(47.0-75.0) FW; (181.0-290.0) DW	58
Shrimp, *Penaeus japonica*; muscle	63.0 DW	41
Shrimp, *Penaeus monodon*; muscle	(21.5-70.0) DW	37
White shrimp, *Peneus setiferus*; muscle; Gulf of Mexico	107.0 DW	62
Shrimp, *Penaeus* spp.		
Whole	62.0 DW	42
Whole	9.6 FW	11
Prawn, *Penaeus indicus*; juveniles		
Hepatopancreas		
Whole	150.3 DW	70
Soluble fraction	25.1 DW	70
Insoluble fraction	125.2 DW	70
Gills		
Whole	78.4 DW	70
Soluble fraction	35.0 DW	70
Insoluble fraction	43.4 DW	70
Tiger shrimp, *Penaeus monodon*; Vietnam; 2003-2005; juveniles; max. values		
Muscle	51.0 DW	74
Exoskeleton	25.8 DW	74
Hepatopancreas	165.0 DW	74
Crab, *Porcellana platycheles*; whole	54.0 FW	3
Crab, *Portunus depurator*		
Whole	21.0 FW	3
Blood	1.8 FW	3
Muscle	15.0 FW	3
Stomach fluid	13.0 FW	3
Hepatopancreas	24.0 FW	3
Urine	0.3 FW	3

(Continues)

Table 7.21: Cont'd

Organism	Concentration	Reference[a]
Excretory organ	8.0 FW	3
Gill	25.0 FW	3
Shell	3.0 FW	3
Crab, *Portunus holsatus*; whole	34.1 FW	43
Crab, *Portunus pelagicus*; muscle	(73.7-80.2) DW	37
Crab, *Portunus puber*		
Whole	27.0 FW	3
Blood	7.0 FW	3
Muscle	28.0 FW	3
Stomach fluid	15.0 FW	3
Hepatopancreas	42.0 FW	3
Urine	0.2 FW	3
Excretory organ	10.0 FW	3
Gill	23.0 FW	3
Shell	13.0 FW	3
Vas deferens	20.0 FW	3
Ovary	87.0 FW	3
Crab, *Pugettia producta*; fecal pellets	(39.0-119.0) AW	44
Shrimp		
Muscle	56.3 DW	10
Exoskeleton	10.3 DW	10
Muscle	(12.0-36.0) FW; (44.0-130.0) DW; (610.0-1100.0) AW	22
Carapace	(4.0-39.0) FW; (17.0-110.0) DW; (190.0-710.0) AW	22
Whole	790.0 AW	22
Molts	76.0 DW	45
Stomatopod, *Squilla oratoria*; muscle	110.0 DW	41
Amphipods, *Themisto* spp.; whole	76.0 (72.0-81.0) DW	48
Fiddler crab, *Uca pugnax*; New Jersey		
Intermolt; contaminated site		
Soft tissue	165.8 DW	65
Carapace	46.6 DW	65
Intermolt; reference site		
Soft tissue	149.8 DW	65
Carapace	46.6 DW	65
Immediate postmolt; contaminated site		

(Continues)

Table 7.21: Cont'd

Organism	Concentration	Reference[a]
Soft tissue	133.3 DW	65
Exoskeleton	79.6 DW	65
Immediate postmolt; reference site		
Soft tissue	108.8 DW	65
Exoskeleton	25.6 DW	65
Crab, *Xantho incisus*; whole	26.0 FW	3
Zooplankton; mostly copepods; whole; N.W. Mediterranean Sea	268.0 (52.0-742.0) DW	46

Values are in mg Zn/kg fresh weight (FW), dry weight (DW), or ash weight (AW).
[a]1, Zafiropoulos and Grimanis, 1977; 2, Bohn and McElroy, 1976; 3, Bryan, 1968; 4, Sastry and Bhatt, 1965; 5, Ireland, 1973; 6, Stenner and Nickless, 1975; 7, Walker et al., 1975a; 8, Walker et al., 1975b; 9, Sims and Presley, 1976; 10, Horowitz and Presley, 1977; 11, Wolfe et al., 1973; 12, Greig, 1975; 13, Greig et al., 1977a; 14, Greig et al., 1977b; 15, Topping, 1973; 16, Bryan, 1966; 17, Wharfe and Van Den Broek, 1977; 18, Martin et al., 1977; 19, Martin and Knauer, 1973; 20, Vinogradova and Koual'skiy, 1962; 21, Tijoe et al., 1977; 22, Lowman et al., 1970; 23, Wright, 1976; 24, Stickney et al., 1975; 25, Stenner and Nickless, 1974b; 26, Hall et al., 1978; 27, Cutshall et al., 1977; 28, Robertson et al., 1972; 29, Robertson, 1967; 30, Greig and Wenzloff, 1977; 31, Bryan, 1967; 32, Bryan, 1964; 33, Van As et al., 1973; 34, Van As et al., 1975; 35, Leatherland et al., 1973; 36, Fowler, 1977; 37, Zingde et al., 1976; 38, Bender, 1975; 39, Lowman et al., 1966; 40, Ray et al., 1980; 41, Ishii et al., 1978; 42, Knauer, 1970; 43, DeClerck et al., 1979; 44, Boothe and Knauer, 1972; 45, Bertine and Goldberg, 1972; 46, Hardstedt-Romeo and Laumond, 1980; 47, Weeks and Moore, 1991; 48, Rainbow, 1989; 49, Anil and Wagh, 1988; 50, Ramelow et al., 1989; 51, Ridout et al., 1989; 52, White and Rainbow, 1985; 53, Sprague, 1986; 54, Ward et al., 1986; 55, Waiwood et al., 1987; 56, Khan et al., 1989; 57, Nugegoda and Rainbow, 1988b; 58, Shrestha and Morales, 1987; 59, Palmer et al., 2006; 60, Santos et al., 2006; 61, Frias-Espericueta et al., 2007; 62, Vazquez et al., 2001; 63, Turkmen et al., 2006; 64, Kadar et al., 2006; 65, Bergey and Weis, 2007; 66, Feng et al., 2006; 67, Silva et al., 2006; 68, MacFarlane et al., 2006; 69, Karouna-Renier et al., 2007; 70, Nunez-Nogueira et al., 2006a; 71, Zumholz et al., 2006; 72, Colaco et al., 2006; 73, Ruelas-Inzunza and Paez-Osuna, 2008; 74, Tu et al., 2008; 75, D'Adamo et al., 2008; 76, Zauke and Schmalenbach, 2006.

Zinc is toxic to the copepod *A. tonsa* when fed a zinc-contaminated algal diet (Bielmyer et al., 2006). In that study, diatoms (*T. pseudonana*) were held in seawater containing 0.0003 mg Zn/L for 7 days, reaching a zinc concentration of 0.55 mg Zn/kg DW diatom; diatoms, in turn, were fed to *Acartia* for 7 days, adversely affecting copepod reproduction. Authors note that the current proposed zinc criterion for marine life protection is 0.086 mg Zn/L, which exceeds the 0.0003 mg Zn/L used in this study to produce a toxic diet (Bielmyer et al., 2006). Crustaceans can accumulate zinc from both the medium and the diet (USEPA, 1987b). In uncontaminated waters, diet is the major source of zinc. Absorption from the stomach is efficient and occurs, in part, via the hepatopancreas. When a large pulse of zinc reaches the blood from the stomach, some is excreted, but much is resorbed and stored in hepatopancreas in a relatively nonlabile form. Ultimately, stored zinc is also excreted, although removal via the gut is unimportant (Bryan et al., 1986). Zinc absorption occurs initially at the gill surface, followed by transport on a saturable carrier in the cell wall, and is most efficient at low dissolved ambient zinc concentrations. Urinary excretion is an important body removal pathway, especially at high dissolved ambient concentrations when it can account for 70-80% of total zinc excretion (Bryan et al., 1986). Excess zinc in crustaceans is usually sequestered by metal-binding proteins and subsequently transported to storage or detoxification sites; soluble proteins and amino acids may contain 20-70% zinc (Sprague, 1986). Half-time persistence of zinc in the prawn, *P. elegans*, is about 17 days (Nugegoda and Rainbow, 1988b) and between 30 and 270 days for five other crustacean species (USNAS, 1989). Differences in half-time persistence are linked to differences in excretion rates of ionic zinc and complexed zinc. In general, ionic zinc is turned over faster than internally adsorbed zinc; molting accounts for 22-50% loss of the total zinc body burden in crabs (Bergey and Weis, 2007; Eisler, 1981).

Crustaceans can regulate body concentrations of zinc against fluctuations in intake, although the ways in which regulation is achieved vary among species (Bryan et al., 1986). Regulation of whole-body zinc to a constant level is reported for many crustaceans, including intertidal prawns (*Palaemon* spp.), sublittoral prawns (*P. montagui*), green crab (*C. maenas*), lobster (*Homarus gammarus*), isopods (*Asellus communis*), and others (Bryan et al., 1986; Devineau and Amiard-Triquet, 1985; Johnson and Jones, 1989; Lewis and McIntosh, 1986; Nugegoda and Rainbow, 1988b; Rainbow and White, 1989). The body zinc concentration at which zinc is regulated in crustaceans usually increases with temperature, salinity, molting frequency, bioavailability of the uncomplexed free metal ions, and chelators in the medium (Nugegoda and Rainbow, 1987, 1988a, 1989a,b). Lobsters (*H. gammarus*) are able to equilibrate over a 30-day period with seawater containing between 0.002 and 0.505 mg Zn/L. In response to a 100-fold increase in seawater concentrations (0.005-0.5 mg/L), zinc levels in whole body, blood, hepatopancreas, excretory organs, and gills approximately doubled, but changed little in muscle. Shell zinc concentration increased about 12 times, largely through adsorption (Bryan et al., 1986). Regulation of zinc in lobster blood is achieved by balancing uptake through the gills against urinary excretion and loss over the body surface, including the gills (Bryan et al., 1986). The sublittoral prawn, *P. montagui*, can regulate total body zinc concentration to a constant level (75.0 mg/kg DW) and dissolved zinc concentrations up to 0.022 mg/L, beyond which there is net accumulation of body zinc. This threshold of zinc regulation breakdown is lower than that in *P. elegans* (0.093 mg/L) and *Palaemonetes varians* (0.190 mg Zn/L) under the same physicochemical conditions (Nugegoda and Rainbow, 1987, 1988a,b, 1989a,b,c; Rainbow and White, 1989). The authors conclude that regulation of body zinc concentration is most efficient in decapods adapted to the fluctuating environments of littoral habitats, possibly as a result of changes in permeability of uptake surfaces in combination with improved zinc excretion systems. Metallothioneins are actively involved in zinc regulation during normal growth processes in the blue crab, *C. sapidus*, as judged by a decrease in zinc content in hemolymph and digestive gland during molting (Engel, 1987). Unlike decapod crustaceans, marine amphipods are unable to regulate body zinc concentrations; amphipod body burdens of zinc may reflect sediment total zinc levels and suggest that certain groups may be suitable bioindicators (Rainbow et al., 1989). Molting had no effect on body zinc concentration in four species of adult marine amphipods (Weeks and Moore, 1991), and this forces a re-examination of the role of cast exuviae in zinc transport.

In marsh crabs, *P. erythodactyla*, soft tissues reflected zinc concentrations in the ambient sediments; this pattern held for both males and females (MacFarlane et al., 2006). At higher salinities, males—but not females—showed significant increases in soft tissue zinc levels. In general, crabs with elevated zinc burdens had increased glutathione peroxidase activity; a similar case is made for lead and copper (MacFarlane et al., 2006).

Bioconcentration factors for whole amphipods, *P. walkeri*, from the Australian Antarctic Territory in 1999-2000 ranged from 670 to 2800 when seawater contained 0.0004-0.0009 mg Zn/L; BCFs decreased as seawater zinc concentrations increased (Clason et al., 2003). Increased sensitivity to zinc occurred when seawater concentrations were in the range 0.009-0.026 mg Zn/L (Clason et al., 2003). Bioconcentration factors for zinc and representative crustaceans range widely from 85 to 1500,000 (Eisler, 1980). Numerous factors account for variations in zinc content between and among crustaceans, of which industrial contamination is one of the most important (Bloom and Ayling, 1977; Eisler, 2000h). In the case of barnacles, *Balanus* sp., the high (>3300.0 mg/kg DW soft parts) zinc levels recorded were attributed to inorganic granules that contained up to 38% zinc and which accumulated in tissues surrounding the midgut (Eisler, 1980, 1981). The granules consist of phosphorus, zinc, potassium, sulfur, and chlorine, in that order (Thomas and Ritz, 1986). These insoluble, membrane-limited, form in response to high zinc levels in the ambient seawater within 12 days of exposure, and concentrate in specialized cells around the gut, the stratum perintestinale (Sprague, 1986; Thomas and Ritz, 1986; Walker et al., 1975a,b). Zinc granules in barnacles represent a detoxification mechanism for surplus zinc (Thomas and Ritz, 1986). Older barnacles have greater whole-body zinc accumulations than younger stages, and accumulations change seasonally (Anil and Wagh, 1988). Zinc detoxification mechanisms of the stored zinc in the barnacle *E. modestus* includes production of metabolically inert zinc phosphate granules (Rainbow and White, 1989); however, *E. modestus* transplanted from an area of high-ambient zinc (0.101 mg/L) to a low-ambient zinc (0.004 mg/L) environment lost zinc slowly (0.3% body burden daily) over an 11-week period. Whole-body zinc burden declined from 1554.0 to 125.0 mg/kg DW or about 4.1 mg/kg DW daily (Thomas and Ritz, 1986). In the crab, *C. maenas*, two heavy metal-binding proteins of about 27,000 and 11,500 MW, respectively, were isolated from the midgut gland (Rainbow and Scott, 1979). About 1.6% of the total soluble zinc in midgut gland was associated with the higher weight protein, and 6.3% with the lower MW protein. Both metal-binding proteins also bound to measurable amounts of copper and cadmium, but not lead (Rainbow and Scott, 1979).

Acquired zinc tolerance is reported in amphipods from zinc-contaminated sewage wastes and in fiddler crabs (*Uca* spp.) from a metals-contaminated area. *Uca* from zinc-contaminated areas were more resistant to zinc than were conspecifics from pristine areas, as judged by increased survival and lower tissue zinc burdens (Devi, 1987; Devi and Rao, 1989a,b). Zinc-tolerant populations of the amphipod *Gammarus duebeni* had 50% mortality in 14 days at 1.0 mg Zn/L and 10 ppt salinity and 33% dead in 14 days at 1.0 mg/L and 30 ppt salinity versus nonresistant populations showing reduced survival at more than 0.1 mg Zn/L in 7 days, and 100% mortality in 7 days at 10 ppt salinity and 84% dead in 7 days at 1.0 mg/l and 30 ppt salinity (Johnson and Jones, 1989).

Chemical species of zinc need to be considered in risk assessment. In seawater at pH 8.1, the dominant species of soluble zinc are zinc hydroxide (62.0%), free ion (17.0%), monochloride ion (6.4%), and zinc carbonate (5.8%) (USEPA, 1987b). In estuaries, the relative abundance of zinc species changes with increasing salinity. At low salinities, $ZnSO_4$ and $ZnCl^+$ predominate; at higher salinities, the highly toxic zinc aquo ion $[(Zn(H_2O)_6)^{2+}]$ predominates (Spear, 1981). But as salinity decreases, the concentration of free zinc ion increases and the concentration of zinc-chloro complexes decreases, resulting in increased bioavailability of the free metal ion and increased bioconcentration by resident organisms (Nugegoda and Rainbow, 1989b).

The biocidal properties of zinc, and presumably accumulation rates, were more pronounced for crustacean adults than for other groups of adult invertebrates tested (Eisler, 1981; Eisler and Hennekey, 1977). In general, toxicity was greatest to larvae (Ahsanullah and Arnott, 1978; Brown and Ahsanullah, 1971; Eisler, 1980; Kobayashi, 1979) at elevated temperatures (McLusky and Hagerman, 1987; Spear, 1981; Sprague, 1986) during extended exposures (USEPA, 1980e, 1987a), under conditions of starvation (USNAS, 1989; Verriopoulos and Moraitu-Apostolopoulu, 1989), at salinity extremes above and below the isosmotic point (McLusky and Hagerman, 1987), in summer (Eisler, 1980), in proximity to anthropogenic discharges (Eisler, 2000h), and at low-sediment particulate loadings (Memmert, 1987). Zinc was fatal to sensitive crustaceans under the following conditions: at 0.01 mg/L to the copepod, *Tisbe holothuriae* in the fourth generation of exposure (Verriopoulos and Hardouvelis, 1988); at 0.07 mg/L to *T. holothuriae* in a single generation (Verriopoulos and Hardouvelis, 1988); at 0.001-0.09 mg/L to hermit crab in a dose-dependent manner (delayed mortality) (Ajmalkhan et al., 1986); at <0.10 mg/L to embryos of the horseshoe crab, *L. polyphemus* (an arachnoid, not a true crustacean) (Botton et al., 1998); at 0.10 mg/L to larvae of the hermit crab, *Clibanarius olivaceous* (Ajmalkhan et al., 1986); at 0.13-0.38 mg Zn/L to larvae of the American lobster, *H. americanus* (Spear, 1981; USEPA, 1987a); at 0.2 mg/L to adults of a hermit crab, *P. longicarpus* (Eisler and Hennekey, 1977); and at 0.294 mg/L to a copepod, *A. tonsa* (USEPA, 1987a). Among amphipods, juveniles were more sensitive to zinc than were adults, and epibenthic species were more sensitive than infaunal tube-dwelling species (King et al., 2006). At 5 ppt salinity, zinc was more toxic to juveniles of the euryhaline crab *C. granulatus*, as judged by LC50 (96 h) values (7.1 mg Zn/L) than at 25 ppt salinity (11.4 mg/L) (Beltrame et al., 2008).

Acute toxicity bioassays of 96-h duration with adults of *T. japonicus* copepods—a relatively resistant species that is proposed for routine use in ecotoxicity testing in the western Pacific—showed that all survived 2.5 mg Zn/L, 10% died at 4.7 mg/L, and 50% died at 7.8 mg Zn/L (Lee et al., 2007).

Adverse sublethal effects of zinc insult to crustaceans include gill histopathology (Patel and Kaliwal, 1989); increased tissue total proteins, decreased glycogen, and decreased acid phosphatase activity in the crab *Portunus pelagicus* (Hilmy et al., 1988); retardation of limb regeneration of fiddler crab, *U. pugilator* (Weis, 1980); altered enzyme activity in the green crab, *C. maenas* (Elumalai et al., 2007); and elevated tissue residues in American lobster, *H. americanus* (Waiwood et al., 1987). For example, tissue zinc concentrations in *H. americanus* exposed for 4 days to 25.0 mg Zn/L were especially high in gill (2570.0 mg Zn/kg DW versus 126.0 at start), hepatopancreas (734.0 versus 135.0), and green gland (1032.0 versus 148.0) (Waiwood et al., 1987). After 7 days in uncontaminated seawater, tissue zinc residues remained elevated in gill (675.0 mg Zn/kg DW), hepatopancreas (603.0), green gland (286.0), and other tissues (Waiwood et al., 1987). Zinc concentrations in crustacean soft tissues usually range between 50.0 and 208.0 mg/kg DW, and exceed soft tissue zinc enzymatic requirements by factors of 1.4-6.0 (Depledge, 1989). Metabolic derangement is also associated with zinc insult. Thus, amylase and trypsin activities were disrupted in *A. salina* during immersion in sublethal concentrations of zinc. Trypsin activity was disturbed within 24 h by 5.0 mg Zn/L; amylase activity was significantly different from controls after 72 h in 5.0 mg Zn/L (Alayse-Danet et al., 1979). However, at very low sublethal zinc concentrations *P. montagui*, a coldwater shrimp, showed no significant accumulations during exposure to 0.065 mg Zn/L solutions for 14 days (Ray et al., 1980), although zinc concentrations increased in eggs and hepatopancreas during the first 6 days of depuration (Ray et al., 1980).

Seasonal fluctuations in zinc body burdens of copepods, barnacles, and decapod crustaceans are documented. Summer maxima in *Acartia* were associated with termination of active vertical transport (Pearcy and Osterberg, 1967; Zafiropoulos and Grimanis, 1977), and proximity to anthropogenic discharges (Pearcy and Osterberg, 1967). In barnacles, *B. balanoides*, and some decapods, seasonal variations in zinc content were correlated with river flow rates, tidal flow, and primary productivity (Bender, 1975; Ireland, 1974; Tennant and Forster, 1969). Among euphausiids, there were no seasonal variations in zinc content from collections made at 550 m; however, some variation was evident among conspecifics at 150-500 m depth (Pearcy and Osterberg, 1967).

Zinc content of euphausiids and pelagic shrimp was inversely correlated with body weight. Depending on species, zinc concentrations of smaller individuals were 1.2-4.1 times greater than larger conspecifics (Fowler, 1974). A similar pattern was observed in *Orchestoidea* sp.

(Bender, 1975). Other factors known to impact uptake, retention, and translocation of zinc by crustaceans include water temperature (Cross et al., 1969; Fowler et al., 1971; Jones, 1975; McKenney and Neff, 1979), salinity (Jones, 1975; McKenney and Neff, 1979), sediment composition, diet, and general health of the organism (Cross et al., 1969).

Zinc interacts with salts of cadmium, copper, zinc, nickel, and chromium, but results were not always predictable. Waterborne solutions of zinc-cadmium mixtures were usually additive in toxicity to copepods (Verriopoulos and Dimas, 1988) and amphipods (Ahsanullah et al., 1988). Zinc counteracted adverse effects of cadmium on limb regeneration and growth of the fiddler crab *U. pugilator* (Weis, 1980). Zinc concentrations in larvae of shrimp *Palaemon serratus*, within its threshold regulation range of 0.075-0.525 mg Zn/L, were not affected by the addition of 0.1 mg Cd/L (Devineau and Amiard Triquet, 1985). Zinc-copper mixtures are more than additive in toxicity to marine copepods (Sunda et al., 1990; Verriopoulos and Dimas, 1988); however, zinc-copper mixtures were less than additive in toxicity to the marine amphipod *A. compressa* (Ahsanullah et al., 1988). Zinc added to seawater depressed copper accumulations in barnacles *E. modestus*; however, simultaneous exposure to copper and zinc resulted in enhanced uptake of both metals (Elliott et al., 1985). Antifouling paints containing mixtures of zinc pyrithione and copper (as a replacement for tributyltin) were more than additive in toxicity to amphipods, *Elasmopus rapax*, during 96-h tests, and this should be considered when developing zinc water quality criteria (Bao et al., 2008). Lead-zinc mixtures were more than additive in toxicity to copepods (Verriopoulos and Dimas, 1988) and significantly delayed larval development of mud crab, *Rithropanopeus harrissii* (USEPA, 1987a). Nickel-zinc mixtures were additive in toxicity to marine copepods (Verriopoulos and Dimas, 1988). Chromium-zinc mixtures were more than additive in toxicity to *T. holothuriae*, a marine copepod. Zinc in combination with chromium was more toxic to copepods than were mixtures of zinc with copper, lead, nickel, or cadmium (Verriopoulos and Dimas, 1988).

Zinc exchange with the environment is reportedly slow. Zinc pools within adult aquatic organisms exchange only slowly, if at all, with zinc atoms in the organism's food or surrounding water (Renfro et al., 1975), although not all investigators agree. For example, among euphausiids, there is a major loss of ^{65}Zn due to isotopic exchange with seawater, with 96% of the radiozinc body burden eliminated over a period of 5 months (Fowler et al., 1971). In any event, diet is considered more important than the medium as a pathway for zinc accumulation among decapod crustaceans (Bender, 1975; Renfro et al., 1975). Among shrimp, *P. serratus*, different zinc excretion rates occur depending on the size of the animal. In general, ionic zinc is excreted first and then zinc complexes (Small et al., 1974). There is also the possibility that ionic particulate zinc and dissolved organic compounds may be excreted separately with subsequent combination in water to yield zinc complexes (Small et al., 1974). Surface-adsorbed zinc is turned over faster than internally adsorbed zinc in the dungeness crab, *C. magister* (Tennant and Forster, 1969). Zooplankters, including copepods and euphausiids, all eliminate accumulated zinc—presumably surface-adsorbed zinc—at rates of 1-5%/h (Kuenzler, 1969). In the well-studied euphausiid, *M. norvegica*, about half the whole-body zinc content is shed through molting when accumulated through the medium, but only 33% when accumulated by grazing on brine shrimp, *A. salina* (Fowler et al., 1971, 1972); molts constitute an important vehicle of zinc transfer into marine ecosystems (Fowler, 1977; Fowler et al., 1972). Finally, the participatory turnover time (i.e., the time required to cycle an element in a system through a given material in that system) for ionic zinc in the Ligurian Sea by adult euphausiids is lengthy. This interval is estimated at 498-1234 years depending upon available food and considering food chain as the only route of zinc accumulation (Small and Fowler, 1973). Fecal pellet deposition represented over 90% of zinc flux in this model (Fowler, 1977; Small et al., 1973). In general, molting results in a 33-50% loss of total zinc in marine crustaceans; molts, together with fecal pellets, constitute an important vehicle of zinc transfer in marine ecosystems (Eisler, 1981).

For marine life protection, the United States Environmental Protection Agency recommends that total recoverable zinc in seawater should average less than 0.058 mg/L and never exceed 0.17 mg/L; for acid-soluble zinc, these values are less than 0.086 and 0.095 mg/L, respectively (USEPA, 1980e, 1987a). Most crustaceans can tolerate up to 0.23 mg Zn/L without adverse effects (Sprague, 1986); however, as shown earlier, adverse effects occur at 0.01 mg Zn/L (Verriopoulos and Hardouvelis, 1988). Zinc is important in pH regulation of sperm of the horseshoe crab; zinc reduction to <0.0065 mg/L in semen adversely affected sperm pH and motility (Clapper et al., 1985a,b). To protect against zinc deficiency, whole crustaceans should contain at least 34.5 mg Zn/kg DW (White and Rainbow, 1985) and the medium should contain at least 0.0065 mg Zn/L (Clapper et al., 1985a,b). Zinc deficiency in natural aquatic ecosystems, however, has not yet been credibly documented.

7.60 Zirconium

Planktonic copepods from the Black Sea reportedly contain 70.0 mg Zr/kg AW (Vinogradova and Koual'skiy, 1962). Whole shrimp, *Crangon* sp., had 0.0096 mg Zr/kg DW (Zumholz et al., 2006).

7.61 Literature Cited

Aarkrog, A., 1971. Radioecological investigations of plutonium in an arctic marine environment. Health Phys. 20, 31–47.
Aarkrog, A., 1990. Environmental radiation and radiation releases. Int. J. Radiat. Biol. 57, 619–631.
Abbott, O.J., 1977. The toxicity of ammonium molybdate to marine invertebrates. Mar. Pollut. Bull. 8, 204–205.
Ahsanullah, M., 1982. Acute toxicity of chromium, mercury, molybdenum and nickel to the amphipod *Allorchestes compressa*. Aust. J. Mar. Freshw. Res. 33, 465–474.
Ahsanullah, M., Arnott, G.H., 1978. Acute toxicity of copper, cadmium, and zinc to larvae of the crab *Paragrapsus quadridentatus* (H. Milne Edwards), and implications for water quality criteria. Aust. J. Mar. Freshw. Res. 29, 1–8.

Ahsanullah, M., Williams, A.R., 1991. Sublethal effects and bioaccumulation of cadmium, chromium, copper and zinc in the marine amphipod *Allorchestes compressa*. Mar. Biol. 108, 59–65.

Ahsanullah, M., Mobley, M.C., Rankin, P., 1988. Individual and combined effects of zinc, cadmium and copper on the marine amphipod *Allorchestes compressa*. Aust. J. Mar. Freshw. Res. 39, 33–37.

Ajmalkhan, S., Rajendran, K., Natarajan, R., 1986. Effect of zinc on zoeal development of the estuarine hermit crab *Clibanarius olivaceus* (Henderson). Proc. Indian Acad. Sci. 95, 515–524.

Alayse-Danet, A.M., Charlou, J.L., Jezequel, M., Samain, J.F., 1979. Modele de detection rapide des effets subletaux des pollutants: Modification faux d'amylase et de trypsine d'*Artemia salina* contaminees par le cuivre ou le zinc. Mar. Biol. 51, 41–46.

Alliot, A., Frenet-Piron, M., 1990. Relationship between metals in sea-water and metal accumulation in shrimps. Mar. Pollut. Bull. 21, 30–33.

Alquezar, R., Markich, S.J., Twining, J.R., 2007. Uptake and loss of dissolved ^{109}Cd and ^{75}Se in estuarine macroinvertebrates. Chemosphere 67, 1201–1210.

Amiard-Triquet, C., Amiard, J.C., 1974. Contamination de chaines trophiques par le cobalt 60. Rev. Int. Oceanogr. Med. 33, 49–59.

Amiard-Triquet, C., Amiard, J.D., 1975. Etude experimentale du transfert du cobalt 60 dans une chaine trophique marine benthique. Helg. Wiss. Meeres. 27, 283–297.

Amiard-Triquet, C., Amiard, J.C., 1976. L'organotropisme du ^{60}Co chez *Scrobicularia plana* et *Carcinus maenas* en fonction du vecteur de contamination. Oikos 27, 122–126.

Ancellin, J., Bovard, P., Vilquin, A., 1967. New studies on experimental contamination of marine species by ruthenium-106. In: Actes du Congres Int. sur la Radioprotection du Milieu Soc. Franc. de Radioprot, Toulouse, France, January 14-16, 1967, pp. 213–234.

Anderson, E.P., Mackas, D.L., 1986. Lethal and sublethal effects of a molybdenum mine tailing on marine zooplankton: Mortality, respiration, feeding and swimming behavior in *Calanus marshallae*, *Metridia pacifica* and *Euphausia pacifica*. Mar. Environ. Res. 19, 131–155.

Anil, A.C., Wagh, A.B., 1988. Accumulation of copper and zinc by *Balanus amphitrite* in a tropical estuary. Mar. Pollut. Bull. 19, 177–180.

Anonymous, 1978. Selected pollution profiles: North Atlantic, North Sea, Baltic Sea, and Mediterranean Sea. Ambio 7, 75–78.

Antonini-Kane, J., Fowler, S.W., Heyraud, M., Keckes, S., Small, L.F., Beglia, A., 1972. Accumulation and loss of radionuclides by *Meganyctiphanes norvegica* M. Sars. Rapp. Comm. Int. Mer Medit. 21, 289–290.

Aono, A., Takeuchi, I., 2008. Effects of tributyltin at concentrations below ambient levels in seawater on *Caprella danilevskii* (crustacea: amphipoda: caprellidae). Mar. Pollut. Bull. 57, 515–523.

Argese, E., Bettiol, C., Ghirardini, A.V., Fasolo, M., Giurin, G., Ghetti, P.F., 1998. Comparison of in vitro submitochondrial particle and Microtox assays for determining the toxicity of organotin compounds. Environ. Toxicol. Chem. 17, 1005–1012.

Bao, V.W.W., Leung, K.M.Y., Kwok, K.W.H., Zhang, A.Q., Lui, G.C.S., 2008. Synergistic toxic effects of zinc pyrithione and copper to three marine species: Implications on setting appropriate water quality criteria. Mar. Pollut. Bull. 57, 616–623.

Barbaro, A., Francescon, A., Polo, B., Bilio, M., 1978. *Balanus amphitrite* (Cirripedia: Thoracica): A potential indicator of fluoride, copper, lead, chromium, and mercury in North Atlantic lagoons. Mar. Biol. 46, 247–257.

Barka, S., 2007. Insoluble detoxification of trace metals in a marine copepod *Tigriopus brevicornis* (Muller) exposed to copper, zinc, nickel, cadmium, silver and mercury. Ecotoxicology 16, 491–502.

Barnes, H., Stanbury, F.A., 1948. The toxic action of copper and mercury salts both separately and when mixed on the harpacticoid copepod, *Nitroca spinipes* (Boeck). J. Exp. Biol. 25, 270–275.

Beasley, T.M., Held, E.E., 1971. Silver-108m in biota and sediments at Bikini and Eniwetok Atolls. Nature 230 (5294), 450–451.

Bechmann, R.K., 1994. Use of life tables and LC50 tests to evaluate chronic and acute toxicity effects of copper on the marine amphipod *Tisbe furcata* (Baird). Environ. Toxicol. Chem. 13, 1509–1517.

Beltrame, M.O., De Marco, S.G., Marcovecchio, J.E., 2008. Cadmium and zinc in Mar Chiquita coastal lagoon (Argentina): Salinity effects on lethal toxicity in juveniles of the burrowing crab *Chasmagnathus granulatus*. Arch. Environ. Contam. Toxicol. 55, 78–85.

Benayoun, G., Fowler, S.W., Oregioni, B., 1975. Flux of cadmium through euphausiids. Mar. Biol. 27, 205–212.

Bender, J.A., 1975. Trace metal levels in beach dipterans and amphipods. Bull. Environ. Contam. Toxicol. 14, 187–192.

Bengtsson, B.E., 1977. Accumulation of cadmium in some aquatic animals from the Baltic Sea. Ambio Spec. Rep. 5, 69–73.

Benijts-Claus, C., Benijts, F., 1975. The effect of low lead and zinc concentrations on the larval development of the mudcrab, *Rithropanopeus harrisii* Gould. In: Koeman, J.H., Strik, J.J.T.W.A. (Eds.), Sublethal Effects of Toxic Chemicals on Aquatic Animals. Elsevier, Amsterdam, pp. 43–52.

Bergey, L.L., Weis, J.S., 2007. Molting as a mechanism of depuration of metals in the fiddler crab, *Uca pugnax*. Mar. Environ. Res. 64, 556–562.

Bernard, F.J., Lane, C.E., 1961. Absorption and excretion of copper ion during settlement and metamorphosis of the barnacle, *Balanus amphitrite nireus*. Biol. Bull. 121, 438–448.

Bernard, F.J., Lane, C.E., 1963. Effects of copper ion in oxygen uptake to planktonic cyprids of the barnacle, *Balanus amphitrite nireus*. Proc. Soc. Exp. Biol. Med. 113, 418–420.

Bernhard, M., Zattera, A., 1975. Major pollutants in the marine environment. In: Pearson, E.A., Frangipane, E.D. (Eds.), Marine Pollution and Marine Waste Disposal. Pergamon, Elmsford, NY, pp. 195–300.

Berry, W.J., Cantwell, M.G., Edwards, P.A., Serbst, J.R., Hansen, D.J., 1999. Predicting toxicity of sediments spiked with silver. Environ. Toxicol. Chem. 18, 40–48.

Bertine, K.K., Goldberg, E.D., 1972. Trace elements in clams, mussels, and shrimp. Limnol. Oceanol. 17, 877–884.

Bianchini, A., Playle, R.C., Wood, C.M., Walsh, P.J., 2007. Short-term silver accumulation in tissues of three marine invertebrates: Shrimp *Penaeus duorarum*, sea hare *Aplysia californica*, and sea urchin *Diadema antillarum*. Aquat. Toxicol. 84, 182–189.

Bielmyer, G.K., Grosell, M., Brix, K.V., 2006. Toxicity of silver, zinc, copper, and nickel to the copepod *Acartia tonsa* exposed via a phytoplankton diet. Environ. Sci. Technol. 40, 2063–2068.

Bjerragaard, P., 1982. Accumulation of cadmium and selenium and their mutual interaction in the shore crab *Carcinus maenus* (L.). Aquat. Toxicol. 2, 113–125.

Bloom, H., Ayling, G.M., 1977. Heavy metals in the Derwent Estuary. Environ. Geol. 2, 3–22.

Blunden, S.J., Chapman, A.H., 1982. The environmental degradation of organotin compounds: A review. Environ. Technol. Lett. 3, 267–272.

Blunden, S.J., Chapman, A.H., 1986. Organotin compounds in the environment. In: Craig, P.J. (Ed.), Organometallic Compounds in the Environment. Longman Group, London, pp. 111–159.

Blunden, S.J., Cusack, P.A., Hill, R., 1985. The Industrial Uses of Tin Chemicals. Royal Society of Chemistry, London, 337 pp.

Blust, R., Fontaine, A., Decleir, W., 1991. Effect of hydrogen ion and inorganic complexing on the uptake of copper by the brine shrimp *Artemia franciscana*. Mar. Ecol. Prog. Ser. 76, 273–282.

Bohn, A., 1975. Arsenic in marine organisms from West Greenland. Mar. Pollut. Bull. 6, 87–89.

Bohn, A., McElroy, R.O., 1976. Trace metals (As, Cd, Cu, Fe, and Zn) in Arctic cod, *Boreogadus saida*, and selected zooplankton from Strathcona Sound, Northern Baffin Island. J. Fish. Res. Board Can. 33, 2836–2840.

Boney, A.D., Corner, E.D.S., Sparrow, B.W.P., 1959. The effects of various poisons on the growth and viability of sporelings of the red alga *Plumaria elegans* (Bonnem.). Biochem. Pharmacol. 2, 37–49.

Boon, D.D., 1973. Iron, zinc, magnesium, and copper concentrations in body meat of the blue crab, *Callinectes sapidus*. Chesapeake Sci. 14, 143–144.

Boothe, P.N., Knauer, G.A., 1972. The possible importance of fecal material in the biological amplification of trace and heavy metals. Limnol. Oceanol. 17, 270–274.

Botton, M.L., Johnson, K., Helleby, L., 1998. Effects of copper and zinc on embryos and larvae of the horseshoe crab, *Limulus polyphemus*. Arch. Environ. Contam. Toxicol. 35, 25–32.

Brown, B., Ahsanullah, M., 1971. Effects of heavy metals on mortality and growth. Mar. Pollut. Bull. 2, 182–187.

Bryan, G.W., 1961. The accumulation of radioactive caesium in crabs. J. Mar. Biol. Assoc. UK 41, 551–575.

Bryan, G.W., 1963. The accumulation of ^{137}Cs by brackish water invertebrates and its relation to the regulation of potassium and sodium. J. Mar. Biol. Assoc. UK 43, 541–565.

Bryan, G.W., 1964. Zinc regulation in the lobster *Homarus americanus*. I. Tissue zinc and copper concentrations. J. Mar. Biol. Assoc. UK 44, 549–563.

Bryan, G.W., 1965. Ionic regulation in the squat lobster *Galathea squamifera*, with special reference to the relationship between potassium metabolism and the accumulation of radioactive caesium. J. Mar. Biol. Assoc. UK 45, 97–113.

Bryan, G.W., 1966. The metabolism of Zn and ^{65}Zn in crabs, lobsters and freshwater crayfish. In: Radioecological Concentration Processes. Pergamon, Elmsford, NY, pp. 1005–1016.

Bryan, G.W., 1967. Zinc concentrations of fast and slow contracting muscles in the lobster. Nature 213, 1043–1044.

Bryan, G.W., 1968. Concentrations of zinc and copper in the tissues of decapod crustaceans. J. Mar. Biol. Assoc. UK 48, 303–321.

Bryan, G.W., Ward, E., 1965. The absorption and loss of radioactive and non-radioactive manganese by the lobster, *Homarus vulgaris*. J. Mar. Biol. Assoc. UK 45, 65–95.

Bryan, G.W., Hummerstone, L.W., Ward, E., 1986. Zinc regulation in the lobster *Homarus gammarus*: Importance of different pathways and excretion. J. Mar. Biol. Assoc. UK. 66, 175–199.

Bryant, V., Newbery, D.M., McLuskey, D.S., Campbell, R., 1985. Effect of temperature and salinity on the toxicity of arsenic to three estuarine invertebrates (*Corophium volutator*, *Macoma balthica*, *Tubifex costatus*). Mar. Ecol. Prog. Ser. 24, 129–137.

Burger, J., Cooper, K., Saliva, J., Gochfeld, D., Lipsky, D., Gochfeld, M., 1992. Mercury bioaccumulation in organisms from three Puerto Rican estuaries. Environ. Monit. Assess. 22, 181–197.

Burger, J., Stern, A.H., Gochfeld, M., 2005. Mercury in commercial fish: Optimizing individual choices to reduce risk. Environ. Health Perspect. 113, 266–270.

Burger, J., Fossi, C., McClellan-Green, P., Orlando, E.F., 2007a. Methodologies, bioindicators, and biomarkers for assessing gender-related differences in wildlife exposed to environmental chemicals. Environ. Res. 104, 135–152.

Burger, J., Gochfeld, M., Jeitner, C., Burke, S., Stamm, T., Snigaroff, R., et al., 2007b. Mercury levels and potential risk from subsistence foods from the Aleutians. Sci. Total Environ. 384, 93–105.

Canli, M., Furness, R.W., 1993. Toxicity of heavy metals dissolved in sea water and influences of sex and size on metal accumulation and tissue distribution in the Norway lobster *Nephrops norvegicus*. Mar. Environ. Res. 36, 217–236.

Cappon, C.J., Smith, J.C., 1982. Chemical form and distribution of selenium in edible seafood. J. Anal. Toxicol. 6, 10–21.

Cardwell, R.D., Sheldon, A.W., 1986. A risk assessment concerning the fate and effects of tributyltins in the aquatic environment. In: Maton, G.L. (Ed.), Proceedings Oceans 86 Conference, Washington, DC, September 23-25, 1986, vol. 4, Organotin Symposium. Available from Marine Technology Society, 2000 Florida Avenue, New York, Washington, DC, pp. 1117–1129.

Carr, R.S., McCullough, W.L., Neff, J.M., 1982. Bioavailability of chromium from a used chrome lignosulphonate drilling mud to five species of marine invertebrates. Mar. Environ. Res. 6, 189–203.

Chakravarti, D., Eisler, R., 1961. Strontium-90 and gross beta activity in the fat and nonfat fractions of the liver of the coconut crab (*Birgus latro*) collected at Rongelap Atoll during March 1958. Pacific Sci. 15, 155–159.

Champ, M.A., 1986. Organotin symposium: Introduction and review. In: Maton, G.L. (Ed.), Proceedings Oceans 86 Conference, Washingon, DC, September 23-25, 1986, vol. 4, Organotin Symposium. Available from Marine Technology Society, 2000 Florida Avenue, New York, Washington, DC, pp. 1093–1100.

Chau, Y.K., Maguire, R.J., Wong, P.T.S., Glen, B.A., Bengert, G.A., Tkacz, R.J., 1984. Occurrence of methyltin and butyltin species in environmental samples in Ontario. Natl. Water Res. Inst., Rep. 8401. Available from Department of Environment, Canada Centre for Inland Waters, Burlington, Ontario, Canada L7R 4A6, pp. 1–25.

Cheevaparanapivat, V., Menasveta, P., 1979. Total and organic mercury in marine fish of the upper Gulf of Thailand. Bull. Environ. Contam. Toxicol. 23, 291–299.

Chou, C.L., Uthe, J.F., Zook, E.G., 1978. Polarographic studies on the nature of cadmium in scallop, oyster, and lobster. J. Fish. Res. Board Can. 35, 409–413.

Clapper, D.L., Davis, J.A., Lamothe, P.J., Patton, C., Epel, D., 1985a. Involvement of zinc in the regulation of ph, motility, and acrosome reactions in sea urchin sperm. J. Cell Biol. 100, 1817–1824.

Clapper, D.L., Lamothe, P.J., Davis, J.A., Epel, D., 1985b. Sperm motility in the horseshoe crab. V: Zinc removal mediates chelator initiation of motility. J. Exp. Zool. 236, 83–91.

Clark, J.R., Patrick Jr., J.M., Moore, J.C., Lores, E.M., 1987. Waterborne and sediment-source toxicities of six organic chemicals to grass shrimp (*Palaemonetes pugio*) and amphioxus (*Branchiostoma caribaeum*). Arch. Environ. Contam. Toxicol. 16, 401–407.

Clason, B., Duquesne, S., Liess, M., Schulz, R., Zauke, G.P., 2003. Bioaccumulation of trace metals in the Antarctic amphipod *Paramoera walkeri* (Stebbing 1906): Comparison of two-compartment and hyperbolic toxicokinetic models. Aquat. Toxicol. 65, 117–140.

Coelho, J.P., Reis, A.T., Ventura, S., Pereira, M.E., Duarte, A.C., Pardal, M.A., 2008. Pattern and pathways for mercury lifespan bioaccumulation in *Carcinus maenas*. Mar. Pollut. Bull. 56, 1104–1110.

Colaco, A., Bustamante, P., Fouquet, Y., Sarradin, P.M., Serrao-Santos, R., 2006. Bioaccumulation of Hg, Cu, and Zn in the Azores triple junction hydrothermal vent fields food web. Chemosphere 65, 2260–2267.

Conklin, P.J., Drysdale, D., Doughti, D.G., Rao, K.R., Kakareka, J.P., Gilbert, T.R., et al., 1983. Comparative toxicity of drilling muds: Role of chromium and petroleum hydrocarbons. Mar. Environ. Res. 10, 105–125.

Connell, D.B., Sanders, J.G., Riedel, G.F., Abbe, G.R., 1991. Pathways of silver uptake and trophic transfer in estuarine organisms. Environ. Sci. Technol. 25, 921–924.

Corner, E.D.S., 1959. The poisoning of *Maia squinado* (Herbst) by certain compounds of mercury. Biochem. Pharmacol. 2, 121–132.

Corner, E.D.S., Rigler, F.H., 1958. The modes of action of toxic agents. III. Mercuric chloride and n-amylmercuric chloride on crustaceans. J. Mar. Biol. Assoc. UK 37, 85–96.

Correa, J.D., da Silva, M.R., da Silva, A.C.B., de Lima, S.M.A., Malm, O., Allodi, S., 2005. Tissue distribution, subcellular localization and endocrine disruption patterns induced by Cr and Mn in the crab *Ucides cordatus*. Aquat. Toxicol. 73, 139–154.

Correia, A.D., Costa, M.H., Ryan, K.P., Nott, J.A., 2002. Studies on biomarkers of copper exposure and toxicity in the marine amphipod *Gammarus locusta* (Crustacea): I. Copper-containing granules within the midgut gland. J. Mar. Biol. Assoc. UK 82, 827–834.

Costa, L.G., 1985. Inhibition of gamma: [^{3}H] aminobutyric acid uptake by organotin compounds *in vitro*. Toxicol. Appl. Pharmacol. 79, 471–479.

Costa, M.R.M., da Fonseca, M.I.C., 1967. Tero de arsenico en mariscos. Rev. Port. Farm. 17 (1), 1–19.

Cross, F.A., Dean, J.M., Osterberg, C.L., 1969. The effect of temperature, sediment, and feeding on the behavior of four radionuclides in a marine benthic amphipod. In: Proceedings of the Second National Symposium on Radioecology. U.S. Atomic Energy Commission, Conf. 67053, pp. 450–461.

Culkin, F., Riley, J.P., 1958. The occurrence of gallium in marine organisms. J. Mar. Biol. Assoc. UK 37, 607–615.

Cumont, G., Gilles, G., Bernard, F., Briand, M.B., Stephan, G., Ramonda, G., et al., 1975. Bilan de la contamination des poissons de mer par le mercure a l'occasion d'un controle portant sur 3 annees. Ann. Hyg. L. Fr. Med. et Nut. 11 (1), 17–25.

Custer, T.W., Mitchell, C.A., 1993. Trace elements and organochlorines in the shoalgrass community of the lower Laguna Madre, Texas. Environ. Monit. Assess. 25, 235–246.

Cutshall, N.H., Naidu, J.R., Pearcy, W.G., 1977. Zinc and cadmium in the Pacific hake, *Merluccius productus* off the western U.S. coast. Mar. Biol. 44, 195–202.

D'Adamo, R., Di Stasio, M., Fabbrocini, A., Petitto, F., Roselli, L., Volpe, M.G., 2008. Migratory crustaceans as biomonitors of metal pollution in their nursery areas. The Lesina lagoon (SE Italy) as a case study. Environ. Monit. Assess. 143, 15–24.

D'Agostino, A., Finney, C., 1974. The effect of copper and cadmium on the development of *Tigriopus japonicus*. In: Vernberg, F.J., Vernberg, W.B. (Eds.), Pollution and Physiology of Marine Organisms. Academic Press, New York, pp. 445–463.

Dahab, O.A., Khalil, A.N., Halim, Y., 1990. Chromium fluxes through Mex Bay inshore waters. Mar. Pollut. Bull. 21, 68–73.

Davies, A.G., Smith, P.J., 1982. Tin. In: Wilkinson, G. (Ed.), Comprehensive Organometallic Chemistry. Pergamon, New York, pp. 519–627.

DeClerck, R., Vanderstappen, R., Vyncke, W., 1974. Mercury content of fish and shrimps caught off the Belgian coast. Ocean Manage. 2, 117–126.

DeClerck, R., Vanderstappen, R., Vyncke, W., Van Hoeyweghen, P., 1979. La teneur en metaux lourds dans les organismes marins provenant de la capture accessoire de la peche coteiere belge. Rev. de l'Agric. 3 (32), 793–801.

De Gieter, M., Leermakers, M., Van Ryssen, R., Noyen, J., Goeyens, L., Baeyens, W., 2002. Total and toxic arsenic levels in North Sea fish. Arch. Environ. Contam. Toxicol. 43, 406–417.

Denton, G.R.W., Concepcion, L.C., Wood, H.R., Morrison, R.J., 2006. Trace metals in marine organisms from four harbours in Guam. Mar. Pollut. Bull. 52, 1784–1804.

Depledge, M.H., 1989. Re-evaluation of metabolic requirements for copper and zinc in decapod crustaceans. Mar. Environ. Res. 27, 115–126.

Depledge, M.H., Forbes, T.L., Forkes, V.E., 1993. Evaluation of cadmium, copper, zinc, and iron concentrations and tissue distributions in the benthic crab, *Dorippe granulata* (De Haan, 1841) from Tolo Harbour, Hong Kong. Environ. Pollut. 81, 15–19.

Devescovi, M., Lucu, C., 1995. Seasonal changes of the copper level in shore crabs *Carcinus mediterraneus*. Mar. Ecol. Prog. Ser. 120, 169–174.

Devi, V.U., 1987. Heavy metal toxicity to fiddler crabs, *Uca annulipes* Latreille and *Uca triangularis* (Milne Edwards): Tolerance to copper, mercury, cadmium, and zinc. Bull. Environ. Contam. Toxicol. 39, 1010–1027.

Devi, V.U., Rao, Y.P., 1989a. Heavy metal toxicity to fiddler crabs, *Uca annulipes* Latreille and *Uca triangularis* (Milne Edwards): Respiration on exposure to copper, mercury, cadmium, and zinc. Bull. Environ. Contam. Toxicol. 43, 165–172.

Devi, V.U., Rao, Y.P., 1989b. Zinc accumulation in fiddler crabs *Uca annulipes* Latreille and *Uca triangularis* (Milne Edwards). Ecotoxicol. Environ. Safety 18, 129–140.

Devineau, J., Amiard Triquet, C., 1985. Patterns of bioaccumulation of an essential trace element (zinc) and a pollutant metal (cadmium) in larvae of the prawn *Palaemon serratus*. Mar. Biol. 86, 139–143.

Dietz, R., Riget, F., Johansen, P., 1996. Lead, cadmium, mercury and selenium in Greenland marine animals. Sci. Total Environ. 186, 67–93.

Doi, R., Ui, J., 1975. The distribution of mercury in fish and its form of occurrence. In: Krenkel, P.A. (Ed.), Heavy Metals in the Aquatic Environment. Pergamon Press, New York, pp. 197–221.

Doughti, D.G., Conklin, P.J., Rao, K.R., 1983. Cuticular lesions induced in grass shrimp exposed to hexavalent chromium. J. Invert. Pathol. 42, 249–258.

Drifmeyer, J.E., Odum, W.E., 1975. Lead, zinc and manganese in dredge-spoil pond ecosystems. Environ. Conserv. 2, 39–43.

Duke, T.W., Baptist, J.P., Hoss, D.E., 1966. Bioaccumulation of radioactive gold used as a sediment tracer in the estuarine environment. US Fish Wildl. Serv. Fish. Bull. 65, 427–436.

Duncan, J., 1980. The toxicology of molluscicides. The organotins. Pharmacol. Ther. 10, 407–429.

Edmonds, J.S., Francesconi, K.A., Stick, R.V., 1993. Arsenic compounds from marine organisms. Nat. Prod. Rep. 10, 421–428.

Eftekhari, M., 1975. Teneur en mercure de quelques crevettes du Golfe Persique. Science Peche. Bull. Inst. Peches Marit. 250, 9–10.

Eganhouse, R.P., Young, D.R., 1978. Total and organic mercury in benthic organisms near a major submarine wastewater outfall system. Bull. Environ. Contam. Toxicol. 19, 758–766.

Eisler, R., 1980. Accumulation of zinc by marine biota. In: Nriagu, J.O. (Ed.), Zinc in the Environment. Part II. Health Effects. John Wiley, New York, pp. 259–351.

Eisler, R., 1981. Trace Metal Concentrations in Marine Organisms. Pergamon, Elmsford, New York, 687 pp.

Eisler, R., 1995. Ecological and toxicological aspects of the partial meltdown of the Chernobyl nuclear plant reactor. In: Hoffman, D.J., Rattner, B.A., Burton Jr., G.A., Cairns Jr., A.J. (Eds.), Handbook of Ecotoxicology. Lewis Publishers, Boca Raton, FL, pp. 549–564.

Eisler, R., 2000a. Arsenic. In: Handbook of Chemical Risk Assessment, vol. 3. Lewis Publishers, Boca Raton, FL, pp. 1501–1566.

Eisler, R., 2000b. Boron. In: Handbook of Chemical Risk Assessment, vol. 3. Lewis Publishers, Boca Raton, FL, pp. 1567–1612.

Eisler, R., 2000c. Selenium. In: Handbook of Chemical Risk Assessment, vol. 3. Lewis Publishers, Boca Raton, FL, pp. 1649–1705.

Eisler, R., 2000d. Cadmium. In: Handbook of Chemical Risk Assessment, vol. 1. Lewis Publishers, Boca Raton, FL, pp. 1–44.

Eisler, R., 2000e. Chromium. In: Handbook of Chemical Risk Assessment, vol. 1. Lewis Publishers, Boca Raton, FL, pp. 45–92.

Eisler, R., 2000f. Copper. In: Handbook of Chemical Risk Assessment, vol. 1. Lewis Publishers, Boca Raton, FL, pp. 93–200.

Eisler, R., 2000g. Tin. In: Handbook of Chemical Risk Assessment, vol. 1. Lewis Publishers, Boca Raton, FL, pp. 551–603.

Eisler, R., 2000h. Zinc. In: Handbook of Chemical Risk Assessment, vol. 1. Lewis Publishers, Boca Raton, FL, pp. 605–714.

Eisler, R., 2003. The Chernobyl nuclear power plant reactor accident: Ecotoxicological update. In: Hoffman, D.J., Rattner, B.A., Burton Jr., G.A., Cairns Jr., J. (Eds.), Handbook of Ecotoxicology, second ed. Lewis Publishers, Boca Raton, FL, pp. 703–736.

Eisler, R., 2006. Mercury Hazards to Living Organisms. CRC Press, Boca, Raton, FL, 312 pp.

Eisler, R., Hennekey, R.J., 1977. Acute toxicities of Cd^{2+}, Cr^{+6}, Hg^{2+}, Ni^{2+}, and Zn^{2+} to estuarine macrofauna. Arch. Environ. Contam. Toxicol. 6, 315–323.

Eisler, R., Zaroogian, G., Hennekey, R.J., 1972. Cadmium uptake by marine organisms. J. Fish. Res. Board Can. 29, 1367–1369.

Elferink, J.G.R., Deierkauf, M., Stevenick, J.V., 1986. Toxicity of organotin compounds for polymorphonuclear leukocytes. The effect on phagocytosis and exocytosis. Biochem. Pharmacol. 35, 3727–3732.

Elliott, N.G., Ritz, D.A., Swain, R., 1985. Interaction between copper and zinc accumulation in the barnacle *Elminius modestus* Darwin. Mar. Environ. Res. 17, 13–17.

Elumalai, M., Antunes, C., Guilhermino, L., 2007. Enzymatic biomarkers in the crab *Carcinus maenas* from the Minho River estuary (NW Portugal) exposed to zinc and mercury. Chemosphere 66, 1249–1255.

Engel, D.W., 1987. Metal regulation, and molting in the blue crab, *Callinectes sapidus*: Copper, zinc, and metallothionein. Biol. Bull. 172, 69–82.

Ericksson, S.P., Weeks, J.M., 1994. Effects of copper and hypoxia on two populations of the benthic amphipod, *Corophium volutator* (Pallas). Aquat. Toxicol. 29, 73–81.

Erk, M., Muyssen, B.T.A., Ghekiere, A., Janssen, C.R., 2008. Metallothioneins and cytosolic metals in *Neomysis integer* exposed to cadmium at different salinities. Mar. Environ. Res. 65, 437–444.

Establier, R., 1977. Estudio de la contaminacion marine por metals pesados y sus effectos biologicos. Inf. Tec. Inst. Invest. Pesq. 47, 1–36.

Eustace, I.J., 1974. Zinc, cadmium, copper, and manganese in species of finfish and shellfish caught in the Derwent Estuary, Tasmania. Aust. J. Mar. Freshw. Res. 25, 209–220.

Evans, D.W., Laughlin, R.B. Jr. 1984. Accumulation of bis(tributyltin) oxide by the mud crab. *Rithropanopeus harrisii*. Chemosphere 13, 213–219.

Eversole, A.G., 1978. Marking clams with rubidium. Proc. Nat. Shellfish. Assoc. 68, 78.

Fang, J., Wang, K.X., Tang, J.L., Wang, Y.M., Ren, S.J., Wu, H.Y., et al., 2004. Copper, lead, zinc, cadmium, mercury, and arsenic in marine products of commerce from Zhejiang coastal area, China, May, 1998. Bull. Environ. Contam. Toxicol. 73, 583–590.

Feng, T.H., Hwang, J.S., Hsiao, S.H., Chen, H.Y., 2006. Trace metals in seawater and copepods in the ocean outfall area of the northern Taiwan coast. Mar. Environ. Res. 61, 224–243.

Fleeger, J.W., Gust, K.A., Marlborough, S.J., Tita, G., 2007. Mixtures of metals and polynuclear aromatic hydrocarbons elicit complex nonadditive toxicological interactions in meiobenthic copepods. Environ. Toxicol. Chem. 26, 1677–1685.

Fowler, 1974. The effect of organism size on the content of certain trace metals in marine zooplankton. Rapp. Comm. Int. Mer Medit. 22 (9), 145–146.

Fowler, S.W., 1977. Trace elements in zooplankton particulate products. Nature 269, 51–53.

Fowler, S.W., Benayoun, G., 1976a. Selenium kinetics in marine zooplankton. Mar. Sci. Comm. 2, 43–67.

Fowler, S.W., Benayoun, G., 1976b. Influence of environmental factors on selenium flux in two marine invertebrates. Mar. Biol. 37, 59–68.

Fowler, S.W., Benayoun, G., 1976c. Accumulation and distribution of selenium in mussel and shrimp tissues. Bull. Environ. Contam. Toxicol. 16, 339–346.

Fowler, S.W., Unlu, M.Y., 1978. Factors affecting bioaccumulation and elimination of arsenic in the shrimp *Lysmata seticaudata*. Chemosphere 9, 711–720.

Fowler, S.W., Small, L.F., Dean, J.M., 1971. Experimental studies on elimination of zinc-65, cesium-137, and cerium-144 by euphausiids. Mar. Biol. 8, 224–231.

Fowler, S.W., Small, L.F., La Rosa, J., 1972. The role of euphausiid molts in the transport of radionuclides in the sea. Rapp. Comm. Int. Mer Medit. 21 (6), 291–292.

Fowler, S.W., Heyraud, M., Small, L.F., Benayoun, G., 1973. Flux of ^{141}Ce through euphausiid crustacean. Mar. Biol. 21, 317–325.

Fowler, S.W., Heyraud, M., La Rosa, J., 1978. Factors affecting methyl and inorganic mercury dynamics in mussels and shrimp. Mar. Biol. 46, 267–276.

Fowler, S.W., Buat-Menard, P., Yokoyama, Y., Ballestra, S., Holm, E., Van Nguyen, H., 1987. Rapid removal of Chernobyl fallout from Mediterranean surface waters by biological activity. Nature 329, 56–58.

Francesconi, K.A., Micks, P., Stockton, R.A., Irgolic, K.J., 1985. Quantitative determination of arsenobetaine, the major water-soluble arsenical in three species of crab, using high pressure liquid chromatography and an inductively coupled argon plasma emission spectrometer as the arsenic-specific detector. Chemosphere 14, 1443–1453.

Frias-Espericueta, M.G., Izaguirre-Fierro, G., Valenzuela-Quinonez, F., Osuna-Lopez, J.L., Voltolina, D., Lopez-Lopez, G., et al., 2007. Metal content of California blue shrimp *Litopenaeus stylirostris* (Stimpson). Bull. Environ. Contam. Toxicol. 79, 214–217.

Fujita, T., Yamamoto, T., Yamazi, I., Shigematsu, T., 1969. The contents of ash, iron and manganese in marine plankton. J. Chem. Soc. Jpn. 90, 680–686.

Fukai, R., 1965. Analysis of trace amounts of chromium in marine organisms by the isotope dilution of Cr-51. In: Radiochemical Methods of Analysis. IAEA, Vienna, Austria, pp. 335–351.

Fukai, R., Broquet, D., 1965. Distribution of chromium in marine organisms. Bull. Inst. Oceanol. 65 (1336), 1–19.

Fukai, R., Meinke, W.W., 1959. Some activation analyses of six trace elements in marine biological ashes. Nature 184, 815–816.

Fukai, R., Meinke, W.W., 1962. Activation analyses of vanadium, arsenic, molybdenum, tungsten, rhenium, and gold in marine organisms. Limnol. Oceanol. 7, 186–200.

Fukai, R., Suzuki, H., Watankae, K., 1962. Strontium-90 in marine organisms during the period 1957-1961. Bull. Inst. Oceanogr. Monaco 1251, 1–16.

Gardner, W.S., Windom, H.L., Stephens, J.A., Taylor, F.A., Stickney, R.R., 1975. Concentrations of total mercury in fish and other coastal organisms: Implications to mercury cycling. In: Howell, F.G., Gentry, J.B., Smith, M.H. (Eds.), Mineral Cycling in Southeastern Ecosystem, pp. 268–278. U.S. Ener. Res. Dev. Admin. CONF-740513.

Gentile, S.M., Gentile, J.H., Walker, J., Heltshe, J.F., 1982. Chronic effects of cadmium on two species of mysid shrimp: *Mysidopsis bahia* and *Mysidopsis bigelowi*. Hydrobiologia 93, 195–204.

Gibbs, P.E., Bryan, G.W., 1972. A study of strontium, magnesium, and calcium in the environment and exoskeleton of decapod crustaceans with special reference to *Uca burgersi* on Barbuda, West Indies. J. Exp. Mar. Biol. Ecol. 9, 97–110.

Gibson, C.I., Thatcher, T.O., Apts, C.W., 1976. Some effects of temperature, chlorine, and copper on the survival and growth of the coon stripe shrimp. In: Esch, G.W., McFarlane, R.W. (Eds.), Thermal Ecology II, pp. 88–92 U.S. Ener. Res. Dev. Admin. CONF-750425.

Glickstein, N., 1978. Acute toxicity of mercury and selenium to *Crassostrea gigas* embryos and *Cancer magister* larvae. Mar. Biol. 49, 113–117.

Gordon, C.M., Carr, R.A., Larson, R.E., 1970. The influence of environmental factors on the sodium and manganese content of barnacle shells. Limnol. Oceanol. 15, 461–466.

Gould, E., Greig, R.A., 1983. Short-term low-salinity response in lead-exposed lobsters, *Homarus americanus* (Milne Edwards). J. Exp. Mar. Biol. Ecol. 69, 283–295.

Green Jr., F.A., Anderson, J.W., Petrocelli, S.R., Presley, B.J., Sims, R., 1976. Effect of mercury on the survival, respiration, and growth of postlarval white shrimp, *Penaeus setiferus*. Mar. Biol. 37, 75–81.

Greig, R.A., 1975. Comparison of atomic absorption and neutron activation analyses for determination of silver, chromium, and zinc in various marine organisms. Anal. Chem. 47, 1682–1684.

Greig, R., Wenzloff, D., 1977. Final report on heavy metals in small pelagic finfish, euphausiid crustaceans, and apex predators, including sharks, as well as on heavy metals and hydrocarbons (C_{15+}) in sediments collected at stations in and near Deepwater Dumpsite 106, Vol. III, Contaminant inputs and chemical characteristics. In: Baseline Report of the Environmental Conditions in Deepwater Dumpsite 106. U.S. Department of Commerce, NOAA, pp. 547–564.

Greig, R.A., Wenzloff, D., Shelpuk, C., 1975. Mercury concentrations in fish, North Atlantic offshore waters—1971. Pestic. Monit. J. 9, 15–20.

Greig, R.A., Adams, A., Wenzloff, D.R., 1977a. Trace metal content of plankton and zooplankton collected from the New York Bight and Long Island Sound. Bull. Environ. Contam. Toxicol. 18, 3–8.

Greig, R.A., Wenzloff, D.R., Adams, A., Nelson, B., Shelpuk, C., 1977b. Trace metals in organisms from ocean disposal sites of the middle eastern United States. Arch. Environ. Contam. Toxicol. 6, 395–409.

Guary, J.C., Fowler, S.W., 1978. Uptake from water and tissue distribution of neptunium-237 in crabs, shrimp, and mussels. Mar. Pollut. Bull. 9, 331–334.

Guary, J.C., Negrel, R., 1980. Plutonium and iron association with metal-binding proteins in the crab *Cancer pagurus* (L.). J. Exp. Mar. Biol. Ecol. 42, 87–98.

Guary, J.C., Masson, M., Fraizier, A., 1976. Etude preliminaire, *in situ*, de la distribution du plutonium dans differents tissues et organes de *Cancer pagurus* (crustacea: decapoda) et de *Pleuronectes platessa* (pisces: pleuronectidae). Mar. Biol. 36, 13–17.

Guerin, T., Sirot, V., Volatier, J.L., Leblanc, J.C., 2007. Organotin levels in seafood and its implications for health risk in high-seafood consumers. Sci. Total Environ. 388, 66–77.

Hall Jr., L.W., Pinkney, A.E., 1985. Acute and sublethal effects of organotin compounds on aquatic biota: An interpretive literature evaluation. CRC Crit. Rev. Toxicol. 14, 159–209.

Hall, W.S., Pulliam, G.W., 1995. An assessment of metals in an estuarine wetlands ecosystem. Arch. Environ. Contam. Toxicol. 29, 164–173.

Hall, R.A., Zook, E.G., Meaburn, G.M., 1978. National Marine Fisheries Service survey of trace elements in the fishery resource. U.S. Dept. Commerce NOAA Tech. Rept., NMFS SSRF-721, pp. 1–313.

Hall Jr., L.W., Bushong, S.J., Hall, W.S., Johnson, W.E., 1988a. Acute and chronic effects of tributyltin to a Chesapeake Bay copepod. Environ. Toxicol. Chem. 7, 41–46.

Hall, W.S., Bushong, S.J., Hall Jr., L.W., Lenkevich, M.S., Pinkney, A.E., 1988b. Monitoring dissolved copper concentrations in Chesapeake Bay, U.S.A. Environ. Monit. Assess. 11, 33–42.

Hall Jr., L.W., Zeigenfuss, M.C., Anderson, R.D., Lewis, B.L., 1995. The effect of salinity on the acute toxicity of total and free cadmium to a Chesapeake Bay copepod and fish. Mar. Pollut. Bull. 30, 376–384.

Hall Jr., L.W., Anderson, R.D., Lewis, B.L., Arnold, W.R., 2008. The influence of salinity and dissolved organic carbon on the toxicity of copper to the estuarine copepod, *Eurytemora affinis*. Arch. Environ. Contam. Toxicol. 54, 44–56.

Han, B.C., Jeng, W.L., Chen, R.Y., Fang, G.T., Hung, T.C., Tseng, R.J., 1998. Estimation of target hazard quotients and potential health risks for metals by consumption of seafood in Taiwan. Arch. Environ. Contam. Toxicol. 35, 711–720.

Hanaoka, K., Tagawa, S., 1985a. Isolation and identification of arsenobetaine as a major water soluble arsenic compound from muscle of blue pointer *Isurus oxyrhincus* and whitetip shark *Carcharhinus longimanus*. Bull. Jpn. Soc. Sci. Fish. 51, 681–685.

Hanaoka, K., Tagawa, S., 1985b. Identification of arsenobetaine in muscle of roundnose flounder *Eopsetta grigorjewi*. Bull. Jpn. Soc. Sci. Fish. 51, 1203.

Hansen, J.I., Mustafa, T., Depledge, M., 1992a. Mechanisms of copper toxicity in the shore crab, *Carcinus maenas*. I. Effects on Na, K, ATPase activity, haemolymph electrolyte concentrations and tissue water contents. Mar. Biol. 114, 253–257.

Hansen, J.I., Mustafa, T., Depledge, M., 1992b. Mechanisms of copper toxicity in the shore crab, *Carcinus maenas*. II. Effects on key metabolic enzymes, metabolites and energy charge potential. Mar. Biol. 114, 259–264.

Hardstedt-Romeo, M., Laumond, F., 1980. Zinc, copper and cadmium in zooplankton from the N.W. Mediterranean. Mar. Pollut. Bull. 11, 133–138.

Harino, H., Fukushima, M., Yamamoto, Y., Kawai, S., Miyazaki, N., 1998. Organotin compounds in water, sediment, and biological samples from the Port of Osaka, Japan. Arch. Environ. Contam. Toxicol. 35, 558–564.

Harrington, J.M., 1991. Tributyltin residues in Lake Tahoe and San Diego Bay, California, 1988, State of California Dept. Fish Game. Environ. Serv. Div. Admin. Rep. 91-1, pp. 1–29.

Harvey, J., Harwell, L., Summers, J.K., 2008. Contaminant concentrations in whole-body fish and shellfish from US estuaries. Environ. Monit. Assess. 137, 403–412.

Heyraud, M., Cherry, R.D., 1979. Polonium-210 and lead-210 in marine food chains. Mar. Biol. 52, 227–236.

Hilmy, A.M., El-Hamid, N.F.A., Ghazaly, K.S., 1988. Biochemical and physiological changes in the tissues and serum of both sexes in *Portunus pelagicus* (L.) following acute exposures to zinc and copper. Folia Morphol. 36, 79–94.

Hiyama, Y., Shimizu, M., 1964. On the concentration factors of radioactive Cs, Sr, Cd, Zn and Ce in marine organisms. Rec. Oceanogr. Works Jpn. 7 (2), 43–77.

Holden, A.V., Topping, G., 1972. XIV: Occurrence of specific pollutants in fish in the Forth and Tay estuaries. Proc. R. Soc. Edin. 71B, 189–194.

Hoover, W.L., Melton, J.R., Howard, P.A., Bassett Jr., J.W., 1974. Atomic absorption spectrometric determination of arsenic. JAOAC 57, 18–21.

Horowitz, A., Presley, B.J., 1977. Trace metal concentrations and partitioning in zooplankton, neuston, and benthos from the south Texas Outer Continental Shelf. Arch. Environ. Contam. Toxicol. 5, 241–255.

Hu, J., Zhen, H., Wan, Y., Gao, J., An, W., An, L., et al., 2006. Trophic magnification of triphenyltin in a marine food web of Bohai Bay, North China: Comparison to tributyltin. Environ. Sci. Technol. 40, 3142–3147.

Huang, Y., Zhu, L., Liu, G., 2006. The effects of bis(tributyltin) oxide on the development, reproduction and sex ratio of calanoid copepod *Pseudodiaptomus marinus*. Estuar. Coast. Shelf Sci. 69, 147–152.

Huggett, R.J., Unger, M.A., Seligman, P.F., Valkirs, A.O., 1992. The marine biocide tributyltin. Environ. Sci. Technol. 26, 232–237.

Hui, C.A., Rudnick, D., Williams, E., 2005. Mercury burdens in Chinese mitten crabs (*Eriocheir sinensis*) in three tributaries of southern San Francisco Bay, California, USA. Environ. Pollut. 133, 481–487.

Hutcheson, M.S., 1974. The effect of temperature and salinity on cadmium uptake by the blue crab, *Callinectes sapidus*. Chesapeake Sci. 15, 237–241.

Ichikawa, R., 1961. On the concentration factors of some important radionuclides in marine food organisms. Bull. Jpn. Soc. Sci. Fish. 27, 66–74.

Ikebe, K., Tanaka, R., 1979. Determination of vanadium and nickel in marine samples by flameless and flame atomic absorption spectrophotometry. Bull. Environ. Contam. Toxicol. 21, 526–532.

Ireland, M.P., 1973. Result of fluvial zinc pollution on the zinc content of littoral and sublittoral organisms in Cardigan Bay, Wales. Environ. Pollut. 4, 27–35.

Ireland, M.P., 1974. Variations in the zinc, copper, manganese, and lead content of *Balanus balanoides* in Cardigan Bay, Wales. Environ. Pollut. 7, 65–75.

Ishii, T., Suzuki, H., Koyanagi, T., 1978. Determination of trace elements in marine organisms: I. Factors for variation of concentration of trace element. Bull. Jpn. Soc. Sci. Fish. 44, 155–162.

Jenkins, D.W., 1980. Biological monitoring of toxic trace metals. Vol. 2. Toxic trace metals in plants and animals of the world. Part III. U.S. Environ. Protect. Agen. Rep. 600/3-80-092, pp. 1130–1148.

Jennings, J.R., Rainbow, P.S., 1979a. Studies on the uptake of cadmium by the crab *Carcinus maenas* in the laboratory. I. Accumulation from seawater and a food source. Mar. Biol. 50, 131–139.

Jennings, J.R., Rainbow, P.S., 1979b. Accumulation of cadmium by *Artemia salina*. Mar. Biol. 51, 47–53.

Jennings, J.R., Rainbow, P.S., Scott, A.G., 1979. Studies on the uptake of cadmium by the crab *Carcinus maenas* in the laboratory. II. Preliminary investigation of cadmium-binding proteins. Mar. Biol. 50, 141–149.

Johnson, D.L., Braman, R.S., 1975. The speciation of arsenic and the content of germanium and mercury in members of the pelagic *Sargassum* community. Deep Sea Res. 22, 503–507.

Johnson, I., Jones, M.B., 1989. Effects of zinc/salinity combinations on zinc regulation in *Gammarus duebeni* from the estuary and the sewage treatment works at Looe, Cornwall. J. Mar. Biol. Assoc. UK 69, 249–260.

Jones, M.B., 1975. Effects of copper on survival and osmoregulation in marine and brackish water isopods (Crustacea). In: Proceedings of Ninth European Marine Biology Symposium, pp. 419–431.

Jung, K., Zauke, G.P., 2008. Bioaccumulation of trace metals in the brown shrimp *Crangon crangon* (Linnaeus, 1758) from the German Wadden Sea. Aquat. Toxicol. 88, 243–249.

Kadar, E., Costa, V., Santos, R.S., 2006. Distribution of micro-essential (Fe, Cu, Zn) and toxic (Hg) metals in tissue of two nutritionally distinct hydrothermal shrimps. Sci. Total Environ. 358, 143–150.

Kaise, T., Fukui, S., 1992. The chemical form and acute toxicity of arsenic compounds in marine organisms. Appl. Organomet. Chem. 6, 155–160.

Karouna-Renier, N.K., Snyder, R.A., Allison, J.G., Wagner, M.G., Rao, K.R., 2007. Accumulation of organic and inorganic contaminants in shellfish collected in estuarine waters near Pensacola, Florida: Contamination profiles and risks to human consumers. Environ. Pollut. 145, 474–488.

Keckes, S., Fowler, S.W., Small, L.F., 1972. Flux of different forms of ^{106}Ru through a marine zooplankter. Mar. Biol. 13, 94–99.

Kennedy, V.S., 1976. Arsenic concentrations in some coexisting marine organisms from Newfoundland and Labrador. J. Fish. Res. Board Can. 33, 1388–1393.

Ketchum, B.M., Bowen, V.T., 1958. Biological factors determining the distribution of radioisotopes in the sea. In: Proceedings of the Second Annual United Nations International Congress on the Peaceful Uses of Atomic Energy: Vol. 18, Waste Treatment and Environmental Aspects of Atomic Energy. IAEA, Geneva, pp. 429–433.

Khan, A.T., Weis, J.S., D'Andrea, L., 1989. Bioaccumulation of four heavy metals in two populations of grass shrimp, *Palaemonetes pugio*. Bull. Environ. Contam. Toxicol. 42, 339–343.

Kharkar, D.P., Thomson, J., Turekian, K.K., Forster, W.O., 1976. Uranium and thorium decay series nuclides in plankton from the Caribbean. Limnol. Oceanol. 21, 294–299.

King, C.K., Gale, S.A., Hyne, R.V., Stauber, J.L., Simpson, S.L., Hickey, C.W., 2006a. Sensitivities of Australian and New Zealand amphipods to copper and zinc in waters and metal-spiked sediments. Chemosphere 63, 1466–1476.

King, C.K., Gale, S.A., Stauber, J.L., 2006b. Acute toxicity and bioaccumulation of aqueous and sediment-bound metals in the estuarine amphipod *Melita plumulosa*. Environ. Toxicol. 21, 489–504.

Klemmer, H.W., Unninayer, C.S., Ukubo, W.I., 1976. Mercury content of biota in coastal waters in Hawaii. Bull. Environ. Contam. Toxicol. 15, 454–457.

Knauer, G.A., 1970. The determination of magnesium, manganese, iron, copper, and zinc in marine shrimp. Analyst 93, 476–480.

Knothe, D.W., Van Riper, G.G., 1988. Acute toxicity of sodium molybdate dihydrate (Molhibit 100) to selected saltwater organisms. Bull. Environ. Contam. Toxicol. 40, 785–790.

Kobayashi, N., 1979. Barnacle larvae and rotifers as indicatory materials for marine pollution bioassay, preliminary experiments. Sci. Eng. Rev. Doshisha Univ. 19, 61–67.

Kuenzler, E.J., 1969. Elimination and transport of cobalt by marine zooplankton. In: Proceedings of the 2nd National Symposium on Radioecology. U.S. Atom. Ener. Comm., Conf. 670503, pp. 483–492.

Kumagai, H., Saeki, K., 1978. Contents of total mercury, alkyl mercury and methylmercury in some coastal fish and shells. Bull. Jpn. Soc. Sci. Fish. 44, 807–811.

Laughlin, R., Nordlund, K., Linden, O., 1984. Long-term effects of tributyltin compounds on the Baltic amphipod, *Gammarus oceanicus*. Mar. Environ. Res. 12, 243–271.

Laughlin Jr., R.B., Johannesen, R.B., French, W., Guard, H., Brinckman, F.E., 1985. Structure-activity relationships for organotin compounds. Environ. Toxicol. Chem. 4, 343–351.

Laughlin Jr., R.B., French, W., Guard, H.E., 1986. Accumulation of bis(tributyltin)oxide by the marine mussel *Mytilus edulis*. Environ. Sci. Technol. 20, 884–890.

Leatherland, T.M., Burton, J.D., 1974. The occurrence of some trace metals in coastal organisms with particular reference to the Solent region. J. Mar. Biol. Assoc. UK 54, 457–468.

Leatherland, T.M., Burton, J.D., Culkin, F., McCartney, M.J., Morris, R.J., 1973. Concentrations of some trace metals in pelagic organisms and of mercury in Northeast Atlantic Ocean water. Deep Sea Res. 20, 679–685.

LeBlanc, P.J., Jackson, A.L., 1973. Arsenic in marine fish and invertebrates. Mar. Pollut. Bull. 4, 88–90.

Lee, K.M., Johnston, E.L., 2007. Low levels of copper reduce the reproductive success of a mobile invertebrate predator. Mar. Environ. Res. 64, 336–346.

Lee, K.W., Raisuddin, S., Hwang, D.S., Park, H.G., Lee, J.S., 2007. Acute toxicities of trace metals and common xenobiotics to the marine copepod *Tigriopus japonicus*: Evaluation of its use as a benchmark species for routine ecotoxicity tests in western Pacific coastal regions. Environ. Toxicol. 22, 532–538.

Lee, K.W., Raisuddin, S., Hwang, D.S., Park, H.G., Dahms, H.U., Ahn, I.Y., et al., 2008. Two-generation toxicity study on the copepod model species *Tigriopus japonicus*. Chemosphere 72, 1359–1365.

Lewis, T.E., McIntosh, A.W., 1986. Uptake of sediment-bound lead and zinc by the freshwater isopod *Asellus communis* at three different pH levels. Arch. Environ. Contam. Toxicol. 15, 495–504.

Lewis, A.G., Whitfield, P.H., Ramnarine, A., 1972. Some particulate and soluble agents affecting the relationship between metal toxicity and organism survival in the calanoid copepod *Euchaeta japonica*. Mar. Biol. 17, 215–221.

Lewis, A.G., Whitfield, P., Ramnarine, A., 1973. The reduction of copper toxicity in a marine copepod by sediment extract. Limnol. Oceanol. 18, 324–326.

Lindsay, D.M., Sanders, J.G., 1990. Arsenic uptake and transfer in a simplified estuarine food chain. Environ. Toxicol. Chem. 9, 391–395.

Lowman, F.G., Phelps, D.K., Ting, R.Y., Escalera, R.M., 1966. Progress Summary Report No. 4, Marine Biology Program June 1965-June 1966. Puerto Rico Nucl. Cen Rep. PRNC, 85, pp. 1–57 + Appendices A-F.

Lowman, F.G., Martin, J.H., Ting, R.Y., Barnes, S.S., Swift, D.J.P., Seiglei, G.A., et al., 1970. Bioenvironmental and Radiological-Safety Feasibility Studies, Atlantic-Pacific Interoceanic Canal, vol. I-IV, Contract AT (26-1)-171. Battelle Memorial Institute, Columbus, OH.

Lucu, C., Jelisavcic, O., Lucic, S., Strohal, P., 1969. Interactions of ^{233}Pa with tissues of *Mytilus galloprovincialis* and *Carcinus mediterraneus*. Mar. Biol. 2, 103–104.

Luoma, S.N., 1976. The uptake and interorgan distribution of mercury in a carnivorous crab. Bull. Environ. Contam. Toxicol. 16, 719–723.

Luoma, S.N., 1977. The dynamics of biologically available mercury in a small estuary. Estuar. Coast. Mar. Sci. 5, 643–652.

MacFarlane, G.R., Schreider, M., McLennan, B., 2006. Biomarkers of heavy metal contamination in the red fingered marsh crab, *Parasesarma erythodactyla*. Arch. Environ. Contam. Toxicol. 51, 584–593.

Mackay, D.W., Halcrow, W., Thornton, I., 1972. Sludge dumping in the Firth of Clyde. Mar. Pollut. Bull. 3, 7–10.

Maddock, B.G., Taylor, D., 1980. The acute toxicity and bioaccumulation of some lead alkyl compounds in marine animals. In: Branica, M., Konrad, Z. (Eds.), Lead in the Marine Environment. Pergamon, Oxford, UK, pp. 233–261.

Maher, W.A., 1983. Selenium in marine organisms from St. Vincent's Gulf, South Australia. Mar. Pollut. Bull. 14, 35–36.

Maher, W.A., 1985. The presence of arsenobetaine in marine animals. Biochem. Physiol. 80C, 199–201.

Manyin, T., Rowe, C.L., 2008. Modeling effects of cadmium on population growth of *Palaemonetes pugio*: Results of a full life cycle exposure. Aquat. Toxicol. 88, 111–120.

Martin, J.H., 1970. The possible transport of trace metals via molted copepod exoskeletons. Limnol. Oceanol. 15, 756–761.

Martin, J.L.M., 1972. Study of the uptake, concentration, and metabolism of strontium 85 by the decapod crustacean *Carcinus maenas*. Mar. Biol. 12, 154–158.

Martin, J.L.M., 1973. Iron metabolism in *Cancer irroratus* (crustacea: decapoda) during the intermoult cycle, with special reference to iron in the gills. Comp. Biochem. Physiol. 46A, 123–129.

Martin, J.H., Knauer, G.A., 1973. The elemental composition of plankton. Geochim. Cosmochim. Acta 37, 1639–1653.

Martin, J.L.M., Van Wormhoudt, A., Ceccaldi, H.H., 1977. Zinc-hemocyanin binding to the hemolymph of *Carcinus maenas* (Crustacea: Decapoda). Comp. Biochem. Physiol. 58A, 193–195.

Matida, Y., Kumada, H., 1969. Distribution of mercury in water, bottom mud and aquatic organisms of Minamata Bay, the River Agano and other water bodies in Japan. Bull. Freshw. Fish. Res. Lab. Tokyo 19 (2), 73–93.

Matsuto, S., Stockton, R.A., Irgolic, K.J., 1986. Arsenobetaine in the red crab, *Chionoecetes opilio*. Sci. Total Environ. 48, 133–140.

Mauchline, J., Templeton, W.L., 1966. Strontium, calcium, and barium in marine organisms from the Irish Sea. J. Cons. Perm. Int. Explor. Mer 30, 161–170.

McGee, B.L., Wright, D.A., Fisher, D.J., 1998. Biotic factors modifying acute toxicity of aqueous cadmium to estuarine amphipod *Leptocheirus plumulosus*. Arch. Environ. Contam. Toxicol. 34, 34–40.

McKenney Jr., C.L., Neff, J.M., 1979. Individual effects and interactions of salinity, temperature, and zinc on larval development of the grass shrimp *Palaemonetes pugio*. I. Survival and developmental duration through metamorphosis. Mar. Biol. 52, 177–188.

McLeese, D.W., 1974. Toxicity of copper at two temperatures and three salinities to the American lobster (*Homarus americanus*). J. Fish. Res. Board Can. 31, 1949–1952.

McLusky, D.S., Hagerman, L., 1987. The toxicity of chromium, nickel and zinc: Effects of salinity and temperature, and the osmoregulatory consequences in the mysid *Praunus flexuosus*. Aquat. Toxicol. 10, 225–238.

Memmert, U., 1987. Bioaccumulation of zinc in two freshwater organisms (*Daphnia magna*, crustacea and *Brachydanio rerio*, pisces). Water Res. 21, 99–106.

Miller, C.D., Nelson, D.M., Guillard, R.R.L., Woodward, B.L., 1980. Effects of media with low silicic acid concentrations on tooth formation in *Acartia tonsa* Dana (Copepoda, Calanoida). Biol. Bull. 159, 349–363.

Moore, P.G., Rainbow, P.S., Hayes, E., 1991. The beach hopper *Orchestia gammarellus* (Crustacea: Amphipoda) as a biomonitor for copper and zinc: North Sea trials. Sci. Total Environ. 106, 221–238.

Morris, R.J., Law, R.J., Allchin, C.R., Kelly, C.A., Fileman, C.F., 1989. Metals and organochlorines in dolphins and porpoises of Cardigan Bay, West Wales. Mar. Pollut. Bull. 20, 512–523.

Mushak, P., 1980. Metabolism and systemic toxicity of nickel. In: Nriagu, J.O. (Ed.), Nickel in the Environment. John Wiley, New York, pp. 499–523.

Neff, J.M., Anderson, J.W., 1977. The effects of copper (II) on molting and growth of juvenile lesser blue crabs *Callinectes similis*. In: Giam, C.S. (Ed.), Pollutant Effects on Marine Organisms. D.C. Heath, Lexington, MA, pp. 155–165.

Neff, J.M., Foster, R.S., Slowey, J.F., 1978. Availability of sediment-absorbed heavy metals o benthos with particular emphasis on deposit-feeding infauna. Tech. Rep. D-78-42. U.S. Army Waterways Exp. Sta., Vicksburg, MS, 286 pp.

Neff, J.M., Carr, R.S., McCullough, W.L., 1981. Acute toxicity of a used chrome lignosulfonate drilling mud to several species of marine invertebrates. Mar. Environ. Res. 4, 251–266.

Nicholls, G.D., Curl Jr., H., Bowen, V.T., 1959. Spectrographic analyses of marine plankton. Limnol. Oceanol. 4, 472–476.

Nimmo, D.R., Lightner, D.V., Bahner, L.H., 1977. Effects of cadmium on the shrimps, *Penaeus duorarum*, *Palaemonetes pugio*, and *Palaemonetes vulgaris*. In: Vernberg, F.J., Calabrese, A., Thurberg, F.P., Vernberg, W.B. (Eds.), Physiological Responses of Marine Biota to Pollutants. Academic Press, New York, pp. 131–183.

Nonnotte, L., Boitel, F., Truchot, J.P., 1993. Waterborne copper causes gill damage and hemolymph hypoxia in the shore crab *Carcinus maenas*. Can. J. Zool. 71, 1569–1576.

Norin, H., Vahter, M., Christakopoulos, A., Sandstrom, M., 1985. Concentration of inorganic and total arsenic in fish from industrially polluted water. Chemosphere 14, 1125–1134.

Nugegoda, D., Rainbow, P.S., 1987. The effect of temperature on zinc regulation by the decapod crustacean *Palaemon elegans* Rathke. Ophelia 27, 17–30.

Nugegoda, D., Rainbow, P.S., 1988a. Effect of a chelating agent (EDTA) on zinc uptake and regulation by *Palaemon elegans* (crustacea: decapoda). J. Mar. Biol. Assoc. UK 68, 25–40.

Nugegoda, D., Rainbow, P.S., 1988b. Zinc uptake and regulation by the sublittoral prawn *Pandalus montagui* (crustacea: decapoda). Estuar. Coast. Shelf Sci. 26, 619–632.

Nugegoda, D., Rainbow, P.S., 1989a. Effects of salinity changes on zinc uptake and regulation by the decapod crustaceans *Palaemon elegans* and *Palaemonetes varians*. Mar. Ecol. Prog. Ser. 51, 57–75.

Nugegoda, D., Rainbow, P.S., 1989b. Salinity, osmolality, and zinc uptake in *Palaemon elegans* (crustacea: decapoda). Mar. Ecol. Prog. Ser. 55, 149–157.

Nugegoda, D., Rainbow, P.S., 1989c. Zinc uptake and regulation breakdown in the decapod crustacean *Palaemon elegans* Rathke. Ophelia 30, 199–212.

Nunez-Nogueira, G., Mouneyrac, C., Amiard, J.C., Rainbow, P.S., 2006a. Subcellular distribution of zinc and cadmium in the hepatopancreas and gills of the decapod crustacean *Penaeus indicus*. Mar. Biol. 150, 197–211.

Nunez-Nogueira, G., Rainbow, P.S., Smith, B.D., 2006b. Assimilation efficiency of zinc and cadmium in the decapod crustacean *Penaeus indicus*. J. Exp. Mar. Biol. Ecol. 332, 75–83.

Nuorteva, P., Hasanen, E., 1975. Bioaccumulation of mercury in *Myoxocephalus quadricornis* (L.) (Teleostei, Cottidae) in an unpolluted area of the Baltic. Ann. Zool. Fennici 12, 247–254.

O'Hara, J., 1973a. The influence of temperature and salinity on the toxicity of cadmium to the fiddler crab, *Uca pugilator*. US Dept. Comm. Fish. Bull. 71 (1), 149–153.

O'Hara, J., 1973b. Cadmium uptake by fiddler crabs exposed to temperature and salinity stress. J. Fish. Res. Board Can. 30, 846–848.

Oweson, C., Baden, S., Henroth, B., 2006. Manganese induced apoptosis in haematopoietic cells of the lobster, *Nephrops norvegicus*. Aquat. Toxicol. 77, 322–328.

Ozretic, B., Krajinovic-Ozretic, N., Santin, J., Medjugorac, B., Kras, M., 1990. As, Cd, Pd, and Hg in benthic animals from the Kvarner-Rijeka Bay region, Yugoslavia. Mar. Pollut. Bull. 21, 595–597.

Palmer, A.S., Snape, I., Stark, J.S., Johnstone, G.J., Townsend, A.T., 2006. Baseline metal concentrations in *Paramoera walkeri* from East Antarctica. Mar. Pollut. Bull. 52, 1441–1449.

Parrish, K.M., Carr, R.A., 1976. Transport of mercury through a laboratory two-level marine food chain. Mar. Pollut. Bull. 7, 90–91.

Parveneh, V., 1977. A survey of the mercury content of the Persian Gulf shrimp. Bull. Environ. Contam. Toxicol. 18, 778–782.

Patel, H.S., Kaliwal, M.B., 1989. Histopathological effects of zinc on the gills of prawn *Macrobrachium hendersodyanum*. Zeit. Ange. Zool. 76, 505–509.

Pearcy, W.G., Osterberg, C.L., 1967. Depth, diel, seasonal, and geographic variations in zinc-65 of midwater animals off Oregon. Int. J. Oceanol. Limnol. 1, 103–116.

Peden, J.D., Crothers, J.H., Waterfall, C.W., Beasley, J., 1973. Heavy metals in Somerset marine organisms. Mar. Pollut. Bull. 4, 7–10.

Pedroso, M.S., Bersano, J.G.F., Bianchini, A., 2007a. Acute silver toxicity in the euryhaline copepod *Acartia tonsa*: Influence of salinity and food. Environ. Toxicol. Chem. 26, 2158–2165.

Pedroso, M.S., Pinho, C.L.L., Rodrigues, S.C., Bianchini, A., 2007b. Mechanism of acute silver toxicity in the euryhaline copepod *Acartia tonsa*. Aquat. Toxicol. 82, 173–180.

Pempkowiak, J., Walkusz-Miotk, J., Beldowski, J., Walkusz, W., 2006. Heavy metals in zooplankton from the southern Baltic. Chemosphere 62, 1697–1708.

Penrose, W.R., Black, R., Hayward, M.J., 1975. Limited arsenic dispersion in seawater, sediments, and biota near a continuous source. J. Fish. Res. Board Can. 32, 1275–1281.

Pequegnat, J.E., Fowler, S.W., Small, L.F., 1969. Estimates of the zinc requirements of marine organisms. J. Fish. Res. Board Can. 26, 145–150.

Pereira, E., Abreu, S.N., Coelho, J.P., Lopes, C.B., Pardal, M.A., Vale, C., et al., 2006. Seasonal fluctuations of tissue mercury contents in the European shore crab *Carcinus maenas* from low and high contamination areas (Ria de Aveiro, Portugal). Mar. Pollut. Bull. 52, 1450–1457.

Pesch, G.G., Stewart, N.E., 1980. Cadmium toxicity to three species of invertebrates. Mar. Environ. Res. 3, 145–146.

Peshut, P.J., Morrison, R.J., Brooks, B.A., 2008. Arsenic speciation in marine fish and shellfish from American Samoa. Chemosphere 71, 484–492.

Petri, G., Zauke, G.P., 1993. Trace metals in crustaceans in the Antarctic Ocean. Ambio 22, 529–536.

Phillips, D.J.H., 1990. Arsenic in aquatic organisms: A review, emphasizing chemical speciation. Aquat. Toxicol. 16, 151–186.

Phillips, D.J.H., Depledge, M.H., 1986. Chemical forms of arsenic in marine organisms, with emphasis on *Hemifusus* species. Water Sci. Technol. 18, 213–222.

Phillips, D.J.H., Thompson, G.B., Gabuji, K.M., Ho, C.T., 1982. Trace metals of toxicological significance to man in Hong Kong seafood. Environ. Pollut. 3B, 27–45.

Pinho, G.L.L., Pedroso, M.S., Rodrigues, S.C., d Souza, S.S., Bianchini, A., 2007. Physiological effects of copper in the euryhaline copepod *Acartia tonsa*: Waterborne versus waterborne plus dietborne exposure. Aquat. Toxicol. 84, 62–70.

Pinkney, A.E., Matteson, L.L., Wright, D.A., 1990. Effects of tributyltin on survival, growth, morphometry, and RNA-DNA ratio of larval striped bass, *Morone saxatilis*. Arch. Environ. Contam. Toxicol. 19, 235–240.

Polikarpov, G.G., Oregioni, B., Parchevskaya, D.S., Benayoun, G., 1979. Body burden of chromium, copper, cadmium and lead in the neustonic copepod *Anomalocera patersoni* (Pontellidae) collected from the Mediterranean Sea. Mar. Biol. 53, 79–82.

Pouvreau, B., Amiard, J.C., 1974. Etude experimentale de l'accumulation de l'argent 110 a chez divers organismes marins. Comm. Ener. Atom. France, Rept. CEA-R-4571, pp. 1–19.

Powell, M.L., White, K.N., 1990. Heavy metal accumulation by barnacles and its implications for their use as biological monitors. Mar. Environ. Res. 30, 91–118.

Prange, A., Kremling, K., 1985. Distribution of dissolved molybdenum, uranium and vanadium in Baltic Sea waters. Mar. Chem. 16, 259–274.

Pullen, J.S.H., Rainbow, P.S., 1991. The composition of pyrophosphate heavy metal detoxification granules in barnacles. J. Exp. Mar. Biol. Ecol. 150, 249–266.

Pyefinch, K.A., Mott, J.C., 1948. The sensitivity of barnacles and their larvae to copper and mercury. J. Exp. Biol. 25, 296–298.

Rabitsch, W.B., 1995. Metal accumulation in arthropods near a lead/zinc mine in Arnoldstein, Austria. Environ. Pollut. 90, 221–227.

Rainbow, P.S., 1989. Copper, cadmium and zinc concentrations in oceanic amphipod and euphausiid crustaceans, as a source of heavy metals to pelagic seabirds. Mar. Biol. 103, 513–518.

Rainbow, P.S., Scott, A.G., 1979. Two heavy metal-binding proteins in the midgut gland of the crab *Carcinus maenas*. Mar. Biol. 55, 143–150.

Rainbow, P.S., White, S.L., 1989. Comparative strategies of heavy metal accumulation by crustaceans: Zinc, copper and cadmium in a decapod, an amphipod and a barnacle. Hydrobiologia 174, 245–262.

Rainbow, P.S., Scott, A.G., Wiggins, E.A., Jackson, R.W., 1980. Effects of chelating agents on the accumulation of cadmium by the barnacle *Semibalanus balanoides*, and complexation of soluble Cd, Zn and Cu. Mar. Ecol. Prog. Ser. 2, 143–152.

Rainbow, P.S., Moore, P.G., Watson, D., 1989. Talitrid amphipods (crustacea) as biomonitors for copper and zinc. Estuar. Coast. Shelf Sci. 28, 567–582.

Ramelow, G.J., Webre, C.L., Mueller, C.S., Beck, J.N., Young, J.C., Langley, M.P., 1989. Variations of heavy metals and arsenic in fish and other organisms from the Calcasieu River and Lake, Louisiana. Arch. Environ. Contam. Toxicol. 18, 804–818.

Ramos, A., de Campos, M., Olszyna-Marzys, A.E., 1979. Mercury contamination of fish in Guatemala. Bull. Environ. Contam. Toxicol. 22, 488–491.

Ray, S., McLeese, D.W., Waiwood, B.A., Pezzack, D., 1980. The disposition of cadmium and zinc in *Pandalus montagui*. Arch. Environ. Contam. Toxicol. 9, 675–681.

Rees, H.L., Waldock, R., Matthiessen, P., Pendle, M.A., 1999. Surveys of the epibenthos of the Crouch Estuary (UK) in relation to TBT contamination. J. Mar. Biol. Assoc. UK 79, 209–223.

Reeve, M.R., Walter, M.A., Darcy, K., Ikeda, T., 1977. Evaluation of potential indicators of sub-lethal toxic stress on marine zooplankton (feeding, fecundity, respiration, and excretion): Controlled ecosystem pollution experiment. Bull. Mar. Sci. 27, 105–113.

Reimer, A.A., Reimer, R.D., 1975. Total mercury in some fish and shellfish along the Mexican coast. Bull. Environ. Contam. Toxicol. 14, 105–111.

Reinke, J., Uthe, J.F., Freeman, H.C., Johnston, J.R., 1975. The determination of arsenite and arsenate ions in fish and shellfish by selective extraction and polarography. Environ. Lett. 8, 371–380.

Renfro, W.C., Fowler, S.W., Heyraud, M., LaRosa, J., 1975. Relative importance of food and water in long-term-zinc-65 accumulation by marine biota. J. Fish. Res. Board Can. 31, 1339–1345.

Renzoni, A., Bacci, E., Falciai, L., 1973. Mercury concentration in the water, sediments, and fauna of an area of the Tyrrhenian Coast. In: Sixth International Symposium Medic. Ocean., Portoroz, Yugoslavia, 26-30 September 1973, pp. 17–45.

Ridout, P.S., Rainbow, P.S., Roe, H.S.J., Jones, H.R., 1989. Concentrations of V, Cr, Mn, Fe, Ni, Co, Cu, Zn, As and Cd in mesopelagic crustaceans from the north east Atlantic Ocean. Mar. Biol. 100, 465–471.

Ringenary, M.J., Molof, A.H., Tancredi, J.T., Schreibman, M.P., Kostarelos, K., 2007. Long-term sediment bioassay of lead toxicity in two generations of the marine amphipod *Elasmopus laevis*, S.I. Smith, 1873. Environ. Toxicol. Chem. 26, 1700–1710.

Roberts, D., Maguire, C., 1976. Interactions of lead with sediment and meiofauna. Mar. Pollut. Bull. 7, 211–213.

Roberts, D.A., Poore, A.G.B., Johnston, E.L., 2006. Ecological consequences of copper contamination in macroalgae: Effects on epifauna and associated herbivores. Environ. Toxicol. Chem. 25, 2470–2479.

Robertson, D.E., 1967. Trace elements in marine organisms. Rapp. Amer., BNWL 481-2, pp. 56–59.

Robertson, D.E., Rancitelli, L.A., Langford, J.C., Perkins, R.W., 1972. Battelle-Northwest contribution to the IDOE base-line study. Battelle Pac. Northwest Lab., Richland, Washington, 1–46.

Roesijadi, G., Petrocelli, S.E., Anderson, J.W., Presley, B.J., Sims, R., 1974. Survival and chloride ion regulation of the porcelain crab *Petrolisthes armatus* exposed to mercury. Mar. Biol. 17, 213–217.

Rosemarin, A., Notini, N., Holmgren, K., 1985. The fate of arsenic in the Baltic Sea *Fucus vesiculosus* ecosystem. Ambio 14, 342–345.

Ruelas-Inzunza, J., Paez-Osuna, F., 2008. Trophic distribution of Cd, Pb, and Zn in a food web from Altata-Ensenada del Pebellon subtropical lagoon, SE Gulf of California. Arch. Environ. Contam. Toxicol. 54, 584–596.

Saliba, L.J., Ahsanullah, M., 1973. Acclimation and tolerance of *Artemia salina* and *Ophryotrocha labronica* to copper sulphate. Mar. Biol. 12, 297–302.

Saliba, L.J., Krzyz, R.M., 1976. Acclimation and tolerance of *Artemia salina* to copper salts. Mar. Biol. 38, 231–238.

Sanders, J.G., 1986. Direct and indirect effects of arsenic on the survival and fecundity of estuarine zooplankton. Can. J. Fish. Aquat. Sci. 43, 694–699.

Sankar, T.V., Zynudheen, A.A., Anandan, R., Nair, P.G.V. 2006. Distribution of organochlorine pesticides and heavy metal residues in fish and shellfish from Calicut region, Kerala, India. Chemosphere 65, 583–590.

Santos, I.R., Silva-Filho, E.V., Schaefer, C., Sella, S.M., Silva, C.A., Gomes, V., et al., 2006. Baseline mercury and zinc concentrations in terrestrial and coastal organisms of Admiralty Bay, Antarctica. Environ. Pollut. 140, 304–311.

Sastry, V.W., Bhatt, Y.M., 1965. Zinc content of some marine bivalves and barnacles from Bombay shores. J. Indian Chem. Soc. 42, 121–122.

Sather, B.T., 1967. Chromium absorption and metabolism by the crab, *Podophthalmus vigil*. In: Proceedings of the International Symposium on Radioecological Concentration Processes, Stockholm, 1966, pp. 943–976.

Schreiber, W., 1983. Mercury content of fishery products: Data from the last decade. J. Am. Med. Assoc. 289, 1667–1674.

Schuhmacher, M., Bosque, M.A., Domingo, J.L., Corbella, J., 1990. Lead and cadmium concentrations in marine organisms from the Tarragona coastal waters, Spain. Bull. Environ. Contam. Toxicol. 44, 784–789.

Schuhmacher, M., Domingo, J.L., Bosque, M.A., Corbella, J., 1992. Heavy metals in marine species from the Tarragona Coast, Spain. J. Environ. Sci. Health 27A, 1939–1948.

Schuhmacher, M., Batiste, J., Bosque, M.A., Domingo, J.L., Corbella, J., 1994. Mercury concentrations in marine species from the coastal area of Tarragona Province, Spain. Dietary intake of mercury through fish and seafood consumption. Sci. Total Environ. 156, 269–273.

Scott-Fordsmand, J.J., Depledge, M.H., 1993. The influence of starvation and copper exposure on the composition of the dorsal carapace and distribution of trace metals in the shore crab *Carcinus maenas* (L.). Comp. Biochem. Physiol. 106C, 537–543.

Seebaugh, D.R., Estephan, A., Wallace, W.G., 2006. Relationship between dietary cadmium absorption by grass shrimp (*Palaemonetes pugio*) and trophically available cadmium in amphipod (*Gammarus lawrencianus*) prey. Bull. Environ. Contam. Toxicol. 76, 16–23.

Shealy, M.H., Sandifer, P.A., 1975. Effects of mercury on survival and development of the larval grass shrimp, *Palaemonetes vulgaris*. Mar. Biol. 33, 7–16.

Shimizu, M., Kajihara, T., Hiyama, Y., 1970. Uptake of ^{60}Co by marine animals. Rec. Oceanogr. Works Jpn. 10, 137–145.

Shiomi, K., Shinagawa, A., Hirota, K., Yamanaka, H., Kikuchi, T., 1984a. Identification of arsenobetaine as a major arsenic compound in the ivory shell *Buccinum striatissimum*. Agric. Biol. Chem. 48, 2863–2864.

Shiomi, K., Shinigawa, A., Igarashi, T., Yamanaka, H., Kikuchi, T., 1984b. Evidence for the presence of arsenobetaine as a major arsenic compound in the shrimp *Sergestes lucens*. Experientia 40, 1247–1248.

Shrestha, K.P., Morales, E., 1987. Seasonal variations of iron, copper and zinc in *Penaeus brasiliensis* from two areas of the Caribbean Sea. Sci. Total Environ. 65, 175–180.

Side, J., 1987. Organotins: Not so good relations. Mar. Pollut. Bull. 18, 205–206.

Silva, C.A.R., Smith, B.D., Rainbow, P.S., 2006. Comparative biomonitors of coastal trace metal contamination in tropical South America (N. Brazil). Mar. Environ. Res. 61, 439–455.

Sims Jr., R.R., Presley, B.J., 1976. Heavy metal concentrations in organisms from an actively dredged Texas bay. Bull. Environ. Contam. Toxicol. 16, 520–527.

Sirota, G.R., Uthe, J.P., 1977. Determination of tetraalkyl lead compounds in biological materials. Anal. Chem. 49, 823–825.

Small, L.F., Fowler, S.W., 1973. Turnover and vertical transport of zinc by the euphausiid *Meganyctiphanes norvegica* in the Ligurian Sea. Mar. Biol. 18, 284–290.

Small, L.F., Fowler, S.W., Keckes, S., 1973. Flux of zinc through a macroplanktonic crustacean. In: Radioactive Contamination of the Marine Environment. IAEA, Vienna, Austria, pp. 437–452.

Small, L.F., Keckes, S., Fowler, S.W., 1974. Excretion of different forms of zinc by the prawn, *Palaemon serratus* (Pennant). Limnol. Oceanol. 18, 789–793.

Smith, P.J., 1978. Structure/Activity Relationships for Di- and Triorganotin Compounds. I.T.R.I Rep. 569. Available from International Tin Research Institute, Greenford, Middlesex, UK, 16 pp.

Smith, B.S., 1981. Tributyltin compounds induce male characteristics on female mud snails *Nassarius obsoletus* = *Ilyanassa obsolete*. J. Appl. Toxicol. 1, 141–144.

Sorentino, C., 1979. Mercury in marine and freshwater fish of Papua, New Guinea. Aust. J. Mar. Freshw. Res. 30, 617–623.

Sparling, D.W., Lowe, T.P., 1996. Environmental hazards of aluminum to plants, invertebrates, fish, and wildlife. Rev. Environ. Contam. Toxicol. 145, 1–127.

Spear, P.A., 1981. Zinc in the aquatic environment: Chemistry, distribution, and toxicology. Natl. Res. Coun. Canada, Publ. NRCC 17589, 145 pp.

Sprague, J.G., 1986. Toxicity and Tissue Concentrations of Lead, Zinc, and Cadmium for Marine Molluscs and Crustaceans. Available from International Lead Zinc Research Organization, 2525 Meridian Parkway, Research Triangle Park, North Carolina 27709-2036, 215 pp.

Srinivasin, M., Mahajan, B.A., 1989. Mercury pollution in an estuarine region and its effects on a coastal population. Int. J. Environ. Stud. 35, 63–69.

Stenner, R.D., Nickless, G., 1974a. Absorption of cadmium, copper, and zinc by dog whelks in the Bristol Channel. Nature 247, 198–199.

Stenner, R.D., Nickless, G., 1974b. Distribution of some heavy metals in organisms in Hardangerfjord and Skjerstadfjord, Norway. Water Air Soil Pollut. 3, 279–291.

Stenner, R.D., Nickless, G., 1975. Heavy metals in organisms of the Atlantic coast of S.W. Spain and Portugal. Mar. Pollut. Bull. 6, 89–92.

Stickney, R.R., Windom, H.L., White, D.B., Taylor, E.F., 1975. Heavy-metal concentrations in selected Georgia estuarine organisms with comparative food-habit data. In: Howell, F.G., Gentry, J.B., Smith, M.H. (Eds.), Mineral Cycling in South-Eastern Ecosystems. U.S. ERDA. Available as CONF 740513 from NTIS, Springfield, Virginia, pp. 257–267.

Sunda, W.G., Tester, P.A., Huntsman, S.A., 1990. Toxicity of trace metals to *Acartia tonsa* in the Elizabeth River and southern Chesapeake Bay. Estuar. Coast. Shelf Sci. 30, 207–221.

Sunderland, E.M., 2007. Mercury exposure from domestic and imported estuarine and marine fish in the U.S. seafood market. Environ. Health Perspect. 115, 235–242.

Sundlein, B., 1983. Effects of cadmium on *Pontoporeia affinis* (Crustacea: Amphipoda) in laboratory soft bottom microcosms. Mar. Biol. 74, 203–212.

Szefer, P., Pempkowiak, J., Skwarzec, B., Bojanowski, F., Holm, E., 1993. Concentration of selected metals in penguins and other representative fauna of the Antarctica. Sci. Total Environ. 138, 281–288.

Tariq, J., Jaffar, M., Ashraf, M., Moazzam, M., 1993. Heavy metal concentrations in fish, shrimp, seaweed, sediment, and water from the Arabian Sea, Pakistan. Mar. Pollut. Bull. 26, 644–647.

Tennant, D.A., Forster, W.D., 1969. Seasonal variation and distribution of 65-Zn, 54-Mn, and 51-Cr in tissues of the crab *Cancer magister* Dana. Health Phys. 18, 649–659.

Thain, J.E., Waldock, M.J., 1986. The impact of tributyl tin (TBT) antifouling paints on molluscan fisheries. Water Sci. Technol. 18, 193–202.

Thomas, P.G., Ritz, D.A., 1986. Growth of zinc granules in the barnacle *Elminius modestus*. Mar. Biol. 90, 255–260.

Thompson, J.A.J., Davis, J.C., Drew, R.E., 1976. Toxicity, uptake, and survey studies of boron in the marine environment. Water Res. 10, 869–875.

Thompson, J.A.J., Sheffer, M.G., Pierce, R.C., Chau, Y.K., Cooney, J.J., Cullen, W.R., et al., 1985. Organotin compounds in the aquatic environment: Scientific criteria for assessing their effects on environmental quality. Natl. Res. Coun. Canada, Publ. NRCC 22494. Available from Publications, NRCC/NRC, Ottawa, Canada, K1A OR6, 284 pp.

Thorsson, M.H., Hedman, J.E., Bradshaw, C., Gunnarsson, J.S., Gilek, M., 2008. Effects of settling organic matter on the bioaccumulation of cadmium and BDE-99 by Baltic Sea invertebrates. Mar. Environ. Res. 65, 264–281.

Thurberg, F.P., Calabrese, A., Gould, E., Greig, R.A., Dawson, M.A., Tucker, R.K., 1977. Response of the lobster *Homarus americanus*, to sublethal levels of cadmium and mercury. In: Vernberg, F.J., Calabrese, A.,

Thurberg, W.B., Vernberg, W.B. (Eds.), Physiological Responses of Marine Biota to Pollutants. Academic Press, New York, pp. 185–197.

Tijoe, P.S., de Goeij, J.J.M., de Bruin, M., 1977. Determination of trace elements in dried sea plant homogenate (SP-M-1) and in dried copepod homogenate (MA-A-1) by means of neutron activation analysis. Interuniv. Reactor Inst. (Delft, Netherlands), Rept. 133-77-05, pp. 1–14.

Topping, G., 1973. Heavy metals in shellfish from Scottish waters. Aquaculture 1, 379–384.

Tu, N.P.C., Ha, N.N., Ikemoto, T., Tuyen, B.C., Tanabe, S., Takeuchi, I., 2008. Regional variations in trace element concentrations in tissues of black tiger shrimp *Penaeus monodon* (decapod: penaeidae) from South Vietnam. Mar. Pollut. Bull. 57, 858–866.

Tuncel, G., Ramelow, G., Balkas, T.I., 1980. Mercury in water, organisms and sediments from a section of the Mediterranean coast. Mar. Pollut. Bull. 11, 18–22.

Turkmen, A., Turkmen, M., Tepe, Y., Mazlum, Y., Oymael, S., 2006. Metal concentrations in blue crab (*Callinectes sapidus*) and mullet (*Mugil cephalus*) in Iskenderun Bay, northern east Mediterranean, Turkey. Bull. Environ. Contam. Toxicol. 77, 186–193.

U.S. Environmental Protection Agency (USEPA), 1980a. Ambient water quality criteria for selenium. USEPA Rep., 440/5-80-070, pp. 1–123.

USEPA, 1980b. Ambient water quality criteria for copper. USEPA Rep., 440/5-80-036, pp. 1–162.

USEPA, 1980c. Ambient water quality criteria for nickel. USEPA Rep., 440/5-80-060, pp. 1–206.

USEPA, 1980d. Ambient water quality criteria for silver. USEPA Rep., 440/5-80-071, pp. 1–212.

USEPA, 1980e. Ambient water quality criteria for zinc. USEPA Rep., 440/5-80-079, pp. 1–158.

USEPA, 1985a. Ambient water quality criteria for arsenic—1984. USEPA Rep., 440/5-84-033, pp. 1–66.

USEPA, 1985b. Ambient water quality criteria for lead—1984. USEPA Rep., 440/5-84-027, pp. 1–81.

USEPA, 1987a. Ambient water quality criteria for selenium—1987. USEPA Rep., 440/5-87-006, pp. 1–121.

USEPA, 1987b. Ambient water quality criteria for zinc—1987. USEPA Rep., 440/5-87-003, pp. 1–207.

U.S. National Academy of Sciences (USNAS), 1989. Zinc. USNAS, Natl. Res. Coun., Subcomm. Zinc, Univ. Park Press, Baltimore, MD, 471 pp.

U'ren, S.C., 1983. Acute toxicity of bis(tributyltin)oxide to a marine copepod. Mar. Pollut. Bull. 14, 303–306.

Urrutia, C., Rudolph, A., Lermanda, M.P., Ahumada, R., 2008. Assessment of EDTA in chromium (III-VI) toxicity on marine intertidal crab (*Petrolisthes laevigatus*). Bull. Environ. Contam. Toxicol. 80, 1–3.

Van As, D., Fourie, H.O., Vleggaar, C.M., 1973. Accumulation of certain trace elements in the marine organisms from the sea around the Cape of Good Hope. In: Radioactive Contamination of the Marine Environment. IAEA, Vienna, Austria, pp. 615–624.

Van As, D., Fourie, H.O., Vleggaar, C.M., 1975. Trace element concentrations in marine organisms from the Cape West Coast. South Afr. J. Sci. 71, 151–154.

Van Weerelt, M., Preifer, W.C., Fiszman, M., 1984. Uptake and release of ^{51}Cr (VI) and ^{51}Cr (III) by barnacles (*Balanus* sp.). Mar. Environ. Res. 11, 201–211.

Vaskofsky, V.E., Korotchenko, O.D., Kosheleva, L.P., Levin, V.S., 1972. Arsenic in the lipid extracts of marine invertebrates. Comp. Biochem. Physiol. 41B (4), 777–784.

Vattuone, G.M., Griggs, K.S., McIntyre, D.R., Littlepage, J.L., Harrison, F.L., 1976. Cadmium concentrations in rock scallops in comparison with some other species. U.S. Ener. Res. Dev. Admin, UCRL 52022, pp. 1–11.

Vazquez, F.G., Sharma, V.K., Mendoza, Q.A., Hernandez, R., 2001. Metals in fish and shrimp of the Campeche Sound, Gulf of Mexico. Bull. Environ. Contam. Toxicol. 67, 756–762.

Vernberg, W.B., DeCoursey, P.J., 1977. Effect of sublethal metal pollutants on the fiddler crab *Uca pugilator*. U.S. Environ. Protect. Agen., Rept 600/3-77-024, pp. 1–59.

Vernberg, W.B., O'Hara, J., 1972. Temperature-salinity stress and mercury uptake in the fiddler crab, *Uca pugilator*. J. Fish. Res. Board Can. 29, 1491–1494.

Vernberg, W.B., Vernberg, F.J., 1972. The synergistic effects of temperature, salinity, and mercury on survival and metabolism of the adult fiddler crab, *Uca pugilator*. US Dept. Comm. Fish. Bull. 70 (2), 415–420.

Vernberg, W.B., DeCoursey, P.J., O'Hara, J., 1974. Multiple environmental factor effects on physiology and behavior of the fiddler crab, *Uca pugilator*. In: Vernberg, F.J., Vernberg, W.B. (Eds.), Pollution and Physiology of Marine Organisms. Academic Press, New York, pp. 381–425.

Vernberg, W.B., DeCoursey, P.J., Kelly, M., Johns, D.M., 1977. Effects of sublethal concentrations of cadmium on adult *Palaemonetes pugio* under static and flow-through conditions. Bull. Environ. Contam. Toxicol. 17, 16–24.

Verriopoulos, G., Dimas, S., 1988. Combined toxicity of copper, cadmium, zinc, lead, nickel, and chrome in the copepod *Tisbe holothuriae*. Bull. Environ. Contam. Toxicol. 41, 378–384.

Verriopoulos, G., Hardouvelis, D., 1988. Effects of sublethal concentrations of zinc on survival and fertility in four successive generations of *Tisbe*. Mar. Pollut. Bull. 19, 162–166.

Verriopoulos, G., Moraitu-Apostolopoulou, M., 1989. Toxicity of zinc to the marine copepod *Tisbe holothuriae*; the importance of the food factor. Arch. Hydrobiol. 114, 457–463.

Vighi, M., Calamari, D., 1985. QSARs for organotin compounds on *Daphnia magna*. Chemosphere 14, 1925–1932.

Vinogradova, Z.A., Koual'skiy, V.V., 1962. Elemental composition of the Black Sea plankton. Doklady Acad. Sci. USSR Earth Sci. Sec. 147, 217–219.

Vos, G., Hovens, J.P.C., 1986. Chromium, nickel, copper, zinc, arsenic, selenium, cadmium, mercury and lead in Dutch fishery products 1977-1984. Sci. Total Environ. 52, 25–40.

Waiwood, B.A., Zitko, V., Haya, K., Burridge, L.E., McLeese, D.W., 1987. Uptake and excretion of zinc in several tissues of the lobster (*Homarus americanus*). Environ. Toxicol. Chem. 6, 27–32.

Waldock, R., Rees, H.L., Matthiessen, P., Pendle, M.A., 1999. Surveys of the benthic infauna of the Crouch Estuary (UK) in relation to TBT contamination. J. Mar. Biol. Assoc. UK 79, 225–232.

Walker, G., 1977. "Copper" granules in the barnacle *Balanus balanoides*. Mar. Biol. 39, 343–349.

Walker, G., Rainbow, P.S., Foster, P., Crisp, D.J., 1975a. Barnacles: Possible indicators of zinc pollution? Mar. Biol. 30, 57–65.

Walker, G., Rainbow, P.S., Foster, P., Holland, D.L., 1975b. Zinc phosphate granules in tissues surrounding the midgut of the barnacle *Balanus balanoides*. Mar. Biol. 33, 161–166.

Wang, Z., Kong, H., Wu, D., 2007a. Acute and chronic copper toxicity to a saltwater cladoceran *Moina monogolica* Daday. Arch. Environ. Contam. Toxicol. 53, 50–56.

Wang, Z.S., Kong, H.N., Wu, D.Y., 2007b. Reproductive toxicity of dietary copper to a saltwater cladoceran, *Moina monogolica* Daday. Environ. Toxicol. Chem. 36, 126–131.

Ward, G.S., Hollister, T.A., Heitmuller, P.T., Parrish, P.R., 1981. Acute and chronic toxicity of selenium to estuarine organisms. Northeast Gulf Sci. 4, 73–78.

Ward, T.J., Correll, R.L., Anderson, R.B., 1986. Distribution of cadmium, lead and zinc amongst the marine sediments, sea grasses and fauna, and the selection of sentinel accumulators, near a lead smelter in South Australia. Aust. J. Mar. Freshw. Res. 37, 567–585.

Ward, T.J., Boeri, R.L., Hogstrand, C., Kramer, J.R., Lussier, S.M., Stubblefield, W.A., et al., 2006. Influence of salinity and organic carbon on the chronic toxicity of silver to mysids (*Americamysis bahia*) and silversides (*Menidia beryllina*). Environ. Toxicol. Chem. 25, 1809–1816.

Weeks, J.M., 1992. Copper-rich granules in the ventral caeca of talitrid amphipods (crustacea: amphipoda: talitridae). Ophelia 36, 119–133.

Weeks, J.M., Moore, P.G., 1991. The effect of synchronous moulting on body copper and zinc concentrations in four species of talitrid amphipods (crustacea). J. Mar. Biol. Assoc. UK 71, 481–488.

Weeks, J.M., Rainbow, P.S., 1991. The uptake and accumulation of zinc and copper from solution by two species of talitrid amphipods (Crustacea). J. Mar. Biol. Assoc. UK 71, 811–826.

Weeks, J.M., Rainbow, P.S., 1993. The relative importance of food and seawater as sources of copper and zinc to talitrid amphipods (crustacea; amphipoda; talitridae). J. Appl. Ecol. 30, 722–735.

Weimin, Y., Batley, G.E., Ahsanullah, M., 1994. Metal bioavailability to the soldier crab *Mictyris longicarpus*. Sci. Total Environ. 141, 27–44.

Weis, J.S., 1978. Interactions of methylmercury, cadmium and salinity on regeneration in the fiddler crabs *Uca pugilator*, *U. Pugnax*, and *U. Minax*. Mar. Biol. 49, 119–124.

Weis, J., 1980. Effect of zinc on regeneration in the fiddler crab *Uca pugilator* and its interactions with methylmercury and cadmium. Mar. Environ. Res. 3, 249–255.

Weis, J.S., Kim, K., 1988. Tributyltin is a teratogen in producing deformities in limbs of the fiddler crab, *Uca pugilator*. Arch. Environ. Contam. Toxicol. 17, 583–587.

Weis, J.S., Weis, P., 1992. Transfer of contaminants from CCA-treated lumber to aquatic biota. J. Exp. Mar. Biol. Ecol. 161, 189–199.

Weis, J.S., Gottlieb, J., Kwiatkowski, J., 1987. Tributyltin retards regeneration and produces deformities of limbs in the fiddler crab, *Uca pugilator*. Arch. Environ. Contam. Toxicol. 16, 321–326.

Welander, A.D., 1969. Distribution of radionuclides in the environment of Eniwetok and Bikini Atolls, August 1964. In: Nelson, D.J., Evans, F.C. (Eds.), Symposium on Radioecology. Proceedings of the Second National Symposium. Available as CONF-370503 from NTIS, Springfield, Virginia, pp. 346–354.

Wharfe, J.R., Van Den Broek, W.L.F., 1977. Heavy metals in macroinvertebrates and fish from the lower Medway Estuary, Kent. Mar. Pollut. Bull. 8, 31–34.

White, S.L., Rainbow, P.S., 1985. On the metabolic requirements for copper and zinc in molluscs and crustaceans. Mar. Environ. Res. 16, 215–229.

Wieser, W., 1967. Conquering terra firma: The copper problem from the isopod's point of view. Helg. wiss. Meeres. 15, 282–293.

Wilson, K.W., Connor, P.M., 1971. The use of a continuous-flow apparatus in the study of longer-term toxicity of heavy metals. Int. Coun. Explor. Sea C.M., 1971/E8, pp. 343–347.

Windom, H., Gardner, W., Stephens, J., Taylor, F., 1976. The role of methylmercury production in the transfer of mercury in a salt marsh ecosystem. Estuar. Coast. Mar. Sci. 4, 579–583.

Wolfe, D.A., Cross, F.A., Jennings, C.D., 1973. The flux of Mn, Fe, and Zn in an estuarine ecosystem. In: Radioactive Contamination of the Marine Environment. IAEA, Vienna, Austria, pp. 159–175.

Won, J.H., 1973. The concentrations of mercury, cadmium, lead, and copper in fish and shellfish of Korea. Bull. Korean Fish. Soc. 6, 1–19.

Wong, C.K., Chu, K.H., Tang, K.W., Tam, T.W., Wong, J.L., 1993. Effects of chromium, copper and nickel on survival and feeding behavior of *Metapenaeus ensis* larvae and postlarvae (Decapoda: Penaeidae). Mar. Environ. Res. 36, 63–78.

World Health Organization (WHO), 1980. Tin and organotin compounds: A preliminary review. Environ. Health Crit. 15, 1–109.

Wrench, J., Fowler, S.W., Unlu, M.Y., 1979. Arsenic metabolism in a marine food chain. Mar. Pollut. Bull. 10, 18–20.

Wright, D.A., 1976. Heavy metals in animals from the north east coast. Mar. Pollut. Bull. 7, 36–38.

Wright, D.A., 1977a. The uptake of cadmium into the haemolymph of the shore crab *Carcinus maenas*: The relationship with copper and other divalent cations. J. Exp. Biol. 67, 147–161.

Wright, D.A., 1977b. The effect of salinity on cadmium uptake by the tissues of the shore crab *Carcinus maenas*. J. Exp. Biol. 67, 136–146.

Wright, D.A., 1977c. The effect of calcium on cadmium uptake by the shore crab *Carcinus maenas*. J. Exp. Biol. 67, 163–173.

Wright, D.A., Brewer, C.C., 1979. Cadmium turnover in the shore crab *Carcinus maenas*. Mar. Biol. 50, 151–156.

Yang, Z.B., Zhao, Y.L., Li, N., Yang, J., 2007. Effect of waterborne copper on the microstructures of gill and hepatopancreas in *Eriocheir sinensis* and its induction of metallothionein synthesis. Arch. Environ. Contam. Toxicol. 52, 222–228.

Yang, Z.B., Zhao, Y.L., Li, N., Yang, J., 2008. Effect of waterborne copper on the microstructures and ultrastructure of the X-organ sinus gland complex in *Eriocheir sinensis*. Bull. Environ. Contam. Toxicol. 80, 68–73.

Yannai, S., Sachs, K., 1978. Mercury compounds in some eastern Mediterranean fishes, invertebrates, and their habitats. Environ. Res. 16, 408–418.

Yoshinari, T., Subramanian, V., 1976. Adsorption of metals by chitin. In: Nriagu, J.O. (Ed.), Environmental Biogeochemistry: Vol. 2. Metals Transfer and Ecological Mass Balances. Ann Arbor Science Publishers, Ann Arbor, MI, pp. 542–555.

Zafiropoulos, D., Grimanis, A.P., 1977. Trace elements in *Acartia clausi* from Elefsis Bay of the upper Saronikos Gulf, Greece. Mar. Pollut. Bull. 8, 79–81.

Zauke, G.P., 1977. Mercury in benthic invertebrates of the Elbe estuary. Helgol. Wiss. Meers. 19, 358–374.

Zauke, G.P., Schmalenbach, I., 2006. Heavy metals in zooplankton and decapod crustaceans from the Barents Sea. Sci. Total Environ. 359, 283–294.

Zingde, M.D., Singbal, S.Y.S., Moraes, C.F., Reddy, C.V.G., 1976. Arsenic, copper, zinc, and manganese in the marine flora and fauna of coastal and estuarine waters around Goa. Indian J. Mar. Sci. 5, 212–217.

Zodle, B., Whitman, K.J., 2003. Effects of sampling, preparation and defecation on metal concentrations in selected invertebrates at urban sites. Chemosphere 52, 1095–1103.

Zuckerman, J.J., Reisdorf, R.P., Ellis III, H.V., Wilkinson, R.R., 1978. Organotins in biology and the environment. In: Brinckman, F.E., Bellama, J.M. (Eds.), Organometals and Organometalloids, Occurrence and Fate in the Environment. Amer. Chem. Soc. Sympos. Series 82. American Chemical Society, Washington, DC, pp. 388–424.

Zumholz, K., Hansteen, T.H., Klugel, A., Piatkowski, U., 2006. Food effects on statolith composition of the common cuttlefish (*Sepia officinalis*). Mar. Biol. 150, 237–244.

CHAPTER 8
Insects

Sea-skaters of the genus *Halobates* are the only known insects found in the open ocean and the only marine invertebrates which live exclusively at the air-water interface. Other marine insects, including flies, occupy inshore niches. Attempts have been made to relate environmental contamination with concentrations of cadmium and other trace metals in marine insects; however, data are insufficient to warrant firm conclusions. Variables known to influence metal concentrations in marine insects include life stage and gender (Table 8.1).

8.1 Cadmium

Mean cadmium concentrations in whole *Halobates* spp. ranged from 6.0 to 152.0 mg Cd/kg DW (dry weight), although maxima as high as 309.0 mg Cd/kg DW are recorded (Table 8.1).

Halobates can accumulate cadmium from diet as well as from the medium; however, diet was less important as a route than the medium, in this case the surface microlayers (Cheng et al., 1979). Since these insects normally feed on fluidized soft tissues of prey zooplankton, it was probable that the high concentrations of cadmium and other metals in zooplankton exoskeletons were not biologically available. *Halobates*, in turn, are preyed upon by at least two species of Pacific seabirds (Bull et al., 1977), but transfer rates are unknown.

Laboratory studies with *Halobates robustus* and radiocadmium demonstrate cadmium accumulation in tissues, but levels were below those in the ambient medium (Schulz-Baldes and Cheng, 1979). It was concluded that cadmium was incorporated into body tissues, not merely adsorbed onto surfaces. Uptake, as well as depuration, followed a biphasic pattern, with the biological half-life of cadmium ranging from 0.6 h to 19.7 days for uptake into the first compartment, and depuration in the second compartment, respectively (Schulz-Baldes and Cheng, 1979), suggesting that the chemical form or site of bound cadmium had changed.

8.2 Chromium

Mean chromium concentrations in *H. robustus* ranged from 1.0 to 3.0 mg Cr/kg DW, being highest in females, lowest in males, and intermediate in nymphs (Table 8.1).

Table 8.1: Trace Element Concentrations in Field Collections of Marine Insects

Element and Organism	Value	Reference[a]
Cadmium		
Halobates micans; whole	22.7 (0.0-309.0) DW	1
Halobates robustus; whole		
Males	6.0 DW	2
Females	8.0 DW	2
Nymphs	7.0 DW	2
Halobates sericeus; whole	40.0 DW	3
Halobates sobrinus; whole	152.0 DW	3
Marine flies; whole; 2 spp.	40.0-150.0 DW	4
Rheumobates aestuarius; whole	>5.0 DW	3
Chromium		
Halobates robustus; whole		
Males	1.0 DW	2
Females	3.0 DW	2
Nymphs	2.0 DW	2
Copper		
Halobates robustus; whole		
Males	129.4 DW	2
Females	162.3 DW	2
Nymphs	134.6 DW	2
Halobates sericeus; whole	45.0 DW	3
Halobates sobrinus; whole	50.0 DW	3
Marine flies; whole; 2 spp.	10.0-125.0 DW	4
Rheumobates aestuarius; whole	64.0 DW	3
Iron		
Halobates sericeus; whole	178.0 DW	3
Halobates sobrinus; whole	289.0 DW	3
Marine flies; whole; 2 spp.	100.0-225.0 DW	4
Rheumobates aestuarius; whole	204.0 DW	3

(*Continues*)

Table 8.1: Cont'd

Element and Organism	Value	Reference[a]
Lead		
Halobates robustus; whole		
Males	<1.0 DW	2
Females	<1.0 DW	2
Nymphs	<1.0 DW	
Halobates sericeus; whole	7.0 DW	3
Halobates sobrinus; whole	10.0 DW	3
Rheumobates aestuarius; whole	>2.0 DW	3
Manganese		
Marine flies; whole; 2 spp.	<10.0 DW	4
Nickel		
Halobates sericeus; whole	7.0 DW	3
Halobates sobrinus; whole	18.0 DW	3
Rheumobates aestuarius; whole	6.0 DW	3
Zinc		
Halobates robustus; whole		
Males	115.4 DW	2
Females	145.7 DW	2
Nymphs	109.8 DW	2
Halobates sericeus; whole	176.0 DW	3
Halobates sobrinus; whole	176.0 DW	3
Marine flies; whole; 2 spp.	150.0 to >300.0 DW	4
Rheumobates aestuarius; whole	197.0 DW	3

Values are in mg element/kg dry weight (DW).
[a] 1, Bull et al., 1977; 2, Cheng et al., 1979; 3, Cheng et al., 1976; 4, Bender, 1975.

8.3 Copper

The highest mean copper concentration on record for marine insects is 162.3 mg Cu/kg DW whole female *H. robustus* (Table 8.1).

8.4 Iron

Mean iron concentrations in whole marine insects ranged from 100.0 to 289.0 mg Fe/kg DW (Table 8.1).

8.5 Lead

The highest lead concentration recorded is 10.0 mg Pb/kg DW in whole *Halobates sobrinus* (Table 8.1).

8.6 Manganese

A single datum is available for manganese, being less than 10.0 mg Mn/kg DW whole fly (Table 8.1).

8.7 Nickel

Mean nickel concentrations in three species of marine insects ranged from 6.0 to 18.0 mg Ni/kg DW whole organism (Table 8.1).

8.8 Zinc

Mean zinc concentrations in a variety of marine insects ranged from a low of 109.8 mg Zn/kg DW in whole *H. robustus* nymphs to >300.0 mg Zn/kg DW in whole marine flies (Table 8.1).

8.9 Literature Cited

Bender, J.A., 1975. Trace metal levels in beach dipterans and amphipods. Bull. Environ. Contam. Toxicol. 14, 187–192.

Bull, K.E., Murton, R.K., Osborn, D., Ward, P., Cheng, L., 1977. High levels of cadmium in Atlantic seabirds and sea-skaters. Nature 269, 507–509.

Cheng, L., Alexander, G.V., Franco, P.J., 1976. Cadmium and other heavy metals in seaskaters (Gerridae: *Halobates, Rheumobates*). Water Air Soil Pollut. 6, 33–38.

Cheng, L.P., Franco, P.J., Schulz-Baldes, M., 1979. Heavy metals in the sea-skater *Halobates robustus* from the Galapagos Islands: Concentrations in nature and uptake experiments with special reference to cadmium. Mar. Biol. 54, 201–206.

Schulz-Baldes, M., Cheng, L., 1979. Uptake and loss of radioactive cadmium by the sea-skater *Halobates robustus* (Heteroptera: Gerridae). Mar. Biol. 52, 253–258.

CHAPTER 9
Chaetognaths

The phylum Chaetognatha is a small group of pelagic, mostly oceanic, invertebrates consisting of about 50 species. The most studied genus is *Sagitta* with about 30 species. Chaetognaths seldom exceed 3 cm in length and each contains a distinct head, trunk, and tail region. Typically, the head contains a pair of eyes and two pairs of sickle-shaped setae, the body a pair of lateral fins, and the tail a lateral caudal fin. These hermaphroditic carnivores frequently occur in large schools and are preyed upon by pelagic fishes; in turn, they readily devour fish fry, some as large as themselves.

Most of what is known about trace metal composition of chaetognaths is based upon analysis of collections from three different geographic locales: the mid-Pacific Ocean (Nicholls et al., 1959); the Black Sea (Vinogradova and Koual'skiy, 1962); and Strathcona Sound, Baffin Island, near the Arctic Circle (Bohn and McElroy, 1976). A strict comparison between locales was not possible for several reasons. First, the observations of Vinogradova and Koual'skiy (1962) represent maximum values recorded, whereas those of Nicholls et al. (1959) and Bohn and McElroy (1976) are means. Second, different species were collected from different areas with no species in common; it is probable that closely related species of chaetognaths differ substantially in their ability to accumulate trace metals, as is the case for other phyla. Finally, different analytical techniques were used including atomic absorption spectrophotometry and emission spectrographic analysis, as well as different sample preparation techniques. Until these and other sample preparation techniques and methodologies are cross-validated by interlaboratory comparisons, there will always be some doubt regarding the accuracy and precision of elemental data in chaetognaths and other groups.

9.1 Arsenic

Chaetognaths collected near the Arctic circle contain 7.5-7.7 mg As/kg DW (Table 9.1).

9.2 Barium

The single datum available shows that *Sagitta* sp. from the Black Sea contain up to 1500.0 mg Ba/kg AW (Table 9.1).

Table 9.1: Trace Element Concentrations in Field Collections of Chaetognaths

Element and Organism	Concentration	Reference[a]
Arsenic		
Chaetognaths	7.5-7.7 DW	1
Barium		
Sagitta sp.	1500.0 AW	2
Boron		
Sagitta elegans	130.0 AW	3
Cadmium		
Chaetognaths	1.2-1.3 DW	1
Eukronia hamata		
Barents Sea	0.8 DW	4
Greenland Sea	1.1 DW	5
Chromium		
Sagitta elegans	<1.0 AW	3
Sagitta sp.	80.0 AW	2
Cobalt		
Sagitta elegans	110.0 AW	3
Copper		
Chaetognaths	(5.6-6.3) DW	1
Eukronia hamata		
Barents Sea	4.0 DW	4
Greenland Sea	3.0 DW	5
Sagitta elegans	1100.0 AW	3
Sagitta sp.	30,000.0 AW	2
Gallium		
Sagitta sp.	10.0 AW	2
Iron		
Chaetognaths	(32.0-33.0) DW	1
Sagitta sp.	100,000.0 AW	2

(Continues)

Table 9.1: Cont'd

Element and Organism	Concentration	Reference[a]
Lead		
Eukronia hamata; Greenland Sea	<0.3 DW	5
Sagitta elegans	300.0 AW	3
Sagitta sp.	1500.0 AW	2
Lithium		
Sagitta sp.	500.0 AW	2
Molybdenum		
Sagitta elegans	3.0 AW	3
Nickel		
Sagitta elegans	480.0 AW	3
Sagitta sp.	300.0 AW	2
Silver		
Sagitta sp.	60.0 AW	2
Strontium		
Sagitta sp.	1500.0 AW	2
Tin		
Sagitta elegans	20.0 AW	3
Sagitta sp.	400.0 AW	2
Titanium		
Sagitta elegans	3.0 AW	3
Sagitta sp.	500.0 AW	2
Vanadium		
Sagitta elegans	13.0 AW	3
Sagitta sp.	20.0 AW	2
Zinc		
Chaetognaths; Arctic Circle	76.0-90.0 DW	1

Eukronia hamata
   Barents Sea
   Greenland Sea
	
	 124.0 DW
	 69.0 DW
		
		 4
		 5

(Continues)

Table 9.1: Cont'd

Element and Organism	Concentration	Reference[a]
Sagitta sp.	20,000.0 AW	2
Zirconium		
Sagitta sp.	70.0 AW	2

Values are in mg element/kg whole organism dry weight (DW), or ash weight (AW).
[a] 1, Bohn and McElroy, 1976; 2, Vinogradova and Koual'skiy, 1962; 3, Nicholls et al., 1959; 4, Zauke and Schmalenbach, 2006; 5, Ritterhoff and Zauke, 1997.

9.3 Boron

The single datum available shows that *Sagitta elegans* from the mid-Pacific Ocean contain 130.0 mg B/kg AW (Table 9.1).

9.4 Cadmium

Chaetognaths from the Arctic Circle contain 1.2-1.3 mg Cd/kg DW; concentrations in chaetognaths from the Barents Sea are 0.8 mg Cd/kg DW, and for those from the Greenland Sea it is 1.1 mg Cd/kg DW (Table 9.1).

9.5 Chromium

Sagitta spp. contain up to 80.0 mg Cr/kg AW (Table 9.1).

9.6 Cobalt

S. elegans from the mid-Pacific Ocean contain 110.0 mg Co/kg AW (Table 9.1). Ketchum and Bowen (1958) report that *Sagitta* sp. can concentrate cobalt isotopes from seawater by a factor of 1000. *Sagitta enflata* eliminates radiocobalt at the rate of 11%/h (Kuenzler, 1969).

9.7 Copper

The highest copper concentrations recorded in field collections of chaetognaths were 6.3 mg Cu/kg on a DW (dry weight) basis, and 30,000.0 mg Cu/kg on an AW (ash weight) basis (Table 9.1).

9.8 Gallium

Sagitta sp. from the Black Sea contain up to 10.0 mg Ga/kg AW (Table 9.1).

9.9 Iron

Chaetognaths near the Arctic Circle contain 32.0-33.0 mg Fe/kg DW, and *Sagitta* sp. from the Black Sea contain up to 100,000.0 mg Fe/kg AW (Table 9.1). Iron isotopes in seawater are accumulated by *Sagitta* sp. by a factor approaching 5000 (Ketchum and Bowen, 1958). Much of the radioiron eliminated by *S. enflata* was particulate, with significant amounts nonexchangeable, that is, not biologically available (Kuenzler, 1969).

9.10 Lead

Chaetognaths from the mid-Pacific Ocean contain 300.0 mg Pb/kg AW and those from the Black Sea contain up to 1500.0 mg Pb/kg AW (Table 9.1).

9.11 Lithium

Chaetognaths from the Black Sea contain up to 500.0 mg Li/kg AW (Table 9.1).

9.12 Molybdenum

The single datum available indicates that *S. elegans* from the mid-Pacific Ocean contain an average of 3.0 mg Mo/kg AW (Table 9.1).

9.13 Nickel

Sagitta spp. from two locales contain 300.0 to 480.0 mg Ni/kg AW (Table 9.1).

9.14 Silver

Sagitta sp. from the Black Sea contain up to 60.0 mg Ag/kg AW (Table 9.1).

9.15 Strontium

Sagitta sp. from the Black Sea contain up to 1500.0 mg Sr/kg AW whole organism (Table 9.1). *Sagitta* sp. can accumulate strontium isotopes from seawater by a factor of 70 (Ketchum and Bowen, 1958).

9.16 Tin

The maximum mean tin concentration recorded in field collections of chaetognaths is 400.0 mg Sn/kg AW (Table 9.1).

9.17 Titanium

Sagitta spp. from two locales contained 3.0-500.0 mg Ti/kg AW (Table 9.1). The role of titanium in crustacean metabolism is unknown.

9.18 Vanadium

Sagitta spp. from two locales contained between 13.0 and 20.0 mg V/kg AW (Table 9.1).

9.19 Zinc

Chaetognaths collected near the Arctic Circle contain an average of 76.0-90.0 mg Zn/kg dry weight; *Sagitta* sp. from the Black Sea had a maximum of 20,000.0 mg Zn/kg AW, and *Eukronia hamata* contained 69.0-124.0 mg Zn/kg DW (Table 9.1). Radiozinc is eliminated from *S. enflata* at a rate of 3%/h (Kuenzler, 1969).

9.20 Zirconium

Sagitta sp. from the Black Sea contain up to 70.0 mg Zr/kg AW (Table 9.1).

9.21 Literature Cited

Bohn, A., McElroy, R.O., 1976. Trace metals (As, Cd, Cu, Fe, and Zn) in Arctic cod, *Boreogadus saida*, and selected zooplankton from Strathcona Sound, Northern Baffin Island. J. Fish. Res. Board Can. 33, 2836–2840.

Ketchum, B.M., Bowen, V.T., 1958. Biological factors determining the distribution of radioisotopes in the sea. In: Proceedings of the Second Annual United Nations Conference on Peaceful Uses of Atomic Energy, Vol. 18, Geneva, pp. 429–433.

Kuenzler, E.J., 1969. Elimination and transport of cobalt by marine zooplankton. In: Proceedings of the 2nd National Symposium on Radioecology. U.S. Atom. Ener. Comm. Conf. 670503, pp. 483–492.

Nicholls, G.D., Curl Jr., H., Bowen, V.T., 1959. Spectrographic analyses of marine plankton. Limnol. Oceanol. 4, 472–476.

Ritterhoff, J., Zauke, G.P., 1997. Trace metals in field samples of zooplankton from the Fram Strait and the Greenland Sea. Sci. Total Environ. 199, 255–270.

Vinogradova, Z.A., Koual'skiy, V.V., 1962. Elemental composition of the Black Sea plankton. Doklady Acad. Sci. USSR Earth Sci. Sec. 147, 217–219.

Zauke, G.P., Schmalenbach, I., 2006. Heavy metals in zooplankton and decapod crustaceans from the Barents Sea. Sci. Total Environ. 359, 283–294.

CHAPTER 10
Annelids

This phylum of segmented worms includes earthworms and leeches as well as a large number of free-living or tube-building species belonging to the class Polychaeta. The name Polychaeta refers to the numerous tufts of setae at the sides of the body. Class Polychaeta is the largest in this phylum with more than 10,000 species recorded, and widely distributed in marine habitats—although some species are found in brackish and fresh water and in moist earth. Polychaetes are most common in coastal zones of all oceans in depths up to 600 m, but some have been recorded from the abyss. Several groups of polychaetes are dominant in tropical ecosystems, others in frigid waters, and many others with limited geographic ranges. Some are entirely pelagic, but others are pelagic only through larval stages. Some are mainly sand dwelling and others rock borers, some commensal and a few parasitic. Polychaetes are used extensively by several investigators as indicators of metal-contaminated sediments and as experimental animals to assess the comparative toxicity and physiological effects of metals and metalloids.

10.1 Aluminum

The polychaete *Branchipolynoe seepensis,* a commensal parasite on gills of a hydrothermal vent mussel, *Bathymodiolus azoricus,* contained 11.9 mg Al/kg DW (dry weight) whole animal (Kadar et al., 2006).

Eggs of *Hydroides elegans* were subjected to various concentrations of aluminum from fertilization through the 24 h trocophore stage. Arrested development was evident at 1.1 mg Al/L for the 2 h blastula, and at 0.21 mg Al/L for trocophore larvae (Gopalakrishnan et al., 2007).

10.2 Americium

Beasley and Fowler (1976) report that americium concentrations in whole sandworms, *Nereis diversicolor*, reflect sediment americium concentrations; *Nereis* contain about 0.5% of the americium burden of the surrounding sediments.

10.3 Antimony

Limited data indicate that whole-body concentrations in field populations of polychaetes do not exceed 0.03 mg Sb/kg DW (Table 10.1).

10.4 Arsenic

Arsenic appears to concentrate in lipids, reaching 46,000.0 mg As/kg extract on a fresh weight (FW) basis (Table 10.1).

The Mediterranean fan worm, *Sabella spallanzanii*, normally contains about 1400.0 mg As/kg DW in branchial crowns and about 30.0 mg As/kg DW in body portions, including high levels (84%) of dimethylarsinic acid (DMA), a relatively toxic compound with a possible antipredatory role (Notti et al., 2007). *Sabella* exposed under laboratory conditions to arsenate (0.02 mg As/L), DMA (0.06 mg As/L), trimethylarsine (TMA; 0.06 mg As/L),

Table 10.1: Antimony, Arsenic, Barium, and Cadmium Concentrations in Field Collections of Annelids

Element and Organism	Concentration	Reference[a]
Antimony		
Sandworm, *Nereis diversicolor*; whole	0.03 DW	1
Arsenic		
Polychaete, *Chaetopterus variopedatus*; lipid extract	46,000.0 FW	2
Sandworm, *Nereis diversicolor*; whole	5.2 DW	1
Barium		
Annelids	0.16 FW	3
Cadmium		
Polychaete, *Branchipolynoe seepensis*; from gill of deep-sea-hydrothermal vent mussel; 2002; whole	13.6 DW	7

Polychaete, *Nephtys hombergi*; whole
    March
    October
	
	9.0 FW
	89.0 FW
		
		4
		4

Sandworm, *Nereis diversicolor*; whole	0.53 (0.08-3.6) FW	5, 6

Values are in mg element/kg fresh weight (FW) or dry weight (DW).
[a] 1, Leatherland and Burton, 1974; 2, Vaskovsky et al., 1972; 3, Mauchline and Templeton, 1966; 5, Rosenberg, 1977; 5, Bryan and Hummerstone, 1973b; 6, Bryan and Hummerstone, 1977; 7, Kadar et al., 2006.

or arsenobetaine (0.06 mg As/L) for 20 days showed the highest increases in branchial crowns treated with arsenate, lower increases with DMA and TMA, and no increase of arsenobetaine. Accumulated arsenic, except for arsenobetaine, could be chemically transformed to DMA, accounting for the elevated DMA levels typical of this species (Notti et al., 2007).

Water-soluble arsenic compounds in the polychaete *Arenicola marina* from Odense Fjord, Denmark differ significantly from most other marine animals in that arsenobetaine was present only as a minor (6%) constituent (Geiszinger et al., 2002). Water-soluble inorganic forms included arsenite (58%) and arsenate (16%). Other arsenic compounds were dimethylarsinate (4%), two arsenosugars (at 1% and 3%, respectively), tetramethylarsonium ion (1.5%), arsenocholine (<1%), trimethylarsoniopropionate (<1%), and an unknown anionic arsenical (*ca*. 10%). Polychaetes subjected to 0.01-1.0 mg As/L, as arsenate, in the laboratory took up arsenic in a dose-dependent nonlinear manner. There were no deaths during the 12 day exposure, and the highest whole-body concentration of 270.0 mg As/kg DW was recorded at 1.0 mg As/L. Most of the accumulated arsenic was biotransformed to arsenite and dimethylarsinate, with the rest being accumulated as unchanged arsenate. Other arsenic compounds found in *A. marina* did not increase as a result of arsenate exposure (Geiszinger et al., 2002).

10.5 Barium

Barium concentrations in whole annelids average 0.16 mg Ba/kg FW (Table 10.1).

10.6 Cadmium

The abnormally high concentration of 89.0 mg Cd/kg FW whole *Nephtys hombergi* (Table 10.1) is attributed to contamination from sediments as a result of dredging operations (Rosenberg, 1977). Cadmium resistance has been documented in marine annelids (Wallace et al., 1998). Oligochaetes (*Limnodrilus hoffmeisteri*) from Foundry Cove, New York—a severely cadmium-contaminated site—were more tolerant of cadmium than conspecifics from a reference site, surviving twice as long in a 7 day acute toxicity study (1.0 mg Cd/L) and with bioconcentration factors of 2020 versus 577 for the controls (radiocadmium-109). The cadmium-resistant worms produced metal-rich granules and metallothioneins for cadmium storage and detoxification, whereas nonresistant worms only produced metallothioneins. Grass shrimp (*Palaemonetes pugio*) fed with cadmium-resistant worms absorbed 21% of the ingested cadmium versus 75% for shrimp fed with nonresistant worms (Wallace et al., 1998).

Settling organic matter is a major food source for heterotrophic benthic fauna (Thorsson et al., 2008). In one study, the polychaete *Marenzellaria* spp. fed with ^{109}Cd-labeled green microalgae, *Tetraselmis* spp. for 34 days had a whole-body concentration factor of 12 over the diet (Thorsson et al., 2008). Uptake and retention of radiocadmium by a polychaete, *Eunice* sp., was studied (Alquezar et al., 2007). In that study *Eunice* was held for 385 h in seawater spiked with ^{109}Cd followed by 189 h in ^{109}Cd-free media. During uptake, radiocadmium was concentrated by a factor of 41 over seawater; during depuration 62% was lost. A biphasic depuration pattern was evident: the biological half-life of the short-lived component was 11 h, and for the long-lived component it was 5950 h (Alquezar et al., 2007).

Cadmium interacts with zinc (Goto and Wallace, 2007). Polychaetes, *Capitella capitata* were coexposed to 0.05 mg Cd/L plus 0.086 mg Zn/L with ^{109}Cd and ^{65}Zn radiotracers for 1 week, followed by depuration for 1 week. Uptake and loss of cadmium and zinc were similar, with worms depurating most of the cadmium (75%) and zinc (64%). Cadmium and zinc behaved similarly at the subcellular level; worms partitioned most of the cadmium (65%) and zinc (55%) to the heat stable proteins and organelles fraction. The heat denaturable proteins and metal-rich granules contained less than 6% of both metals. Authors conclude that cadmium-zinc interactions reduce partitioning of cadmium and zinc to heat stable proteins, indicating that metallothioneins and other metal-binding proteins are key to these interactions (Goto and Wallace, 2007). Cadmium also interacts with petroleum hydrocarbons (Sun and Zhou, 2007). A polychaete, *Hediste japonica*, exposed to 0.5 mg Cd/L or to 0.5 mg Cd/L plus 0.1 mg petroleum hydrocarbons/L for 3 days had BCF values of 25 for cadmium alone and 81 for the mixture, indicating significant interaction. Net accumulation rates, in mg Cd/kg body weight daily, were 4.3 for cadmium alone and 8.5 for the mixture (Sun and Zhou, 2007).

Three laboratory studies with *Nereis* spp. and cadmium are of interest. Ueda et al. (1976) found that *Nereis japonica*, when compared to controls, contained six times more cadmium when subjected to cadmium-contaminated sediments for 8 days, and up to 200 times more cadmium from cadmium-containing seawater in a similar period; they concluded that the main cadmium uptake route for this species is via the medium. *N. japonica* excreted the accumulated cadmium in multicompartmental phases: the major short-lived component had a computed biological half-life of 4 days. In a second study, *Nereis virens* of 1-2 g in weight, immersed in seawater containing 0.03 mg Cd/L for 14 days, accumulated cadmium at the rate of 0.01 mg Cd/kg DW whole animal per hour, with a final concentration of 7.5 mg Cd/kg DW whole worm (Ray et al., 1980). At a seawater concentration of 9.2 mg Cd/L, these values were 2.127 mg Cd/kg DW/h and a final whole-body content of 755.5 mg Cd/kg DW. Intermediate values were obtained with seawater cadmium concentrations of 0.09, 0.90, 1.90, and 4.4 mg/L. When the study was repeated with larger worms of 5-7 g wet weight, similar—but slightly lower—values were obtained in each case (Ray et al., 1980). Further studies indicated that cadmium uptake by *N. virens* is primarily from the aqueous phase rather than from cadmium-laden sediments (Ray et al., 1980). In the third study, Bryan and Hummerstone (1973b) demonstrated that survival and cadmium accumulation rates over a 17 day period for *N. diversicolor* was related to the organism's previous history of exposure to cadmium. Nereids from a low-cadmium sediment environment of 0.22 mg Cd/kg DW had shorter survival times, higher daily absorption rates, and higher whole-body concentrations at any given cadmium concentration in the medium than did nereids from a high-cadmium sediment environment of 3.1 mg Cd/kg DW. At 1.0 mg Cd/L, for example, *N. diversicolor* from the low-cadmium sediment environment had a daily absorption rate of 6.0 mg Cd/kg on a DW basis and a final concentration of 208.0 mg Cd/kg whole-body DW after 17 days; at 10.0 mg Cd/L, these values were 55.0 and 1860.0, respectively. Nereids from the high-cadmium sediment environment, when held in 1.0 mg Cd/L, had a daily absorption rate of only 4.0 mg Cd/kg whole-body DW and a final whole-body content of 140.0 mg Cd/kg DW; at 10.0 mg Cd/L, these values were 18.0 and 629.0, respectively. This pattern held at all cadmium concentrations tested, namely, 1.0, 2.5, 10.0, 25.0, and 100.0 mg Cd/L. Bryan and Hummerstone (1973b) suggest that long-time contamination from mining operations produced the elevated levels of cadmium in sediments, and that the polychaete populations from this area could withstand metals-induced stress to a higher degree than polychaete populations from less-contaminated areas. It is not known if differences between the two populations of nereids are genetic or the result of physiological adaptation. Bryan and Hummerstone (1973b) state that cadmium is not regulated by worms, and the resultant body concentrations appear to reflect sediment cadmium burdens.

A study with glycerid polychaetes showed that bloodworms, *Glycera dibranchiata*, accumulate cadmium through the body surface and the intestine (Rice and Chien, 1979). Cadmium absorbed through the gut bound rapidly to coelomic proteins, including hemoglobin. Ionic cadmium injected into the coelom at 7.0 mg Cd/L increased proline incorporation rates into hemoglobin by a factor of 15 in 72 h (Rice and Chien, 1979), and this may adversely affect survival over time. Sandrini et al. (2006) exposed the polychaete *Laeonereis acuta* to 0.1 mg Cd/L for up to 24 h. After 12 h, the mean whole-body cadmium content was 1.0 mg Cd/kg DW; after 24 h, it was about 2.3 mg Cd/kg DW. Cadmium was stored mainly in the cytosolic fractions. Cadmium generates oxidative stress in *Laeonereis* as indicated by altered enzyme activity (Sandrini et al., 2006). Polychaetes, *Perinereis aibuhitensis*, exposed to 0.05 mg Cd/L for 1 week contained 0.35 mg Cd/kg FW versus 0.04 mg Cd/kg FW for controls (Ng et al., 2008). However, cadmium exposure did not affect overall cadmium uptake rates, efflux rates, and metallothionein concentrations. During exposure, cadmium concentrations increased in the cytosolic fraction of *Perinereis*; it also increased during depuration, suggesting that the insoluble fraction of cadmium was lost at a faster rate than cytosolic cadmium (Ng et al., 2008).

Embryos and sperm of the polychaete *H. elegans* were the two most sensitive stages to cadmium insult with toxic effects apparent to sperm at 0.09 mg Cd/L (20 min exposure) and 0.18 mg Cd/L to embryos (2 h exposure); by contrast, the LC-50 (96 h) value for adults was 0.23 mg Cd/L (Gopalakrishnan et al., 2008). To protect annelids and other marine benthos against acute cadmium intoxication, a maximum water concentration of 0.023 mg Cd/L is recommended; however, to protect against chronic cadmium poisoning, a water concentration of <0.00025 mg Cd/L is recommended (Hall et al., 1998). Eisler (2000) states that seawater concentrations in excess of 0.0045 mg Cd/L should be considered potentially hazardous to marine life until additional data prove otherwise.

10.7 Cesium

The principal isotope of radiocesium found in the muscle of juvenile flounders, *Pleuronectes platessa*, was ^{137}Cs; the major source of cesium to *Pleuronectes* was diet, especially the *N. hombergi* fraction of the diet, despite its relatively low ^{137}Cs content (Pentreath and Jefferies, 1971). Some evidence exists demonstrating that various species of polychaetes accumulate ^{137}Cs from the surrounding geophysical environment. Ueda et al. (1977) report a concentration factor of 6 after 11 days by *N. japonica* for ^{137}Cs; Bryan (1963) found that *N. diversicolor* also had a sixfold accumulation factor and that *Perinereis* sp. had a whole-body concentration factor of 7.5, with individual *Perinereis* tissues showing concentration factors of 11.2 for pharynx, 12.2 for body wall, and 13.2 for gut.

10.8 Chromium

The highest concentration of chromium in field collections of annelids is 38.0 mg Cr/kg DW whole organism from urban coastal areas considered grossly contaminated by metals wastes (Table 10.2; Phelps et al., 1975).

Chipman (1967) conducted laboratory studies on uptake and excretion of Cr^{3+} and Cr^{6+} by *Hermione hystrix*. It was determined that trivalent chromium was not readily accumulated by *Hermione* when present in seawater owing to the formation of particles and adsorption to the surface; moreover, little accumulation of Cr^{3+} was evident on contact with contaminated sediments. In the case of Cr^{6+} in the medium, however, Cr^{6+} was readily accumulated by *Hermione*; the process was slow and only small amounts (0.01-0.03 mg Cr^{6+}/kg FW whole organism) was taken up from media containing 0.003-0.01 mg Cr^{6+}/L over a period of 19 days. Higher body burdens of 0.49-1.1 mg Cr/kg fresh body weight were reported at 0.1-0.5 mg Cr^{6+}/L, but some deaths were noted at these concentrations. Chromium accumulation by *Hermione* is a passive process and directly related to Cr^{6+} concentration of the medium. At least two rates of biological loss are involved, one of 8 days, another of 123 days. Chipman (1967) concluded that the greater part of chromium accumulated by *Hermione* from a long exposure would be bound in a body component having a slow turnover rate with an estimated biological half-life of 123 days.

Uptake of Cr^{6+} from seawater is recorded for *Neanthes arenaceodentata*. Whole *Neanthes* contained 30.0 mg Cr/kg DW after exposure for 150 days in 0.03 mg Cr/L (Mearns and Young, 1977) and 0.5-1.6 mg Cr/kg FW following immersion for 440 days (Oshida et al., 1976); both of these observations are similar to those of Chipman (1967) after correction for wet/DWs.

Table 10.2: Chromium, Cobalt, and Copper Concentrations in Field Collections of Annelids

Element and Organism	Concentration	Reference[a]
Chromium		
Polychaetes; 3 spp.; whole	8.1-14.7 DW; 16.9-20.0 FW	1, 3
Sandworm, *Nereis diversicolor*; whole	0.55 (<0.1-2.4) DW	2
Polychaetes; whole	23.8-38.0 DW	4
Cobalt		
Polychaete, *Branchipolynoe seepensis*; from gill of deep-sea mussel; 2002; whole	0.7 DW	12
Sandworm, *Nereis diversicolor*; whole	4.7 (1.6-7.9) DW	2
Copper		
Polychaete, *Branchipolynoe seepensis*; mid-Atlantic Ridge; hydrothermal vent field; whole; 2001		
1700 m deep	29.0 DW	13
2300 m deep	45.0 DW	13
Bloodworm, *Glycera gigantea*		
Jaws	14,340.0-16,320.0 DW	5
Whole	43.0-60.0 DW	5
Polychaetes; lugworm; whole; Wales, coastal areas; 1989	3.9 FW	11
Polychaete, *Lycastis ouanaryensis*; whole; India; 1984-1985; contaminated site vs. reference site	32.0-95.0 DW vs. 4.0-27.0 DW	10
Polychaete, *Neanthes arenaceodentata*; whole; held for 85 days on various sediment types		
No sediments	31.6 DW	6
Sand	25.1 DW	6
Sand-mud mixture	23.3 DW	6
Mud	18.7 DW	6
Sandworm, *Nereis diversicolor*		
Jaws	10.0 DW	7
Basal section jaws	2.0 DW	7
Distal section jaws	15.0 DW	7
Whole	44.0 (22.0-78.0) DW	2
Whole	1.6-5.1 FW	8

(*Continues*)

Element and Organism	Concentration	Reference[a]
Whole	28.0-1142.0 DW	9
Parapodia	779.0-902.0 DW	9
Body wall	48.0-543.0 DW	9
Gut	1678.0-1980.0 DW	9
Segment	494.0-638.0 DW	9

Values are in mg element/kg fresh weight (FW) or dry weight (DW).
[a]1, Bernhard and Zattera, 1975; 2, Bryan and Hummerstone, 1977; 3, Fukai and Broquet, 1965; 4, Phelps et al., 1975; 5, Gibbs and Bryan, 1980; 6, Pesch, 1979; 7, Bryan and Gibbs, 1979; 8, Wharfe and Van Den Broek, 1977; 9, Bryan and Hummerstone, 1971; 10, Athalye and Gokhole, 1991; 11, Morris et al., 1989; 12, Kadar et al., 2006; 13, Colaco et al., 2006.

Concentrations as low as 0.0125 mg Cr^{6+}/L can decrease brood size in *N. arenaceodentata* (Mearns et al., 1976; Oshida et al., 1976), although chromium body burdens in experimental subjects were the same as in controls. Uptake of Cr^{6+} by *Neanthes* was related to dose at low ambient chromium concentrations. Worms subjected to 0.0026, 0.0045, 0.0098, and 0.0166 mg Cr/L for 309 days contained 0.5, 0.7, 2.2, and 2.5 mg Cr/kg whole fresh organism, respectively (Oshida and Word, 1982). There was no direct relation between tissue concentration and brood size, suggesting that chromium in *Neanthes* attaches to proteins in the body wall, gut, and parapodial regions (Oshida and Word, 1982).

N. arenaceodentata is the most sensitive marine organism yet tested to hexavalent chromium and for which data are available. In worms exposed to sublethal concentrations, feeding was disrupted after 14 days at 0.079 mg/L (USEPA, 1980a), reproduction ceased after 440 days (three generations) at 0.1 mg/L (Oshida and Word, 1982; Oshida et al., 1981), and abnormalities in larval development increased after 5 months in 0.025 mg/L (Reish, 1977). However, exposure for 293 days (two generations) in 50.4 mg trivalent chromium/L caused no adverse effects on survival, maturation time required for spawning, or brood size (Oshida et al., 1981). The polychaete *C. capitata* was more resistant to hexavalent chromium than *Neanthes*; a decrease in brood size was noted only after exposure for 5 months to 0.05 and 0.1 mg Cr^{6+}/L (USEPA, 1980a).

10.9 Cobalt

Limited data indicate that the highest concentration of cobalt recorded in field collections of annelids is 7.9 mg Co/kg DW in sandworms, *N. diversicolor* (Table 10.2).

Cobalt uptake and excretion by *N. japonica* is reported by Ueda et al. (1977). They found that whole *Nereis* contain six times more cobalt than the medium after exposure for 11 days. Half the accumulated cobalt was excreted in 37 days after transfer to cobalt-free media; this period was slightly longer in unfed worms. Triquet (1973) reports that cobalt accumulation in lugworm, *A. marina* is mainly via the medium and not from cobalt-contaminated sediments; blood and digestive tract contained the highest cobalt concentrations.

10.10 Copper

The highest concentration of copper measured in whole marine polychaetes was 1142.0 mg Cu/kg DW in organisms collected from coastal areas contaminated by metals wastes (Table 10.2). Copper tends to concentrate in the jaws of some polychaetes, and this needs to be considered when assessing risk on the basis of whole organism. In fact, copper is the principal metal in jaws of *Glycera gigantas*, a glycerid polychaete, accounting for about 1.5% of the DW and up to 67% of the total body burden of copper (Gibbs and Bryan, 1980). The copper is concentrated in the distal tip of the jaw where it accounts for up to 13% of the copper body content; high copper levels were measured in jaws of other species of glycerid polychaetes, and this seems characteristic of that group (Gibbs and Bryan, 1980).

Copper accumulations in marine polychaetes are influenced mainly by the copper content and nature of the surrounding substrate. In one case, body content of copper in *N. diversicolor* directly reflected sediment copper concentrations (Bryan, 1974; Bryan and Hummerstone, 1971). However, in *Capitella* and *Paraprionspio*, highest concentrations of copper (and also samarium, scandium, and iron) were found in annelids collected from organic silt substrates, with consistently lower values from conspecifics collected from gravelly sand or clayey silts (Phelps et al., 1969). Pesch and Morgan (1978) clearly demonstrate the influence of presence or absence of sediment on whole-body burden of copper in *N. arenaceodentata*; at any given copper concentration in the medium, whole-body burden was significantly lower when sand was the substrate versus no substrate except glass. In follow-up studies, Pesch (1979) showed that sediment type influences the survival and accumulation of copper in *N. arenaceodentata* immersed in seawater containing 0.01 mg Cu/L for 85 days. In general, *Neanthes* survival was highest in mud, mud-sand mixtures, sand, and no sediment, in that order. Whole-body copper burdens after death, in mg Cu/kg DW were 270.0 without sediment, 994.0 in sand, 1047.0 in the mixture, and 1464.0 in mud. It was concluded that high organic content and small grain size were the two most important sediment parameters positively influencing survival of copper-stressed polychaetes (Pesch, 1979).

Polychaetes, and other invertebrates, reared or found in high-copper environments may tolerate additional copper stress more readily than conspecifics from low-copper environments. In one case, polychaetes from metal-polluted estuaries had elevated body burdens of copper and were more resistant to copper lethality than were polychaetes from low-copper environments and with low body burdens of copper (Bryan, 1974). In another case, *Ophryotrocha labronica* had greater tolerance to 1.0 mg Cu/L solutions after exposure to two to three generations in 0.025 to 0.1 mg Cu/L than did controls; however, this tolerance diminished in successive generations (Saliba and Ahsanullah, 1973).

Naturally occurring chelators can reduce copper toxicity to marine annelids. For example, fulvic acid—a major component in dissolved organic matter—complexes with copper, effectively reducing copper toxicity to larvae of the polychaete *H. elegans* (Qiu et al., 2007). Copper alone causes 50% abnormal development in 48 h at 0.056 mg Cu/L; however, the addition of 20.0 mg fulvic acid/L resulted in an increase to 0.137 mg Cu/L in the same period (Qiu et al., 2007). Polychaetes, *H. japonica*, exposed to 0.1 mg Cu/L for 6 days or to 0.1 mg Cu/L plus 0.1 mg petroleum hydrocarbons/L had BCF values of 83 for copper alone and 133 for the mixture, indicating significant interaction. Net accumulation rates, in mg Cu/kg body weight daily, were 2.1 for copper alone and 4.4 for the mixture (Sun and Zhou, 2007).

Copper was the most toxic metal tested to trocophore larvae of *H. elegans*. Arrested development occurred at 0.122 mg Cu/L, with order of toxicity being copper > aluminum > lead > nickel > zinc (Gopalakrishnan et al., 2007). Laboratory and field studies with nereid polychaetes suggest that body wall and gut are the most useful indicators of environmental copper as these tissues show the greatest concentration potential (Bryan and Hummerstone, 1971; Raymont and Shields, 1963). Age of the organism and salinity of the medium are additional variables for consideration in the interpretation of copper residues. Juveniles of *N. arenaceodentata* are more sensitive than adults to copper insult and show greater accumulations during 28 day tests (Reish et al., 1976). Salinity affects sensitivity of *N. diversicolor* to copper stress, with adverse effects most pronounced at comparatively high and low salinities (Jones et al., 1976). The synergistic effect of increased copper toxicity under conditions of extreme salinities is more important as a lethal factor than sediment copper levels or amount of copper accumulated in *Nereis* tissues (Jones et al., 1976). Larvae of the sandworm (*N. diversicolor*) are more resistant to copper with increasing organism age and with increasing temperature and salinity of the medium (Ozoh and Jones, 1990). In adult sandworms, whole-body loadings of copper usually increase with increasing temperature in the range of 12-22 °C and with decreasing salinity in the 7-31 ppt salinity range (Ozoh, 1992a,b); copper-temperature-salinity interactions are significant and complex in this species (Ozoh, 1994). Adult lugworms (*A. marina*) living in sediments containing 182.0-204.0 mg Cu/kg DW sediment had inhibited digestive processes; it was concluded that the digestive system of lugworms, and perhaps other deposit feeders, are vulnerable to copper-contaminated sediments (Chen and Mayer, 1998).

Polychaetes, *Hediste diversicolor*, exposed to increasing concentrations of copper showed behaviors indicative of metal stress, including inconsistent burrowing, eversion of the proboscis, and abnormal crawling. Behavioral endpoints at 10 ppt salinity and copper concentrations as low as 0.2 mg Cu/L were considered predictors of survival in acute toxicity bioassays (Burlinson and Lawrence, 2007). However, in formulation of marine water quality criteria for metals, behavioral responses should be accompanied by tissue concentrations of the stressor (Eisler, 1979). Antifouling paints containing mixtures of zinc pyrithione and copper (tributyltin replacement) were more than additive in toxicity to larval polychaetes, *H. elegans*, in 96 h survival tests; this should be considered in developing copper and zinc water quality criteria (Bao et al., 2008).

10.11 Iron

High variability was evident in concentrations of iron from polychaetes at different collection locales (Table 10.3). Highest values (>22,000.0 mg Fe/kg DW whole organism) were recorded in select deposit feeders, that is, those polychaetes which feed primarily at the sediment-water interface (Phelps, 1967).

10.12 Lead

Limited data indicate that lead content in whole *N. diversicolor* ranges from 2.1 to 261.0 mg Pb/kg DW whole organism (Table 10.3). Studies with the polychaete *H. elegans* show that the two most sensitive stages to lead insult are sperm at 0.38 mg Pb/L (20-min exposure) and egg at 0.69 mg Pb/L (20 min exposure); by contrast, the LC50 (96 h) value for adults is 0.95 mg Pb/L (Gopalakrishnan et al., 2008).

10.13 Manganese

Manganese concentrations in field collections of annelids were variable (Table 10.3). Manganese burdens in *N. diversicolor* are influenced by salinity of the medium, with concentrations greatest in worms collected from low salinity regimes (Bryan and Hummerstone, 1973a), and this, in part, may account for some of the variability.

Laboratory studies with *N. diversicolor* show high accumulations of manganese in whole worms after exposure for 2 weeks (Bryan and Hummerstone, 1973a). At 1.0 mg Mn/L, body burdens were 27.0 mg Mn/kg DW versus 4.0 for controls; at 5.0 mg Mn/L, this value was 81.0. On transfer to manganese-free media for 7 days, whole *Nereis* body burdens were 2.0 mg Mn/kg DW (controls), 9.0 (1.0 mg Mn/L), and 20.0 (5.0 mg Mn/L), respectively (Bryan and Hummerstone, 1973a); authors suggest that manganese occurs in *Nereis* in two distinct pools, one exchanging slowly, the other more rapidly.

10.14 Mercury

The highest mercury values recorded in field collections of annelids were 0.15 mg Hg/kg FW and 0.6 mg Hg/kg DW (Table 10.4). Periodic dredging to enlarge existing ship channels and for other purposes was associated with increasing body concentrations of mercury in annelids. For example, *N. hombergi* from a recently dredged area contained up to 0.15 mg Hg/kg FW, or three times more mercury than conspecifics collected from a control area

Table 10.3: Iron, Lead, and Manganese in Field Collections of Annelids

Element and Organism	Concentration	Reference[a]
Iron		
Polychaete, *Aglophamus* sp.; whole	390.0-1500.0 FW	1
Polychaete, *Armandia maculata*; whole	30,600.0 DW	2
Polychaete, *Capitella* sp.; whole	22,100.0 DW	2
Bloodworm, *Glycera gigantea*; jaws	34.0-122.0 DW	3

Sandworm, *Nereis diversicolor*
    Whole
    Jaws, whole
    Jaws, basal section
    Jaws, distal section
	
	 362.0 (260.0-462.0) DW
	 586.0 DW
	 95.0 DW
	 1089.0 DW
		
		 5
		 4
		 4
		 4

Polychaete, *Paraprionspio* sp.; whole	24,600.0 DW	2
Polychaetes; 6 spp., whole	500.0-4000.0 (300.0-5300.0) DW	6
Lead		
Sandworm, *Nereis diversicolor*; whole	45.0 (2.1-261.0) DW	5
Manganese		
Polychaete, *Branchipolynoe seepensis*; parasite on gill of deep-sea mussel; 2002; whole	10.8 DW	9
Bloodworm, *Glycera gigantea*; jaws	Max. 5.0 DW	3

Polychaete, *Hermione hystrix*
    Whole
    Body
    Cuticle
    Bristles
	
	 1510.0 AW
	 1970.0 AW
	 3980.0 AW
	 1360.0 AW
		
		 7
		 7
		 7
		 7

Sandworm, *Nereis diversicolor*
    Whole
    Jaws, whole
    Jaws, basal section
    Jaws, distal section
	
	 5.0-12.5 DW
	 806.0 DW
	 380.0 DW
	 1079.0 DW
		
		 5, 8
		 4
		 4
		 4

Polychaetes, 6 spp.; whole	<10.0-58.0 (<10.0-110.0) DW	6

Polychaete, *Spirographis spallanzini*
    Body
    Tube
	
	 102.0 AW
	 337.0 AW
		
		 7
		 7

Values are in mg element/kg fresh weight (FW), dry weight (DW), or ash weight (AW).

[a] 1, Phelps et al., 1969; 2, Phelps, 1967; 3, Gibbs and Bryan, 1980; 4, Bryan and Gibbs, 1979; 5, Bryan and Hummerstone, 1977; 6, Cross et al., 1970; 7, Chipman and Thommeret, 1970; 8, Bryan and Hummerstone, 1973a; 9, Kadar et al., 2006.

Table 10.4: Mercury and Nickel Concentrations in Field Collections of Annelids

Element and Organism	Concentration	Reference[a]
Mercury		
Lugworm, *Arenicola marina*; whole	0.01-0.07 FW	1
Polychaete, *Branchipolynoe seepensis*; whole		
From gill of deep-sea mussel; 2002 Mid-Atlantic Ridge; hydrothermal vent field; 2001	0.6 DW	7
1700 m deep	0.4 DW	8
2300 m deep	0.2 DW	8
Polychaete, *Nephtys hombergi*; whole	0.01 DW; 0.15 FW	1, 4
Sandworm, *Nereis diversicolor*		
Whole	0.03-0.08 FW	1
Whole	0.35 DW	2
Polychaete, *Nereis succinea*; whole	0.008-0.13 FW	3
Nickel		
Sandworm, *Nereis diversicolor*; whole		
Britain; coastal waters	2.2-5.2 DW	4
Bidasoa estuary; French-Spanish border; April 1993	5.4 (3.2-8.5) DW	5
Polychaetes; 3 spp.; whole; California	3.8-18.7 DW	6

Values are in mg Hg/kg fresh weight (FW) or dry weight (DW).
[a] 1, Zauke, 1977; 2, Leatherland and Burton, 1974; 3, Luoma, 1977; Rosenberg, 1977; 4, Bryan and Hummerstone, 1977; 5, Saiz-Salinas et al., 1996; 6, Jenkins, 1980; 7, Kadar et al., 2006; 8, Colaco et al., 2006.

(Table 10.4; Rosenberg, 1977); about 55% of the accumulated mercury was excreted over the next 18 months when the area was free from dredging (Rosenberg, 1977).

Luoma (1977) reports that whole *Nereis succinea* accumulated up to 1.8 mg Hg/kg FW when fed terrigenous sediment; this accumulation exceeded mercury concentrations in solution by a factor of 350. The increased suspended solids in seawater, to which most of the mercury compounds readily sorbed, was the probable source of increased mercury in *N. succinea*. Luoma (1977) also noted that mercury uptake by *N. succinea* was markedly reduced at salinities of 16 ppt and lower. The presence of sediments in assay containers strongly affects mercury accumulation and elimination by adult *Eurythoe complanata*, a polychaete annelid (Vazquez-Nunez et al., 2007). Inorganic mercury was strongly accumulated during exposure for 8 days to 0.0015-0.011 mg Hg^{2+}/L, but was much lower in the presence of sediments. Mercury elimination of 25-36% occurred during an 8 day depuration period, but only in the presence of sediments (Vazquez-Nunez et al., 2007).

In laboratory studies, methylmercury was rapidly accumulated by whole *G. dibranchiata*; elimination was rapid, with 30% loss in 8 h following immersion for 8 days in 0.01 mg methylmercury/L (Medeiros et al., 1980). The addition of cysteine to the depuration media accelerated loss, with only 14% of the methylmercury remaining after 8 h (Medeiros et al., 1980). Absorption of glycine from ambient seawater across the body surface by the oligochaete *Enchytraeus albidus* and the polychaete *N. diversicolor* was significantly reduced in the presence of 0.1-0.15 mg Hg/L (Siebers and Ehlers, 1979). Mercury was more effective in preventing glycine absorption than were the following: copper, silver, and cadmium at 0.25-2.0 mg/L; aluminum, chromium, iron, lead, molybdenum, vanadium, and zinc up to 10.0 mg/L; and nickel, manganese, cobalt, and selenium up to 150.0 mg/L (Siebers and Ehlers, 1979). Early life stages of the polychaete *H. elegans* were most sensitive to inorganic mercury. Toxicity was evident to sperm at 0.032 mg Hg/L (20-min exposure) and to embryos at 0.054 mg Hg/L (2 h exposure); by contrast, the LC-50 (96 h) value for adults was 0.1 mg Hg/L (Gopalakrishnan et al., 2008). In general, mercury was most toxic to early life stages of *H. elegans*, followed by cadmium, lead, nickel, and zinc, in that order (Gopalakrishnan et al., 2008).

10.15 Nickel

Sandworms, *N. diversicolor*, from British coastal waters contained between 2.2 and 5.2 mg Ni/kg DW whole organism (Table 10.4); sandworms from the Bidasoa estuary along the French-Spanish border contained up to 8.5 mg Ni/kg DW whole organism (Table 10.4). Concentrations up to 18.7 mg Ni/kg DW were recorded in whole polychaetes collected from the California coast (Table 10.4).

Polychaete annelids were comparatively resistant to nickel. No deaths occurred in 168 h among sandworm adults immersed in seawater containing 10.0 mg Ni/L (Eisler and Hennekey, 1977), and LC50(96 h) values for three species of polychaetes ranged from 17.0 to 49.0 mg Ni/L (USEPA, 1980b). However, early life stages of marine polychaetes were more sensitive to nickel insult than were adults. Toxicity was evident to sperm of *H. elegans* at 0.73 mg Ni/L after 20 min exposure, and to eggs at 1.18 mg Ni/L after 20 min exposure (Gopalakrishnan et al., 2008).

10.16 Plutonium

Accidental contamination of the marine environment near Greenland from the crash of a United States military aircraft transporting plutonium-containing materials raised median background levels of plutonium by three orders of magnitude (Aarkrog, 1971). From 1968 to 1970, the $^{239+240}$Pu concentrations in benthic organisms, including annelids and molluscs, decreased by an order of magnitude; since 1970, the decline is less evident (Aarkrog, 1977).

Laboratory studies with *N. diversicolor* show that whole-body content of plutonium is about 0.5% of the plutonium content of the surrounding sediment, and that no preferential uptake of ^{238}Pu over $^{239+240}$Pu is evident (Beasley and Fowler, 1976).

10.17 Ruthenium

Over a period of 11 days, *N. japonica* concentrated ^{106}Ru from seawater by a factor of about 6; the time for 50% excretion was 35 days in fed worms and was slightly longer in starved worms (Ueda et al., 1977). Uptake factor for ^{106}Ru by a gobiid fish fed with *N. japonica* was only 0.002; thus, transfer of ^{106}Ru from worm to fish was relatively small in comparison with accumulation from the surrounding seawater (Kimura and Ichikawa, 1969).

10.18 Samarium

The highest concentration of samarium recorded in field collections of whole annelids is 3.6 mg Sm/kg DW (Table 10.5). Samarium concentrations in polychaetes seem to reflect the levels in surrounding sediments (Phelps, 1967).

Table 10.5: Samarium, Scandium, Silver, and Strontium in Field Collections of Annelids

Element and Organism	Concentration	Reference[a]
Samarium		
Polychaete, *Aglophamus* sp.; whole	0.14 FW	1
Polychaetes; benthic; whole	0.04-3.6 DW	2
Scandium		
Polychaetes; whole; 9 spp.		
3 spp.	0.7-3.0 DW	2
3 spp.	0.3-6.9 DW	2
3 spp.	2.9-26.4 DW	2
Silver		
Sandworm, *Nereis diversicolor*; whole	5.2 (0.7-30.0) DW	3
Polychaete, *Marphysa sanguinea*; whole; San Francisco Bay; 1982	Max. 5.5 DW	5
Strontium		
Annelids; whole	2.1 FW	4

Values are in mg element/kg fresh weight (FW) or dry weight (DW).
[a] 1, Phelps et al., 1969; 2, Phelps, 1967; 3, Bryan and Hummerstone, 1977; 4, Mauchline and Templeton, 1966; 5, Luoma and Phillips, 1988.

10.19 Scandium

The highest scandium concentration recorded in whole marine annelids is 26.4 mg Sc/kg DW (Table 10.5), although this requires verification; concentrations seem to reflect scandium concentrations in the surrounding sediments (Phelps, 1967). Scandium is concentrated by a few species more markedly than by ecological groups; however, there is a tendency for reduced accumulations as the organisms move up the food chain from deposit feeders (Phelps, 1967).

10.20 Selenium

Uptake and retention of radioselenium was determined in a polychaete, *Eunice* sp. (Alquezar et al., 2007). *Eunice* was held for 385 h in seawater spiked with ^{75}Se followed by 189 h in ^{75}Se-free media. During uptake, whole *Eunice* concentrated radioselenium from the medium by a factor of 45; about 71% was lost over the next 189 h. During depuration, the short-lived component had a biological half-time of 34 h; for the long lived component, it was 555 h (Alquezar et al., 2007).

10.21 Silver

The maximum silver concentration recorded in annelids of 30.0 mg Ag/kg DW was in whole *N. diversicolor* (Table 10.5) from a metals-contaminated estuary (Bryan and Hummerstone, 1977).

Uptake and retention of silver by adults of *Sabella pavonia* held in seawater containing 0.05 mg Ag/L for 8 weeks, then transferred to silver-free media for an additional 8 weeks is reported (Koechlin and Grasset, 1988). During immersion, the maximum whole-body silver burden was 22.1 mg/kg DW versus 0.8 in controls; main sites of accumulation were the connective tissues of nephridia and gut. There was no detectable histopathology. A constant elimination of silver in urine occurs simultaneously with silver accumulation. During depuration, new connective tissues were formed and silver concentrations were reduced by 88%. Ionic or free silver interferes with calcium metabolism of marine polychaetes contaminated with silver, that is, the calcium content of nephridial cells was reduced although silver was not detected in the calcium vesicles. Silver binds with protein sulfhydryl groups, and this process protects the annelid against silver intoxication (Koechlin and Grasset, 1988).

10.22 Strontium

Whole marine annelids contain an average of 2.1 mg Sr/kg FW (Table 10.5).

10.23 Tin

No deaths were documented in larvae of the lugworm, *Arenicola cristata*, in 168 h at 0.002 mg tributyltin (TBT)/L (Walsh et al., 1986); however, all larvae died at 0.004 mg TBT/L in 96 h (Maguire et al., 1984). At sublethal TBT concentrations, teratogenic effects were observed in lugworm larvae (Walsh et al., 1986). Laboratory studies with *H. elegans* show that eggs were the most sensitive stage tested in 48 h assays: LC-50 was 0.00018 mg TBT/L (Lau et al., 2007). Results of 17 day exposure from egg to juvenile demonstrate that TBT concentrations >0.00001 mg/L adversely affect survival and growth of *Hydroides*. In longer exposures of 44 days, only juveniles exposed to <0.0001 mg TBT/L reached maturity; and in 60 day fecundity tests, egg production, fertilization success, and egg development were all reduced at >0.0001 mg TBT/L (Lau et al., 2007).

Annelids accumulate organotins from sediments. For example, sediments spiked with 0.98 mg Sn (as TBT)/kg DW resulted in concentrations of 4.41 mg/kg DW whole body in oligochaete annelids after 22 weeks, up from 0.38 at the start (Maguire and Tkacz, 1985). Annelids in sediments containing less than 0.61 mg TBT/kg DW had normal growth and survival (Cardwell and Sheldon, 1986). However, survival decreased when annelids were exposed simultaneously to organotins through both water and sediments (Clark et al., 1987).

10.24 Zinc

Whole-body concentrations of zinc in annelids from different collections range from 22.0 to 1554.0 mg Zn/kg on a DW basis (Table 10.6). It is noteworthy that jaws of nereid polychaetes analyzed contained 2880.0-34,950.0 mg Zn/kg DW, and this accounted for up to 40% of the total body zinc burden (Bryan and Gibbs, 1979). In *N. diversicolor*, zinc content was not related to zinc content of surrounding sediments; a high zinc content seems to be a structural characteristic of nereid jaws (Bryan and Gibbs, 1979).

Among the polychaetes, the composition of the feeding community based on feeding types is an important factor in the partitioning of stable zinc; the less dependent an organism is on the sediment as a source of food, the lower its concentration of zinc (Phelps, 1967). Polychaetes that feed mainly from the sediment-water interface (selective deposit feeders), concentrate iron to the apparent exclusion of zinc, and polychaetes that feed mainly below the sediment water interface (nonselective deposit feeders, omnivores, carnivores) preferentially concentrate zinc over iron (Phelps, 1967). *N. diversicolor*, as one example, could remove up to 4% of the ^{65}Zn in the upper 2 cm of a hypothetical radioactive estuary (Renfro, 1973). Qualitative and quantitative structure of the benthic faunal community has a direct effect on zinc distribution: a shift in preponderance from polychaete annelids to bivalve molluscs could result in a decrease in the amount of zinc incorporated into the biological community (Phelps, 1967).

Table 10.6: Zinc Concentrations in Field Collections of Annelids

Organism	Concentration	Reference[a]
Polychaete, *Branchipolynoe seepensis*; whole; mid-Atlantic Ridge; hydrothermal vent field; 2001		
1700 m deep	219.0 DW	10
2300 m deep	304.0 DW	10
Sandworm, *Nereis diversicolor*		
Whole	130.0-315.0 DW	4, 5
Whole	22.0-37.0 FW	6
Jaws	14,000.0 DW	7
Jaws, basal section	1790.0 DW	7
Jaws, distal section	34,950.0 DW	7
Head	843.0-995.0 DW	9
Parapodia	216.0-413.0 DW	9
Trunk	158.0-218.0 DW	9
Polychaetes; 9 spp.; jaws	4971.0-24,050.0 DW	7
Polychaetes; whole; 5 spp.	78.0-165.0 DW	1
Polychaetes; whole		
Lugworm, *Arenicola marina*	1.8 FW	8
Polynoid spp.	55.0-59.0 DW	2
Ceratonereis sp.	123.0 DW	2
Protoaricia sp.	370.0-1554.0 DW	2
Bloodworm, *Glycera gigantea*		
Jaws	2880.0-3300.0 DW	3
Whole	2210.0-4070.0 DW	3

Values are in mg Zn/kg fresh weight (FW) or dry weight (DW).
[a] 1, Cross et al., 1970.; 2, Phelps, 1967; 3, Gibbs and Bryan, 1980; 4, Bryan and Hummerstone, 1973b; 5, Bryan and Hummerstone, 1977; 6, Wharfe and Van Den Broek, 1977; 7, Bryan and Gibbs, 1979. 8, Morris et al., 1989; 9, Fernandez and Jones, 1989; 10, Colaco et al., 2006.

In the sandworm, *N. diversicolor*, zinc is localized in the gut wall, epidermis, nephridia, and blood vessels; most of the body zinc is present in wandering amoebocyte cells of excretory organs (Fernandez and Jones, 1989). Zinc in *Nereis* may be present as insoluble granules in membrane-bound vesicles. Excretion is via exocytosis with the aid of amoebocytes (Fernandez and Jones, 1989). Unlike the insoluble zinc phosphate granules of molluscs and crustacean, zinc granules in *Nereis* were very soluble and retained only by sulfide precipitation (Pirie et al., 1985). Zinc uptake in *N. diversicolor* increased with increasing sediment zinc concentrations, at lower salinities (Eisler, 1980), and at elevated temperatures (Fernandez and Jones, 1987, 1989). Zinc had no measurable effect on burrowing behavior of

Nereis, even at acutely lethal concentrations (Fernandez and Jones, 1987). Sandworms from zinc-contaminated sediments were more resistant to waterborne zinc insult by 10-100 times than those form clean sediments (USEPA, 1987). Tolerance to zinc in sandworms may be a result of acclimatization or genetic adaptation. In either event, the degree of metal tolerance decreases rapidly as the level of zinc contamination declines, suggesting that some zinc-tolerant sandworms may be competitively inferior to normal individuals in clean environments (Grant et al., 1989). Body zinc burdens of nereid worms tend to reflect zinc burdens in the surrounding sediments (Bryan, 1974). Physiological adaptation of *N. diversicolor* to zinc-impacted sediments is proposed by Bryan and Hummerstone (1973b), owing, in part, to decreased body surface permeability (Bryan, 1974). Under laboratory conditions Bryan and Hummerstone (1973b) found that *Nereis* from high-zinc sediments, when compared to those collected from low-zinc sediments, accumulated less zinc per unit body weight daily and had higher survival; this pattern appeared to hold over a range of salinities. It is emphasized that the rate of net absorption, in mg Zn/kg DW daily, in whole *N. diversicolor* was 2230.0—the highest reported value for any group of invertebrates (Bryan and Hummerstone, 1973b). That study was conducted for 35 days in seawater solutions containing 250.0 mg Zn/L with *Nereis* collected from sediments of low zinc content. Rate of net absorption decreased with increasing zinc concentration until it was 55.0 mg Zn/kg DW daily in 10.0 mg Zn/L (Bryan and Hummerstone, 1973b). Sometimes, however, trace metal content of polychaetes did not differ markedly when collected from sediments that contained substantially different quantities of iron, manganese, or zinc. The order of enrichment in six species of polychaetes over sediments is zinc > iron > manganese. This order of enrichment agrees with the hypothesis that enrichment of divalent metal ions by the marine biosphere follows the order of stability of metal-ligand complexes (Cross et al., 1970).

There was no difference in body burden of zinc and other metals analyzed in polychaetes from two stations which had markedly different concentrations of trace metals in the sediments, suggesting that these organisms may be able to regulate trace metal content or that much of the metal associated with the sediment is in an unacceptable chemical form for biological uptake (Cross et al., 1970). In another study, zinc levels in benthic communities—mainly annelids—show no relation to proximity of the land mass or to changes in substrate, although there is a trend for higher zinc values to occur in organisms found in the offshore silts and clays (Phelps et al., 1969). Time to reach equilibrium with the sediment is another variable; for example, the polychaete *Hermione hystrix* required 60 days to approach a steady state with zinc in the sediments (Renfro, 1973). In many species of marine biota, zinc is bound to metallothioneins; however, not all zinc-binding proteins are metallothioneins (Andersen et al., 1989; Eriksen et al., 1990). Low-molecular-weight metal-binding proteins—not metallothioneins—were induced in polychaete annelids in metals-contaminated environments (Andersen et al., 1989). In less contaminated environments, zinc was mainly associated with high-molecular-weight proteins in four species of sediment-feeding polychaetes, suggesting that metallothionein-like proteins may not be satisfactory for monitoring purposes and that other cytosolic components should be studied (Eriksen et al., 1990).

Unlike other major groups of marine benthic organisms, polychaete annelids, as judged by *N. arenaceodentata*, have a limited capacity to regulate zinc (Mason et al., 1988). Uptake in *Neanthes* occurs from the free ionic pool of zinc, whereas EDTA and EDTA-zinc complexes are largely excluded. Zinc accumulates linearly over time (350 h), and the rate decreases with increasing temperature in the range 4-21 °C. Mason et al. (1988) conclude that uptake and accumulation of zinc is passive in *Neanthes* and does not require metabolic energy. Zinc transfer across the plasma membrane is by way of diffusion. Within the cell, zinc binds to a variety of excising ligands which maintain an inwardly directed diffusion gradient, preventing zinc efflux. Accumulation rate is determined by the number and binding characteristics of the available ligands and their accessibility to zinc. After 50 h of exposure, worms selectively accumulated zinc over cadmium from the medium by a process requiring metabolic energy, and this is attributed to a change in the turnover rate and the size and nature of the zinc-binding ligands (Mason et al., 1988).

Life stage of the organism, possible discrimination against various chemical species of zinc, normal interspecies differences, and dredging practices should also be considered when interpreting zinc residues in annelids. In regard to life stage, juveniles of *Neanthes* and *Capitella* were more sensitive to zinc effects than were adults (Reish et al., 1976); a similar case is made for *H. elegans* (Gopalakrishnan et al., 2008). *H. hystrix* is apparently capable of distinguishing radiozinc species from stable zinc species: the biological half-life of ^{65}Zn in *Hermione* is 52-197 days but only 14-17 days for stable zinc (Renfro, 1973). Variability in zinc content of closely related species of polychaetes is considerable; for six species, the zinc content differed by an eightfold factor (Cross et al., 1970). The influence of dredging on zinc concentrations in *N. hombergi* is considerable (Rosenberg, 1977). *Nephtys* from a recently dredged area contained elevated levels of zinc when compared to conspecifics from a reference area; however, 95% to 98% was excreted during the next 18 months while the area was free from dredging.

Abnormal larval development of the polychaete *C. capitata* was documented during exposure for ten days to 0.05-0.1 mg Zn/L (USEPA, 1987). Zinc was lethal to adult *C. capitata* at 1.25 mg Zn/L in 28 days (USEPA, 1987), to adult *N. arenaceodentata* at 0.9 mg/L in 28 days (Spear, 1981), and to adults of *N. diversicolor* at 2.6 mg/L in seven days (Eisler and Hennekey, 1977). Adults of *N. diversicolor* held in 10.0 mg Zn/L for 96 h at 6 °C contained zinc burdens (in mg/kg DW) of 1031.0 in head versus 843.0 in controls, 366.0 versus 158.0 in trunk, and 455.0 versus 275.0 in parapodia; uptake was higher at 12 °C and 20 °C (Fernandez and Jones, 1989). Surviving adults of *N. diversicolor* subjected to zinc concentrations of 10.0 mg/L for 34 days had whole-body burdens of 2500.0 mg/kg DW versus 180.0 in controls (as quoted in Eisler, 1980).

10.25 Zirconium

Uptake studies with radiozirconium-95 and *N. japonica* indicate a concentration factor of 4 over the medium, a biological half-life (time for 50% excretion) of 32 days for accumulated ^{95}Zr in fed worms, and a slightly longer half-life for starved worms (Ueda et al., 1977).

10.26 Literature Cited

Aarkrog, A., 1971. Radioecological investigations of plutonium in an Arctic marine environment. Health Phys. 20, 31–47.

Aarkrog, A., 1977. Environmental behaviour of plutonium accidentally released at Thule, Greenland. Health Phys. 32, 271–284.

Alquezar, R., Markich, S.J., Twining, J.R., 2007. Uptake and loss of dissolved ^{109}Cd and ^{75}Se in estuarine macroinvertebrates. Chemosphere 67, 1201–1210.

Andersen, R.A., Eriksen, K.D.H., Bakke, T., 1989. Evidence of presence of a low molecular weight, non-metallothionein-like metal-binding protein in the marine gastropod *Nassarius reticulatus* L. Comp. Biochem. Physiol. 94B, 285–291.

Athalye, R.P., Gokhole, K.S., 1991. Heavy metals in the polychaete *Lycastis ouanaryensis* from Thane Creek, India. Mar. Pollut. Bull. 22, 233–236.

Bao, V.W.W., Leung, K.M.Y., Kwok, K.W.H., Zhang, A.Q., Lui, G.C.S., 2008. Synergistic toxic effects of zinc pyrithione and copper to three marine species: Implications on setting appropriate water quality criteria. Mar. Pollut. Bull. 57, 616–623.

Beasley, T.M., Fowler, S.W., 1976. Plutonium and americium: Uptake from contaminated sediments by the polychaete *Nereis diversicolor*. Mar. Biol. 38, 95–100.

Bernhard, M., Zattera, A., 1975. Major pollutants in the marine environment. In: Pearson, E.A., Frangipane, E.D. (Eds.), Marine Pollution and Marine Waste Disposal. Pergamon, New York, pp. 195–300.

Bryan, G.W., 1963. The accumulation of radioactive caesium by marine invertebrates. J. Mar. Biol. Assoc. UK 43, 519–539.

Bryan, G.W., 1974. Adaptation of an estuarine polychaete to sediments containing high concentrations of heavy metals. In: Vernberg, F.J., Vernberg, W.B. (Eds.), Pollution and Physiology of Marine Organisms. Academic Press, New York, pp. 123–135.

Bryan, G.W., Gibbs, P.E., 1979. Zinc—a major inorganic component of nereid polychaete jaws. J. Mar. Biol. Assoc. UK 59, 969–973.

Bryan, G.W., Hummerstone, L.G., 1971. Adaptation of the polychaete *Nereis diversicolor* to estuarine sediments containing high concentrations of heavy metals. I. General observations and adaptation to copper. J. Mar. Biol. Assoc. UK 51, 845–863.

Bryan, G.W., Hummerstone, L.G., 1973a. Adaptation of the polychaete *Nereis diversicolor* to manganese in estuarine sediments. J. Mar. Biol. Assoc. UK 53, 859–872.

Bryan, G.W., Hummerstone, L.G., 1973b. Adaptation of the polychaete *Nereis diversicolor* to estuarine sediments containing high concentrations of zinc and cadmium. J. Mar. Biol. Assoc. UK 53, 839–857.

Bryan, G.W., Hummerstone, L.G., 1977. Indicators of heavy metal contamination in the Looe estuary (Cornwall) with particular regard to silver and lead. J. Mar. Biol. Assoc. UK 57, 75–92.

Burlinson, F.C., Lawrence, A.J., 2007. Development and validation of a behavioural assay to measure the tolerance of Hediste diversicolor to copper. Environ. Pollut. 145, 274–278.

Cardwell, R.D., Sheldon, A.W., 1986. A risk assessment concerning the fate and effects of tributyltins in the aquatic environment. In: Maton, G.L. (Ed.), Proceedings Oceans 86 Conference, Washington, DC, 23-25 September 1986, vol. 4. Organotin symposium. Available from Marine Technology Society, 2000 Florida Ave, New York, Washington, DC, pp. 1117–1129.

Chen, Z., Mayer, L.M., 1998. Digestive processes of the lugworm (*Arenicola marina*) inhibited by Cu from contaminated sediments. Environ. Toxicol. Chem. 17, 433–438.

Chipman, W.A., 1967. Some aspects of the accumulation of chromium-51 by marine organisms. In: Proceedings of an International Symposium on Radioecological Concentration Processes, Stockholm, 1966, pp. 931–994.

Chipman, W., Thommeret, J., 1970. Manganese content and the occurrence of fallout ^{54}Mn in some marine benthos of the Mediterranean. Bull. Inst. Oceanogr. 69(1402), 1–15.

Clark, J.R., Patrick Jr., J.M., Moore, J.C., Lores, E.M., 1987. Waterborne and sediment-source toxicities of six organic chemicals to grass shrimp (*Palaemonetes piugio*) and amphioxus (*Branchiostoma caribaeum*). Arch. Environ. Contam. Toxicol. 16, 401–407.

Colaco, A., Bustamante, P., Fouquet, Y., Sarradin, P.M., Serraos-Santos, R., 2006. Bioaccumulation of Hg, Cu, and Zn in the Azores triple junction hydrothermal vent fields food web. Chemosphere 65, 2260–2267.

Cross, F.A., Duke, T.W., Willis, J.N., 1970. Biogeochemistry of trace elements in a coastal plain estuary: Distribution of manganese, iron, and zinc in sediments, water and polychaetous worms. Chesapeake Sci. 11, 221–234.

Eisler, R., 1979. Behavioural responses of marine poikilotherms to pollutants. Phil. Trans. R. Soc. Lond. 286B, 507–521.

Eisler, R., 1980. Accumulation of zinc by marine biota. In: Nriagu, J.O. (Ed.), Zinc in the Environment. Part II. Health Effects. John Wiley, New York, pp. 259–351.

Eisler, R., 2000. Cadmium. In: Handbook of Chemical Risk Assessment, vol. 1. Lewis Publishers, Boca Raton, FL, pp. 1–43.

Eisler, R., Hennekey, R.J., 1977. Acute toxicities of Cd^{2+}, Cr^{+6}, Hg^{2+}, Ni^{2+} and Zn^{2+} to estuarine macrofauna. Arch. Environ. Contam. Toxicol. 6, 315–323.

Eriksen, K.D.H., Andersen, T., Stenersen, J., Andersen, R.A., 1990. Cytosolic binding of Cd, Cu, Zn, Zn and Ni in four polychaete species. Comp. Biochem. Physiol. 95C, 111–115.

Fernandez, T.V., Jones, N.V., 1987. Some studies on the effect of zinc on *Nereis diversicolor* (polychaete-annelida). Trop. Ecol. 28, 9–21.

Fernandez, T.V., Jones, N.V., 1989. The distribution of zinc in the body of *Nereis diversicolor*. Trop. Ecol. 30, 285–293.

Fukai, R., Broquet, D., 1965. Distribution of chromium in marine organisms. Bull. Inst. Oceanogr. 65(1336), 1–19.

Geiszinger, A.E., Goessler, W., Francesconi, K.A., 2002. The marine polychaete *Arenicola marina*: Its unusual arsenic compound pattern and its uptake of arsenate from seawater. Mar. Environ. Res. 53, 37–50.

Gibbs, P.E., Bryan, G.W., 1980. Copper—The major metal component of glycerid polychaete jaws. J. Mar. Biol. Assoc. UK 60, 205–214.

Gopalakrishnan, S., Thilagam, H., Raja, P.V., 2007. Toxicity of heavy metals on embryogenesis and larvae of the marine sedentary polychaete *Hydroides elegans*. Arch. Environ. Contam. Toxicol. 52, 171–178.

Gopalakrishnan, S., Thilagam, H., Raja, P.V., 2008. Comparison of heavy metal toxicity in life stages (spermotoxicity, egg toxicity, and larval toxicity) of *Hydroides elegans*. Chemosphere 71, 515–528.

Goto, D., Wallace, W.G., 2007. Interaction of Cd and Zn during uptake and loss in the polychaete *Capitella capitata*: Whole body and subcellular perspectives. J. Exp. Mar. Biol. Ecol. 352, 65–77.

Grant, A., Hately, J.G., Jones, N.V., 1989. Mapping the ecological impact of heavy metals on the estuarine polychaete *Nereis diversicolor* using inherited metal tolerance. Mar. Pollut. Bull. 20, 235–238.

Hall Jr., L.W., Scott, M.C., Killen, W.D., 1998. Ecological risk assessment of copper and cadmium in surface waters of Chesapeake Bay watershed. Environ. Toxicol. Chem. 17, 1172–1189.

Jenkins, D.W., 1980. Biological monitoring of trace metals. Volume 2. Toxic trace metals in plants and animals of the world. Part III. 1–290 U.S. Environ. Protect. Agent. Rep. 600/3-80-092, pp. 1-290.

Jones, L.H., Jones, N.V., Radlett, A.J., 1976. Some effects of salinity on the toxicity of copper to the polychaete *Nereis diversicolor*. Estuar. Coast. Mar. Sci. 4, 107–111.

Kadar, E., Santos, R.S., Powell, J.J., 2006. Biological factors influencing tissue compartmentalization of trace metals in the deep-sea hydrothermal vent bivalve. *Bathymodiolus azoricus* at geochemically distinct vent sites of the mid-Atlantic Ridge. Environ. Res. 101, 221–229.

Kimura, K., Ichikawa, R., 1969. Accumulation and retention of ingested ruthenium-106 by genuine goby. Bull. Jpn. Soc. Sci. Fish. 35, 435–440.

Koechlin, N., Grasset, M., 1988. Silver contamination in the marine polychaete annelid *Sabella pavonina* S.: A cytological and analytical study. Mar. Environ. Res. 26, 249–263.

Lau, K.C., Chan, K.M., Leung, K.M.Y., Luan, T.G., Yang, M.S., Qiu, J.W., 2007. Acute and chronic toxicities of tributyltin to various life stages of the marine polychaete *Hydroides elegans*. Chemosphere 69, 135–144.

Leatherland, T.M., Burton, J.D., 1974. The occurrence of some trace metals in coastal organisms with particular reference to the Solent Region. J. Mar. Biol. Assoc. UK 54, 457–468.

Luoma, S.N., 1977. Physiological characteristics of mercury uptake by two estuarine species. Mar. Biol. 41, 269–273.

Luoma, S.N., Phillips, D.J.H., 1988. Distribution, variability, and impacts of trace elements in San Francisco Bay. Mar. Pollut. Bull. 19, 413–425.

Maguire, R.J., Tkacz, R.J., 1985. Degradation of the tri-*n*-butyltin species in water and sediment from Toronto Harbor. J. Agric. Food Chem. 33, 947–953.

Maguire, R.J., Wong, P.T.S., Rhamey, J.S., 1984. Accumulation and metabolism of tri-*n*-butyltin cation by a green alga, *Ankistrodesmus falcatus*. Can. J. Fish. Aquat. Sci. 41, 537–540.

Mason, A.Z., Jenkins, K.D., Sullivan, P.A., 1988. Mechanisms of trace metal accumulation in the polychaete *Neanthes arenaceodentata*. J. Mar. Biol. Assoc. UK 68, 61–80.

Mauchline, J., Templeton, W.L., 1966. Strontium, calcium, and barium in marine organisms from the Irish Sea. J. Cons. Int. Explor. Mer. 30, 161–170.

Mearns, A.J., Young, D.R., 1977. Chromium in the southern California marine environment. In: Giam, C.S. (Ed.), Pollutant Effects on Marine Organisms. D.C. Heath, Lexington, MA, pp. 125–142.

Mearns, A.J., Oshida, P.S., Sherwood, M.J., Young, D.R., Reish, D.J., 1976. Chromium effects on coastal organisms. J. Water Pollut. Control Fed. 48, 1929–1939.

Medeiros, D.M., Caldwell, L.L., Preston, R.L., 1980. A possible physiological uptake mechanism of methylmercury by the marine bloodworm (*Glycera dibranchiata*). Bull. Environ. Contam. Toxicol. 24, 97–101.

Morris, R.J., Law, R.J., Allchin, C.R., Kelly, C.A., Fileman, C.F., 1989. Metals and organochlorines in dolphins and porpoises of Cardigan Bay, West Wales. Mar. Pollut. Bull. 20, 512–523.

Ng, T.Y.T., Rainbow, P.S., Amiard-Triquet, C., Amiard, J.C., Wang, W.X., 2008. Decoupling of cadmium biokinetics and metallothionein turnover in a marine polychaete after metal exposure. Aquat. Toxicol. 89, 47–54.

Notti, A., Fattorini, D., Razzetti, E.M., Regoli, F., 2007. Bioaccumulation and biotransformation of arsenic in the Mediterranean polychaete *Sabella spallanzanii*: Experimental observations. Environ. Toxicol. Chem. 26, 1186–1191.

Oshida, P.S., Word, L.S., 1982. Bioaccumulation of chromium and its effects on reproduction in *Neanthes arenaceodentata* (Polychaeta). Mar. Environ. Res. 7, 167–174.

Oshida, P.S., Mearns, A.J., Reish, D.J., Word, L.S., 1976. The effects of hexavalent and trivalent chromium on *Neanthes arenaceodentata* (Polychaeta: Annelida). SCCWRP (1500 E. Imperial Highway, El Segundo, California) TM 225, 1–58.

Oshida, P.S., Word, L.S., Mearns, A.J., 1981. Effects of hexavalent and trivalent chromium on the reproduction of *Neanthes arenaceodentata* (Polychaeta). Mar. Environ. Res. 5, 41–49.

Ozoh, P.T.E., 1992a. The effects of salinity, temperature and sediment on the toxicity of copper to juvenile *Hediste* (*Nereis*) *diversicolor* (O.F. Muller). Environ. Monit. Assess. 21, 1–10.

Ozoh, P.T.E., 1992b. The effect of temperature and salinity on copper body-burden and copper toxicity to *Hediste* (*Nereis*) *diversicolor*. Environ. Monit. Assess. 21, 11–17.

Ozoh, P.T.E., 1994. The effect of salinity, temperature and time on the accumulation and depuration of copper in ragworm, *Hediste* (*Nereis*) *diversicolor* (O.F. Muller). Environ. Monit. Assess. 29, 155–164.

Ozoh, P.T.E., Jones, N.V., 1990. The effects of salinity and temperature on the toxicity of copper to 1-day and 7-day-old larvae of *Hediste* (*Nereis*) *diversicolor* (O.F. Muller). Ecotoxicol. Environ. Safety 19, 24–32.

Pentreath, R.J., Jefferies, D.F., 1971. The uptake of radionuclides by I-group plaice (*Pleuronectes platessa*) off the Cumberland Coast, Irish Sea. J. Mar. Biol. Assoc. UK 51, 963–976.

Pesch, C.E., Morgan, D., 1978. Influence of sediment in copper toxicity tests with the polychaete *Neanthes arenaceodentata*. Water Res. 12, 747–751.

Pesch, C.E., 1979. Influence of three sediment types on copper toxicity to the polychaete *Neanthes arenaceodentata*. Mar. Biol. 52, 237–245.

Phelps, D.K., 1967. Partitioning of the stable elements Fe, Zn, Sc, and Sm within the benthic community, Anasco Bay, Puerto Rico. In: Proceedings of an International Symposium on Radioecological Concentration Processes, Stockholm, 1966, pp. 721–734.

Phelps, D.K., Santiago, R.J., Luciano, D., Irizarry, N., 1969. Trace element composition of inshore and offshore benthic populations. In: Proceedings of the 2nd National Symposium on Radioecology. USAEC Conf. 67053, pp. 509–526.

Phelps, D.K., Telek, G., Lapan Jr., R.L., 1975. Assessment of heavy metal distribution within the food web. In: Pearson, E.A., Frangipane, E.D. (Eds.), Marine Pollution and Marine Waste Disposal. Pergamon, Elmsford, NY, pp. 341–348.

Pirie, B., Fayi, L., George, S., 1985. Ultrastructural localisation of copper and zinc in the polychaete, *Nereis diversicolor*, from a highly contaminated estuary. Mar. Environ. Res. 17, 197–198.

Qiu, J.W., Tang, X., Zheng, C., Li, Y., Huang, H., 2007. Copper complexation by fulvic acid affects copper toxicity to the larvae of the polychaete *Hydroides elegans*. Mar. Environ. Res. 64, 563–573.

Ray, S., McLeese, D., Pezzack, D., 1980. Accumulation of cadmium by *Nereis virens*. Arch. Environ. Contam. Toxicol. 9, 1–8.

Raymont, J.E.G., Shields, J., 1963. Toxicity of copper and chromium in the marine environment. Int. J. Air Water Pollut. 7, 435–443.

Reish, D.J., 1977. Effects of chromium on the life history of *Capitella capitata* (Annelida: Polychaeta). In: Vernberg, F.J., Calabrese, A., Thurberg, F.P., Vernberg, W.B. (Eds.), Physiological Responses of Marine Biota to Pollutants. Academic Press, New York, pp. 199–207.

Reish, D.J., Martin, J.M., Piltz, F.M., Word, J.Q., 1976. The effect of heavy metals on laboratory populations of two polychaetes with comparisons to other water quality conditions and standards in southern California marine waters. Water Res. 10, 299–302.

Renfro, W.C., 1973. Transfer of ^{65}Zn from sediments by marine polychaete worms. Mar. Biol. 21, 305–316.

Rice, M.A., Chien, P.K., 1979. Uptake, binding and clearance of divalent cadmium in *Glycera dibranchiata* (Annelida: Polychaeta). Mar. Biol. 53, 33–39.

Rosenberg, R., 1977. Effects of dredging operations on estuarine benthic macrofauna. Mar. Pollut. Bull. 8, 102–104.

Saiz-Salinas, J.I., Ruiz, J.M., Frances-Zubillaga, G., 1996. Heavy metal levels in intertidal sediments and biota from the Bidasoa estuary. Mar. Pollut. Bull. 32, 69–71.

Saliba, L.J., Ahsanullah, M., 1973. Acclimation and tolerance of *Artemia salina* and *Ophryotrocha labronica* to copper sulphate. Mar. Biol. 23, 297–302.

Sandrini, J.Z., Regoli, F., Fattorini, D., Notti, A., Inacio, A.F., Linde-Arias, A.R., et al., 2006. Short-term responses to cadmium exposure in the estuarine polychaete *Laeonereis acuta* (polychaeta, nereididae): Subcellular distribution and oxidative stress generation. Environ. Toxicol. Chem. 25, 1337–1344.

Siebers, D., Ehlers, U., 1979. Heavy metal action on trans-integumentary absorption of glycine in two annelid species. Mar. Biol. 50, 175–179.

Spear, P.A., 1981. Zinc in the aquatic environment: Chemistry, distribution, and toxicology. Natl. Res. Coun. Can. Publ. NRCC 17589. pp. 1–145.

Sun, F.H., Zhou, Q.X., 2007. Metal accumulation in the polychaete *Hediste japonica* with emphasis on interaction between heavy metals and petroleum hydrocarbons. Environ. Pollut. 149, 92–98.

Thorsson, M.H., Hedman, J.E., Bradshaw, C., Gunnarsson, J.S., Gilek, M., 2008. Effects of settling organic matter on the bioaccumulation of cadmium and BDE-99 by Baltic Sea benthic invertebrates. Mar. Environ. Res. 65, 264–281.

Triquet, C., 1973. Study of the contamination of *Arenicola marina* L. by cobalt-60. Comptes Rend. Acad. Sci. 276 (Ser. D, No. 4), 645–648.

Ueda, T., Nakamura, R., Suzuki, Y., 1976. Comparison of ^{115m}cd accumulation from sediments and sea water by polychaete worms. Bull. Jpn. Soc. Sci. Fish. 42, 299–306.

Ueda, T., Nakamura, R., Suzuki, Y., 1977. Comparison of influences of sediments and sea water on accumulation of radionuclides by worms. J. Radiat. Res. 18, 84–92.

U.S. Environmental Protection Agency (USEPA), 1980a. Ambient water quality criteria for chromium. USEPA Rep. 440/5-80-035.

USEPA, 1980b. Ambient water quality criteria for nickel. USEPA Rep. 440/5-80-060. pp. 1–206.

USEPA, 1987. Ambient water quality criteria for zinc-1987. USEPA Rep. 440/5-87-003. pp. 1–207.

Vaskovsky, V.E., Korotchenko, O.D., Kosheleva, L.P., Levin, V.S., 1972. Arsenic in the lipid extracts of marine invertebrates. Comp. Biochem. Physiol. 41B (4), 777–784.

Vazquez-Nunez, R.V., Mendez, N., Green-Ritz, C., 2007. Bioaccumulation and elimination of Hg in the fireworm *Eurythoe complanata* (annelida: polychaeta) from, Mazatlan, Mexico. Arch. Environ. Contam. Toxicol. 52, 541–548.

Wallace, W.G., Lopez, G.R., Levinton, J.S., 1998. Cadmium resistance in an oligochaete and its effect on cadmium trophic transfer to an omnivorous shrimp. Mar. Ecol. Prog. Ser. 172, 225–237.

Walsh, G.E., Louie, M.K., McLaughlin, L., Lores, E.M., 1986. Lugworm (*Arenicola cristata*) larvae in toxicity tests: Survival and development when exposed to organotins. Environ. Toxicol. Chem. 5, 749–754.

Wharfe, J.R., Van Den Broek, W.L.F., 1977. Heavy metals in macroinvertebrates and fish from the lower Medway Estuary, Kent. Mar. Pollut. Bull. 8, 31–34.

Zauke, G.P., 1977. Mercury in benthic invertebrates of the Elbe estuary. Helgol. Wiss. Meeres. 29, 358–374.

CHAPTER 11
Echinoderms

This widely distributed group includes sea urchins, sea cucumbers, starfishes, brittle stars, sand dollars, and sea lilies. These metazoan coelomates have a unique water vascular system, and all show a division of most of the body structures into five sectors commonly produced as arms or rays. Although some species tolerate a change in salinity better than others, only a few species can tolerate brackish waters. Within the sea, however, echinoderms occur anywhere between the littoral zone to depths of 6000 m or more, with the greatest number and variety between 1000 and 2000 m. Echinoderms are cosmopolitan and range from cold seas, where they are the most abundant, to the tropics where numerous colorful varieties abound. All are comparatively sluggish and can remain immobile for extended periods. Almost all are bottom dwellers, although a few highly modified species are free floating. Echinoderms are prime movers of sediments and detritus in the sea and are probably quite important in the cycling of trace metals.

11.1 Aluminum

Maximum concentrations of aluminum in echinoderm field collections of 2100.0 mg Al/kg dry weight (DW) in intestine and 1000.0 mg Al/kg DW in gonad are recorded (Table 11.1). The significance of these elevated levels is imperfectly understood, but in the case of intestine this may reflect environmental sediment levels (Riley and Segar, 1970).

11.2 Antimony

Antimony concentrations in whole echinoderms usually ranged between 0.01 and 0.07 mg Sb/kg DW, although one sea urchin, *Arbacia lixula*, inexplicably contained 5.5 mg Sb/kg DW whole animal (Table 11.2).

11.3 Arsenic

It is clear from Table 11.3 that arsenic accumulates in lipids of echinoderms and that gonadal tissues of sea urchins may be useful in assessing anthropogenic arsenic contamination. Vaskovsky et al. (1972) found that arsenic content of

Table 11.1: Aluminum Concentrations in Field Collections of Echinoderms

Organism	Concentration	Reference[a]
Starfish, *Asterias rubens*		
Whole	<100.0 DW	1
Oral skin	230.0 DW	1
Aboral skin	<95.0 DW	1
Pyloric ceca	<100.0 DW	1
Gonad	<100.0 DW	1
Echinoderms; body and shell	159.0-183.0 DW	2
Echinoid, *Echinus esculentus*		
Oral shell	120.0 DW	1
Aboral shell	140.0 DW	1
Aristotle's lantern	160.0 DW	1
Spines	130.0 DW	1
Intestines	2100.0 DW	1
Gonad	95.0 DW	1
Starfish, *Henricia sanguinolenta*; whole	90.0 DW	1
Sea cucumber, *Holothuria* sp.; whole	73.0 FW	3
Starfish, *Solaster papposus*; whole	240.0-500.0 DW	1
Sea urchin, *Spatangus purpureus*		
Test and spines	160.0 DW	1
Gonad	1000.0 DW	1

Values are in mg Al/kg fresh weight (FW) or dry weight (DW).
[a]1, Riley and Segar, 1970; 2, Culkin and Riley, 1958; 3, Matsumoto et al., 1964.

Table 11.2: Antimony Concentrations in Field Collections of Echinoderms

Organism	Concentration	Reference[a]
Sea urchin, *Arbacia lixula*; whole	5.5 DW	1
Starfish, *Asterias rubens*; whole	0.01 DW	2
Starfish, *Echinaster sepositus*; whole	0.02 DW	1
Sea cucumber, *Holothuria tubulosa*; whole	0.05 DW	1
Starfish, *Marthasterias glacialis*; whole	0.02-0.03 DW	1, 2
Brittle star, *Ophioderma longicauda*; whole	0.01 DW	1
Sea urchin, *Paracentrotus lividus*; whole	0.07 DW	1
Echinoid, *Sphaerechinus granularis*; whole	0.13 DW	1

Values are in mg Sb/kg dry weight.
[a]1, Papadopoulu et al., 1976; 2, Leatherland and Burton, 1974.

Table 11.3: Arsenic Concentrations in Field Collections of Echinoderms

Organism	Concentration	Reference[a]
Starfish, *Asterias amurensis*; lipid extract, soft tissues	51,200.0 FW	1
Starfish, *Asterias rubens*; whole	10.0 DW	2
Sea cucumber, *Cucumaria fraudatrix*; lipid extract, soft tissues	10,200.0 FW	1
Starfish, *Distolasterias nippon*; lipid extract, soft tissues	41,800.0 (23,500.0-59,200.0) FW	1
Echinoid, *Echinocardium cardatum*; lipid extract, soft tissues	7900.0 FW	1
Guam; 1998-1999; 4 locations; sea cucumbers; dorsal body wall vs. hemal system; maximum concentration		
Bohadschia argus	17.7 DW vs. 42.9 DW	6
Holothuria atra	23.2 DW vs. 28.3 DW	6
Starfish, *Marthasterias glacialis*; whole	5.8 DW	2
Sea urchin, *Paracentrotus lividus*; whole; 1991-2005; western Mediterranean Sea; Balearic Islands; Port of Mahon	1.6 (0.7-4.1) DW	7
Starfish, *Patira pectinifera*; lipid extract, soft tissues	17,400.0 FW	1
Starfish, *Pisaster ochraceus*; whole	1.3 DW	3
Sea cucumber, *Stichopus japonicus*; lipid extract, soft tissues	1400.0 FW	1
Sea urchin, *Strongylocentrotus droebachiensis* Gonads; distance in meters from arsenic mining wastes outfall		
0	6.0 DW	4
35	5.6 DW	4
157	3.8 DW	4
189	1.9 DW	4
Whole	2.7-4.5 DW	5
Gonads	6.8-16.0 DW	5
Sea urchin, *Strongylocentrotus* spp.; lipid extract, soft tissues	13,000.0-16,700.0 FW	1

Values are in mg As/kg fresh weight (FW) or dry weight (DW).
[a] 1, Vaskovsky et al., 1972; 2, Leatherland and Burton, 1974; 3, Gorgy et al., 1948; 4, Penrose et al., 1975; 5, Bohn, 1979; 6, Denton et al., 2006; 7, Deudero et al., 2007.

echinoderms collected from different localities varied substantially. For example, arsenic content in lipids of sea stars, *Distolasterias nippon*, ranged from 23,500.0 to 59,200.0 mg As/kg fresh weight (FW), and concluded that reasons for this are essentially unknown. In addition, Vaskovsky and his colleagues report that lipids of echinoderms of the same species, but occupying different habitats, had different arsenic contents.

Laboratory studies of 7 week duration on uptake and retention of arsenic by the sea urchin *Strongylocentrotus droebachiensis* fed arsenic-loaded algae, *Fucus vesiculosus*, showed that each animal, on average, consumed 0.203 mg arsenic, excreted 0.036 mg, and converted most of the arsenic in tissues to an organic form (Penrose et al., 1977). Other studies by Penrose et al. (1977) with ^{74}As confirmed that more than half the inorganic arsenic was converted to an organic form in as little as 72 h, with up to 40% conversion in 15 min. Embryos of the sea urchin *Anthocidaris crassispina* developed normally in media containing 2.1 mg pentavalent As/L, but 4.2 mg/L was fatal (Kobayashi, 1971). Subsequent studies showed that seawater solutions containing 3.0 mg As/L were associated with retarded development and abnormal pluteus formation (Kobayashi, 1977).

11.4 Barium

Mauchline and Templeton (1966) report that echinoderms from the Irish Sea contain between 0.2 and 1.3 mg Ba/kg FW whole organism.

11.5 Boron

Embryos of a sea urchin, *A. crassispina*, are relatively resistant to boron as boric acid (H_3BO_3). Normal development occurs at 37.0 mg B/L; however, 75.0 mg B/L is fatal (Kobayashi, 1971). No data wee available on boron concentrations in field collections of echinoderms.

11.6 Cadmium

Most cadmium values recorded in echinoderm tissues were less than 1.0 mg Cd/kg DW (Table 11.4). However, some species, such as the starfish *Asterias rubens*, frequently exceed this value, and some tissues including intestine and pyloric ceca have comparatively elevated cadmium burdens (Table 11.4). This latter observation tends to confirm the findings of Hiyama and Shimisu (1964), who demonstrate that digestive tract of the sea urchin *Strongylocentrotus* sp. had the highest cadmium residues of all tissues examined.

Table 11.4: Cadmium Concentrations in Field Collections of Echinoderms

Organism	Concentration	Reference[a]
Starfish, *Asterias rubens*		
Whole	3.7 DW	1
Oral skin	1.7 DW	1
Aboral skin	3.6 DW	1
Pyloric ceca	8.7 DW	1
Gonad	1.5 DW	1
Whole	1.6 DW	2
Whole	0.2 DW	3
Starfish, *Coscinasterias calamaria*; whole	0.6 FW	4
Echinoderms; whole	Max. 18.0 DW	5
Echinoid, *Echinus esculentus*		
Oral shell	0.7 DW	1
Aboral shell	0.3 DW	1
Aristotle's lantern	0.2 DW	1
Spines	0.7 DW	1
Intestines	8.9 DW	1
Gonad	0.7 DW	1
Starfish, *Henricia sanguinolenta*; whole	3.5 DW	1
Guam; 1998-1999; sea cucumbers; 4 stations; dorsal body wall vs. hemal system; max. concentrations		
Bohadschia argus	0.11 DW vs. 0.39 DW	9
Holothuria atra	0.07 DW vs. 0.25 DW	9
Sea urchin, *Molpadia intermedia*; muscle	1.7 DW	6
Sea urchin, *Paracentrotus lividus*; whole; 1991-2005; Balearic Islands; Port of Mahon; western Mediterranean Sea	0.67 (0.14-1.6) DW	10
Starfish, *Patirella regularis*; whole	0.7 FW	4
Starfish, *Solaster papposus*; whole	4.5-5.3 DW	1
Echinoid, *Spatangus purpureus*		
Test and spines	0.1 DW	1
Gonad	1.1 DW	1

(*Continues*)

Table 11.4: Cont'd

Organism	Concentration	Reference[a]
Sea urchin, *Strongylocentrotus droebachiensis*		
Whole	0.6-1.8 DW	7
Gonads	0.9-1.5 DW	7
Sea urchin, *Strongylocentrotus transicanus*; gonad	0.57 FW	8

Values are in mg Cd/kg fresh weight (FW) or dry weight (DW).
[a]1, Riley and Segar, 1970; 2, Leatherland and Burton, 1974; 3, De Clerck et al., 1979; 4, Eustace, 1974; 5, Stenner and Nickless, 1974; 6, Thompson and Paton, 1978; 7, Bohn, 1979; 8, Vattuone et al., 1976; 9, Denton et al., 2006; 10, Deudero et al., 2007.

Laboratory studies show that seawater containing 0.35-1.41 mg Cd/L resulted in developmental abnormalities in embryos of the sea urchins *A. crassispina* and *Hemicentrotus* sp., including decreased rates of fertilization, decreased cell division during gastrulation, and increased rate of polyspermy (Kobayashi and Fujinaga, 1976). Both species developed normally in 0.18 mg Cd/L, which is substantially in excess of cadmium concentrations normally encountered in coastal waters. Death of adult starfish, *A. rubens*, is recorded at 0.70 mg Cd/L in starfish during exposure for 7 days (Eisler and Hennekey, 1977), and of *A. crassispina* embryos at 1.6 mg Cd/L within 48 h (Kobayashi, 1971).

11.7 Cerium

Laboratory studies with radiocerium and the sea urchin *Strongylocentrotus pulcherrimus* demonstrate that cerium was concentrated in digestive tract by factors in excess of 250-fold over the ambient medium; for gonad and test, these values were 3 and 15, respectively (Hiyama and Shimisu, 1964). Addition of the chelating agent EDTA to the medium was associated with lower radiocerium values in these tissues by an order of magnitude in most cases (Hiyama and Shimisu, 1964). The fate and effects of chelated cerium are not known with certainty.

11.8 Cesium

Concentrations of stable cesium in five species of representative coastal echinoderms from Greece ranged between 0.003 and 0.050 mg Cs/kg DW whole organism (Papadopoulu et al., 1976). A sixth species, the echinoid *Sphaerechinus granularis*, contained 0.13 mg

Cs/kg DW whole animal (Papadopoulu et al., 1976). Radiocesium-137 levels were below the minimum detection limit in the green sea urchin, *Strongylocentrotus polyacanthus*, collected from the Aleutian Islands in summer 2004: the site of nuclear tests between 1965 and 1971 (Burger et al., 2007b).

Accumulation of radiocesium under controlled conditions was investigated for a sea urchin, *S. pulcherrimus* (Hiyama and Shimisu, 1964), and an echinoid *Psammechinus miliaris* (Bryan, 1963). The sea urchin accumulated radiocesium from the medium in digestive tract by a factor of 9.3 in 2 days; values for gonad and test were lower: 2 for gonad and 0.4 for test. The echinoid also demonstrated radiocesium uptake over the medium in gut (28.1), followed by gonad (17.7), muscle (9.1), shell plus tube feet (4.0), teeth (1.9), and no uptake in coelomic fluid (1.0) (Bryan, 1963).

11.9 Chromium

Chromium concentrations in echinoderm tissues were usually less than 1.0 mg Cr/kg DW, with the exception of samples from Puerto Rico (Table 11.5). The elevated levels (24.2-43.2 mg Cr/kg FW whole organism) seen in *Echinometra* and *Ophiothrix* from Puerto Rico were not reflected in *Holothuria* from the same vicinity (Table 11.5), and the former values should be interpreted with caution pending verification. Echinoderms from the United Kingdom and environs were comparatively low in chromium, with values of less than 0.46 mg Cr/kg DW whole organism recorded for *Asterias*, *Echinus*, *Solaster*, *Porania*, and *Spatangus* (Riley and Segar, 1970).

In laboratory tests, embryos of the sea urchin, *Anthocidaris* sp., developed normally in seawater containing 3.2-4.2 mg Cr/L, but failed to develop at 8.4-10.0 mg Cr/L (Kobayashi, 1971; Okubo and Okubo, 1962). Larvae of *Hemicentrotus* sp. were more sensitive, showing abnormal development or death within 24 h at concentrations less than 1.0 mg Cr/L (Okubo and Okubo, 1962). Hexavalent chromium at 6.0 mg/L was associated with abnormal development of *A. crassispina* embryos (Kobayashi, 1977). Results of 7-day tests with adult starfish, *A. rubens*, and hexavalent chromium show that all survived at 5.0 mg/L, 50% died at 8.0 mg/L, and all died at 20.0 mg/L (Eisler and Hennekey, 1977).

11.10 Cobalt

Whole sea cucumbers, *Stichopus tremulus*, accumulated cobalt up to 240 times over ambient seawater concentrations (Ichikawa, 1961); however, cobalt body residues were low, and probably reflected the low environmental levels. In fact, cobalt concentrations in most field collections of echinoderms were near or below analytical detection limits for cobalt, with the

Table 11.5: Chromium Concentrations in Field Collections of Echinoderms

Organism	Concentration	Reference[a]
Sea urchin, *Arbacia lixula*; whole	13.0 DW	1
Starfish, *Asterias rubens*; whole	0.46 FW	2
Echinoid, *Echinaster sepositus*; whole	0.83 DW	1
Echinoderms; whole	0.03-1.2 DW	3
Sea urchin, *Echinometra lacunter*; whole	43.2 FW	4
Guam; 1998-1999; 4 stations; sea cucumbers; max. concentrations; dorsal body wall vs. hemal system *Bohadschia argus* *Holothuria atra*	0.43 DW vs. 31.9 DW 0.25 DW vs. 8.6 DW	8 8
Starfish, *Henricia sanguinolenta*; whole	0.46 DW	5
Sea cucumber, *Holothuria forksalii*; muscle	0.28 DW	6
Sea cucumber, *Holothuria tubulosa*; whole	0.8 DW	1
Starfish, *Marthasterias glacialis*; whole	1.6 DW	1
Sea urchin, *Molpadia intermedia*; muscle	2.2 DW	7
Brittle star, *Ophioderma longicauda*; whole	0.46 DW	1
Brittle star, *Ophiothrix suensoni*; whole	24.2 FW	4
Brittle star, *Ophiura texturata*; whole	0.8 DW	4
Sea urchin, *Paracentrotus lividus* Whole Whole	4.8 DW 1.6 (0.4-3.2) DW	1 9
Echinoid, *Sphaerechinus granularis*; whole	6.4 DW	1
Sea cucumber, *Stichopus regalis*; whole less gut	1.1 DW	4

Values are in mg Cr/kg fresh weight (FW) or dry weight (DW).
[a]1, Papadopoulu et al., 1976; 2, De Clerck et al., 1979; 3, Fukai and Broquet, 1965; 4, Bernhard and Zattera, 1975; 5, Riley and Segar, 1970; 6, Fukai, 1965; 7, Thompson and Paton, 1978; 8, Denton et al., 2006; 9, Deudero et al., 2007.

possible exceptions of gonad and pyloric ceca tissues (Table 11.6). Radiocobalt-60 levels were not detectable in sea urchins collected from the Aleutian Islands during 2004 near sites used in nuclear tests from 1965 to 1971 (Burger et al., 2007b).

In the laboratory, it was shown that embryos of a sea urchin *A. crassispina* developed normally in seawater solutions containing 17.0 mg Co/L as cobalt acetate; however, 170.0 mg Co/L was fatal (Kobayashi, 1971).

Table 11.6: Cobalt Concentrations in Field Collections of Echinoderms

Organism	Concentration	Reference[a]
Starfish, *Asterias rubens*		
Whole	<0.3 DW	1
Oral skin	<0.3 DW	1
Aboral skin	<0.3 DW	1
Pyloric ceca	2.1 DW	1
Gonad	0.6 DW	1
Echinoderms; whole		
3 spp.	<0.2-1.0 DW	1
7 spp.	0.09-0.66 DW	2
Echinoid, *Echinus esculentus*		
Gonad	0.4 DW	1
Other tissues	<0.3 DW	1
Echinoid, *Spatangus purpureus*		
Test and spines	<0.2 DW	1
Gonad	0.5 DW	1

Values are in mg Co/kg dry weight (DW).
[a]1, Riley and Segar, 1970; 2, Papadopoulu et al., 1976.

11.11 Copper

Maximum copper concentrations measured in echinoderm body parts, in mg/kg, were 31.4 in whole starfish on a FW basis, 90.0 DW in body and shell, and 324.0 ash weight (AW) in sea urchin sperm (Table 11.7). Copper concentrations in echinoderms, when compared to other marine invertebrate groups examined (except coelenterates) were low, suggesting sensitivity to this metal (Eisler, 1979).

Laboratory studies with eggs of sea urchins, *Hemicentrotus pulcherrimus* and *A. crassispina*, showed that development was normal at 0.04-0.08 mg Cu/L, but higher levels of 0.08-0.16 mg/L inhibited development (Kobayashi and Fujinaga, 1976). Sensitivity of sea urchin sperm and eggs to copper was investigated by Kobayashi (1980); sperm activity was inhibited at low sublethal copper concentrations, and fertilization and gastrulation stages were more sensitive to copper than were first cleavage, blastulation, and pluteus stages. Abnormal development was observed at concentrations as low as 0.06 mg Cu/L (Kobayashi, 1977). Copper was more toxic to sperm of the sea urchin, *Paracentrotus lividus* (EC50 of 0.018 mg Cu/L) than embryos (0.046 mg Cu/L); however, mixtures of copper, triazines, and dimethylureas interact significantly on spermotoxicity and embryotoxicity (Manzo et al., 2008). Mixtures of copper and zinc or copper and cadmium produced more than additive

Table 11.7: Copper Concentrations in Field Collections of Echinoderms

Organism	Concentration	Reference[a]
Starfish, *Asterias rubens*		
Whole	18.8 FW	1
Whole	4.3 DW	2
Oral skin	2.5 DW	2
Aboral skin	2.0 DW	2
Pyloric ceca	16.0 DW	2
Gonad	4.1 DW	2
Starfish, *Coscinasterias calamaria*; whole	10.4 FW	3
Echinoderms		
Body and shell	20.0-90.0 DW	4
Whole	Max. 18.0 DW	3, 5
Echinoid, *Echinus esculentus*		
Oral shell	1.8 DW	2
Aboral shell	0.9 DW	2
Aristotle's lantern	0.4 DW	2
Spines	1.6 DW	2
Intestines	5.9 DW	2
Gonad	16.0 DW	2
Starfish, *Henricia sanguinolenta*; whole	8.2 DW	2
Sea cucumber, *Holothuria* sp.; whole	1.9 FW	6
Sand dollar, *Mellita lata*; whole	6.0-8.2 FW; 10.0-14.0 DW; 11.0-15.0 AW	7
Sand dollar, *Mellita sexiesperforata*; whole	5.0-20.0 FW; 9.4-38.0 DW; 10.0-40.0 AW	7
Sea urchin, *Molpadia intermedia*; muscle	26.0 DW	8
Sea urchin, *Paracentrotus lividus*; whole; 1991-2005; Balearic Islands, Port of Mahon; western Mediterranean Sea	4.9 (1.6-7.5) DW	12
Asteroid, *Pisaster brevispinus*; California		
Gonad	2.0-10.0 DW	11
Hepatic caecum	18.0-36.0 DW	11
Stomach	5.0-40.0 DW	11
Starfish, *Patiriella regularis*; whole	31.4 FW	3
Sea urchin, *Pseudocentrotus depressus*		
Spermatozoa homogenate	17.0-19.0 AW	9
Spermatozoa head	0.0 AW	9

(*Continues*)

Table 11.7: Cont'd

Organism	Concentration	Reference[a]
Spermatozoa mid-piece	257.0-324.0 AW	9
Spermatozoa tail	0.0 AW	9
Starfish, *Solaster papposus*; whole	6.1-11.0 DW	2
Echinoid, *Spatangus purpureus*		
Test and spines	1.2 DW	2
Gonad	8.7 DW	2
Sea urchin, *Strongylocentrotus droebachiensis*		
Whole	2.3-4.0 DW	10
Gonad	3.3-9.0 DW	10

Values are in mg Cu/kg fresh weight (FW), dry weight (DW), or ash weight (AW).
[a]1, De Clerck et al., 1979; 2, Riley and Segar, 1970; 3, Eustace, 1974; 4, Culkin and Riley, 1958; 5, Stenner and Nickless, 1974; 6, Matsumoto et al., 1964; 7, Lowman et al., 1966; 8, Thompson and Paton, 1978; 9, Morisawa and Mohri, 1972; 10, Bohn, 1979; 11, Jenkins, 1980; 12, Deudero et al., 2007.

effects in suppressing embryonic development of sea urchins than could be predicted on the basis of individual metals; however, copper-nickel mixtures acted additively (Kobayashi and Fujinaga, 1976). In another study with sea urchins, adults of the comparatively sensitive *Strongylocentrotus nudus* were held in seawater containing 0.025 mg Cu/L for 30 days and the resultant embryos reared in uncontaminated seawater for 30 days (Durkina and Evtushenko, 1991). Food consumption of adults decreased after day 16; there was accelerated development of pluteal stages; and all developmental stages had increased activities of acid phosphatase (Durkina and Evtushenko, 1991).

Recent studies (Rosen et al., 2008) with eggs and larvae of the purple sea urchin, *Strongylocentrotus purpuratus*, related whole body residues to effects. For example, controls exposed for up to 96 h to 0.0025 mg Cu/L contained up to 3.7 mg Cu/kg DW; higher concentrations up to 0.0091 mg Cu/L produced whole body residues up to 23.8 mg Cu/kg DW but with no observable adverse effects. Adverse effects on normal larval development were associated with 0.014-0.021 mg Cu/L and whole body residues of 108.0-192.0 mg Cu/kg DW (Rosen et al., 2008).

11.12 Europium

Radioeuropium-152 levels in sea urchins from the Aleutian Islands in summer 2004 were below detection limits; collection locales were near the site of three nuclear tests conducted between 1965 and 1971 (Burger et al., 2007b).

11.13 Gallium

Culkin and Riley (1958) report values of 0.02-0.35 mg Ga/kg DW in body and shell of echinoderms from the Irish Sea and environs.

11.14 Iron

Echinoderms accumulate radioiron to relatively high levels over ambient seawater: concentration factors were 1000 in intestine of the sea cucumber *Stichopus japonicus*, 78,000 in whole *Stichopus*, and 10,000 in ovary of the sea urchin *A. crassispina* (Ichikawa, 1961). Concentrations of stable iron in echinoderms are quite variable (Table 11.8). It is possible that the grossly elevated iron burdens reported in intestine and whole body of some organisms may reflect anthropogenic contamination of sediments. Iron, as was true of copper, is comparatively abundant in the mid-piece section of spermatozoa from the sea urchin *Pseudocentrotus depressus* when compared to other portions of the spermatozoan (Table 11.8), but the significance of this observation is not clear. Embryonic development of larval sea urchins proceeds normally at 3.2 mg Fe/L; however, developmental abnormalities occur at 10.0 mg Fe/L and higher (Okubo and Okubo, 1962).

Table 11.8: Iron Concentrations in Field Collections of Echinoderms

Organism	Concentration	Reference[a]
Sea urchin, *Arbacia lixula*; whole	620.0 DW	1
Starfish, *Asterias rubens*		
Whole	37.0 DW	2
Oral skin	18.0 DW	2
Aboral skin	21.0 DW	2
Pyloric ceca	87.0 DW	2
Gonad	19.0 DW	2
Echinoid, *Echinaster sepositus*; whole	170.0 DW	1
Echinoid, *Echinus esculentus*		
Oral shell	6.9 DW	2
Aboral shell	2.5 DW	2
Aristotle's lantern	1.6 DW	2
Spines	16.0 DW	2
Intestines	22,000.0 DW	2
Gonad	15.0 DW	2
Echinoderms; body and shell	149.0-1700.0 DW	3

(Continues)

Table 11.8: Cont'd

Organism	Concentration	Reference[a]
Sea urchin, *Echinometra lucunter*; skeleton	35.0-38.0 DW	4
Starfish, *Henricia sanguinolenta*; whole	1800.0 DW	2
Sea cucumber, *Holothuria* sp.; whole	50.0 FW	5
Sea cucumber, *Holothuria tubulosa*; whole	74.0 DW	1
Starfish, *Marthasterias glacialis*; whole	110.0 DW	1
Brittle star, *Ophioderma longicauda*; whole	130.0 DW	1
Sea urchin, *Paracentrotus lividus*; whole	600.0 DW	1
Sea urchin, *Pseudocentrotus depressus*; spermatozoa		
Homogenate	194.0-219.0 AW	6
Head	201.0-222.0 AW	6
Mid-piece	1171.0-1465.0 AW	6
Tail	88.0-140.0 AW	6
Starfish, *Solaster papposus*; whole	170.0-200.0 DW	2
Echinoid, *Spatangus purpureus*		
Test and spines	53.8 DW	2
Gonad	19.2 DW	2
Echinoid, *Sphaerechinus granularis*; whole	1100.0 DW	1
Sea urchin, *Strongylocentrotus droebachiensis*; collected various distances from Zn/Pb mine		
1 km; whole	1310.0 DW	7
15 km; whole vs. gonads	495.0 DW vs. 217.0 DW	7
Sea urchin, *Tripneustes esculentus*; skeleton	53.0-81.0 DW	4

Values are in mg Fe/kg fresh weight (FW), dry weight (DW), or ash weight (AW).
[a]1, Papadopoulu et al., 1976; 2, Riley and Segar, 1970; 3, Culkin and Riley, 1958; 4, Stevenson and Ufret, 1966; 5, Matsumoto et al., 1964; 6, Morisawa and Mohri, 1972; 7, Bohn, 1979.

11.15 Lead

The elevated concentrations of lead in echinoderms from the western Norway coast (up to 460.0 mg Pb/kg DW; Stenner and Nickless, 1974) and Japan (14.4 mg/kg FW; Matsumoto et al., 1964) were probably associated with human activities. In contrast, echinoderms from the United Kingdom showed much lower lead burdens (Table 11.9).

Table 11.9: Lead Concentrations in Field Collections of Echinoderms

Organism	Concentration	Reference[a]
Starfish, *Asterias rubens*		
Whole	2.3 DW	1
Oral skin	1.1 DW	1
Aboral skin	1.1 DW	1
Pyloric ceca	4.0 DW	1
Gonad	0.5 DW	1
Whole	1.6 FW	5
Echinoderms; whole; West Norway coast	Max. 460.0 DW	2
Starfish, *Henricia sanguinolenta*; whole	<0.5 DW	1
Sea cucumber, *Holothuria* sp.; whole	14.4 FW	3
Sea urchin, *Molpadia intermedia*; muscle	1.4 DW	4
Sea urchin, *Paracentrotus lividus* Italy; soft parts		
Reference site	20.0 DW	6
Polluted site	42.0 DW	6
Western Mediterranean Sea; whole; 1991-2005	3.0 (0.8-6.7) DW	7
Starfish, *Solaster papposus*; whole	5.2-6.8 DW	1

Values are in mg Pb/kg fresh weight (FW) or dry weight (DW).
[a] 1, Riley and Segar, 1970; 2, Stenner and Nickless, 1974; 3, Matsumoto et al., 1964; 4, Thompson and Paton, 1978; 5, De Clerck et al., 1979; 6, Sheppard and Bellamy, 1974; 7, Deudero et al., 2007.

Kobayashi (1971) reports that embryos of a sea urchin, *A. crassispina*, develop normally in seawater containing 1.1 mg Pb/L as lead acetate, but do not develop at 2.2 mg Pb/L.

11.16 Manganese

Most values were less than 57.0 mg Mn/kg on a DW basis (Table 11.10). Where interspecies comparisons were possible, elevated levels were recorded in gonad, aboral skin, and Aristotle's lantern (Table 11.10). A sea cucumber, *Stichopus regalis*, took up radiomanganese from seawater by a factor of about 200-fold (Ichikawa, 1961), suggesting a need for this metal by *Stichopus*. On the other hand, high burdens of manganese inhibited motility of sperm of the sea urchin *Arbacia punctulata*; sperm motility was restored with the addition of

Table 11.10: Manganese Concentrations in Field Collections of Echinoderms

Organism	Concentration	Reference[a]
Starfish, *Asterias rubens*		
Whole	6.5 DW	1
Oral skin	8.2 DW	1
Aboral skin	11.0 DW	1
Pyloric ceca	2.3 DW	1
Gonad	3.7 DW	1
Starfish, *Coscinasterias calamaria*; whole	6.9 FW	2
Sea urchin, *Echinometra lucunter*; skeleton	20.0-21.0 DW	3
Echinoid, *Echinus esculentus*		
Oral shell	56.0 DW	1
Aboral shell	57.0 DW	1
Aristotle's lantern	88.0 DW	1
Spines	4.5 DW	1
Intestines	11.0 DW	1
Gonad	0.5 DW	1
Starfish, *Henricia sanguinolenta*; whole	34.0 DW	1
Sea cucumber, *Holothuria tubulosa*; whole	8.9 AW	4
Starfish, *Patiriella regularis*; whole	51.7 FW	2
Starfish, *Solaster papposus*; whole	31.0-43.0 DW	1
Echinoid, *Spatangus purpureus*		
Test and spines	27.0 DW	1
Gonad	82.0 DW	1
Sea urchin, *Tripneustes esculentus*; skeleton	13.0-14.0 DW	3

Values are in mg Mn/kg fresh weight (FW), dry weight (DW), or ash weight (AW).
[a]1, Riley and Segar, 1970; 2, Eustace, 1974; 3, Stevenson and Ufret, 1966; 4, Chipman and Thommeret, 1970.

EDTA, a chelating agent (Young and Nelson, 1974). Comparatively high concentrations of manganese, that is 6.6-10.0 mg/L, do not interfere with embryonic development of sea urchins, *Anthocidaris* spp.; however, higher concentrations of 13.2-32.0 mg Mn/L adversely affect growth and survival (Kobayashi, 1971; Okubo and Okubo, 1962).

Manganese occurs naturally in marine sediments as manganese dioxide (MnO_2); however, during hypoxic conditions it is converted to the bioavailable Mn^{2+} and can be accumulated by starfish, *A. rubens* (Oweson et al., 2008). Starfish exposed to 5.5 mg Mn/L in the laboratory for up to 25 days showed cytotoxicity and adverse effects on immune mechanisms that suppress defense against pathogens (Oweson et al., 2008).

11.17 Mercury

Mercury concentrations in echinoderms from nonpolluted areas never exceed 0.40 mg Hg/kg on a FW basis or 0.92 mg Hg/kg on a DW basis (Eisler, 1981; Table 11.11). Mercury contamination from a chloralkali plant in Cuba was reflected in elevated soil mercury concentrations and its terrestrial vegetation, and in sediments and gonads of benthic fauna from these sediments (Table 11.11). Mercury contamination is measurable for at least 5 km from the chloralkali plant (Gonzalez, 1991). Asteroids from a mercury-contaminated site in

Table 11.11: Mercury Concentrations in Field Collections of Echinoderms

Organism	Concentration	Reference[a]
Starfish, *Asterias rubens*		
Whole	0.22 DW	1
Whole	0.12 FW	2
Asteroids; 4 spp.; Golfo Triste, Venezuela; 1981; mercury-contaminated in 1980		
Stomach contents	0.17-2.2 FW; 0.53-7.3 DW	8
Exoskeleton	0.59-2.0 FW; 2.2-7.3 DW	8
Gonads	0.88-1.6 FW; 3.8-8.7 DW	8
Echinoderms; whole	0.28-0.40 FW	3
Sea urchin, *Lytechinus variegatus*; gonads; Cuba; 1985-1987; near chloralkali plant		
Near discharge	0.38 DW	5
Transition area	0.29 DW	5
Reference site	0.07 DW	5
Starfish, *Marthasterias glacialis*; whole	0.92 FW	1
Sea urchin, *Paracentrotus lividus*; whole; western Mediterranean Sea; Balearic Islands; Port of Mahon	0.14 (0.03-0.38) DW	6
Sea urchin, *Strongylocentrotus fragilis*; gonads; total mercury vs. organic mercury	0.020-0.034 FW vs. 0.003 FW	4
Green sea urchin, *Strongylocentrotus polyacanthus*; edible portions; Aleutian Islands; 2004	0.005 FW; max. 0.016 FW	7

Values are in mg Hg/kg fresh weight (FW) or dry weight (DW).
[a]1, Leatherland and Burton, 1974; 2, De Clerck et al., 1979; 3, Williams and Weiss, 1973; 4, Eganhouse and Young, 1978; 5, Gonzalez, 1991; 6, Deudero et al., 2007; 7, Burger et al., 2007a; 8, Iglesias and Penchaszadeh, 1983.

Venezuela had comparatively high concentrations in stomach when compared to exoskeleton, and elevated concentrations in tissues of inshore specimens when compared to offshore conspecifics (Iglesias and Penchaszadeh, 1983).

Sea urchin development is arrested at concentrations exceeding 0.010-0.023 mg inorganic Hg/L (Kobayashi, 1971; Okubo and Okubo, 1962), with a high proportion of abnormal plutei at 0.03 mg Hg/L (Kobayashi, 1977). Effects were more pronounced at similar concentrations of organic mercury salts (Kobayashi, 1971). Mercury inhibits the motility of spermatozoa of a sea urchin, *A. punctulata* (Young and Nelson, 1974). This effect was reversed by EDTA. EDTA alone can block motility by complexing essential seawater cations, but may also deplete intracellular regulatory cations, which otherwise may exist in a state of dynamic equilibrium with seawater (Young and Nelson, 1974).

11.18 Molybdenum

The starfish, *A. rubens*, is relatively resistant to molybdenum salts. Concentrations from 127.0 to 254.0 mg Mo/L were necessary to kill 50% in 24 h, with mortality higher at the lower end of the pH range tested of 5.8-8.2 (Abbott, 1977).

11.19 Nickel

Echinoderms collected from Puerto Rico, including *Echinometra* and *Tripneustes*, contained up to 10 times more nickel than did asteroids and echinoids from Britain (Table 11.12). It is not known if these differences are real, or to differences in sample collection, digestion, and analysis procedures.

Laboratory studies with Japanese sea urchins suggest that ambient seawater concentrations up to 0.18 mg Ni/L do not affect embryonic development; however, 0.37-1.47 mg Ni/L inhibited development in as little as 15 h (Kobayashi and Fujinaga, 1976). Research on other species demonstrate that concentrations as low as 0.058 mg Ni/L adversely affects development of embryos of the sea urchin, *Lytechinus pictus* (Timourian and Watchmaker, 1972; USEPA, 1980a), that 0.059 mg Ni/L inhibits motility of sperm of the purple sea urchin *Strongylocentrotus purpurea* (Timourian and Watchmaker, 1977), and that death of adults of the starfish, *Asterias forbesi*, and the sea urchin *A. punctulata* occur at comparatively elevated concentrations of 13.0-17.0 mg Ni/L (Eisler and Hennekey, 1977; USEPA, 1980a).

11.20 Plutonium

Echinoderms from Bylot Sound, Greenland, had elevated levels of radioplutonium following accidental release of that element from military nuclear missiles when a bomber crashed at that site (Aarkrog, 1971). Half-time retention of $^{239+240}$Pu in brittle stars is estimated at 50 years (Aarkrog, 1990).

Table 11.12: Nickel Concentrations in Field Collections of Echinoderms

Organism	Concentration	Reference[a]
Starfish, *Asterias rubens*		
Whole	1.5 DW	1
Oral skin	0.8 DW	1
Aboral skin	0.7 DW	1
Pyloric ceca	4.1 DW	1
Gonad	2.4 DW	1
Rock boring sea urchin, *Echinometra lucunter*		
Skeleton	49.0-59.0 DW	2
Puerto Rico; skeleton vs. whole	51.0 (42.0-78.0) DW vs. 37.0 DW	3
Echinoid, *Echinus esculentus*		
Spines	1.6 DW	1
Gonad	7.7 DW	1
Other tissues	<0.2 DW	1
Florida; near sewage outfall; exposure for 120 days		
Sea urchin, *Lytechinus variegatus*; whole	30.0 DW	4
Sea cucumber, *Holothuria mexicana*; whole	40.0 DW	4
Starfish, *Henricia sanguinolenta*; whole	3.7 DW	1
Sea urchin, *Paracentrotus lividus*; whole; 1991-2005; western Mediterranean Sea; Balearic Islands; Port of Mahon	1.7 (0.8-2.7) DW	5
Starfish, *Solaster papposus*; whole	2.3-4.0 DW	1
Echinoid, *Spatangus purpureus*		
Test and spines	<0.2 DW	1
Gonad	1.4 DW	1
Sea urchin, *Tripneustes esculentus*		
Skeleton	35.0 (18.0-54.0) DW	2, 3
Ovary	1.4 FW	3
Testes	22.0 FW	3

Values are in mg Ni/kg fresh weight (FW) or dry weight (DW).
[a] 1, Riley and Segar, 1970; 2, Stevenson and Ufret, 1966; Jenkins, 1980; 4, Montgomery et al., 1978; 5, Deudero et al., 2007.

11.21 Rubidium

Representative species of adult echinoderms from the Aegean Sea contained between 0.9 and 4.7 mg Rb/kg DW whole organism (Table 11.13).

11.22 Scandium

Adults of six species of echinoderms found in the Aegean Sea contained 0.005-0.051 mg Sc/kg DW whole organism (Papadopoulu et al., 1976). A seventh species, *S. granularis*, contained 0.171 mg Sc/kg DW whole adult (Papadopoulu et al., 1976), and this may be related to the comparatively high concentrations of iron, chromium, and other metals found in that same sample.

11.23 Selenium

Concentrations of selenium in whole adults of seven species from coastal waters of Greece ranged from 0.8 to 4.4 mg Se/kg DW (Table 11.14).

11.24 Silver

Silver concentrations in almost all species of echinoderms examined were either low, nondetectable, or near the limits of analytical detection (Table 11.15). The comparatively high value of 0.6 mg Ag/kg DW in whole *Henricia* is indicative of the high body burdens of other metals reported in this species (Riley and Segar, 1970). Silver concentrations of 7.3 mg Ag/kg DW in whole *Marthasterias* and 4.9 mg/kg DW in hemal system of *Holothuria atra* (Table 11.15) requires confirmation.

Table 11.13: Rubidium Concentrations in Field Collections of Echinoderms

Organism	Concentration	Reference[a]
Sea urchin, *Arbacia lixula*; whole	2.6 DW	1
Echinoid, *Echinaster sepositus*; whole	0.9 DW	1
Sea cucumber, *Holothuria tubulosa*; whole	2.0 DW	1
Starfish, *Marthasterias glacialis*; whole	2.1 DW	1
Brittle star, *Ophioderma longicauda*; whole	1.8 DW	1
Sea urchin, *Paracentrotus lividus*; whole	4.7 DW	1
Echinoid, *Sphaerechinus granularis*; whole	3.5 DW	1

Values are in mg Rb/kg dry weight (DW).
[a]1, Papadopoulu et al., 1976.

Table 11.14: Selenium Concentrations in Field Collections of Echinoderms

Organism	Concentration	Reference[a]
Sea urchin, *Arbacia lixula*; whole	2.4 DW	1
Echinoid, *Echinaster sepositus*; whole	4.4 DW	1
Sea cucumber, *Holothuria tubulosa*; whole	3.4 DW	1
Starfish, *Marthasterias glacialis*; whole	3.0 DW	1
Brittle star, *Ophioderma longicauda*; whole	1.9 DW	1
Sea urchin, *Paracentrotus lividus*; whole	1.1 DW	1
Echinoid, *Sphaerechinus granularis*; whole	0.8 DW	1

Values are in mg Se/kg dry weight (DW).
[a]1, Papadopoulu et al., 1976.

Table 11.15: Silver Concentrations in Field Collections of Echinoderms

Organism	Concentration	Reference[a]
Sea urchin, *Arbacia lixula*; whole	0.30 DW	1
Starfish, *Asterias rubens* All tissues	Not detectable	2
Echinoid, *Echinaster sepositus*; whole	0.25 DW	1
Echinoid, *Echinus esculentus*		
Oral shell	0.09 DW	2
Aboral shell	<0.05 DW	2
Aristotle's lantern	0.53 DW	2
Spines	<0.03 DW	2
Intestine	<0.8 DW	2
Gonad	<0.03 DW	2
Guam; 1998-1999; 4 stations; sea cucumbers; max. concentrations; dorsal body wall vs. hemal system		
Bohadschia argus	<0.14 DW vs. <0.14 DW	4
Holothuria atra	0.24 DW vs. 4.9 DW	4
Starfish, *Henricia sanguinolenta*; whole	0.6 DW	2
Sea cucumber, *Holothuria tubulosa*; whole	0.05 DW	1
Starfish, *Luidia clathrata*; Tampa Bay, Florida vs. Gulf of Mexico; 1982		
Body wall	0.26-0.84 DW vs. 0.67 DW	3
Pyloric ceca	0.4-1.1 DW vs. 0.17 DW	3

(Continues)

Table 11.15: Cont'd

Organism	Concentration	Reference[a]
Starfish, *Marthasterias glacialis*; whole	7.3 DW	1
Brittle star, *Ophioderma longicauda*; whole	0.07 DW	1

Sea urchin, *Paracentrotus lividus*
  Whole
  Whole; 1991-2005; western Mediterranean Sea; Balearic Islands; Port of Mahon
	0.32 DW
	1.7 (0.2-6.8) DW
		1
		5

Echinoid, *Sphaerechinus granularis*; whole	0.32 DW	1

Values are in mg Ag/kg dry weight (DW).
[a]1, Papadopoulu et al., 1976; 2, Riley and Segar, 1970; 3, Lawrence et al., 1993; 4, Denton et al., 2006; 5, Deudero et al., 2007.

Silver accumulates in eggs of the sea urchin *Diadema antillarum* (Bianchini et al., 2007). In that study, sea urchins were exposed to 0.001 or 0.01 mg Ag/L for 48 h. Silver burdens were highest in eggs (20.0-30.0 mg Ag/kg FW), tube feet (5.0 mg/kg FW in the high-dose group), and less than 2.0 mg Ag/kg FW in other tissues, including muscle and hemolymph. The effects of these concentrations on reproduction are unknown (Bianchini et al., 2007). Ionic silver at 0.0005 mg/L reportedly inhibits embryo development of the sea urchin, *A. lixula*, after 52 h (USEPA, 1980b). In 30-day tests, all adults survived 0.019 mg/L of dissolved silver (about 0.000029 ionic Ag/L); only 75% survived 0.033 mg/L of dissolved silver (about 0.000049 mg ionic Ag/L; Ward et al., 2006). Sea urchin sperm (0.000014 mg/L) and embryos (0.000028 mg/L) were more sensitive to ionic silver in 1 h and 48 h tests, respectively, than were adults in 30-day tests (Ward et al., 2006).

11.25 Strontium

Strontium was especially abundant in shell and other hard tissues of echinoderms, and comparatively low in soft tissues (Table 11.16). In this respect, it probably follows calcium levels with which it is closely and positively associated. Variability in strontium accumulations among echinoderm species may be associated with variability between species in ability to accumulate strontium from seawater; interspecies echinoderm strontium concentration factors from seawater range from 0.4 to 2.0 (Ichikawa, 1961; Hiyama and Shimisu, 1964).

Strontium is the most abundant trace element in disk and arm tissue of brittle stars, *Ophiothrix spiculata*, collected from San Diego Bay, California, in summer 2001 (Deheyn and Latz, 2006). In order of decreasing concentration in disc, strontium was followed by iron, aluminum, zinc, tin, nickel, lead, manganese, arsenic, selenium, titanium, copper, silver, chromium, and cadmium; for arm, this order was strontium, iron, zinc, aluminum, tin,

Table 11.16: Strontium and Tin Concentrations in Field Collections of Echinoderms

Element and Organism	Concentration	Reference[a]
Strontium		
Starfish, *Asterias rubens*		
Whole	390.0 DW	1
Oral skin	600.0 DW	1
Aboral skin	520.0 DW	1
Pyloric ceca	<17.0 DW	1
Gonad	18.0 DW	1
Echinoderms		
Whole, 4 spp.	700.0-2500.0 AW	2
Body and shell	2.3-8.8 FW	3
Echinoid, *Echinus esculentus*		
Oral shell	930.0 DW	1
Aboral shell	820.0 DW	1
Aristotle's lantern	920.0 DW	1
Spines	760.0 DW	1
Intestines	150.0 DW	1
Gonad	20.0 DW	1
Starfish, *Henricia sanguinolenta*; whole	200.0 DW	1
Starfish, *Solaster papposus*; whole	260.0-490.0 DW	1
Echinoid, *Spatangus purpureus*		
Test and spines	1100.0 DW	1
Gonad	<19.0 DW	1
Tin		
Sea urchin; edible portions; France; 2005; organotins		
Total organotins	0.0053 FW	4
Butyltins	0.0047 FW	4
Phenyltins	0.0003 FW	4
Octyltins	0.0003 FW	3

Values are in mg element/kg fresh weight (FW), dry weight (DW), or ash weight (AW).
[a] 1, Riley and Segar, 1970; 2, Fukai et al., 1962; 3, Mauchline and Templeton, 1966; 4, Guerin et al., 2007.

manganese, lead, arsenic, copper, nickel, silver, titanium, selenium, cadmium, and chromium (Deheyn and Latz, 2006).

Radiostrontium-90 levels were below detection limits in green sea urchins, *S. polyacanthus*, collected in summer 2004 from the Aleutian Islands near the sites of nuclear tests conducted between 1965 and 1971 (Burger et al., 2007b).

11.26 Technetium

No detectable levels of radiotechnetium-99 were found in sea urchins collected during June-July 2004 in the Aleutian Islands from the vicinity of nuclear tests conducted in 1965-1971 (Burger et al., 2007b).

11.27 Tin

Edible portions of sea urchins sold for human consumption in France in 2005 had low organotin content (0.0053 mg/kg FW), of which 89% was butyltins (Table 11.16); tributyltins comprised 62% of all organotins (Guerin et al., 2007).

Suppression of regeneration in echinoderms by organotins is due mainly to the neurotoxicological action of organotins, and secondarily due to direct action on tissue at the breakage point (Walsh et al., 1986). Arm regeneration in a brittle star, *Ophioderma brevispina*, was inhibited at 0.001 mg Sn (as tributyltin)/L (Walsh et al., 1986). Tributyltin chloride, at 0.007 mg Sn/L, resulted in 50% immobilization of sperm of the sea urchin, *A. crassispina* (Shim et al., 2006). Tributyltin chloride was the most toxic of all organotins tested to sea urchin sperm, followed by triphenyltin chloride, dibutyltin chloride, monophenyltin chloride, trimethyltin chloride, diphenyltin chloride, monobutyltin chloride, and dimethyltin dichloride, in that order; the same general order of toxicity was observed for egg development (Shim et al., 2006).

11.28 Zinc

Zinc concentrations greater than 100.0 mg/kg DW in various tissues were not unusual (Table 11.17), supporting a conclusion that zinc was not limiting among echinoderms and probably accumulated in excess of the organism's immediate needs.

Despite the comparatively high accumulations of zinc in echinoderms, several investigators have demonstrated that relatively low seawater concentrations of <0.0065 mg Zn/L were associated with developmental abnormalities and other adverse effects, including reduced semen production in sea urchins and starfish and reduced sperm pH and motility (Clapper et al., 1985a,b). At 0.023 mg Zn/L for 5 days, embryos of the purple sea urchin, *S. purpuratus*, experienced 50% inhibition of development (USEPA, 1987); similar results were observed at 0.097-0.107 mg/L in only 96 h (Phillips et al., 1998). At 0.028 mg Zn/L for 60 min, sperm of the sand dollar, *Dendraster excentricus*, had a 50% reduction in fertilization success (USEPA, 1987). This may be a case of nutritional zinc deficiency, which has been induced experimentally in echinoderms (Eisler, 2000). Concentrations of zinc, in mg/L, which inhibited echinoderm embryogenesis were 0.065-0.65 in *P. miliaris* (Cleland, 1953); 0.06-0.41 in *A. crassispina* (Okubo and Okubo, 1962; Kobayashi, 1977; Kobayashi and

Table 11.17: Zinc Concentrations in Field Collections of Echinoderms

Organism	Concentration	Reference[a]
Starfish, *Asterias rubens*		
Whole	190.0 DW	1
Oral skin	110.0 DW	1
Aboral skin	110.0 DW	1
Pyloric ceca	160.0 DW	1
Gonad	360.0 DW	1
Whole	220.0 DW	2
Whole	56.6 FW	3
Starfish, *Coscinasterias calamaria*; whole	40.6 FW	4
Echinoderms		
Whole	40.6 FW	5
Whole, 7 spp.	16.0-120.0 DW	6
Various species; whole	Usually 100.0 DW or lower; frequently >100.0 DW; max. 245.0 FW, 1500.0 DW	12
Echinoid, *Echinus esculentus*		
Oral shell	110.0 DW	1
Aboral shell	12.0 DW	1
Aristotle's lantern	22.0 DW	1
Spines	28.0 DW	1
Intestines	550.0 DW	1
Gonad	110.0 DW	1
Starfish, *Henricia sanguinolenta*; whole	240.0 DW	1
Sea cucumber, *Holothuria* sp.; whole	8.7 FW	7
Sand dollar, *Mellita lata*; whole	31.0-130.0 FW; 54.0-220.0 DW; 57.0-240.0 AW	8
Sand dollar, *Mellita sexiesperforata*; whole	24.0-56.0 FW; 43.0-83.0 DW; 46.0-88.0 AW	8
Sea urchin, *Molpadia intermedia*; whole	171.0 DW	9
Sea urchin, *Paracentrotus lividus*; whole; 1991-2005; western Mediterranean Sea; Balearic Islands; Port of Mahon	40.3 (5.7-74.5) DW	13
Starfish, *Patiriella regularis*; whole	245.0 FW	4
Sea urchin, *Pseudocentrotus depressus*; spermatozoa		
Homogenate	66.0-79.0 AW	10
Head	34.0-41.0 AW	10

(*Continues*)

Table 11.17: Cont'd

Organism	Concentration	Reference[a]
Mid-piece	281.0–337.0 AW	10
Tail	172.0–186.0 AW	10
Microtubules	118.0–142.0 AW	10
Starfish, *Solaster papposus*; whole	120.0–130.0 DW	1
Echinoid, *Spatangus purpureus*		
Test and spines	35.0 DW	1
Gonad	170.0 DW	1
Sea urchin, *Strongylocentrotus droebachiensis*		
Whole; 1 km from zinc ore tailings vs. 15 km	57.0 DW vs. 34.0 DW	11
Gonads; 15 km from zinc ore tailings	153.0 DW	11

Values are in mg Zn/kg fresh weight (FW), dry weight (DW), or ash weight (AW).
[a]1, Riley and Segar, 1970; 2, Leatherland and Burton, 1974; 3, De Clerck et al., 1979; 4, Eustace, 1974; 5, Stenner and Nickless, 1974; 6, Papadopoulu et al., 1976; 7, Matsumoto et al., 1964; 8, Lowman et al., 1966; 9, Thompson and Paton, 1978; 10, Morisawa and Mohri, 1972; 11, Bohn, 1979; 12, Young et al., 1980; 13, Deudero et al., 2007.

Fujinaga, 1976; Nakamura et al., 1989); and 0.2 for *Henricia pulcherrimus* (Kobayashi and Fujinaga, 1976). Zinc inhibits the formation of the fertilization membrane in sea urchin eggs by interfering with cortical granule-derived proteases and proteins (Nakamura et al., 1989). At 0.24 mg/L, female adults of the starfish, *A. rubens*, showed increased steroid metabolism in pyloric ceca after 21 days (Voogt et al., 1987). Adults of the starfish, *A. rubens*, survived exposure to 1.0 mg Zn/L for 168 h; however, 50% died at 2.3 mg/L in 168 h (Eisler and Hennekey, 1977).

Bioconcentration factors for stable zinc and echinoderms were comparatively low, and usually ranged between 15 and 500 (Eisler, 1980). Sea cucumbers, *S. tremulus*, took up radiozinc-65 from solution by a factor of 1400 (Ichikawa, 1961), suggesting that the ambient medium is an important route of zinc intake. However, concentration factors based on ^{65}Zn accumulation data should be viewed with reservation as the addition of stable zinc substantially reduced radiozinc burdens in echinoderm viscera, sometimes by an order of magnitude (Hiyama and Shimisu, 1964). Diet and water depth influence zinc accumulations in echinoderms. Carey (1970) stated that certain detrital feeders contain more zinc than carnivores, and that species which feed near the surface contain more zinc per unit weight than sediment feeders. Further, Carey (1970) found lower zinc concentrations in echinoderms collected offshore in deeper waters when compared to conspecifics collected inshore.

To protect important species of marine animals, the United States Environmental Protection Agency (USEPA) proposed that total recoverable zinc in seawater should average less than 0.058 mg/L and never exceed 0.17 mg/L; for acid-soluble zinc, these values are <0.086 and 0.095 mg/L (USEPA, 1987). As shown earlier, adverse effects of zinc to echinoderms are documented at 0.023-0.028 mg Zn/L, suggesting re-examination of these proposed criteria.

11.29 Literature Cited

Aarkrog, A., 1971. Radiological investigations of plutonium in an Arctic marine environment. Health Phys. 20, 31–47.
Aarkrog, A., 1990. Environmental radiation and radiation releases. Int. J. Radiat. Biol. 57, 619–631.
Abbott, O.J., 1977. The toxicity of ammonium molybdate to marine invertebrates. Mar. Pollut. Bull. 8, 204–205.
Bernhard, M., Zattera, A., 1975. Major pollutants in the marine environment. In: Pearson, E.A., Frangipane, R. (Eds.), Marine Pollution and Marine Waste Disposal. Pergamon, Elmsford, NY, pp. 195–300.
Bianchini, A., Playle, R.C., Wood, C.M., Walsh, P.J., 2007. Short-term silver accumulation in tissues of three marine invertebrates: shrimp *Penaeus duorarum*, sea hare *Aplysia californica*, and sea urchin *Diadema antillarum*. Aquat. Toxicol. 84, 182–189.
Bohn, A., 1979. Trace metals in fucoid algae and purple sea urchins near a high Arctic lead/zinc ore deposit. Mar. Pollut. Bull. 10, 325–327.
Bryan, G.W., 1963. The accumulation of radioactive caesium by marine invertebrates. J. Mar. Biol. Assoc UK 43, 519–539.
Burger, J., Gochfeld, M., Jeitner, C., Burke, S., Stamm, T., Snigaroff, R., et al., 2007a. Mercury levels and potential risk from subsistence foods from the Aleutians. Sci. Total Environ. 384, 93–105.
Burger, J., Gochfeld, M., Jewett, S.C., 2007b. Radionuclide concentrations in benthic invertebrates from Amchitka and Kiska Islands in the Aleutian Chain, Alaska. Environ. Monit. Assess. 128, 329–341.
Carey Jr., A.G., 1970. Zn65 in benthic invertebrates off the Oregon coast. Available from as RLO-1750-55, NTIS, Springfield, VA, as RLO-1750-55, 27 pp.
Chipman, W., Thommeret, J., 1979. Manganese content and the occurrence of fallout 54Mn in some marine benthos of the Mediterranean. Bull. Inst. Ocean. 69(1402), 1–15.
Clapper, D.L., Davis, J.A., Lamothe, P.J., Patton, C., Epel, D., 1985a. Involvement of zinc in the regulation of pH, motility, and acrosome reactions in sea urchin sperm. J. Cell Biol. 100, 1817–1824.
Clapper, D.L., Lamothe, P.L., Davis, J.A., Epel, D., 1985b. Sperm motility in the horseshoe crab. V: zinc removal mediates chelator initiation of motility. J. Exp. Zool. 236, 83–91.
Cleland, K.W., 1953. Heavy metals, fertilization and cleavage in the eggs of *Psammechinus miliaris*. Exp. Cell Res. 4, 246–248.
Culkin, F., Riley, J.P., 1958. The occurrence of gallium in marine organisms. J. Mar. Biol. Assoc. UK 37, 607–615.
De Clerck, R., Vanderstappen, R., Vyncke, W., Van Hoeyweghen, P., 1979. La teneur en metaux lourds dans les organismes marins provenant de la capture accessoire de la peche coterie belge. Rev. de l'Agric. 3 (32), 793–801.
Deheyn, D.D., Latz, M.I., 2006. Bioavailability of metals along a contamination gradient in San Diego Bay (California, USA). Chemosphere 63, 818–834.
Denton, G.R.W., Concepcion, L.R., Wood, H.R., Morrison, R.J., 2006. Trace metals in marine organisms from four harbours in Guam. Mar. Pollut. Bull. 52, 1784–1804.
Deudero, S., Box, A., March, D., Valencia, J.M., Grau, A.M., Tintora, J., et al., 2007. Temporal trends in benthic invertebrate species from the Balearic Islands, Western Mediterranean. Mar. Pollut. Bull. 54, 1545–1558.
Durkina, V.B., Evtushenko, Z.S., 1991. Changes in activity of certain enzymes in sea urchin embryos and larvae after exposure of adult organisms to heavy metals. Mar. Ecol. Prog. Ser. 72, 111–115.
Eganhouse, R.P., Young, D.R., 1978. Total and organic mercury in benthic organisms near a major submarine wastewater outfall system. Bull. Environ. Contam. Toxicol. 19, 758–766.
Eisler, R., 1979. Copper accumulations in coastal and marine biota. In: Nriagu, J.O. (Ed.), Copper in the Environment, Part 1: Ecological Cycling. John Wiley, New York, pp. 383–449.

Eisler, R., 1980. Accumulation of zinc by marine biota. In: Nriagu, J.O. (Ed.), Zinc in the Environment. Part II: Health Effects. John Wiley, New York, pp. 259–351.

Eisler, R., 1981. Trace Metal Concentrations in Marine Organisms. Pergamon, Elmsford, NY, 687 pp.

Eisler, R., 2000. Zinc. In: Handbook of Chemical Risk Assessment. Volume 1. Metals. Lewis Publishers, Boca Raton, FL, pp. 605–714.

Eisler, R., Hennekey, R.J., 1977. Acute toxicities of Cd^{2+}, Cr^{+6}, Hg^{2+}, Ni^{2+}, and Zn^{2+} to estuarine macrofauna. Arch. Environ. Contam. Toxicol. 6, 315–323.

Eustace, I.J., 1974. Zinc, cadmium, copper and manganese in species of finfish and shellfish caught in the Derwent Estuary, Tasmania. Aust. J. Mar. Freshw. Res. 25, 209–220.

Fukai, R., 1965. Analysis of trace amounts of chromium in marine organisms by the isotope dilution of Cr-51. In: Radiochemical Methods of Analysis. IAEA, Vienna, pp. 335–351.

Fukai, R., Broquet, D., 1965. Distribution of chromium in marine organisms. Bull. Inst. Oceanogr. 65 (1336), 1–19.

Fukai, R., Suzuki, H., Watanake, K., 1962. Strontium-90 in marine organisms during the period 1957-1961. Bull. Inst. Oceanogr. Monaco 1251, 1–16.

Gonzalez, H., 1991. Mercury pollution caused by a chloralkali plant. Water Air Soil Pollut. 56, 219–233.

Gorgy, S., Rakestraw, N.W., Fox, D.L., 1948. Arsenic in the sea. J. Mar. Res. 7, 22–32.

Guerin, T., Sirot, V., Volatier, J.L., Leblanc, J.C., 2007. Organotin levels in seafood and its implications for health risk in high-seafood consumers. Sci. Total Environ. 388, 66–77.

Hiyama, Y., Shimisu, M., 1964. On the concentration factors of radioactive Cs, Sr, Cd, Zn, and Ce in marine organisms. Rec. Oceanogr. Works Jpn. 7 (2), 43–77.

Ichikawa, R., 1961. On the concentration factors of some important radionuclides in marine food organisms. Bull. Jpn. Soc. Sci. Fish. 27, 66–74.

Iglesias, N., Penchaszadeh, P.E., 1983. Mercury in sea stars from Golfo Triste, Venezuela. Mar. Pollut. Bull. 14, 396–398.

Jenkins, D.W., 1980. Biological monitoring of toxic trace metals. Volume 2, Toxic trace metals in plants and animals of the world. Part II. U.S. Environ. Prot. Agen. Rep. 600/3-80-091, 505–618 pp.

Kobayashi, N., 1971. Fertilized sea urchin eggs as an indicatory material for marine pollution bioassay, preliminary experiments. Publ. Seto Mar. Biol. Lab. 18, 379–406.

Kobayashi, N., 1977. Preliminary experiments with sea urchin pluteus and metamorphosis in marine pollution bioassay. Publ. Seto Mar. Biol. Lab. 24, 9–21.

Kobayashi, N., 1980. Comparative sensitivity of various developmental stages of sea urchins to some chemicals. Mar. Biol. 58, 163–171.

Kobayashi, N., Fujinaga, K., 1976. Synergism of inhibiting actions of heavy metals upon the fertilization and development of sea urchin eggs. Sci. Eng. Rev. Doshisha Univ. 17, 54–69.

Lawrence, J.M., Mahon, W.D., Avery, W., Lares, M., 1993. Concentrations of metals in *Luidia clathrata* and *Luidia senegalensis* (Echinodermata: Asteroidea) in Tampa Bay and the nearshore Gulf of Mexico, Florida. Comp. Biochem. Physiol. 105C, 203–206.

Leatherland, T.M., Burton, J.D., 1974. The occurrence of some trace metals in coastal organisms with particular reference to the Solent Region. J. Mar. Biol. Assoc. UK 54, 457–468.

Lowman, F.G., Phelps, D.K., Ting, R.Y., Escalera, R.M., 1966. Progress Summary Report No. 4, Marine Biology Program June 1965-June 1966. Puerto Rico Nucl. Cen. Rep., PRNC 85, pp. 1–57.

Manzo, S., Buono, S., Cremisini, C., 2008. Predictability of copper, irgarol, and diuron combined effects on sea urchin *Paracentrotus lividus*. Arch. Environ. Contam. Toxicol. 54, 57–68.

Matsumoto, T., Satake, M., Yamamoto, J., Haruna, S., 1964. On the micro constituent elements in marine invertebrates. J. Oceanogr. Soc. Jpn. 20 (3), 15–19.

Mauchline, J., Templeton, W.L., 1966. Strontium, calcium and barium in marine organisms from the Irish Sea. J. Cons. Perm. Int. Explor. Mer 30, 161–170.

Montgomery Jr., M., Price, M., Thurston, J., de Castro, G.L., Cruz, L.L., Zimmerman, D.D., 1978. Biological availability of pollutants to marine organisms. US Environ. Prot. Agen. Rep. 600/3-78-035, pp. 1–134.

Morisawa, M., Mohri, H., 1972. Heavy metals and spermatozoan motility. I. Distribution of iron, zinc, and copper in sea urchin spermatozoa. Contrib. Tamano Mar. Lab. 187, 1–6.

Nakamura, S., Ohmi, C., Kojima, M.K., 1989. Effect of zinc ion on formation of the fertilization membrane in sea urchin eggs. Zool. Sci. (Jpn.) 6, 329–333.

Okubo, K., Okubo, T., 1962. Study on the bioassay method for the evaluation of water pollution—II. Use of the fertilized eggs of sea urchins and bivalves. Bull. Tokai Reg. Fish. Res. Lab. 32, 131–140.

Oweson, C., Skold, H., Pinsino, A., Matranga, V., Hernroth, B., 2008. Manganese effects on haematopoietic cells and circulating coelomocytes of *Asterias rubens* (Linnaeus). Aquat. Toxicol. 89, 75–81.

Papadopoulu, C., Kanias, G.D., Kassinati, E.M., 1976. Stable elements of radioecological importance in certain echinoderm species. Mar. Pollut. Bull. 7, 143–144.

Penrose, W.R., Black, R., Hayward, M.J., 1975. Limited arsenic dispersion in seawater, sediments, and biota near a continuous source. J. Fish. Res. Board Can. 32, 1275–1281.

Penrose, W.R., Conacher, H.B.S., Black, R., Meranger, J.C., Miles, W., Cunningham, H.M., et al., 1977. Implications of inorganic/organic interconversion on fluxes of arsenic in marine food webs. Environ. Health Perspect. 19, 53–59.

Phillips, B.M., Anderson, B.S., Hunt, J.W., 1998. Spatial and temporal variation in results of purple urchin (*Strongylocentrotus purpuratus*) toxicity tests with zinc. Environ. Toxicol. Chem. 17, 453–459.

Riley, J.P., Segar, D.A., 1970. The distribution of the major and some minor elements in marine animals. I. Echinoderms and coelenterates. J. Mar. Biol. Assoc. UK 50, 721–730.

Rosen, G., Rivera-Duarte, I., Chadwick, B.C., Ryan, A., Santore, R.C., Paquin, P.R., 2008. Critical tissue copper residues for marine bivalve (*Mytilus galloprovincialis*) and echinoderm (*Strongylocentrotus purpuratus*) embryonic development: conceptual, regulatory and environmental implications. Mar. Environ. Res. 66, 327–336.

Sheppard, C.R.C., Bellamy, D.J., 1974. Pollution of the Mediterranean around Naples. Mar. Pollut. Bull. 5, 42–44.

Shim, W.J., Hong, S.H., Agafonova, I.G., Aminin, D.L., 2006. Comparative toxicities of organotin compounds on fertilization and development of sea urchin (*Anthocidaris crassispina*). Bull. Environ. Contam. Toxicol. 77, 755–763.

Stenner, R.D., Nickless, G., 1974. Distribution of some heavy metals in organisms in Hardangerfjord and Skjerstadfjord, Norway. Water Air Soil Pollut. 3, 279–291.

Stevenson, R.A., Ufret, S.L., 1966. Iron, manganese and nickel in skeletons and food of the sea urchins *Tripneustes esculentus* and *Echinometra lucunter*. Limnol. Oceanogr. 11, 11–17.

Thompson, J.A.J., Paton, D.W., 1978. Heavy metals in benthic organisms from Point Grey dumpsite—Vancouver, B.C. A preliminary report. Inst. Ocean Sci. Sidney B.C., Canada, PMCR 78-11, pp. 1–18.

Timourian, H., Watchmaker, G., 1972. Nickel uptake by sea urchin embryos and their subsequent development. J. Exp. Zool. 182, 379–387.

Timourian, H., Watchmaker, G., 1977. Assay of sperm motility to study the effects of metal ions. In: Drucker, H., Wildung, R.E. (Eds.), Biological Implications of Metals in the Environment. Proceedings of the Fifteenth Annual Hanford Life Sciences Symposium. Richland, WA, September 29-October 1, 1975, Ener. Res. Develop. Admin. Sympos., Ser 42, CONF-750929, pp. 523–536.

U.S. Environmental Protection Agency (USEPA), 1980a. Ambient water quality criteria for nickel USEPA Rep., 440/5-80-060, pp. 1–206.

USEPA, 1980b. Ambient water quality criteria for silver. USEPA Rep., 440/5-80-071, pp. 1–212.

USEPA, 1987. Ambient water quality criteria for zinc—1987. USEPA Rep., 440/5-87-003, pp. 1–207.

Vaskovsky, V.E., Korotchenko, O.D., Kosheleva, L.P., Levin, V.S., 1972. Arsenic in the lipid extracts of marine invertebrates. Comp. Biochem. Physiol. 41B (4), 777–784.

Vattuone, G.M., Griggs, K.S., McIntyre, D.R., Littlepage, J.L., Harrison, F.L., 1976. Cadmium concentrations in rock scallops in comparison with some other species. U.S. Ener. Res. Dev. Admin. UCRL 52020, pp. 1–11.

Voogt, P.A., Besten, P.J.D., Kusters, G.C.M., Messing, M.W.J., 1987. Effects of cadmium and zinc on steroid metabolism and steroid level in the sea star *Asterias rubens* L. Comp. Biochem. Physiol. 86C, 83–89.

Walsh, G.E., McLaughlin, L.L., Louie, M.K., Deans, C.H., Lores, E.M., 1986. Inhibition of arm regeneration by *Ophioderma brevispina* (Echinodermata, Ophiuroidea) by tributyltin oxide and triphenyltin oxide. Ecotoxicol. Environ. Safety 12, 95–100.

Ward, T.J., Kramer, J.R., Boeri, R.L., Gorsuch, J.W., 2006. Chronic toxicity of silver to the sea urchin (*Arbacia punctulata*). Environ. Toxicol. Chem. 25, 1568–1573.

Williams, P.M., Weiss, H.V., 1973. Mercury in the marine environment: concentration in sea water and in pelagic food chain. J. Fish. Res. Board Can. 30, 293–295.

Young, L.G., Nelson, L., 1974. The effects of heavy metal ions on the motility of sea urchin spermatozoa. Biol. Bull. 147, 236–246.

Young, D.E., Jan, T.K., Hershelman, G.P., 1980. Cycling of zinc in the nearshore marine environment. In: Nriagu, J.O. (Ed.), Zinc in the Environment. Part I: Ecological Cycling. John Wiley, New York, pp. 297–335.

CHAPTER 12
Tunicates

Tunicates are common, cosmopolitan, frequently seen but seldom noticed, marine organisms related to vertebrates. Tunicates possess distinctive gills and a unique dorsal tubular spinal cord found in no other invertebrate animals. In general, mature tunicates are permanently attached forms with no capacity for locomotion, using an enlarged and elaborate gill system to filter large volumes of seawater to procure its planktonic diet. A tunic is secreted by the skin which helps anchor the organism to a solid structure and give rigidity to its outer form, allowing the fragile inner organism to feed undisturbed. Some tunicates are solitary creatures growing to the size of an orange, and some are colonial. Tunicates are universally distributed throughout the seas of the world from the intertidal zone to the deep abyss. Some groups have given up a fixed existence for a drifting floating life. The more primitive group, the ascidians or sea squirts, are associated with the sea bottom and may be found along or near the shore of all seas, growing on seaweeds, on or under rocks, in submerged sand flats, on coral reefs, ship bottoms, wharf piles, and attached to lobsters and sea snails. They are common from the polar regions to the tropics. Some species are cosmopolitan, others so limited in range as to be known only from isolated locations. Alone among animals it secretes its coat of tunicin, a substance similar to that of cellulose in plants. Its blood is unique. Although tunicate blood has no respiratory pigment and contains no more oxygen than a similar amount of seawater, it is frequently rich in both sulfuric acid and vanadium.

12.1 Antimony

Whole body concentrations of antimony in field collections of tunicates ranged from 0.02 to 0.26 mg Sb/kg DW (dry weight) (Table 12.1).

12.2 Arsenic

Arsenic concentrations in tunicate tissues did not exceed 6.5 mg As/kg on a fresh weight (FW) basis or 34.0 on a DW basis, except for lipids (Table 12.1). Arsenic concentrations in tunicate lipids reached a spectacular 23,700.0 mg/kg FW (Table 12.1), and this corroborates the findings of other investigators who have analyzed arsenic in lipid extracts of marine organisms other than tunicates.

Table 12.1: Antimony, Arsenic, Barium, Boron, and Cadmium Concentrations in Field Collections of Tunicates

Element and Organism	Concentration	Reference[a]
Antimony		
Botryllus schlosseri; whole	0.26 DW	1
Ciona intestinalis		
Whole	0.16 DW; 0.007 FW	2
Tunic	0.15 DW	2
Whole less tunic	0.40 DW	2
Cynthia claudicans; soft parts	0.17 DW; 0.032 FW	3
Microcosmus sulcatus; whole	0.1 DW	2
Pyrosoma sp.; whole	0.02 DW	4
Styella clava; whole	0.15 DW	1
Arsenic		
Botryllus schlosseri; whole	6.6 DW	1
Cynthia claudicans; soft parts	6.5 FW; 34.0 DW	3
Guam; 1998-1999; 4 locales; whole		
Ascidia sp.	2.7-3.9 DW	9
Rhopalaea sp.	2.3-3.6 DW	9
Halocynthia aurantium; lipid extract, soft tissues	23,700.0 FW	5
Pyrosoma sp.; whole	1.5 DW	4
Styella clava; whole	4.8 DW	1
Barium		
Cynthia claudicans; soft parts	0.5 FW	3
Tunicates, whole	0.07-0.1 FW	6
Boron		
Salpa fusiformis; whole	50.0 AW	7
Cadmium		
Ascidacea sp.; whole	0.2 FW	8
Botryllus schlosseri; whole	2.7 DW	1

(Continues)

Table 12.1: Cont'd

Element and Organism	Concentration	Reference[a]
Guam; 1998-1999; 4 locations; whole		
Ascidia sp.	0.08-0.36 DW	9
Rhopalaea sp.	0.13-0.44 DW	9
Pyrosoma sp.; whole	0.44 DW	4

Values are in mg element/kg fresh weight (FW), dry weight (DW), or ash weight (AW).
[a]1, Leatherland and Burton, 1974; 2, Papadopoulu and Kanias, 1977; 3, Papadopoulu et al., 1972; 4, Leatherland et al., 1973; 5, Vaskovsky et al., 1972; 6, Mauchline and Templeton, 1966; 7, Nicholls et al., 1959; 8, Eustace, 1974; 9, Denton et al., 2006.

12.3 Barium

Barium levels in tunicates do not exceed 0.1 mg Ba/kg FW whole organism or 0.5 mg Ba/kg FW soft parts (Table 12.1).

12.4 Boron

A single datum indicates that whole salp contains 50.0 mg B/kg on an ash weight (AW) basis (Table 12.1).

12.5 Cadmium

Cadmium burdens in whole tunicates range from 0.08 to 2.7 mg Cd/kg DW (Table 12.1).

12.6 Cesium

Concentrations of stable cesium in whole tunicates did not exceed 0.071 mg/kg DW or 0.008 mg/kg FW (Table 12.2).

Results of studies by Bryan (1963) with *Ciona intestinalis* and ^{137}Cs show that cesium was always absorbed more slowly than potassium. At equilibrium, however, concentration factors for cesium usually exceeded those of potassium. *Ciona* accumulated cesium over environmental levels by factors of 1.01 for blood, 1.11 for test, 4.6 for gut, and 6.5 for pharynx plus body wall muscles (Bryan, 1963).

Table 12.2: Cesium, Chromium, and Cobalt Concentrations in Field Collections of Tunicates

Element and Organism	Concentration	Reference[a]
Cesium		
Ciona intestinalis; Greece		
Whole	0.071 DW; 0.003 FW	1
Tunic	0.021 DW	1
Microcosmus sulcatus; whole; Greece	0.047 DW; 0.008 FW	1
Chromium		
Ciona intestinalis; Greece		
Whole	5.5 DW; 0.24 FW	1
Tunic	1.9 DW; 0.075 FW	1
Whole less tunic	6.1 DW; 0.25 FW	1
Guam; 1998-1999; 4 locations; whole		
Ascidia sp.	1.0-5.1 DW	3
Rhopalaea sp.	1.8-9.7 DW	3
Microcosmus sulcatus; whole; Greece	6.6 DW; 1.1 FW	1
Podoclavella moluccensis; blood		
Plasma, whole	0.35 DW	2
Plasma fractions 9-12	940.0 DW	2
EtOH/HCl extract	58.0 DW	2
Cell residue	22.0 DW	2
Cobalt		
Ciona intestinalis; Greece		
Whole	0.52 DW; 0.022 FW	1
Tunic	0.44 DW; 0.017 FW	1
Microcosmus sulcatus; whole; Greece	1.9 DW	1

Values are in mg element/kg fresh weight (FW), or dry weight.
[a]1, Papadopoulu and Kanias, 1977; 2, Hawkins et al., 1980; 3, Denton et al., 2006.

12.7 Chromium

Stable chromium in whole tunicates from Greece did not exceed 6.6 mg/kg DW, with tunic containing disproportionately low burdens (Table 12.2). Elevated accumulations of 940.0 mg Cr/kg DW is recorded in selected plasma fractions of *Podoclavella moluccensis* (Hawkins et al., 1980; Table 12.2).

Eggs of the ascidian, *Pyura momus*, reportedly accumulated radiochromium-51 from the medium (Townsley et al., 1962), but the extent of accumulation or its significance is unknown.

12.8 Cobalt

The highest concentration of stable cobalt in tunicate tissues is 1.9 mg/kg DW (Table 12.2).

Tunicates are important in cobalt cycling. Kuenzler (1969) reports that diurnally migrating zooplankton, including salps, are significant agents in cobalt cycling from the sea surface and in its transport through the thermocline. Elimination rates of accumulated cobalt by *Salpa cylindrica* ranged from 8% to 19%/h; some of the eliminated cobalt was particulate, but most was positively charged (Kuenzler, 1969). *Pyrosoma verticillatum*, one of the luminous species of tunicates, has a comparatively low cobalt elimination rate of 2-3%/h (Kuenzler, 1969). *Styella plicata* accumulated radiocobalt-60 from seawater over a 30-day period by factors of 850 in viscera, and 400 in test (Shimizu et al., 1970). A concentration factor of 60 for cobalt was reported in whole *Salpa* sp. (Ketchum and Bowen, 1958).

12.9 Copper

The highest copper concentrations reported in field collections of tunicates were 8.3 mg/kg in whole *Ascidacea* on a FW basis, 130.0 mg/kg DW in blood of *Podoclavella*, and 500.0 mg/kg AW in whole *Salpa* (Table 12.3).

Table 12.3: Copper, Iron, and Lead Concentrations in Field Collections of Tunicates

Element and Organism	Concentration	Reference[a]
Copper		
Ascidacea sp.; whole	8.3 FW	1
Podoclavella moluccensis; blood		
Plasma; whole	0.66 DW	2
Plasma fractions 9-12	130.0 DW	2
EtOH extract	0.0 DW	2
EtOH/HCl extract	18.0 DW	2
Cell residue	16.0 DW	2
Sea squirt, *Ciona intestinalis*		
California; tunic vs. viscera	55.0 DW vs. 73.0 DW	8
Sweden; whole	13.0 DW	8
Guam; 1998-1999; 4 locales; whole		
Ascidia sp.	3.1-5.6 DW	10
Rhopalaea sp.	6.4-9.9 DW	10
Pyura microcosmus; soft parts	7.3-7.8 DW	3

(*Continues*)

Table 12.3: Cont'd

Element and Organism	Concentration	Reference[a]
Salpa fusiformis; whole	500.0 AW	4
Various species; Wales; 1989; whole; coastal area	2.6 FW	9
Iron		
Ascidia ceratodes; whole	1900.0 DW	5
Sea squirt, *Ciona intestinalis*		
Blood	50.0 AW	5
Whole	880.0 DW; 39.0 FW	6
Tunic	610.0 DW; 24.0 FW	6
Whole less tunic	3000.0 DW; 120.0 FW	6
Clavelina huntsmani; whole	900.0 DW	5
Cynthia claudicans; soft parts	330.0 FW; 1740.0 DW	7
Distaplia occidentalis; whole	500.0-2200.0 DW	5
Eudistoma molle; whole	400.0 DW	5
Eudistoma ritteri; whole	2400.0-2600.0 DW	5
Microcosmus sulcatus; whole	840.0 DW; 140.0 FW	6
Molgula manhattensis; whole less tunic		
Intestines cleared	900.0-1700.0 DW	5
Intestines not cleared	16,400.0-26,600.0 DW	5
Perophora annectens; whole	940.0 DW	5
Podoclavella moluccensis; blood		
Plasma, whole	1.3 DW	2
Plasma fractions 9-12	4530.0 DW	2
EtOH/HCl extract	624.0 DW	2
Cell residue	964.0 DW	2
Synoieum sp.; whole	600.0 DW	5
Lead		
Cynthia claudicans; soft parts	3.4 FW; 18.0 DW	7
Salpa fusiformis; whole	3.0 AW	4

Values are in mg element/kg fresh weight (FW), dry weight (DW), or ash weight (AW).

[a]1, Eustace, 1974; 2, Hawkins et al., 1980; 3, Papadopoulu et al., 1967; 4, Nicholls et al., 1959; 5, Swinehart et al., 1974; 6, Papadopoulu and Kanias, 1977; 7, Papadopoulu et al., 1972; 8, Jenkins, 1980; 9, Morris et al., 1989; 10, Denton et al., 2006.

12.10 Iron

Elevated iron concentrations up to 2600.0 mg/kg DW are recorded in whole tunicates (Table 12.3). The high iron levels may have been associated with iron-containing sediments in the intestines. For example, *Molgula* with intestines cleared had significantly lower total iron than did freshly collected specimens (Swinehart et al., 1974). Age of the tunicate and tissue specificity should also be considered when interpreting the significance of iron concentrations. Experiments with *C. intestinalis* and radioiron-59 show that accumulation was maximal after 2-3 days, especially in the tunic; young *Ciona* accumulated greater ^{59}Fe concentrations than did adults (Battani et al., 1968).

12.11 Lead

The highest lead concentration recorded in tunicates is 18.0 mg Pb/kg DW soft parts (Table 12.3).

12.12 Manganese

Manganese concentrations up to 80,000.0 mg/kg on an AW basis and 112.6 mg/kg on a FW basis are recorded in tunicate tissues; moreover, manganese concentrations are similar in viscera and body wall (Table 12.4), suggesting uniform distribution.

12.13 Mercury

Mercury concentrations in tunicates were relatively low, with the exception of the specimen collected from Minamata Bay, Japan (Table 12.4). Minamata Bay—a site of notorious mercury pollution—is discussed elsewhere in this volume.

12.14 Nickel

Whole tunicates had nickel maxima of 0.13 mg/kg FW and 60.0 mg/kg AW (Table 12.4).

12.15 Niobium

Niobium concentrations in tunicates were either negligible, that is, <0.001 mg Nb/kg DW, or considerable, ranging up to 56.0 mg Nb/kg DW (Table 12.4). Carlisle (1958) shows a direct inverse correlation with niobium and vanadium; high vanadium levels being associated with low niobium concentrations and the reverse. If, as has been suggested, the vanadium present in petroleum deposits is derived from ascidian tunicates, it also seems likely that tunicates are the source of the niobium present on petroleum deposits.

Table 12.4: Manganese, Mercury, Nickel, and Niobium Concentrations in Field Collections of Tunicates

Element and Organism	Concentration	Reference[a]
Manganese		
Ascidacea sp.; whole	112.6 FW	1
Cynthia claudicans; soft parts	44.0 FW	2
Halocynthia papillosa; whole	80,000.0 AW	3
Phallusia mammilata		
Viscera	24.0 AW	3
Body wall	23.1 AW	3
Podoclavella moluccensis; blood plasma		
Whole	0.45 DW	4
Fractions 9-12	45.0 DW	4
Cell residue	7.0 DW	4
Mercury		
Botryllus schlosseri; whole	0.57 DW	5
Ciona intestinalis; whole	0.03-0.12 FW	6
Cynthia claudicans; soft parts	0.049 FW; 0.26 DW	5
Styella plicata; whole; Minamata Bay, Japan	35.0 DW	7
Nickel		
Halocynthia roretzi; whole	0.13 FW	8
Salpa fusiformis; whole	60.0 AW	9
Niobium		
Molgula manhattensis		
Whole	26.5 DW	10
Test	1.9 DW	10
Flesh without test	56.0 DW	10
Phallusia mammillata; whole	<0.001 DW	10

Values are in mg element/kg fresh weight (FW), dry weight (DW), or ash weight (AW).
[a] 1, Eustace, 1974; 2, Papadopoulu et al., 1972; 3, Chipman and Thommeret, 1970; 4, Hawkins et al., 1980; 5, Leatherland and Burton, 1974; 6, Yannai and Sachs, 1978; 7, Matida and Kumada, 1969; 8, Ikebe and Tanaka, 1979; 9, Nicholls et al., 1959; 10, Carlisle, 1958.

12.16 Rubidium

Mean rubidium concentrations in tissues from two species of tunicates from the Aegean Sea ranged from 2.6 to 7.0 mg/kg DW (Table 12.5).

Table 12.5: Rubidium, Scandium, Selenium, and Silver Concentrations in Field Collections of Tunicates

Element and Organism	Concentration	Reference[a]
Rubidium		
Ciona intestinalis		
Whole	2.6 DW; 0.11 FW	1
Tunic	3.7 DW; 0.14 FW	1
Whole less tunic	7.0 DW; 0.29 FW	1
Microcosmus sulcatus; whole	2.6 DW; 0.43 FW	1
Scandium		
Ciona intestinalis		
Whole	0.15 DW; 0.0065 FW	1
Tunic	0.0410 DW; 0.0046 FW	1
Whole less tunic	0.520 DW; 0.021 FW	1
Microcosmus sulcatus; whole	0.078 DW; 0.012 FW	1
Selenium		
Ciona intestinalis		
Whole	1.2 DW; 0.051 FW	1
Tunic	1.0 DW; 0.04 FW	1
Whole less tunic	2.7 DW; 0.11 FW	1
Microcosmus sulcatus; whole	5.1 DW; 2.8 DW	1
Silver		
Ciona intestinalis		
Whole	0.021 DW; 0.0009 FW	1
Tunic	0.011 DW; 0.0004 FW	1
Whole less tunic	0.067 DW; 0.0027 FW	1
Cynthia claudicans; soft parts	0.9 FW; 4.8 DW	2
Guam; 1998-1999; 4 locales; whole		
Ascidia sp.	0.13-0.33 DW	3
Rhopalaea sp.	<0.27-0.27 DW	3
Microcosmus sulcatus; whole	0.031 DW; 0.005 FW	1

Values are in mg element/kg fresh weight (FW) or dry weight (DW).
[a] 1, Papadopoulu and Kanias, 1977; 2, Papadopoulu et al., 1972; 3, Denton et al., 2006.

12.17 Ruthenium

Of all tunicates studied, *Dendroda grassularia* was the most active concentrator of radioruthenium-106 (Ancellin et al., 1967). The chemical form of ruthenium is important: insoluble ruthenium compounds were accumulated about 10 times greater than dissolved ruthenium compounds (Ancellin and Bovard, 1971).

12.18 Scandium

Concentrations of scandium in two species of whole tunicates from the same collection area were 0.078 and 0.15 mg Sc/kg DW and 0.0065 and 0.012 mg Sc/kg on a FW basis (Table 12.5).

12.19 Selenium

Average selenium concentrations in two species of tunicates from the same collection area were 1.2 and 5.1 mg Se/kg on a DW basis and 0.051 and 2.8 mg Se/kg on a FW basis (Table 12.5), suggesting significant interspecies differences in selenium burdens.

12.20 Silver

The maximum silver concentration recorded in whole tunicates was 0.33 mg Ag/kg on a DW basis; for soft parts, it was 4.8 mg Ag/kg DW (Table 12.5).

12.21 Strontium

Strontium is present in tunicates at concentrations up to 7.3 mg Sr/kg FW (Table 12.6).

Table 12.6: Strontium, Tin, and Titanium Concentrations in Field Collections of Tunicates

Element and Organism	Concentration	Reference[a]
Strontium		
Cynthia claudicans; soft parts	7.2 FW	1
Tunicates; whole	3.7-7.3 FW	2
Tin		
Salpa fusiformis; whole	8.0 AW	3
Titanium		
Salpa fusiformis; whole	1.0 AW	3

Values are in mg element/kg fresh weigh (FW) or dry weight (DW).
[a] 1, Papadopoulu et al., 1972; 2, Mauchline and Templeton, 1966; 3, Nicholls et al., 1959.

12.22 Tin

Tin is present at 8.0 mg Sn/kg AW in whole salps (Table 12.6).

12.23 Titanium

A single datum indicates that whole salps contain 1.0 mg Ti/kg on an AW basis (Table 12.6).

12.24 Vanadium

Vanadium is accumulated by many groups of marine organisms, but concentrations in some species of ascidian tunicates are spectacular (Table 12.7).

Table 12.7: Vanadium Concentrations in Field Collections of Tunicates

Organism	Concentration	Reference[a]
Ascidia ceratodes		
Whole	1300.0 DW	1
Blood	6200.0 AW	1
Outer fluid	8000.0 AW	1
Ascidia nigra		
Whole blood	51.0 FW	2
Centrifuged blood cells	14,600.0 FW	3
Amaroucium sp.; whole	<10.0 AW	1
Ciona intestinalis		
Whole	100.0 DW	4
Ovary	76.0 DW	4
Incurrent siphon	109.0 DW	4
Anterior gut	911.0 DW	4
Posterior gut	178.0 DW	4
Stomach	248.0 DW	4
Blood	1.2-1.8 DW	5
Whole	400.0-7000.0 AW	6
Blood	160.0 AW	1
Clavelina huntsmani; whole	30.0 DW	1
Cynthia claudicans; soft parts	0.03 FW	7
Cystodes lobatus; whole	50.0 DW	1
Distaplia occidentalis; whole	600.0-1200.0 DW	1
Eudistoma diaphanes; whole	25.0 DW	1

(Continues)

Table 12.7: Cont'd

Organism	Concentration	Reference[a]
Eudistoma molle		
Whole	1000.0 DW	1
Zooids	560.0-1200.0 DW; 5000.0-6000.0 AW	1
Test	1000.0 AW	1
Eudistoma psammion		
Whole	80.0-130.0 DW	1
Zooids	610.0-614.0 DW	1
Eudistoma ritteri; whole		
Large colony	320.0 AW	1
Small colony	270.0 AW	1
Euherdmania sp.; whole	475.0 DW	4
Halocynthia roretzi; whole	0.12 FW	8
Molgula manhattensis		
Test	16.0 DW	9
Flesh without test	101.0 DW	9
Whole	54.0 DW	9
Whole	<20.0 AW	1
Perophora annectens; whole	700.0-9000.0 DW	1
Phallusia mammillata		
Whole	1260.0 DW	9
Eggs	170.0 DW	3
Phlebobranchia; whole; 12 species	206.0-4123.0 DW	3
Podoclavella moluccensis; blood		
Whole plasma	4.1 DW	10
Plasma fractions 9-12	96.0 DW	10
EtOH extract	0.0 DW	10
EtOH/HCl extract	480.0 DW	10
Cell residue	681.0 DW	10
Polyclinum planum; whole	<10.0 AW	1
Polyclitoridae; whole	0.0 FW	3
Pyura microcosmus; soft tissues	0.95-1.04 DW	11
Rhopalaea abdominalis; whole	1800.0 DW	1
Salpa fusiformis; whole	7.0 AW	12
Salpa spp.; whole	0.0 FW	3

(*Continues*)

Table 12.7: Cont'd

Organism	Concentration	Reference[a]
Styella spp.; whole	0.0 DW	4
Styelidae; whole	0.0 FW	3
Synocidae	0.0 FW	3

Values are in mg V/kg fresh weight (FW), dry weight (DW), or ash weight (AW).
[a]1, Swinehart et al., 1974; 2, Kustin et al., 1976; 3, Ciereszko et al., 1963; 4, Goldberg et al., 1951; 5, Rummel et al., 1967; 6. Elroi and Komarovsky, 1961; 7, Papadopoulu et al., 1972; 8, Ikebe and Tanaka, 1979; 9, Carlisle, 1958; 10, Hawkins et al., 1980; 11, Papadopoulu et al., 1967; 12, Nicholls et al., 1959.

It is clear that not all groups of tunicates readily take up vanadium. For example, members of the family Ascidiidae and Perophoridae from the order Phlebobranchia and several species in the order Aplausobranchia have comparatively large vanadium burdens, while other groups contain little or none. It is generally agreed that vanadium accumulates in tunicate blood cells, or vanadocytes, and forms a steady-state concentration in membranes of the branchial net and gastrointestinal tract (Kustin et al., 1975, 1976; Rummel et al., 1967). However, there are at least five morphologically distinct types of vanodocytes in tunicates, and the distribution and valence of vanadium in these cells is uncertain (Kustin et al., 1976). Thus, tunicates may accumulate vanadium from seawater as V^+ (Kustin et al., 1976), and through transformation show up as V^{4+} in order Aplausobranchia or V^{3+} in the order Phlebobranchia (Swinehart et al., 1974).

Studies by Hawkins et al. (1980) with *Pyura stolonifera*, in which iron is the major plasma metal, and *Ascidia ceratodes*, in which vanadium is the major plasma metal, indicates differential uptake and binding of radioiron and radiovanadium. In the case of *Pyura*, iron uptake is independent of iron concentration, and vanadium is not strongly bound to plasma components. With *Ascidia*, iron uptake is inhibited at lower iron concentrations but vanadium is strongly bound and retained by plasma components.

The interaction of niobium with vanadium has been documented earlier by Carlisle (1958). In addition, vanadium uptake by tunicates is allegedly inhibited by high phosphate levels in seawater (Goldberg et al., 1951). These last two observations assume increasing significance in coastal waters impacted by wastes from the fossil fuel industry and phosphate loadings of household cleansers.

12.25 Zinc

Highest zinc concentrations reported in various ascidian tissues were 720.0 mg/kg on a DW basis and 64.0 mg/kg on a FW basis (Table 12.8). It is probable that tunicates reflect environmental levels of zinc, and this could account for some of the variability observed.

Table 12.8: Zinc Concentrations in Field Collections of Tunicates

Organism	Concentration	Reference[a]
Ascidacea sp.; whole	64.0 FW	1
Botryllus schlosseri		
Whole	179.0-251.0 DW	2
Whole	135.0 DW	3
Ciona intestinalis		
Whole	100.0 DW	4
Tunic	110.0 DW	4
Microcosmus sulcatus; whole	180.0 DW	4
Podoclavella moluccensis; blood		
Whole plasma	4.3 DW	5
Plasma fractions 9-12	370.0 DW	5
Cell residue	49.0 DW	5
Polycarpa pediculata; whole; near lead smelter outfall, Australia; 2.5-5.2 km from outfall vs. 18-18.8 km	153.0 DW; max. 345.0 DW vs. 98.0 DW	9
Pyrosoma sp.; whole	105.0 DW	6
Pyura microcosmus; soft parts	684.0-720.0 DW	7
Tunicates; whole	200.0 DW; max. 370.0 DW; max. 64.0 FW	8

Values are in mg Zn/kg fresh weight (FW) or dry weight (DW).
[a] 1, Eustace, 1974; 2, Ireland, 1973; 3, Leatherland and Burton, 1974; 4, Papadopoulu and Kanias, 1977; 5, Hawkins et al., 1980; 6, Leatherland et al., 1973; 7, Papadopoulu et al., 1967; 8, Young et al., 1980; 9, Ward et al., 1986.

For example, *Salpa* spp. showed decreasing concentrations of radiozinc with increasing distance from the mouth of the Columbia River (United States), suggesting that the river is the major source of ^{65}Zn (Osterberg, 1962). Accumulated radiozinc is eliminated at the rate of about 1%/h by *P. verticillatum*; for radiocobalt and radioiron, elimination rates were 2-3%/h, suggesting a need for zinc by this species (Kuenzler, 1969).

12.26 Literature Cited

Ancellin, J., Bovard, P., 1971. Observations concernant les contaminations experimentales et les contaminations "in situ" d'especes marines par le ruthenium 106. Rev. Int. Oceanogr. Med. 21, 85–92.

Ancellin, J., Bovard, P., Vilquin, A., 1967. New studies on experimental contamination of marine species by ruthenium-106. In: Actes du Congress Int. sur le radioprotection du mileu Soc. Franc. de radioprot, Toulouse, France, January 14-16, 1967, pp. 213–234.

Battani, M., Chambost, M.D., Leandri, M., 1968. Study of the contamination by iodine-131 and iron-59 of *Ciona intestinalis* Linn. In: Proc. Third Inter. Coll. Medical Ocean., Fifth and Sixth Sess., pp. 71–86.

Bryan, G.W., 1963. The accumulation of radioactive caesium by marine invertebrates. J. Mar. Biol. Assoc. UK 43, 519–539.
Carlisle, D.B., 1958. Niobium in ascidians. Nature 181, 933.
Chipman, W., Thommeret, J., 1970. Manganese content and the occurrence of 54Mn in some marine benthos of the Mediterranean. Bull. Inst. Oceanogr. 69(1402), 1–15.
Ciereszko, L.S., Ciereszko, E.M., Harris, E.R., Lane, C.A., 1963. Vanadium content of some tunicates. Comp. Biochem. Physiol. 8, 137–140.
Denton, G.R.W., Concepcion, L.P., Wood, H.R., Morrison, R.J., 2006. Trace metals in marine organisms from four harbours in Guam. Mar. Pollut. Bull. 52, 1784–1804.
Elroi, D., Komarovsky, B., 1961. On the possible use of the fouling ascidian *Ciona intestinalis* as a source of vanadium, cellulose and other products. Proc. Gen. Fish. Counc. Medit. 6, 261–267.
Eustace, I.J., 1974. Zinc, cadmium, copper and manganese in species of finfish and shellfish caught in the Derwent Estuary, Tasmania. Aust. J. Mar. Freshw. Res. 25, 209–220.
Goldberg, E.D., McBlair, W., Taylor, K.M., 1951. The uptake of vanadium by tunicates. Biol. Bull. 101, 84–94.
Hawkins, C.J., Merefield, P.M., Parry, D.L., Biggs, W.R., Swinehart, J.H., 1980. Comparative study of the blood plasma of the ascidians *Pyura stolonifera* and *Ascidia ceratodes*. Biol. Bull. 159, 656–668.
Ikebe, K., Tanaka, R., 1979. Determination of vanadium and nickel in marine samples by flameless and flame atomic absorption spectrophotometry. Bull. Environ. Contam. Toxicol. 21, 526–532.
Ireland, M.P., 1973. Result of fluvial zinc pollution on the zinc content of littoral and sublittoral organisms in Cardigan Bay, Wales. Environ. Pollut. 4, 27–35.
Jenkins, D.W., 1980. Biological monitoring of toxic trace metals. Volume 2. Toxic trace metals in plants and animals of the world. Part II. U.S. Environ. Protect. Agen, Rep. 600/3-80-091, pp. 505–618.
Ketchum, B.M., Bowen, V.T., 1958. Biological factors determining the distribution of radioisotopes in the sea. In: Proceedings of the Second Annual United Nations Conference on Peaceful Uses of Atomic Energy, Geneva, vol. 18, pp. 429–433.
Kuenzler, E.J., 1969. Elimination of iodine, cobalt, iron, and zinc by marine zooplankton. In: Proceedings of the Second National Symposium on Radioecology, USAEC Conf. 670503, pp. 462–473.
Kustin, K., Ladd, K.V., McLeod, G.C., 1975. Site and rate of vanadium assimilation in the tunicate *Ciona intestinalis*. J. Gen. Physiol. 615, 315–328.
Kustin, K., Levine, D.S., McLeod, G.S., Curby, W.A., 1976. The blood of *Ascidia nigra*: blood cell frequency distribution, morphology, and the distribution and valence of vanadium in living blood cells. Biol. Bull. 150, 426–441.
Leatherland, T.M., Burton, J.D., Culkin, F., McCartney, M.J., Morris, R.J., 1973. Concentrations of some trace metals in pelagic organisms and of mercury in Northeast Atlantic Ocean water. Deep Sea Res. 20, 679–685.
Leatherland, T.M., Burton, J.D., 1974. The occurrence of some trace metals in coastal organisms with particular reference to the Solent region. J. Mar. Biol. Assoc. UK 54, 457–468.
Matida, Y., Kumada, H., 1969. Distribution of mercury in water, bottom mud and aquatic organisms of Minamata Bay, the River Agano and other water bodies in Japan. Bull. Freshw. Fish. Res. Lab. Tokyo 19(2), 73–93.
Mauchline, J., Templeton, W.L., 1966. Strontium, calcium and barium in marine organisms from the Irish Sea. Cons. Perm. Int. Explor. Mer. 30, 161–170.
Morris, R.J., Law, R.J., Allchin, C.R., Kelly, C.A., Fileman, C.F., 1989. Metals and organochlorines in dolphins and porpoises of Cardigan Bay, West Wales. Mar. Pollut. Bull. 20, 512–523.
Nicholls, H.D., Curl Jr., H., Bowen, V.T., 1959. Spectrographic analyses of marine plankton. Limnol. Oceanogr. 4, 472–476.
Osterberg, C., 1962. Zinc-65 content of salps and euphausids. Limnol. Oceanogr. 7, 478–479.
Papadopoulu, C., Kanias, G.D., 1977. Tunicate species as marine pollution indicators. Mar. Pollut. Bull. 8, 229–231.
Papadopoulu, C.P., Cazianis, C.T., Grimanis, A.P., 1967. Neutron activation analysis of vanadium, copper, zinc, bromine and iodine in *Pyura microcosmus*. In: Proceedings of a Symposium. Nuclear Activation Techniques in the Life Sciences. IAEA, Vienna, pp. 365–377.

Papadopoulu, C., Hadzistelios, I., Grimanis, A.P., 1972. Trace element uptake by *Cynthia claudicans* (Savigny). Greek Limnol. Oceanogr. 11, 651–653.

Rummel, W., Beilig, H.J., Forth, W., Pfleger, K., Rudiger, W., Seifen, E., 1967. Absorption and accumulation of vanadium by tunicates. Protides Biol. Fluids Proc. Colloq. Bruges. 14, 205–210.

Shimizu, M., Kajihara, T., Hiyama, Y., 1970. Uptake of ^{60}Co by marine animals. Rec. Oceanogr. Works Jpn. 10, 137–145.

Swinehart, J.H., Biggs, W.R., Halko, D.J., Schroeder, N.C., 1974. The vanadium and selected metal contents of some ascidians. Biol. Bull. 146, 302–312.

Townsley, S.J., Reid, D.F., Ego, W.T., 1962. Theaccumulation of radioactive isotopes by tropical marine organisms. Annual Rep. 1960-61. USAEC, TID 14420, pp. 1–36.

Vaskovsky, V.E., Korotchenko, O.D., Kosheleva, L.P., Levin, V.S., 1972. Arsenic in the lipid extracts of marine invertebrates. Comp. Biochem. Physiol. 41B(4), 777–784.

Yannai, S., Sachs, K., 1978. Mercury compounds in some eastern Mediterranean fishes, invertebrates, and their habitats. Environ. Res. 16, 408–418.

Young, D.R., Jan, T.K., Hershelman, G.P., 1980. Cycling of zinc in the nearshore marine environment. In: Nriagu, J.O. (Ed.), Zinc in the Environment. Part I: Ecological Cycling. John Wiley, New York, pp. 297–335.

Ward, T.J., Correll, R.L., Anderson, R.B., 1986. Distribution of cadmium, lead and zinc amongst the marine sediments, sea grasses and fauna, and the selection of sentinel accumulators, near a lead smelter in South Australia. Aust. J. Mar. Freshw. Res. 37, 567–585.

CHAPTER 13
Concluding Remarks

13.1 General

Much of the variability in elemental content of marine plants and invertebrates—as documented herein—is now attributed to tissue analyzed, age and gender of the organism, proximity to anthropogenic and natural point sources, to variations in seawater pH, temperature, and salinity, and to interactions with other chemicals in the environment, especially organic chemicals introduced as a result of human activities. Diet, sediment composition, inherent species differences, and metabolism of different chemical forms of metals and metalloids also influence elemental content. Additional studies are recommended on interaction effects of contaminant mixtures on metal contents of representative marine flora and fauna and on their modes of action.

Basic research on most trace metals is now lacking or deficient in a number of areas including essentiality, metabolism, uptake, and retention mechanisms; frequency of carcinogenicity, mutagenicity, teratogenicity, and histopathology; acclimatization mechanisms; and effects of long-term exposure to low sublethal levels under changing thermosaline regimes typical of areas now subjected to global warming effects. Also needed is a vastly improved technology for prevention or removal of comparatively toxic metals originating from anthropogenic sources into the marine biosphere, and repeated monitoring over time using selected indicator species native to the region for assessment of progress.

Since the publication of *Trace Metal Concentrations in Marine Organisms* (Eisler, 1981), there has been a steady output in the trace metal and metalloid technical literature on marine plants and animals from scientists in western Europe, North America, Australia, New Zealand, and Japan, with emphasis on polar fauna, uptake mechanisms, and metabolism. At the same time there has been a disproportionate increase in the technical literature on trace metal concentrations in marine flora and fauna from China, India, Mexico, South America, Africa, and numerous other regions. The reasons for this are unclear, but it is probable that nations that have benefited from the recent redistribution of global wealth now have the resources to train and equip marine scientists with the dual goals of protecting the dwindling marine resources, and the health of human consumers of seafood products of commerce. Environmental awareness seems high in these formerly impoverished regions and their political leaders have provided the long-range funding and the expensive equipment necessary for establishment of metal and metalloid standards based on concentrations in tissues, sediments, diet, and seawater.

Throughout this volume, and the companion volume on vertebrates (Eisler, 2009), proposed criteria for resource protection and human consumers of seafood are abundant and based mainly on recommendations of national and international regulatory organizations. These criteria are almost always based on maximum permissible concentrations in seawater, tissues, sediments, chemical species (most notably, arsenic, chromium, mercury, and tin), and food chain transfer. Criteria, unlike legislatively administered standards, are virtually unenforceable. Legislatively administered standards exist in isolated geographic regions; however, enforcement of these standards are lax, except for the most egregious violations. Chemical risk assessment needs to be based—at a minimum—on sound science, public support, and meaningful and enforceable penalties for violators.

13.2 Breadth of Coverage

Breadth of coverage—as indicated by citations of metals and metalloids for each of the marine groups examined, namely, algae and macrophytes, annelids, chaetognaths, coelenterates, crustaceans, echinoderms, insects, molluscs, protists, sponges, and tunicates—was most extensive for cadmium, copper, iron, nickel, and zinc and least extensive for dysprosium, erbium, gadolinium, hafnium, holmium, lutetium, neodymium, praseodymium, tantalum, tellurium, and thulium (Table 13.1). Additional research is merited on the less extensively studied metals and metalloids, with emphasis on acquisition of baseline data.

Similar data for marine vertebrates, namely, elasmobranchs, fishes, reptiles, birds, and mammals, was most extensive for arsenic, cadmium, chromium, copper, iron, lead, manganese, mercury, selenium, silver, strontium, and zinc, and least extensive for bismuth, europium, germanium, lanthanum, neptunium, niobium, palladium, platinum, radium, rhenium, scandium, technetium, tellurium, thorium, yttrium, and zirconium (Eisler, 2009).

13.3 Depth of Coverage

Of the approximately 1980 references cited (Table 13.2), molluscs—especially the bivalves, viz., clams, oysters, mussels, and scallops—continue to be the dominant group researched (38.2% of the total references) by virtue of their economic importance, their comparatively

Table 13.1: Trace Metals and Marine Biota: Breadth of Coverage

Trace Element	Breadth of Coverage[a] (%)
Cadmium, copper, iron, nickel, zinc	100
Arsenic, barium, chromium, cobalt, lead, manganese, strontium	91
Mercury, silver, tin	82
Aluminum, antimony, cesium, gallium	73
Boron, selenium, titanium, ruthenium	64
Molybdenum, plutonium, rubidium, scandium, vanadium	55
Zirconium, lithium	46
Bismuth, cerium, europium, radium, silicon, uranium	37
Americium, gold, niobium, rhenium, samarium, technetium, thorium, tungsten,	28
Beryllium, germanium, lanthanum, neptunium, polonium, protactinium, terbium, thallium, ytterbium, yttrium	19
Dysprosium, erbium, gadolinium, hafnium, holmium, lutetium, neodymium, praseodymium, tantalum, tellurium, thulium	10

[a] A rating of 100% means that the eleven major groups covered herein (algae and macrophytes, annelids, chaetognaths, coelenterates, crustaceans, echinoderms, insects, molluscs, protists, sponges, and tunicates) are listed in the technical literature. Ratings of 91%, 82%, 73%, 64%, 55%, 46%, 37%, 28%, 19%, and 10% represent ten, nine, eight, seven, six, five, four, three, two, and one major taxonomic groups, respectively.

sessile existence, and their ability to reflect trace metals in their immediate biosphere. The crustacean literature is also abundant (23.0% of total references) and reflects their economic importance, although trace metal interpretation is confounded by molting frequency. Algae and other plants that form the basis of most marine food chains assume an increasingly important role in the trace metal literature with 19.9% of the total references (Table 13.2). Collectively, algae and macrophytes, molluscs, and crustaceans account for about 81.1% of the total research effort. Comparatively neglected are the annelids (4.8% of all references), protists (4.7%), echinoderms (3.6%), coelenterates (2.0%), tunicates (1.7%), sponges (1.4%), chaetognaths (0.4%), and insects (0.3%); acquisition of additional baseline data on these groups is strongly recommended.

Similar data for marine vertebrates, based on 1376 references, showed that 45.9% of all references were devoted to fishes, as expected, due to their commercial importance (Eisler, 2009). Marine birds accounted for 24.8%, mammals 20.5%, elasmobranchs 5.0%, and reptiles 3.8% (Eisler, 2009). It is clear that acquisition of baseline data on elasmobranches and marine reptiles should be given a high priority.

Table 13.2: Trace Metals and Marine Biota: Depth of Coverage

Taxonomic Group	Approximate Number of References (Percent of Total)
Algae and macrophytes	395 (19.9)
Annelids	96 (4.8)
Chaetognaths	7 (0.4)
Coelenterates	40 (2.0)
Crustaceans	455 (23.0)
Echinoderms	71 (3.6)
Insects	5 (0.3)
Molluscs	755 (38.2)
Protists	94 (4.7)
Sponges	28 (1.4)
Tunicates	34 (1.7)
Total	1980 (100.0)

13.4 Literature Cited

Eisler, R., 1981. Trace Metal Concentrations in Marine Organisms. Pergamon, Elmsford, NY, 687 pp.
Eisler, R., 2009. Compendium of Trace Metals and Marine Biota. Volume 2: Vertebrates. Elsevier, Amsterdam.

Index

A

Acetaldehyde, 267
Acid phosphatase, 493, 563
Acid volatile sulfide, 470
Acushnet estuary, 422
Adak Island, 153, 162, 181, 243, 260, 275, 297
Admiralty Bay, 38, 52, 72
Adriatic Sea, 9, 10, 153–4, 156, 163, 181–2, 189, 205, 210, 231, 244, 261, 267, 275, 287, 297, 320, 330, 343
AE, 169–70, 310, 360, 412, 478
Aegean Sea, 148, 163, 176, 206, 224–5, 244, 343, 571, 591
Alabama, 198
ALAD, 448
Alanine, 126
Alaska, 151, 153, 162, 181, 243, 260, 275, 281, 297, 402, 404, 453
Aleutian Islands, 52, 270, 312, 453, 559–60, 563, 568, 574–5
Alexandria, 24, 316, 318
Alginates, 416
Alginic acid, 23, 71
Algulose, 23, 28, 71
Alkaline phosphatase, 356, 359
Aluminum, 1, 2, 7–8, 69, 99–100, 115–6, 123–4, 144–7, 269, 328, 399–400, 527, 536, 540, 553–4, 573, 601
Amchitka Island, 147, 176, 192, 224, 292, 312, 329
American Samoa, 151, 404
Americium, 1, 2, 8, 147–8, 293, 527, 601
Ammonium chloride, 108
Amylase, 437, 493
Anaheim Bay, 242
Anasco Bay, 463
Anglesey, 345
Antarctica, 38, 52, 72, 231, 276, 344, 401, 409–10, 417, 421, 428, 431, 437–45, 448, 450, 454, 461, 462, 464–5, 467, 472–3, 476, 481–2, 486
Antarctic Ocean, 425, 428, 432, 437
Antarctic region, 9–10
Antimony, 1, 2, 8–9, 35, 115–6, 123–4, 148–9, 399, 401, 528, 553–4, 583–4, 601
Antimony trioxide, 8
Arabian Sea, 64, 126, 128, 130, 131–3, 135, 140, 467
Arcachon Bay, 197, 219, 317, 328
Arctic Circle, 521, 523–6
Arctic Ocean, 59, 425, 428, 432, 437
Argentina, 10, 15, 17, 24, 27, 29, 38, 43, 49, 55, 57, 63, 71, 73
Arsenate, 9, 12–14, 33, 99, 150, 152, 199, 218, 405, 435, 528–9
Arsenic, 1, 2, 9–14, 35, 99–101, 115–8, 123–4, 150–5, 269, 316, 325, 399–406, 443, 521, 522, 528–9, 553–6, 573–4, 583–5, 600–1
Arsenite, 9, 12–14, 150, 152, 405, 529
Arsenobetaine, 13, 99, 150–2, 400–1, 405, 529
Arsenocholine, 13, 400, 529
Arsenolipids, 13, 400
Arsenosugars, 9, 12–13, 400, 529
Ascorbic acid, 435
Assimilation efficiency, 169–70, 310, 360, 412, 478
Athens, 150
Atlantic Ocean, 52, 57, 60, 72, 99, 198, 237, 239, 271, 273, 290, 428, 432, 452, 481, 485
Atlantic Provinces, 27, 29, 49
ATP, 104
Australia, 18–19, 24, 29, 30, 38–9, 43, 45, 49, 50, 58, 71, 73, 75–7, 158, 162, 171, 173, 213, 219, 237–8, 242, 296, 306, 320, 333–4, 354, 404, 407, 411, 420, 431, 434, 437, 445, 447, 466, 482, 486, 491, 596, 599
Australian Antarctic Territory, 407, 437, 447, 491
Aveiro, 108, 322
AVS, 470
Ayamonte, 196, 336
Azores, 164, 175, 189, 209, 232, 246, 262, 277, 294, 297, 314, 346

B

Baffin Island, 521
Baja California, 19, 32, 77, 203, 303
Balearic Islands, 154, 164, 181, 183, 205, 208, 275, 288, 555, 557, 562, 568, 570, 573, 576
Baltic Sea, 17, 30, 44, 49, 57, 74, 146, 162, 180, 194, 203, 230, 242, 255, 260, 266, 275, 286, 341, 412, 421, 423, 433, 447, 460, 463

603

Baltimore, 337
Barbate, 196, 336
Barbuda, 67
Barcelona, 163, 181, 205, 244, 276, 343
Barents Sea, 408, 425, 479, 522, 523, 524, 526
Barium, 1, 2, 14, 28, 71, 100–1, 116, 118, 124–5, 155–7, 406, 421, 521–4, 528–9, 556, 584–5, 601
Barium sulfonate, 421
Bay of Biscay, 151, 158, 178, 187, 193, 227, 238, 271, 300, 330, 333
Beer, 171, 344, 346–7
Belgium, 303, 401, 457
Bentonite, 172, 421
Benzo(a)anthracene, 119
Benzo(a)pyrene, 119
Benzo(b)fluoranthene, 119
Benzo(e)pyrene, 119
Benzo(j)fluoranthene, 119
Benzo(k)fluoranthene, 119
Bergen Harbor, 280
Bermuda, 132
Beryllium, 1, 2, 14–15, 406–7, 601
Bidaron estuary, 290
Bidasoa estuary, 539, 540
Bioconcentration factors (BCFs), 46, 48, 65, 67, 77–9, 147–8, 167, 174, 177, 192, 223, 266, 282, 293, 298, 306, 315, 326, 360–1, 407, 437, 447, 491, 530, 536
Biomagnification, 26, 33, 60, 68, 108, 282
Bismuth, 1, 2, 15, 125, 156–7, 359, 407, 600–1
Black Sea, 36, 57, 64, 66, 70–1, 79, 176, 205, 343, 448, 476, 495, 521, 524–6
Block Island Sound, 194
Bodega Bay, 206, 303
Bohai Bay, 68, 317, 472–3
Bohai Sea, 184, 212, 248, 271, 319, 349
Boric acid, 556

Boron, 1, 2, 15–16, 125, 156–7, 406–7, 522, 524, 556, 584–5, 601
Boston Harbor, 106, 109
Brazil, 19, 31, 40, 50, 58, 75, 158–60, 199, 228, 257, 284, 336–7, 408, 425, 441, 449, 462, 479
Bristol Channel, 168, 171, 210, 344, 347
British Columbia, 55, 271, 320
Brunswick, 53
Bryozoans, 469, 470
Butyltins, 325, 327, 472
Bylot Sound, 293, 569

C

Cadmium, 1, 2, 16–21, 78, 101, 109, 118, 126–7, 157–74, 308, 311, 358–9, 407–16, 434, 448, 468, 491, 517, 522, 524, 528–32, 540, 546, 556–8, 574, 585, 601
Calcasieu River Estuary, 140, 335, 467, 479
Calcium, 1, 2, 21, 61, 66–7, 108, 118–9, 125, 134, 167, 172, 222, 293, 312, 327, 355, 359, 361, 407, 416, 436, 478, 542, 573
Calcium carbonate, 61, 172
California, 8, 18, 19, 30–2, 39, 41, 44–6, 50–1, 69, 75, 77, 159, 161, 180, 186, 201, 206, 239, 268, 302–3, 308, 354, 357, 400, 409–10, 429, 445, 455, 484–5, 539–40, 562, 573, 587
Camel estuary, 166, 184, 191, 213, 234–5, 248, 264, 290, 306, 350
Canada, 27, 29, 49, 55, 57, 72, 204, 271, 321, 327
Canary Islands, 164, 246
Carbon, 2, 13, 15, 37, 105, 109, 170, 222, 356, 435, 470, 475
Carbonates, 251, 359, 435
Carbonic anhydrase, 359, 478
Carbonyl, 37

Carboxylic, 60
Carboxypeptidase, 359
Carcinogenicity, 599
Caribbean Sea, 464
Catalonia, 116–7, 278, 454
Cellulose, 583
Cerium, 1, 2, 22, 119, 175, 417–8, 558, 601
Cesium, 1, 2, 22–3, 102, 127, 175–7, 417–8, 532, 558–9, 585–6, 601
Charlotte Harbor, 12
Chelators, 26, 34, 217, 359, 422, 433, 435, 490, 536
Chernobyl, 176, 295, 417–8, 465
Chertsey, 148, 270
Chesapeake Bay, 8, 44, 52, 65, 105, 108, 145, 194, 198, 218, 239, 258, 301, 308, 316, 337, 340, 357, 424
Chile, 9, 16, 29, 44, 52, 58, 64, 67, 71, 75, 161, 202, 230, 259, 286, 297, 341
China, 29, 31, 43, 45, 68, 73, 76, 132, 151, 158–9, 166, 171, 184, 195, 197, 212, 238–9, 248, 264, 271, 279, 317, 319, 334, 349, 457, 472–3, 599
Chincoteague inlet, 449
Chinese National Water Quality Standard for Fisheries, 438
Chitin, 434, 436, 461
Chlorine, 2, 356, 491
Chlorophyll, 8, 14, 37, 41, 79
Chromated copper arsenate, 33, 199, 218, 435
Chrome lignosulfonate, 421
Chromium, 1, 2, 23–6, 102, 127–8, 177–86, 291, 308, 418–24, 517–9, 524, 532–4, 559–60, 571, 586
Citrate, 34, 225, 237
Ciudad del Carmen, 456
Coatzacoalcos estuary, 240
Cobalt, 1, 2, 26–8, 42, 78, 102, 119, 127–8, 186–92, 423–4, 524, 534–5, 559–61, 587, 601
Cobalt acetate, 560
Cobo Bay, 200

Columbia, 72
Columbia River, 596
Connecticut, 301, 303
Cook Inlet, 151
Copper, 1, 2, 28–36, 102–3, 119, 127–9, 192–224, 279, 424–38, 519, 524, 535–7, 561–3, 587–8
Copper acetate, 435
Copper chloride, 435
Copper sulfate, 434–5, 438
Copper sulfide, 222
Corio Bay, 162
Cornwall, 208, 345
Corpus Christi Harbor, 101
Corsica, 19, 25, 27, 45, 53, 58
Croatia, 275, 318
Cuba, 568
Cupric ion, 34, 435
Cysteine, 540
Cytochrome P450, 326

D

DDT, 78
Deficiency, 56, 78, 103, 107, 362, 436–7, 495, 575
Delaware, 13, 249–50, 437, 449
Denmark, 32, 529
Dibutyltin, 68, 108, 317–21, 473, 475, 575
Dimethylarsenosoribosyl, 12
Dimethylarsinate, 12, 529
Dimethylarsinic acid, 99, 151, 152, 528
Dimethyltin, 68, 108, 319, 322, 475, 575
Dimethylureas, 561
Dinitrophenol, 61
Diphenyltin, 318–20, 322, 475, 575
Dissolved organic matter, 536
DNA, 63, 295, 298, 325
Dorset, 18, 171, 344, 346
Douro River estuary, 16
Dulas Bay, 17–18, 38–9, 43, 44, 49, 57–8, 73, 74
Dysprosium, 2, 438, 601

E

East Antarctica, 417, 439, 442, 461, 464–5, 472–3, 486

East China Sea, 151, 195, 238, 271
East Indies, 49, 72
East Japan Sea, 293, 298, 315
EDTA, 26, 47, 72, 217, 359, 417, 422, 435, 546, 558, 567, 569
Egypt, 24, 316, 318, 418–9
Eikhamrane, 17, 43, 52, 73
Elizabeth River estuary, 424
England, 24, 29, 52, 57–8, 66, 130, 157, 160, 184, 190–1, 197, 204, 208, 213, 234, 240, 242–3, 248, 258, 260, 264, 283, 285, 290, 323, 325, 338, 349–50, 354
English Channel, 200, 214, 301
Eniwetok Atoll, 407
Erbium, 1, 3, 438, 600–1
Europium, 3, 224–5, 439, 563, 600–1

F

Fal estuary, 354
Faroe Islands, 271
Fermain Bay, 200
Ferric hydroxide, 61, 226
Ferric oxide, 41, 103
Ferritin, 356, 443
Ferrochrome, 421
Ferromanganese, 104
Firth of Clyde, 485
Firth of Tay, 52–3
Flak, 17, 43, 52, 73
Flatworms, 434–5
Florida, 12, 54, 57, 152, 160, 179, 198–9, 239, 243, 272, 284, 296, 318, 337, 402, 404, 408, 419, 444, 453, 466, 473, 570, 572
Florida Everglades, 54
Fluoranthrene, 416
Former Soviet Union, 57
Foundry Cove, 529
France, 19, 25, 27, 45, 51, 53, 58, 181, 197, 205, 219, 244, 276, 290, 299, 316–8, 328, 463, 472
Fucoidan, 28, 71
Fulvic acid, 536

G

Gadolinium, 1, 3, 439, 600–1
Gallium, 1, 3, 36, 100, 103, 117, 119, 129–30, 224–5, 439, 522, 524, 564, 601
Galveston Bay, 239, 337
Gannel estuary, 166, 184, 190–1, 213, 234–5, 248, 264, 290, 306, 349–50
Genova, 163, 181, 205, 244, 276, 343
Georges Bank, 194
Georges River, 213
Georgia, 53, 54, 145, 152, 156, 157, 160, 179, 198, 228, 239, 258, 272, 281, 284, 299, 301, 337
Germanium, 1, 3, 36–7, 439, 440, 600–1
Germany, 243
Glucose–6-phosphate dehydrogenase, 16
Glutamate, 126
Glutathione, 21, 170, 295, 447, 491
Glutathione peroxidase, 170, 295, 447, 491
Glycine, 540
Glycogen, 174, 311, 493
Glycoprotein, 173
Glycyl-*L*-histidine, 107
Goa, 27, 29, 38, 43, 49, 57, 72
Gold, 1, 3, 36–7, 224–5, 439, 440, 601
Golfo Triste, 568
Greece, 150, 155, 163, 181, 192, 206, 231, 244, 280, 287, 312, 329, 558, 571, 586
Greenland, 58, 74, 160, 240, 243, 251, 272, 293, 342, 354, 409, 445, 454, 464, 466, 540, 569
Greenland Sea, 522–4
Guam, 11, 18, 24, 64, 116–7, 120, 124, 126, 128, 130, 136, 154, 163, 182, 207, 276, 304, 403, 409, 420, 429, 467, 555, 557, 560, 572, 584–7, 591
Gulf coast, 198, 208, 239, 245, 258, 301, 338, 345

Gulf of California, 18–19, 30–2, 39, 41, 44–6, 50, 75, 77, 159, 239, 336, 410, 445, 485
Gulf of Gabes, 159
Gulf of Gdansk, 17, 30, 44, 49, 57, 74, 201, 255, 340
Gulf of Gemlik, 163, 182, 189, 205, 231, 244, 261, 287, 343
Gulf of Mannar, 118, 120
Gulf of Mexico, 37, 79, 240, 308, 318, 324, 337, 411, 431, 446, 457, 487, 572

H

Hackensack River, 24, 419
Hafnium, 1, 3, 225–6, 600–1
Hainan Island, 132
Halifax, 204
Harbour Island, 27, 57, 58, 75
Hardangerfjord, 17, 43, 52, 73
Hawaii, 99, 127, 129, 137, 139
Hemocyanin, 195, 222, 424, 436
Hemoglobin, 531
Hemolymph, 171–2, 207–8, 221, 232, 309, 345, 360, 415, 422, 426, 428, 436, 443, 463–4, 469, 480, 490, 573
Hexavalent chromium, 23, 26, 102, 185, 186, 422, 534, 559
Histidine, 107
Holmium, 1, 3, 440, 600–1
Hong Kong, 29, 43, 73, 101, 109, 132, 153, 165, 183, 197, 211, 213, 219, 234, 247, 263, 289, 322, 327, 348–9, 428
Huangshi Sea, 271
Humic acid, 172, 222, 435
Hydrogen peroxide, 107
Hydroxycarboxylic, 60
4-Hydroxy–3-nitrobenzene arsonic acid, 99
Hyperion sewer outfall, 268

I

Iberian Peninsula, 326
Iceland, 27, 29, 49, 72
Imposex, 325–7
India, 18–9, 25–7, 29–31, 38–9, 43–4, 46, 49, 52, 57, 66, 68, 72, 75, 78, 112, 118, 120, 165, 180, 183, 202, 211, 233, 247, 259, 263, 278, 289, 326, 341, 348, 533, 599
Indonesia, 132, 326
Ireland, 17, 24, 27, 77, 140, 181, 303, 448, 489, 493, 596
Irish Sea, 18, 27, 29, 36, 38, 44, 49, 58, 64, 66, 74, 99, 195, 400, 556, 564
Iron, 1, 3, 26, 34, 38–42, 100, 103, 117, 119, 131, 226–37, 440–3, 518, 520, 525, 537, 538, 564–5, 588–9
Iron oxide, 172, 192, 299, 356
Iskenderun Bay, 400, 408, 419, 423, 444
Israel, 17, 29, 57, 288, 305
Italy, 19, 31, 40, 45, 53, 163, 181, 205, 206, 231, 244, 273, 276, 321, 343, 410, 420, 445, 451, 566
Izmir Bay, 162, 182, 205, 231, 244, 261, 275, 343

J

Jamaica, 63, 66
James River, 340
Japan, 15, 19, 22, 24–6, 32, 36, 40, 42, 53, 55, 59, 61, 63, 74, 131–2, 134, 204, 208, 267, 273, 297, 319, 328, 452, 589, 599
Japan Sea, 293, 297, 298, 315
Jiaozhou Bay, 166, 171, 184, 212, 248, 264, 349

K

Korea, 19, 25, 27, 32, 40, 51, 52, 166, 183, 190, 212, 241, 248, 264, 290, 315, 318, 321, 349, 457
Kyushu, 267, 319, 328

L

La Haque, 463
Lagoon of Venice, 326
Laguna Madre, 11, 24, 63, 405, 420
Lanthanum, 1, 3, 237, 443, 600–1
Laverbread, 60
Lead, 1, 3, 35, 42–8, 103–4, 117, 120, 131–2, 144, 170–1, 174, 218–9, 237–54, 269, 316, 325, 334, 342, 354, 358–9, 443–8, 491, 494, 520, 523, 525, 537–8, 540, 565–6, 573–4, 588–9, 600–1
Lead acetate, 447, 566
Lebanon, 285
Lesina lagoon, 420, 430, 445, 451, 485
Leucine, 103
Lisbon, 319
Lithium, 1, 3, 48, 125, 132–3, 255, 448, 523, 525, 601
LOAEL (lower observed adverse effect level), 469
Lofoten, 17, 52
Long Island, 196, 204, 283, 290, 449
Long Island Sound, 196, 204, 449
Looe estuary, 162, 184, 191, 204, 213, 234, 242, 260, 264, 290, 302–3
Los Angeles, 268
Louisiana, 301, 335, 467, 479
Lutetium, 1, 3, 448, 600–1

M

Magnesium, 1, 3, 21, 60, 67, 102, 107, 108, 222, 356, 407, 436, 478
Maine, 320
Malaysia, 52, 193, 209, 322
Malic dehydrogenase, 359
Manchester, 199
Manganese, 1, 3, 42, 48–51, 69, 99, 100, 103–4, 120, 133, 144, 170–2, 192, 218, 255–66, 291, 355, 359, 443, 449–52, 519–20, 537–8, 540, 545, 566–7, 573–4, 589–90, 600–1
Manganese dioxide, 452, 567
Manganese oxide, 172, 192, 356
Mannoside, 359
Marmara Sea, 160, 162, 200, 205, 272, 275
Marseille, 51, 53, 163, 181, 205, 244, 276, 343

Maryland, 8, 52, 145, 198, 218, 301, 308, 316, 340, 357, 400, 402, 424
Massachusetts, 106, 109, 249, 250, 422
Maximum acceptable toxicant concentration (MATC), 438
Mediterranean Sea, 11, 17, 19, 25, 26, 27, 29, 30, 40, 45, 57, 58, 61, 66, 78, 118–20, 151, 154, 158, 163–4, 169, 176, 178, 180–1, 183, 187, 189, 193, 205, 208, 227, 238, 244, 245, 261, 273, 275–7, 287–8, 300, 321, 330, 333, 343, 345, 357, 399, 417–9, 432, 443, 448, 464, 471, 489, 555, 557, 562, 566, 568, 570, 573, 576
Melanin, 147
Menai Straits, 17–18, 38–9, 43–4, 49, 57–8, 73, 74, 440
Mercurials, 54–5, 106, 280
Mercuric chloride, 106, 269, 279, 458
Mercuric reductase, 106
Mercury, 4, 51–5, 101, 104–107, 117, 120, 133–4, 266–80, 295, 452–9, 537–40, 568–9, 589–90, 601
Mersin coast, 158, 169, 178, 238
Metallothioneins, 168, 170–1, 174, 223, 311, 356, 412, 438, 490, 529–31, 545–6
Methane, 99, 105–6
Methylarsonic acid, 12
Methylcobalamin, 105
Methylmercury, 104–6, 275, 279–80, 416, 452, 455, 540
Mex Bay, 24
Mexico, 18–9, 30–2, 36–7, 39, 41, 44–6, 50–1, 59, 75, 77, 79, 161, 169, 202–3, 219, 240–1, 272, 303, 308, 318, 324, 337, 340, 410–1, 429, 431, 445–6, 456–7, 484–5, 487, 572, 599
Mid-Atlantic ridge, 145, 158, 187, 194, 227, 256, 270, 300, 313, 333, 427–8, 441, 454–5, 482, 484, 533, 539, 544
Minamata Bay, 52–3, 105–6, 134, 267, 273, 452, 454, 589
Mission Point, 327
Mississippi, 239, 272, 404
Molting, 399, 405, 413, 418, 433, 443, 447, 452, 465, 490, 495, 601
Molybdate, 55, 56
Molybdenum, 1, 4, 55–6, 134–5, 280–2, 460, 495, 523, 525, 540, 569, 601
Monaco, 24
Monobutyltin, 68, 108, 317–22, 324–5, 473, 575
Monomethylarsonate, 400
Monomethylarsonic acid, 99, 152
Monophenyltin, 319, 320, 322, 575
Monterey Bay, 50, 69, 75
Morocco, 161–2, 169, 180, 203, 242, 259, 274, 286, 340–1
Mutagenicity, 599

N

Narragansett Bay, 23, 24, 206, 303, 354
Natal, 19, 31, 40, 50, 58, 75, 160, 199, 337, 408, 425, 441, 449, 462, 479
Neodymium, 4, 461, 600–1
Neptunium, 4, 282, 461, 600–1
Netherlands, 162, 201, 242, 260, 286, 342, 361, 402
New Bedford, 422
New Caledonia, 151, 159, 179, 188, 196, 228, 257, 284, 300, 336
New England, 130, 258, 283
New Jersey, 24, 31, 43, 46, 276, 419, 432, 437, 446, 457, 488
New South Wales, 158, 171, 238, 296, 333, 404, 411, 420, 431, 445, 466, 486
New York, 196, 204, 283, 290, 529
New York Bight, 194, 422, 449
New Zealand, 178, 194, 204, 242, 256, 283, 304, 305, 333, 599
Newfoundland, 341
Newport River, 198, 337
Newport River estuary, 73
Nickel, 1, 4, 35, 56–60, 101, 109, 117, 120, 134–5, 144, 218, 255, 269, 282–9, 461–3, 494, 519–20, 523, 525, 536, 539, 540, 569–70, 573–4, 589–90, 600–1
Nickel-cadmium battery plant, 56
Nickel chloride, 107
Niobium, 1, 4, 292, 463, 589, 595, 600–1
Nitrates, 15–6
Nitrogen, 4, 23, 56, 61
NOAEL (no observed adverse effect level), 469
Nobeoka Bay, 208
North America, 148, 150, 153, 177, 241, 316, 402, 421, 427, 444, 450, 454, 460, 462, 466–7, 472–3, 477, 599
North Carolina, 73, 77, 198–9, 201, 337–8
North River, 337
North Sea, 154, 303, 403, 430, 457
North Wales, 345
Norway, 17, 18, 24, 27, 29, 31–2, 38–40, 43, 44, 46, 51–2, 55, 57–8, 63, 72–4, 76, 161, 165, 202, 204, 241, 243, 246, 275, 280, 286, 288, 303, 341, 437, 565–6
Nova Scotia, 10, 57–8, 153

O

Octyltins, 318, 322, 326, 473, 574
Odense Fjord, 529
Ogasawara Island, 131
Okhotsk Sea, 160
Oregon, 11, 19, 25, 32, 41, 59, 77, 99
Organoarsenicals, 11–4, 99, 150, 400, 405
Organoethylleads, 47
Organoiron, 41
Organolead, 47, 103
Organomercurial lyase, 106

Organomercury, 54, 268–9, 271, 276, 279–80, 452, 459
Organomethylleads, 47
Organoselenium, 62
Organotins, 68, 69, 108–9, 316–8, 323–8, 472–3, 475–6, 543, 574–5
Orkney, 12
Osaka Bay, 319, 474
Otago, 194
Outer continental shelf, 59, 291, 463

P

Pacific Ocean, 42, 61, 99, 130, 413, 435, 475, 521, 524, 525
PAH (polycyclic aromatic hydrocarbon), 104, 119
Pakistan, 64
Palk Bay, 55
Panama, 72
Paralytic shellfish poisoning (PSP), 147
Pensacola, 152, 160, 179, 198, 239, 272, 284, 296, 318, 337, 402, 408, 419, 425, 444, 453, 462, 466, 473, 479
Persian Gulf, 457
Perth, 320
Petroleum hydrocarbons, 421, 530, 536
Phenanthrene, 416
Phenyltins, 318, 472–3, 574
Phosphate, 9, 12, 16, 42, 103, 107, 359, 491, 544, 595
Phosphatidylcholine, 400
Phosphorus, 4, 9, 355, 361, 407, 491
Photosynthesis, 15, 21, 26, 37, 42, 47, 54, 60, 69, 78, 79
Phytochelatins, 21, 35
Plutonium, 1, 4, 8, 60–1, 120, 292–3, 443, 463–4, 540–1, 569, 601
Polonium, 1, 4, 293, 464, 601
Polychlorinated biphenyls (PCBs), 144
Polycyclic aromatic hydrocarbon, 104

Polysaccharides, 21, 61
Port of Mahon, 154, 163–4, 181, 183, 205, 208, 244–5, 275, 277, 288, 345, 555, 557, 565, 568, 570, 573, 576
Port Phillip Bay, 162, 237, 242
Portland, 18
Portugal, 16, 20, 32–3, 47, 52, 108, 154, 163, 204, 207, 232, 245, 261, 276, 281, 287, 294, 297, 319, 322, 331, 344, 453
Potassium, 1, 4, 67, 102, 108, 356, 491, 585
Praseodymium, 1, 4, 464, 600–1
Proline, 531
Protactinium, 1, 4, 293, 464, 601
Proteases, 577
Puerto Rico, 23, 24, 50–1, 57, 76, 119, 457, 463, 559, 569–70
Puget Sound, 46, 53, 58, 242
Purdy, 198

Q

Queensland, 24, 38, 71

R

Radium, 1, 4, 61, 134, 195, 221, 293, 464, 600–1
Rappahannock River estuary, 198
Raritan Bay, 31, 43, 44, 46
Ras Beirut, 285
Red tide, 79
Rhenium, 1, 4, 61, 294, 464, 600–1
Rhode Island, 23–4, 99, 198, 206, 303, 354
Rio de Aveiro, 322
Rio Pedras, 336
Rio Tinto estuary, 17, 29, 43, 72, 202, 341
River Calligan, 181
River Crouch, 321, 328, 476
River Tyne, 320–1
Ross Sea, 193
Rotifers, 434
Rubidium, 1, 4, 16, 61–2, 134, 294, 465, 571, 591, 601
Ruthenium, 1, 4, 61, 119–20, 134, 175, 295, 465, 541, 592, 601

S

Saanach Inlet, 55
Sacca Sessola, 163, 181, 206, 231, 244, 261, 343
St. Croix, 132
Samarium, 1, 4, 295, 465, 535, 541, 601
Samoa, 127–8, 136, 139, 151, 404
San Antonio Bay, 239
Sancti-Petri, 196, 336
San Diego, 204, 243, 268, 342
San Diego Bay, 303, 320, 573
San Diego Harbor, 204, 243, 320, 342
San Francisco Bay, 8, 145, 219, 301–3, 306, 340, 400, 455, 541
San Giuliano, 163, 181, 206, 231, 244, 261, 343
Sanlucar, 196, 336
Santa Marie Bay, 151, 159, 179, 284, 300, 336
Sardinia, 163, 205, 244, 287
Sargasso Sea, 99
Saronikos Gulf of Greece, 181, 206, 244
Savannah, 145, 152, 156–7, 160, 179, 198, 228, 239, 258, 272, 281, 284, 299, 337
Scandium, 1, 5, 127, 135–6, 295–6, 465, 535, 541–2, 571, 591–2, 600–1
Scheldt estuary, 146, 153, 162, 181, 189, 230, 242, 260, 286, 303, 342
Scotland, 12, 31, 52–3, 57–8, 195, 242, 462–3
Sea of Japan, 24, 27, 29, 38, 49, 57, 70–1, 160
Seal Harbour, 10, 153
Selenite, 62, 295–6, 298, 466
Selenium, 1, 5, 35, 62–3, 117, 120, 268–9, 295–9, 459, 466, 468–9, 540, 542, 571–2, 574, 591–2, 600–1
Sellafield, 67
Sepetiba Bay, 159, 336
Severn estuary, 18, 44, 74

Shenzhen, 29, 43, 73
Silica, 37
Silicates, 15, 37, 63, 469
Silicic acid, 37, 63
Silicon, 1, 5, 62–3, 135–6, 299, 407, 469, 478, 601
Silver, 1, 5, 64–5, 107–8, 121, 135, 172, 218, 269, 299–312, 325, 466–7, 469–70, 523, 525, 540–2, 571–4, 591–2, 600–1
Silver sulfide, 310–1
Sinaloa, 161, 202, 219, 241, 340
Sodium metavanadate, 70
Solent region, 52
Sorfjorden, 17, 18, 43–4, 46, 73, 74, 76
South Africa, 27, 40, 51, 76, 157, 197, 261, 284
South Australia, 482
South Wales, 158, 171, 238, 296, 333, 404, 411, 420, 431, 445, 466, 486
Southampton, 148, 236, 270
Soviet Union, 24, 57
Spain, 11, 17, 29, 32, 40, 43, 52, 72, 77, 116–8, 120, 163, 164, 180–1, 196, 202, 204–5, 209, 243–4, 246, 249, 256, 276, 278, 283, 290, 296, 300, 333, 336, 341, 343, 347, 419, 446, 454
Spitsbergen, 59
Strait of Magellan, 9, 64, 161, 202, 230, 286, 297, 341
Strathcona Sound, 521
Strontium, 1, 5, 8, 14, 21, 40, 42, 50, 66–7, 69–70, 72, 101, 108, 121, 125, 137–9, 155, 312–3, 470–2, 523, 525, 541–2, 573–4, 592, 600–1
Succinate, 324
Sulfate, 34, 56, 105, 107, 108, 416, 434–6, 438
Sulfide, 222, 310–1, 470, 544
Sulfur, 5, 99, 356, 438, 491
Sulfuric acid, 583
Superoxide dismutase, 170
Svalbard, 59

Sweden, 61, 117, 120, 130, 201, 255, 587
Sydney, 18, 24, 30, 39, 45, 50, 58, 75, 159, 213, 336

T

Tagus estuary, 20, 33, 47
Tahiti, 127–9, 136, 137, 139
Taiwan, 150–2, 155, 197, 219, 399–400, 405, 408, 418–9, 426, 432, 440–1, 444, 450, 457, 481
Tamar estuary, 18, 39, 44, 49, 58, 64, 74, 184, 191, 213, 234, 248, 264, 290, 350
Tamil Nadu, 30, 39, 44, 49, 52, 67–8, 74
Tampa Bay, 572
Tampico, 456
Tantalum, 1, 5, 312, 600–1
Tanzania, 145, 160, 179, 201, 229, 240, 258, 339
Tarragona coast, 249, 446
Tasmania, 162, 197, 260, 353
TBT, 35, 67–9, 108–9, 219, 316, 320–8, 474–6, 543
Technetium, 1, 5, 67, 312, 575, 600–1
Tellurium, 1, 5, 315, 600–1
Teratogenicity, 599
Terbium, 1, 5, 315, 472, 601
Terra Nova Bay, 52
Tetrachloroaurate, 37
Tetraethyllead, 42, 104, 252, 254, 448
Tetramethylarsonium, 529
Tetramethyllead, 42, 104, 254, 448
Tetramethyltins, 68
Tetravalent tin, 108
Texas, 11, 24, 27, 32, 45, 50–1, 55, 57–9, 63, 75, 101, 198, 239, 272, 291, 337, 405, 418, 420, 457, 463, 467
Texas Continental Shelf, 50, 51, 75
Thailand, 272, 322
Thallium, 1, 5, 315–6, 472, 601
Thiol, 37

Thorium, 1, 5, 61, 137–8, 316, 317, 472, 600–1
Thulium, 1, 5, 472, 600–1
Tin, 1, 5, 67–9, 108–9, 137, 138, 219, 316–28, 472–6, 523, 525, 543, 573–5, 592–3, 600–1
Titanium, 1, 5, 8, 14, 40, 42, 50, 66, 69, 70, 101, 109, 127, 137, 317, 322, 329, 473–4, 476, 523, 526, 573–4, 592, 593, 601
Tokyo Bay, 319
Triazines, 561
Tributyltin, 35, 68, 108, 137, 219, 317–22, 325, 472–5, 494, 536, 543, 575
Triethyllead, 47, 254, 448
Trimethylarsine, 528
Trimethylarsoniopropionate, 529
Trimethyllead, 47, 254, 448
Trimethyltins, 68–9, 319, 322, 475
Tripentyltins, 69
Triphenyltins, 68–9, 317–8, 320, 322, 327, 472–3
Tripropyltins, 69, 109
Trivalent chromium, 23, 186, 422, 532, 534
Trivalent gold, 37
Trondheimsfjord, 17, 38, 52, 57, 73
Trypsin, 437, 493
Tungsten, 1, 5, 69, 317, 323, 329, 473–4, 476, 601
Tunicin, 583
Tunisia, 25, 30, 76, 78, 159, 238
Turkey, 158, 160, 162–3, 169, 178, 182, 189, 195, 200, 205–6, 231, 238, 240, 244, 261, 272, 275, 287, 334, 343, 354, 400, 408, 419, 423, 425, 441, 444, 449, 462, 479
Tuscany, 53
Tyrrhenian Sea, 456

U

United Kingdom, 17–8, 24–5, 27, 29, 32, 38–9, 43–4, 49, 51, 55, 57–60, 62, 64, 68, 70–1, 73–4, 148, 162, 166, 168, 171, 179, 200, 208, 210, 270, 284–6,

288, 303, 317, 320–1, 328, 346, 440, 476, 559, 565
United States, 23, 31, 43–5, 53–5, 73, 77, 79, 162, 197–8, 204, 206, 208, 239, 242, 245, 249, 258, 260, 276, 284–5, 288, 301, 303, 320, 323, 328, 337–8, 342, 345, 353–4, 401, 407, 452–3, 457, 476, 495, 540, 577, 596
United States Environmental Protection Agency, 79, 495, 577
United States Food and Drug Administration, 452
United States Navy, 323, 328
Unleaded gasoline, 131–2
Uranium, 1, 6, 61, 69–70, 139, 293, 317, 323, 329, 476, 601

V

Vanadium, 1, 6, 70–1, 118, 121, 125, 139–40, 172, 329–30, 476–7, 523, 526, 540, 589, 593–5, 601
Vanadocytes, 595
Vancouver, 327
Venezuela, 130, 167, 185, 216, 250, 291, 332, 352, 568–9
Venice lagoon, 25, 32, 58, 76, 163, 181, 206, 231, 244, 261, 318, 321, 343
Vera Cruz, 456
Victoria, 327
Vietnam, 401, 404, 406–7, 411, 417, 420, 423, 431, 446, 451, 456, 460, 465–7, 471–2, 474, 477, 487
Virginia, 196, 198, 328, 338, 445, 449
Vistula Lagoon, 17, 30, 44, 49, 57, 74

W

Wakayama Prefecture, 15
Wales, 67, 74, 130, 203, 345, 427, 533, 588
Washington, 46, 53, 58, 198–9, 242
West Indies, 67
Western Port Bay, 162
Windscale, 60–1
World Health Organization, 237

Y

Yangtze River estuary, 31, 45, 76
Yingluo Bay, 31, 45, 76
York River, 340
Ytterbium, 1, 6, 329, 476, 601
Yttrium, 1, 6, 71–2, 477, 600–1
Yugoslavia, 153

Z

Zhejiang, 271, 457
Zinc, 1, 21, 34, 35, 42, 72–9, 109, 121, 140, 170–1, 218, 252, 269, 308, 329–63, 477–95, 520, 526, 540, 543–6, 573, 575–8, 595–6, 600
Zinc aquo ion, 492
Zinc hydroxide, 492
Zinc phosphate, 491, 544
Zinc pyrithione, 35, 494, 536
Zirconium, 1, 6, 79, 119, 363, 495, 524, 526, 600–1